Bacterial Disease Mechanisms
An introduction to cellular microbiology

Because of the rapid increase in antibiotic resistance, the past decade has seen an unprecedented increase in research into the mechanisms used by bacteria to cause human disease. This renaissance of bacteriological research has spawned a new scientific discipline, Cellular Microbiology, a fusion of Bacteriology, Molecular Biology and Cell Biology, with major inputs from Structural Biology and Immunology. The result of this research – the subject of this textbook – is an awareness of the intimacy of the interactions that exist between bacteria and our own cells both in health and during infections. The central premise of this textbook is that both bacteria and their multicellular hosts have co-evolved molecular and cellular mechanisms to evade and overcome the offence/defence mechanisms of one another.

This textbook describes, from the standpoint of human bacterial disease, the molecular and cellular interactions that occur between ourselves and the bacteria that live with us and introduces the reader to the latest techniques being used to unravel bacteria–host interactions.

Bacterial Disease Mechanisms is a core textbook for students taking courses in Microbiology, Medical Microbiology, Biotechnology and Pathology and will be of interest to those taking courses in Biology, Cell Biology, Molecular Biology, Medicine and Biochemistry.

MICHAEL WILSON is Professor of Microbiology in the Faculty of Clinical Sciences, University College London, and Head of the Department of Microbiology at the Eastman Dental Institute, University College London. His main research interests are bacterial virulence factors, biofilms, antibiotic resistance and the development of new antimicrobial strategies.

ROD McNAB is Lecturer in Molecular Microbiology at the Eastman Dental Institute, University College London, and works on streptococcal adhesion and colonisation factors, biofilms and bacterial cell–cell communication.

BRIAN HENDERSON is Professor of Cell Biology and Head of the Cellular Microbiology Research Group at the Eastman Dental Institute, University College London. His research centres around cytokine biology and the interactions of bacteria with myeloid and lymphoid cells.

Bacterial Disease Mechanisms

An introduction to cellular microbiology

Michael Wilson, Rod McNab
and Brian Henderson
Eastman Dental Institute, University College London

CAMBRIDGE
UNIVERSITY PRESS

PUBLISHED BY THE PRESS SYNDICATE OF THE UNIVERSITY OF CAMBRIDGE
The Pitt Building, Trumpington Street, Cambridge, United Kingdom

CAMBRIDGE UNIVERSITY PRESS
The Edinburgh Building, Cambridge CB2 2RU, UK
40 West 20th Street, New York, NY 10011-4211, USA
477 Williamstown Road, Port Melbourne, VIC 3207, Australia
Ruiz de Alarcón 13, 28014 Madrid, Spain
Dock House, The Waterfront, Cape Town 8001, South Africa

http://www.cambridge.org

First published 2002

Printed in the United Kingdom at the University Press, Cambridge

Typeface Swift 9/12.25pt (Monotype). System 3B2

A catalogue record for this book is available from the British Library

ISBN 0 521 79250 9   hardback
ISBN 0 521 79689 X   paperback

Contents

Chapter 10 | Bacterial evasion of host defence mechanisms — 514

Preface

It is unlikely that even Steven Spielberg could do justice to the history of Bacteriology, with its fascinating story of ignorance, blinding enlightenment, industrial success and an ultimate denouement that the reader is living through. This story is populated with some of the most fascinating characters in science such as Antonie van Leeuwenhoek, Louis Pasteur, Robert Koch, Élie Metchnikoff, Paul Ehrlich, Alexander Fleming and Selman Waksman, to name but a few.

With some 5000 years of recorded human history, it is stunning to realise that it was only in the 17th century that we saw bacteria for the first time, through the 'magic' of Antonie van Leeuwenhoek's single lens microscopes. It was only 150 years ago that Pasteur firmly established the concept of the microorganism. In the 30 years spanning the last two decades of the 19th century and the first decade of the 20th century, in what is the proverbial 'twinkling of an eye', the major human bacterial pathogens were identified. No sooner had they been identified than vaccination for bacterial diseases (e.g. anthrax) was developed. Ehrlich developed the first antibacterial, Salvarsan, in 1910 and penicillin was discovered by Fleming in 1929, although introduced into clinical practice only in the 1940s. In the 1940s and 50s, Selman Waksman and others developed new antibiotics such as streptomycin from soil organisms. However, antibiotics proved to be the death-knell of bacteriological research and by 1969 the American Surgeon General felt able to say 'we can close the book on infectious diseases'.

Thirty or so years later, the writers of this book, and its readers, are facing a crisis – antibiotic resistance. It is the rapid rise in resistance to antibiotics that has forced humanity to take another look at bacteria and this has led to a renaissance in research in Bacteriology during the past 10–15 years. Such research is revealing the enormous complexity of the interactions occurring between bacteria and their hosts and is dispelling for ever the belief that bacteria are simple organisms easily defeated by our 'wonderdrugs'. With pride-coming-before-a-fall we need to remember that antibiotics in current clinical use are compounds (or derivatives of these) made by microbes to kill or block the growth of other microbes.

In the past two decades our understanding of the mechanisms of bacterial infections has increased enormously. In the process, that branch of bacteriology dealing with bacterial pathogenesis has evolved into a new science by melding with Cell Biology, Molecular Biology and Structural Biology to become a multiskill discipline, which has been termed Cellular Microbiology. Current studies of the mechanisms used by bacteria to cause disease utilise tools such as the Human Genome Project database, prokaryotic genome sequence databases, transgenic knockout animals, transposon mutagenesis, *in vivo* expression technology, high level expression systems and X-ray crystal-lography. It is this new science, and its application to the problem of bacterial infection, that is the subject of this textbook.

This textbook is designed for students (both undergraduate and post-graduate) of the biological and medical sciences who need to under-stand the complex cellular and molecular processes that are involved when bacteria infect us and cause disease. The first two chapters describe bacteria and the diseases they cause and adopt a generic approach to these areas. Chapter 3 reviews the growing number of genetic methods for analysing bacterial virulence mechanisms. Chapters 4, 5 and 6 introduce the reader to the host response to bac-teria and describe the very large number of strategies, from the purely physical to the molecular, which have evolved to prevent or inhibit infection. Chapters 7 and 8 describe the systems used by bacteria to adhere to host cells and to invade these cells. Adhesion and invasion are now perceived to be central activities in bacterial infection and survival and the study of these subjects is revealing unexpected complexity. Chapter 9 deals with a topic that has been the subject of extensive research and which strikes fear into most hearts – toxins. Probing the mechanisms of bacterial toxins has long been associated with new discoveries in Cell Biology. Chapter 10 considers how bacteria respond to the machinations of our immune systems and describes the mechanisms that have evolved to enable bacteria to evade our 'best shots' in immunity. The final chapter gives a brief introduction to the very latest areas in bacterial virulence and attempts to gain insight from the crystal ball. Throughout the book we have emphasised a key feature of bacteria host–cell interactions that has emerged only during the last decade the amazing way in which bacteria manipulate normal host cell functions for their own ends.

This book seeks to involve the reader and utilises a number of teach-ing aids such as 'Chapter outline', 'Aims', 'Concept check' and 'What's next?' in each chapter to allow the reader to see where s/he is going and what s/he has covered having been there. Each chapter includes a sprinkling of 'micro-asides' to illustrate what the writers consider to be particularly important or fascinating aspects of the subject and also contains a number of questions to test the student in what s/he has learned. While the book deals with the general aspects of bacterial virulence and does not concentrate on any single organism, we have chosen two 'paradigm' organisms – *Escherichia coli* and *Streptococcus pyogenes* – and will describe them within the context of the individual chapters. An extensive reading list is included at the end of each chapter and includes appropriate books, review articles and some key papers. Finally, like it or loathe it, the Internet cannot be ignored in this day and age. We have, therefore, included the addresses of useful websites at the end of each chapter. One of the problems with many websites is that they are often set up with great enthusiasm but then abandoned soon after. We have, therefore, been very selective in our choice and have included only those sites that we believe are likely to remain active over the next few years.

We hope that you, the reader, will have as much fun with this book as we have had in writing it and wish you well as you explore one of the most fascinating stories ever written by evolution.

General reading

Cossart, P., Boquet, P., Normark, S. & Rappouli. R. (2000). *Cellular Microbiology*. Washington, DC: ASM Press.

Henderson, B., Wilson, M., McNab, R. & Lax, A. (1999). *Cellular Microbiology: Bacteria–Host Interactions in Health and Disease*. Chichester: John Wiley Ltd.

Salyers, A. A. & Whitt, D.D. (2001). *Bacterial Pathogenesis: A Molecular Approach*, 2nd edition. Washington, DC: ASM Press.

Abbreviations used

Bacterial genera

Actinobacillus	*A.*
Aeromonas	*Aer.*
Bacillus	*B.*
Bacteroides	*Bac.*
Bordetella	*Bord.*
Borrelia	*Bor.*
Brucella	*Br.*
Campylobacter	*Camp.*
Chlamydia	*Chl.*
Clostridium	*Cl.*
Corynebacterium	*C.*
Coxiella	*Cox.*
Enterococcus	*Ent.*
Epulopiscium	*Epul.*
Escherichia	*E.*
Francisella	*Fran.*
Haemophilus	*H.*
Helicobacter	*Hel.*
Klebsiella	*K.*
Lactobacillus	*Lac.*
Legionella	*Leg.*
Listeria	*Lis.*
Micrococcus	*Mic.*
Mycobacterium	*M.*
Mycoplasma	*Myc.*
Neisseria	*N.*
Pasteurella	*P.*
Peptostreptococcus	*Pep.*
Porphyromonas	*Por.*
Propionibacterium	*Prop.*
Pseudomonas	*Ps.*
Rickettsia	*R.*
Salmonella	*Sal.*
Shigella	*Shig.*
Staphylococcus	*Staph.*
Streptococcus	*Strep.*
Treponema	*T.*
Vibrio	*V.*
Wolbachia	*W.*
Yersinia	*Y.*

Other abbreviations

A/E	attaching and effacing
ABC	ATP binding cassette
AHL	acyl-homoserine lactone
AI	auto-inducer
AIDS	acquired immunodeficiency syndrome
ail	attachment–invasion locus
APAF-1	apoptotic protease-activating factor-1
APC	antigen-presenting cell
ATP	adenosine triphosphate
Avr	avirulence protein
Bax	Bcl-2 associated protein
Bcl	B cell lymphoma proto-oncogene product
bfp	bundle-forming pili
BMEC	brain microvascular endothelial cell
BoNT	botulinum neurotoxin
bp	base-pair(s)
BPI	bacterial permeability-inducing protein
bvg	*Bordetella* virulence gene
C4BP	C4-binding protein
cag	cytotoxin-associated gene (in *Helicobacter pylori*)
CAK	Cdc2-activating kinase
c-AMP	cyclic adenosine monophosphate
CAP	cationic antimicrobial protein
CAPK	ceramide-activated protein kinase
CAPP	ceramide-activated protein phosphatase
CCPs	complement control protein repeats
CD	cluster of differentiation
CDEC	cell-detaching *Escherichia coli*
Cdk	cyclin-dependent kinase
cDNA	complementary DNA
CDT	cytolethal distending toxin
CEACAM	carcinoembryonic-antigen-related cellular adhesion molecule
CF	cystic fibrosis
CFTR	cystic fibrosis transmembrane conductance regulator
CFA	colonisation factor antigen
CFT	channel-forming toxin
CFTR	cystic fibrosis transmembrane conductance regulator
CFU	colony-forming units
CGD	chronic granulomatous disease
c-GMP	cyclic guanosine monophosphate
CHD	coronary heart disease
CHO	Chinese hamster ovary
CIP	Cdk inhibitory protein

CKIs	cyclin kinase inhibitors
CLDTEC	cytolethal distending toxin-producing *Escherichia coli*
CLIP	class II-associated invariant chain peptide
CMR	comprehensive microbial resource
CNF	cytotoxic necrotising factor
COGs	clusters of orthologous groups
ConA	concanavalin A
cox	cyclo-oxygenase gene
CR	complement receptor
CRE	c-AMP response element
CREB	CRE-binding
CRP	C-reactive protein
CSF	colony-stimulating factor
csp	control of signal production gene
Csr	capsule synthesis regulator
CT	cholera toxin
DAF	decay-accelerating factor
DAG	diacylglycerol
DALPC	direct analysis of large protein complexes
DC	dendritic cell
DcR	decoy receptor
DD	death domain
DED	death effector domain
DFI	differential fluorescence induction
DICE	disseminated insertion of class I epitopes
DISC	death-inducing signalling complex
DNT	dermonecrotic toxin
DR3	death receptor 3
DT	diphtheria toxin
DTH	delayed-type hypersensitivity
EAEC	enteroaggregative *Escherichia coli*
EAST	EAEC heat-stable enterotoxin
EB	elementary body
ECM	extracellular matrix
ECV	endosomal carrier vesicle
EF	oedema factor
EF2	elongation factor 2
EGF	epidermal growth factor
EHEC	enterohaemorrhagic *Escherichia coli*
EIEC	enteroinvasive *Escherichia coli*
EP	endogenous pyrogen
EPEC	enteropathogenic *Escherichia coli*
ER	endoplasmic reticulum
ERAD	endoplasmic reticulum-associated protein degradation
ERK	extracellular signal-regulated kinase (a synonym for MAP)
ES	embryonic stem cell

ESAT-6	6 kDa early secretory antigenic target
ESI	electronspray ionisation
EspB	EPEC-secreted protein B
EST	expressed signature tag
ETEC	enterotoxigenic *Escherichia coli*
ExoA	exotoxin A
ExoS	exoenzyme S
ExoT	exoenzyme T
FAB	fast atom bombardment
FACS	fluorescence-activated cell sorter
FADD	Fas-associated death domain
FAE	follicle-associated epithelium
FAK	focal adhesion kinase
FAS	fatty acid synthase
FasL	Fas ligand
FDA	Food and Drug Administration
ffh	fifty-four homologue
FGF	fibroblast growth factor
FGFR	fibroblast growth factor receptor
FH	factor H
FHL	factor H-like protein
fM	femtomolar (10^{-15} M)
FMLP	*N*-formyl-methionyl-leucyl-phenylalanine
fucP	L-fucose permease gene
GALT	gut-associated lymphoid tissue
GAMBIT	genomic analysis and mapping by *in vitro* transposition
GAP	GTPase-activating protein
GAPDH	glyceraldehyde-3-phosphate dehydrogenase
GCF	gingival crevicular fluid
G-CSF	granulocyte colony-stimulating factor
GDI	GDP dissociation inhibitor
GEF	guanine nucleotide exchange factor
GFP	green fluorescent protein
GM-CSF	granulocyte-macrophage colony-stimulating factor
GOD	generation of diversity
GPCR	G protein-coupled receptor
GPI	glycosylphosphatidylinositol
GSP	general secretory pathway
GTP	guanosine triphosphate
HBMEC	human brain microvascular endothelial cell
HC	heavy chain
HEV	high endothelial venule
HFF	human foreskin fibroblast
HGF	human gingival fibroblast
HHV	human herpes virus
hil	hyperinvasive locus
HIV	human immunodeficiency virus

HLA	human leukocyte antigen
HPLC	high pressure liquid chromatography
Hsp	heat shock protein
HUS	haemolytic uraemic syndrome
HUVEC	human umbilical vascular endothelial cell
ICAM	intercellular adhesion molecule
ICAT	isotope-coded affinity tag
ICE	pro-IL-1β converting enzyme
ID$_{50}$	dose infective in 50% of experimental animals
IEL	intra-epithelial lymphocyte
IFN	interferon
IFN$_\gamma$	interferon γ
Ig	immunoglobulin
IKK	IκB kinase
IL	interleukin
IL-1	interleukin-1
IL-1RI	type I IL-1 receptor
IL-1RII	type II IL-1 receptor
IL-1ra	IL-1 receptor antagonist
IL-1RacP	IL-1 receptor accessory protein
InlA	internalin
iNOS	inducible nitric oxide synthase
IP$_3$	inositol 1,4,5-trisphosphate
ipa	invasion plasmid antigen gene
ipg	invasion plasmid gene
IRAK	IL-1 receptor-associated kinase
IS	insertion sequence
IVET	*in vivo* expression technology
IVIAT	*in vivo*-induced antigen technology
IVIF	*in vivo* induction of fluorescence
JAK	Janus kinase
JNK	c-Jun N-terminal kinase
kb	kilobase-pair(s)
KDO	2-keto-3-deoxyoctulonic acid
Ksr	kinase suppressor of Ras
LAM	lipoarabinomannan
LAMP	lysosome-associated membrane protein
LBP	LPS-binding protein
LC	light chain
LD$_{50}$	lethal dose in 50% of experimental animals
LEE	locus of enterocyte effacement
LEM	leukocyte endogenous mediator
Leu-Cam	leukocyte cell adhesion molecule
LF	lethal factor
LFA	lymphocyte-function-associated antigen
Lgp	lysosomal glycoprotein
LIF	leukaemia-inhibitory factor
LJP	localised juvenile periodontitis
LMP	low molecular mass polypeptide

LPS	lipopolysaccharide
LT	heat-labile toxin
LTA	lipoteichoic acid
mAb	monoclonal antibody
MAC	membrane attack complex
MADD	MAP kinase-activating death domain
MALDI	matrix-assisted laser desorption ionisation
MALT	mucosa-associated lymphoid tissue
MAMP	microbe-associated molecular pattern
MAP	mitogen-activated protein (a synonym for ERK)
MAPK	mitogen-activated protein kinase
MASP	mannose binding protein-associated serine protease
Mb	megabase-pairs
MBL	mannose-binding lectin
MBP	mannose-binding protein
MCP	monocyte chemotactic protein (Chapters 4 and 8)
MCP	membrane cofactor protein (Chapters 6 and 10)
MCSF	macrophage colony-stimulating factor
MEK	this acronym comes from a combination of MAP and ERK
MHC	major histocompatibility complex
mig	macrophage-induced gene
MIP	macrophage inflammatory protein
MOMP	major outer membrane protein
MP-R	mannose 6-phosphate receptor
mRNA	messenger ribonucleic acid
MS	mass spectrometry
MSCRAMMs	microbial surface components recognising adhesive matrix molecules
msr	methionine sulphoxide reductase gene
MTB	main terminal branch
mxi-spa	membrane expression of Ipas-surface presentation of antigens locus
MyD88	myeloid differentiation protein
NAK	NF-κB-activating kinase
NCBI	National Center for Biotechnology Information
NF	nuclear factor
NGFR	nerve growth factor receptor
NIK	NF-κB-inducing kinase
NK	natural killer
nM	nanomolar (10^{-9} M)
NO	nitric oxide
Nramp	natural resistance-associated macrophage protein
NTEC	necrotoxic *Escherichia coli*
OHHL	*N*-3-oxohexanoyl-L-homoserine lactone

OMP	outer membrane protein
Opa	opacity-associated protein
Opp	oligopeptide permease
org	oxygen-related gene
ORF	open reading frame
oriC	origin of replication
OSM	oncostatin M
PA	protective antigen
PAS	the *Drosophila* period clock protein, the vertebrate aryl hydrocarbon-receptor nuclear translocator and the *Drosophila* single-minded protein
PAF	platelet-activating factor
PAGE	polyacrylamide gel electrophoresis
PAI	pathogenicity island
PAMP	pathogen-associated molecular pattern
PAP	pylonephritis-associated pilus
PCD	programmed cell death
PCR	polymerase chain reaction
PCV	post-capillary venule
PDE	phosphodiesterase
PDGF	platelet-derived growth factor
PE proteins	a family of acidic, glycine-rich mycobacterial proteins containing proline-glutamate (i.e. PE) near the N-terminus
PEEC	pathogen-elicited epithelial chemoattractant
PG	prostaglandin
PHA	phytohaemagglutinin
PHD	pleckstrin homology domain
PI	phosphatidylinositol
PI3K	phosphoinositide 3-kinase
PIP_2	phosphatidylinositol 4,5-bisphosphate
PI-PLC	phosphatidylinositol-specific phospholipase C
PKA	protein kinase A
PKC	protein kinase(s) C
PLA_2	phospholipase A_2
PLC	phospholipase C
PLD	phospholipase D
PMN	polymorphonuclear neutrophil
PMT	*Pasteurella multocida* toxin
PPE proteins	a family of acidic, glycine-rich mycobacterial proteins containing proline-proline-glutamate (i.e. PPE) near the N-terminus
prg	*phoP*-repressed gene
PRR	pattern recognition receptor
PS	phosphatidylserine
PT	pertussis toxin
PTK	protein tyrosine kinase
ptl	pertussis toxin liberation gene

RAG	recombination-activating genes
RAIDD	RIP-associated ICH-1/CED-3-homologous protein with a death domain
RAP-PCR	RNA arbitrary primed PCR
RB	reticulate body
RCA	regulators of complement activation
RIP	receptor-interacting protein
RIVET	recombinase-based IVET system
RNS/RNI	reactive nitrogen species/intermediates
ROS/ROI	reactive oxygen species/intermediates
Rpf	resuscitation promoting factor
rRNA	ribosomal RNA
RT	reverse transcriptase
RTK	receptor tyrosine kinase
RTX	repeat in toxin
SAA	serum amyloid A
SAgs	superantigens
SALT	skin-associated lymphoid tissue
SCID-hu	humanised severe combined immunodeficiency
SCR	short consensus repeat
SDH	surface dehydrogenase
SDS-PAGE	sodium dodecyl sulphate-polyacrylamide gel electrophoresis
SEA	staphylococcal enterotoxin A
SEB	staphylococcal enterotoxin B
Sec	secretory protein
SfaS	S fimbria adhesin S
Sfbl	streptococcal fibronectin-binding protein I
SH2	src homology region
Sif	*Salmonella*-induced filament
Sip	*Salmonella* invasion protein
sir	*Salmonella* invasion regulator locus
SLA	streptolysin A
SLO	streptolysin O
SLPI	secretory leukocyte protease inhibitor
SLS	streptolysin S
SLT	Shiga-like toxin
SOD	superoxide dismutase
SODD	silencer of death domains
Sop	*Salmonella* outer membrane protein
SOS	son of sevenless
sp.	species (singular)
Spa	streptococcal protective antigen
SP-A and D	surfactant proteins A and D
SPE	streptococcal pyrogenic exotoxin
SPI	*Salmonella* pathogenicity island
spp.	species (plural)
SptP	secreted protein tyrosine phosphatase
SR	scavenger receptor

SRP	signal recognition particle
sst1	susceptibility to tuberculosis 1 gene
ST	stable toxin
STATs	signal transducers and activators of transcription
STEC	Shiga-toxin producing *Escherichia coli*
STI	*Salmonella typhimurium* inhibitor
STM	signature-tagged mutagenesis
Stx	Shiga toxin
TACO	tryptophan aspartate-containing coat protein
TAP	transporter associated with antigen processing
Tat	twin-arginine translocation
TCP	toxin co-regulated pilus
TCR	T cell receptor
TCSTS	two-component signal transduction system
TeNT	tetanus neurotoxin
TFF	trefoil factor family
TGF	transforming growth factor
Th	T helper cell
TI	thymus-independent
TIGR	The Institute for Genomic Research
TIR	Toll/IL-1R
Tir	translocated intimin receptor
TLR	Toll-like receptor
TNF	tumour necrosis factor
TNFR	tumour necrosis factor receptor
TOF-MS	time-of-flight mass spectrometry
TRADD	TNF receptor-associated death domain
TRAF	TNF receptor-associated factor
TRAIL	TNF-related apoptosis-inducing ligand
TRAMP	TNFR-related apoptosis-mediating protein
tRNA	transfer RNA
TSH	thyroid-stimulating hormone
TSS	toxic shock syndrome
TSST	toxic shock syndrome toxin
Tus	termination utilisation substance
UEA	*Ulex europeaeus* agglutinin
UTI	urinary tract infection
VASP	vasodilator-stimulated phosphoprotein
VCAM	vascular cell adhesion molecule
VLA	very late antigen
vMIP	viral macrophage inflammatory protein
VPI	*Vibrio cholerae* pathogenicity island
VSG	variant surface glycoprotein
WASP	Wiskott–Aldrich syndrome protein
WWW	World Wide Web
yad	*Yersinia* adherence gene
Yop	Yersinia outer membrane protein
Ysc	Yop secretion protein

An introduction to bacterial diseases

Chapter outline

Aims

The principal aims of this chapter are:

- to describe the normal microflora of humans
- to discuss the role of the normal microflora in health and disease
- to describe the types of interaction that exist between humans and bacteria
- to outline some of the diseases caused by members of the normal microflora
- to give examples of diseases caused by exogenous pathogens
- to describe a classification system for infectious diseases based on the presence or absence of key virulence factors of the causative organism
- to introduce the paradigm organisms – *Streptococcus pyogenes* and *Escherichia coli*

1.1 | Introduction

Human beings display a tremendous range of attitudes to bacteria, ranging from a paranoid loathing accompanied by an ultimately futile determination to extinguish all 'germs' from the immediate environment to an admiring appreciation of their tremendously important role in the functioning of the biosphere. While a well-educated biologist will, one hopes, tend towards the latter attitude, when it comes to a consideration of infectious diseases, even s/he will often exhibit a response (bacteriophobia) more characteristic of the former. This retrenchment to a 'them' against 'us' confrontational attitude, while understandable, is not particularly helpful in trying to unravel the complexities of the origin of, and the mechanisms underlying, infectious diseases. What many people fail to realise is that, when it comes to viewing the bacterial world, it is not a case of 'us' and 'them' because we are, in fact, more 'them' than 'us'. Hence, we do not exist as germ-free (i.e. 'gnotobiotic') animals, but each of us is an ecosystem comprising approximately 10^{14} microbes (from more than 1000 different species, of which most are bacteria) but only 10^{13} mammalian cells. For most of our lives, the mammalian component of this association lives in harmony with its microbial 'burden', which is known as the 'normal microflora'. Some of our closest friends, therefore, are microbes. However, many of these organisms can, under certain circumstances (to be described later), turn against their host and induce an infectious process (known as an 'endogenous' or 'opportunistic' infection). Furthermore, some bacteria that are not normally members of the normal microflora can also cause disease (these are known as 'exogenous' infections) and such organisms include the familiar 'pathogens' of humans such as *Salmonella typhi*, *Vibrio cholerae*, *Neisseria gonorrhoeae*, *Bordetella pertussis* and *Mycobacterium tuberculosis*. Examples of the causative organisms of important endogenous and exogenous infections, together with their sites of action, are given in Figures 1.1 and 1.2, respectively.

While much of the emphasis in this book will be on diseases caused by exogenous pathogens, it is appropriate, first of all, to learn more about our close friends in the microbial world – the normal microflora.

1.2 | The normal bacterial flora of humans

Only during the first nine months of our existence (i.e. *in utero*) are humans composed entirely of mammalian cells. Once we leave the protection of our mother's womb we are exposed to microbes and begin a life-long association with our normal microflora. Ultimately, of course, the microbes take over completely as, once we have died, our normal microflora rapidly takes advantage of the collapse of our defence systems (which are described in Chapters 5 and 6) and avails itself of the resulting enormous supply of nutrients.

The normal microflora consists of those microbes (mainly bacteria, but it also includes fungi, viruses and protozoans) that are present on body surfaces exposed to the external environment, i.e. the skin, oral cavity, respiratory tract, gastrointestinal tract, vagina and urinary tract

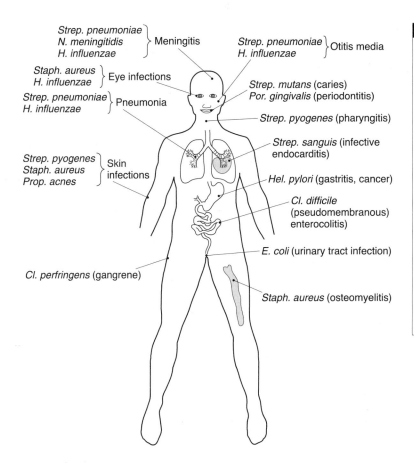

Figure 1.1 Some of the bacteria that are able to cause infections at the anatomical sites indicated. As all of these organisms can usually, or often, be isolated from healthy individuals, the diseases are regarded as being 'endogenous' or 'opportunistic' infections due to 'opportunistic' pathogens. In most cases, the organisms cause disease at the anatomical site that they usually inhabit. However, some members of the normal microflora cause disease when they gain access to a site that is normally sterile. For example, *Streptococcus sanguis* may cause infective endocarditis when it gains access to the bloodstream. For genus abbreviations, see list on p. xxi.

Beneficial aspects of the normal microflora

Some of the great 19th century pioneers of bacteriology and immunology, such as Louis Pasteur and Élie Metchnikoff, were firm believers that the normal microflora was essential for the well-being of humankind. This very enlightened attitude is surprising in view of the fact that the existence, and all-pervading presence, of invisible, disease-inducing organisms had only recently been discovered. To regard these newly discovered organisms as potential allies certainly required considerable insight. The normal microflora is now recognised as being beneficial to the host from several points of view. In the first instance, these organisms bind to specific receptors on host cells (see Section 7.3), so blocking these receptors and preventing them from serving as sites of attachment for exogenous pathogenic organisms – if the latter cannot adhere to the host then, generally, they will be expelled from the host's body. As well as physically preventing adhesion of pathogens, members of the normal microflora often produce antimicrobial factors that will help to kill, or limit the growth of, pathogenic organisms. The complex interactions between members of the oral microflora involving food webs, antagonistic agents, competition for nutrients, etc. also serve to control the relative proportions of the constituent organisms, so preventing the 'overgrowth' of species with disease-inducing potential. The consequences of disrupting the normal microflora are dramatically illustrated by the ills that can befall patients given broad-spectrum antibiotics or antibiotic combinations. These include enterocolitis due to the overgrowth of *Clostridium difficile* in the colon, candidosis due to proliferation of *Candida albicans* (a yeast) in the mouth or vagina and infections with environmental organisms such as *Pseudomonas aeruginosa*. As well as serving a protective role, the normal microflora also carries out a range of biochemical reactions that benefit the host. Intestinal organisms, for example, produce proteolytic enzymes that can break down proteins in food, thereby aiding digestion. They can also break down bile acids to products that are important in the emulsification of fats. Furthermore, a whole range of intestinal species produce vitamin K, which is needed for the synthesis of prothrombin, an enzyme involved in blood clotting. More recently, it has been established that the intestinal microflora plays an important role in the development of the intestinal epithelium and the gut-associated lymphoid tissue (see Chapters 5 and 11).

(Figure 1.3). Although the term 'normal microflora' is well established, its use is questioned by many who point out that 'normal' is very difficult to define and that 'flora' is inappropriate as microbes are not plants! The term 'indigenous biota' has, therefore, been proposed as a replacement. As pointed out previously, there are enormous

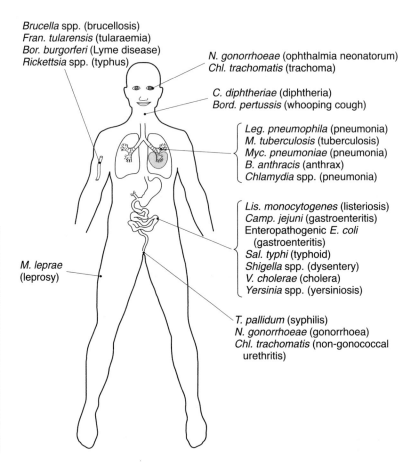

Brucella spp. (brucellosis)
Fran. tularensis (tularaemia)
Bor. burgorferi (Lyme disease)
Rickettsia spp. (typhus)

N. gonorrhoeae (ophthalmia neonatorum)
Chl. trachomatis (trachoma)

C. diphtheriae (diphtheria)
Bord. pertussis (whooping cough)

Leg. pneumophila (pneumonia)
M. tuberculosis (tuberculosis)
Myc. pneumoniae (pneumonia)
B. anthracis (anthrax)
Chlamydia spp. (pneumonia)

Lis. monocytogenes (listeriosis)
Camp. jejuni (gastroenteritis)
Enteropathogenic *E. coli*
  (gastroenteritis)
Sal. typhi (typhoid)
Shigella spp. (dysentery)
V. cholerae (cholera)
Yersinia spp. (yersiniosis)

M. leprae
(leprosy)

T. pallidum (syphilis)
N. gonorrhoeae (gonorrhoea)
Chl. trachomatis (non-gonococcal
  urethritis)

Figure 1.2 Some of the bacteria that are able to cause infections at the sites indicated. As these organisms are rarely present in healthy individuals, the diseases are considered to be 'exogenous' infections due to 'classical' pathogens. Although most of the organisms have been designated as causing an infection at a particular anatomical site (usually the site of entry), many of them become disseminated throughout the body. For genus abbreviations, see list on p. xxi.

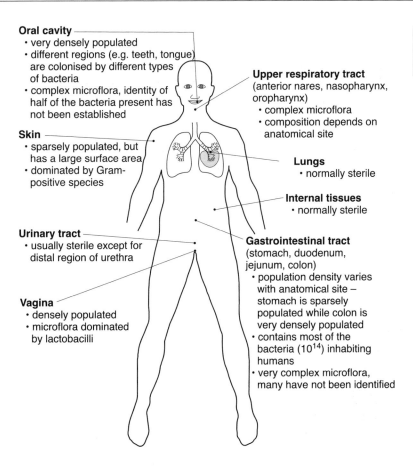

Oral cavity
- very densely populated
- different regions (e.g. teeth, tongue) are colonised by different types of bacteria
- complex microflora, identity of half of the bacteria present has not been established

Skin
- sparsely populated, but has a large surface area
- dominated by Gram-positive species

Urinary tract
- usually sterile except for distal region of urethra

Vagina
- densely populated
- microflora dominated by lactobacilli

Upper respiratory tract (anterior nares, nasopharynx, oropharynx)
- complex microflora
- composition depends on anatomical site

Lungs
- normally sterile

Internal tissues
- normally sterile

Gastrointestinal tract (stomach, duodenum, jejunum, colon)
- population density varies with anatomical site – stomach is sparsely populated while colon is very densely populated
- contains most of the bacteria (10^{14}) inhabiting humans
- very complex microflora, many have not been identified

Figure 1.3 Diagram showing, in general terms, the distribution of bacteria at different body sites.

numbers of bacteria inhabiting these surfaces; they outnumber mammalian cells by a factor of 10 and comprise an appreciable proportion of the total mass of our bodies (Table 1.1).

While our knowledge of the identity of the bacteria present on our body surfaces is considerable, we know very little about the interactions that occur between these organisms and the cells and tissues to which they are attached. Important questions to be addressed include: 'How is a healthy balance between microbes and host cells

Table 1.1. Mass of the microflora associated with various body sites

Organ	Associated microflora (grams wet weight)
Eyes	1
Nose	10
Mouth	20
Upper respiratory tract	20
Vagina	20
Skin	200
Intestines	1000

maintained?' and 'How do these very different types of cell communicate?' Unfortunately, relatively little research has been conducted in these areas – what is known will be described in Section 4.4.

The types of bacteria found associated with an individual vary enormously from site to site within that individual so that it is essential to speak in terms of the normal microflora of a particular anatomical region, for example the skin, oral cavity, etc. Even within a particular region, for example the oral cavity, there are large differences between the types of bacteria found at particular locations such as the teeth, the tongue, the gums, etc. These variations arise as a result of the differing environmental selection factors operating at individual sites (Table 1.2). Hence the particular chemical, physical, biological and mechanical factors at the site combine to produce a unique environment that limits the types of bacteria that can survive and grow at that particular location. Variations between individuals with regard to these factors complicate any attempt to define what constitutes the normal microflora of a particular anatomical region. For example, it is well established that the diet, hygiene practices, gender, age, etc. of an individual can all markedly affect the types of bacteria to be found at a particular anatomical site, particularly within the gastrointestinal tract. The task of defining the normal microflora is rendered even more difficult because most bacteriological research has focused on disease-inducing bacteria rather than the 'harmless' species with which we are colonised, so less is known about the latter. Furthermore, these microbial communities are extremely complex and this leads to problems with the isolation and identification of the species present. It has also

Table 1.2.	Environmental selection factors that influence the survival of a particular organism within a particular habitat
Factor	**Examples**
Chemical	Types and quantities of macronutrients available
	Presence of essential vitamins and/or growth factors
	Presence of trace elements
	Composition of the gaseous phase (O_2, CO_2, etc.)
	pH
Physical	Temperature
	Redox potential
	Osmotic pressure
	Hydrostatic pressure
Biological	Host secretions
	Host defence systems
	Presence of other bacteria (that could produce growth inhibitors or whose waste products could serve as nutrients)
Mechanical	Mechanical and fluid shear forces

been estimated that fewer than half of the microbes present in the gut, oral cavity and skin have ever been cultivated in the laboratory. The application of molecular techniques to ascertaining which organisms are present in a particular habitat is revealing the considerable diversity of the normal microflora. Although only a limited number of studies have been carried out to date, the results of these suggest that the number of different bacterial species inhabiting humans is likely to be of the order of 1000. Bearing in mind the limitations discussed, this section will summarise what is known of the nature of the bacteria with which we, usually, live in harmony. For each anatomical site, a brief description of the major environmental selection factors operating there will be given in order to provide some insight into why particular organisms are to be found there.

1.2.1 The skin

The skin is a readily accessible region for bacterial colonisation and comes into regular contact with a large number, and variety, of bacteria from the environment and from other anatomical sites, for example the respiratory and gastrointestinal tracts. Many of these (known as 'transients') cannot survive for long on the skin surface while others ('residents') are able to grow and multiply there and become established members of the community. The skin is rather sparsely populated in comparison with other body surfaces and the reasons for this will be described later. However, because of its large surface area (approximately $2\,m^2$), the total number of skin bacteria is considerable and amounts to approximately 10^{12} per person. So what is the nature of the environment offered to bacteria attempting to colonise the skin? It must be recognised that the various anatomical sites provide very different environments. As a result of the distribution of hair and sebaceous glands, as well as the gross anatomy, both the moisture content and temperature vary enormously between sites. For example, the armpits, which are enclosed, hairy regions with many sebaceous glands, provide a warmer, more moist environment than the back. The armpits, therefore, support a denser population (several million per square centimetre) than the back (several hundred per square centimetre). Even within these distinct sites, the distribution of bacteria shows considerable variation. Hence, the actual surface of the skin (the epidermis) is generally dry and so supports fewer organisms than do the moister sweat glands and hair follicles (see Figure 5.4). The principal sources of nutrients for skin bacteria are sweat and sebum.

Sweat consists mainly of Na^+ (0.13%), Cl^- and K^+, together with small quantities of fatty acids, steroids and vitamins (mainly B_1 and C), and the average adult produces approximately 500 ml/day. Sebum consists of a mixture of the esters of a range of fatty acids. The ways in which both of these secretions can act to reduce or prevent bacterial colonisation (as well as other skin-associated antibacterial mechanisms) are described in Section 5.4.

Bacteria and under-arm odour

Bacteria are responsible for many of the unpleasant odours emanating from the oral cavity, the gastrointestinal tract and various regions of the skin. With regard to the skin, the most powerful and distinctive odours are those produced by the axillae. In these regions, bacteria metabolise steroidal components of sweat (mainly 16-androstene compounds such as androstadienone and androstadienol) to a mixture of 3α- and 3β-androstenols, which have urine- and musk-like odours and, together, are suggestive of under-arm odour. The microflora of the axillae is dominated by *Corynebacterium* spp., *Propionibacterium* spp., *Micrococcus* spp. and *Staphylococcus* spp. Of these, *Corynebacterium* spp., particularly *Corynebacterium xerosis*, appear to be mainly responsible for odour generation. Hence, in a study involving 34 adult males, those with the highest odour intensity had microfloras dominated by *Corynebacterium* spp., whereas those with a low odour intensity had lower proportions of these organisms, but higher proportions of *Staphylococcus* spp. Furthermore, of the bacteria isolated, only the *Corynebacterium* spp. were capable of odour production when incubated with material obtained (by swabbing with ether) from the axillae of the volunteers.

Skin shedders

It has been estimated that the average person sheds approximately 5×10^8 skin scales (squames) per day, of which approximately 10% have bacteria attached to them. Not all regions of the skin contribute equally to the shedding process – the nose and the perineal area are particularly active in this respect. As the latter two areas are the preferred habitats of *Staphylococcus aureus*, the shedding of squames constitutes an effective means of disseminating this organism – particularly by males, who have been shown to shed more staphylococci than females. Indeed, the number of staphylococci in the air can be increased 10-fold (to approximately $360\,\text{cells}/\text{m}^3$) by a 'good' skin shedder. The squames are dispersed into the air when the individual moves or undresses and when contaminated articles such as clothes or bedding are disturbed. In hospitals, transmission of *Staph. aureus* via squames may be a major route by which other patients become infected. The number of squames in the air may be reduced by dusting with damp cloths and by preventing air circulating between infected and susceptible patients.

Bacteria on the skin have ready access to oxygen so that most of the indigenous species are either obligate aerobes such as *Micrococcus* spp., or facultative anaerobes including staphylococci and *Corynebacterium* spp. However, more anaerobic conditions prevail within hair follicles and the ducts of sebaceous glands and this enables the survival of anaerobic organisms such as *Propionibacterium* spp. and *Peptostreptococcus* spp. Successful colonisers of the skin ought, therefore, to have the following characteristics: be aerobic or facultatively anaerobic (or anaerobic, if inhabiting hair follicles or sebaceous glands), be able to adhere to keratinised epithelial cells, be able to utilise lipids as a carbon and energy source and be able to tolerate high salt concentrations. In general, the skin microflora is dominated by bacteria with these characteristics – the predominant organisms being species belonging to the genera *Staphylococcus*, *Micrococcus*, *Propionibacterium* and *Corynebacterium*. These are all Gram-positive organisms. The only Gram-negative organism regularly found on the skin is the obligate aerobe *Acinetobacter calcoaceticus*.

Several members of the normal skin microflora are able to induce disease under certain circumstances. The most important of these are *Staph. aureus* (which is responsible for boils, wound infections, food poisoning), *Staph. epidermidis* (infective endocarditis, infections of prosthetic devices) and *Propionibacterium acnes* (acne).

1.2.2 The oral cavity

The oral cavity contains such a varied set of habitats, each with its own characteristic microflora, that it is meaningless to speak of a 'normal oral microflora' without specifying which particular oral habitat (Figure 1.4). More than 500 different bacterial species have been isolated from the human oral cavity and it is likely that a similar number of, as yet, uncultivated species may be present. The total number of bacteria in the oral cavity is approximately 10^{10}. The main habitats are the teeth, the buccal mucosa, the tongue and the gingival crevice (i.e. the gap between a tooth and the adjacent gums). Each of these provides an environment differing in terms of nutrients, oxygen content, redox potential and pH. One of the unique features of the oral cavity is that it contains the only non-shedding surfaces in the body, teeth, thus enabling bacteria to accumulate to form dense aggregates (or biofilms, see Sections 2.7 and 7.2) known as dental plaques. These biofilms typically contain as many as 10^{11} bacteria per gram wet weight. Bacteria in the oral cavity are subjected to powerful removal forces, including the constant flow of saliva, swallowing, tongue movements and chewing. Consequently, the ability to adhere to oral surfaces (or to already

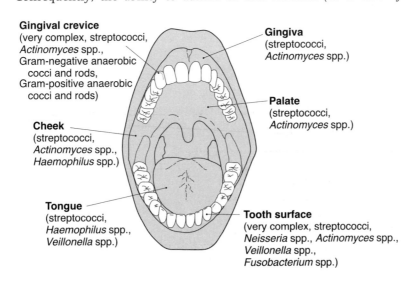

Gingival crevice
(very complex, streptococci, *Actinomyces* spp., Gram-negative anaerobic cocci and rods, Gram-positive anaerobic cocci and rods)

Gingiva
(streptococci, *Actinomyces* spp.)

Palate
(streptococci, *Actinomyces* spp.)

Cheek
(streptococci, *Actinomyces* spp., *Haemophilus* spp.)

Tongue
(streptococci, *Haemophilus* spp., *Veillonella* spp.)

Tooth surface
(very complex, streptococci, *Neisseria* spp., *Actinomyces* spp., *Veillonella* spp., *Fusobacterium* spp.)

Figure 1.4 The predominant microfloras of the main anatomical regions within the human oral cavity. This is a gross oversimplification of the situation *in vivo* as the microflora of each site (particularly the tooth surface, gingival crevice and tongue) is far more complex than that suggested by the diagram, which lists only those organisms generally predominating at each site.

adherent bacteria) is an essential prerequisite for an organism to colonise the oral cavity. While ostensibly an aerobic environment, most organisms in the oral cavity are facultative or obligate anaerobes, with the relative proportions of each varying with the nature of the microhabitat. The gingival crevice, for example, because of its low redox potential, restricted access to oxygen and plentiful supply of nutrients from a serum-like exudate (gingival crevicular fluid), harbours high proportions of Gram-negative obligate anaerobes such as *Fusobacterium* spp., *Veillonella* spp. and spirochaetes. In contrast, the microflora of a freshly colonised tooth surface is dominated by aerobic and facultatively anaerobic bacteria such as *Neisseria* spp. and streptococci.

The microflora of dental plaque is extremely complex, varies with the nature of the anatomical site and also changes with time. Nevertheless, some generalisations can be made. The numerically dominant organisms are invariably streptococci and *Actinomyces* spp., with members of the following genera usually being present in smaller numbers – *Veillonella*, *Haemophilus*, *Neisseria*, *Fusobacterium* and *Propionibacterium*. The genus *Streptococcus* comprises a group of facultatively anaerobic, Gram-positive, chain-forming cocci. Traditionally this genus is subdivided on the basis of their effects on blood-containing agar into three groups: non-haemolytic, β-haemolytic (complete destruction of red blood cells) and α-haemolytic (partial destruction of red blood cells accompanied by a green coloration). α-Haemolytic (also known as 'viridans' streptococci) and non-haemolytic species are the most prevalent streptococci in the oral cavity and these can give rise to endocarditis when they gain access to the bloodstream of individuals with damaged heart valves.

Saliva has a very high content of bacteria (approximately 10^8 per millilitre) but these are derived mainly from the teeth and oral mucosal surfaces as a result of mechanical abrasion caused by chewing, talking, swallowing, etc.

1.2.3 The respiratory tract

Because of the regular inhalation of air, the respiratory tract comes into regular contact with a large number of bacteria, either freely suspended or associated with particulate matter. Although it has been estimated that the average individual inhales approximately 10 000 bacteria per day, a number of mechanisms exist for reducing colonisation by these organisms. The first barrier, hairs in the anterior nares, serves to remove large particles. Any bacteria or particles overcoming this defence system are expelled by the 'mucociliary escalator' (see Section 5.4). This very effective cleansing mechanism depends on the existence of the ciliated epithelial cells and mucus-secreting cells that line the respiratory epithelium. The mucus-secreting cells provide a layer of mucus in which microbes become trapped. The mucus is continually propelled along by the cilia towards the larynx where, together with the entrapped microbes, it is either swallowed or coughed up. It has been estimated that the time required to clear inhaled particles from the bronchi is approximately 30 minutes. These systems serve to reduce microbial colonisation of the respiratory tract and maintain the lower regions (bronchi and bronchioles) almost microbe-free. Nevertheless, the anterior nares, nasopharynx and oropharynx have a characteristic microflora.

In order to establish themselves as part of the normal microflora, bacteria have to resist expulsion by the above mechanisms and must be able to adhere to the epithelium lining the respiratory tract, as well as overcome antibacterial factors present in mucus. The latter include lysozyme, lactoferrin (a protein that binds iron thereby limiting bacterial growth), secretory IgA antibodies (which prevent attachment of bacteria to epithelial cells), complement components that can kill bacteria (see Sections 6.4.2 and 6.6.1.3) and superoxide radicals generated by lactoperoxidase. Bacteria able to attach to the epithelium and survive the host defences listed above (the means by which bacteria evade such defences are described in Chapter 10) must then be able to grow and reproduce in this aerobic environment in which the only nutrients are those molecules secreted by host cells. The microflora is dominated by aerobic and facultatively anaerobic species such as streptococci, staphylococci, *Corynebacterium* spp. and Gram-negative cocci (Figure 1.5).

The anterior nares have an epithelium, and an associated microflora, essentially similar to those of the skin. The dominant organisms are *Staph. epidermidis*, *Micrococcus* spp. and *Corynebacterium* spp. The opportunistic pathogen *Staph. aureus* is usually present in 20–40% of the population.

The nasopharynx has a more complex microflora than that found in the anterior nares containing, in addition to staphylococci and corynebacteria, a number of Gram-negative species such as *Haemophilus influenzae*, *Moraxella catarrhalis* and *Neisseria* spp. Some strains of *H. influenzae* (those possessing particular types of capsule) may cause life-threatening infections such as meningitis, acute epiglotitis and pneumonia. Such strains are present in approximately 4% of the population. However, most strains are of low virulence and are generally associated with less serious diseases such as otitis media and sinusitis. While most of the *Neisseria* spp. found in the nasopharynx are unable to cause disease in healthy individuals, as many as 10% of the population harbour *Neisseria meningitidis*, which can cause a life-threatening meningitis.

The microflora of the oropharynx exhibits a greater diversity than that of the nasopharynx, possibly because of the availability of a wider range of nutrients from food ingested by the host. Streptococci (mainly α-haemolytic species) are the dominant organisms but the microflora also includes species belonging to the genera *Haemophilus*, *Neisseria*, *Mycoplasma* and *Moraxella*. As many as 10% of the population harbour *Streptococcus pyogenes*, a β-haemolytic organism (strains of which are also known as 'group A streptococci'). This is the most frequent bacterial cause of pharyngitis, which, in some cases, can be followed by

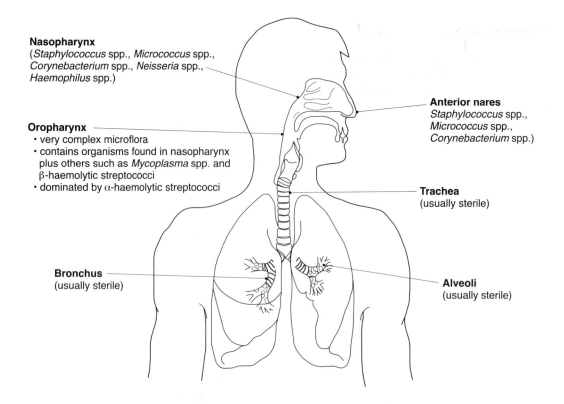

Nasopharynx
(*Staphylococcus* spp., *Micrococcus* spp., *Corynebacterium* spp., *Neisseria* spp., *Haemophilus* spp.)

Anterior nares
Staphylococcus spp.,
Micrococcus spp.,
Corynebacterium spp.)

Oropharynx
• very complex microflora
• contains organisms found in nasopharynx plus others such as *Mycoplasma* spp. and β-haemolytic streptococci
• dominated by α-haemolytic streptococci

Trachea
(usually sterile)

Bronchus
(usually sterile)

Alveoli
(usually sterile)

Figure 1.5 Predominant bacteria colonising the various regions of the human respiratory tract. While the upper regions have a complex microflora, the lower regions (trachea, bronchi and alveoli) are usually free from bacteria.

rheumatic fever or acute glomerulonephritis. It is also the causative agent of impetigo, cellulitis, lymphangitis and erysipelas. Up to 70% of the population may have *Streptococcus pneumoniae* (an α-haemolytic species) in their oropharynx and this is responsible for a wide range of infections, including pneumonia, meningitis, otitis media and sinusitis.

The lower respiratory tract (trachea, bronchi and alveoli) is usually sterile owing to the efficient removal of particles and microbes by specialised structures in the upper regions. Any organisms that do reach the alveoli are usually destroyed by phagocytic macrophages.

1.2.4 The gastrointestinal tract

The gastrointestinal tract contains most of the bacteria inhabiting humans (approximately 10^{14}, with a total mass of approximately 1000 g) and has a large surface area amounting to approximately 200 m^2. It consists of several distinct regions (Figure 1.6), which have in common a number of important environmental selection factors that differ markedly from those affecting the anatomical regions discussed so far. Therefore (i) as there is very little ingress of air, these ecosystems are predominantly anaerobic and have a low redox potential, (ii) bacteria inhabiting these regions are not dependent on host secretions for nutrients as there are an enormous variety of nutrients available from food ingested by the host, (iii) extremes of pH are

Parties are bad for your health – or are they?

In 1996 there was an outbreak of meningococcal meningitis in a secondary school in Wales. This consisted of two cases in a class of 15 to 16 year olds and one other case in a different class – the causative organism was a group B strain of N. meningitidis. In order to assess the risk factors for carriage of the organism, throat swabs were taken from most (744) of the 760 pupils at the school and cultured for the presence of N. meningitidis. At the same time, the pupils were asked to complete a questionnaire about various aspects of their home, lifestyle and behaviour. Neisseria meningitidis was recovered from 60 (7.9%) of the pupils, of whom 33 (55%) were in the same class as those who had contracted meningitis. The main factors associated with carriage of the organism were: increasing age (>14 years), smoking (both active and passive), receiving bad news, being in the same class as the two pupils with meningitis and attending the end-of-year party. When data concerning carriage of the particular meningococcal strain actually responsible for the two cases of meningitis were analysed, it was found that the four main factors governing carriage of this organism were being in the same class as the pupils who had contracted the disease, being older than 14, being male, and having attended the end-of-year party.

Do these findings support the views of most parents that their children should stay at home and study rather than go to parties? Not necessarily. Pupils who go to lots of parties and social events (particularly those who smoke) are more likely to carry meningococci of all types. This would increase their long-term risk of contracting meningitis but may protect them from infection by a virulent strain (by having 'primed' the body's defence systems, for example antibody production, against the organism) during an outbreak of the disease. In contrast, those who spend their nights studying quietly at home for most of the year and then attend the end-of-year party may be easy prey for a virulent strain of the meningococcus. Life certainly isn't fair!

encountered, particularly in the stomach, and (iv) the tract consists of a number of fluid-filled cavities so that the ability of an organism to adhere to the mucosal surface is not an essential prerequisite for it becoming established as a member of the normal microflora. There is some difference of opinion as to which of the organisms found in these cavities can be regarded as members of the microflora. Some argue that only those organisms attached to the mucosal surfaces are truly part of the microflora of these regions and any other organisms (i.e. those in the intestinal lumen) should be regarded as being transients. The latter spend only between 12 and 18 hours (the normal intestinal transit time) inside the host.

As well as having the normal antibacterial mechanisms associated with mucosal surfaces, the gastrointestinal tract also produces proteolytic enzymes and bile salts that exert an antibacterial effect.

Because of its acidity (the pH is as low as 2) and the presence of proteolytic enzymes, the stomach functions as a protective barrier to the ingress of bacteria into the rest of the intestinal tract. Very few bacteria (approximately 10^3 bacteria/ml) are found in the stomach contents and these are mainly members of the aciduric

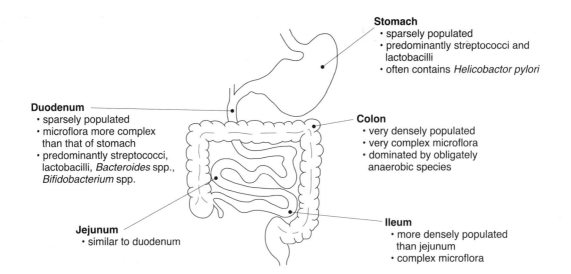

Stomach
- sparsely populated
- predominantly streptococci and lactobacilli
- often contains *Helicobactor pylori*

Duodenum
- sparsely populated
- microflora more complex than that of stomach
- predominantly streptococci, lactobacilli, *Bacteroides* spp., *Bifidobacterium* spp.

Colon
- very densely populated
- very complex microflora
- dominated by obligately anaerobic species

Jejunum
- similar to duodenum

Ileum
- more densely populated than jejunum
- complex microflora

Figure 1.6 Predominant microfloras of the main regions of the human gastrointestinal tract. In view of the fact that at least 500 bacterial species have been identified from the gastrointestinal tract (with the certainty that a great many more will be detected in the future), this diagram is a very simplified generalisation of the situation *in vivo*.

genera *Streptococcus* and *Lactobacillus* (Table 1.3). Some of these are also found attached to the gastric mucosa and so may be regarded as true residents rather than as transients. More controversial is the situation regarding *Helicobacter pylori*, a Gram-negative, spiral-shaped, motile bacterium that attaches itself to the mucosal wall or embeds itself in the mucus associated with the wall. This organism causes gastritis and peptic and duodenal ulcers but is also found in the stomach of healthy individuals and it has been reported that, by the age of 10 years, it is present in up to 80% of the population. Interestingly, *Hel. pylori* has been shown to produce antibacterial peptides similar in structure to cecropins – antibacterial peptides produced by mammalian cells (see Section 5.4.3.11). The organism is itself resistant to these peptides so that their production may help to eliminate other, competing, organisms from this habitat.

The duodenum and jejunum, which are still quite acidic (with a pH between 4 and 5), also have a sparse microflora (approximately 10^5 bacteria/ml), although it is more complex than that of the stomach. In addition to those organisms found in the stomach, members of the genera *Bacteroides* and *Bifidobacterium* are also present. The ileum, the next region of the small intestine, has a more substantial (up to 10^9 bacteria/ml) and diverse bacterial flora, consisting of *Lactobacillus* spp., *Bifidobacterium* spp., *Enterococcus faecalis*, *Bacteroides* spp., *Veillonella* spp., *Clostridium* spp. and Gram-negative facultative rods such as *Escherichia coli*.

Large numbers of bacteria are attached to the mucosal surface of the colon and are present in the lumen, where they constitute approximately 55% of the solids. The pH of this region is approximately neutral and very little oxygen is present. Nearly 500 bacterial species have been isolated from the colon, although only approximately 40 are regularly encountered, with one species, *Bacteroides vulgatus*, regularly constituting 10% of the microflora. Obligate anaerobes comprise >90% (approximately 10^{10} cells/g of intestinal contents) of the microflora,

Table 1.3. Normal microflora of the gastrointestinal tract

Anatomical region	Dominant environmental features	Predominant organisms found
Stomach	Very acidic, aerobic, pepsin (a protease), lipase proteolytic enzymes	Sparse microflora; *Streptococcus* spp., *Lactobacillus* spp., *Helicobacter pylori*
Duodenum jejunum	Acidic, little oxygen present, contains bile and a range of digestive enzymes (peptidases, nucleases, amylase, disaccharidases, lipase)	Sparse microflora; *Streptococcus* spp., *Lactobacillus* spp., *Bacteroides* spp., *Bifidobacterium* spp.
Ileum	pH almost neutral, anaerobic, contains bile and digestive enzymes	Sparse, but more diverse, microflora than in stomach, duodenum and jejunum: *Lactobacillus* spp., *Bifidobacterium* spp., *Enterococcus faecalis*, *Bacteroides* spp., *Veillonella* spp., *Clostridium* spp. and Gram-negative facultative rods such as *Escherichia coli*
Colon	pH almost neutral, anaerobic, very low redox potential, very little oxygen	Very dense and diverse microflora, at least 40 species regularly isolated: >90% of the microflora are obligate anaerobes; predominantly species of *Bacteroides*, *Eubacterium*, *Bifidobacterium*, *Peptostreptococcus*, *Fusobacterium* followed by *Ruminococcus*, *Streptococcus*, *Clostridium* and *Enterococcus*

with five genera accounting for most of these – *Bacteroides*, *Eubacterium*, *Bifidobacterium*, *Peptostreptococcus* and *Fusobacterium*. Other numerically important genera include *Ruminococcus*, *Streptococcus*, *Clostridium* and *Enterococcus*. Species belonging to the following genera are regularly isolated, although they comprise smaller proportions of the microflora: *Escherichia*, *Enterobacter*, *Proteus*, *Lactobacillus*, *Veillonella*.

Common inhabitants of the colon that are known to cause infections include *Cl. perfringens* (gas gangrene), *Bacteroides* spp. (peritonitis, intra-abdominal abscesses), *Cl. difficile* (pseudomembranous colitis), *E. coli* (diarrhoeal diseases, urinary tract infections, neonatal meningitis).

1.2.5 The urogenital tract

The urinary tract consists of the urethra and the bladder and, apart from the distal portion of the urethra, does not have a normal micro-flora as it is regularly flushed by sterile urine. Bacteria from the skin, anus and vagina often colonise the distal region of the urethra of females so that streptococci, coagulase-negative staphylococci, lactoba-cilli and *E. coli* are often found there.

Prior to puberty and after the menopause, vaginal secretions are alkaline and the vaginal microflora is dominated by staphylococci and streptococci. In contrast, between puberty and the menopause, the secretions are acidic (pH 4.7). This is the result of the fermentation of glycogen (by host cells or bacteria), which accumulates in epithelial cells due to the action of oestrogens. The low pH encourages colonisa-tion by aciduric species such as lactobacilli (particularly *Lactobacillus acidophilus*, *Lactobacillus fermentum*, *Lactobacillus casei* and *Lactobacillus cellobiosus*) that constitute the dominant members of the vaginal microflora. This is a good example of the way in which host factors (in this case oestrogens) can have a major effect on the composition of the normal microflora. A variety of other bacteria have also been isolated from vaginal secretions including staphylococci, corynebacteria, streptococci, *Bifidobacterium* spp., *Peptostreptococcus* spp., *Peptococcus* spp., *Eubacterium* spp., *Propionibacterium* spp. and anaerobic Gram-negative bacilli.

1.3 | Bacteria and disease

1.3.1 Types of host–bacteria interaction

The relationship between the prokaryotes and eukaryotes comprising *Homo sapiens* is harmonious and constitutes an example of mutualism, i.e. where both partners benefit and neither suffers. Hence the host provides a warm, moist, nutrient-rich environment for the bacteria, which in turn provide protection against colonisation by pathogenic species and also synthesise a range of vitamins (niacin, thiamine, riboflavin, vitamin K) that may be utilised by the host. In the case of many organisms inhabiting our body, however, we do not appear to derive any benefit. The bacteria, on the other hand, are provided with nutrients, moisture, etc. – such an association (while appearing rather unfair, although ultimately not harmful to us) is termed 'commensalism'. However, for a variety of reasons, these harmless relationships break down and can lead to a situation in which the prokaryotes gain at the expense of the eukaryotes: such an association is described as 'parasitism'. This constitutes what is often termed an endogenous or opportunistic infection and will be examined in more detail later in Section 1.3.2. Furthermore, there are certain bacteria with which we do not form such pleasant associations and when these organisms are encountered (they do not usually constitute part of our normal microflora) the result is a parasitic relationship in which the prokaryote (known as an 'exogenous pathogen') causes considerable damage to its host. This constitutes what most readers will recognise as the classical (exogenous) infectious disease (see Table 1.4). It must be remembered that the most successful parasites are not those that kill their host but those that can exploit the latter to ensure their continued carriage, viability and dissemination to other hosts. On one hand, a dead host presents the parasite with a huge problem – once this source of nutrients has been exhausted it must gain access to a new host. A successful parasite, on the other hand, will do minimal damage to its host (thereby ensuring a long-term supply of nutrients) and use its host's respiratory (coughing, exhalation of aerosols), excretory (faeces), reproductive (mucosa–mucosa contact) activities to optimise its chances of reaching other hosts. Before describing which organisms are responsible for the various infectious diseases of humankind, it is important to realise that it is often very difficult to establish the

Table 1.4.	Examples of exogenous pathogens and their natural reservoirs

Organism	Disease	Natural reservoir
Clostridium tetani	Tetanus	Soil
Legionella pneumophila	Pneumonia	Water
Borrelia burgdorferi	Lyme disease	Deer, mice, other rodents, birds
Bacillus anthracis	Anthrax	Animals, soil
Yersinia pestis	Plague	Rodents
Vibrio cholerae	Cholera	Humans, water
Salmonella typhi	Typhoid fever	Humans
Brucella spp.	Brucellosis	Cattle, dogs, pigs, goats
Treponema pallidum	Syphilis	Humans
Mycobacterium tuberculosis	Tuberculosis	Humans
Corynebacterium diphtheriae	Diphtheria	Humans
Rickettsia rickettsii	Rocky Mountain spotted fever	Ticks
Neisseria gonorrhoeae	Gonorrhoea	Humans
Shigella spp.	Dysentery	Humans
Campylobacter jejuni	Enteritis	Poultry
Listeria monocytogenes	Listeriosis	Domestic mammals, rodents, birds

aetiological agent of a particular disease. Fortunately, this task has been greatly aided by Robert Koch, who, in 1890, proposed a set of criteria necessary for establishing the causative agent of a disease (Table 1.5). These are known as Koch's postulates and they have proved invaluable in identifying the aetiological agents of many infectious diseases. However, there are a number of problems associated with applying these postulates. Firstly, not all pathogenic bacteria can be cultivated on artificial media, for example *M. leprae*, *Treponema pallidum* and all viruses. Secondly, for some diseases, for example meningococcal meningitis, there are no suitable animal models, as the organisms responsible are only pathogenic for humans. Technological advances, however, have enabled these problems to be circumvented, and a set of molecular guidelines now exists that can help to determine whether or not a particular organism is responsible for a disease (Table 1.6).

Regardless of the nature of either the disease or the organism responsible, several stages can be identified in the infectious process that eventually lead to a recognisable pathology in the host and each of these presents the organism with a series of, sometimes formidable, problems. Once the organism has come into contact with its host, it must establish itself at that site. This usually involves adhesion of the bacterium to a host cell or extracellular component (see Chapter 7). It must then avoid being destroyed, or expelled, by one or more of the many host defence mechanisms that will usually come into play once the organism has been 'detected' by the host (see Chapter 10). At the same time, it must obtain from its immediate environment all of the nutrients essential for its growth and be able to compete for these with

---

Table 1.5. Koch's postulates for determining whether or not an organism is the aetiological agent of a particular disease

---

- The organism should be found in all cases of the disease and should be present in the lesions of the disease
- It should be grown in pure culture on artificial media
- Inoculation of the pure culture into a susceptible animal should result in a similar disease
- It should be possible to recover the organism from the lesions of the diseased animal

---

Table 1.6. Molecular Koch's postulates

(i) The nucleic acid sequence of a putative pathogen should be present in most cases of the disease. The sequences should be found preferentially in those organs or sites known to be diseased but not in those sites that lack pathology

(ii) The nucleic acid sequences should be absent (or in lower copy number) in healthy individuals without disease

(iii) Resolution of the disease (e.g. following effective treatment) should result in a decrease in the copy number of pathogen-associated nucleic acid sequences

(iv) The presence in a healthy subject of a nucleic acid sequence characteristic of a pathogen should predict subsequent development of the associated disease in that individual

(v) The nature of the microorganism inferred from the available sequence should be consistent with the known biological characteristics of that group of organisms

(vi) These sequence-based findings should be reproducible

---

members of the normal microflora at that site. These bacterial activities will often result in some form of direct damage to the host or, as is often the case, the response of the host can be overzealous so that damage is self-inflicted. Some organisms may then go on to actually invade the host tissues, thereby causing more extensive damage (see Chapter 8). Finally, the organism must escape from its host and ensure that it is transmitted to another individual. The means by which an organism accomplishes the above tasks are designated as its 'virulence factors' (Table 1.7) and will be discussed in great detail in subsequent chapters.

1.3.2 Diseases caused by members of the normal microflora

First, a few points about terminology. The dictionary definition of the word 'disease' is 'an unhealthy condition of the mind, body or some part thereof'. The meaning of 'infectious disease' is, therefore, quite obvious – a disease that is caused by a microbe. Care must be taken, however, with the term 'infection'. Some consider it to be synonymous with 'infectious disease'. Others, however, use it to mean merely the presence of a particular organism in the host regardless of whether it is causing any damage. Difficulties in the use of the term 'infection' are encountered particularly when one speaks of diseases caused by

The bacteriologist as guinea pig

Barry Marshall, who did much to bring *Hel. pylori* to the attention of the medical and scientific communities is one of the few brave enough to use himself as a guinea pig to test out his own hypothesis. He had noticed the association between what was then known as 'pyloric campylobacter' and gastritis and was convinced that there was a causal relationship between the two. Evidence had accumulated to satisfy the first two of Koch's postulates but the remaining two needed to be tested – and this he set out to do. His first step was to establish that he had no inflammation or abnormality of his stomach by having biopsies taken from a number of regions. He then swallowed a culture of *Hel. pylori* (10^9 colony-forming units), which had been isolated from an individual suffering from gastritis, and the subsequent effects were monitored. Eight days later he vomited a little mucus and then suffered from headaches during the second week. He also became irritable and his colleagues noticed that his breath had become putrid! Biopsy samples from his stomach revealed the presence of helical bacteria adhering to the epithelial cells, which appeared abnormal. The latter contained very little mucus and were devoid of microvilli. Many polymorphonuclear cells were present and 'pyloric campylobacters' were cultured from the biopsy specimens. The actions of this highly committed researcher provided data to satisfy the remaining two of Koch's postulates – that the inoculation of the organism into a susceptible animal (i.e. a human) produced disease in that animal and the organism could be found in the disease lesions of the inoculated animal.

organisms that are members of the normal microflora. Hence, in one sense, most healthy individuals are 'infected' with *Strep. sanguis*, a member of the normal microflora of the oral cavity, but most of us are not suffering from an infectious disease (e.g. bacterial endocarditis) caused by this organism.

In the previous sections, the types of organisms associated with particular anatomical sites were described and the point was made that many of these organisms (known as 'opportunistic pathogens') are capable of initiating an infectious process (termed an 'opportunistic infection') but, in general, do not do so (Table 1.8). In considering such diseases, therefore, one of the first questions to ask is: 'what has upset the pre-existing balanced state of affairs to precipitate the disease?' The

Table 1.7.	Bacterial virulence factors
Virulence factor	Function
Adhesin	Enables binding of the organism to a host tissue
Invasin	Enables the organism to invade a host cell/tissue
Impedin	Enables the organism to avoid one or more of the host's defence mechanisms
Aggressin	Causes damage to the host directly, e.g. exotoxins, enzymes
Modulin	Induces damage in the host indirectly by perturbing cytokine networks, e.g. LPS

LPS, lipopolysaccharide.

Table 1.8. | Examples of infections caused by members of the normal microflora

Infection	Causative organism(s)
Urinary tract infection	*Escherichia coli, Staphylococcus epidermidis, Proteus* spp.
Infective endocarditis	α-Haemolytic streptococci, *Staphylococcus* spp.
Periodontal diseases	*Porphyromonas gingivalis, Bacteroides forsythus,* spirochaetes, *Actinobacillus actinomycetemcomitans*
Caries	*Streptococcus mutans, Lactobacillus* spp., *Actinomyces* spp.
Aspiration pneumonia	Oral anaerobes, *Staphylococcus* spp.
Pneumonia following viral infection	*Streptococcus pneumoniae, Haemophilus influenzae, Staphylococcus aureus*
Otitis media	*Haemophilus influenzae, Streptococcus pneumoniae*
Pharyngitis	*Streptococcus pyogenes*
Meningitis	*Haemophilus influenzae, Neisseria meningitidis, Streptococcus pneumoniae*
Burn and wound infections	*Staphylococcus aureus, Streptococcus pyogenes*
Acne	*Propionibacterium acnes*
Toxic shock syndrome	*Staphylococcus aureus*
Boils	*Staphylococcus aureus*
Actinomycosis	*Actinomyces* spp.
Pseudomembranous colitis	*Clostridium difficile*
Gas gangrene	*Clostridium perfringens*
Peritonitis	*Bacteroides* spp.
Infections associated with prosthetic devices	*Staphylococcus epidermidis, Staphylococcus aureus*

changes resulting in a switch from a mutualistic or commensal association to a disease-inducing parasitic one have, in some cases, been identified, although many remain obscure. Some of the most important means by which opportunistic infections arise are described in the following sections.

1.3.2.1 Damage to the epithelium

One of the most important functions of the epithelium is to exclude the enormous number of bacteria that inhabit our body surfaces from the underlying sterile tissues. The thickness of the epithelium varies with the anatomical site but can be as thin as a single layer of cells. Damage to this barrier allows ingress of the normal microflora, some members of which are able to initiate a disease process. Burns and wounds, for example, invariably result in some impairment of the epithelium and the skin organism *Staph. aureus* often colonises, and causes disease in, the damaged sites. Environmental organisms such as *Ps. aeruginosa*, which are notoriously resistant to antibiotics, are also frequent causes of burn and wound infections. A dreaded, life-threatening, complication of bowel surgery or rupture of the appendix is peritonitis due to gut organisms such as *Bacteroides fragilis*. Puncture wounds (i.e. penetration of the epithelium by a sharp object such as a knife or thorn) can enable access of a wide range of organisms (from the skin or the environment) to the bloodstream, resulting in a bacteraemia and, possibly, a septicaemia. Puncture wounds may also become colonised by anaerobic, toxin-producing organisms such as *Cl. perfringens* – resulting in gangrene, a life-threatening condition. Using a toothbrush may also damage the junction between the teeth and gums, enabling the access of oral bacteria to underlying tissues and, possibly, resulting in a transient bacteraemia. This does not appear to be a problem for most people. However, in individuals who have congenital cardiac abnormalities or prosthetic heart valves, or who have had rheumatic fever, the organisms may colonise the cardiac tissue or prosthetic device and so cause endocarditis. This is a serious condition that can be very difficult to treat.

1.3.2.2 The presence of a foreign body

Advances in surgery and the science of biomaterials have provided us with a wide range of artificial body parts (prostheses) such as heart valves and joints that can be used to replace their damaged natural counterparts. A number of medical and surgical procedures also involve the insertion of a variety of tubes (catheters) into body orifices, or through the skin, where they remain for varying periods of time. Unlike the epithelium, such devices do not have shedding surfaces and so enable the accumulation of bacteria, resulting in the formation of biofilms. Biofilms (see Sections 2.7 and 7.2) are less susceptible to phagocytosis and to the bactericidal effects of serum and blood because of the altered physiological state of the bacteria and the presence of large amounts of extracellular material. Such devices may also interfere with blood and lymphatic flow in neighbouring tissues, hence rendering the host less able to deal with the adherent organisms. They can also interfere with mechanisms designed to remove bacteria from these sites, for example the flushing action of urine and the mucociliary escalator in the respiratory tract. The types of organism associated with such infections will depend on the anatomical site involved but often include members of the normal microflora such as *Staph. epidermidis*, *Staph. aureus*, and *Candida* spp. (yeasts) or environmental organisms such as *Pseudomonas* spp. Diseases that are the result of some medical procedure are termed iatrogenic.

1.3.2.3 The transfer of bacteria to sites where they are not part of the normal microflora

The close proximity of the colon to the urethra in females facilitates the colonisation of the peri-urethral area by members of the normal microflora of the colon such as *E. coli*, *Proteus* spp. and *Klebsiella* spp. Such organisms can then ascend the urethra to the bladder resulting in a urinary tract infection. Females between the ages of 20 and 40 years are particularly susceptible to urinary tract infections, with *E. coli* being the most frequent causative organism. It has been shown that certain strains of *E. coli* predominate as the causative agents in a particular geographical area, suggesting that the organism can be transmitted from person to person (e.g. via food or water) and then become established in the gastrointestinal tract.

As described previously, the lower regions of the respiratory tract are usually sterile. However, members of the oral microflora, or other organisms, can gain access to these regions when (i) an individual loses consciousness, (ii) tubes are inserted, and (iii) food or gastric fluid is inhaled. The presence of such organisms, particularly anaerobic members of the oral microflora, in the lower respiratory tract can lead to aspiration pneumonia – one of the most common causes of death in the elderly. The disease is usually polymicrobial and a wide range of organisms may be implicated, including anaerobes from the oral cavity, Gram-negative bacilli (particularly *Klebsiella* spp. and *Pseudomonas* spp.) and staphylococci.

1.3.2.4 Suppression of the immune system by drugs or radiation

Cancer therapy involves the use of radiation or drugs that are designed to kill rapidly growing cells. Unfortunately, this includes polymorphonuclear neutrophils, which constitute one of the major host defences against bacteria (see Section 6.3.2). Other side effects of these procedures include depressed antibody production and impaired complement function, both of which will weaken the individual's ability to deal with infecting organisms. The immune response is deliberately suppressed in patients undergoing organ (liver, kidney, heart, etc.) transplantation so as to decrease the chance of organ rejection. These treatments, therefore, render the patient prone to infection by a wide variety of bacteria that normally do not cause problems in individuals with an intact defence system. Gram-negative bacilli (e.g. *E. coli*, *Klebsiella* spp. and *Pseudomonas* spp.), *Staph. aureus* and the yeast *Candida albicans* are the most frequent causes of infections in such patients. Many of these infections are acquired while the patient is in hospital (termed a 'nosocomial infection') and, as many of the organisms present in hospitals are antibiotic resistant, the treatment of such infections can be very difficult.

1.3.2.5 Impairment of host defences due to infection by an exogenous pathogen

Influenza is a common infection caused by orthomyxoviruses. These viruses destroy cells lining the upper and lower respiratory tracts and so impair the ability of the epithelium to exclude bacteria. They also inhibit phagocytosis of bacteria by alveolar macrophages, so enabling the survival of members of the normal microflora such as *Staph. aureus*, *Strep. pneumoniae* and *H. influenzae*. These organisms can then cause an often fatal pneumonia.

Infection with the human immunodeficiency virus (HIV), the causative agent of acquired immunodeficiency syndrome (AIDS), destroys one of the key cells (CD4 T lymphocytes, see Section 6.7) of the immune response, leaving patients vulnerable to opportunistic infections. Many of the organisms responsible for such infections are members of the normal microflora such as *Candida albicans*, *Pneumocystis carinii*, *Strep. pneumoniae*, *Staph. aureus* and *Corynebacterium* spp.

1.3.2.6 Disruption of the normal microflora by antibiotics

The microflora inhabiting a particular anatomical site consists of a complex community in which the relative proportions of the different species are controlled by, amongst other factors, positive and negative interactions between the organisms present. These include competition for adhesion sites and nutrients, interdependence as a result of food webs and the production of bacteriocins and other antagonistic chemicals.

The administration of an antibiotic to an individual can have a dramatic effect on these communities and can encourage the proliferation of those species that are intrinsically resistant to the particular antibiotic. Consequently, an organism that may have been present in only small numbers prior to administration of the antibiotic could emerge as a dominant member of the new community and may be able to initiate an infectious disease: this is often referred to as a 'superinfection'. This scenario is well documented in the case of broad-spectrum agents such as tetracyclines, ampicillin and cephalosporins. For example, administration of tetracycline can result in an overgrowth of *Candida albicans* in the oral cavity and this is often accompanied by thrush (candidosis). One of the most frightening, potentially life-threatening, adverse consequences of the administration of antibiotics is pseudomembranous colitis. This is a diarrhoeal disease caused by a member of the normal flora of the colon, *Cl. difficile*. The organism is usually present in small numbers, being held in check by complex interactions involving other members of the microflora. Administration of antibiotics, especially ampicillin, clindamycin and cephalosporins, can disrupt this balance, allowing the proliferation of the organism and secretion of the potent toxins that are responsible for the diarrhoea (for a discussion of these toxins, see Section 9.5.3.1).

1.3.2.7 Unknown precipitating factor

The commonest infectious diseases of humans are the inflammatory periodontal diseases, which include gingivitis (when only the gingivae, i.e. gums, are involved) and periodontitis (when tooth-supporting structures such as the alveolar bone are involved). These diseases, which are caused by members of the normal oral microflora, affect everyone at one time or another and are the major cause of tooth loss in adults. The causative organisms (known as periodontopathogens) are mainly, though not exclusively, anaerobes such as *Porphyromonas gingivalis*, *Prevotella intermedia*, spirochaetes and *Eubacterium* spp. These may comprise part of the dental plaque community in the gingival crevice (the gap between the tooth and gums) and, in the absence of disease, may be present in low numbers. Some environmental perturbation may then occur that could favour the proliferation of one (or more) of these periodontopathogens to disease-inducing levels. One possible event leading to such a scenario would be an increased flow of nutrient-rich, antibody-containing gingival crevicular fluid (GCF, an exudate from serum) due to an inflammatory response to the accumulation of dental plaque. The GCF may supply a key nutrient, the absence of which may have been limiting the growth of one of the pathogens, or may provide antibodies against a non-periodontopathogenic species whose elimination may serve to allow pathogens to proliferate.

1.3.3 Exogenous infections

Bacteria that are capable of inducing disease in individuals with intact specific and non-specific defence systems are usually referred to as 'pathogens'. The ability of a particular pathogen to damage its host (its 'virulence') is expressed in terms of the number of cells required to elicit a particular pathogenic response in its host (Table 1.9). Although the list of bacteria recognised as being pathogens continues to grow steadily (owing, mainly, to advances in detection methods), it is important to realise that the number of such species is minute in comparison to the number of species with which we live in harmony. Many of the most dangerous pathogens are not usually part of the normal human microflora, although humans may be the natural host, for example *Bord. pertussis* (the causative agent of whooping cough), *Shigella* spp. (cause dysentery), *M. tuberculosis* (causes tuberculosis), *T. pallidum* (causes syphilis). Others are primarily inhabitants of other animals (e.g. *M. bovis*, *Bacillus anthracis*, *Yersinia pestis*) or are found in the soil (e.g. *Cl. tetani*) or in water (e.g. *V. cholerae*). There are, however, some exceptions to this generalisation as some organisms (e.g. *Sal. typhi*, *N. meningitidis*, *C. diphtheriae*) can colonise certain individuals (known as 'carriers') without causing disease and can then be transmitted to susceptible individuals. Furthermore, patients recovering from an infection may harbour the causative organism for an appreciable length of time and so can act as a source of infection for other individuals. The most notorious example of the latter was 'Typhoid Mary', a cook who lived in New York state in the early 1900s. This lady was a carrier of *Sal. typhi* and, while working for a number of different families, unwittingly infected approximately 200 individuals.

One complication associated with designating a particular bacterial species as being a pathogen is that different strains of a particular species can exhibit considerable differences in virulence. Such differences are attributable to the presence or absence in the strain of those virulence factors that are considered to be important in the disease process. Until relatively recently, the recognition of different strains within a species was based on phenotypic characteristics, for example the ability to ferment particular sugars or to produce particular enzymes. However, because of advances in molecular genetic

Table 1.9. The terms pathogen, pathogenicity and virulence have been defined in a number of ways over the years and these, together with a proposal for a new set of definitions, are shown below

Term	Previous definitions	Proposed definition
Pathogen	A microbe capable of causing disease	A microbe capable of causing host damage; the definition can encompass classical pathogens and opportunistic pathogens; host damage can result from either direct microbial action or the host immune response
	A microbe that can increase in living tissue and produce disease	
	A microbe whose survival depends upon its capacity to replicate and persist on or within another species by actively breaching or destroying a cellular or humoral host barrier that ordinarily restricts or inhibits other microbes	
	A parasite capable of causing or producing some disturbance in the host	
Pathogenicity	Capacity of a microbe to produce disease	The capacity of a microbe to cause damage in a host
Virulence	Degree of pathogenicity	The relative capacity of a microbe to cause damage in a host
	Virulence is proportional to the reciprocal of resistance	
	Strength of the pathogenic activity	
	Relative capacity to overcome available defences	
	Disease severity as assessed by reductions in host fitness following infection	
	Percentage of death per infection	
	A synonym for pathogenicity	
	Property of invasive power	
	Measure of the capacity of a microbe to infect or damage a host	
	Relative capacity to enter and multiply in a given host	
Virulence factor or determinant	A component of a pathogen that, when deleted, specifically impairs virulence but not viability	A component of a pathogen that damages the host; this can also include components essential for viability
	Microbial products that permit a pathogen to cause disease	

techniques, there is now much greater interest in identifying strain variation by analysing the genetic make-up of the organisms (i.e. 'geno-typing'). Such analyses have established the existence of large numbers of clonal types in some pathogenic species. Some clones of a particular organism may constitute part of the normal human microflora whereas others are highly virulent and colonisation by these often results in disease. A good example of the wide variation in virulence displayed by strains of a particular species is *H. influenzae*, which exists as six serotypes distinguishable on the basis of the antigenicity of their capsular polysaccharides. Serotypes a, c, d, e and f are present in the upper respiratory tract of most individuals, where they rarely cause disease. In contrast, serotype b of this organism is highly virulent and is rarely found in healthy individuals. It has been shown that in North America 104 clonal types of *H. influenzae* serotype b exist. Further-more, genetic analysis of serotype b strains has shown that only 6 of the 104 clonal types are responsible for 81% of disease outbreaks. At the other extreme is the enteric pathogen *Shigella sonnei*, in which only one clonal type has been identified. A common method of quantifying the virulence of different strains of a pathogen is to determine the number of cells (lethal dose, LD) necessary to kill 50% of a population of suscep-tible animals – this number is known as the LD_{50}. It is important to remember that, just as the virulence of members of a particular bacter-ial species can differ, the susceptibility of the host to a particular organism can vary markedly among individuals. The outcome (disease versus no disease) of the interaction between a bacterium and its host depends, therefore, not only on the virulence of the infect-ing organism but also on the susceptibility of that particular indi-vidual. Factors that are known to affect host susceptibility include age, nutritional status, whether or not the individual has been exposed previously to the organism and, as we are only now coming to appreciate, a variety of genetic factors.

1.4 | The spectrum of bacterial diseases

The number of different diseases caused by bacteria is very large and, when attempting to review the whole range of such diseases, it is con-venient to classify them into a smaller number of groups, each of which has certain features in common. The classification systems used by authors of textbooks are, inevitably, biased towards the target audience of the particular book. Hence, books written primarily for scientists have tended to use the type of bacteria involved in the disease process as the basis for their classification system, for example diseases due to staphylococci, streptococci, etc. Books written for clinicians have tended to regard the target organ or organ system as the basis of their classification scheme, for example diseases affecting the respiratory tract, skin, etc. As this book is concerned with disease mechanisms, we have based our classification system on the bacterial virulence

Effects of host genotype on susceptibility to infectious diseases

A number of studies, involving a variety of approaches, have been carried out to determine whether host genetic factors affect an individual's susceptibility to infectious diseases. One approach has been to compare the susceptibility of a sibling of an infected individual to a particular infection to that of the general population. However, any increased susceptibility in such individuals may be attributable, in part, to environmental rather than genetic factors. This problem can be overcome in studies involving identical and non-identical twins. The susceptibility to infection with *M. tuberculosis*, *M. leprae* and *Hel. pylori* has been shown to be more similar in identical than in non-identical twins. As there is general agreement that genetic factors do affect susceptibility to infection, the next question to ask is: which genes are involved? To answer this question, the following approach is necessary: (i) identify the range of variations (polymorphisms) found in a particular gene which is likely to be involved, for example one of the many genes involved in host defence systems, (ii) determine whether susceptibility to a particular infection can be correlated with the presence or absence of a particular gene polymorphism. Most of these candidate genes chosen for study have been those involved in immune mechanisms, for example genes of the major histocompatibility complex (also known as human leukocyte antigens, HLA) and genes encoding cytokines and their receptors (see Sections 4.2.6 and 6.7.2). It has been found that the class II antigen HLA-DR2 is frequently, but not always, associated with susceptibility to both leprosy and tuberculosis. The susceptibilities of individuals to leprosy and meningococcal meningitis have also been shown to be associated with polymorphisms in the tumour necrosis factor α (TNFα) gene while variations in the genes encoding the interferon γ (IFNγ) receptor or the interleukin 12 (IL-12) receptor contribute to susceptibility to mycobacterial infections.

Host genetic factors also appear to influence the chances of developing gastric cancer following infection with *Hel. pylori*. Individuals with particular genotypes that increase production of the cytokine interleukin 1 (IL-1) show an increased risk of *Hel. pylori*-associated gastric cancer. Among individuals infected with *Hel. pylori*, those with the IL-1B-31T or the IL-1RN*2 allele of the IL-1β gene have a higher risk of developing gastric cancer.

factors involved in the disease process. A remarkable feature of many infectious diseases is the predictability of the course of the disease. This gives rise to a characteristic set of features in the patient that is very useful to the clinician who is trying to establish what is wrong with that individual. Hence, a particular organism will usually infect a certain tissue or organ, will either be limited to that tissue or spread throughout the body, and will produce a characteristic pattern of tissue destruction. Such predictable behaviour stems from the fact that a particular bacterium possesses a defined set of virulence factors that determine to which tissue an organism can adhere, whether it can invade that tissue and what damage it causes. A useful classification system for bacterial infections, which emphasises the underlying virulence mechanisms, is one based on the following criteria: (i) their site of action, (ii) whether or not the organisms invade the tissue and reach other parts of the body, (iii) whether the organisms secrete toxins, and (iv), in the case of those species that invade the host, whether the bacteria take up residence inside host cells (intracellular lifestyle) or not (extracellular lifestyle) (Table 1.10). The scheme is not, of course, ideal, as it largely ignores the role of the host in the infectious process. Hence, an organism that is unable to induce a disseminated infection in one individual may readily do so in another, particularly if the latter is very young, very old, malnourished or immunocompromised in some way. Another problem is defining what is meant by an 'invasive' pathogen. Should any organism capable of invading an epithelial surface be classed as 'invasive' or should the term be restricted to those organisms that are habitually disseminated (by the bloodstream or lymphatic system, or by carriage within phagocytic cells) to many parts of the body? In our scheme, we have restricted the term to mean the latter rather than the former as organisms capable of causing a disseminated infection must possess a distinctive array of virulence factors (Table 1.7), particularly impedins, to enable them to withstand the full onslaught of the many facets of the host defence system that can be mounted against them. Examples of each of the major types of infection, based on this classification system, are given below.

1.4.1 Infections that are not usually accompanied by tissue invasion and dissemination

In order to maintain themselves in their host, bacteria must adhere to some tissue (see Chapter 7). Once they have adhered, many bacterial species are then able to harm their host. Such organisms do not generally invade the tissue to which they have adhered or, if they do, they rarely reach the bloodstream and so are not carried to other parts of the body. Examples of such diseases include diphtheria, pharyngitis, cholera, whooping cough and some forms of gastroenteritis.

1.4.1.1 Infections confined to mucosal surfaces

1.4.1.1.1 Causative organism produces an exotoxin

Diphtheria is a disease that originates in the pharynx and is invariably confined to the upper respiratory tract. It is caused by *C. diphtheriae*, which is transmitted via a bacteria-laden aerosol. The organism causes a very severe pharyngitis that can actually block the air passage resulting in suffocation of the infected individual. The bacteria also secrete a toxin (the gene for which is carried by a bacteriophage) that enters the bloodstream and is able to affect nearly all human tissues. Heart failure is the usual cause of death, although many organs are damaged by the toxin. On recovering from the infection, an individual will often harbour the organism in the throat or nose for several weeks and so

Table 1.10. Classification of bacterial infections. This scheme is very microbe centred and basically ignores the role of the host! It may not be valid when the host is very young, very old, malnourished or immunocompromised

Involving tissue invasion and dissemination throughout the body				Generally not accompanied by dissemination throughout the body			
Exotoxins involved		No exotoxins involved		Confined to mucosa		Confined to skin	
Infecting organism is intracellular	Infecting organism is extracellular	Infecting organism is intracellular	Infecting organism is extracellular	Exotoxins involved	No exotoxins involved	Exotoxins involved	No exotoxins involved
Dysentery (*Shigella* spp.)	Gangrene (*Clostridium perfringens*)	Tuberculosis (*Mycobacterium tuberculosis*)	Meningitis (*Haemophilus influenzae*)	Diphtheria (*Corynebacterium diphtheriae*)	Gonorrhoea (*Neisseria gonorrhoeae*)	Tetanus (*Clostridium tetani*)	Acne (*Propionibacterium acnes*)
Listeriosis (*Listeria monocytogenes*)	Pneumonia (*Streptococcus pneumoniae*)	Typhoid fever (*Salmonella typhi*)	Meningitis (*Neisseria meningitidis*)	Cholera (*Vibrio cholerae*)	Diarrhoea (enteropathogenic *Escherichia coli*)	Impetigo (*Staphylococcus aureus* or *Streptococcus pyogenes*)	Erythrasma (*Corynebacterium minutissimum*)
Legionnaires' disease (*Legionella pneumophila*)	Puerperal fever (*Streptococcus pyogenes*)	Brucellosis (*Brucella* spp.)		Whooping cough (*Bordetella pertussis*)	Vincent's stomatitis (*Fusobacterium* spp. and *Treponema* spp.)	Scalded skin syndrome (*Staphylococcus aureus*)	
	Meningitis (*Streptococcus pneumoniae*)	Lyme disease (*Borrelia burgdorferi*)		Some urinary tract infections	Otitis media (*Haemophilus influenzae*)	Boils (*Staphylococcus aureus*)	
	Plague (*Yersinia pestis*)	Trachoma (*Chlamydia trachomatis*)		Gastritis and peptic ulcers (*Helicobacter pylori*)	Periodontitis (*Porphyromonas gingivalis*)	Erysipelas (*Streptococcus pyogenes*)	
	Anthrax (*Bacillus anthracis*)	Pneumonia (*Chlamydia pneumoniae*)		Tonsillitis (*Streptococcus pyogenes*)			
	Gastroenteritis (*Yersinia enterocolitica*)	Typhus (*Rickettsia typhi*)		Gastroenteritis (*Campylobacter* spp., enterotoxigenic *Escherichia coli*)			
		Q fever (*Coxiella burnetii*)		Pseudomembranous colitis (*Clostridium difficile*) Staphylococcal toxic shock syndrome			

can transmit the organism to others via aerosols or direct contact with respiratory secretions.

The causative agent of cholera, *V. cholerae*, is transmitted in faeces-contaminated water or food. Generally, quite high numbers (10^8–10^9 cells) of the organism need to be ingested to induce the disease. The organism colonises the mucosa of the small intestine and produces an enterotoxin that stimulates epithelial cells to secrete chloride, bicarbonate and water; this results in a profuse, watery diarrhoea that can amount to more than 1 litre per hour. The infected individual usually dies (the case fatality rate can be as high as 14%) unless the water and electrolytes are replaced.

Whooping cough is a disease predominantly affecting young children and is caused by *Bord. pertussis*, which is spread in aerosols created by infected individuals. It is highly infectious, with attack rates greater than 50% in susceptible individuals and it kills approximately 600 000 children worldwide each year. The organism adheres to the cilia of the respiratory epithelia and secretes several toxins that kill epithelial cells, cause dysfunction of the central nervous system and induce hypoglycaemia. The prolonged paroxysms of coughing result in apnoea and, eventually, death.

Infections of the urinary tract are more common in females than in males and involve the colonisation of the urethra by organisms such as *E. coli* and *Proteus* spp., which are normal inhabitants of the colon. In order to resist the flushing action of urine, one of the prime requirements of a urinary pathogen is an ability to adhere strongly to mucosal cells and in *E. coli* this is mediated by pili. Many uropathogenic strains of *E. coli* also produce an exotoxin that can kill host cells and stimulate an inflammatory reaction.

It is now recognised that the main cause of gastric and duodenal ulceration is infection with *Hel. pylori*. This organism has been isolated from as many as 80% of patients with gastric ulcers and from 95% with duodenal ulcers. It is able to penetrate the protective mucus layer of the stomach and attach to mucus-secreting cells, where it produces a urease that liberates ammonia, so providing the buffering power needed to enable it to survive at the low pH of the stomach. It also secretes a vacuolating cytotoxin that plays a role in ulcer development. The natural host for the organism is humans but its mode of transmission is uncertain. Gastric fluid containing the bacterium could reach the oral cavity and so be transmitted via aerosols, or the organism could be excreted in faeces.

1.4.1.1.2 Causative organism does not produce an exotoxin

Some non-invasive organisms induce disease without secreting any toxins. For example, *Mycoplasma pneumoniae*, an aerobic, pleomorphic bacterium without a cell wall, is one of the most common causes of pneumonia in children and adolescents. It has been estimated that up to 18% of pneumonias in non-hospitalised patients are due to this organism. The normal habitat of *Myc. pneumoniae* is the nasopharynx

Corynebacterium diphtheriae toxin

The main virulence factor of *C. diphtheriae* is its toxin, which, when injected into an appropriate animal, can induce the characteristic features of the disease. Remarkably, however, it has been shown that the gene (*tox*) encoding this toxin is provided by a phage that infects the organism. Strains that are not infected with a toxin-encoding phage are unable to cause diphtheria and, in fact, most strains of the organism that are isolated from the throats of individuals outside epidemic areas are non-toxinogenic. The expression of the *tox* gene is controlled by a number of factors including pH, amino acid concentration, type of carbon and nitrogen sources available and the concentration of iron. The effects of variations in iron concentration on toxin production have been the most extensively investigated and it has been shown that high concentrations ($>0.5 \,\mu g/ml$) inhibit production of the toxin.

of humans and it is transmitted between individuals by aerosols. The organism adheres to epithelial cells of the upper respiratory tract, where it initially inhibits ciliary action, so enabling colonisation of the lower regions of the tract, and then kills the cells. The mechanisms underlying these processes have not yet been elucidated, although there is evidence that in some *Mycoplasma* spp. the toxic effects are due to their ability to produce hydrogen peroxide.

Gonorrhoea is caused by *N. gonorrhoeae*, an organism whose only host is the human. It is a Gram-negative, aerobic, non-motile coccus that does not survive for long outside its host, so that transmission of the infection is invariably by direct person-to-person contact. It has been estimated that approximately 1000 viable bacteria are needed to initiate an infection in the male urethra. The bacteria adhere to columnar epithelial cells of the cervix and urethra by means of pili and induce a strong inflammatory response, giving rise to a characteristic purulent discharge. This is presumably due to the induction of cytokine release from host cells, although the bacterial constituents responsible for this have not yet been identified. Following firm attachment, bacteria are taken up by the epithelial cells and transported to the basement membrane. However, in only approximately 1% of all cases does the infection become systemic, resulting in arthritis and, less commonly, endocarditis.

1.4.1.2 Infections confined to the skin

1.4.1.2.1 Causative organism produces an exotoxin

Tetanus is caused by *Cl. tetani*, which is widely distributed in the environment, being found in soil, in the faeces of humans (up to 40% of the population) and other animals and in the dust of houses and hospitals. The disease results from the contamination of minor puncture wounds, but only if the conditions there are suitable for spore germination and toxin production, and it has been estimated that approximately one million cases of tetanus occur each year throughout

Bacteria can cause cancer

Helicobacter pylori was first described only in 1982 but since then has been recognised as a common bacterial pathogen and is thought to be present in the stomach of up to 80% of adults. Once infected, an individual generally carries the organism for life, unless it is eradicated by antibiotics. It was soon shown that *Hel. pylori* was responsible for gastritis but, surprisingly, it was also found to be the main cause of gastric ulcers and duodenal ulcers – which, in developed countries, occur in approximately 15% of individuals infected with the organism. For many years, it had been assumed that gastric and duodenal ulcers were idiopathic in nature or attributable to the use of aspirin or non-steroidal anti-inflammatory drugs. Even more astounding revelations were to follow. For a number of years chronic gastritis had been known to be linked to the development of adenocarcinoma of the stomach, this being the most important gastric malignancy and the second most common cause of death due to cancer. Increasing evidence then accumulated to support an association between *Hel. pylori* and gastric carcinoma until in 1994 the International Agency for Cancer Research declared that *Hel. pylori* was a carcinogen for humans. The organism is also thought to be responsible for at least two other forms of cancer – gastric non-Hodgkin's lymphomas and gastric mucosa-associated lymphoid tissue lymphoma. Such is the interest in this organism, and closely related species, that nearly 8% of all research publications in microbiology between 1991 and 1997 concerned members of the *Helicobacter* genus.

The finding that *Hel. pylori* is a carcinogen has spurred research into determining the possible involvement of bacteria in other carcinomas and in chronic diseases such as arthritis, heart disease and even schizophrenia.

the world. The bacteria remain at the site of the wound but the toxin binds to nerve terminal membranes, is internalised and then migrates to the central nervous system. Its principal effect is to block the normal inhibition of spinal motor neurones, which results in characteristic muscular spasms. Estimates of the case fatality rate vary enormously, but in the United Kingdom it is thought to be approximately 10%.

Cutaneous abscesses are invariably caused by *Staph. aureus*, a member of the normal skin microflora. This organism produces a wide range of toxins and extracellular enzymes of which the most likely to be involved in abscess formation are the Panton–Valentine leukocidins and α-haemolysin (which kill polymorphonuclear leukocytes), coagulase (which walls off the lesion and protects bacteria from phagocytes) and lipases (which release inflammatory fatty acids).

1.4.1.2.2 Causative organism does not produce an exotoxin

Acne is a disease caused by *Prop. acnes*, which is a member of the normal skin microflora and, indeed, is probably the most common organism found on human skin. These bacteria inhabit sebaceous glands and can hydrolyse the lipids present in sebum to produce fatty acids that induce the inflammation characteristic of the condition.

Corynebacterium minutissimum infects the stratum corneum of moist areas of the skin such as the toe webs and axillae, producing red scaly patches – a condition known as erythrasma.

1.4.2 Infections accompanied by tissue invasion and dissemination

Those organisms that are able to invade and traverse the protective epithelial layers characteristically adopt either an intracellular or extracellular lifestyle (see Section 8.4) and this is a useful criterion for classification.

1.4.2.1 Causative organism produces an exotoxin

1.4.2.1.1 Infecting agent adopts an extracellular lifestyle

Clostridial myonecrosis (gas gangrene) is an aggresive, often lethal, infection of muscle due mainly to *Cl. perfringens* following contamination of deep wounds or after abdominal surgery. The organism is present in soil and dust and is a member of the normal intestinal microflora. Under appropriate conditions, the spores present in a wound may germinate and the resulting vegetative organisms multiply and then spread into adjacent groups of muscle, releasing a number of potent toxins that cause extensive tissue destruction. The most important environmental condition affecting the initiation of the infection is the redox potential of the tissues. Clostridia cannot multiply in normal tissues because they need a highly reducing environment (with a redox potential below approximately $+75\,\mathrm{mV}$). Normal tissues have a much higher redox potential $(>+120\,\mathrm{mV})$. However, tissue damage often results in a decreased supply of (oxygenated) blood to the damaged area, giving rise to tissue anoxia and hence a low redox potential and thereby creating ideal conditions for the growth of anaerobic bacteria. Gangrene can also be caused by other members of the genus *Clostridium*, including *Cl. novyi*, *Cl. septicum* and *Cl. histolyticum*.

Streptococcus pneumoniae, an α-haemolytic streptococcus, is present in the nasopharynx of at least one-third of the human population. However, it can also invade the respiratory epithelium and enter the bloodstream to reach the meninges, where it is responsible for a life-threatening meningitis. When the organism lyses it releases a potent toxin (pneumolysin) that can kill host cells and can also act as a very effective inducer of cytokine synthesis. *Streptococcus pneumoniae* is also the most frequent cause of bacterial pneumonia acquired in the community (i.e. as opposed to pneumonia contracted in hospitals), accounting for between 30% and 50% of cases of the disease. Other diseases due to this organism include otitis media, sinusitis and chronic bronchitis.

1.4.2.1.2 Infecting agent adopts an intracellular lifestyle

Bacterial dysentery is a gastrointestinal infection in which blood is often present in the accompanying diarrhoea. Any of the four species of the genus *Shigella* may be responsible for the infection.

These bacteria are essentially human pathogens and are usually transmitted via contaminated food or water, although, as the infectious dose is so low (a few hundred organisms), it can also be transmitted by person-to-person contact. The bacteria cross the intestinal mucosa via M cells and then invade epithelial cells via their basolateral faces (see Section 8.2.1.1.3). They then spread to adjacent cells by forcing their way through the intervening cytoplasmic membranes, killing the cells as they go. They generally invade only the gut wall and the mesenteric lymph glands – bacteraemia is not common. All four *Shigella* spp. secrete toxins but *Shig. dysenteriae* also releases an intracellular Shiga toxin on lysis that is enterotoxic, neurotoxic and cytotoxic and is thought to be responsible for an, often fatal, acute kidney failure. Patients recovering from the infection often continue to excrete the organism in their stools for several weeks and some become persistent carriers of the organism.

Legionella pneumophila is an organism that is ubiquitous in aquatic environments – both natural and artificial. Inhalation of aerosols (e.g. from air-conditioning systems, showers, etc.) can lead to a potentially fatal form of pneumonia known as Legionnaires' disease. Once they are inhaled, the bacteria invade and colonise predominantly the macrophages present in the lung alveoli but they can also invade, and survive within, neutrophils and tissue monocytes. Tissue damage results from the secretion of a number of proteases and exotoxins including a heat-stable cytotoxin and a haemolysin. The bacteria are often transported within phagocytes to other parts of the body, where they may be released and so infect other cells.

1.4.2.2 Causative organism does not produce an exotoxin

1.4.2.2.1 Infecting agent adopts an extracellular lifestyle

Haemophilus influenzae forms part of the normal microflora of the upper respiratory tract. Most of these normal residents are strains of low virulence but colonisation by serotype b strains can result in a number of serious infections including meningitis. Colonisation of the nasopharynx by the organism can result in nasopharyngitis, which is sometimes accompanied by a bacteraemia. However, little is known concerning how the organism then invades the meninges to cause meningitis. The organism is also responsible for a range of respiratory tract infections including otitis media, sinusitis, epiglottitis, pneumonia and chronic bronchitis.

Neisseria meningitidis is found in the nasopharynx of as many as 10% of the human population. It is responsible for a potentially fatal form of meningitis, particularly in infants but also in older children and adults, in which death may occur within 24 hours of the appearance of symptoms. These include headache, fever, a skin rash and neck stiffness. The disease is acquired by inhaling aerosols produced by an individual who is either infected with, or a carrier of, the organism. The bacteria then colonise the nasopharynx and invade the respiratory epithelium, passing through this into the bloodstream. They are then carried to the meninges, which are invaded and become inflamed. If the condition is left untreated, death occurs in approximately 85% of cases.

1.4.2.2.2 Infecting agent adopts an intracellular lifestyle

Tuberculosis is a chronic infection that originates in the lungs following inhalation of the causative organism, *M. tuberculosis*. This organism is able to survive, and multiply, in alveolar macrophages and the complex mechanisms allowing survival are described in Section 10.4.2.2.2. Macrophages containing the organism are able to fuse to form giant cells. Eventually a walled-off lesion, known as a granuloma, is formed which consists of a fibroblast covering together with bacteria, T cells, myeloid cells and giant cells. This usually limits the primary infection, as most of the mycobacteria cannot tolerate the low oxygen content and pH at the centre of the granuloma. In some cases, the organism spreads to infect the bones, joints, kidneys and meninges. Some organisms also survive in a dormant form and, when reactivated, can give rise to post-primary tuberculosis, resulting in severe damage to the lungs and possibly to other regions of the respiratory tract, the intestinal tract, bladder and kidneys.

Typhoid fever results from the ingestion of *Sal. typhi* (approximately 10^7 cells) in contaminated water or, less commonly, in food. The organisms invade the outermost cells (M cells) of the Peyer's patches of the intestinal tract and lymphoid follicles, multiply in the submucosal layers and then enter the bloodstream. They then multiply in the spleen and liver and re-enter the bloodstream and exit through the gut epithelium, which becomes severely ulcerated. During an attack of the disease, the bacteria can be isolated from the faeces, blood and urine of the patient. Most of the symptoms of typhoid fever are attributable to cytokine induction by lipopolysaccharide (LPS) and other components of the organism's cell wall.

Brucella spp. normally infect domesticated animals such as cattle, goats and pigs. Some species, including *Brucella abortus* (from cows), *Br. suis* (from pigs) and *Br. melitensis* (from goats and sheep), can cause infections (brucellosis) in humans. Brucellosis is, therefore, an example of a zoonosis. The infection is transmitted via direct contact with infected milk, vaginal secretions or carcases. It is a chronic disease characterised by recurrent episodes of fever, sweating, aches and pains and hence is also known as undulant fever. The bacteria can penetrate mucous membranes or skin abrasions and are then disseminated widely to become intracellular parasites of cells of the mononuclear phagocyte system (see Section 6.3). A number of other animal pathogens can give rise to disease in humans and examples are given in Table 1.11.

Table 1.11. | Examples of diseases of humans (zoonoses) that are caused by organisms whose normal host is another animal

Disease	Causative organism	Normal host	Mode of transmission to humans
Anthrax	*Bacillus anthracis*	Domestic mammals (sheep, cattle, horses)	Inhalation of spores or entry via skin abrasions
Brucellosis	*Brucella* spp.	Domesticated mammals (pigs, goats, sheep, cattle, dogs)	Milk, cheese, inhalation, occupational contact (vets, farmers, slaughterhouse workers)
Leptospirosis	*Leptospira interrogans*	Rodents, foxes, domestic mammals	Ingestion of contaminated water or food, skin abrasions
Pasteurellosis	*Pasteurella multocida*	Dogs, cats, birds, wild mammals	Animal bites
Rat bite fever	*Streptobacillus moniliformis*	Rats, cats	Animal bites
Q fever	*Coxiella burnetii*	Domestic animals	Inhalation
Psittacosis	*Chlamydia psittaci*	Parrots, turkeys, other birds	Inhalation
Salmonellosis	*Salmonella* spp.	Fowl, domestic mammals	Ingestion in contaminated food
Listeriosis	*Listeria monocytogenes*	Domestic mammals, rodents, birds	Ingestion in contaminated food
Lyme disease	*Borrelia burgdorferi*	Deer, rodents	Bite of a tick that has previously fed on an infected animal
Plague	*Yersinia pestis*	Rats, other rodents	Bite of a flea that has previously fed on an infected animal
Typhus	*Rickettsia typhi*	Rats	Bite of a flea that has previously fed on an infected animal
Relapsing fever	*Borrelia* spp.	Rats, wild mammals	Bite of a flea that has previously fed on an infected animal
Tularaemia	*Francisella tularensis*	Rabbits, ticks	Direct contact with infected rabbits, tick bite

1.5 | Concept check

- All body surfaces with access to the external environment are colonised by bacteria and other microbes – such organisms are known collectively as the 'normal microflora'.
- The composition of the normal microflora is extremely complex.
- In humans, bacterial cells outnumber mammalian cells by 10 to 1.
- The microflora at a particular anatomical site has a characteristic composition dictated by local environment factors and is stabilised by bacteria–bacteria and host–bacteria interactions.
- A healthy individual and his/her normal microflora are engaged in a mutualistic relationship.
- Most of the species comprising the normal microflora do not induce disease under normal circumstances.
- The normal microflora is essential for human well-being.
- Medical and surgical interventions can render individuals susceptible to infection by members of the normal microflora.
- The 'virulence' of an organism is a measure of its ability to cause disease in a healthy host.
- Members of the normal microflora able to cause disease are known as 'opportunistic pathogens' and their virulence is low.

- Exogenous pathogens are highly virulent and are able to cause disease in a healthy individual.
- Identification of the causative agent of an infection is often difficult and requires the application of a set of criteria such as Koch's postulates or their molecular equivalent.
- In order to cause a disease, an organism must possess one or more virulence factors.
- The virulence factors possessed by a particular organism result in a disease with a characteristic pathology and this can form the basis of a useful classification scheme for infectious diseases.

1.6 | Introduction to the paradigm organisms

In each chapter of this book two organisms, *Strep. pyogenes* and *E. coli*, will be used as paradigms to exemplify the particular ways in which specific organisms have evolved to deal with issues raised in that chapter.

1.6.1 *Streptococcus pyogenes*

Streptococcus pyogenes is a Gram-positive, facultatively anaerobic, non-motile coccus that typically forms chains of cells (Figure 1.7a). The organism is found only in humans and is present in the throats of as many as 20% of healthy children. Carriage rates are somewhat lower in adults. It is usually transmitted from person to person via aerosols created from saliva or respiratory secretions. *Streptococcus pyogenes* produces a wide range of extracellular enzymes and toxins (more than 20) that account for its versatility as a pathogen. Some of these toxins are haemolysins, which give rise to characteristic zones of haemolysis (see Section 1.2.2) when the organism is grown on blood-containing media (Figure 1.7b). The organism is responsible for a wide range of diseases including pharyngitis, tonsillitis and impetigo (all three being localised, non-invasive infections) as well as puerperal fever, meningitis, endocarditis, necrotising fasciitis and toxic shock syndrome, all of which involve invasion by the organism. A characteristic feature of infections with this organism is that they can lead to a number of potentially life-threatening complications once the acute illness has passed. Rheumatic fever, the most serious of these, can affect as many as 3% of infected individuals. It occurs two to three weeks after the infection and involves damage to the muscles and valves of the heart but also affects the joints and central nervous system. The mechanisms underlying this disease have not been fully established but it is known that certain antigens of the organism have structures similar to that of antigens present in heart tissue – damage may result when these cardiac antigens bind to antibodies that the body has raised to protect itself against the organism. Recurrent infections lead to progressive damage and therefore the patient must be protected against the organism with antibiotics such as penicillin. Acute glomerulonephritis

Figure 1.7 (a) Transmission electron micrograph of *Streptococcus pyogenes*, showing characteristic chain formation. Magnification 4640×. (Reproduced with permission from V. A. Fischetti, the Laboratory of Bacterial Pathogenesis and Immunology, the Rockefeller University, New York, USA. Website: http://www.rockefeller.edu/vaf) (b) *Streptococcus pyogenes* growing on a blood-containing agar plate, showing characteristic zone of haemolysis around each colony.

(a)

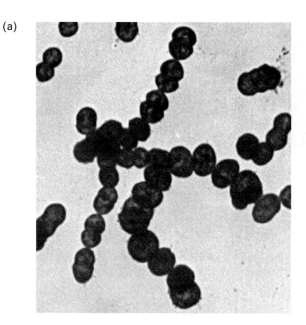

(b)

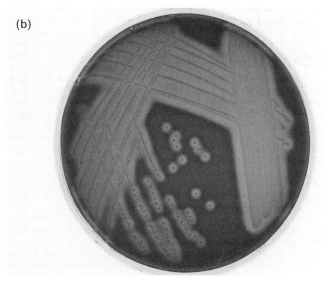

may also occur several weeks after infection with *Strep. pyogenes*. The damage to the kidneys that accompanies this condition is the result of inflammation induced by the deposition of antigen–antibody complexes in the blood vessels (glomeruli) of this organ. Only a limited number of strains of the organism (nephritogenic strains) are able to cause glomerulonephritis and, unlike rheumatic fever, recurrence of this complication is rare and requires reinfection by a nephritogenic strain. It has been estimated that as many as 15% of individuals infected with a nephritogenic strain will develop acute glomerulonephritis.

Since the mid-1980s, there has been a resurgence of infections due to this organism. Hence, there has been an increase in the number of outbreaks of rheumatic fever, streptococcal toxic shock syndrome and

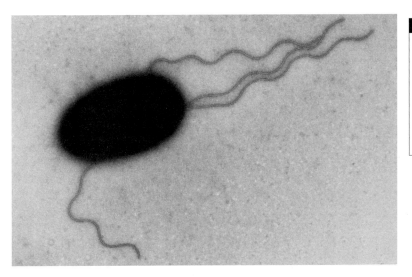

Figure 1.8 Transmission electron micrograph of *Escherichia coli*, showing its characteristic rod shape. This particular cell has four flagella and numerous (much shorter) hair-like projections (fimbriae). Magnification 15 770×. (Copyright Dennis Kunkel Microscopy.)

severe invasive skin and soft tissue infections. The reasons for this remain unknown but may be due to the emergence of certain serotypes of the organism, for example, M protein serotype 1 (see Section 2.9.1) is very often associated with invasive disease.

1.6.2 *Escherichia coli*

Escherichia coli is a Gram-negative, facultatively anaerobic, motile bacillus that constitutes part of the normal microflora of humans and most warm-blooded animals (Figure 1.8). In humans, it is the predominant facultative organism of the large bowel, where, for many years, it was considered to be a harmless commensal. It was known, however, that outside its normal habitat E. *coli* could induce disease, for example in the urinary tract. The first indication of its ability to cause disease within its normal habitat was in the 1940s, when it was shown to be responsible for outbreaks of 'summer diarrhoea' in infants. Since then, the organism has been shown to be a major cause of a number of intestinal tract infections in adults – these range from a mild diarrhoea to a life-threatening condition. It is now known that specific strains of the organism are responsible for the various diseases and these 'enterovirulent' strains have been grouped into six major classes, each of which induces disease by a characteristic mechanism (Table 1.12). The nomenclature of such strains can be confusing, as some researchers use the term 'enteropathogenic' to mean any E. *coli* strain capable of causing gastroenteritis (i.e. synonymous with enterovirulent), while others reserve the term for one of the six main classes of diarrhoea-inducing strains. Furthermore, as not all **S**higa **t**oxin-producing E. *coli* (STEC) strains are pathogenic for humans, the term **e**ntero**h**aemorrhagic **E.** *coli* (EHEC) has been introduced to denote a subset of strains of STEC that cause disease in humans. To further complicate the matter, two other enteropathogenic classes of the organism have recently been proposed – **c**ell-**d**etaching **E.** *coli* (CDEC) and **c**ytolethal **d**istending

Table 1.12. The principal types of enterovirulent strains of *Escherichia coli*

Type of *E. coli*	Pathogenic mechanism	Clinical symptoms	Source of infection
Enteropathogenic (EPEC)	Forms microcolonies at a limited number of sites on the epithelial cell ('localised' adherence), induces an attaching/effacing lesion	Watery diarrhoea, fever	Water, possibly fresh fruit and vegetables
Diffusely adherent (DAEC)	Does not form microcolonies, induces an attaching/effacing lesion?	Diarrhoea?, vomiting common	?
Enteroaggregative (EAEC)	Forms a thick, mucoid biofilm, secretes enterotoxins	Watery and mucoid diarrhoea, fever	Water, possibly fresh fruit and vegetables
Shiga toxin-producing (STEC)	Secretes enterotoxins belonging to the Shiga toxin family, forms an attaching/effacing lesion	Watery and bloody diarrhoea. Induction of haemolytic uraemic syndrome	Water, raw/undercooked hamburgers, sausages, chicken, vegetables, milk
Enteroinvasive (EIEC)	Invasive, secretes enterotoxins?	Watery diarrhoea sometimes with blood, sometimes fever	Water, cheese, guacamole
Enterotoxigenic (ETEC)	Secretes a heat-labile enterotoxin, a heat-stable enterotoxin or both	Watery diarrhoea without blood, no fever	Water, fresh fruit and vegetables, scallops, soft cheeses
Cell detaching (CDEC)	Produces an α-haemolysin	?	?
Cytolethal distending toxin-producing (CLDTEC)	Produces cytolethal distending toxin	?	?

toxin-producing **E. coli** (CLDTEC). Although both CDEC and CLDTEC have been isolated from children with diarrhoea, their association with the disease process has not been definitively established.

Escherichia coli constitutes a good example of the difficulty of designating a particular species as being a pathogen or a commensal – many strains inhabiting the gastrointestinal tract do not cause disease. It is thought that many enterovirulent strains of the organism acquired their virulence-associated genes from more well-known enteric pathogens such as *V. cholerae* and *Shig. dysenteriae*.

1.7 | What's next?

This chapter has served as a general introduction to bacteria as agents of disease. Most of the remaining chapters are devoted to describing how bacteria cause disease, the defence systems that have evolved for preventing bacterial infections and the ways in which bacteria overcome these systems. A central theme will be the many ways in which bacteria manipulate normal cellular functions for their own benefit and these are summarised diagrammatically in Figure 1.9 to whet the

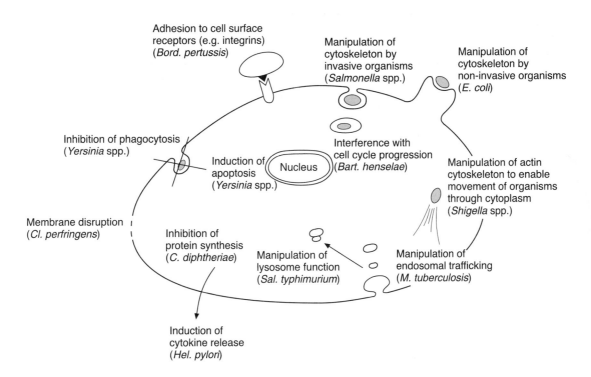

reader's appetite for what is to come. However, before describing the mechanisms involved in the initiation and progression of an infectious process, it is important to familiarise the reader with key aspects of the biology of bacteria. Chapter 2, therefore, will describe those aspects of the structure, genetics and physiology of bacteria that are essential for an understanding of the mechanisms underlying the initiation and progression of infectious diseases.

Figure 1.9 Diagram showing some of the ways in which bacteria can interfere with the normal functions of the host cell for their own benefit. In each case, only one example of a bacterial species with the stated ability is given – many other examples will be described in subsequent chapters. For genus abbreviations, see list on p. xxi.

1.8 | Questions

1. Define the terms pathogen, parasite, commensal, opportunistic infection, symbiosis, mutualism, superinfection, virulence. Give specific examples of each, explaining how they conform to your definitions.
2. The normal microflora is essential to the well-being of its host: discuss.
3. Describe and compare the normal microfloras of the various regions of the human gastrointestinal tract. How do such differences in microfloras arise?
4. Describe how modern medical and surgical procedures have rendered patients vulnerable to a broader range of infectious diseases.
5. Antibiotics do more harm than good: discuss.
6. *Staphylococcus aureus* is a more successful parasite than *Salmonella typhi*: discuss.
7. Describe the range of diseases caused by *Streptococcus pyogenes*.

8. *Escherichia coli* is a harmless commensal of the human gastrointestinal tract: discuss.
9. Explain why certain body sites with access to the external environment (e.g. the alveoli, the stomach) are largely free from bacteria while others (e.g. the pharynx, the colon) support a large bacterial population.
10. Why is it thought that modern, molecular methods of bacterial identification will revolutionise our knowledge of the normal human microflora?

1.9 | Further reading

Books

Fischetti, V. A., Novick, R. P., Ferretti, J. F., Portnoy, D. A. & Rood, J. I. (2000). *Gram-Positive Pathogens*. Washington, DC: ASM Press.

Murray, P. R., Baron, E. J. & Pfaller, M. A. (1999). *Manual of Clinical Microbiology*. Washington, DC: ASM Press.

Nataro, J. P., Blaser, M. J. & Cunningham-Rundles, S. (2000). *Persistent Bacterial Infections*. Washington, DC: ASM Press.

Scheld, W. M., Craig, W. M. & Hughes, J. M. (2000). *Emerging Infections*. Washington, DC: ASM Press.

Tannock, G. W. (1995). *Normal Microflora*. London: Chapman & Hall.

Tannock, G. W. (1999). *Medical Importance of the Normal Microflora*. Dordrecht: Kluwer Academic Publishers.

Review articles

Adamsson, I., Edlund, C. & Nord, C. E. (2000). Microbial ecology and treatment of *Helicobacter pylori* infections. *J Chemother* **2**, 5–16.

Ashley, E. A., Johnson, M. A. & Lipman, M. C. (2000). Human immunodeficiency virus and respiratory infection. *Curr Opin Pulm Med* **6**, 240–245.

Bateson, M. C. (2000). *Helicobacter pylori*. *Postgrad Med J* **76**, 141–144.

Bellamy, R. (1999). Host factors in genetic susceptibility to infectious diseases. *Rev Med Microbiol* **10**, 175–183.

Besser, R. E., Griffin, P. M. & Slutsker, L. (1999). *Escherichia coli* O157:H7 gastroenteritis and the haemolytic uremic syndrome: an emerging infectious disease. *Annu Rev Med* **50**, 355–367.

Boman, H. G. (2000). Innate immunity and the normal microflora. *Immunol Rev* **173**, 5–16.

Brook I. (1998). Microbiology of common infections in the upper respiratory tract. *Prim Care* **25**, 633–648.

Brook, I. (1999). Bacterial interference. *Crit Rev Microbiol* **25**, 155–172.

Casadevall, A. & Pirofski, L. (1999). Host–pathogen interactions: redefining the basic concepts of virulence and pathogenicity. *Infect Immun* **67**, 3703–3713.

Casadevall, A. & Pirofski, L. (2000). Host–pathogen interactions: basic concepts of microbial commensalism, colonization, infection, and disease. *Infect Immun* **68**, 6511–6518.

Cohen, M. L. (2000). Changing patterns of infectious disease. *Nature* **406**, 762–767.

Collins, M. D. & Gibson, G. R. (1999). Probiotics, prebiotics, and synbiotics: approaches for modulating the microbial ecology of the gut. *Am J Clin Nutr* **69**, 1052S-1057S.

Cunningham, M. W. (2000). Pathogenesis of group A streptococcal infections. *Clin Microbiol Rev* **13**, 470-511.

Davies, H. D. & Schwartz, B. (1999). Invasive group A streptococcal infections in children. *Adv Pediatr Infect Dis* **14**, 129-145.

Desselberger, U. (2000). Emerging and re-emerging infectious diseases. *J Infect* **40**, 3-15.

Ernst, P. (2000). The disease spectrum of *Helicobacter pylori* – ulcers to cancer. *Annu Rev Microbiol* **54**, 615-640.

Fein, A. M. (1999). Pneumonia in the elderly: overview of diagnostic and therapeutic approaches. *Clin Infect Dis* **28**, 726-729.

Fredericks, D. N. & Relman, D. A. (1996). Sequence-based identification of microbial pathogens: a reconsideration of Koch's postulates. *Clin Microbiol Rev* **9**, 18-33.

Gaydos, C. A. & Quinn, T. C. (2000). The role of *Chlamydia pneumoniae* in cardiovascular disease. *Adv Intern Med* **45**, 139-173.

Gilligan, P. H. (1999). *Escherichia coli*. EAEC, EHEC, EIEC, ETEC. *Clin Lab Med* **19**, 505-521.

Gold, R. (1999). Epidemiology of bacterial meningitis. *Infect Dis Clin North Am* **3**, 515-525.

Graham, D. Y. (2000). *Helicobacter pylori* infection is the primary cause of gastric cancer. *J Gastroenterol* **35**, Suppl 12, 90-97.

Graham, M. R., Smoot, L. M., Lei, B. & Musser, J. M. (2001). Toward a genome-scale understanding of group A streptococcus pathogenesis. *Curr Opin Microbiol* **4**, 65-70.

Greenberg, D., Leibovitz, E., Shinnwell, E. S., Yagupsky, P. & Dagan, R. (1999). Neonatal sepsis caused by *Streptococcus pyogenes*: resurgence of an old etiology? *Pediatr Infect Dis J* **18**, 479-481.

Guerrant, R. L., Steiner, T. S., Lima, A. A. M. & Bobak, D. A. (1999). How intestinal bacteria cause disease. *J Infect Dis* **179**, Suppl 2, S331-S337.

Habash, M. & Reid, G. (1999). Microbial biofilms: their development and significance for medical device-related infections. *J Clin Pharmacol* **39**, 887-898.

Hentschel, U., Steinert, M. & Hacker, J. (2000). Common molecular mechanisms of symbiosis and pathogenesis. *Trends Microbiol* **8**, 226-231.

Hill, A. V. (1998). The immunogenetics of human infectious diseases. *Annu Rev Immunol* **16**, 593-617.

Huebner, J. & Goldmann, D. A. (1999). Coagulase-negative staphylococci: role as pathogens. *Annu Rev Med* **50**, 223-236.

Kerr, J. R. & Matthews, R. C. (2000). *Bordetella pertussis* infection: pathogenesis, diagnosis, management, and the role of protective immunity. *Eur J Clin Microbiol Infect Dis* **19**, 77-88.

Kononen, E. (2000). Development of oral bacterial flora in young children. *Ann Med* **32**, 107-112.

Lyczak, J. B., Cannon, C. L. & Pier, G. B. (2000). Establishment of *Pseudomonas aeruginosa* infection: lessons from a versatile opportunist. *Microbes Infect* **2**, 1051-1060.

McNicholl, J. M., Downer, M.V., Udhayakumar, V., Alper, C. A. & Swerdlow, D. L. (2000). Host-pathogen interactions in emerging and re-emerging infectious diseases: a genomic perspective of tuberculosis, malaria, human immunodeficiency virus infection, hepatitis B, and cholera. *Annu Rev Pub Health* **21**, 15–46.

Murphy, T. F. (2000). *Haemophilus influenzae* in chronic bronchitis. *Semin Respir Infect* **15**, 41–51.

Nataro, J. P. & Kaper, J. B. (1998). Diarrheagenic *Escherichia coli*. *Clin Microbiol Rev* **11**, 142–201.

Prier, R. & Solnick, J. V. (2000). Foodborne and waterborne infectious diseases. Contributing factors and solutions to new and re-emerging pathogens. *Postgrad Med* **107**, 245–252.

Qureshi, S. T., Skamene, E. & Malo, D. (1999). Comparative genomics and host resistance against infectious diseases. *Emerg Infect Dis* **5**, 36–47.

Schneider, R. F. (1999). Bacterial pneumonia. *Semin Respir Infect* **14**, 327–332.

Smith, H. (2000). Host factors that influence the behaviour of bacterial pathogens *in vivo*. *Int. J. Med. Microbiol* **290**, 207–213.

Stevens, D. L. (2000). Streptococcal toxic shock syndrome associated with necrotizing fasciitis. *Annu Rev Med* **51**, 271–288.

Tannock, G. W. (1999). Analysis of the intestinal microflora: a renaissance. *Antonie Van Leeuwenhoek* **76**, 265–278.

Papers

Fitzpatrick, P. E., Salmon, R. L., Hunter, P. R., Roberts, R. J. & Palmer, S. R. (2000). Risk factors for carriage of *Neisseria meningitidis* during an outbreak in Wales. *Emerg Infect Dis* **6**, 65–69.

Marshall, B. J., Armstrong, J. A., McGechie, D. B. & Glancy, R. J. (1985). Attempt to fulfil Koch's postulates for pyloric *Campylobacter*. *Med J Aust* **142**, 436–439.

Rennie, P. J., Gower, D. B. & Holland, K. T. (1991). *In-vitro* and *in-vivo* studies of human axillary odour and the cutaneous microflora. *Br J Dermatol* **124**, 596–602.

Tai, S. P., Krafft, A. E., Nootheti, P. & Holmes, R. K. (1990). Coordinate regulation of siderophore and diphtheria toxin production by iron in *Corynebacterium diphtheriae*. *Microb Pathogen* **9**, 267–273.

1.10 | Internet links

1. http://www.asm.org

   The website of the American Society for Microbiology has a great deal of information and links to other relevant sites. It can be accessed by non-members but many sections are available only to members.

2. http://www.socgenmicrobiol.org.uk

   The website of the Society for General Microbiology. Primarily about the activities of the society but also has links to other relevant sites.

3. http://www.sfam.org.uk

   The website of the Society for Applied Microbiology. Primarily about the activities of the society but also has links to other relevant sites.

4. http://www.amm.co.uk

The website of the Association of Medical Microbiologists. Not much accessible information here but a useful list of links to relevant sites.

5. http://gsbs.utmb.edu/microbook/toc.htm
   A complete Medical Microbiology textbook on line – easy to use and search.

6. http://www.bact.wisc.edu/microtextbook/toc.html
   A complete set of lecture notes from the undergraduate microbiology course at the University of Wisconsin-Madison.

7. http://www.cdc.gov/mmwr
   Free access to *Morbidity and Mortality Weekly Reports*, which documents data on notifiable infectious diseases.

8. http://pages.prodigy.net/pdeziel/
   Infectious disease weblink. A wealth of information plus many links to other sites.

9. http://www.phls.co.uk
   Public Health Laboratory Service of the United Kingdom. Data on infectious diseases in the United Kingdom and a list of useful links.

10. http://omni.ac.uk/subject-listing/WC100.html
    The communicable diseases section of OMNI (Organising Medical Networked Information). A list of well over 100 links to sites devoted to infectious diseases.

11. http://web.bham.ac.uk/bcm4ght6/res.html
    A website devoted entirely to one of our paradigm organisms – *E. coli*. Contains much useful information about this organism and has many links to related internet sites.

12. http://www.vet.uga.edu/mmb/pathogenic/WEBFILES/margie/Intro.htm
    A course on pathogenic microbiology from the University of Georgia College of Veterinary Medicine. Key concepts are well illustrated with photomicrographs and animated images.

13. http://www.mic.ki.se/Diseases/c1.html
    A jump-station (compiled by the Karolinska Institute, Sweden) to a huge number of websites covering all aspects of infectious diseases.

14. http://microbiol.org/vlmicro/index.htm
    The section of the WWW Virtual Library that deals with Microbiology and Virology. Basically, a jump-station divided into sections on culture collections, databases, educational sites, journals, government and regulatory sites, images, medical microbiology, university departments, virology, protista, and references to news groups, email lists, and other resources for the microbiologist.

15. http://www.cdc.gov/ncidod/index.htm
    The website of the National Center for Infectious Diseases of the USA. A wealth of information on all aspects of infectious diseases arranged according to the Center's major programmes – AIDS, STD, and TB Laboratory Research; Bacterial and Mycotic Diseases; Hospital Infections; Vector-Borne Infectious Diseases; Viral and Rickettsial Diseases. Particularly useful are the pages dealing (in great detail) with more than 200 different infectious diseases. (http://www.cdc.gov/ncidod/diseases/index.htm).

2

Bacterial cell biology

Chapter outline

Aims

The principle aims of this chapter are:

- to provide a basic understanding of those aspects of bacterial structure and function that are particularly relevant to disease pathogenesis
- to describe the various mechanisms used by bacteria to export virulence factors and other proteins
- to provide an introduction to bacterial genetics, with emphasis on virulence mechanisms and the evolution of pathogens
- to introduce the concept of bacterial biofilms

2.1 | Introduction

In the succeeding chapters we will, time and time again, refer to the means by which bacteria damage their host, interfere with host cell functions and avoid the defence systems mounted by the host. All of these features of the lifestyle of a pathogenic bacterium are attributable either to the molecules of which the organism is composed (mainly those located on its surface) or to those molecules that it secretes into its environment. A knowledge of bacterial ultrastructure and the means by which bacteria export molecules is, therefore, central to understanding bacterial virulence mechanisms. However, this is only part of the story. New developments in molecular and genetic techniques are enabling us to make rapid strides in determining how bacteria have acquired those genes that encode virulence factors as well as how these genes are regulated. Such information not only contributes to our knowledge and understanding of these fascinating organisms but also could form the basis of new approaches to controlling bacterial diseases. In this chapter, therefore, we will describe those bacterial structures that contribute to bacterial virulence, the means by which bacteria export molecules involved in disease pathogenesis and the genetic basis of bacterial virulence.

2.2 | Bacterial ultrastructure

Due to their extremely small size, little can be discerned of bacterial ultrastructure by light microscopy. Nevertheless, some important information can be obtained, both directly and with the use of certain stains (Gram stain, dyes to reveal capsule, flagella, etc.), such as bacterial shape and organisation, and this can provide some very important and rapid clues to the diagnostic microbiologist regarding the identity of the organism.

Electron microscopy has provided us with a more detailed view of bacterial ultrastructure (Table 2.1, Figure 2.1) and this has revealed the absence of a nuclear membrane enclosing the chromosomal DNA as well as the absence of other membrane-bound organelles. Outside the complex and rigid cell wall, there may be an additional layer – the capsule or glycocalyx – as well as one or more of a variety of surface appendages (fimbriae, flagella, fibrils). Some bacteria, notably the mycoplasmas, do not have a cell wall.

2.2.1 Size, shape and organisation

Most bacteria appear as rods (bacilli), as spherical cocci or as spiral-shaped cells. Cocci may be seen as single units. Often, however, they appear in arrangements that are characteristic of the species, for example pairs (*Neisseria gonorrhoeae*), chains (*Streptococcus* spp.), bunches (*Staphylococcus* spp.). Rod-shaped bacteria vary greatly in their length : width ratio; coccobacilli such as *Haemophilus* spp. are very short and stubby, whereas others (e.g. *Fusobacterium* spp.) grow as long filaments. Vibrios are curved bacteria, while spirochaetes have a tightly coiled morphology. In some instances there is no characteristic shape associated with an organism and these bacteria are described as pleiomorphic as is the case for mycoplasmas, which lack a rigid and shape-defining cell wall.

Table 2.1. | Bacterial ultrastructure

Structure	Function
Cytoplasmic membrane	Selectively permeable barrier, location of some important metabolic and biosynthetic activities, energy generation, environmental sensing
Ribosomes	Protein synthesis
Inclusion bodies	Storage, e.g. of carbon or phosphate
Nucleoid	Tightly coiled chromosomal DNA
Cell wall	Shape and mechanical strength
Periplasmic space	Nutrient processing and uptake
Capsule, slime layer, S layer	Adhesion, protection from phagocytosis and desiccation
Fimbriae, pili, fibrils	Adhesion, protection from phagocytosis, horizontal DNA transfer
Flagella	Motility
Endospore	Survival under harsh environmental conditions

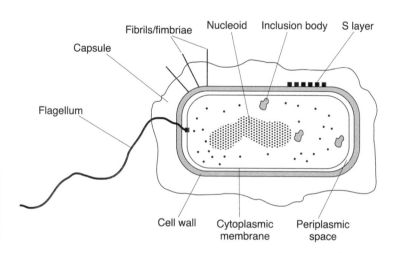

Figure 2.1 Schematic showing the major structures that may be present in a bacterial cell. Not all bacterial species possess all of these structures.

The size of bacteria is limited primarily by two key constraints. Firstly, the volume has to be sufficient to allow packaging of the bacterial chromosome and ribosomes. Secondly, the upper limit is determined by diffusion limitation; with a larger size comes a smaller surface area : volume ratio that restricts the diffusion of nutrients. Bacteria vary in size, with diameters ranging from 0.1 to 50 μm – the diameter of the 'average' bacterium being approximately 1.0 μm, although there have been recent reports of bacteria that lie outwith the limits set above.

Size matters

Epulopiscium fishelsoni is truly a monster among bacteria. This cigar-shaped organism lives in the intestines of Red Sea brown surgeonfish (*Acanthurus nigrofuscus*) and has a diameter of some 50 μm and a length ranging from 200 to 600 μm (i.e. up to 0.6 mm long – almost visible to the naked eye). Initially reported as a protist, DNA sequencing of its 16 S rRNA confirmed that *Epul. fishelsoni* is a bacterium related to the Gram-positive genus *Clostridium*. Diffusion limitation problems posed by its large size are overcome by having a highly convoluted plasma membrane that greatly increases its surface area. Decreasing in size from the behemoth *Epul. fishelsoni*, we pass through the size range occupied by most bacteria (diameter 1 μm). At the lower end of the range lie the mycoplasmas – pleomorphic, spherical or pear-shaped bacteria with a diameter of 0.3 to 0.8 μm. The smallest coccoid forms capable of independent growth are 0.2 to 0.3 μm in diameter, around the size of pox-viruses. *Mycoplasma genitalium* is an obligate intracellular parasite that has lost many biochemical functions and consequently has a reduced genome size (0.6 Mb compared with *Escherichia coli* at 4.6 Mb and *Streptococcus pneumoniae* at 2.0 Mb), which may account for its reduced size. Other limitations to size include packaging sufficient ribosomes for synthesis of the structural and biochemical components required by bacteria. Recently, the lower limit to bacterial size has been challenged by the description of tiny (diameter 0.05 to 0.5 μm) bacterium-like entities in human blood and urine. These so-called nanobacteria are capable of forming a calcium phosphate shell that renders them very resistant to heat and chemicals. There has been speculation that these are the equivalent of the fossilised structures found recently on a Martian meteorite. The occurrence of nanobacteria in kidney stones has led to the suggestion that the calcium carbonate shells formed by nano-bacteria act as foci for crystallisation and stone growth. However, the subject is highly controversial and many scientists suggest that the particles are not life forms but are merely unusual crystal structures. Nevertheless, rRNA genes have been amplified from them and DNA sequencing of these has placed nanobacteria in a group of bacteria that includes the genera *Brucella* and *Bartonella*, species of which are pathogenic to humans and animals. Watch this space (see Section 11.4)!

2.2.2 Cytoplasmic membrane

The innermost layer of the cell envelope in all bacteria is the cytoplasmic membrane. The unit membrane comprises a phospholipid bilayer in which integral membrane proteins are embedded. Peripheral membrane proteins may be associated with both the inner and outer leaflets. The bacterial membrane differs from that of eukaryotic cells in that it lacks cholesterol, although some bacterial membranes may contain hopanoids, cholesterol-like molecules that probably help to stabilise the membrane. However, the cytoplasmic membrane of *Mycoplasma* spp., which lack cell walls, may contain cholesterol. As well as being a semipermeable barrier, the cytoplasmic membrane is the site of respiration and also of the synthesis of cell wall components.

2.2.3 Cytoplasm

Contained within the cytoplasmic membrane is the cytoplasm itself. In contrast to eukaryotic cells, the prokaryotic cytoplasm is devoid of membrane-bound organelles, although it has abundant ribosomes. Ribosomes may also be associated with the inner surface of the cytoplasmic membrane. The cytoplasm is thought to have a gel-like consistency due to its high protein content. The cytoplasm of some bacteria may contain inclusion bodies – granules of glycogen, poly-β-hydroxybutyric acid or polyphosphate – for storage of carbon and phosphate, respectively.

The single circular bacterial chromosome is not delineated by a nuclear membrane, although it is often visible as a compact structure known as the nucleoid. The DNA in the nucleoid is, by necessity, tightly packed: the *E. coli* chromosome of 4.6 Mb has a length of approximately 1 mm, some 500 times the length of the bacterium itself. The DNA is tightly looped and coiled, probably with the aid of nucleoid proteins, while the negative charge of the molecule is partially neutralised by polyamines and Mg^{2+}.

Many bacteria also contain plasmids, i.e. small, usually circular, molecules of DNA that can replicate independently of the chromosome. Although not generally required for the viability of the organism, many plasmids carry antibiotic resistance genes or encode accessory functions related to virulence. Some plasmids, including those of the spirochaete *Borrelia burgdorferi* (which causes Lyme disease), are linear molecules.

2.2.4 Bacterial cell wall

Lying outside the cytoplasmic membrane is the cell wall, which provides mechanical strength and also defines the shape of the cell. On the basis of a simple staining procedure developed over 100 years ago by Christian Gram, bacterial cells can be divided into two major groups: Gram positive and Gram negative. Electron microscopy studies have revealed that this differential staining reflects quite marked differences in wall structure and organisation.

The structural polymer giving strength to cell walls is peptidoglycan and is found in both Gram-positive and Gram-negative walls, though it is present in greater proportions in the former. The peptidoglycan backbone is composed of alternating *N*-acetylglucosamine –

−G−M−G−M−G−[M−G]$_n$
|
L-Ala
|
D-Glu
|
−DAP   D-Ala −
|
D-Ala −DAP
|
D-Glu
|
D-Ala
|
−G−M−G−M−G−[M−G]$_n$

(a)   The side chains of *E. coli* peptidoglycan are cross-linked directly

−G−M−G−M−G−[M−G]$_n$
|
L-Ala
|
D-Glu
|
−[Gly]$_5$−L-Lys   D-Ala −[Gly]$_5$ −
|
D-Ala −[Gly]$_5$−L-Lys
|
D-Glu
|
D-Ala
|
−G−M−G−M−G−[M−G]$_n$

(b)   The side chains of *Staph. aureus* peptidoglycan are linked via a pentaglycine interbridge

Figure 2.2   Peptidoglycan structure. (a) The peptide side chains in the peptidoglycan of Gram-negative bacteria (e.g. *Escherichia coli*) are cross-linked directly. In contrast, the peptide side chains in the peptidoglycan of Gram-positive species are linked by a peptide. In (b) the structure of the peptidoglycan of *Staphylococcus aureus* is shown; in this case the cross-linking peptide is pentaglycine. G, *N*-acetylglucosamine; M, *N*-acetylmuramic acid; DAP, diaminopimelic acid.

N-acetylmuramic acid residues (Figure 2.2). Attached to the carboxyl group of *N*-acetylmuramic acid is a peptide side chain of four amino acid residues, some of which (D-glutamic acid, D-alanine, diaminopimelic acid) are not found in proteins. Peptidoglycan strands are interconnected either by direct cross-links between peptide side chains, or else by means of a peptide interbridge (Figure 2.2). The exact nature and composition of the side chains and cross-bridges is species dependent. Peptidoglycan synthesis is the target for a number of commonly used antibiotics such as the penicillins, cephalosporins and vancomycin.

Many of the wall components of disease-causing bacteria, including peptidoglycan, have important roles in host–bacteria interactions and in the induction of pathology.

2.2.4.1 Gram-positive cell wall

Peptidoglycan is the main component of the Gram-positive bacterial cell wall, being approximately 20 to 40 layers thick (Figure 2.3). The other major components of the wall are anionic polymers such as teichoic acid and lipoteichoic acid. Teichoic acids are polymers of glycerol or ribitol linked by phosphodiester bonds, and are covalently attached to *N*-acetylmuramic acid residues of the peptidoglycan. The amino acids D-alanine and L-lysine are common substituents of teichoic

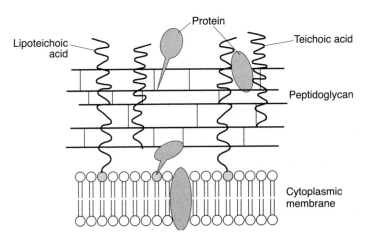

Lipoteichoic acid

Protein

Teichoic acid

Peptidoglycan

Cytoplasmic membrane

Figure 2.3   Cross-section through the wall of a typical Gram-positive bacterium. Peptidoglycan is the main constituent of the wall of such organisms

acids. Lipoteichoic acids are polymers of glycerol phosphate and are substituted with D-alanine and sugars. They are covalently attached not to peptidoglycan but to membrane lipids. The function of these polymers remains unclear, although they can function as adhesins (see Section 7.3.3) and may help to maintain cell wall structure. Additionally, a number of surface proteins of certain pathogens are held at the cell surface by interaction with teichoic acid and lipoteichoic acids. Thus pneumococcal surface protein A is anchored at the cell surface by interaction with choline residues attached to the teichoic acid and lipoteichoic acid of this bacterium. Similarly, InlB of *Listeria monocytogenes*, required for invasion, is associated with lipoteichoic acid.

Teichoic acids and lipoteichoic acids appear to extend to the surface of the peptidoglycan layer and may contribute to the net negative charge of Gram-positive bacteria. Additionally, both are antigenic and often constitute the major somatic antigens. Lipoteichoic acids and peptidoglycan may be recognised by the host through Toll-like receptor 2 (TLR2) and scavenger receptors and this initiates an innate immune defence against Gram-positive bacteria (see Section 6.6.2.1). In infections due to Gram-positive organisms, these molecules may contribute to inflammatory pathology due to their ability to stimulate the release of pro-inflammatory cytokines (Section 6.4.4).

Cell-wall-anchored proteins of Gram-positive bacteria can also play an important role in pathogenesis. These proteins may be involved in evasion of host defences, adhesion, host tissue destruction and induction of cytokine release. Important examples include the anti-phagocytic M proteins of *Strep. pyogenes*, immunoglobulin-binding protein A of *Staph. aureus*, and the invasion-associated InlA protein of *Lis. monocytogenes* (see Chapter 10). Many cell wall-associated proteins of Gram-positive bacteria are covalently linked to the amino acid side chains of peptidoglycan by an only recently defined mechanism (see Microaside, below). Others, as described above, interact with teichoic acid and/or lipoteichoic acid.

Sortase as a target for novel antimicrobials

Gram-positive pathogenic bacteria display an array of proteins on their surfaces that may interact with host cells and tissues and contribute to virulence. Many of these are anchored by a common mechanism that involves covalent attachment of the C-terminus of the protein to the cross-bridge of peptidoglycan, a reaction catalysed by the enzyme sortase. Mutants of *Staph. aureus* no longer expressing sortase were found to be defective in the assembly and display of surface proteins, and were greatly attenuated in virulence in a mouse model of infection. Thus the LD_{50} increased 100-fold. Sortase has been detected in many Gram-positive bacteria and displays a high degree of sequence conservation, particularly around the catalytic domain. This fact, along with the surface location (and thus accessibility) of sortase, makes it an ideal target for the development of novel therapeutics against Gram-positive bacteria (for further discussion of future antimicrobial agents, see Section 11.3).

2.2.4.2 Gram-negative cell wall

The cell wall of Gram-negative bacteria has a more complex multi-layered structure than that of Gram-positive species (Figure 2.4). The peptidoglycan layer is very much thinner (possibly only a single layer thick) and lies between the cytoplasmic membrane and a second membrane known as the outer membrane. The peptidoglycan is highly hydrated, with a low degree of cross-linking and is attached to the outer membrane by the so-called Braun lipoprotein. The fluid-filled region lying between the cytoplasmic and outer membranes is known as the periplasm or periplasmic space. The inner leaflet of the outer membrane consists mainly of phospholipids. In contrast, the outer leaflet is composed of lipopolysaccharide (LPS). LPS is an amphipathic molecule comprising three regions: (i) a glycolipid (lipid A) that is responsible for most of the biological activities of LPS, (ii) a core oligosaccharide containing a characteristic sugar acid, 2-keto-3-deoxy-octulonic acid (KDO), and other sugars including a heptose, and (iii) an antigenic polysaccharide composed of a chain of repeating oligosaccharide subunits (O antigenic side chain) (Figure 2.5). The lipid A component is embedded in the outer leaflet of the outer membrane and the remainder of the LPS molecule projects out from the cell surface, renders the bacterium highly hydrophilic, and may also contribute to the net negative charge of Gram-negative bacteria.

LPS is the main somatic antigen of Gram-negative bacteria and its structure is strain dependent. Although readily recognised by the immune system, some bacteria can rapidly change the nature of their O antigen side chains to avoid host immune defences. LPS can provoke a wide range of responses in animals and many of these are due to its ability, even at extremely low levels (pg/ml), to induce the release of cytokines by a variety of host cell types (see Sections 6.6.1.1

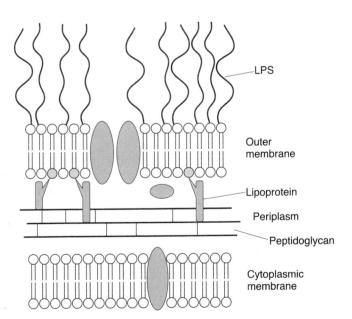

LPS

Outer membrane

Lipoprotein

Periplasm

Peptidoglycan

Cytoplasmic membrane

Figure 2.4 Cross-section through the wall of a typical Gram-negative bacterium. In these organisms, the wall is a multilayered structure consisting of peptidoglycan (a thinner layer than in the case of Gram-positive species) and an outer membrane. LPS, lipopolysaccharide.

Figure 2.5   Structure of LPS showing the three major regions of the molecule: lipid A, a core polysaccharide and an O antigenic side chain. The precise chemistry of the lipid A and polysaccharide components of LPS from a *Salmonella* spp. is provided. GlcN, glucosamine; KDO, 2-keto-3-deoxyoctulonic acid; Eth, ethanolamine; Hep, heptulose; Glc, glucose; Gal, galactose; NAG, *N*-acetylglucosamine; Rha, rhamnose; Man, mannose; Abe, abequose; P, phosphate.

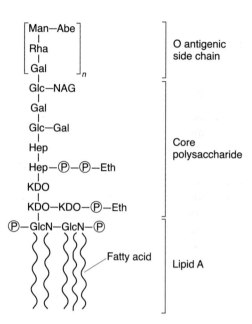

and 10.3). The receptor on cells responsible for the inflammatory effects of LPS is TLR4 and such signalling will be described in more detail in Sections 4.2.6.6.2 and 6.6.1.2.

As well as LPS and phospholipids, the outer membrane contains a large number of proteins comprising some 50% of the dry mass. These **outer membrane proteins (OMPs)** include porins that form trimers and span the outer membrane to create a narrow channel with a diameter of approximately 1 nm, allowing the passage of small, relatively hydrophilic molecules up to around 700 Da in size. Larger molecules may have their own transport proteins. Porins, which can also be potent inducers of cytokine release from monocytes and lymphocytes, are tightly associated with LPS, which is involved in assembling and maintaining porin structure.

2.2.4.3 Cell wall of *Mycobacterium* spp.

Members of the genus *Mycobacterium* are important pathogens of humans and animals and cause chronic diseases, including leprosy and tuberculosis. The peptidoglycan of mycobacterial cell walls is covered by a complex, lipid-rich layer comprising some 60% of the dry mass of the wall. A variety of lipids, glycolipids and lipoproteins are present and several of these contain mycolic acids, unique to the genera *Mycobacterium*, *Nocardia* and *Corynebacterium*. This layer renders the bacterial cell surface waxy and highly hydrophobic and, as a result, cells are inherently more resistant to many antimicrobial agents (and also to a variety of stains) due to the inability of these agents to penetrate the cell. A major component of the cell surface is lipoarabinomannan, a phosphatidylinositol-linked lipoglycan containing arabinan and mannose sugars. Lipoarabinomannan is essential for the survival of mycobacteria in the host, and is a potent inducer of cytokine release from certain immune cells.

2.2.5 Surface appendages

Many bacteria are motile and, in most cases, motility is due to the presence of one or more flagella. Flagella are long, thin appendages that are clearly visible by electron microscopy but cannot be seen by light microscopy unless stained. Their arrangement on the cell surface is species specific and varies from a single polar flagellum to multiple flagella arranged in tufts (lophotrichous), or distributed evenly over the cell surface (peritrichous). Motility permits the organism to move towards, or away from, beneficial or adverse environmental conditions, respectively, and there are bacterial species that display phototaxis, chemotaxis, aerotaxis and magnetotaxis.

Flagella are composed of multiple flagellin subunits polymerised in a helical arrangement to form a long (up to 20 μm) hollow filament that terminates in a complex structure – the basal body – which is anchored in the cytoplasmic membrane and cell wall. More than 40 genes are required for the structure, assembly and function of flagella in *Salmonella* and *Escherichia* spp. Other important pathogenic bacteria that express flagella include *Clostridium* spp., *Vibrio cholerae*, *Pseudomonas* spp., *Campylobacter* spp. and *Helicobacter pylori*. Flagellins are highly antigenic and constitute the H antigens of motile bacteria.

Fimbriae (also, confusingly, termed pili) are generally shorter (0.5 to 10 μm) and thinner than flagella, and are not involved in motility. Instead, many fimbriae appear to be involved in mediating the adhesion of bacteria to various surfaces, including host cells, and to other bacteria. Fimbriae are composed of helically-arranged identical protein subunits (fimbrillins or pilins) that range in size from 14 to 30 kDa. Some fimbriae are known to carry adhesins as minor components, either along their sides or at their tip. At least 34 different types of fimbriae have been identified on different types of *E. coli* strains, many of which provide different host cell specificities for attachment. A detailed description of the role of fimbriae/pili in adhesion is provided in Section 7.3.3. Fimbriae are expressed predominantly by Gram-negative bacteria, but are also found on some Gram-positive organisms including *Streptococcus* spp. and *Actinomyces* spp.

Conjugative (sex) pili are similar in structure to fimbriae, although longer and thicker, and there are generally only 1 to 10 per cell. Pili are involved in the horizontal transfer of DNA during bacterial conjugation. Additionally, however, they may act as receptors for certain viruses. For example, the toxin co-regulated pilus of *V. cholerae* is the receptor for the filamentous phage CTXφ, which carries the cholera toxin genes (see Micro-aside, p. 491).

Fibrils are shorter (up to 0.5 μm) and much thinner than fimbriae, and may be present in high numbers on the bacterial surface. Less is known about fibrils, though some from Gram-positive bacteria are known to be composed of large (about 300 kDa), wall-anchored proteins and are involved in adhesion.

2.2.6 Cell-surface-associated components

Many bacteria, especially when first isolated from their natural environment, have additional layers external to the normal wall structure. These include capsules, slime layers and crystalline S layers, and they often contribute to the virulence of the organism.

The capsule of a bacterium is a highly hydrated, gelatinous matrix composed of polysaccharide and/or protein which is loosely associated with the cell surface and is often shed by the bacterium. Capsules are highly antigenic and constitute the K antigens of bacteria. They may function in attachment to surfaces, resistance to phagocytosis, desiccation and hydrophobic toxic agents such as detergents, and also in the sequestration of nutrients from the environment. Capsules play an important role in virulence for many pathogenic bacteria including *Strep. pneumoniae* (see Micro-aside, below), *Pseudomonas* spp., *Klebsiella* spp., *Neisseria* spp. and *Haemophilus influenzae*. Slime layers, produced notably by *Staphylococcus* spp., are more hydrated and less well organised than capsules and are generally less firmly attached to the bacterial surface.

Some bacteria have a cell surface layer (an S layer) composed of a highly ordered two-dimensional array of protein or glycoprotein subunits, and this can constitute up to 15% of the total cellular protein of the bacterium. The major function of these crystalline S layers is not fully understood, but in pathogenic bacteria they may function like capsules in resistance to phagocytosis and in adhesion.

The capsule of *Streptococcus pneumoniae*

Streptococcus pneumoniae is a common human commensal organism that normally colonises the nasopharynx but is, nevertheless, an important cause of invasive diseases associated with high morbidity and mortality. The capsule of *Strep. pneumoniae* is a key virulence factor and is believed to provide protection against non-specific host defences including phagocytosis – strains lacking a capsule are severely attenuated in animal models of infection.

There are some 90 structurally distinct capsular serotypes of *Strep. pneumoniae*. Some serotypes are more strongly associated with invasive disease than others, and it is thought that the capsule structure may influence virulence. Antibody against capsular polysaccharide provides strain-specific protection from disease, and the current vaccine against the disease consists of a combination of purified capsular polysaccharide from the 23 serotypes most frequently associated with serious disease. Unfortunately, this vaccine is only poorly effective in the young and elderly, whose immune systems do not respond well to polysaccharide antigens. The next generation of vaccines for pneumococci is being developed in which the immune response to polysaccharide antigens is improved greatly by coupling them to specific pneumococcal surface proteins.

Approximately 19 genes are required for capsule synthesis and assembly and these are clustered in a 'capsule cassette' located in the same position on the chromosome of all pneumococcal strains. Interestingly, the genes that define the different chemical composition of each capsule are often flanked by genes that are common to most serotypes. This cassette-like arrangement allows shuffling of capsule-type-specific genes by homologous recombination (Figure 2.6): *Strep. pneumoniae* is naturally competent for transformation and can take up naked DNA from its environment, including the chromosomal region encoding a different capsular polysaccharide. There is disturbing recent evidence that a multiply antibiotic-resistant *Strep. pneumoniae* strain has given rise to several different serotype strains that are otherwise very closely related, if not indistinguishable, in overall genotype, including multidrug resistance.

2.3 | Bacterial cell cycle

For most bacteria, growth of an individual cell continues until the cell undergoes binary fission to produce two daughter cells of equal size. Four main phases can be recognised to comprise the bacterial cell cycle: bacterial growth, DNA replication, DNA segregation, and cell division. These processes are subject to tight temporal and spatial regulation to ensure that each daughter cell receives a copy of the chromosome, together with sufficient numbers of other macromolecules (e.g. ribosomes), monomers and inorganic ions to survive as an independent cell. The whole cycle can take as little as 20 minutes for *Escherichia coli* growing under ideal conditions, but is considerably longer in many other bacteria. The cell cycle of *E. coli* remains the best studied, although the asymmetric cell division by *Bacillus subtilis* during endospore formation (see Section 2.4) is also well characterised.

DNA replication is a complex process that involves over 30 different proteins. Initiation of replication of the circular bacterial chromosome occurs at a specific site, the **ori**gin of replication (*oriC*). This short

Figure 2.6 Serotype switching in *Streptococcus pneumoniae*. The genes defining capsule serotype specificity are flanked by conserved *cps* genes allowing homologous recombination. *Streptococcus pneumoniae* is naturally competent for transformation and can take up foreign DNA from the environment.

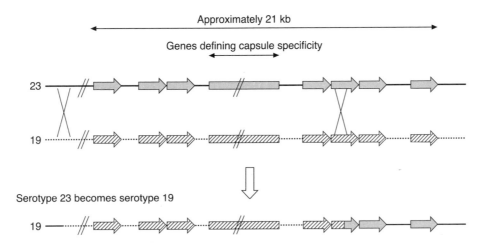

sequence is the site of binding of multiple copies of a protein (DnaA) that promotes unwinding of the DNA helix to allow assembly of the replication machinery that includes DnaB (a helicase to unwind the DNA in front of the replication fork) and DNA polymerases. DNA replication is bi-directional away from *oriC*, and DNA is synthesised at a rate of 1000 nucleotides per second. Replication terminates within a 450 bp region (*ter*) directly opposite *oriC*. This region contains two 'traps' that prevent continued movement of the two replication origins. The traps are unidirectional to allow full replication of the chromosome: the clockwise replication fork passes through one Ter site, designed to stop the counter-clockwise fork, before being trapped by the second Ter site, and vice versa. Termination by the traps requires Tus (termination **u**tilisation **s**ubstance), which acts as a contra-helicase and blocks further unwinding of the template DNA.

Once the replication forks meet in the *ter* region, separation of the daughter chromosomes occurs, a process requiring DNA gyrase and topoisomerase IV. This takes place within a few minutes of completion of DNA synthesis, and thus there is essentially no equivalent of the eukaryotic G_2 phase in bacteria (see Section 10.6.1.1). The DNA molecules are then segregated into those regions of the mother cell that are destined to become daughter cells. Partitioning of the chromosomes involves a number of Muk (Japanese mukaku = anucleate) proteins. MukB, which forms a complex with MukE and MukF, has a domain structure that resembles those of the eukaryotic motor proteins, kinesin and myosin, and can bind DNA as well as ATP and GTP. MukB is proposed to function as a motor protein driving the separation of chromosomes, possibly binding to, and using, the FtsZ ring (see below) to provide leverage.

Cell division involves the formation of a septum at the mid-point of the cell. The septum is an invagination of the cell wall and membrane and ultimately divides the parent cell into two daughter cells. The formation of the septum is directed by a cytoskeletal element known as the FtsZ or Z ring (fts is from temperature-sensitive filamentation). The Z ring is formed by polymerisation of tubulin-like FtsZ protein monomers and their assembly as an annular structure on the internal face of the cytoplasmic membrane at the mid-point of the cell. A whole series of other proteins are recruited to the Z ring. The functions of many of these are not fully understood, although FtsI is a septum-specific penicillin-binding protein required for the synthesis of peptidoglycan, while FtsK may be involved in resolution of chromosome dimers. Soon after cell division, a new Z ring appears and marks the future cell division site. Septum formation results in a double layer of peptidoglycan across the middle of the dividing cell, and this is cleaved by the action of specific peptidoglycan hydrolases. This is the end of the process for Gram-positive bacteria. In contrast, Gram-negative bacteria have an additional layer, the outer membrane, and invagination of this structure occurs as the separation of the double peptidoglycan layer takes place. This process requires EnvA (also known as LpxC), an enzyme involved in lipid A biosynthesis.

2.4 | Sporulation

Sporulation in bacteria describes the formation of a specialised structure – the endospore – which is very resistant to heat, desiccation, ultraviolet radiation and harsh chemicals. The endospore is a survival structure produced under adverse conditions, and spore-forming bacteria are found most commonly in the soil. Because of their remarkable heat resistance, and because many sporulating bacteria are dangerous pathogens (see Micro-aside, p. 60), sporulation is a considerable problem in food preservation and medical microbiology. Endospores can also remain viable for very long periods of time, certainly for decades and possibly even for many millions of years. For example, they have been found preserved in the guts of extinct bees embedded in amber from Dominica.

The endospore structure is complex (Figure 2.7). The outermost layer, the exosporium, is a thin proteinaceous covering within which lies the spore coat, comprising layers of spore-specific proteins. Underneath the spore coat is the cortex, which is composed of peptidoglycan with less cross-linking than that found in vegetative cells. Within the cortex, is the core, which comprises a normal cell wall (core wall), cytoplasmic membrane, cytoplasm, nucleoid, etc. The resistance shown by spores to extreme conditions is not fully understood, but may arise by several mechanisms. The spore protoplast contains only 10–30% of the water content of the vegetative cell, and this contributes to heat resistance. Dipicolinic acid (a spore-specific compound), complexed with calcium ions, may make up some 15% of the endospore dry weight and may stabilise spore nucleic acids.

Spore formation, best studied in *Bacillus* and *Clostridium* spp., involves as many as 200 genes encoding spore-specific structural proteins and regulatory proteins involved in a complex cascade of signalling. Entry into sporulation is established, in part, by changes in cell division and chromosome partitioning. The symmetrical cell division of vegetative growth is replaced by an asymmetric division to generate the smaller forespore and larger mother cell necessary for the production of the mature endospore. The bacterial cell signalling pathways regulating

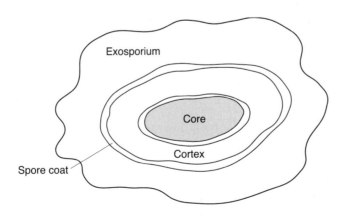

Figure 2.7 Cross-section through a bacterial endospore.

this switch have received intense study. There are a number of spore-specific genes switched on very early in the proceedings by alternative σ factors and other spore-specific transcriptional factors, and some of these are involved in the relocation of the Z ring to a polar position.

The conversion of an endospore back to an actively growing vegetative cell is a three-stage process involving activation, germination and outgrowth. Activation of spores, at least in the laboratory, is most easily achieved by heating the spores briefly. Many normal nutrients such as sugars and amino acids can then stimulate germination of the activated spore. Germination is relatively rapid and involves a loss of resistance to heat and chemicals, together with the degradation, or release, of spore components. The germinating spore swells due to water uptake and active synthesis of nucleic acids and proteins, and the cell emerges from the remains of the spore coat and eventually begins to divide in the normal vegetative fashion.

Spore-forming pathogenic bacteria include members of the genera *Clostridium* and *Bacillus*. The histotoxic clostridia, for example *Clostridium perfringens* and *Cl. septicum*, cause necrotising infections of skeletal muscle (clostridial myonecrosis) when spores contaminating tissue injuries germinate and the vegetative bacteria produce a barrage of toxins (see Chapter 9). Similarly, *Cl. tetani* spores can germinate in skin wounds and the vegetative cells produce tetanus toxin. Botulism is a form of food poisoning caused by the toxins of *Cl. botulinum*. The most common source of infection is home-canned food that has not been heated sufficiently to kill the spores. Within the can, the spores germinate and *Cl. botulinum* grows and produces a potent neurotoxin (BoNT; see Section 9.5.3.3.1). If the contents of the can are consumed without adequate heating to inactivate the toxin, disease (flaccid paralysis) ensues.

Inhalation anthrax and the germination of spores in lung macrophages

Anthrax is caused by *B. anthracis*, which infects mainly herbivores such as cattle and sheep but whose spores remain viable in soil and animal products for many decades. *Bacillus anthracis* produces an unusual three-component AB toxin responsible for the symptoms of anthrax (see Section 9.4.1.1.3). Infection is most commonly via a break in the skin and cutaneous anthrax ensues. If the spores reach the intestines, gastrointestinal anthrax may result. However, disease arising from inhalation of anthrax spores (pulmonary anthrax, or woolsorter's disease), though not common, has elicited much interest due to the spectre of biological weapons. The course of inhalation anthrax is dramatic and the mortality rate is high. The onset of disease may be influenza-like, but progression to hypotension and haemorrhage may occur within 1 to 2 days. A rapid decline culminating in septic shock, respiratory distress and death within 24 hours is not uncommon. The high mortality rate, rapid disease progression, the robust nature of the spores and the ease of dissemination mean that the potential use of anthrax as a biological weapon is taken very seriously.

When anthrax spores are inhaled they are rapidly phagocytosed by lung macrophages and can be carried via the lymphatic system to local lymph nodes. In a mouse model of inhalation anthrax, spores germinate within the phagolysosome of alveolar macrophages as early as 30 minutes after infection. There is also a rapid onset of expression of the anthrax toxin genes, carried by a large plasmid, pXO1, which also encodes genes required for germination. *Bacillus anthracis* is thus capable of responding rapidly to changes in environmental conditions, and it may be that the acid environment of the phagolysosome acts as the spore activation trigger. Once in the phagolysosome, spores can germinate and grow to fully formed, encapsulated vegetative cells within 30 minutes.

2.5 | Bacterial protein secretion systems

In order to survive and proliferate, bacteria must be able to sense, and interact with, their external environment. Hence, they must be able to respond appropriately to environmental changes, obtain nutrients from their surroundings and, in the case of those organisms that live in or on animals and plants, interact with host cells and tissues. The ways in which bacteria sense, and respond to, their environment are described in Section 4.3 and later in this chapter (Section 2.6.1). An important means by which bacteria obtain nutrients from their environment is by secreting enzymes (proteases, nucleases, polysaccharidases, lipases) to degrade macromolecules to their small molecular mass constituents, which are then taken up by the cell by diffusion or active transport. Interaction with host cells (described in greater detail in subsequent chapters – particularly Chapters 7, 8 and 9) is often mediated by proteins secreted by the bacterium. Both nutrient sequestration and interaction with host cells, therefore, require the export of proteins from the bacterium. It has been estimated that approximately 20% of the polypeptides synthesised by bacteria are destined for a location outside of the cytosol. In addition to secreted toxins and extracellular enzymes, such molecules will include proteinaceous surface structures (such as flagella) and other cell-surface-associated proteins, and integral cytoplasmic membrane and cell wall proteins. How, then, do bacteria orchestrate the translocation and correct targeting of proteins from their site of synthesis (the cytoplasm) to their ultimate destination? Although the ultrastructure of bacteria may seem quite primitive compared with that of eukaryotic cells (with their internal membrane-delineated organelles), Gram-negative bacteria in particular possess several distinct compartments. Thus there is the cytosol, cytoplasmic membrane, periplasm, outer membrane and, finally, the external environment. Secreted proteins must, therefore, be transported across all of the intervening barriers. Gram-positive bacteria are simpler as they lack an outer membrane and, consequently, a membrane-defined periplasmic space. Collectively, bacteria have a number of pathways available for the export of proteins to their final destination, and these include the general secretory

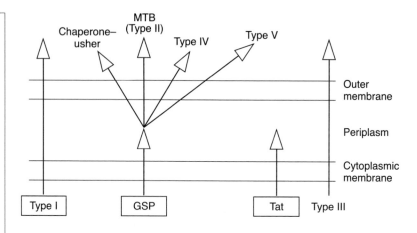

Figure 2.8 Overview of the principle protein translocation pathways of bacteria. Boxed systems are common to both Gram-positive and Gram-negative species. In Gram-positive bacteria (which lack an outer membrane) translocation results in the secretion or surface location of the transported protein. In contrast, the outer membrane of Gram-negative bacteria represents an additional barrier that proteins have to cross in order to be secreted (or surface-located). Note that the type IV system of *Helicobacter pylori* translocates the CagA protein directly into the target cell cytoplasm (see Section 2.5.3.3). GSP, general secretary pathway; MTB, main terminal branch; Tat, twin-arginine translocation.

pathway (GSP) and the type I to type IV secretion systems (summarised in Figure 2.8).

2.5.1 The general secretory pathway

The GSP is the major system by which proteins are translocated into, or across, the bacterial cytoplasmic membrane. This system is sufficient for the export of proteins of Gram-positive organisms. However, proteins destined for beyond the periplasm of Gram-negative bacteria need to be transported across the periplasm and inserted into, or transported across, the outer membrane by means of the so-called terminal branches of the GSP (see below).

Protein substrates of the GSP system (i.e. pre-secretory proteins) have a characteristic N-terminal signal sequence, or leader peptide, that is removed by an enzyme, known as signal peptidase, during translocation across the cytoplasmic membrane. The signal sequence is typically 18 to 26 amino acid residues long (often longer, up to 45 residues, in Gram-positive bacteria) and has a characteristic modular design (Figure 2.9). The central hydrophobic domain is important, and this tends to be more hydrophobic in integral membrane proteins. The primary amino acid sequence within the polar C-terminal region of the signal sequence is important for signal sequence recognition and cleavage by signal peptidase. For some periplasmic proteins of *E. coli*, the signal sequence can be dispensed with, although the translocation efficiency is reduced, indicating that sequence information in the mature portion of some proteins may also be involved in translocation signalling.

The GSP of *E. coli* (the best-studied organism) comprises a membrane-associated Sec (**sec**retory) translocase and one or more cytoplasmic molecular chaperones. The Sec translocase comprises at least five integral membrane proteins, SecD, E, F, G and Y. SecA is a peripheral membrane protein that binds as a homodimer to the SecYEG complex. Cytoplasmic SecB is a chaperone that is specific for exported proteins. SecB binds as a tetramer to a subset of nascent pre-secretory

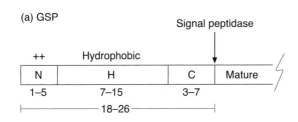

(a) GSP

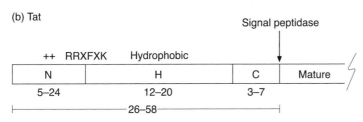

(b) Tat

Figure 2.9 Characteristics of signal sequences for (a) GSP-exported or (b) Tat-exported proteins (see Section 2.5.2). The numbers refer to the number of amino acid residues associated with each region. The twin arginine motif Arg-Arg-X-Phe-X-Lys (where X is any amino acid residue) straddles the N and H domains of the signal sequence of Tat-exported proteins. Signal peptidase cleaves at the points indicated in the diagrams to release the mature proteins.

proteins and maintains them in an unfolded and translocation-competent state. SecB is thought to interact with a large region (>180 amino acid residues) of the target protein. SecB then delivers the pre-secretory protein to the translocase by binding to membrane-associated SecA (Figure 2.10). Each SecA protein has two nucleotide-binding sites and undergoes significant nucleotide-induced conformational changes during protein translocation. The current working model hypothesises that pre-secretory proteins are translocated across the membrane in a stepwise fashion (approximately 2.5 kDa

Figure 2.10 The general secretory pathway. Polypeptides are delivered to Sec translocase either by SecB or via the FtsY-SRP system (for details, see the text). SRP, signal recognition particle.

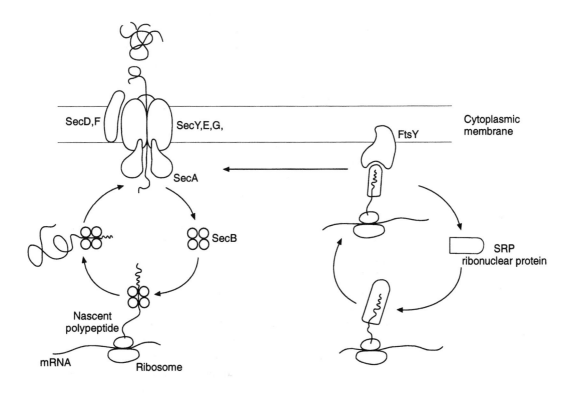

per step) by repeated nucleotide-induced cycles of SecA membrane insertion and de-insertion. SecD and SecF may function to prevent backsliding of the target protein in the translocation channel and may also be involved in release of the protein from the channel. No clear homologue of SecB has been found yet in Gram-positive bacteria.

SecB mutant strains of *E. coli* suffer a relatively mild translocation defect that affects only a subset of pre-secretory proteins, indicating that there are additional routes to Sec targeting. A second general targeting system was identified by searching for homologues of the eukaryotic signal recognition particle (SRP) system. The eukaryotic SRP is a ribonucleoprotein complex comprising a 7 S RNA molecule associated with six different polypeptides. A 54 kDa protein (SRP54) interacts with signal sequences. The SRP receptor comprises a peripheral membrane protein, SRα, and an integral membrane component, SRβ. *Escherichia coli* has a simplified version of the eukaryotic system that has three components: Ffh, a homologue of SRP54 (Ffh stands for fifty four homologue) and 4.5 S RNA comprise the prokaryotic SRP; and FtsY, a homologue of SRα, comprises the receptor for SRP (Figure. 2.10). Ffh, 4.5 S RNA and FtsY can partially substitute for the corresponding component in eukaryotic systems, and vice versa. SRP binds to nascent pre-secretory proteins after the first 70 to 150 amino acid residues have been incorporated, and hydrophobic sequences within the signal sequence are important for Ffh binding. Binding of SRP arrests further protein synthesis, and the SRP–ribosome complex is then targeted at the membrane-associated FtsY. Once bound at the membrane, SRP is released from the pre-secretory protein, thus relieving the translational arrest, and the pre-secretory protein is delivered to the Sec translocase (Figure 2.10). SRP-dependent translocation is thus co-translational, as opposed to SecB-dependent post-translational translocation. The primary substrates for the SRP targeting system are integral membrane proteins, which may explain the preference of SRP for highly hydrophobic signal sequences. By arresting translation, the SRP system may prevent misfolding and aggregation in the cytoplasm of highly hydrophobic integral membrane proteins.

Thus SecB and SRP form two different targeting routes that converge at the Sec translocase. When the pre-secretory protein is not recognised by SecB or SRP, the signal sequence of that protein may target it directly at membrane-associated SecA. In *E. coli*, other chaperones such as DnaK and its associated proteins, DnaJ and GrpE, might also contribute to translocation in the absence of SecB or SRP chaperones.

2.5.2 The Tat export pathway

Before we move on to deal with the terminal branches of the GSP and with other specific export pathways, mention should be made of a relatively recently described distinct second general pathway. This pathway has been termed the twin-arginine translocation (Tat) system as substrate proteins have a characteristic amino acid motif that includes invariant consecutive arginine residues within the signal sequence (see Figure 2.9). As we have just learned, proteins secreted by the Sec system are generally bound by cytoplasmic chaperones to prevent folding prior to export and are 'threaded' through the Sec translocase in an extended conformation. In contrast, proteins targeted to the Tat system are already folded. In most cases, the substrates of the pathway are proteins that must bind co-factors in the cytoplasm and hence fold prior to translocation. These proteins function predominantly in the respiratory and photosynthetic electron transport chains. Some translocated proteins even pre-assemble into oligomeric complexes prior to transport, and a Tat-specific signal sequence on a single subunit may be sufficient to mediate translocation of the whole complex. Other proteins lacking co-factors may fold too fast, or else require cytoplasmic 'foldases', to obtain the correct conformation, and thus must use the Tat system rather than Sec. Analysis of the *E. coli* genome sequence indicates that 22 gene products may be Tat-exported. Apart from *E. coli*, many other bacteria, both Gram-positive and Gram-negative, have a Tat export system. Notable exceptions include obligate intracellular parasites, methanogens and organisms that are wholly fermentative and therefore do not possess membrane-associated respiratory chains. Like SRP, the Tat system has its eukaryotic counterpart, being closely related to the ΔpH-dependent protein import pathway of the plant chloroplast.

The signal sequence of Tat-exported proteins has a similar tripartite structure to Sec-dependent proteins. However, the Tat signal contains the twin-arginine motif and the H domain is generally less hydrophobic. The lower hydrophobicity of the signal sequence, in conjunction with the occurrence of basic residues within the C region, is thought to function as a Sec-avoidance mechanism. The twin arginines are absolutely required for translocation, and replacing a Sec-specific signal sequence with a Tat sequence is sufficient to redirect the protein via the Tat pathway.

What, then, is known about the Tat translocase machinery? Clearly, any pore formed in the membrane to allow translocation of folded proteins and protein complexes must be relatively large – probably up to 7 nm in diameter. Such a pore would require careful control to prevent full movement of ions across the membrane, and this might be achieved using a translocase capable of dilating and contracting to fit snugly around the translocating protein as it crosses the membrane; however, this is still purely speculative. In *E. coli*, four integral membrane proteins are involved in Tat translocation. Three of these, TatA, TatB and TatC are encoded by a four-gene operon that also encodes the cytoplasmic protein TatD. The fourth transmembrane protein, TatE, is encoded at a separate chromosomal location. TatC has six transmembrane domains and is essential for translocation. TatA, B and E each have a single transmembrane domain and share some sequence homology. Each may interact with a different subset of folded proteins. TatD does not appear to be essential for Tat-mediated translocation. How these components assemble to form the translocation apparatus is, as yet, unknown.

2.5.3 Terminal branches of the GSP

The GSP delivers proteins into or across the cytoplasmic membrane. For Gram-positive organisms this is sufficient, and exported proteins may be released from the cell or else attached covalently, or non-covalently, to various cell wall polymers. In Gram-negative bacteria, however, the GSP delivers the protein only as far as the periplasm. For some proteins, this may be the final location. However, outer membrane proteins and secreted proteins must be inserted into, or translocated across, the outer membrane, respectively.

The insertion of proteins into the outer membrane is only poorly understood and may involve either a periplasmic intermediate or else direct insertion at, as-yet-unconfirmed, contact sites between the inner and outer membrane, the so-called Bayer's bridges. For secretion, there are several minor branches of the GSP. These include autosecretion of proteins, for example the immunoglobulin (Ig) A1 protease of *N. gonorrhoeae*, where C-terminal sequence information is sufficient to direct outer membrane translocation. The list of so-called autotransporter proteins is growing and includes a number of virulence factors of Gram-negative bacterial pathogens (see Micro-aside below). Other proteins have dedicated integral outer membrane components that mediate secretion, as is the case for the *Serratia marcesens* haemolysin ShlA. However, the great majority of secreted proteins of Gram-negative bacteria, including many exoenzymes and exotoxins involved in disease, are translocated across the outer membrane by means of the main terminal branch of the GSP, also known as type II secretion (types I, III and IV secretion will be discussed soon). Another important terminal branch of the GSP is the chaperone–usher pathway, which is important for the assembly of certain structures such as fimbriae and pili.

D-I-Y protein translocation

This terminal branch of the GSP was first described for the IgA1 protease of *N. gonorrhoeae* and was initially termed type IV secretion. Recently, however, the system has been renamed type V secretion, although the term autotransporter is preferred, since, as the name implies, all the information required for translocation across the outer membrane is contained within the precursor of the secreted protein itself. Autotransporter proteins comprise three domains: (i) an N-terminal signal sequence that directs translocation of the protein into the periplasm via the Sec system (some autotransporters have an extended signal sequence, although the significance of this is not yet known); (ii) a surface-located or secreted mature protein (termed the passenger or α-domain); and (iii) a C-terminal domain (termed the β- or pore-forming domain) that mediates translocation across the outer membrane. The β-domain is predicted to form a 14-stranded β-barrel within the outer membrane to create a pore. This is a structural motif that appears also in bacterial pore-forming toxins (see Section 9.4.1.1).

The current model of autotransporter secretion proposes that the protein enters the periplasmic space via the Sec system, with concomitant cleavage of the signal sequence. Once in the periplasm, the β-domain spontaneously inserts into the outer membrane to create a pore. The passenger domain, still attached to the β-barrel, is translocated through the pore to the cell surface. Once on the surface of the bacterium, the autotransporter may be cleaved to release the passenger domain into the external milieu. This is the case for the IgA1 proteases, although whether the processing is autoproteolytic or whether it requires a surface-located accessory protease is not clear. For some autotransporter proteins, cleavage does not result in secretion of the passenger domain, which instead remains associated non-covalently with the β-domain. A third group of autotransporters are not processed and remain anchored at the cell surface by the β-domain. This is the case for the Hsf fibrillar protein that mediates the adhesion of *H. influenzae* to human respiratory epithelial cells. Autotransporter proteins have been shown to function in adhesion, invasion and serum resistance. Secreted autotransporters include the IgA1 proteases, which may function in immune avoidance, and other serine proteases. The vacuolating cytotoxin of *Hel. pylori* (VacA) is one of several autotransporter toxins.

2.5.3.1 The main terminal branch (type II secretion)

The main terminal branch (MTB), or type II secretion system, was first described in *Klebsiella oxytoca* and subsequently shown to be widespread in Gram-negative bacteria. The system is complex and involves some 12 to 14 components, most of which are, perhaps paradoxically, integral cytoplasmic membrane proteins. The protein substrates for translocation require to be folded correctly in the periplasm before translocation. Indeed, some proteins must first be organised into higher-order structures. Thus the B pentamer of cholera toxin (see Section 9.5.1.1) must be assembled prior to secretion, and export of the A subunit also requires the B pentamer. It is not yet clear how the system distinguishes between periplasmic proteins and those destined for secretion, although tertiary structure information is thought to play a role. The secretion apparatus of *Pseudomonas aeruginosa*, responsible for the export of virulence factors such as lipase, elastase, alkaline phosphatase and exotoxin A, is composed of 12 proteins. Eleven of these components are integrated into, or associated with, the cytoplasmic membrane. One component is an integral protein of the outer membrane. It is fascinating to note that five components of the *Pseudomonas* system have signal sequences typical of type IV pilin subunits and also share some sequence homology with the pilins. Other non-pilin components of the type II system also have counterparts in the type IV pilin biosynthetic system. Type IV pili are involved in adhesion and twitching motility of many Gram-negative bacteria. Like pilin subunits, the *Pseudomonas* pseudo-pilins are processed in the periplasmic space by prepilin peptidase. It is very tempting to speculate that the pseudo-pilins constitute a pilus-like structure between the cytoplasmic and outer membranes for translocation of proteins. However, as yet there is little direct evidence that a stable transperiplasmic structure is formed, and it is worth remembering that type II secreted proteins are first delivered to the periplasm by the GSP and are folded in the periplasm, and thus would probably have to enter such an apparatus from the side. Nevertheless, very recent work has demonstrated that, when the pullulanase type II secretion system of *K. oxytoca* was overexpressed in *E. coli*, one pseudo-pilin, PulG, was assembled into pilus-like bundles. The pullulanase system is probably best understood and has two additional components: PulS, an outer membrane lipoprotein, and PulE, a cytoplasmic protein that may associate with the Pul complex at the internal face of the cytoplasmic membrane (Figure 2.11).

The outer membrane component of the type II system is a member of a larger family of outer membrane proteins, the secretins. Secretins function also in type III secretion, type IV pilus assembly and in the extrusion of filamentous phage. The secretins are rich in β-sheet structure and form homomultimers of some 10 to 12 units that define a transmembrane pore of up to 9 nm, sufficient for the export of folded proteins and protein complexes, and for the extrusion of filamentous phage and pili. As with the Tat system, careful control is required to prevent the free movement of ions across the membrane.

In the absence of solid evidence for the existence of a transperiplasmic pseudo-pilus tube, a second mechanism has been proposed for type II secretion. In this model, proteins are delivered to the periplasm by the GSP and fold into their final structure. These then assemble at the, presumably shut, secretin complex in the outer membrane. Transient

Figure 2.11 Terminal branches of the general secretory pathway. Proteins delivered to the periplasmic space by Sec translocase are fully exported by one of three major terminal branches, the chaperone–usher system (a), type II secretion system (main terminal branch) (b) and type IV secretion system (not shown). Other means exist of escaping the periplasmic space, such as dedicated transporters.

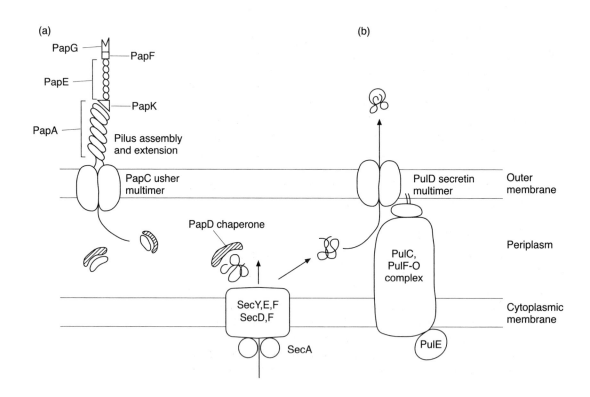

construction of a pseudo-pilus then shunts the proteins through the secretin channel, resulting in their secretion.

In many Gram-negative bacteria, natural competence for DNA uptake requires a subset of the type II secretion/pilus biosynthetic system components. Furthermore, competence in the Gram-positive bacterium *B. subtilis* requires four pre-pilin-like proteins and two further proteins, all of which have homology to components of the above systems.

2.5.3.2 The chaperone–usher pathway

This system is required for the synthesis of a number of adhesive pili or fimbriae, including type I and P pili that are widespread amongst the Enterobacteriaceae. These structures can be relatively complex. P pili, for example, are composite fibres composed of five different proteins. PapE subunits form a short, slim tip fimbrillum that is joined to the end of the main pilus rod (polymerised PapA subunits) by PapK. The PapG adhesin is attached to the tip fimbrillum by PapF accessory protein, and PapH terminates the rod pilus and is thought to anchor the pilus in the membrane (Figure 2.11). Despite such a complex organisation, only two pilus-specific proteins, a chaperone (PapD in the case of P pili) and an usher (PapC), are required for subunit export and pilus assembly.

Sec translocase normally releases its substrates into the periplasmic space. However, the chaperone component is required for release of pilus subunits. The chaperone binds to pilus subunits as they enter the periplasm and facilitates correct folding. Binding of the chaperone also caps the pilin interactive surfaces and prevents the aggregation of subunits within the periplasm. The chaperone–pilin complexes are then targeted to the outer membrane usher assembly sites where the chaperone dissociates from the subunit. Ushers are members of the secretin family of outer membrane proteins that were described above, and form homomultimers (6 to 12 subunits) defining a central transmembrane pore approximately 2 to 3 nm in diameter. This is wide enough to accommodate the tip fimbrillum but not the assembled 6.8 nm diameter helical pilus rod. It is proposed that oligomerised pilus subunits are exported across the outer membrane as a linear fibre, which, upon reaching the outer surface of the bacterium, 'snaps' into the 6.8 nm diameter helical pilus rod.

2.5.3.3 Type IV secretion

Pertussis toxin (PT), produced by *Bordetella pertussis*, is a slightly unusual AB$_5$ toxin that is composed of five different protein subunits (see Section 9.5.2.1). Two copies of subunit S4, together with single copies of subunits S2, S3 and S5, form a pentameric ring that mediates host cell receptor binding and translocation of the toxigenic S1 subunit (an ADP ribosyltransferase). The individual subunits of PT are delivered to the periplasm of *Bord. pertussis* by the Sec pathway. Secretion of fully assembled PT then requires a dedicated type IV system. This secretion apparatus is assembled from nine gene products encoded on the same operon as the *ptx* genes. The *ptl* (pertussis toxin liberation) genes are closely related to the *tra* genes required for conjugal transfer of self-transmissible plasmids, and to the *vir* genes of *Agrobacterium tumefaciens* required for transfer of oncogenic DNA into plant cells. In the latter systems, a transfer pilus is synthesised that traverses the membranes of donor and recipient (bacteria and plant) cell. Like the type II system, many of the type IV constituents are membrane associated and one has homology to the pilin proteins. It is not clear whether a secretion pilus is formed or whether PT is injected directly into target cells.

A full understanding of the type IV secretion system has yet to be obtained, but it does provide an excellent example of the ability of bacteria to appropriate and adapt groups of proteins to fit new needs. The question arises, however, as to why PT is not secreted by the type II system, which efficiently exports other AB toxins, including the structurally related cholera toxin.

More recent work has revealed that other bacteria utilise type IV systems to export proteins and/or DNA. In the case of *Legionella pneumophila* and *Hel. pylori*, the type IV system is used to deliver bacterial proteins into the host cell (see also the type III secretion system described in Section 2.5.5). The protein delivered by the intracellular pathogen *Leg. pneumophila* (LcmX) is essential for the establishment of an intracellular organelle within the host that avoids fusion with endocytic vesicles. *Helicobacter pylori* translocates the CagA protein that is responsible for activation of host cell signal transduction leading to cytoskeletal reorganisation and induction of an inflammatory response. CagA, and its associated type IV secretion system, are encoded on a 40 kb pathogenicity island (*cai* pathogenicity island, see Section 2.6.4). Evidence suggests that CagA is exported directly from the cytoplasm of *Hel. pylori* into the target cell cytoplasm. As such, the *Hel. pylori* type IV system more closely resembles the *vir* system of *Agrobacterium tumefaciens* than the *ptl* system of *Bord. pertussis*. As more bacterial genome sequences are completed, potential type IV systems are appearing with increasing frequency, for example in species belonging to the genera *Rickettsia*, *Actinobacillus* and *Brucella*.

2.5.4 Type I secretion

As we have seen, most secreted proteins have cleaved N-terminal signal sequences and cross the cytoplasmic membrane via the Sec or Tat systems. Many bacteria also possess type I dedicated systems for the export of specific proteins, peptides or, indeed, non-protein substrates. These export systems are part of a larger family of so-called ABC (ATP-binding cassette) transporters, present also in eukaryotes, which mediate import or export of a wide range of substances. The name ABC transporter is derived from the characteristic ATP-binding component that provides the energy for transport via hydrolysis of ATP. Associated with this are transmembrane components (with 6 or 12 transmembrane helices) thought to provide a translocation channel. In importer

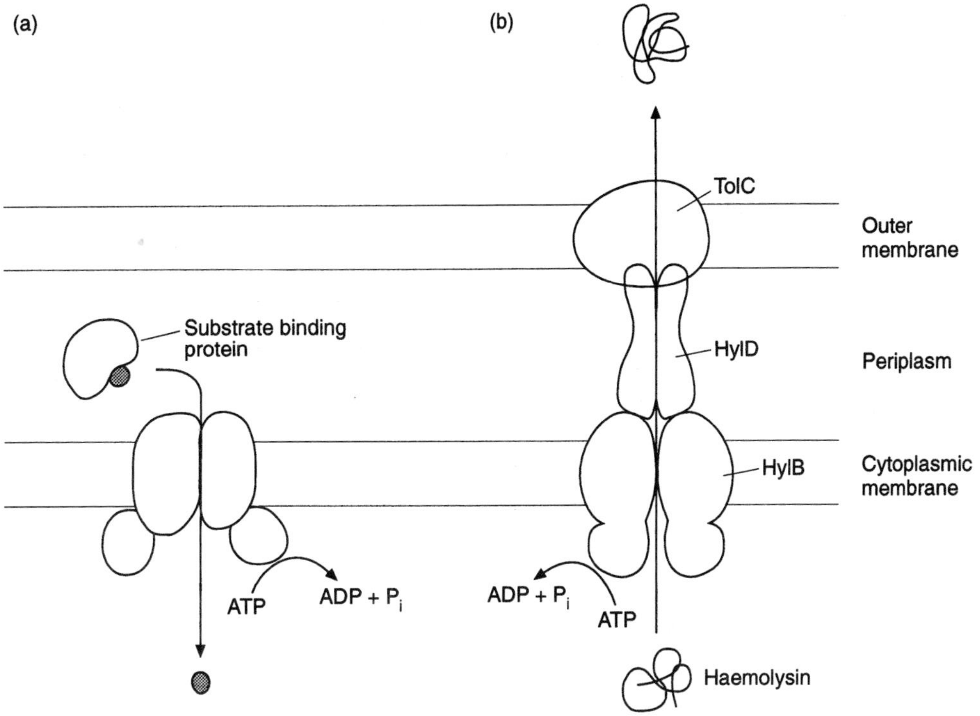

Figure 2.12 ABC transporter systems. (a) In most uptake systems, the ATP-binding cassette components, which provide energy for the translocation process by hydrolysis of ATP, are separate from the transmembrane components. (b) In contrast, for many export systems, these protein components are fused with the transmembrane pore-forming components. Note that in Gram-positive bacteria, which lack an outer membrane, the substrate-binding component of the ABC importer is lipid-modified and thus anchored in the membrane.

systems, these components are generally separate (Figure 2.12). In contrast, for most export systems, these components are fused to produce a single polypeptide (Figure 2.12). In Gram-negative systems, there is an accessory factor that may serve to bridge the periplasmic space, as well as, in some cases, an outer membrane component. The accessory factor is generally part of the operon encoding the ABC transporter. Interestingly, some ABC exporters of Gram-positive bacteria possess homologues of the accessory factor. Given the lack of an outer membrane and associated periplasm, the function of the accessory factor of Gram-positive bacteria remains to be determined.

The best-defined type 1 protein secretion system is that required for export of the RTX (repeat in toxin) toxin, haemolysin, of *E. coli* (Figure 2.12, see also Section 9.4.1.4.1). The gene encoding the outer membrane component, TolC, is not linked to the *hyl* operon. RTX toxins are produced by a range of other Gram-negative bacteria and, in all cases, the toxin is secreted by a dedicated type I transport system. The secretion signal lies in the C-terminus of the HylA toxin and is not removed during secretion. Other proteins secreted by type I systems include an alkaline protease of the opportunistic pathogen *Ps. aeruginosa*. Type 1 secretion systems are also responsible for the export of some antibacterial peptides (bacteriocins, lantibiotics) of both Gram-positive and Gram-negative bacteria. These peptides vary in length (generally less than 11 kDa in size) and often undergo significant processing and unusual post-translational modifications.

2.5.5 Type III secretion

Type III secretion systems mediate the injection of proteins (known as 'effector' proteins) directly into eukaryotic cells. They are not, therefore, secretion systems in the true sense (i.e. they do not export bacterial products to the external environment) but are specialised systems for delivering specific virulence factors directly into the cytoplasm of the target host cell. As protein translocation was initially thought to be triggered by contact of the bacterium with the target cell, the process was also termed 'contact-dependent secretion'. It is now known, however, that environmental signals can also trigger secretion in *Salmonella* and *Shigella* systems. Type III secretion systems are found in a number of Gram-negative pathogens, where they play an important role in virulence, including adhesion, invasion and cytotoxicity (Table 2.2, see also Chapters 7, 8 and 9). As well as being present in the organisms listed in Table 2.2, the system has also been detected (but less extensively studied) in other organisms including *Chlamydia*

| Table 2.2. | Type III secretion systems of human pathogens |

Organism	Secreted proteins	Role in virulence	Genetic location
Yersinia spp.	Yop	Cytotoxicity, inhibition of phagocytosis	Plasmid
Yersinia enterocolitica	Ysp	Initial invasion of Peyer's patches?	Chromosomal
Salmonella typhimurium	Sip	Cell invasion, induction of apoptosis	SPI-1 pathogenicity island[a]
	Sse	Proliferation in macrophages	SPI-2 pathogenicity island[a]
Shigella flexneri	Ipa	Invasion, apoptosis	Plasmid
Enteropathogenic *Escherichia coli*	Esp, Tir	Adhesion, induction of attaching and effacing lesions	LEE pathogenicity island[a]
Pseudomonas aeruginosa	Exo	Cytotoxicity, inhibition of phagocytosis	Chromosomal

[a] See Section 2.6.4.

pneumoniae, Chl. trachomatis, Bord. bronchiseptica, Bord. pertussis and *Burkholderia pseudomallei*.

Like type I secretion, the type III system is Sec-independent (although some components of the secretion apparatus itself are transported by the Sec system), and substrates for translocation lack a cleaved signal sequence and, indeed, a periplasmic intermediate form. One of the first type III systems described was that of pathogenic *Yersinia* spp. (see also Section 9.5.4), and this system will serve as a paradigm. Some six translocated effector proteins (designated 'Yop' – ***Y****ersinia* **o**uter membrane **p**rotein) have been characterised, and these have a range of cellular effects that are mediated, on the whole, by interruption or modification of host cell signalling pathways (see Table 9.9; for details of signalling mechanisms, see also Chapter 4; and, for a fuller description of the interaction of *Yersinia* spp. with host cells see Section 8.2.1.1.1).

The type III secretion apparatus is complex, involving up to 29 gene products and the majority of these are thought to be associated with the cytoplasmic membrane (Figure 2.13). Although the translocated effector proteins differ in sequence and function, many of the apparatus components (designated 'Ysc' – **Y**op **s**e**c**retion) are well conserved among bacteria. Additionally, some components demonstrate significant homology to components of flagellar biosynthesis systems of Gram-positive and Gram-negative bacteria. Nevertheless, a definitive model for type III secretion systems has yet to be obtained. In the

Figure 2.13 Contact-dependent type III secretion. Under normal growth conditions, YopN is thought to act as a plug blocking secretion of Yop effectors. Upon contact with a suitable host cell, YopN is shed into the medium, and the YopB/YopD complex is inserted into the host cell membrane with the help of LcrV to create a pore through which the Yop effectors are translocated. Energy for translocation is provided by ATP hydrolysis mediated by YscN.

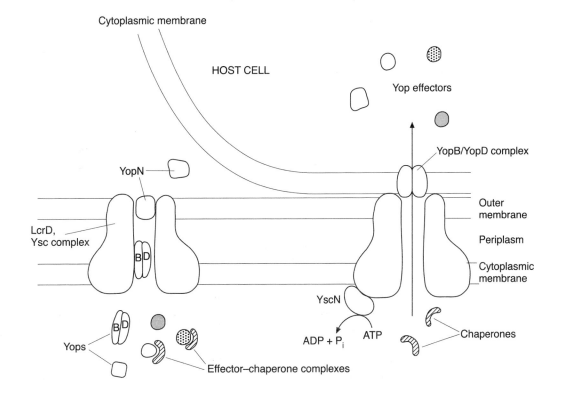

Cytoplasmic membrane

HOST CELL

Yop effectors

YopN

YopB/YopD complex

LcrD, Ysc complex

Outer membrane

Periplasm

Cytoplasmic membrane

YscN

Yops

ADP + P$_i$ ATP

Chaperones

Effector–chaperone complexes

Yersinia system, LcrD (also known as YscV) is an integral membrane protein with six or possibly eight transmembrane helices and may define a channel across the cytoplasmic membrane. Other proteins thought to be involved in the formation of such a channel include YscD, YscR, YscS and YscT. YscU is also in the cytoplasmic membrane and this may interact with the YscN cytosolic ATPase required for energising translocation. YscJ is an outer membrane lipoprotein that may also be anchored via its C-terminus in the cytoplasmic membrane and thus traverse the periplasmic space. A second lipoprotein, YscW, is important for the correct insertion of YscC into the outer membrane and may also be part of the transperiplasmic apparatus. YscC is the only confirmed outer membrane protein and is a member of the secretin family that we have already encountered (type II secretion, chaperone–usher pathway). This protein forms a ring-shaped structure in the outer membrane that has a central pore with a diameter of about 5 nm. The substrates for translocation have no definable amino acid sequence directing them to the translocation apparatus. Instead, the structure of the 5′ end of the mRNA may be involved in targeting via an as yet unknown mechanism. Cytosolic chaperones are required for secretion of some Yops. For example, SycG and SycH serve as chaperones for YopE (cytotoxic factor) and YopH (inhibition of phagocytosis), respectively, and these may also help to target their respective Yop to the secretion apparatus.

An intriguing feature of this system is that, in *Yersinia* spp., it is contact dependent. Thus proteins are not translocated until the bacterium makes contact with a host cell. YopN (possibly in association with TyeA and LcrG) is a regulator of secretion. This protein is associated with the bacterial surface but, on contact with the target cell, or in conditions that are thought to mimic this event (low calcium ion concentrations), YopN is released into the external environment (Figure 2.13). YopN is thought to act as a channel 'plug' preventing early release of Yop effectors. Nevertheless, Yop secretion is also tightly regulated at the level of transcription and translation. The first Yops to be translocated through the secretory apparatus are YopB and YopD. These hydrophobic proteins form a channel in the host cell membrane, thereby enabling the translocation of other Yops into the host cell. Several effector proteins are injected into the host cell and these have a variety of effects such as interfering with cytoskeletal rearrangements and cytokine release as well as inducing apoptosis. These effects will be described in more detail in Sections 8.2.1.1.1, 9.5.4.1, 10.3.3, 10.4.2 and 10.6.2.2.1.

As well as the plasmid-encoded Yop system, *Y. enterocolitica* has two chromosomally encoded type III systems, one of which is required for the secretion of at least eight Ysp effector proteins that are thought to be involved in the early stages of the infection process. Furthermore, the flagellar biosynthetic apparatus has been shown to mediate also the secretion of several proteins including the phospholipase virulence factor YplA.

Although the most intensively studied of the type III systems, the secretion apparatus of *Yersinia* spp. has not yet been visualised. However, the apparatus (termed a 'secreton' or 'injectisome') has been observed in both *Salmonella* spp. and *Shigella* spp. and is shown in Figures 8.12 and 8.16. In both cases, it can be seen to consist of a syringe-like structure through which the bacterial proteins are injected directly into the host cell.

At the time of writing, more than 20 effector proteins have been identified as being injected into host cells by the various organisms with type III secretion systems. These effectors display a variety of activities and include the ability: (i) to mediate adhesion of the organism to the target cell, (ii) to interfere with host cell signalling, (iii) to induce cytoskeletal rearrangements and (iv) to up-regulate or down-regulate cytokine production. Details of the specific activities of the effector proteins of the various organisms are given in Chapters 7, 8 and 10.

As yet, type III secretion systems have been found only in Gram-negative bacterial species. Recently, however, it has been shown that *Strep. pyogenes* is also capable of injecting an effector protein directly into the cytoplasm of a host cell. Delivery of the effector molecule (an NAD-glycohydrolase that can interfere with eukaryotic signal transduction) into the host cell cytoplasm is dependent on streptolysin O, a cholesterol-dependent cytolysin. The streptolysin O secreted by the bacterium forms a pore in the membrane of the host cell through which the effector molecule can pass. As cholesterol-dependent cytolysins are widely distributed among Gram-positive organisms, such cytolysin-mediated translocation may constitute the functional equivalent of the type III secretion system found in Gram-negative species.

2.6 | Genetic aspects of bacterial virulence

It is not within the scope of this book to provide a definitive discourse on bacterial genetics. Indeed, the subject is large and complex enough for a book in its own right and the reader is directed to any number of excellent texts, some of which are listed at the end of this chapter. What this section will try to do, instead, is highlight aspects of bacterial genetics that have a significant role in virulence. These include bacterial sensing of, and responding to, environmental cues, the regulation of gene transcription and the organisation of genes and operons into large, virulence-associated regulons. Additionally, we will look at the impact of mobile genetic elements, including pathogenicity islands, on bacterial virulence and, indeed, on bacterial evolution as a whole.

Life in the developed world is relatively pleasant in terms of the living conditions to which we are exposed. Air conditioning and central heating, for example, take the sting out of the year's temperature extremes. The guesswork is mostly removed from global travel too; via the information superhighway we have immediate access to current weather conditions and forecasts for our destination and so can pack accordingly. We can also keep abreast of environmental (natural disaster) or social (civil unrest) factors that might impact on our trip, and immunise ourselves against diseases endemic in the destination country. In other words, we can foresee the likely conditions we may have to face and we can prepare for most eventualities. On a cellular scale, the majority of eukaryotic cells that make up the human body also live a cosseted life. Their environment is kept relatively constant, as is the supply of oxygen, nutrients and removal of waste products. Spare a thought, then, for the perilous plight of bacteria. For example, oral bacteria present in plaque that forms on teeth must learn to cope with a variety of assaults that include being scrubbed off daily, exposed to very hot and very cold foods and subjected to significant fluctuations in pH and redox conditions. Certain oral bacteria, notably the α-haemolytic streptococci, when introduced into the bloodstream by routine dental procedures, have the ability to cause infective endocarditis. Conditions in blood differ radically from those of the oral cavity, and the bacteria must respond to the new conditions by modifying gene expression accordingly. The likely fate of aquatic *V. cholerae* cells when ingested by humans in contaminated water, should they survive the acid conditions in the stomach, is simply to pass through without pause. Unless, that is, *V. cholerae* cells have been infected with bacteriophage VPIϕ (see Micro-aside, p. 491). In which case, cells have the wherewithal to adhere to intestinal mucosal surfaces and, therefore, to colonise humans, a very different environment for growth and replication. Clearly then, many bacteria are capable of rapid, and often profound, changes in gene expression in response to changing environmental conditions.

2.6.1 Gene regulation

In order to survive in its host, a bacterium must be able to induce the expression of appropriate gene products, i.e. those needed to cope with the new environment in which it finds itself (e.g. to enable adhesion to host cells, to obtain nutrients, etc.) and those essential for avoiding host defence systems (e.g. anti-phagocytic mechanisms, invasion of host cells, antigenic variation, etc.). This requires the involvement of two processes – the sensing of environmental cues and the regulation of gene transcription, i.e. the switching on of some genes and the turning off of others.

2.6.1.1 Sensing changes in the environment

Two-component signal transduction systems (TCSTSs) are probably the most important means by which bacteria sense and respond to their environment. TCSTSs are involved in regulating many diverse cellular functions including sporulation, chemotaxis, quorum sensing, survival within eukaryotic cells, and the expression of substrate uptake systems, toxins and other virulence factors. Despite this myriad of functions, most two-component systems can be stripped down to certain modular protein domains, and these are described in detail in Section 4.3.1.2.

A TCSTS detects an external signal and directs the organism to make a response conducive to its survival. This response is achieved by the induction or repression of the transcription of appropriate genes.

2.6.1.2 Gene transcription

Unlike eukaryotic cells, bacteria lack a nuclear membrane. Consequently, transcription (the synthesis of RNA from a DNA template) and translation (the synthesis of protein from an mRNA template) occur within the same compartment. Indeed, translation of an mRNA molecule can begin as soon as its transcription starts. Ribosomes attach in succession to the ribosome-binding site (also known as the Shine–Dalgarno sequence) of the nascent mRNA chain and move along as the mRNA molecule is synthesised. This gives rise to a polysome that can be visualised by electron microscopy (Figure 2.14). When the expression of a gene is initiated, mRNA molecules appear as soon as 2.5 minutes later and the corresponding protein some 30 seconds afterwards. Bacterial messages are highly unstable and the degradation of an mRNA molecule often follows close behind its translation (Figure 2.14). Thus, coupled transcription and translation, and rapid turnover of messages, enable bacteria to adapt quickly to new environmental conditions.

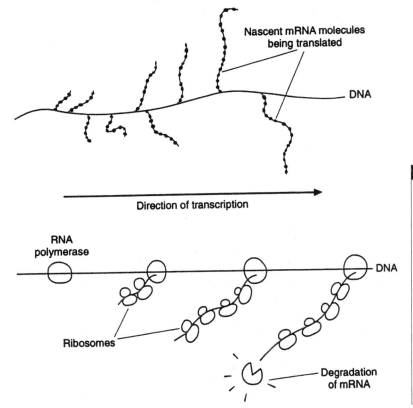

Figure 2.14 Transcription, translation and mRNA degradation can occur concomitantly. The upper diagram is based on an electron microscopic image of coupled transcription and translation. The lengths of the mRNA transcripts, laden with ribosomes, indicate the direction of transcription. The lower diagram shows that, once transcription is initiated, ribosomes bind in succession to the nascent mRNA molecule and begin translation. mRNA degradation follows close behind.

Additionally, a single mRNA molecule may encode two or more proteins that generally have associated functions. The organisation of tandem genes in transcriptional units (known as operons) further speeds and co-ordinates bacterial responses. A good example of this is the *hylCABD* operon of *E. coli*. The *hylA* gene encodes the haemolysin toxin itself. *hylC* encodes an enzyme that activates HylA, while *hylB* and *hylD* encode components of the type I secretion system required for export of the haemolysin toxin (Figure 9.6). Regulons have an additional level of organisation. Regulons are groups of frequently unlinked genes and/or operons that are subject to control by a single transcriptional regulator. The *vir* regulon of *Strep. pyogenes*, for example, is under the control of a transcriptional activator, Mga, which demonstrates homology with response regulators of TCSTSs (see Section 4.3.1.2). Mga controls the expression of multiple virulence-associated genes and operons, including those encoding M and M-related proteins, oligopeptide permease and secreted cysteine protease, in response to CO_2 levels and bacterial growth phase.

Transcription of genes or operons requires RNA polymerase. RNA polymerase core enzyme consists of four subunits, α_2, β, β', and catalyses the incorporation of approximately 40 nucleotides per second, unwinding DNA as it advances along the template. A fifth component, sigma (σ) factor, has no catalytic activity but, instead, helps the core enzyme to recognise promoter elements. Once the RNA polymerase holoenzyme has bound and transcription is initiated, the σ factor falls off and is available for directing other core enzymes (Figure 2.15). Most bacteria can synthesise several distinct σ factors that recognise different promoter sequences. This means that by changing σ factor expression, bacteria can initiate transcription of a whole new set of genes and operons that might be required for adaptation to a particular environment.

The primary σ factor (σ^{70} in *E. coli*) is responsible for the transcription of most genes expressed in growing cells and is essential for cell viability. The main consensus sequences for σ^{70} promoters are the hexanucleotides TTGACA and TATAAT located approximately 35 and 10 nucleotides $5'$ to the transcriptional start site, respectively. Non-essential σ factors resembling σ^{70} include those required for bacterial adaptation to the stationary phase. For example, σ^s of *E. coli* mediates the general stress response. Alternative σ factors control the transcription of specific sets of genes (regulons) under special environmental or physiological conditions. Perhaps the best studied of the alternative σ factors are those involved in sporulation in Gram-positive bacteria. Spore formation in *B. subtilis* requires five different alternative σ factors and these are differentially expressed during sporulation, both temporally and spatially, as cells undergo asymmetric division. Other alternative σ factors may control transcription of genes involved in the biosynthesis of flagella, the response of cells to heat shock and the synthesis of extracellular factors in response to environmental conditions. These σ factors recognise different -10 and -35 promoter sequences. Members of a final group of σ factors, the σ^{54} family, are

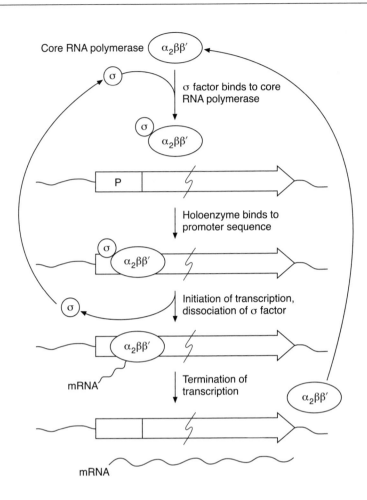

Core RNA polymerase $\alpha_2\beta\beta'$

σ factor binds to core RNA polymerase

$\alpha_2\beta\beta'$

P

Holoenzyme binds to promoter sequence

$\alpha_2\beta\beta'$

Initiation of transcription, dissociation of σ factor

$\alpha_2\beta\beta'$

mRNA

Termination of transcription

$\alpha_2\beta\beta'$

mRNA

Figure 2.15 Initiation of transcription. Core RNA polymerase $(\alpha_2\beta\beta')$ acquires a σ-factor that directs the holoenzyme to bind to a subset of promoters. Once transcription is initiated, the σ-factor is released and can attach to another core polymerase.

unrelated to those described above. These factors are required for nitrogen-regulated genes.

Transcription of a gene or operon is terminated at a specific site that lies just $3'$ to the stop codon. Termination may or may not require the help of a specific factor, rho (ρ), that binds to the core enzyme. All terminator sequences consist of short inverted repeats that in the mRNA form a hydrogen-bonded stem–loop structure that causes RNA polymerase to pause or stop transcribing DNA (Figure 2.16). This allows binding of ρ factor and subsequent dissociation of polymerase from the template. Rho-independent terminators have a run of approximately six uridine residues following the stem–loop and this facilitates release of the core enzyme independently of ρ factor binding.

2.6.1.3 Gene induction and repression

The induction and repression of transcription allows a bacterium to match gene expression to its cellular requirements. This is perhaps best exemplified by looking at the expression of biosynthetic or catabolic pathways regulated by repressor proteins. Repressor proteins bind to an operator site that lies adjacent to, and may overlap, the promoter

Figure 2.16 Structure of transcriptional terminators. Terminator sequences include inverted repeats that form hairpin loops. (a) Rho-independant terminators are enriched for G and C residues within the stem and the stem–loop structure is followed by a run of U residues. (b) Rho-dependent terminators do not exhibit these characteristics.

(a)

```
          A—U
       A         C
        |         |
        C         A
       ⎛  G≡C
       ⎜  C≡G
       ⎜  G≡C
G + C rich ⎨ A=U
       ⎜  C≡G
       ⎜  G≡C
       ⎝  G≡C
```
—A—U—G—A=U—U—U—U—U—U—U—

(b)

```
          U—C
       U         G
        |         |
        A         A
           A=U
           U=A
           C≡G
           U=A
           A=U
```
—A—U—U—C   G—U—A—A—

site. Binding of an active repressor at the operator site interferes with RNA polymerase binding, or function, such that transcription is blocked. Furthermore, binding of accessory molecules to the repressor protein can modify its activity. The end-products of some metabolic pathways act as co-repressors, and binding of these molecules to inactive repressor protein results in a repressor capable of binding the operator site and thus blocking transcription (Figure 2.17). For example, if a particular amino acid is present in the environment at

Figure 2.17 Repressors and activators of gene transcription. (a) The repressor binds to an operator site (O) adjacent to, or overlapping, the promoter (P) of the gene under control. Bound repressor blocks the binding of RNA polymerase or else prevents the movement of RNA polymerase from the promoter region. The binding of an inducer molecule to the repressor leads to a conformational change in the repressor molecule, causing the repressor to fall off the operator site. (b) Native repressor has no affinity for the operator site of regulated genes. Binding of co-repressor results in activation of repressor activity and leads to repression of gene transcription. (c) Activators bind to specific sequences adjacent to promoters of regulated genes. Activators may bind directly to RNA polymerase or else may induce local changes in DNA structure to promote gene transcription.

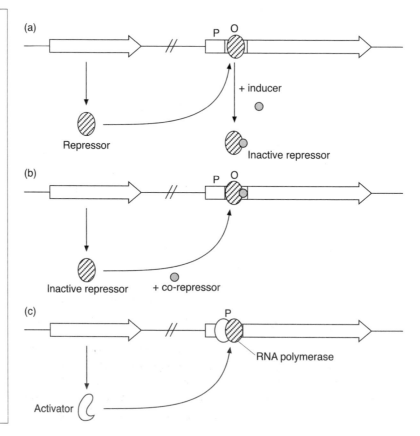

a sufficient level for the bacterium's needs, then the biosynthetic pathway for *de novo* synthesis of that amino acid need not be expressed. In some instances, the repressor protein is expressed in an active form and binds directly to the operator site, thus blocking transcription. The binding of an inducer molecule causes a conformational change in the repressor, resulting in repressor inactivation and thus allowing transcription (Figure 2.17). Perhaps the best example of this is the *lac* operon, whose expression is required for the uptake and breakdown of lactose (a galactose-glucose disaccharide). Allolactose, a modified form of lactose, is the inducer of the *lac* operon. A basal constitutive level of *lac* transcription allows the uptake of lactose, if present in the environment, and its conversion to the inducer allolactose.

The expression of many genes or operons is subject to positive activation. Activator proteins bind to specific sequences adjacent to the promoter site. Activators may function by interacting directly with RNA polymerase core enzyme or with the associated σ factor of the holoenzyme (Figure 2.17). Binding enhances the normal activity of RNA polymerase, promoting the initiation of transcription. Other activators are thought to induce a sharp bend in DNA, altering the promoter conformation and enhancing the initiation of transcription.

Transcriptional attenuation is another means by which bacteria regulate gene expression. Attenuation is involved in the regulation of several amino acid biosynthetic pathways and relies on the close coupling between transcription and translation.

Two-component systems and contact sensing?

Escherichia coli has two stress response systems that sense, and respond, to the presence of misfolded proteins in the bacterial envelope. The first system uses an alternative factor, σ^E, to detect overexpressed or misfolded outer membrane proteins and is activated by general stresses such as heat or ethanol. σ^E is normally held at the cytoplasmic face of the inner membrane by a transmembrane anti-σ factor. Misfolded proteins are somehow detected at the periplasmic face of the inner membrane by the anti-σ factor, which then releases σ^E, allowing it to interact with RNA polymerase (Figure 2.18). The σ^E regulon includes *degP*, encoding a periplasmic protease, and genes encoding envelope-folding factors as well as other, as yet unidentified, genes.

Perhaps more interesting is the CpxA-CpxR TCSTS. Studies on P pilus assembly by uropathogenic *E. coli* have provided much information on this fascinating system. As we have learned, P pilus subunits are delivered to the periplasm by the Sec machinery, whereupon they are capped by chaperone proteins in preparation for assembly into the pilus rod by the usher complex. Overproduction of pilus subunits in the absence of chaperones results in the accumulation of misfolded aggregates associated with the periplasmic face of the inner membrane. The sensor kinase is thought to interact with an inhibitor periplasmic factor (CpxP) that maintains CpxA in an 'off' state. The activating signal, misfolded proteins themselves or other unknown intermediate signalling molecules, may bind to, or titrate out, the inhibitory factor, resulting in autophosphorylation of CpxA and signal transduction to the CpxR response regulator (Figure 2.18). The Cpx regulon also includes *degP*, encoding a periplasmic protease, as well as genes encoding envelope-folding factors, including DsbA, which catalyses the formation of disulphide bonds. DsbA, is required for proper folding of many envelope proteins including P pili subunits and components of the toxin co-regulated pili of *V. cholerae*, for example. Other genes switched on include those encoding virulence factors. Thus, in *Shigella* spp., the Cpx system regulates the expression of the VirF transcriptional activator, which is, in turn, required for expression of Ipa proteins involved in invasion of host cells.

It has been proposed recently that Cpx functions as a sensory system to correctly time the synthesis and assembly of adhesive structures such as P pili. Thus, when an organism makes contact with a suitable host substrate, the outward growth of the extending pilus is halted, resulting in the accumulation of misfolded subunits within the periplasm. These are sensed by the Cpx system, leading to increased expression of factors required for pilus assembly, including DsbA. This results in an increase in the number of pili on the cell surface to strengthen adhesion. Additionally, at least in *Shigella* spp., pilus contact with a suitable host cell may prime the bacteria for invasion by the up-regulation of Ipa genes.

2.6.2 Antigenic and phase variation

The surface of a bacterium is the forum for interactions with its host. Surface components such as pili, OMPs, S layer proteins, flagella, lipopolysaccharide (LPS) and capsules are involved in a number of host–bacteria interactions including adhesion and resistance to phagocytosis, and some may contribute towards the pathogenic effects of the bacterium. Certainly, many of these components are dominant antigens recognised by the host immune system. It is perhaps not surprising, then, that many bacteria, particularly those that normally reside on mucosal surfaces and so are exposed to the mucosal immune system (Chapter 5), have the means to switch on or off the expression of some components (phase variation) or indeed to alter the antigenic nature of these components (antigenic variation).

As well as providing a means of avoiding immune defences, antigenic or phase variation may contribute to the dissemination of bacteria within a colonised host. For example, switching adhesion-mediating OMPs may confer altered adhesion specificity. Similarly, by switching off the expression of pili, bacteria may be able to 'de-camp' from a particular site and spread back into the environment or to another host. Finally, genetic diversity within a population makes it more likely that some cells would survive a sudden environmental change. Bacteria employ numerous means to facilitate antigenic or phase variation and some examples are provided below.

2.6.2.1 DNA inversion

DNA inversion by site-specific recombination is a common means by which bacteria control the expression of genes located within, or adjacent to, the invertible region. One of the best-characterised examples

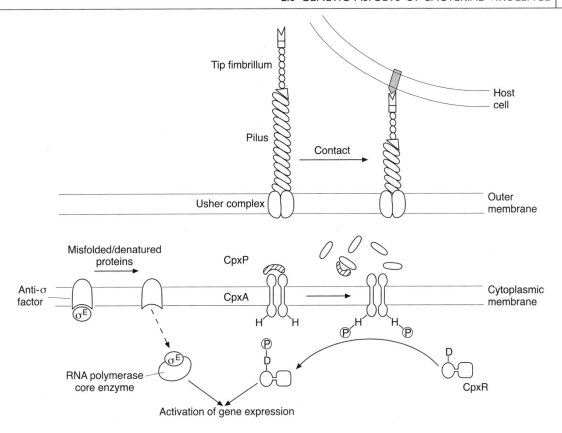

is that of *E. coli* type I fimbriae. *Escherichia coli* uses DNA inversion to switch on, or off, at relatively high frequency, the expression of fimbriae and this is known as phase variation rather than antigenic variation. A 314 bp invertible DNA sequence lies upstream from the gene encoding the structural subunit (*fimA*). In the ON orientation, the promoter is in the correct position to initiate transcription of *fimA* (Figure 2.19). Inversion is mediated by FimB and FimE site-specific recombinases. Other co-factors are required, and these include global regulatory proteins that may bend DNA, thus aligning the recombination sites so that strand exchange can occur.

In a modification of this simple ON↔OFF system, *Salmonella* spp. regulate the expression of H1 or H2 flagellar antigens using an ON_{H1}/OFF_{H2}↔ON_{H2}/OFF_{H1} switch. The invertible 996 bp DNA fragment contains the promoter that drives expression of H2 and also carries the site-specific recombinase itself. Co-transcribed with the *h2* gene is *rh1*, encoding a repressor of the *h1* gene. DNA inversion means that *h2* is no longer transcribed nor is the *rh1* repressor, thus allowing *h1* gene expression. As with the *E. coli* system, other cellular factors are required and Figure 2.19 provides a simplified version of the system.

A similar switch is used to alternate S layer protein expression in *Campylobacter fetus* and in *Lactobacillus acidophilus*, although in the latter example the S protein genes are located within the invertible DNA

Figure 2.18   Contact sensing via the CpxA-CpxR two-component system. *Escherichia coli* has two stress response systems that sense, and respond to, misfolded proteins in the bacterial envelope. In the first system, σ^E is released from its anti-σ factor in response to the presence of heat- or chemically denatured proteins and interacts with RNA polymerase core enzyme. The second system is contact dependent. Once contact is made with a host cell, pilin subunits accumulate within the periplasm and aggregate in the absence of sufficient chaperones. Misfolded or aggregated pilin subunits remove the inhibitory activity of CpxP, allowing activation of the CpxA autokinase domain and autophosphorylation of CpxA and subsequent phospho-transfer to the response regulator CpxR.

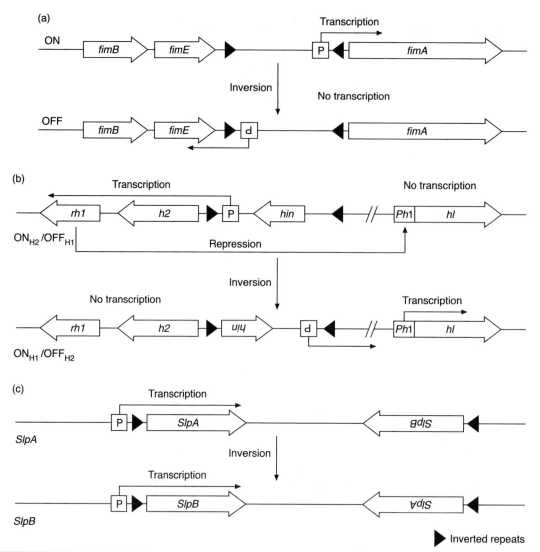

Figure 2.19 Phase and antigenic variation by DNA inversion event. (a) ON↔OFF control of fimbriae expression in *Escherichia coli*. DNA inversion mediated by *fimB* and *fimE* gene products. (b) In the H2 ON orientation, a repressor of *hl* promoter activity is co-transcribed with *h2* and this maintains the *hl* locus silent. DNA inversion means *h2* and *rhl* are no longer transcribed, allowing expression of *hl*. (c) DNA inversion at the *slp* locis of *Lactobacillus acidophilus* moves the position of *slpA* and *slpB* structural genes relative to the *slp* promoter.

fragment and are brought into juxtaposition with a stationary promoter sequence (Figure 2.19).

2.6.2.2 DNA recombination

Neisseria spp. have developed several different mechanisms to permit antigenic variation (Table 2.3). One mechanism uses homologous recombination to shuffle the antigenic make-up of its type IV pilus. The main subunit of the type IV pilus of *N. gonorrhoeae* is expressed at the *pilE* (expressed) locus. There are, however, upwards of 50 silent loci elsewhere in the genome that encode transcriptionally inactive alleles of *pilE* called *pilS* (silent). Recombination occurs frequently and removes the expressed gene and replaces it with a new gene from a *pilS* locus resulting in a different pilin product. Occasionally, recombination errors occur and this results in an ON→OFF phase variation such

Table 2.3. Mechanisms of antigenic or phase variation in pathogenic *Neisseria* spp.

Surface component	Mechanism	Phenotypic change
Pilus	Homologous recombination	Antigenic variation, swapping of expressed pilus subunit
	Post-translational modification	Antigenic variation, altered glycosylation patterns
Opacity proteins	Slipped-strand mispairing during DNA replication	Phase and antigenic variation, switch off or on expression of various Opa proteins
Opc outer membrane protein	Transcriptional slipped-strand mispairing	Phase variation, switch off (low) or on (high) expression of Opa

that no pili are formed. The causative agent of Lyme disease, *Bor. burgdorferi*, uses a similar system for antigenic variation of surface components.

Recombination between homologous sequences need not be confined to DNA from within the same cell. Many mucosal pathogens including *H. influenzae*, *Neisseria* spp. and *Strep. pneumoniae* are competent for DNA transformation. This allows intergenomic recombination, thus widening the repertoire of antigenic variation, and an example was described earlier for the capsule of *Strep. pneumoniae* (see Micro-aside, p. 56).

2.6.2.3 Slipped-strand mispairing

Neisseria gonorrhoeae and *N. meningitidis* use this system for ON/OFF phase variation of some **opa**city (Opa) proteins that are involved in adhesion and invasion. There may be 11 to 12 *opa* loci in *N. gonorrhoeae* and 3 to 4 in *N. meningitidis*, and strains exist that express none (Opa⁻), one, or several opacity proteins.

In *N. gonorrhoeae* the *opa* genes contain a run of several tandem five nucleotide repeat sequences (CTTCT). During DNA replication, transient mispairing can occur, leading to strand slippage. This results in the incorporation of errors – typically the insertion or deletion of whole repeats – in the daughter cell chromosome. The size of the repeat is significant, as removal or addition of a repeat results in a frame shift mutation leading usually to premature termination of *opa* mRNA translation (Figure 2.20). Genome sequencing projects are now revealing the relatively widespread nature of this mechanism for phase variation amongst bacteria, including *Hel. pylori* and *Bord. pertussis*.

Figure 2.20 Phase variation by slipped-strand mispairing. In *Neisseria gonorrhoeae* the *opa* genes contain a run of several tandem five nucleotide repeat sequences (CTTCT). During DNA replication, transient mispairing can occur, leading to strand slippage. This results in the insertion or deletion of whole repeats, causing a frame shift, which, in turn, ultimately leads to premature termination of translation.

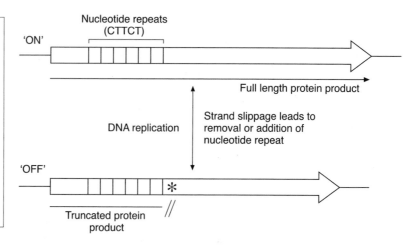

Expression of the Opc outer membrane protein of *N. meningitidis* is modulated also by slipped-strand mispairing, although the control in this case is at the level of transcription. Strand slippage results in the addition or removal of cytosine residues at a poly(C) tract upstream from the opc promoter. Increasing or decreasing the number of cytosine residues in this tract of DNA affects the transcriptional start site of RNA polymerase, resulting in no transcription (C < 10 or C > 15), moderate (11 or 14 C) or high (12 or 13 C) levels of transcription.

2.6.2.4 Epigenetic variation

Epigenetic variation differs from the mechanisms described above. Variations occur in the phenotype of the organism, but not the genotype. In other words, there are no genomic rearrangements or other alterations to gene structure. Post-translational modification of surface structures by the addition of sugar residues, for example, may give rise to antigenic variability. In other cases, such as P fimbriae and antigen 43 OMP of *E. coli*, regulation of gene expression itself is modified by the methylation of DNA. Deoxyadenosine methylase, Dam, of *E. coli* methylates the adenosine residue of most of the 18 000 or so GATC sequences that occur in the chromosome. However, some sites can be protected from methylation. OxyR, the response regulator of a TCSTS, binds upstream from the *agn43* (encoding **antigen 43**) promoter and blocks transcription of the gene. The OxyR binding site contains three GATC sequences and if any of these are methylated on both strands then OxyR cannot bind and so *agn43* is transcribed. A new round of DNA replication generates hemimethylated binding sites (where the adenosine residue on the template strand, but not the new strand, is methylated), allowing OxyR to bind and thus block transcription, possibly by interfering with RNA polymerase binding. OxyR binding also blocks methylation of the three GATC sites.

2.6.2.5 Point mutations

Simple point mutations within the reading frame of genes encoding surface components can give rise to antigenic variants no longer recognised by the host immune system. The best examples of organisms that use this strategy include the human immunodeficiency virus (HIV) and influenza A virus. However, point mutations are also used by bacterial pathogens to generate antigenic variation. There are now over 100 recognised M protein serotypes of *Strep. pyogenes* (described in Section 2.9.1). The M protein is an important adhesin and serves also to block phagocytosis. Antibodies to a particular M protein provide protection against only those organisms expressing that particular M protein and not against other *Strep. pyogenes* serotypes.

2.6.3 Mobile genetic elements

A number of different types of mobile genetic element are found in bacteria. These include plasmids, transposable elements and bacteriophages and they may play an important role in the virulence of bacterial pathogens. Furthermore, since genomes evolve by rearranging existing DNA and by acquiring new sequences, these elements, as well as pathogenicity islands (discussed in Section 2.6.4) may have contributed to the evolution of pathogens as we know them today.

2.6.3.1 Plasmids

The vast majority of plasmids are circular double-stranded DNA molecules that are capable of independent replication within the bacterial cell. Some plasmids, for example those of the spirochaete *Bor. burgdorferi*, are linear. Sizes vary from around 1 kb to over 100 kb, and the number per cell also varies from 1 to more than 40. Often the plasmids are not essential for bacterial viability, at least in the laboratory, and can be cured from cells. An episome is a plasmid that either can exist independently or else can integrate into the bacterial chromosome and be replicated with it. Other plasmids are capable of conjugal transfer between bacterial cells. These plasmids carry genes for the synthesis of conjugative pili or other cell–cell contact mechanisms, and for plasmid mobilisation to recipient cells. Many conjugative plasmids have a broad host range and plasmid transfer can occur between Gram-positive and Gram-negative bacterial cells.

Over 300 different naturally occurring plasmids have been isolated from strains of *E. coli* alone. Plasmids can be organised broadly into several different categories, some of which have relevance to bacterial pathogenicity and are described here (Table 2.4).

The resistance factors, or R plasmids, may carry multiple genes encoding antibiotic resistance mechanisms. Often the resistance genes are carried on transposons, and this facilitates the rapid development of plasmids with multiple resistance determinants. Furthermore, many R plasmids are conjugative, or else can be mobilised by co-resident conjugative plasmids, and this promotes widespread dissemination of resistance genes.

Virulence plasmids are also widespread. These may be relatively small plasmids carrying only one or a few genes encoding, for

Table 2.4.	Examples of bacterial plasmids		
Type	Plasmid (size)	Organism	Phenotype
Resistance	RP4 (54 kb)	*Pseudomonas* spp. other Gram-negative bacteria	Resistance to ampicillin, kanamycin, neomycin, streptomycin
			Conjugative
Virulence	ColV-K30 (2 kb)	*Escherichia coli*	Siderophore for iron uptake
	pO157 (93 kb)	*Escherichia coli* O157 : H7	Type III secretion, RTX toxin
	pCD1 (70 kb)	*Yersinia pestis*	Type III secretion system and Yop effectors
	pXO1 (180 kb)	*Bacillus anthracis*	Germination, A_2B toxin, adenylyl cyclase

example, a toxin or iron acquisition protein, but they may also be very large and encode a complex virulence phenotype. Amongst this group of large virulence plasmids are three that encode functions described elsewhere in this book. *Bacillus anthracis* plasmid pOX1 is approximately 180 kb in size and carries some 140 or so predicted open reading frames (ORFs). These include genes encoding the A_2B components of anthrax toxin (Section 9.4.1.1.3) and genes required for germination of *B. anthracis* spores. Spread throughout the plasmid are multiple insertion sequences and transposons encoding associated transposase proteins and there are many other cryptic (i.e. not-expressed) transposase genes and gene fragments. This is a feature of many of these large plasmids. The *Y. pestis* plasmid pCD1 (70 kb) has approximately 100 putative ORFs and some 50 of these encode a type III secretion system with associated chaperones and effector proteins (Figure 2.21). Enterohaemorrhagic *E. coli* strain O157 : H7 possesses a 93 kb plasmid, pO157, that encodes a variant RTX toxin (see Section 9.5.1.2) and also a type III secretion system amongst the 80 or so proposed ORFs.

2.6.3.2 Bacteriophages

Although bacteriophages have their own agenda, many of them can be considered as mobile genetic elements that impinge on bacterial virulence. Many bacteriophages are lysogenic and can therefore integrate into the chromosome of the host bacterium, often at specific sites. Some lysogenic bacteriophages carry toxin genes, and insertion of the phage DNA into the host chromosome results in phage conversion of bacteria from a non-toxigenic to a toxigenic phenotype. For example, Shiga toxins Stx1 and Stx2 are encoded in the genome of lambdoid phages (see Section 9.5.1.2) and, possibly as a result of this, over 100 serotypes of *E. coli* can produce Shiga toxins. Similarly, the genes encoding cholera toxin are present on the genome of filamentous

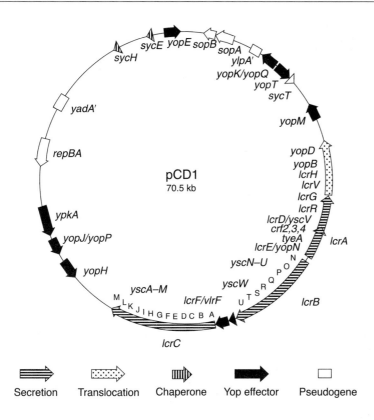

Figure 2.21 The virulence plasmid pCD1 of *Yersinia pestis*. Similar plasmids are found in *Yersinia enterocolitica* (pYV227) and *Yersinia pseudotuberculosis* (pIB1).

prophage CTXφ. Furthermore, the receptor for phage CTXφ binding, toxin co-regulated pilus (TCP), is itself encoded by another filamentous phage, VPIφ.

Bacteriophages can also mediate the horizontal movement of bacterial DNA between cells by transduction. In generalised transduction, during the lytic cycle of the phage, host bacterial chromosomal DNA can be fragmented and pieces can be incorporated into the phage particle. On infection of the next host, the transported fragment of bacterial DNA may recombine with the new host chromosomal DNA producing a transduced cell (Figure 2.22). Specialised transduction requires that the phage DNA is integrated into the host chromosome. On induction of the lytic cycle, the phage genome excises from the bacterial chromosome. Under rare conditions the phage genome is excised incorrectly and can carry with it host DNA lying on either side of the integration site and this bacterial DNA is carried to the new host (Figure 2.22).

2.6.3.3 Transposable elements

Transposable elements are sequences of DNA that can move from one site to another within a bacterial genome (Table 2.5). Transposition does not require extensive regions of DNA homology, and this differentiates transposition from other forms of genetic recombination. Instead, a transposable element carries the gene(s) encoding protein(s) responsible for facilitating transposition and is delineated by short

Figure 2.22 Transduction. (a) In generalised transduction, fragments of host bacterial DNA may, at low frequency, be packaged along with viral DNA in a phage particle. (b) Specialised transduction involves the imprecise excision of integrated prophage from the bacterial chromosome. DNA flanking the phage integration site may also be excised, and this is replicated along with the bacteriophage DNA. Note that if excision is imprecise, all progeny phage will be capable of mediating transduction.

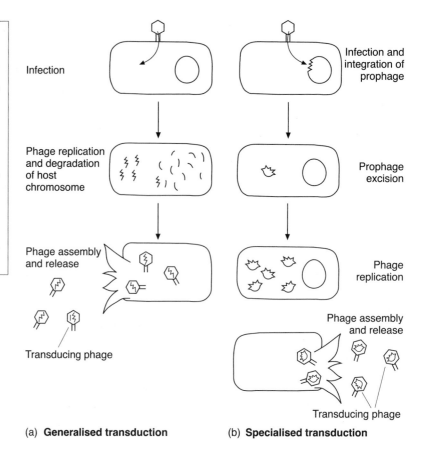

Infection

Phage replication and degradation of host chromosome

Phage assembly and release

Transducing phage

Infection and integration of prophage

Prophage excision

Phage replication

Phage assembly and release

Transducing phage

(a) **Generalised transduction**   (b) **Specialised transduction**

inverted repeats (typically 5 to 40 bp, depending on the element) that are recognised by the transposase. Nevertheless, transposable elements can allow homologous recombination to occur by acting as portable duplicate sequences upon which the cellular recombination system can act. Many transposable elements carry antibiotic resistance

Table 2.5. Examples of transposable elements of bacteria

Element	Element type	Size	Genetic markers
IS1	Insertion sequence	0.77 kb	None
Tn5	Composite transposon	5.7 kb, delimited by IS50 elements	Kanamycin resistance
Tn3	Type II transposon	5.0 kb	Ampicillin resistance
Tn916	Conjugative transposon	18.0 kb	Tetracycline resistance

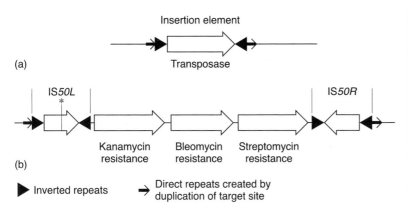

(a)

(b)

Inverted repeats

Direct repeats created by duplication of target site

Figure 2.23 Structure of (a) an insertion element and (b) a composite transposon. Transposon Tn5 comprises two copies of IS50 flanking kanamycin/neomycin, bleomycin and streptomycin resistance markers. The transposase of IS50L is non-functional owing to a point mutation within the transposase gene (indicated by *) resulting in premature termination of translation.

markers and may also affect gene expression by bacteria. Thus transposition of an element within a gene or operon can disrupt its expression. Conversely, it is now known that many elements have promoter DNA sequences near either end that can result in activation of genes lying adjacent to the site of insertion.

Insertion sequences (IS elements) are the simplest form of transposable element and carry no information other than that required for transposition itself (Figure 2.23). IS elements are approximately 750 to 1500 bp in length and can be found on the chromosome and on plasmids. Several hundred distinct IS elements have been described. They may be present in multiple copies, and a given bacterium may have more than one type of IS element. Transposition events are relatively rare, occurring at a frequency of less than 10^{-5} per generation, comparable to spontaneous mutation rates. The target site for IS element insertion varies from 5 to 9 bp in length, and this is duplicated during transposition, with one copy lying on either side of the IS sequence.

Transposons are larger than IS elements and may carry additional genetic information besides the transposase, such as gene(s) encoding antibiotic resistance, toxins, or catabolic enzymes for the biodegradation of organic pollutants. There are several different types of transposon. Composite transposons comprise two identical, or nearly identical, IS elements either side of a gene or group of genes that are mobilised along with the IS elements. Commonly, these IS modules are inverted with respect to each other and, since the transposase gene of only one IS module is required for movement, there is no selective pressure to preserve both transposase genes, and frequently one is inactive due to spontaneous mutation. Tn5 (5.7 kb) provides an example of this and comprises a kanamycin resistance marker flanked by copies of IS50 insertion sequence (Figure 2.23). The IS50 module on the left-hand end of Tn5 (IS50L) has a single base-pair change rendering its transposase inactive.

IS elements and composite transposons move by non-replicative (conservative) transposition. The transposase catalyses the excision of the element and its insertion at the new target site. Thus transposition does not increase the number of copies of that element within the cell. In contrast, type II elements move by replicative transposition. Consequently, a transposition event results in an increase in the number of copies of that transposon. Type II transposons are not bound by IS elements but are nevertheless flanked by short inverted repeats and generate short direct repeats of the translocation target sequence. Replicative transposition involves the formation of a transient co-integrate molecule (where the original and replicated elements are joined). Consequently, type II transposons require, in addition to the translocase, a resolvase activity to mediate recombination between transposons to resolve the co-integrate molecule and release the donor and target elements.

Neither replicative nor conservative transposition involves the generation of a free transposon molecule in the cell. However, members of a third group of transposable element, conjugative transposons, form a covalently closed circular intermediate. This circular intermediate can reinsert into host genomic DNA or can transfer by conjugation to a new host and integrate into the recipient's genome. There is no duplication of the target site sequence. Conjugative transposons are more closely related to bacteriophages and the site-specific recombinases associated with transposition are members of the phage integrase family. Conjugative transposons, such as Tn916, have a very broad host range and probably contribute as much as plasmids to the spread of antibiotic resistance genes. Furthermore, the conjugation functions of this group of transposons can also mobilise co-resident plasmids.

2.6.4 Pathogenicity islands

Pathogenicity islands (abbreviated to PAIs or, when one is referring to a specific, named pathogenicity island, the abbreviation used is Pai) are large elements (frequently greater than 30 kb in length) that are integrated into the chromosome of pathogens and carry one or more virulence determinants (Table 2.6). PAIs are often located at tRNA loci, the site of integration of many bacteriophages and conjugative transposons. This feature of PAIs, plus the observations that many carry features characteristic of mobile elements and that the $G + C$ content of the PAI often differs markedly from that of the host genome, suggest that PAIs may have been acquired by horizontal gene transfer.

PAIs were first studied in detail in uropathogenic and enteropathogenic *E. coli* strains. Pai I (70 kb) of uropathogenic *E. coli* strain 536 encodes a haemolysin. The PAI is flanked by direct repeats of length 16 bp and is inserted at the *selC* locus, which encodes a selenocysteine-specific tRNA. Pai II (190 kb) encodes P fimbriae as well as haemolysin, and is located at the *leuX* locus, which encodes a minor leucine-specific tRNA and is flanked by 18 bp long direct repeats (Table 2.7). Both Pai I and Pai II are lost from the chromosome at a frequency of 10^{-4} to 10^{-5} per generation, and both have a $G + C$ content of only 41% as compared with the host genome content of 51%. The tRNA target sites for

Table 2.6.	Characteristics of pathogenicity islands

- Large elements (frequently > 30 kb) that carry virulence genes
- Present in pathogenic strains but absent, or present sporadically, in related non-pathogenic isolates
- Different G + C content as compared with the rest of the host genome
- Associated with tRNA genes and/or IS elements
- Carry (often cryptic) mobility genes
- Unstable

Pai I and II also serve as the attachment sites for bacteriophages φR73 and P4, respectively, and both PAIs carry cryptic (i.e. not-expressed) phage-like integrase genes. In contrast, the tRNA target sites for Pai IV (170 kb) and Pai V (110 kb) of uropathogenic *E. coli* strain J96, *pheV* and *pheR* respectively, are the sites of integration of conjugative transposons in *Salmonella* and *E. coli* strains. Pai IV encodes haemolysin and adhesion-associated Pap pili whereas Pai V encodes cytotoxic necrotising factor (CNF) in addition to haemolysin and both have a G + C content of 41% as compared with the host genome content of 51% (Table 2.7). DNA showing homology to the origins of replication of plasmids, and to IS elements, are also present on many PAIs.

Enteropathogenic *E. coli* strains carry a PAI that is required for the induction of attaching and effacing lesions on enterocytes (see Section 7.5.2.1.2.1). Pai III or LEE (locus of enterocyte effacement) encodes a type III secretion system and a number of secreted proteins involved in host signalling and pedestal formation. Pai III is located at the *selC* locus and has a G + C content of 39%. Introduction of Pai III into non-pathogenic laboratory strains of *E. coli* renders them capable of inducing lesions in enterocytes.

PAIs are also found in a number of other Gram-negative pathogens (Table 2.7). For example, five large PAIs have been found in strains of *Sal. typhimurium*. *Salmonella* pathogenicity island (SPI)-1 is 40 kb in length and carries some 25 genes. Many of these comprise a type III secretion system (Inv/Spa system) and secreted proteins that are responsible for invasion of enterocytes (see Section 8.2.1.1.4). Induction of apoptosis in *Salmonella*-infected macrophages also involves SPI-1. SPI-1 is not located at a tRNA locus and appears to be stable within the *Salmonella* chromosome. Nevertheless, it has a G + C content significantly different from that of the *Salmonella* chromosome (42% compared with 52%), suggesting SPI-1 was acquired by horizontal gene transfer. Interestingly, the organisation of genes within SPI-1 is broadly similar to that of the invasion genes of the *Shigella* virulence plasmid. SPI-2 is a 40 kb element that encodes some 17 genes. These include a TCSTS and a distinct type III secretion system (Spi/Ssa) that is required for survival within macrophages and is important for systemic disease; mutants in SPI-2 are severely attenuated in virulence. SPI-2 is located at the *valV*

Table 2.7. | Pathogenicity islands (PAIs) of some human pathogens

Pathogen	Pathogenicity island	Size (kb)	G + C content (PAI/host)	tRNA locus	Phenotype
Escherichia coli	Pai I	70	40/51	*selC*	Haemolysin production
	Pai II	190	40/51	*leuX*	Haemolysin, P fimbriae production
	Pai III (LEE)	35	39/51	*selC*	Type III secretion system, induction of attaching and effacing lesions
	Pai IV	170	41/51	*pheV*	Haemolysin, pap pili production
	Pai V	110	41/51	*pheR*	Haemolysin, CNF production
Salmonella typhimurium	SPI-1	40	42/52	–	Type III secretion system, invasion of enterocytes, apoptosis
	SPI-2	40	45/52	*valV*	Survival in macrophages
	SPI-3	17	47.5/52	*selC*	Survival in macrophages
	SPI-4	25	44.5/52	–	Type I secretion system, survival in macrophages
Salmonella dublin	SPI-5	9.5	44/52	*serT*	Enteropathogenicity
Shigella flexneri	*she* PAI	47	49/51	*pheV*	Enterotoxin, autotransporter, proteases
	SHI-2	24	48.6/51	*selC*	Iron acquisition, colicin, immunity
Yersinia enterocolitica	HPI	45	?	*asn-tRNA*	Iron acquisition
Helicobacter pylori	*cag*-PAI	40	36/39	–	Type IV secretion system, pedestal formation, inflammation, cancer?

? Not currently known; HPI, high pathogenicity island; for other abbreviations, see the text.

locus, encoding a valine-specific tRNA, and has a G + C content of 45%, again marking it as having been acquired horizontally. A third SPI was found by looking specifically at the *selC* locus, which in *E. coli* harbours either Pai I or LEE. Sure enough, a 17 kb element, SPI-3, was found that has a reduced G + C content (47.5%). SPI-3 encodes 10 proteins including a putative autotransported protein, a putative regulatory protein and a magnesium transporter, and is required for intra-macrophage survival. SPI-4 encodes 18 putative ORFs organised as a singe operon. Although not inserted at a tRNA gene, SPI-4 does contain a tRNA-like gene upstream from the operon. Some proteins may comprise a type I secretion system like that used to secrete RTX toxins (see Section 9.4.1.4). Unfortunately, the functions of other ORFs are as yet unknown or else the ORF shows homology only to hypothetical proteins uncovered by genome sequencing projects. Gene inactivation experiments indicate that SPI-4 is involved in intra-macrophage survival. A fifth SPI was located at the *serT* locus of *Sal. dublin* encoding a serine-specific tRNA. SPI-5, which is present also in *Sal. typhimurium*, is approximately 9.5 kb and has a G + C content of only 44% and has direct repeats at the left-hand and right-hand boundaries of the element. SPI-5 carries six genes whose expression is linked to the enteropathogenicity of *Sal. dublin*, but which may not be involved in systemic disease. One protein encoded by SPI-5, SopB, is translocated into host cells and promotes fluid secretion and an inflammatory response. Translocation of SopB is dependent on Sip (*Salmonella* invasion protein) proteins (see Section 8.2.1.1.4) that are themselves secreted by the Inv/Spa type III system encoded by SPI-1.

Apart from SPI-1 to SPI-5, numerous other genes or operons are scattered throughout the *Salmonella* chromosome, some of which are flanked by direct repeats and/or exhibit atypical G + C content providing evidence of horizontal acquisition. Collectively, these regions can be considered to comprise a 'pathogenicity archipelago' within the *Salmonella* chromosome, and are required for full virulence. Many of these genes are regulated by the PhoP/PhoQ two-component system that governs the adaptation of *Salmonella* to environments with a low magnesium concentration. PhoP/PhoQ is present in both pathogenic and non-pathogenic isolates. It is clear that after acquisition of PAIs by *Salmonella* (and presumably by other organisms) the control of expression of PAI-encoded virulence factors has been integrated within the regulatory networks of the host organism to ensure their correct expression in terms of spatial and temporal cues.

PAIs have also been found in other human pathogens including *Hel. pylori* (see Micro-aside, following), *Shigella* spp., and *Yersinia* spp. (Table 2.7).

Chromosomal elements that correspond fully with the PAI criteria provided in Table 2.6 have not yet been found in Gram-positive bacteria. There are, however, several examples of clusters of virulence factors that may have been acquired by horizontal DNA transfer. These include the PaLoc (19 kb) of toxigenic *Cl. difficile*, which encodes TcdA enterotoxin and TcdB cytotoxin (see Section 9.5.3.1). This element is not located at a tRNA locus nor is it delineated by direct repeats. Nevertheless, PaLoc is absent from non-toxigenic isolates of *Cl. difficile*. Similarly, a cluster of six virulence genes is present on a 10 kb section of the chromosome of *Lis. monocytogenes*, but is absent, or defective, in non-pathogenic *Listeria* spp. Amongst the genes present at this locus are *hyl*, encoding listeriolysin O, a cholesterol-binding cytolysin that is required for the escape of the organism from host cell vacuoles, and *actA*, encoding a protein required for motility of the organism inside host cells.

Helicobacter pathogenicity island and cancer

Helicobacter pylori is responsible for a spectrum of diseases in humans, ranging from superficial gastritis to duodenal ulceration. Furthermore, patients infected with *Hel. pylori* have a three- to six-fold increased risk of developing gastric cancer. Those strains expressing the 145 kDa CagA cytotoxin-associated protein (see below) are more frequently associated with cancer than are CagA$^-$ strains. CagA is one protein expressed by a 40 kb PAI called *cag*-Pai that carries some 40 genes in total and is flanked by direct repeats of 31 bp long. A number of proteins encoded on *cag*-PAI comprise a type IV secretion system similar to that used by *Bord. pertussis* to secrete pertussis toxin (see above and Section 9.5.2.1). The type IV system is essential for many of the pathological effects of *Hel. pylori* infection, including cytoskeletal rearrangements, a pronounced inflammatory response through the induction of IL-8 expression, and, indeed, cancer.

CagA is translocated directly into gastric epithelial cells by the type IV secretion system, whereupon it is phosphorylated on tyrosine residues by an as yet unidentified host cell kinase. Phosphorylated CagA induces cytoskeletal reorganisation resulting in the formation of a pedestal on which *Hel. pylori* sits in intimate contact with the host cell. CagA has been shown to activate **m**itogen-**a**ctivated **p**rotein (MAP) kinase cascades, leading to the induction of expression of proto-oncogene *c-fos* and phosphorylation of c-Jun that together comprise the transcription factor AP-1 (Figure 2.24). The AP-1 family of transcriptional activators plays a pivotal role in cell proliferation and neoplastic transformation and thus *Hel. pylori* strains carrying *cag*-Pai can induce mitogenic signals and proto-oncogene expression in gastric epithelial cells.

Mutations in the type IV secretion system encoded by *cag*-Pai result in bacteria that are greatly impaired in their ability to induce interleukin (IL-8) secretion by gastric epithelial cells. However, since *cagA*$^-$ cells are relatively unimpaired in IL-8 induction, other *cag*-Pai-encoded effectors may be translocated by the type IV system. The effector(s) are thought to up-regulate the activity of nuclear factor NF-κB, which, together with AP-1, leads to induction of IL-8 expression (Figure 2.24).

2.7 | Bacterial biofilms

In this chapter we have described various aspects of the cell biology, physiology and genetics of bacteria and have done so from the point of view that these organisms exist as isolated, single cells that are independent of their neighbours. However, during the last two decades it has become evident that, in many cases, the natural mode of growth of bacteria is not as single cells suspended in an aqueous environment (i.e. the planktonic mode) but as a community of cells living together in an ordered structure known as a biofilm. The classical definition of a biofilm is 'an accumulation of bacteria and their products on a surface'. Whereas such a definition may have satis-

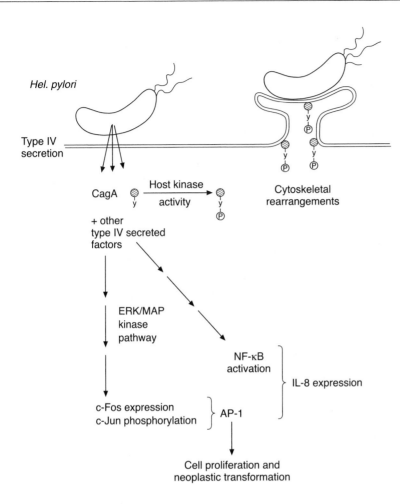

fied many bacteriologists 20 years ago, what has been learned since then about the complexity of biofilms in terms of their structure, organisation and intercellular communication renders this definition woefully inadequate. Although no really adequate definition of the term 'biofilm' exists, we can certainly recognise the main features of a biofilm as being the following (Figure 2.25):

(a) It has a three-dimensional structure containing one or more bacterial species.
(b) It forms at interfaces – solid/liquid, liquid/air, solid/air.
(c) It exhibits spatial heterogeneity due to physicochemical and chemical gradients that develop within it.
(d) It is often permeated by water channels.
(e) The organisms within it display a marked decrease in susceptibility to antimicrobial agents and host defence systems as compared with their planktonic counterparts.

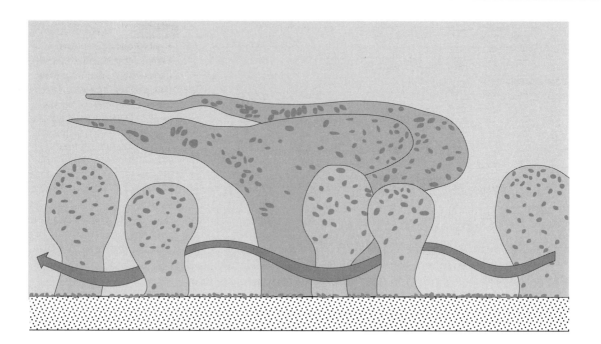

Figure 2.25 Diagrammatic representation of the overall structure of a typical biofilm showing mushroom-like stacks containing bacteria embedded in an extracellular matrix. The stacks are separated by water channels. The arrow shows the direction of fluid flow.

The rapid advances made in our knowledge of biofilms stem from the growing realisation that such films probably represent the natural mode of growth of most bacteria on this planet. Indeed, it has been suggested that broth cultures (i.e. the time-honoured approach to growing bacteria in the laboratory) have outlived their usefulness. While this may be an extreme view, it has to be recognised by medical microbiologists that many infectious diseases are caused by biofilms and, because bacteria display a different phenotype when in a biofilm (see below), investigations of virulence, antibiotic susceptibility and interactions with host defence mechanisms should be carried out using biofilm-grown bacteria, not broth cultures. Examples of biofilm-induced diseases include caries, gingivitis, periodontitis, osteomyelitis, endocarditis, infections associated with prosthetic devices (e.g. catheters, artificial heart valves, implants) and lung infections in individuals with cystic fibrosis. Examples of organisms that commonly form biofilms include *Ps. aeruginosa*, *Staph. epidermidis*, *Staph. aureus*, *E. coli*, *Lactobacillus* spp. and *Streptococcus* spp. Being a relatively new field, there is considerable controversy regarding many aspects of biofilm formation, structure and physiology and our understanding of these bacterial communities is certainly destined to evolve rapidly. During the past decade, for example, our knowledge of the structure of biofilms has changed radically due to the advent of confocal laser scanning microscopy. This has enabled us, for the first time, to view biofilms in their living, hydrated state. This is in contrast to electron microscopy, which requires that the biofilms be dehydrated and examined in a vacuum – such processing destroys the real structure of the biofilm (for electron micrographs of one of the most prevalent biofilms in humans – dental

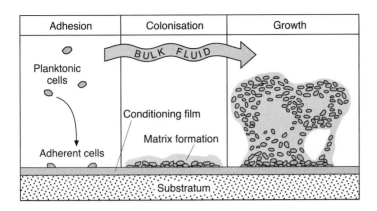

Figure 2.26 Stages involved in biofilm formation. Planktonic bacteria adhere to the conditioning film of the substratum and then grow and synthesise extracellular matrix molecules. Further growth and cell replication lead to the formation of a biofilm.

plaque, see Figure 7.2). We will now describe some features of the formation, growth, structure and physiology of bacterial biofilms.

Three stages can be recognised in biofilm formation on a solid surface (termed a substratum): adhesion, colonisation and growth (Figure 2.26). The first phase involves adhesion of planktonic bacteria to the substratum. As any surface in an aqueous environment is invariably coated with adsorbed molecules, adhesion is usually to this coating (often termed a 'conditioning film') rather than to the material of the substratum itself. It is well established that adhesion to a substratum often leads to changes in gene transcription (as described above), resulting in an alteration in the phenotype of the organism. If an adequate supply of nutrients is available and environmental factors are conducive, adherent bacteria may then grow and reproduce and this stage is known as colonisation. During this stage, the bacteria begin to secrete extracellular polymers (usually polysaccharide) that form a matrix in which the cells become embedded. The community then grows and develops a characteristic structure, the nature of which will depend on many factors including the species involved, the availability of nutrients, the hydrodynamic and mechanical shear forces present, etc. In many cases, the organisms form stacks that are separated by water channels (Figure 2.27). However, this may be characteristic only of biofilms growing in an environment with a high fluid flow rate and/or a low nutrient content. There is some evidence that, in a nutrient-rich environment with a low fluid flow rate, there are fewer water channels and the biofilms have a less open structure. It has been suggested that the water channels function as a primitive circulatory system in that they could deliver fresh nutrients to the stacks of bacteria and remove excretory products. Because of the high density of packing of the bacteria in the stacks, penetration of nutrients, gases, ions, etc. is inhibited to some extent. This results in gradients within the stacks (both horizontally and vertically) in terms of nutrients, oxygen, pH, waste products, etc. The net effect of this is that a vast range of microhabitats is present within the biofilm (Figure 2.28). This has two consequences. Firstly, because of this wide range of microhabitats, survival within the biofilm of organisms with a diverse set of

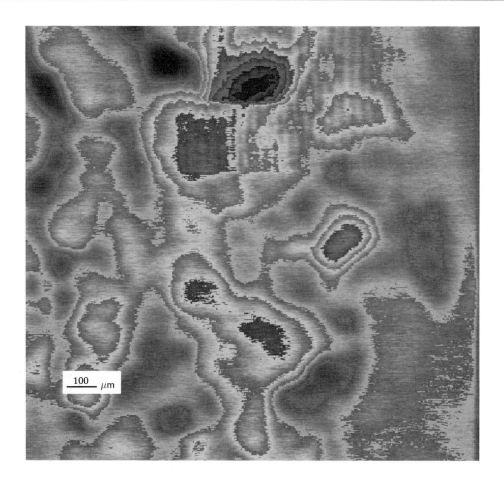

Figure 2.27 Scanning confocal micrograph of a typical bacterial biofilm (viewed from above) showing stacks and water channels. The 'contours' represent different distances from the microscope objective – the darker the colour the shorter the distance. The biofilm was grown in the laboratory from human saliva and so consists of many different bacterial species.

nutrient and physicochemical requirements is possible. Hence, obligate anaerobes can survive in dental plaque, which grows in an essentially aerobic environment – the oral cavity. Secondly, if the biofilm consists of only a single species, the different microhabitats will enable differential gene expression in the different regions of the biofilm. Studies using promotor–reporter constructs (see Section 3.6) have demonstrated that this does occur so that the same organism can display different phenotypes depending on its location within the biofilm.

Following the discovery that gene expression in some bacterial species is regulated by the population density of the organism (quorum sensing, described in Section 4.3.2.1), biofilms were an obvious place to look for examples of this phenomenon. Many organisms forming biofilms are known to utilise quorum sensing, for example *Ps. aeruginosa*, *Staph. aureus*, *Staph. epidermidis*, *E. coli*, *Vibrio* spp. and *Lactobacillus* spp. Finally, an important property of biofilms from the point of view of infectious diseases is that they are remarkably resistant to antibiotics and to host defence systems. Treatment of diseases due to biofilms, therefore, is a major problem. The mechanisms

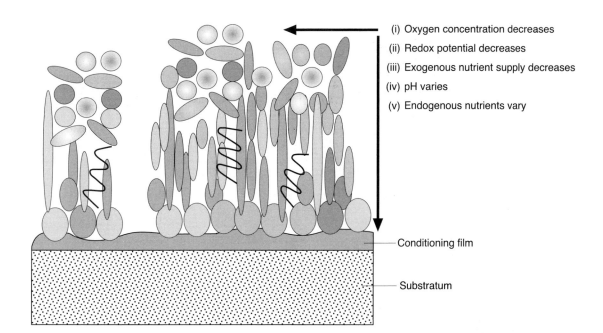

(i) Oxygen concentration decreases

(ii) Redox potential decreases

(iii) Exogenous nutrient supply decreases

(iv) pH varies

(v) Endogenous nutrients vary

Conditioning film

Substratum

underlying such resistance remain to be fully established. However, factors thought to contribute to the ability of biofilms to tolerate high concentrations of antimicrobial agents include: (i) binding of the antimicrobial agent to the biofilm matrix, thereby limiting its penetration; (ii) inactivation of the antimicrobial agent by enzymes trapped in the biofilm matrix; (iii) that the reduced growth rate of bacteria in biofilms renders them less susceptible to the antimicrobial; (iv) that the altered microenvironment within the biofilms (e.g. pH, oxygen content) can reduce the activity of the agent; and (v) that altered gene expression by organisms within the biofilm can result in a phenotype with reduced susceptibility to the agent. With regard to host defences, bacteria within biofilms are less easily phagocytosed and less susceptible to complement than their planktonic counterparts but the reasons for this reduced susceptibility are not understood.

To summarise this brief description of the exciting world of biofilms – these are cellular communities with an ordered structure and a circulatory system, they display different physiologies within different regions, have a form of intercellular communication and can resist noxious chemicals and other threats from their environment. Does this sound familiar? These properties are very similar to those exhibited by multicellular organisms!

Figure 2.28 The range of habitats available within a biofilm because of the formation of nutrient, gaseous and physico-chemical gradients from the biofilm/liquid interface through to the substratum and from the outside of a 'stack' through to its centre. The various shapes represent different species of bacteria.

2.8 | Concept check

- Bacteria come in many different shapes and sizes but can be divided broadly into two groups: Gram-positive and Gram-negative.

These groupings reflect fundamental differences in cell wall structure.

- Many cell surface components such as LPS and lipoteichoic acid may play a central role in the pathology of infection.
- Secreted proteins are important in ensuring the survival of bacteria within the host and bacteria have evolved a variety of pathways for exporting such proteins. Many of these systems are complex and may have been adapted from other biological functions.
- Type III and type IV systems are recently discovered pathways of protein secretion that enable the injection of bacterial proteins directly into host cells and thereby play key roles in the pathogenesis of a variety of infections.
- Bacteria can sense their environmental conditions, and can respond rapidly to changes through alterations in gene expression.
- By altering their cell surfaces (by phase or antigenic variation), bacteria can avoid host immune responses.
- Mobile genetic elements such as plasmids, transposable elements and pathogenicity islands play an important role in bacterial virulence, and may have contributed to the evolution of pathogens as we know them today.
- Many organisms grow as complex communities known as biofilms. These enable bacteria to survive (and cause disease) in hostile environments.

2.9 | Cell biology of the paradigm organisms

As it would be very difficult to summarise what is known of the cell biology of these two paradigm organisms in a reasonable space, we will at this stage review briefly only their surface structures as these contain molecules that play a central role in virulence, for example in adhesion, invasion, evasion of host defence systems and the induction of damage to the host. Other aspects of the cell biology of these organisms (e.g. protein secretion, cell signalling, etc.) will be described subsequently in appropriate chapters.

2.9.1 *Streptococcus pyogenes*

The cell wall of this organism has been shown to be a very complex structure and is shown diagrammatically in Figure 2.29.

A conspicuous feature of the surface of cells of *Strep. pyogenes* is the presence of fibrils (see Figure 7.28) composed of a protein known as the M protein. The structure of this protein has been investigated extensively and has been shown to have several distinct regions (Figure 2.30). At the N-terminus is a signal sequence that directs the M protein for secretion and is removed from the mature protein. Next is a highly variable region that protrudes beyond the cell wall. This is followed by a region rich in proline/glycine and threonine/serine residues

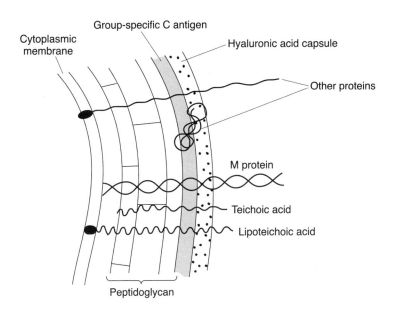

Figure 2.29 The cell envelope of *Streptococcus pyogenes*. Carbohydrate-containing molecules include the group-specific C antigen, the hyaluronic acid capsule, teichoic acid and lipoteichoic acid. Proteinaceous surface components include the M protein and a variety of other wall-anchored or surface-associated proteins (see Table 2.7). The latter include glycolytic enzymes.

that marks the passage of the protein through the peptidoglycan layer. Finally, at the C-terminus is a wall-anchor domain that comprises a Leu-Pro-X-Thr-Gly (where X is any amino acid) sequence, a hydrophobic region that traverses the cytoplasmic membrane, and finally a region of up to seven charged amino acid residues that protrudes into the cytoplasm. The last two regions of the wall-anchor domain serve to pause translocation of the protein through the membrane to allow cleavage of the protein by the enzyme sortase between the Thr and Gly residues of the Leu-Pro-X-Thr-Gly motif. The cleaved molecule is then covalently attached to peptidoglycan via the Thr residue.

The highly variable region of the M protein has enabled the species to be subdivided into a number of M types (approximately 100), which are distinguishable on the basis of the antigenicity of the M protein. M-typing is particularly important in distinguishing between those strains of the organism that are associated with the various non-suppurative sequelae of infection by the organism (see Section 1.6.1) such as glomerulonephritis (M types 2, 49, 57, 59, 60, 61) and rheumatic fever

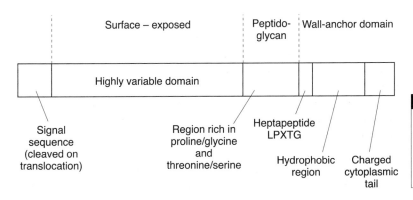

Figure 2.30 Major regions (and their relationship to other surface structures) of a typical M protein of *Streptococcus pyogenes*. The various regions are not drawn to scale.

Table 2.8. | A partial list of the proteins that have been detected on the surface of *Streptococcus pyogenes*

Protein	Function/properties
M protein	Anti-phagocytic, anti-opsonic, adhesion
IgA-binding group A streptococcal M-like protein 2 (ARP2)	IgA-binding
ARP4	IgA-binding
M-related protein 4 (Mrp4)	IgG/fibrinogen-binding
Fibronectin receptor A (FcRA)	Fc-binding
Streptococcal C5a peptidase (SCP)	C5a peptidase
Sfb, PrtF, protein F2, SfBP1, SfbII	Fibronectin-binding
Fibronectin-binding protein 54 (FBP54)	Fibronectin/fibrinogen-binding
Plasminogen-binding group A streptococcal M-like protein (PAM)	Plasmin-binding
Glyceraldehyde-3-phosphate dehydrogenase	Adhesion to fibronectin
Glyceraldehyde-3-phosphate dehydrogenase, phosphoglycerate kinase, phosphoglycerate mutase, α-enolase, triosephosphate isomerase	Generation of ATP?

(M types 1, 3, 5, 6, 14, 18, 19, 24). The M protein is important in evading host defence mechanisms (see Sections 10.4.1.3 and 10.5.1.2), being anti-phagocytic and anti-opsonic.

A variety of other proteins are found on the surface of *Strep. pyogenes* (see Table 2.8), the functions of many of which are unknown, although some are thought to be adhesins. Of particular interest is the presence of a number of enzymes from the glycolytic pathway – specifically five enzymes involved in ATP production. This implies that the organism may be able to generate ATP on its surface. It is possible that this extracellularly produced ATP is able to activate cells via purinoceptors that bind ATP.

Streptococcus pyogenes has two major polysaccharide cell surface components, the capsule and the group-specific C antigen. The capsule is composed of hyaluronic acid, which is chemically indistinguishable from that found in the connective tissue of humans and, therefore, is non-antigenic. The latter property helps the organism to escape the attention of the immune response. The capsule, like the M protein, is an important anti-phagocytic component and is also involved in adhesion to epithelial cells. The group-specific C antigen is an antigenic carbohydrate that can comprise up to 10% of the dry weight of the organism and is used to subdivide the β-haemolytic streptococci into groups (known as Lancefield's groups, labelled A to O); *Strep. pyogenes* belongs to group A of this classification scheme. The carbohydrate consists of a backbone of rhamnose residues to which N-acetylglucosamine side chains are linked. Antibodies to this polysaccharide also react with constituents of heart tissue such as myosin and this is thought to contribute to the pathology of rheumatic fever.

Other carbohydrates that are exposed on the surface of *Strep. pyogenes* include teichoic acid and lipoteichoic acid. Both of these are antigenic and can stimulate the release of cytokines from host cells, the possible role of lipoteichoic acid in adhesion is described in Section 7.8.1.

2.9.2 *Escherichia coli*

The cell wall of *E. coli* has a complex structure typical of that of Gram-negative organisms. In addition, it has a range of surface-associated organelles and molecules that are involved in disease pathogenesis including a capsule, fimbriae, flagella, type III secretion apparatus, lipopolysaccharide, porins and outer membrane proteins (see Figure 2.4).

The capsule is a polysaccharide whose composition is very much strain dependent and may consist of any of the following: it may be one of the 80 distinct polysaccharides (designated as the K antigens), a polymer derived from one of the 170 different O antigens of the LPS molecule, or colanic acid. The first two types of capsule are important virulence factors as they are anti-phagocytic, but colanic acid is unlikely to be a virulence determinant as it is expressed only at temperatures below 30°C.

The fimbriae (or pili) of *E. coli* are undoubtedly important in adhesion of the organism to host cells. Four major types of fimbriae are recognised among bacteria, but most clinical isolates of *E. coli* have type I fimbriae, i.e. they can agglutinate erythrocytes and this is inhibited by mannose. As well as possessing this classical type of fimbrium, some strains can also produce coiled structures (curli) that also mediate attachment to host cells and some (e.g. enteropathogenic strains) produce bundle-forming pili (examples of type IV fimbriae) on contacting host cells (see Section 7.5.2.1.2.1). Production of type I fimbriae is phase variable and oscillates between an expressed and non-expressed state. The suppression of fimbrial production helps the organism to evade the immune response, as fimbrial proteins are highly antigenic.

Escherichia coli has approximately 10 flagella distributed over its surface in an apparently random array. It is tempting to speculate that chemotaxis (dependent, of course, on flagella) may be important in enabling *E. coli* to find a suitable habitat in its host – there is, however, little evidence in support of this. Flagella have been implicated in adhesion and have also been shown to induce cytokine release from host cells. The flagella are antigenic and constitute the H antigen of the organism. More than 50 different types of H antigens have been recognised in *E. coli*.

LPS has long been known to be a key virulence factor of *E. coli*. This is attributable mainly to its ability to stimulate the release of pro-inflammatory cytokines from a variety of host cells (see Chapters 5 and 6). Similarly, a number of OMPs of the organism have also been shown to possess this ability. Both LPS and OMPs may also function as adhesins. The antigenic side chain of the LPS molecule constitutes the O antigen of the organism and more than 170 different types have been recognised in *E. coli*. Serotyping of the organism is usually based on the nature of the O and H antigens of the organism. Hence, one of the most notorious food-poisoning strains of the organism has the O157 antigen and the H7 antigen and is designated *E. coli* O157 : H7.

Another antigenic constituent of the cell wall is the enterobacterial common antigen. This is a glycolipid that is present in all Gram-negative enteric bacteria. It consists of a linear heteropolysaccharide chain covalently linked to phosphoglyceride.

2.10 | What's next?

Now that we have some idea of what makes a bacterium tick, we will now move on to review the new molecular approaches that are being developed to tease out the interplay that occurs between bacteria and the host cells with which they interact. Chapter 3 will, therefore, describe those techniques that are being applied to both bacteria and host cells in order to gain a better understanding of the role each is playing in that intimate and complex relationship that lies at the heart of any bacterial infection.

2.11 | Questions

1. Describe the fundamental differences between Gram-positive and Gram-negative bacterial cell structure. How do these differences impact on basic cell functions such as protein translocation and cell division?
2. Describe, briefly, the means by which Gram-negative bacteria translocate proteins across the outer membrane.
3. Outline the essential features of type III and type IV secretion systems, giving examples of organisms that utilise such systems.
4. How may bacteriophages contribute to bacterial virulence? And to the evolution of pathogens?
5. What are the principle features of pathogenicity islands that mark them as having been acquired by horizontal gene transfer?
6. What is the difference between antigenic variation and phase variation, and what advantages do these systems provide for pathogens?
7. Outline the mechanisms by which *Neisseria* spp. alter their cell surfaces.

8. Discuss the role played by mobile genetic elements in bacterial virulence.

9. Describe the means by which bacteria sense, and respond to, changes in their environment.

10. What do you understand by the term 'biofilm'? Describe what you know of the formation and structure of biofilms. Why do biofilms pose such a problem therapeutically?

2.12 | Further reading

Books

Allison, D. G., Gilbert, P., Lappin-Scott, H. M. & Wilson, M. (eds.) (2000). *Community Structure and Co-operation in Biofilms*. Cambridge: Cambridge University Press.

Baumberg, S. (1999). *Prokaryotic Gene Expression*. Oxford: Oxford University Press.

Kaper, J. B. & Hacker, J. (1999). *Pathogenicity Islands and Other Mobile Virulence Elements*. Washington, DC: ASM Press.

Lewin, B. (1999). *Genes VII*. Oxford: Oxford University Press.

Madigan, M. T., Martinko, J. M. & Parker, J. (2000). *Brock Biology of Microorganisms*, 9th edn, London: Prentice-Hall International.

Wagner, R. (2000). *Transcription Regulation in Prokaryotes*. Oxford: Oxford University Press.

Waldvogel, F. A. & Bisno, A. L. (2000). *Infections Associated with Indwelling Medical Devices*. Washington, DC: ASM Press.

Review articles

Berks, B. C., Sargent, F. & Palmer, T. (2000). The Tat protein export pathway. *Mol Microbiol* **35**, 260–274.

Christie, P. J. & Vogel, J. P. (2000). Bacterial type IV secretion: conjugation systems adapted to deliver effector molecules to host cells. *Trends Microbiol* **8**, 354–360.

Cornelis, G. R., Boland, A., Boyd, A. P., Geuijen, C., Iriarte, M., Neyt, C., Sory, M.-P. & Stainer, I. (1998). The virulence plasmid of *Yersinia*, an antihost genome. *Microbiol Mol Biol Rev* **62**, 1315–1352.

Cornelis, G. & van Gijsegem, F. (2000). Assembly and function of type III secretory systems. *Annu Rev Microbiol* **54**, 735–774.

Cotter P. A. & DiRita, V. J. (2000). Bacterial virulence gene regulation: an evolutionary perspective. *Annu Rev Microbiol* **54**, 519–565.

Costerton, J. W., Stewart, P.S. & Greenberg, E.P. (1999). Bacterial biofilms: a common cause of persistent infections. *Science* **284**, 318–322.

Danese, P. N. & Silhavy, T. J. (1998). Targeting and assembly of periplasmic and outer-membrane proteins in *Escherichia coli*. *Annu Rev Genet* **32**, 59–94.

Davey, M. E. & O'Toole, G. A. (2000). Microbial biofilms: from ecology to molecular genetics. *Microbiol Mol Biol Rev* **64**, 847–867.

Deitsch, K. W., Moxon, E. R. & Wellems, T. E. (1997). Shared themes of antigenic variation and virulence in bacterial, protozoal, and fungal infections. *Microbiol Mol Biol Rev* **61**, 281–293.

Dougan, G., Haque, A., Pickard, D., Frankel, G., O'Goara, P. & Wain J. (2001). The gene pool of *Escherichia coli*. *Curr Opin Microbiol* **4**, 90–94.

Fekkes, P. & Driessen, A. J. M. (1999). Protein targeting to the bacterial cytoplasmic membrane. *Microbiol Mol Biol Rev* **63**, 161–173.

Fernández, L. A. & Berenguer, J. (2000). Secretion and assembly of regular surface structures in Gram-negative bacteria. *FEMS Microbiol Rev* **24**, 21–44.

Groisman, E. A. & Ochman, H. (1997.) How *Salmonella* became a pathogen. *Trends Microbiol* **5**, 343–349.

Hallet, B. (2001). Playing Dr Jekyll and Mr Hyde: combined mechanisms of phase variation in bacteria. *Curr Opin Microbiol* **4**, 570–581.

Henderson, I. R., Cappello, R. & Nataro, J. P. (2000). Autotransporter proteins, evolution and redefining protein secretion. *Trends Microbiol* **8**, 529–532.

Henderson, I. R., Owen, P. & Nataro, J. P. (1999). Molecular switches – the ON and OFF of bacterial phase variation. *Mol Microbiol* **33**, 919–932.

Henderson, I. R. & Nataro, J. P. (2001). Virulence functions of autotransporter proteins. *Infect Immun* **69**, 1231–1243.

Hensel, M. (2000). *Salmonella* pathogenicity island 2. *Mol Microbiol* **36**, 1015–1023.

Hentschel, U & Hacker, J. (2001). Pathogenicity islands: the tip of the iceberg. *Microbes and Infection* **3**, 545–548.

Høiby, N., Johansen, H. K., Moser, C., Song, Z., Ciofu, O. & Kharazmi, A. (2001). *Pseudomonas aeruginosa* and the *in vitro* and *in vivo* biofilm mode of growth. *Microbes Infect* **3**, 23–35.

Hueck, C. J. (1998). Type III protein secretion systems in bacterial pathogens of animals and plants. *Microbiol Mol Biol Rev* **62**, 379–433.

Kaper, J. B. & Hacker, J. (2000). Pathogenicity islands and the evolution of microbes. *Annu Rev Microbiol* **54**, 641–679.

Kim, J. F. (2001). Revisiting the chlamydial type III protein secretion system: clues to the origin of type III protein secretion. *Trends Genet* **17**, 65–69.

Koch, A. L. (2000). The bacterium's way for safe enlargement and division. *Appl Env Microbiol* **66**, 3657–3663.

Koonin, E. V., Makarova, K. S. & Aravind, L. (2001). Horizontal gene transfer in prokaryotes: quantification and classification. *Annu Rev Microbiol* **55**, 709–742.

Lewis, K. (2001). Riddle of biofilm resistance. *Antimicrob Agents Chemother* **45**, 999–1007.

Mah, T. C. & O'Toole, G. A. (2001). Mechanisms of biofilm resistance to antimicrobial agents, *Trends Microbiol* **9**, 34–39.

Mahillon, J. & Chandler, M. (1998). Insertion sequences. *Microbiol Mol Biol Rev* **62**, 725–774.

O'Toole, G., Kaplan, H. B. & Kolter, R. (2000). Biofilm formation as microbial development. *Annu Rev Microbiol* **54**, 49–79.

Raivo, T. L. & Silhavy, T. J. (1999). The σ^E and Cpx regulatory pathways: overlapping but distinct envelope stress responses. *Curr Opin Microbiol* **2**, 159–165.

Saier, M. H. Jr (2000). Families of transmembrane transporters selective for amino acids and their derivatives. *Microbiology* **146**, 1775–1795.

Sutherland, I. W. (2001). Biofilm exopolysaccharides: a strong and sticky framework. *Microbiology* **147**, 3–9.

Thanassi, D. G. & Hultgren S. J. (2000). Multiple pathways allow protein secretion across the bacterial outer membrane. *Curr Opin Cell Biol* **12**, 420–430.

Thanassi, D. G., Saulino, E. T. & Hultgren, S. J. (1998). The chaperone/usher pathway: a major terminal branch of the general secretory pathway. *Curr Opin Microbiol* **1**, 223–231.

van Wely, K. H. M., Swaving, J., Freudl, R. & Driessen, A. J. M. (2001). Translocation of proteins across the cell envelope of Gram-positive bacteria. *FEMS Microbiology Reviews* **25**, 437–454.

Wallis, T. S. & Galyov, E. E. (2000). Molecular basis of *Salmonella*-induced enteritis. *Mol Microbiol* **36**, 997–1005.

Watnick, P. & Kolter R. (2000). Biofilm, city of microbes. *J Bacteriol* **182**, 2675–2679.

Winstanley, C. & Hart, C. A. (2001). Type III secretion systems and pathogenicity islands. *J Med Microbiol* **50**, 116–127.

Wosten, M. M. S. M. (1998). Eubacterial sigma-factors. *FEMS Microbiol Rev* **22**, 127–150.

Young, J. & Holland, I. B. (1999). ABC transporters: bacterial exporters-revisited five years on. *Biochim Biophys Acta* **1461**, 177–200.

Papers

Karaolis, D. K. R., Somara, S., Maneval, D. R. Jr, Johnson, J. A. & Kaper, J. B. (1999). A bacteriophage encoding a pathogenicity island, a type-IV pilus and a phage receptor in cholera bacteria. *Nature* **399**, 375–379.

Madden, J. C., Ruiz, N. & Caparon, M. (2001). Cytolysin-mediated translocation (CMT): a functional equivalent of type III secretion in Gram-positive bacteria. *Cell* **104**, 143–152.

Reid, S. D., Herbelin, C. J., Bumbaugh, A. C., Selander, R. K. & Whittam, T. S. (2000). Parallel evolution of virulence in pathogenic *Escherichia coli*. *Nature* **406**, 64–67.

Stein, M., Rappuoli, R. & Covacci, A. (2000). Tyrosine phosphorylation of the *Helicobacter pylori* CagA antigen after *cag*-driven host cell translocation. *Proc Natl Acad Sci USA* **97**, 1263–1268.

2.13 | Internet links

1. http://www.bact.wisc.edu/microtextbook/index.html
   Excellent background on many aspects of microbiology including bacterial classification, structure, nutrition and metabolism, genetics, and disease. University of Wisconsin-Madison, USA.
2. http://trishul.sci.gu.edu.au/courses/ss12bmi/microbe_structure.html
   Describes the organisation and structure of microbes. Griffith University, Australia.
3. http://gsbs.utmb.edu/microbook/ch002.htm
   Chapter on bacterial structure from an on-line Medical Microbiology textbook.

4. http://www.kcom.edu/faculty/chamberlain/Website/Lects/
Bacteria.htm
A set of lecture notes on bacterial structure.

5. http://gsbs.utmb.edu/microbook/ch005.htm
Chapter on bacterial genetics from an on-line Medical
Microbiology textbook.

6. http://www.med.sc.edu:85/mayer/geneticreg.htm
Section of an on-line series of lecture notes that describes the
regulation of gene expression. Has a good set of explanatory
diagrams.

7. http://www.med.sc.edu:85/mayer/genetic%20ex.htm
Section of an on-line series of lecture notes that describes the
transfer of genetic information between bacteria. Has a good set
of explanatory diagrams.

8. http://instruct1.cit.cornell.edu/Courses/biomi290/Horror/
Biofilmmenu.html
This site provides introductory information on the complex
microbial communities that form biofilms. It provides pictures,
diagrams and films of the structure and formation of biofilms
and the interactions occurring between members of biofilm
communities.

9. http://www.erc.montana.edu
This web page is provided by one of the leading centres of
biofilm research – the Center for Biofilm Engineering at
Montana State University. It has an overview of the institute and
the research going on there and has links to other sites
providing information on biofilms.

10. http://www.denniskunkel.com/
A set of beautiful electron micrographs of a wide range of
bacteria and other organisms.

Molecular analysis of bacterial virulence mechanisms

Chapter outline

Aims

The principal aims of this chapter are:

- to describe the various molecular approaches that have been developed (in both bacteria and eukaryotic cells) to identify bacterial virulence factors
- to highlight the advantages and limitations of each method
- to provide examples of the application of molecular techniques to answer questions about pathogen behaviour during infection

3.1 | Introduction

As described in the Preface to this book, there has been an alarming rise in the incidence of antibiotic resistance amongst bacterial pathogens during the last few years. The boast of the Surgeon General, in his address to the USA Congress in 1969, that 'we can close the book on infectious diseases' sounds very hollow now as we witness the resurgence of diseases caused by pathogens such as *Mycobacterium tuberculosis* and *Streptococcus pyogenes*. We also face a myriad new infectious agents that are emerging as the world shrinks, through international travel and migration, and as population demographics change. Understanding how bacteria cause disease is, consequently, high on the agenda once again.

Some fundamental aspects of bacterial virulence have been investigated in great detail. These include bacterial adhesion, invasion and the production of exotoxins, and these topics are covered in later chapters in this book. However, the results arising from the use of new molecular techniques, described in this chapter, are forcing us to broaden our idea of what we mean by the term 'virulence factor'. Many different screening procedures now exist to identify genes expressed during infection. A significant proportion of the genes identified are so-called 'housekeeping genes' and genes required for survival *in vivo*. A more global definition of the term 'virulence factor' may be 'any gene product that is necessary for survival and persistence within the host'. Hopefully, this will open up new strategies and new targets for intervening in the infectious process. This type of approach to understanding the molecular interactions between bacteria and their host during infection may also advance our understanding of opportunistic infections by commensal bacteria.

The molecular techniques, both old and new, that have been developed to define bacterial virulence mechanisms, fall into a number of categories. Some are protein based, others are DNA based, while the remainder rely on the isolation and analysis of bacterial mRNA molecules. A number of techniques serve as 'promoter traps' to capture gene promoters switched on during the infectious process. Some of these techniques require that quite advanced molecular genetic systems are available for the organism under study. Consequently, there are powerful tools and techniques available for the analysis of bacteria such as *Escherichia coli*, *Salmonella typhimurium*, *Listeria monocytogenes* and *Strep. pneumoniae* that cannot be applied to organisms whose genetics are less well understood. Nevertheless, some molecular techniques are more applicable to bacteria in general and require only that DNA or RNA can be isolated from the bacterium. The advantages and limitations of each of these techniques will be highlighted.

A suitable model system is required to study bacterial gene expression during infection. Experimental animals such as mice and monkeys have long been used as models for human infection. However, their use is limited for several important reasons. In the first instance, there are clear ethical considerations. Furthermore, animal models are frequently costly but, perhaps more importantly, the information they provide may be of limited relevance to human infections. Tissue culture monolayer studies using epithelial cells and cells of the monocyte–macrophage lineage have contributed significantly to our understanding of gene expression *in vivo*, and recent advances in multi-ple-layer tissue culture systems using mixed cell populations (for e.g. epithelial and endothelial cells) have the advantage of more closely mimicking the challenge facing bacterial pathogens at the site of infection.

An understanding of basic molecular biology techniques, such as nucleic acid purification and manipulation, library construction and screening and polymerase chain reaction (PCR), is assumed, and the reader is directed to any number of good molecular biology textbooks and websites, some of which are listed at the end of this chapter.

3.2 | Mutational analysis

The introduction of mutations in the bacterial chromosome in order to disrupt or modify gene expression is a popular approach for investigating bacterial virulence mechanisms. When coupled with an appropriate screening procedure, random mutagenesis can be used to identify new virulence factors. Directed mutagenesis, on the other hand, allows one to test the hypothesis that a specific gene product is involved in virulence.

3.2.1 Directed mutagenesis

In a technique often referred to as reverse genetics, the gene encoding a protein implicated, by other means, as a virulence factor is insertion-ally inactivated. This generates an isogenic mutant that is otherwise identical with the wild-type strain. Thus any reduction in the patho-genicity of the mutant strain, or alteration in its behaviour, can usually be ascribed to the inactivated gene. Consequently, directed mutagen-esis is an important technique to confirm the involvement (or non-involvement) in virulence and pathogenicity of genes and gene prod-ucts identified by other screening methods. In directed mutagenesis, an antibiotic resistance marker is ligated within the coding region of a cloned copy of the gene of interest. The disrupted gene is then intro-duced into the target organism, where it undergoes recombination with the homologous region of the bacterial chromosome (Figure 3.1). This results in the insertion of the antibiotic resistance marker into the chromosome, and recombinant organisms can be selected by plating on medium containing a concentration of the relevant antibiotic that inhibits the growth of the wild-type strain. This method requires a means of introducing DNA into the target organism. It also requires suitable selective markers that are efficiently expressed within the target organism, and an inherent capacity for homologous recombina-tion. Methods for introduction of DNA into bacteria include transfor-mation (with naked DNA), transduction (using bacteriophages), conjugation (bacterial mating) and electroporation (high voltage pulse) (see Chapter 2). If the selective marker on the cloning vector is itself expressed within the target organism, then the construct can be

Figure 3.1 Targeted insertional inactivation. Homologous recombination results in the integration of DNA into the bacterial chromosome. (a) A double cross-over recombination event results in the integration of only the selective marker. (b) The entire recombinant vector is integrated following a single cross-over event.

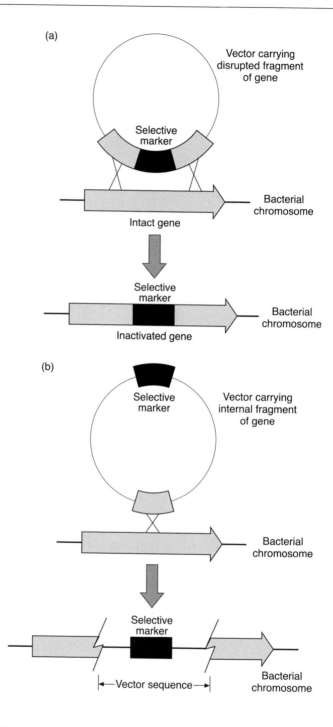

used directly for insertion–duplication mutagenesis (Figure 3.1). A single cross-over recombination event results in integration of the entire plasmid. Since the cloned pathogen DNA is duplicated on the chromosome, if the cloned portion includes DNA 5′ or 3′ to the gene, then a complete gene is regenerated and mutagenesis will be unsuccessful.

Directed mutagenesis can support or refute a hypothesis

The hypothesis has been proposed that the capsules of a number of bacterial pathogens, such as *Haemophilus influenzae* and *Streptococcus* spp., are important virulence determinants. Capsules may mediate adhesion to host surfaces (see Table 7.2), and impair phagocytosis and complement-mediated killing (see Section 10.4.1.1). *Pasteurella multocida* is the causative agent of a wide range of diseases in wild and domestic animals. Experiments with purified capsular material from this organism indicated that it had significant anti-phagocytic activity, thus implicating the capsule as a virulence determinant. In order to test this hypothesis, an isogenic mutant was created by insertion of a tetracycline resistance determinant within the *cexA* gene, encoding a protein required for the export of the capsular polysaccharide. The isogenic mutant was able to synthesise immunoreactive polysaccharide but failed to export it correctly and, consequently, cells were acapsular. When injected into the peritoneum of mice, the 50% infective dose (ID_{50}: number of bacteria required to initiate an infection in 50% of challenged animals) for the wild-type strain was less than 10 colony-forming units (CFU). Indeed, at a dose of 10 CFU only 17% of mice survived the challenge. In contrast, no deaths were recorded for mice challenged with less than 8×10^5 CFU of the acapsular mutants. Clearly, then, capsule production does contribute significantly to the virulence of *P. multocida*. Acapsular mutants, unlike the wild-type strain, were cleared rapidly from the blood and other organs of challenged animals and were up to six times more susceptible to phagocytosis by mouse peritoneal macrophages.

In order to complement the mutant strain, a replicating plasmid was constructed that contained an uninterrupted copy of *cexA*. This plasmid was then introduced back into the mutant strain. The complemented strain was now capable of expressing a capsule and behaved like the wild type in mouse infection studies, thus confirming the role of the capsule in virulence.

Helicobacter pylori causes gastritis and is associated with the development of peptic ulcers and gastric cancer. The urease activity of *Hel. pylori* is central to pathogenesis, as the ammonia produced by the hydrolysis of urea protects this acid-sensitive organism and allows it to grow within the acidic conditions of the stomach. Urea can, in turn, be produced by the hydrolysis of arginine by arginase. It was recently postulated, therefore, that arginase may, like urease, be essential for acid protection *in vivo* by providing the organism with urea. The gene encoding arginase was identified from the complete genome sequence of *Hel. pylori*, and was disrupted by the insertion of a kanamycin resistance gene. The isogenic mutants were devoid of arginase activity but still expressed urease activity as expected. The mutant strain was 1000-fold more sensitive to acid exposure *in vitro* than was the wild-type strain. Nevertheless, addition of urea dramatically increased the survival of both wild-type and mutant *Hel. pylori* cells at pH 2.3, since both strains could produce ammonia from the added urea. The addition of arginine protected the wild-type strain from acid conditions. However, the mutant was not protected by arginine, and was 10 000-fold more sensitive than the wild type in the presence of arginine, as expected. When mice were challenged, however, there was no significant difference in the ability of wild-type and mutant strains to colonise the stomach, although the total bacterial numbers cultivated from samples of the mice infected with arginase-deficient *Hel. pylori* were lower. Clearly, then, although the mutant strain was acid sensitive *in vitro*, gene disruption indicated that arginase was not a significant virulence factor *in vivo*. Presumably other pathways exist to provide *Hel. pylori* with urea for ammonia production.

3.2.2 Random mutagenesis

If no previous knowledge of potential virulence factors is available, then random mutagenesis can be used to identify putative virulence genes. The key to this approach is to have a good selection procedure so that mutants with impaired virulence can be easily identified within a population of mutagenised bacteria; as large numbers of mutant isolates must be examined, screening of mutants individually is impractical. Frequently, the selection method will involve a general biological property that reflects the bacterium's ecological niche. For example, to identify putative virulence genes required for the growth of *Sal. typhimurium* within macrophages, one could screen for acid-sensitive mutants by first plating pools of mutants on normal agar then replica-plating these onto a medium at low pH. Acid-sensitive mutants would be identified by their inability to grow on the latter medium but could be rescued from the original plate. These mutants could then be tested individually in macrophage survival experiments.

Random mutations can be introduced by exposure of organisms to ultraviolet light or to mutagenic chemicals that introduce nucleotide changes within the bacterial chromosome. This approach is little used now, as multiple nucleotide alterations are frequently introduced, and also it is technically quite demanding to identify the location of the single nucleotide change that renders the bacterium impaired in virulence. Instead, transposons are now used routinely. As described in Section 2.6.3.3, transposons are mobile genetic elements that can insert more or less randomly within the genome of a bacterium. Naturally occurring transposons have been tailored to improve their utility for mutagenesis. For example, genes not required have been removed, or else replaced with suitable selective markers, and useful restriction enzyme sites have been introduced. Consequently, large numbers of single-site mutations can be generated with relative ease. Furthermore, since the mutation is marked by the transposon, with its associated antibiotic marker, the DNA flanking the insertion site can readily be recovered for sequencing.

Transposon mutagenesis requires that a suitable transposon is available for use with the bacterium of interest. Often, this is not the case, and so a plasmid library of the genomic DNA can be mutagenised within *E. coli* using one of the many transposons available for this organism. Pooled, mutagenised plasmids can be isolated from *E. coli* and transformed back into the organism under study. The transposon is then introduced into the bacterial chromosome by homologous recombination between the interrupted DNA sequence on the plasmid, and the normal chromosomal DNA. Although this method is ideal for naturally competent bacteria, improved methods for efficient transformation of non-competent bacteria (e.g. electroporation) should improve the utility of this, and other, approaches.

Random mutagenesis can be taken one step further and done, not in the host bacterium nor, indeed, in *E. coli*, but in a test tube, as transposition itself requires only the transposon, the target DNA and the enzyme(s) that catalyse the transposition event. The recently developed GAMBIT (genomic analysis and mapping by *in vitro* transposition) method is based on *in vitro* transposition, and is useful for naturally competent organisms whose genomes have been sequenced. GAMBIT involves the PCR amplification of large, defined regions of the chromosome of the target organism. Each large fragment is subjected to saturation mutagenesis with a transposon. The pool of mutagenised DNA fragments is then introduced back into the host organism, where homologous recombination results in a pool of mutants with transposons randomly inserted within a defined portion of the chromosome. A portion of this pool is held back (pre-selection), while a second portion is grown under selective conditions (for example, at low pH or within a challenged animal). Genomic DNA is then isolated from bacteria in the pre- and post-selection pools and subjected to PCR amplification using a transposon-specific primer and a primer to a known location of the chromosome. A ladder of bands is obtained from amplification of the pre-selection pool DNA, where each unique site of transposon insertion gives a PCR product of unique length (Figure 3.2). Mutagenised bacteria in which the transposon has inserted into a gene essential for growth under the selective conditions will not survive the selective growth conditions. Consequently, there will be no corresponding PCR amplification product. Electrophoretic separation of PCR products in adjacent lanes of an agarose gel reveals 'deleted' clones (for which there is no PCR product) and consequently will identify the genes essential for growth under defined conditions. This procedure can be repeated stepwise around the entire chromosome to generate a 'mutational map'.

3.2.3 Signature-tagged mutagenesis

This is a further development of random mutation which, like GAMBIT, allows the identification of specific mutants from large pools of mutagenised bacteria. The advantage of signature-tagged mutagenesis (STM) is that genomic sequence information is not required in the first instance. STM involves tagging individual transposon molecules with a unique short nucleotide sequence that acts as a 'bar code' identifier. The pool of tagged transposons is used to mutagenise the bacterium under study. Individual mutants are selected and arrayed in 96-well microtitre plates to produce a bank of mutants, each of which has a single, uniquely tagged transposon within the chromosome and so can be distinguished from every other mutant (Figure 3.3). Like GAMBIT, two pools of bacteria are prepared: one pre-selection pool (representing the entire bank) and a post-selection pool that contains only those bacteria able to grow under the selection conditions. DNA is isolated from each pool and the unique tags are amplified by PCR. Labelled tags

Figure 3.2 Genome analysis and mapping by *in vitro* transposition (GAMBIT). (a) A section of the pathogen chromosome is amplified by PCR and subjected to *in vitro* mutagenesis. A pool of chromosomal mutants is then created by transforming the mutant PCR products back into the pathogen. (b) Genomic DNA is isolated from bacteria in pre- and post-selection pools and is subjected to PCR using a fixed chromosomal primer and a transposon-specific primer. Where the transposon has inserted within an essential gene, there are no bacterial representatives within the post-selection pool and consequently no PCR product is obtained.

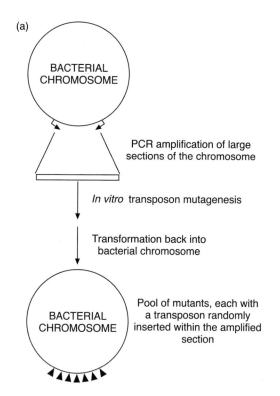

(a)

BACTERIAL CHROMOSOME

PCR amplification of large sections of the chromosome

In vitro transposon mutagenesis

Transformation back into bacterial chromosome

BACTERIAL CHROMOSOME

Pool of mutants, each with a transposon randomly inserted within the amplified section

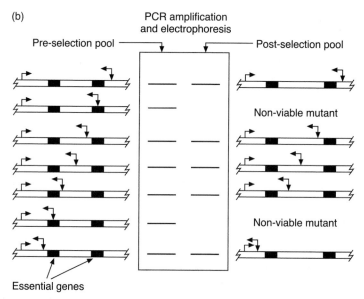

(b)

PCR amplification and electrophoresis

Pre-selection pool

Post-selection pool

Non-viable mutant

Non-viable mutant

Essential genes

are then used to screen, separately, duplicate colony blots of the mutant bank. Mutants that are unable to grow under the selective conditions will be absent, or only poorly represented, in the post-selection pool. Consequently, the signature tag of the transposon that has generated that mutant will be amplified only from the pre-selection

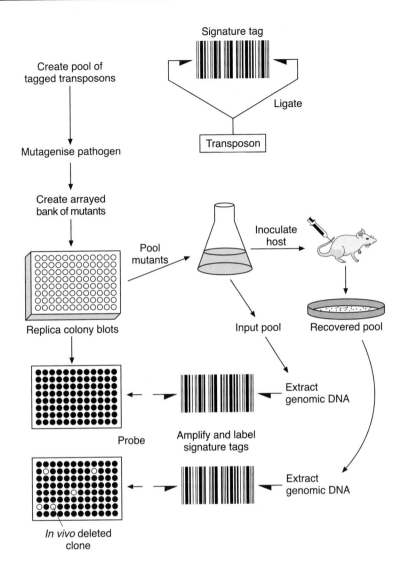

Create pool of
tagged transposons

Mutagenise pathogen

Create arrayed
bank of mutants

Signature tag

Ligate

Transposon

Pool
mutants

Inoculate
host

Input pool

Recovered pool

Replica colony blots

Probe

Amplify and label
signature tags

Extract
genomic DNA

Extract
genomic DNA

In vivo deleted
clone

Figure 3.3 Signature-tagged mutagenesis allows the identification of genes essential for survival and growth within the challenge animal. Two probes are generated by PCR amplification of the unique oligonucleotide sequence tagging each transposon. One probe represents the entire mutant bank (input pool), while the second comprises only those transposon mutants that are recovered from the challenged animal. Differential screening of the arrayed bank of mutants identifies *in vivo*-deleted clones. The 'bar code' represents the unique signature tag that enables the identification of individual transposon mutants within the input and recovered pools.

pool. This means that mutants that are unable to grow under the selective conditions can be identified on the colony blots as those clones hybridising with the pre-selection probe, but not, or only very poorly, with the post-selection probe. As with GAMBIT, the selective conditions can be varied. The mutagenised pool could be used to challenge an experimental animal with the post-selection pool comprising those bacteria that can be isolated from the spleen or other organs, for example, after a specified incubation period. The isolated mutants are subsequently evaluated to rule out the possibility that they possess a general growth defect, and can be tested individually in an animal model of virulence.

STM has been applied now to a wide range of pathogens including both Gram-positive and Gram-negative species (Table 3.1). Very often, however, the results may be difficult to interpret, as the functions of the gene products identified are frequently unknown. Indeed, in a

Table 3.1. | Use of signature-tagged mutagenesis to identify virulence genes

Organism and model	Number of attenuated mutants/total	Genes identified
Mycobacterium tuberculosis, mouse lung	16/1927	Lipid metabolism, transporters, transcriptional regulators. No homology/unknown function, 12.5%
Yersinia enterocolitica, mouse peritoneum	55/2000	Virulence plasmid (type III secretion, YopP), LPS biosynthesis, transporters/nutrient acquisition, shock response. No homology/unknown function, 3.6%
Legionella pneumophila, guinea pig lung	16/1386	dot-icm (intracellular survival, see Section 10.4.2.2.3), transporters/nutrient acquisition. No homology/unknown function, 19%
Streptococcus agalactiae, neonatal rat sepsis	92/1600	Transporters, transcriptional regulators, cell wall metabolism, transposon-related functions. No homology/unknown function, 50%
Escherichia coli K1, infant rat gastrointestinal tract	16/2140	LPS biosynthesis, transporters, transcriptional regulators, adhesins. No homology/unknown function, 25%
Brucella abortus, persistant infection in mice	14/178	Type IV secretion, LPS biosynthesis, transport, amino acid metabolism. No homology/unknown function, 14.3%

LPS, lipopolysaccharide.

recent study of neonatal sepsis caused by *Strep. agalactiae*, some 50% of attenuated mutants were disrupted in genes which either had no homology to genes in the databases, or else showed homology to putative genes of unknown function. The proportion of 'unknowns', often called orphan genes, is lower for those pathogens that have received considerable attention, for example *Y. enterocolitica* (Table 3.1). Nevertheless, this highlights the need for biochemical and physiological studies to augment straightforward genome sequencing exercises. A wide range of gene classes is identified by STM. Chief amongst these are genes encoding proteins involved in the synthesis and metabolism of wall components (lipid, lipopolysaccharide (LPS), peptidoglycan) and genes encoding transporter systems required for nutrient acquisition (Table 3.1).

Brucella abortus uses distinct sets of virulence determinants to establish and maintain chronic infection

Brucella abortus is an intracellular pathogen and is responsible for brucellosis, a chronic infection endemic in Mediterranean countries and in Central and South

America. Bacteria cause a systemic infection and survive and multiply within macrophages. In the mouse model of brucellosis, bacteria divide rapidly within the first two weeks of infection, then numbers stabilise over the next five to six weeks; bacteria have been recovered from the spleens of infected mice some 24 weeks after infection. STM has been used to investigate the virulence determinants required for *Br. abortus* to establish an infection and to set up a chronic infection. Mutants that were defective in *establishing* infection were not recovered either two or eight weeks post infection. In contrast, mutants that were able to establish infection but were unable to maintain chronic infection were recovered from mice two weeks post infection, but not eight weeks post-infection. Four loci were identified that were required for the establishment of infection. Two of these encoded components of a type IV secretion system (see Section 2.5.3.3), while one was involved in LPS biosynthesis and one was involved in amino acid biosynthesis. The type IV secretion mutants were shown to be impaired in their ability to replicate within macrophages. DNA sequence information was obtained for five of the 10 mutants that were unable to sustain chronic infection. One mutant had a transposon insertion in a gene required for glucose and galactose uptake, while two mutants were defective in amino acid metabolism, including glycine dehydrogenase. Of the remaining two mutants, one was disrupted in a gene that had no database homology, and the second was disrupted in a gene that had homology to an open reading frame of unknown function from *Hel. pylori*. Interestingly, in separate studies, it was shown that the gene encoding glycine dehydrogenase was up-regulated 10-fold in *M. tuberculosis* as it enters a state of non-replicative persistence *in vitro*. This suggests that different pathogens responsible for chronic infections may use similar metabolic pathways to set up persistent infections.

3.3 | Protein expression approaches

The analysis of bacterial protein expression under different growth conditions has been a useful means of identifying differential gene expression by pathogenic bacteria. The use of two-dimensional poly-acrylamide gel electrophoresis (PAGE) and N-terminal protein sequencing has been augmented in recent years with a number of powerful mass spectrometric techniques. Proteomics is now a large field in itself, and will be discussed in more detail in Section 3.3.2. There are other protein-based techniques, however, and these are outlined below.

3.3.1 Surface and secreted proteins

Bacterial cell surface and secreted proteins play an important role in infection. The involvement of adhesins, invasins and toxins, for example, is described in other chapters. Additionally, surface and secreted components are involved in nutrient acquisition, environmental sensing, tissue destruction and immune avoidance, for example. Consequently, analysis of bacterial surface and secreted proteins may identify a subset of virulence factors. Two molecular approaches have been devised to identify surface and secreted proteins that might be involved in virulence.

In the first method, a modified transposon containing a gene for an environment-sensitive enzyme is used to identify genes that encode proteins which have a secretion signal and are therefore exported to the cytoplasmic membrane or beyond (see Section 2.5). TnphoA is a modified transposon that carries, near one end, a gene encoding alkaline phosphatase. The alkaline phosphatase enzyme is inactive unless it is secreted from the cell. The *phoA* gene in TnphoA has been truncated so that it lacks DNA for both a promoter and a signal sequence for directing secretion (Figure 3.4). TnphoA inserts randomly within the genome of the target organism. When TnphoA inserts in-frame within a gene encoding an exported polypeptide, then an alkaline phosphatase-exporting clone will be created, and this can be detected readily on growth medium containing a chromogenic substrate of alkaline phosphatase. Selected clones can also be tested directly for their virulence in a suitable model of infection, as TnphoA insertion will have resulted in disruption of the gene that is providing the signal sequence for alkaline phosphatase secretion.

TnphoA mutagenesis is limited to organisms with developed genetic systems. However, there are several ways around this limitation. Firstly, a genomic library can be made in *E. coli* and can be mutagenised with TnphoA within *E. coli*. Naturally, this relies on promoter sequences being recognised by *E. coli* and on the signal sequence functioning in *E. coli*. Nevertheless, this technique has been applied successfully to bacteria with restricted genetic systems, such as *Staphylococcus aureus* and *Mycoplasma fermentans*. A second approach has been used to identify secreted proteins of the human respiratory pathogen *Strep. pneumoniae*, and this relies on the organism's natural competence for DNA transformation and proficiency for homologous recombination. Random fragments of *Strep. pneumoniae* genomic DNA were cloned upstream from a truncated *phoA* gene in a plasmid that was able to replicate in *E. coli* but not in *Strep. pneumoniae*. The plasmid library was then transformed back into *Strep. pneumoniae*, with selection for the plasmid antibiotic resistance marker. Since the plasmid cannot replicate in *Strep. pneumoniae*, resistance arises from homologous recombination that inserted the plasmid into the chromosome by a single cross-over (Figure 3.4). Recombinants were screened for alkaline phosphatase activity, and the plasmids from positive clones were rescued and sequenced to identify the inserted DNA. The method identified a number of distinct clones including those encoding transporters and two-component signal transduction systems (TCSTSs).

The second protein-based approach utilises serum from patients who have experienced disease caused by the pathogen of interest, or else infected experimental animals, to identify proteins that are (i) expressed during the infection and (ii) recognised by the host immune system and able to elicit an antibody response. In this technique, known as IVIAT (**in vivo**-induced antigen technology), sera from patients or experimental animals are used to screen an expression library of the

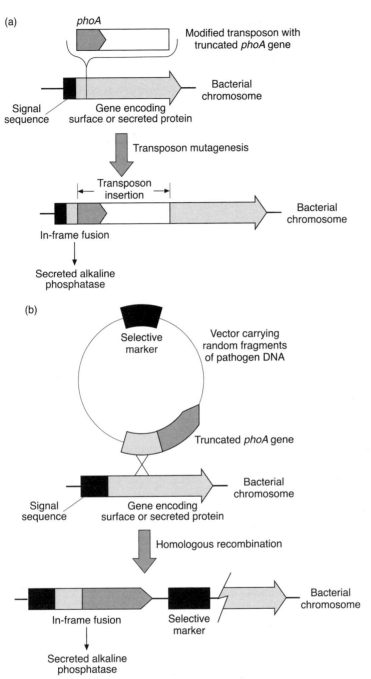

(a)

phoA

Modified transposon with
truncated *phoA* gene

Bacterial
chromosome

Signal
sequence

Gene encoding
surface or secreted protein

Transposon mutagenesis

Transposon
insertion

Bacterial
chromosome

In-frame fusion

Secreted alkaline
phosphatase

(b)

Selective
marker

Vector carrying
random fragments
of pathogen DNA

Truncated *phoA* gene

Bacterial
chromosome

Signal
sequence

Gene encoding
surface or secreted protein

Homologous recombination

Bacterial
chromosome

In-frame fusion

Selective
marker

Secreted alkaline
phosphatase

Figure 3.4 Identification of cell surface and secreted proteins by *phoA* fusion analysis. (a) The transposon *TnphoA* carries a modified *phoA* gene that encodes a truncated alkaline phosphatase lacking the signal sequence that is necessary for secretion of the protein. Insertion of the transposon within a gene encoding a surface or secreted protein, such that an in-frame fusion is created with the truncated *phoA* gene, will result in the generation of a clone that secretes alkaline phosphatase. (b) A modified *phoA* fusion system can be used if a suitable transposon is not available for the organism under study. This system relies on homologous recombination. Random fragments of genomic DNA are cloned into a unique restriction site that lies immediately 5′ to the truncated *phoA* gene, and the pool of recombinant plasmids is introduced into the pathogen. A single cross-over homologous recombination event results in insertion of the truncated *phoA* gene into the chromosome. As before, if the DNA fragment that is cloned in the vector encodes a surface or secreted protein such that an in-frame fusion is created with the truncated *phoA* gene, then alkaline phosphatase is secreted.

pathogen. Reactive clones contain DNA encoding *in vivo*-expressed proteins. In order to identify only *in vivo*-induced antigens, the patient or animal serum can be absorbed with whole cells or cellular extracts prepared from the organism grown *in vitro* to remove antibodies to constitutively expressed proteins. Alternatively, libraries can be differentially screened with patient serum and with serum raised in rabbits to killed whole cells, or cell extracts, of broth-grown bacteria.

Sera from patients suffering from infective endocarditis due to *Enterococcus faecalis* have been used to screen expression libraries of this opportunistic pathogen. Some 38 distinct immunopositive clones were isolated and the expressed genes identified by DNA sequencing. Most of these encoded surface or secreted proteins, as expected, although there was a small group of cytoplasmic proteins, suggesting, perhaps, that cell lysis occurs during infection. A number of transporter components were isolated, including one shown in independent work to be involved in virulence. In a similar approach, sera from neutropenic patients with *Strep. oralis* septicaemia and from patients with infective endocarditis due to *Strep. oralis* and other, related, oral streptococci were used to identify immunodominant antigens. *Streptococcus oralis* genes encoding a heat shock protein (hsp90) and a common oral streptococcal adhesin (PAc protein, also known as antigen I/II) were identified. This adhesin mediates streptococcal attachment to saliva-coated surfaces and so plays a role in oral colonisation by *Strep. oralis*. Nevertheless, the antigen I/II protein from a number of streptococci has been shown also to mediate attachment to human collagen, and this may contribute to the ability of these commensal oral bacteria to cause endocarditis.

3.3.2 Proteomics

In 2001, at the time of writing, we have available a wealth of tools for analysing the genome and its primary product – the transcriptome. This latter term refers to the range of mRNA molecules that can be produced by a cell under a given set of conditions. However, there is a growing realisation that the genome and the transcriptome tell us only what a cell could do, but not what it is doing. It is now well established that the pattern of mRNA in a cell's cytoplasm is not an accurate reflection of the proteins that the cell will produce under a given set of environmental conditions. Thus, if one wants to understand how a cell is working under a given set of conditions, then the best way of doing this is to define what proteins are being produced. The complete set of proteins produced by a cell under a particular set of conditions is called the proteome and the study of the proteome is called 'proteomics'. For clarity, the reader should be aware that the term 'proteome' has also been used to denote the complete complement of proteins that the cell is potentially capable of producing. The relationship between the genome, transcriptome and proteome is shown diagramatically in Figure 3.5. In this section, a brief description of the methodology of proteomics, and some of its uses in microbiology, will be provided.

3.3.2.1 Proteomic methodology

Proteomics seeks to define the pattern of proteins produced by chosen cells under given sets of conditions. Bacteria may produce a few

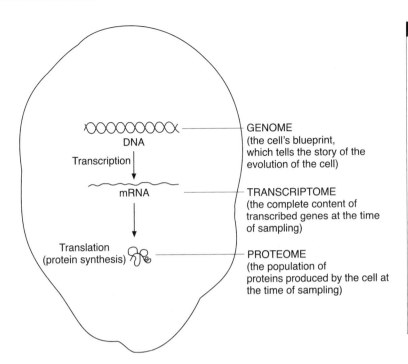

GENOME
(the cell's blueprint, which tells the story of the evolution of the cell)

TRANSCRIPTOME
(the complete content of transcribed genes at the time of sampling)

PROTEOME
(the population of proteins produced by the cell at the time of sampling)

Figure 3.5 Three views of the cell. The genome is analogous to the blueprint, or architect's drawing, and provides information on the evolutionary history of the cell. It tells us what the cell could do, but not necessarily what it is doing. Likewise the transcriptome is the complete collection of mRNA molecules at the time of sampling. Again, its relevance to the current understanding of what the cell is doing is uncertain. The only measure that reveals what a cell is doing at the time of sampling is the proteome – the collection of proteins and their post-translational modifications that the cell is producing.

thousand proteins and eukaryotic cells tens of thousands of proteins under any given set of environmental conditions. The ability to identify the patterns of proteins produced by cells depends on our ability to separate these proteins. A method developed in the 1970s, and refined in subsequent years, is to separate proteins in two dimensions in thin gels using two complementary separation technologies. Proteins are separated in the first dimension by isoelectric focusing, using immobilised pH gradients in narrow strips of polyacrylamide. Isoelectric focusing is a technique by which proteins are separated on the basis of their isoelectric point (pI). Using immobilised gradients of ionophores (which generate the pI gradient), it is possible to obtain reproducible and highly discriminatory separation of proteins. The thin strips containing the isoelectrically separated proteins are then placed into a larger polyacrylamide gel and separated at right angles using conventional sodium dodecyl sulphate–polyacrylamide gel electrophoresis (SDS-PAGE). This separates proteins on the basis of their mass. By this means, and depending on the size of the gel used, it is possible to separate many thousands of proteins (Figure 3.6).

Having separated the proteins, how is it possible to identify them? The standard method of identifying proteins in such gels is by Edman degradation to obtain the N-terminal sequence. Once the sequence is obtained, it can be fed into one of the sequence databases and the protein, if it exists on the database, identified. This technique of N-terminal sequencing is well established but has certain disadvantages. Firstly, it requires relatively large amounts of protein. Secondly, modification at the N-terminus can block the chemistry of the reaction and prevent a sequence being obtained. This can be overcome by

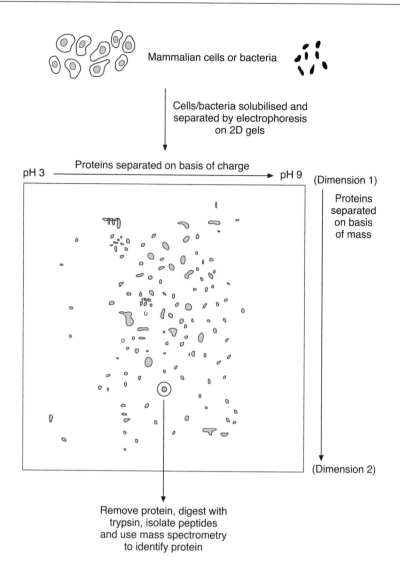

Figure 3.6 The secret to identifying the composition of the proteome lies in being able to separate and identify each protein in a cell. A method that is commonly used is two-dimensional (2D) gel electrophoresis. Here we see a representation of a typical 2D gel showing each protein spot (normally visualised by staining with a dye). This technique can separate complex mixtures of proteins such as those produced by bacteria. Once the proteins are separated, the next step is to trypsin-digest each protein spot, isolate the tryptic peptides and identify them by N-terminal sequencing or, more conveniently, by MALDI-TOF MS (see Figure 3.7).

trypsin-digesting the protein and determining the sequence of internal peptides. However, the current method of choice for identifying proteins in gels is time-of-flight mass spectrometry (TOF MS). This is a technique in which the substance(s) to be analysed is given a charge and is then separated on the basis of its mass:charge (m/z) ratio. A number of methods are available for generating charged particles but of these, the two most commonly used are **matrix-assisted laser desorption ionisation** (MALDI) and **electron spray ionisation** (ESI). In the case of the MALDI-TOF mass spectrometer, the biomolecule(s) to be analysed is mixed with an organic chemical that, when it is illuminated with laser light, ionises the associated biomolecule and gives it a charge. The charged molecule is then allowed to move down a 'drift tube' to a detector. The time taken to move along the drift tube is directly proportional to the mass of the biomolecule and thus the MALDI-TOF can be thought of as an incredibly sensitive balance,

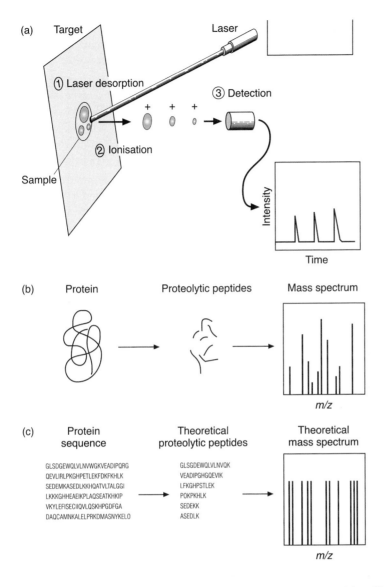

(a) Target Laser

① Laser desorption

③ Detection

+ + +

② Ionisation

Sample

Intensity

Time

(b) Protein Proteolytic peptides Mass spectrum

m/z

(c) Protein sequence Theoretical proteolytic peptides Theoretical mass spectrum

GLSDGEWQLVLNVWGKVEADIPQRG
QEVLIRLPKGHPETLEKFDKFKHLK
SEDEMKASEDLKKHQATVLTALGGI
LKKKGHHEAEIKPLAQSEATKHHIP
VKYLEFISECIIQVLQSKHPGDFGA
DAQCAMNKALELPRKDMASNYKELO

GLSGDEWQLVLNVQK
VEADIPGHGQEVIK
LFKGHPSTLEK
POKPKHLK
SEDEKK
ASEDLK

m/z

Figure 3.7 Matrix-assisted laser desorption ionisation time-of-flight (MALDI-TOF) mass spectrometry (MS) is a commonly used method of identifying proteins. This diagram shows the basics of the instrument and of the analysis. Each protein from the two-dimensional gel shown in Figure 3.6 is cut out and trypsinised to produce a collection of tryptic peptides. These peptides are mixed with an organic chemical such as 4-methoxycinnamic acid and placed in the mass spectrometer. (a) The laser is fired at each specimen and the organic chemical passes the laser energy on to the peptides, thereby ionising them. The peptides are then separated in the drift tube on the basis of their mass, and each peptide is picked up by the detector, giving a mass spectrum, as shown in (b). This spectrum is compared with the theoretical mass spectra of all the proteins encoded by the genome of the organism of interest to see whether there is a match (c).

accurate to hundredths of a Dalton. Now, in order to be able to identify proteins in two-dimensional gels by MALDI-TOF MS, each protein must be removed from the gel and cleaved with a protease. The most common protease used is trypsin. The collection of tryptic peptides is then put into the mass spectrometer and each peptide is 'weighed'. This provides a 'mass fingerprint' for the protein. This information, by itself, is quite useless. However, if the organism that the experimenter is working with has had its genome sequenced then it is possible to produce an 'in silico' digest of each of the proteins encoded by each gene. It is then a relatively simple matter to compare the mass fingerprint of the protein with that of all of the in silico fingerprints to find a match and identify the protein (Figure 3.7).

With a growing number of genome sequences available, it is now possible to ask questions of your favourite organism to see how it responds to its environment. Now, much of the study of bacterial virulence is concerned with understanding how the bacterium responds to changes in environments. Such responses cause the up-regulation of specific genes and the production of a different pattern of proteins. Molecular genetic methods of analysis are described in other sections of this chapter and include differential display, in vivo expression technology (IVET), IVIAT and STM. Proteomics gives a direct readout of the proteome. The disadvantage at the moment is that proteomics cannot identify proteins produced in vivo. However, this is largely a question of sensitivity.

Mass spectrometry for all seasons

MALDI-TOF is simply one of a growing number of mass spectrometric techniques employed in biology. These techniques differ in the methods used to ionise the biomolecules and the nature of the mass analyser use to detect such ions. In addition to laser desorption, the other major ionisation methods used in biology are ESI and fast atom bombardment (FAB). In ESI the biomolecules are sprayed from a fine needle at high voltage towards the inlet of the mass spectrometer (this is under vacuum), which is held at a lower voltage. During this process the fluid droplets dry and gain charge. The spray can come from either a nanospray device or be the output from high performance liquid chromatography (HPLC). In ESI, the mass analyser is generally a quadrupole-type device or an ion trap analyser. As its name implies, the quadrupole mass analyser consists of four cylindrical rods that serve as electrodes. The ability to change the potential on each pair of electrodes allows filtering of ions to occur and it is possible to scan mass spectra with this device. The other ionisation technique used with biomolecules, FAB, uses energetic atoms to bombard and ionise the biomolecules of interest. Other techniques used in biological mass spectrometry include tandem mass spectrometry (often styled MS/MS) in which one mass spectrometer is coupled to another. The first separates the ions and the second can fragment each ion to provide additional structural information.

In addition to straightforward identification of proteins, mass spectrometric techniques are being used increasingly to identify protein–protein interactions within cells. For example, the 60 kDa heat shock protein (chaperonin 60) of E. coli, an oligomeric protein known as GroEL, interacts with proteins and helps their correct folding. Exactly how many proteins in E. coli GroEL bound to was not known until recently. To determine this, use was made of an antibody to immunoprecipitate GroEL from lysates of E. coli. The proteins present in the immunoprecipitates were identified by two-dimensional gel electrophoresis/mass spectrometry. This revealed that GroEL apparently binds to 300 different polypeptides.

Prokaryotic and eukaryotic cells are full of protein and protein–nucleic acid complexes. One of the largest of such complexes is the ribosome. A technique involving multidimensional chromatography and tandem mass spectrometry, linked to comprehensive bioinformatic analysis, has recently been developed and provided with the acronym DALPC (direct analysis of large protein complexes). Application of DALPC to the ribosome of the yeast Saccharomyces cerevisiae was able to identify more than 100 constituent proteins in a single analysis.

Biologists in general, and microbiologists in particular, are attempting to understand how the cell responds to its environment. One of the most powerful methods of identifying cellular changes in response to environmental conditions is the use of DNA microarrays. These will be described in detail in Section 3.8. One of the problems with the use of such microarrays is the uncertainty that the presence of a particular mRNA in a cell actually means that the protein that can be translated from it is actually translated. A new proteomic technique has been introduced to quantify the differences in protein expression between the same cell population maintained within two different environments. This method uses reagents termed isotope-coded affinity tags (ICATs), which are biotin-labelled reagents with a thiol-specific reactive group. A linker joins these two groups, which are of different molecular masses, thus giving rise to a heavy or a light ICAT. To utilise these reagents, cells grown under one environmental condition are labelled, say with the light ICAT, and cells grown under a second environmental state are labelled with the heavy ICAT. The protein mixtures from both cells are then combined and subjected to proteolysis, and the ICAT-labelled peptides are isolated by affinity chromatography on an avidin column. The ICAT-labelled peptides, which differ by eight mass units, can then be separated chromatographically. Both light- and heavy-labelled peptides separate in an identical manner and the differential concentrations of both ICATs can be quantified by ESI-MS using an ion trap. The end result is a measure of the effect of environmental conditions on global protein expression.

3.3.2.2 Proteomics in microbiology

The main application of proteomics in microbiology is in 'comparative proteomics', using either different strains of an organism grown under the same set of environmental conditions or the same strain of organism grown under different conditions. In the former, one major aim is to compare the proteins produced by pathogenic strains and also compare them with those produced by a non-pathogenic strain of the bacterium of interest. Any differences in proteins synthesised are likely to be related to the virulence phenotype. In order to elucidate the effects of the host environment on the production of virulence factors, the same strain of bacterium can be grown under a range of conditions chosen to mimic those likely to be experienced by the bacterium when it infects the host. These include changes in pH, anaerobicity, osmolarity, metal ions (iron, zinc, copper, etc.), host secretions, etc.

Another major use for proteomics is as a rapid means of identifying protein–protein interactions (affinity proteomics). For example, a recent study has used carbohydrate probes to identify adhesins of *Hel. pylori*. This identified the Lewis[b] binding adhesin of this organism. This is a generic technology for identifying bacterial proteins binding to host proteins and vice versa and it promises to be a valuable method for studying host–bacteria interactions in the next decade (Figure 3.8).

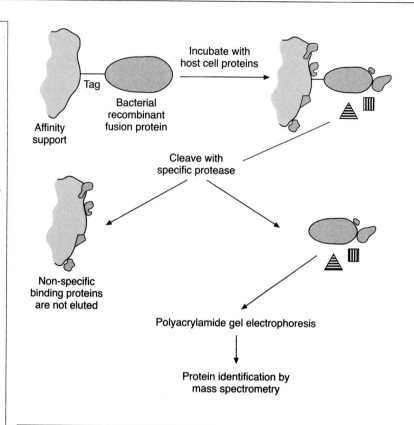

Figure 3.8 Affinity proteomics. This promises to be one of the major uses of proteomics. To identify the host protein(s) that binds to a particular bacterial protein, the gene encoding the bacterial protein is engineered to produce a recombinant fusion protein that has a tag cleavable by a specific protease. The recombinant protein is then attached to a matrix and appropriate mixtures of host cell proteins are added. Some of these human proteins will bind specifically to the bacterial protein. Other proteins may bind non-specifically to the matrix. To collect those human proteins binding to the bacterial protein, the protease site in the tag is cut, thereby releasing the recombinant bacterial protein with any associated human protein(s). These proteins are then separated by SDS-PAGE and each protein band is trypsin-digested and identified by MALDI-TOF MS.

3.4 | Subtractive and differential analysis of mRNA

Subtractive and differential techniques were first used to investigate differential gene expression in eukaryotes. As applied to bacteria, these procedures rely only on being able to isolate mRNA from the organism, and so are applicable to organisms without developed genetic systems. The basic approach is to compare mRNA isolated from bacteria grown under different conditions. To do this, reverse transcriptase (RT), a retroviral enzyme, is first used in conjunction with random primers and PCR amplification (RT-PCR) to synthesise complementary DNA (cDNA) fragments from mRNA templates. The cDNA fragments generated will depend on the mRNA molecules present in the original pool. These then form the basis of the subtractive or differential analysis.

The most straightforward means of differential analysis is to compare cDNA pools visually after PAGE (Figure 3.9). The advantage of this method, known as RNA arbitrary primed (RAP)-PCR, but also sometimes as differential display (DD)-PCR, is that very little starting mRNA is required. Differences in the RAP-PCR profiles represent differences in gene expression, and the cDNA present in bands of interest can be eluted, reamplified and cloned for DNA sequence analysis. Furthermore, specific primers can then be used in RT-PCR to confirm that the identified genes are, in fact, differentially expressed. RAP-PCR has recently been used to identify *Vibrio cholerae* genes differentially

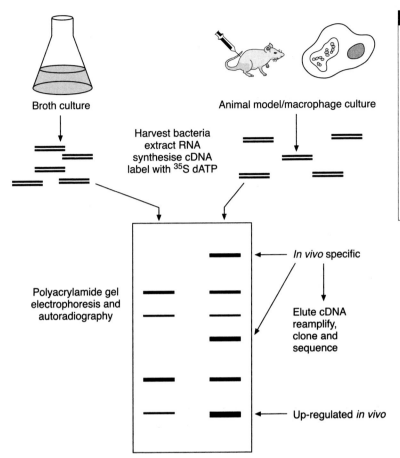

Figure 3.9 RNA arbitrary primed polymerase chain reaction (RAP-PCR). cDNA is synthesised by RT-PCR amplification using, as template, RNA isolated from cells grown in broth culture or the *in vivo* system (e.g. in an animal or in a host cell culture). Labelled cDNA molecules are then separated by PAGE and bands are visualised by autoradiography. cDNA in bands of interest can be eluted from the gel, reamplified and cloned for further study.

expressed in the host infection. Bacteria were grown either to late exponential phase in broth culture, or overnight in ligated rabbit ileal loops, and isolated RNA was subjected to RAP-PCR fingerprinting. Five differentially expressed transcripts were sequenced, three of which were induced *in vivo*, while the remaining two were specific for cells grown *in vitro* (Table 3.2). RAP-PCR has also been used, by two independent groups, to measure differences in gene expression between *M. tuberculosis* strain H37Rv and its avirulent mutant H37Ra (Table 3.3). One group identified four genes encoding proteins of unknown function that were specific to the avirulent strain H37Ra. Two of these proteins belong to the acidic, glycine-rich PPE family of mycobacterial proteins that contain the tripeptide proline-proline-glutamate near the N-terminus. Analysis of the *M. tuberculosis* genome sequence indicates that there are 68 related PPE proteins and, although the subcellular location of PPE proteins is not yet clear, it may be that these proteins have immunological importance, serving to provide *M. tuberculosis* with a form of antigenic variation. The second group obtained quite different results, and described six genes that were expressed in virulent H37Rv, but not in avirulent H37Ra. Amongst the genes identified (Table 3.3), three encoded proteins with

Table 3.2. *In vivo-* and *in vitro-*induced genes of *Vibrio cholerae* identified by RAP-PCR fingerprinting

Clone	Proposed function of gene product	Induction
pRAP1	Peptidoglycan biosynthesis	*In vivo*
pRAP2	Leucyl-tRNA synthase	*In vitro*
pRAP3	Unknown/no homology	*In vivo*
pRAP4	50 S ribosomal protein	*In vitro*
pRAP5	TCA cycle, energy production	*In vivo*

TCA cycle, tricarboxylic acid cycle.

Table 3.3. *Mycobacterium tuberculosis* virulence-associated genes identified by subtractive or differential analysis. RNA was isolated and analysed from *M. tuberculosis* virulent strain H37Rv and avirulent mutant H37Ra

Method	Gene product, function	Virulent- or avirulent-specific?
RAP-PCR	Hypothetical protein, no known function	Virulent
	PPE protein, antigenic variation?	Virulent
	ESAT-6 protein, T cell antigen	Virulent
	Polyketide synthase, lipid synthesis	Virulent
RAP-PCR	PPE protein, antigenic variation?	Virulent
	Hypothetical protein, no known function	Avirulent
Subtractive	Two-component signal transduction system	Virulent

For abbreviations, see list on p. xxii.

no known function, one encoded a PPE protein and one encoded a member of the ESAT-6 (6 kDa early secretory antigenic target) family of potent T cell antigens. The final gene was involved in lipid biosynthesis.

A second approach that utilises cDNA isolated from bacteria grown under different conditions involves library screening. Thus duplicate copies of a library are hybridised independently with labelled cDNA pools. Clones encoding genes expressed *in vivo* but not *in vitro*, for example, are identified by their differential hybridisation with the

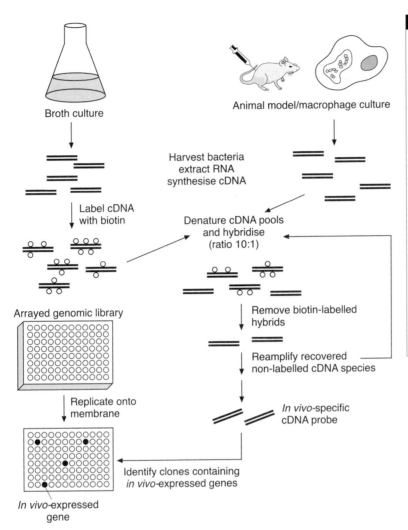

Figure 3.10 Subtractive hybridisation. cDNA is synthesised by RT-PCR amplification using, as template, RNA isolated from cells grown in broth culture or in the *in vivo* system. The cDNA from broth-grown cells is labelled with biotin, denatured by heating and hybridised in solution at a ratio of 10 : 1 with unlabelled, heat-denatured cDNA generated from *in vivo*-grown bacteria. Double-stranded cDNA molecules labelled with biotin are removed using streptavidin-coated magnetic beads, and non-labelled molecules are recovered and reamplified. A further two rounds of subtractive hybridisation are performed to create an *in vivo*-specific cDNA probe that is used to screen an ordered genomic library of the pathogen.

labelled cDNA pools. A major drawback of this procedure is that the majority of RNA isolated from bacteria is not, in fact, mRNA, but is rRNA or tRNA. Furthermore, many of the mRNA molecules will be common to both *in vivo*- and *in vitro*-grown cells, and this can result in a relatively high level of false-positives requiring laborious secondary screening. To circumvent this, a subtractive method is used to remove common, house-keeping, cDNA molecules from the *in vivo* cDNA sample (Figure 3.10). cDNA molecules are first generated for each pool by RT-PCR as normal. The cDNA molecules of the *in vitro* pool are then chemically tagged with biotin. Double-stranded cDNA molecules from both pools are then made single stranded by heat denaturation and are mixed at a ratio of 10 : 1 (*in vitro* cDNA : *in vivo* cDNA) and allowed to hybridise in solution. Biotin-labelled cDNA hybrids (i.e. hybrids with one or two strands of *in vitro*-specific cDNA) are easily removed using magnetic beads coated with streptavidin, which binds to biotin (Figure 3.10). The resultant pool is enriched for *in vivo*-specific cDNA molecules. These are subsequently reamplified and the hybridisation-subtraction process is repeated once or twice more.

Subtractive hybridisation has been used to isolate a gene expressed by *M. avium* growing in macrophages. cDNA, made from bacteria grown in broth culture, was used in three subtractive rounds with cDNA made from bacteria grown in macrophages. The gene identified, *mig* (macrophage induced gene), encodes a secreted protein with an as yet unknown function and, unfortunately, genetic systems in mycobacteria are not yet well enough developed to create a knockout mutant for virulence testing. Interestingly, expression of *mig* could be induced *in vitro* by shifting the culture from pH 7 to pH 6, suggesting that *M. avium* may sense environmental pH to provide clues about its location. Shifts in pH, temperature, etc. have been used in conjunction with subtractive or differential techniques to mimic host-associated environments. Similarly, bacteria can be grown in the presence or absence of specific host factors, such as blood or saliva, and the resulting cDNA pools can be used to identify host factor-induced gene expression.

3.5 | *In vivo* expression technology

IVET comprises a number of related methodologies for monitoring gene expression *in vivo*. The techniques all act as 'promoter traps' to identify genes active *in vivo* but not *in vitro*, and are based on the expression of marker genes required for *in vivo* survival. Relatively well-developed genetic systems are required for IVET and so these techniques have been somewhat restricted in their use.

The first IVET technique to be used required the prior isolation and characterisation of an auxotrophic mutant of the pathogen under study. An auxotroph is a mutant that is deficient in the synthesis of a specific essential nutrient, but that can grow happily if this nutrient is supplied in the growth medium. Deficiency in the *de novo* synthesis of purines (required for nucleic acid synthesis) was commonly used. The IVET plasmid construct comprised a promoterless marker gene, the product of which complements the auxotrophic deficiency. Only if the gene is expressed can the organism grow in the absence of added purines. Random fragments of genomic DNA from the pathogen are cloned immediately upstream from the promoterless marker gene in the IVET construct. These constructs are then introduced back into the chromosome of the pathogen by homologous recombination to create a pool of random fusion clones (Figure 3.11(a) and (b)(i)). Recombinants can be selected using an antibiotic resistance marker present also on the integrated IVET construct. Pooled clones are then inoculated into the experimental animal. After a suitable infection period, bacteria are isolated from the spleen or other organs. Surviving bacteria will be enriched for clones that were able to make purines and thus grow and divide within the animal. In other words, survivors are likely to have an *in vivo*-active promoter fused to the promoterless marker gene.

(a)

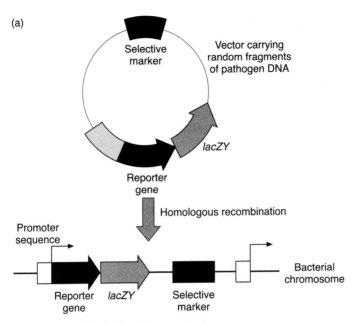

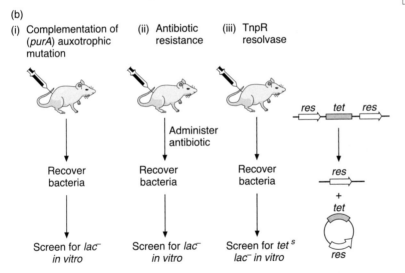

Conversely, if the IVET construct has integrated within DNA that is not expressed *in vivo*, that bacterium cannot synthesise purines and so will not survive.

This system can be further refined if the *lacZ* gene is included as an operon fusion with the promoterless marker gene on the IVET construct. The *lacZ* gene encodes the enzyme β-galactosidase, which can hydrolyse the chromogenic substrate known as X-Gal (5-bromo-4-chloro-3-indolyl-β-D-thiogalactoside) to create a blue-coloured product. Recombinant clones can be plated on rich medium either prior to animal challenge, or else following isolation, to identify clones with marker gene fusions to promoters that are also active *in vitro*; such clones will be blue in colour and can be ignored. An important feature of IVET systems is that recombinant bacteria are generated by insertion–duplication mutagenesis. This means that, unlike transposon mutagenesis, the gene in which the vector has inserted may not have been disrupted. Thus the growth of recombinant bacteria should not be impaired during experimental infection. However, in order to ascertain the importance of *in vivo*-induced genes in the infection process, individual genes must be inactivated by standard procedures.

A second IVET technique that does not require the initial isolation of auxotrophic mutants utilises a promoterless chloramphenicol resistance determinant as the marker gene (Figure 3.11 (b)(ii)). After colonisation of the experimental animal with pools of recombinant clones, the animal is treated with chloramphenicol to provide a level of the drug within the blood and tissues that is sufficient to inhibit growth of the wild-type bacterium. However, if the IVET construct has inserted within an *in vivo*-expressed gene, then the organism expresses the marker gene and is consequently rendered chloramphenicol resistant and is thus isolated in higher numbers when the animal is sacrificed.

Both these initial IVET systems, by their nature, select for genes that are expressed constitutively at relatively high levels throughout the challenge period. Conversely, the systems may miss genes expressed at low levels, or only transiently, but which are nevertheless essential for virulence. A third, recombinase-based **IVET** system, termed RIVET, was developed to overcome this problem. RIVET employs, as the marker, the gene encoding a site-specific recombinase called resolvase. Resolvase catalyses the excision of an unlinked antibiotic resistance marker that is flanked by resolvase recognition sequences (*res*). Thus, if a promoter to which the RIVET construct is fused is active *in vivo*, then the clone loses that antibiotic resistance gene (Figure 3.11(b)(iii)). Bacteria recovered from the experimental animal are replica-plated on medium with or without the antibiotic to determine which have lost the resistance gene. Due to the exquisite sensitivity of this system, even very low levels of expression result in resolvase activity and loss of resistance. Consequently, RIVET is less useful for the identification of *in vivo*-induced genes that are, nevertheless, expressed at low levels during *in vitro* growth. However, RIVET, as well as the original IVET approaches, have been applied successfully to the investigation of *in vivo*-expressed genes in a range of pathogens (Table 3.4).

IVET and endocarditis

Members of the viridans group of streptococci normally reside in the oral cavity where they are considered to be non-pathogenic, except for *Strep. mutans*, which causes dental caries. However, once these bacteria enter a person's bloodstream (e.g. during dental procedures), they may cause serious illness, including streptococcal shock, respiratory distress, and infective endocarditis. How do these normally avirulent organisms cause disease once they gain access to the bloodstream? The answer is likely to be that they respond to the new environmental conditions by switching on the expression of genes encoding virulence factors. Two groups have used IVET to investigate gene expression during infective endocarditis caused by *Strep. gordonii*. The first group has used an experimental animal, while the second group employed a shift in an environmental condition, pH, to mimic the transition from the oral cavity to blood.

The IVET system used in the rabbit model of endocarditis employed a promoter-less *cat* gene, conferring chloramphenicol resistance, as the marker gene. Instead of using *lacZ*, encoding β-galactosidase, as the *in vitro* expression indicator, a gene encoding α-amylase, *amy*, was fused with the marker gene. Expression of α-amylase is detected by growing bacteria on plates containing starch. Recombinant *Strep. gordonii* were inoculated as pools into rabbits prepared for experimental endocarditis, and, starting 6 hours after inoculation, rabbits were given chloramphenicol intravenously twice daily to achieve a serum level sufficient to inhibit the growth of wild-type *Strep. gordonii*. Three days later, rabbits were killed and the heart valves removed. Bacteria growing on these valves were then plated out. Bacteria isolated from the rabbits, and therefore likely to express the *cat* gene *in vivo*, were tested for α-amylase activity and chloramphenicol resistance on plates. Those that were α-amylase negative and chloramphenicol sensitive on plates were selected as probable *in vivo*-specific clones. The promoter regions of those clones that drove marker gene expression were isolated and sequenced. Some 48 clones were obtained and these comprised 13 different genes (some identical clones were isolated several times). Amongst these were 'the usual suspects', including genes encoding transporters and enzymes involved in sugar and amino acid metabolism, as well as regulatory genes.

The pH of the oral cavity is slightly acidic, whereas that of blood is slightly alkaline, and this change in environmental pH may be one of the signals sensed by *Strep. gordonii*. In the second IVET screen, a promoterless gene conferring resistance to spectinomycin was used as the promoter trap to screen for pH-regulated genes. Bacteria were plated on agar at pH 7.3 and containing a spectinomycin concentration sufficient to inhibit the growth of normal *Strep. gordonii*. After 36 hours of incubation, individual colonies growing on the pH 7.3 plates were streaked onto plates at pH 6.2, again containing spectinomycin. A total of five clones that were unable to grow at pH 6.2 were recovered from the pH 7.3 plates. These were shown to have the marker gene fused with a promoter that was active at pH 7.3 (i.e. at blood pH) but not at pH 6.2.

One gene, *msrA*, encoding a methionine sulphoxide reductase, was isolated by both IVET systems. The sulphur groups of methionine residues are highly sensitive to oxidation by oxygen radicals, and modification renders the protein non-functional. The reduction of oxidised methionine residues by methionine sulphoxide reductase may help to restore function. Furthermore, this enzyme may also help to stabilise adhesins on the bacterial cell surface, and therefore contribute to virulence.

3.6 | Reporter systems

It is possible to assess gene expression at the protein level by using reporter genes whose products are easily measured either directly or

Table 3.4. Identification of *in vivo*-induced (*ivi*) genes using IVET methodology

Organism and model	IVET system	Genes identified
Salmonella typhimurium, intraperitoneal infection of mice	Purine auxotrophy complementation	>100 *ivi* genes. Regulatory, metabolic, transporters, adhesion- and invasion-associated. > 50% had no homology/unknown function
Pseudomonas aeruginosa, intraperitoneal infection of mice	Purine auxotrophy complementation	22 *ivi* genes. Regulatory, metabolic, transporters. 6 had no homology/unknown function
Streptococcus gordonii, rabbit endocarditis model	Chloramphenicol resistance	13 *ivi* genes. Regulatory, metabolic, transporters. Methionine sulphoxide reductase.
Staphylococcus aureus, mouse renal abscess model	RIVET	69 *ivi* genes. Regulatory, metabolic, transporters. Capsule synthesis. 45 had no homology/unknown function

ivi, *in vivo* induced; for other abbreviations, see list on p. xxii.

indirectly. Current systems use β-galactosidase, green fluorescent protein (GFP) or luciferase. A promoterless reporter gene is fused with a promoter of interest. Bacteria can then be grown under different conditions to assess the effects on promoter activity. Luciferase genes, isolated from fireflies or from bioluminescent bacteria, have been useful markers for the analysis of gene expression in both Gram-positive and Gram-negative bacteria. The system is very sensitive and, additionally, the proteins have a short half-life and so light production reflects real-time promoter activity. GFP from *Aequoria* jellyfish, and its various mutant forms, is encoded by another popular reporter gene, *gfp*. The advantage of GFP over β-galactosidase or, indeed, luciferase, is that measurement of GFP levels is non-invasive. Thus the protein fluoresces when excited by blue light. In contrast, measurement of both β-galactosidase and luciferase may require bacterial cell permeabilisation to allow the access of enzyme substrate. Mutated versions of GFP have been produced that are more intensely fluorescent than the original protein, and are also more soluble and so remain active even when expressed at high levels. GFP can allow the visualisation of promoter activity of bacterial cells within tissue samples or of internalised bacteria in model systems of infection.

When combined with flow cytometry, the *gfp* marker gene allows the assessment of intracellular fluorescence in individual bacterial cells. Differential fluorescence induction (DFI) is a modification of the IVET system that uses the GFP as the promoter trap. Thus random fragments of bacterial genomic DNA can be cloned upstream from a promoterless *gfp* reporter gene on a plasmid. Recombinant plasmids are then reintroduced back into the pathogen under study to generate a pool of random fusions (Figure 3.12). These are then introduced into the

(a)

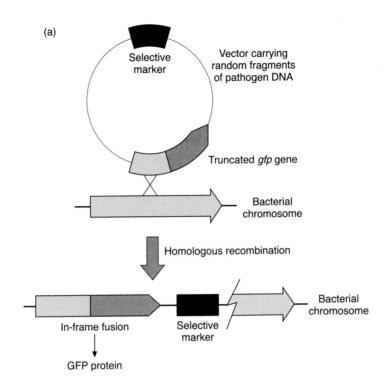

Figure 3.12a Differential fluorescence induction. (a) Random fragments of genomic DNA are cloned into a unique restriction site that lies immediately 5′ to the truncated *gfp* gene, and the pool of recombinant plasmids is introduced into the pathogen. A single cross-over homologous recombination event results in insertion of the truncated *gfp* gene into the chromosome. If the DNA fragment that is cloned in the vector encodes a gene such that an in-frame fusion is created with the truncated *gfp* gene, then green fluorescent protein (GFP) may be expressed.

experimental system, for example tissue culture or animal model. Bacteria that are isolated following a suitable period of time can be sorted using a **fluorescence-activated cell sorter (FACS)**. Highly fluorescent bacteria can be collected and plated onto laboratory media. A second FACS sorting step can then be used to eliminate those clones that remain highly fluorescent after *in vitro* growth. Those with low *in vitro* fluorescence are selected for further analysis.

Salmonella type III secretion systems and infection

The *Sal. typhimurium* pathogenicity island 2 (SPI-2) is necessary for the survival of bacteria within macrophages, and encodes a type III secretion system (see Section 2.5.5). Using DFI, a component of this type III secretory apparatus was shown to be selectively induced in host macrophages. To investigate this more closely, the *gfp* reporter gene was fused, individually, with all the open reading frames (ORFs) within SPI-2. Flow cytometry was then used to analyse *gfp* expression in recombinant strains within infected macrophages. At least four transcriptional units, or operons, were demonstrated and termed regulatory, structural I, secretory and structural II, according to their predicted role (Figure 3.13). The structural operons encoded protein components of the secretory apparatus. The secretory operon comprised secreted effector proteins and their chaperones. The regulatory operon encoded a two-component signal transduction system (TCSTS) that was essential for regulation of all genes in the SPI-2 secretion system. Macrophage-dependent induction of SPI-2 genes required acidification of the phagosome containing the bacteria. However, exposure of bacteria *in vitro* to low pH did not induce gene expression, suggesting that the signal is not simply low pH. Mutants defective in structural, regulatory, or some secretory components were impaired in intracellular replication and survival. Mutants were, nevertheless, still able to colonise mice but were not able to spread beyond the Peyer's patches.

An extension to DFI is *in vivo* induction of fluorescence (IVIF). This allows an analysis of tissue-specific gene expression by sacrificing experimental animals and visualising bacterial fluorescence in specific tissues *in situ* by both scanning laser confocal microscopy and by isolating and sorting individual cells by use of a FACS.

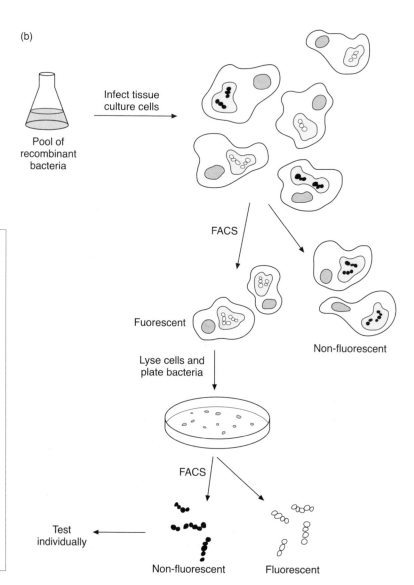

Figure 3.12b Differential fluorescence induction. (b) A fluorescence-activated cell sorter (FACS) is used to identify *in vivo*-specific gene promoters. The pool of recombinant bacteria is used to infect tissue culture cells, and highly fluorescent infected cells are collected by the FACS cytometer. The cells are lysed and the bacteria that are released are grown on laboratory medium. These are themselves sorted by FAC sorting to separate bacteria that are fluorescent from those that are non-fluorescent following *in vitro* growth. Non-fluorescent bacteria potentially carry *gfp* gene fusions with promoters that are expressed only during infection. These strains can then be tested individually in tissue culture.

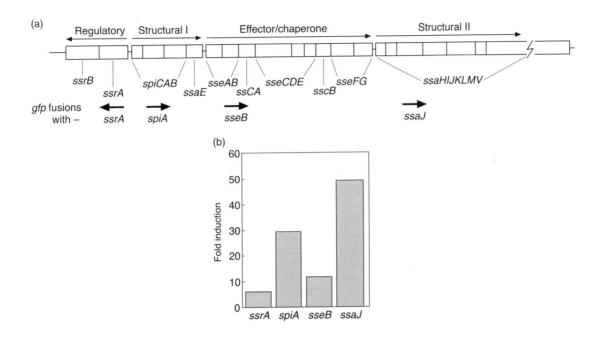

Figure 3.13 Differential fluorescence induction analysis of *in vivo* expression of the *Salmonella* SPI-2 type III secretion system. (a) *gfp* fusions were constructed to representative genes of the type III secretion system. (b) Bacterial fluorescence was measured following infection of tissue culture cells. Six hours post infection all constructs demonstrated fluorescence induction as compared with cells incubated in parallel in tissue culture medium.

3.7 | Genomic approaches

At the time of writing, the genomes of some 39 microbes have been fully sequenced and the information is in the public domain. These include one eukaryotic microbe (*Saccharomyces cerevisiae*, 13 Mb in size), while the remainder are bacteria (31) or archaea (7). The genomes of those prokaryotic organisms that have been sequenced range in size from 6.3 Mb for the opportunistic pathogen *Pseudomonas aeruginosa* down to a remarkably small 0.58 Mb for the obligate intracellular pathogen *Myc. genitalium*. Many of the bacterial species sequenced are important pathogens (Table 3.5). More than 100 other genomes are currently being sequenced, and so any lists rapidly become out of date. This explosion in sequence data has induced the development of ever more powerful tools for analysing the information. The World Wide Web (WWW) has played a central role, allowing access to large, central databases and providing the sheer computing power required for some of the newer tools. A good place to start is The Institute for Genomic Research (TIGR, website http://www.tigr.org), which provides access to most completed genome sequence databases. TIGR has also compiled a comprehensive microbial resource (CMR). The CMR allows researchers to access data from all of the genome sequences completed to date. This so-called 'omniome' is fully annotated with details of the organisms, information on structure and composition of the DNA molecule (e.g. its $G + C$ content, and whether the gene is on a plasmid or on the chromosome). Furthermore, information, such as the molecular mass, or predicted function, of protein sequences derived from the DNA sequence is available. Molecular

Table 3.5.	A list of some bacterial pathogens, and other organisms, that have been sequenced (see also Section 3.14)	
Organism	**Genome size (Mb)**
Bacteria	
Mycoplasma genitalium	0.58
Chlamydia trachomatis	1.07
Chlamydia pneumoniae	1.23
Campylobacter jejuni	1.64
Helicobacter pylori	1.66
Haemophilus influenzae	1.83
Neisseria meningitidis	2.27
Vibrio cholerae	4.00
Mycobacterium tuberculosis	4.40
Escherichia coli K-12	4.60
Escherichia coli O157 : H7	5.60
Other organisms	
Saccharomyces cerevisiae	13
Caenorhabditis elegans	20
Drosophila melanogaster	137

search tools are also available at the National Centre for Biotechnology Information (NCBI, website http://www.ncbi.nlm.nih.gov). NCBI was established in 1988 as a national resource for molecular biology information and creates public databases and conducts research in computational biology to develop software tools for analysing genome data. One tool is the COGs database. COGs (clusters of orthologous groups) is a set of multiple sequence alignments of homologous proteins that have the same function in different organisms.

Using these databases and search tools, a small amount of DNA sequence, obtained from STM or IVET analysis, for example, is often sufficient to obtain information on the entire gene, the proposed function of its protein product (if one is known) and the neighbouring genes. If the gene is an 'orphan', that is to say it has no homology with genes in the database or shares homology with a gene with no known function, then clues may be derived by searching for protein motifs within the derived amino acid sequence. Importantly, however, sequence homology may not necessarily reflect a conserved function and homologous proteins may have different functions in different bacteria. Frequently also, non-homologous proteins can perform the same function. For example, both *Lis. monocytogenes* and *Shigella flexneri* can subvert host actin polymerisation to move within and between cells (see Section 8.4.1.2). The proteins used to initiate actin polymerisation, ActA in *Lis. monocytogenes* and IcsA in *Shig. flexneri*, share no significant sequence similarity. Clearly, biochemical experiments are needed to augment molecular approaches to understanding genomics and life and there are many problems still to be addressed before genomic sequencing is in a position to answer as many questions as it raises.

Another major issue is that of strain-to-strain variation. Consider the two *E. coli* strains listed in Table 3.5, for example. *Escherichia coli* K-12 has been the workhorse of molecular biology almost from the start. This strain has a genome size of 4.6 Mb. Strain O157:H7 has only relatively recently emerged as a serious pathogen in contaminated food. This enterohaemorrhagic strain has a genome of 5.6 Mb, and so has the capacity, theoretically, to encode some 800 to 1000 more genes than K-12. PCR amplification experiments have demonstrated that *E. coli* K-12 lacks many of the virulence factors carried by other *E. coli* isolates, including heat-labile enterotoxin, haemolysin, and colonisation factors. Nevertheless, even amongst pathogenic isolates, we can identify several different *E. coli* strains, such as EPEC, EHEC, ETEC, etc. (see Section 1.6.2 and Table 1.12), each with different lifestyles and, ultimately, different pathogenic outcomes in infection.

Horizontal gene transfer plays a major role in the generation of species diversity, as described in Chapter 2. Thus phages, pathogenicity islands, plasmids and transposable elements all may carry virulence factors that alter the behaviour of the recipient organism. Random mutations and gene duplications also contribute significantly. Nevertheless, for pathogens, the genome sequence contains all of the information about every virulence factor, vaccine candidate and potential drug target. As more and more genomes are completed, we can now start to compare microbial genomes to glean clues about the minimal gene set required for cellular life and, conversely, reveal genes that specify the phenotypic characteristics that make each organism unique.

Mycobacterial comparative genomics

Mycobacterium tuberculosis and *M. leprae* are difficult to work with, owing to their fastidious growth requirements and long generation times. However, the availability of the complete genome sequence of *M. tuberculosis* and most of the *M. leprae* genome has allowed a computational approach to their molecular characterisation.

Mycobacterium tuberculosis H37Rv is a virulent strain that has a genome size of 4.4 Mb. It carries approximately 4000 genes, accounting for some 90% of the potential coding capacity. The genome has a uniform G + C content, which suggests that there has been little or no horizontal acquisition of genes, at least in the relatively recent past. The function of 40% of the protein-coding genes has been tentatively described, leaving approximately 60% as orphans. Around 20% the proteins are involved in fatty acid metabolism or are members of the PE (proline-glutamate) or PPE (proline-proline-glutamate) families of acidic, glycine-rich proteins. The former proteins are required for the synthesis of the large array of lipids, glycolipids and mycolic acids that are present in the mycobacterial cell wall. The PE and PPE proteins have no known function, although they may play a role in antigenic variation.

In contrast to *M. tuberculosis*, only 50–60% of the genome of *M. leprae* consists of transcribable genes. Its genome is also smaller at 2.8 Mb, and may encode only 1600 genes. There is much synergy between the two genomes in terms of common genes and their similar organisation. However, *M. leprae* has numerous large stretches of DNA interrupting conserved stretches that appear to be non-coding or to contain pseudo-genes. These pseudo-genes are non-functional due to the presence of multiple frameshift mutations, caused by small insertions or deletions of bases, and in-frame stop codons. Many genes present in *M. tuberculosis* are lacking in *M. leprae* and this may explain the restricted metabolic activities of *M. leprae* and why it is much slower growing. Nevertheless, the *M. leprae* genome sequence to date has identified some genes that are not present in *M. tuberculosis*. One interesting observation was that the gene encoding catalase-peroxidase in *M. leprae* carried multiple mutations. The catalase-peroxidase of *M. tuberculosis* mediates the toxic effects of the key anti-tubercle drug isoniazid. The mutations in the *M. leprae* gene render it inactive and this explains why *M. leprae* is catalase-peroxidase negative and also highly resistant to isoniazid.

3.8 | Assessing gene expression using DNA microarrays

With a complete genome sequence available, it is now possible to define exactly what genes are expressed under any given set of environmental conditions. A DNA microarray is a highly ordered matrix of many thousands of different DNA sequences. Microarrays can be screened by hybridisation to measure the expression levels of the entire gene complement of a bacterium in a single hybridisation experiment. There are two main approaches to the construction of microarrays. In the first approach, PCR is used to amplify DNA individually from all putative genes in a chromosome. These DNA fragments, generally over 200 bp in length, are then spotted onto a support surface, for example nylon membranes or glass slides, in an ordered array. The second approach is to generate a high density oligonucleotide array (or 'oligo chip') where oligonucleotides are synthesised *in situ* on glass wafers using a photolithotrophic technique. Each gene is represented by some 15 to 20 different oligonucleotides of around 25 bases in length. The use of oligonucleotides, rather than long contiguous stretches of DNA, provides a higher degree of specificity in the hybridisation. Thus the oligonucleotides are specifically selected to differentiate between genes with significant levels of sequence similarity, such as paralogues, i.e. homologous genes in the same cell whose products may perform similar, but not identical, functions.

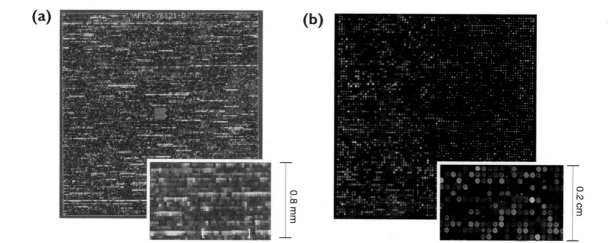

Figure 3.14 DNA microarray technology. Two examples of microarray technology are presented. (a) An oligonucleotide array where multiple (in this case 20) different oligonucleotides for every gene have been synthesised *in situ* using a photolithographic technique. In (b), a cDNA library has been deposited robotically onto a glass slide. A robot could also be used to apply PCR products of every known gene, or plasmids from an ordered genomic library that covers the entire genome. mRNA is isolated from cells grown under two different conditions. cDNA probes are then synthesised from each mRNA pool and labelled with different fluorescent tags. The probes are mixed and hybridised simultaneously. The array is then scanned to generate a quantitative fluorescent image. A pseudo-colour image can then be created to indicate the relative number of transcripts present under two different growth conditions. In this instance: red represents high level expression under condition 1, low level expression under condition 2; green represents low under condition 1, high under condition 2; yellow represents high expression under both conditions; black represents low expression under both conditions. (Reprinted by permission from *Nature* (Lockhart, D.J. & Winzler, E.A., Genomics, gene expression and DNA arrays. *Nature*, 2000, **405**, 827–836) ©2000 Macmillan Magazines Ltd.)

Microarrays can be used to determine the different patterns of gene expression in a bacterium growing under different conditions. RNA is first isolated from the organism grown under the different sets of conditions. RNA can be labelled directly but, commonly, labelled cDNA molecules are made by RT-PCR using random primers and fluorescently labelled nucleotides. Differences in the expression profiles are determined by comparison of identical microarrays probed separately with the sets of labelled probes produced (Figure 3.14, opposite). Transcription levels can be calculated by reference to cDNA samples of known concentration added to the hybridisation mixture. The expression profiles can be compared computationally and qualitative and quantitative differences in gene expression can be identified.

There are, however, problems associated with DNA microarrays. Firstly, the lack of a poly-adenylated tail on bacterial mRNA molecules (unlike eukaryotic mRNA molecules) means that it is not possible to separate mRNA from other species such as rRNA and tRNA that can comprise up to 99% of total RNA. This may have an impact on labelling efficiencies. Secondly, there is a rapid turnover of individual transcripts in bacteria and so it may not be possible to get a valid 'snapshot' at any particular time.

Mycobacterium tuberculosis and drug-induced alterations in gene expression

The global incidence of tuberculosis, caused by *M. tuberculosis*, is around seven million new cases per year, and in recent years multiply resistant strains have emerged. A DNA microarray has been used to characterise the transcriptional response of *M. tuberculosis* to isoniazid, the most commonly prescribed anti-tubercle drug. Isoniazid blocks the biosynthetic pathway that is required for the synthesis of mycolic acid components of the waxy envelope of the organism. PCR was used to amplify DNA fragments of 200 to 1000 bp from 3834 of the predicted 3924 ORFs. The PCR products were arrayed on glass slides using a robotic spotting device. RNA was isolated from *M. tuberculosis* grown in the presence or absence of isoniazid, and differentially labelled cDNA probes were generated by RT-PCR using nucleotides tagged with two different fluorescent labels. These were mixed and simultaneously hybridised with the microarray. The slides were then scanned using a laser-activated confocal scanner and fluorescence ratios were calculated for each spot.

There was a highly induced set of five genes encoding components of the fatty acid synthase (FAS)-II complex, and this correlated with previous biochemical data on the effects of isoniazid exposure. A number of other genes were switched on, including an exported transferase enzyme involved in mycolic acid maturation, an efflux protein, and four orphan genes. The induction of these genes is probably due to biochemical feedback loops detecting either a depletion of mycolic acid molecules in isoniazid-treated cells, or else the accumulation of intermediates in the biochemical pathway prior to the isoniazid-induced block.

The DNA microarray method, therefore, allowed the identification of a number of additional genes whose products are potentially involved in mycolic acid metabolism. These genes might not have been identified by other means and they may encode new drug targets. Isoniazid-resistant strains arise predominantly from mutations in the *katG* gene encoding a catalase-peroxidase required for isoniazid activation. Using the DNA microarray, it was also demonstrated that an isoniazid-resistant (*katG⁻*) strain demonstrated no isoniazid-induced transcriptional alterations, confirming the central role of this gene in isoniazid resistance.

As well as investigating gene transcription, DNA microarrays can be used to help to assign roles for orphan genes. In *E. coli*, for example, some 40% of potential ORFs encode proteins with no known function. Using DNA microarrays it may be possible to look at the co-regulation of orphan genes with genes of known function under various environmental growth conditions to shed some light on possible roles. DNA microarrays can also be used for comparative genomics and for the genotyping of bacterial strains.

3.9 | Eukaryotic molecular methods

Thus far, our discussion has concentrated on manipulation of the bacterial genome to identify bacterial genes encoding virulence determinants. However, bacterial virulence is a game for two players and it is important to be able to identify host genes that encode proteins involved in the overall virulence phenomenon. As an example, in Section 4.2.6 we will describe the key role that the large number of host-produced hormone-like proteins termed cytokines play in both the defence against infection and in the pathology due to infection. There are many cytokines and it is important to identify the roles that they play in the host response to infection with particular bacteria. An obvious way of determining the role of any particular cytokine would be to try to remove it from the experimental situation. In the recent past, investigators have used neutralising antibodies for this purpose. However, it is often inconvenient to administer antibodies on a long-term basis to animals and administration can give rise to untoward side effects. An alternative is to inactivate the gene by homologous recombination to produce animals known as gene knockouts. Cytokine gene knockouts are being used increasingly in the study of bacterial virulence mechanisms.

This section will briefly describe a few techniques that have been applied largely to eukaryotic cells and organisms and which have proved to be useful in identifying genes involved in bacterial pathogenesis.

3.9.1 Yeast two-hybrid screening

3.9.1.1 Description of the technique

One of the major questions asked in bacterial virulence studies is: with what does protein X interact? This is a subset of a general question in

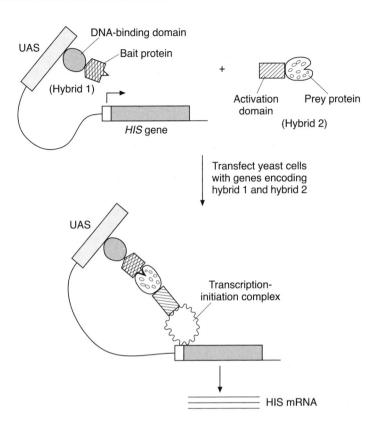

UAS

DNA-binding domain

Bait protein

(Hybrid 1)

+

Activation domain

Prey protein

(Hybrid 2)

HIS gene

Transfect yeast cells with genes encoding hybrid 1 and hybrid 2

UAS

Transcription-initiation complex

HIS mRNA

Figure 3.15 Yeast two-hybrid (bait–prey) cloning relies on having two halves of a transcription factor fused to different proteins. Only if those recombinant proteins interact will the two halves of the transcription protein come together and perform its normal function. The figure shows schematically the two-hybrid recombinant proteins used in this technique. The DNA-binding domain (which binds to a region of the DNA termed the upstream activation sequence (UAS)) is fused to what is normally termed the bait protein, and the activation domain of the transcription protein is fused to the prey protein (sometimes called the 'fish domain'). If the bait and the prey proteins interact within the yeast cell then the DNA-binding and activation domains can interact with the DNA and switch on the transcription of a reporter gene. In the example provided, the gene switched on encodes a protein that complements an amino acid auxotrophic deficiency (histidine deficiency). Cells that can grow in histidine-depleted media contain a prey protein that can interact with the bait protein.

Cell Biology. The complete understanding of how a cell (prokaryotic, archaeal or eukaryotic) behaves must describe all of the interactions between its proteins. It is these interactions that make the cell what it is. One way of assessing protein interactions has already been described and can be termed 'affinity proteomics', in which proteins binding to a particular 'bait' protein can be rapidly identified. In the example given, a human leukocyte receptor was used to fish for bacterial adhesins.

Another, earlier, system for asking questions about interacting proteins was developed by Stanley Fields and Paul Bartel in the late 1980s. This system is variously called the yeast two-hybrid system, two-hybrid system or bait–prey cloning system. Like many good ideas, the basis of this cloning system is simplicity itself. There are certain transcription factors that have two physically separate domains. One of these domains (the DNA-binding domain) enables the transcription factor to target particular DNA (i.e. promoter) sequences. These are termed upstream activation sequences. The other domain (the transcription activation domain) acts to control the assembly of the transcription complex thus enabling transcription of the gene to be initiated (Figure 3.15). The DNA encoding each of these domains can be fused, separately, with the genes of interest and the subsequent hybrid proteins produced then used to determine protein–protein interactions. In practice, if one wishes to know if two proteins (say A and B) interact, then they are individually cloned into vectors which produce a fusion protein consisting of (i) A linked to the DNA-binding domain (generally referred to as the bait) and (ii) B linked to the transcription activation domain (generally referred to as the prey or sometimes the 'fish'). If A and B interact then the two transcriptional domains interact and can induce the transcription of a particular protein. The assay is done in a yeast and, in the original formulation of the technique, the gene transcribed when A and B bound together would have been *lacZ*, with positive colonies being recognised by being blue. In current formulations of the method, the yeast used has defects in genes encoding essential nutrients that are complemented by correct A–B assembly and thus screening is by colony survival.

The yeast two-hybrid methodology is more often used to detect unknown proteins binding to a given 'bait' protein. In this formulation of the screen, a library of cDNAs, or genomic DNA fragments, are fused with the coding region of the transcription activation domain and these are expressed in yeast also expressing the DNA-binding domain fused with the protein of interest (the bait protein).

This technique has undergone constant refinement since its inception in the late 1980s. For example, there is now a three-hybrid system for detecting the interactions of proteins with RNA. A recent advancement has been the development of a bacterial 'two-hybrid' system.

3.9.1.2 Use of the two-hybrid system in investigating bacterial virulence factors

The two-hybrid system is generally used for identifying proteins which interact within the cytoplasm of eukaryotic cells. Given the growing understanding that we have of bacterial invasion of host cells and of direct bacterial secretion of proteins into host cells, the use of two-hybrid screening in investigating bacterial virulence factors is obvious, although it has not yet been used as extensively as it should.

Helicobacter pylori has a major virulence factor termed vacuolating (VacA) toxin. This has specific effects on the formation and maturation of the endosomal compartment of host cells. Using VacA as bait, it has been shown that this protein interacts with a novel eukaryotic 54 kDa protein termed VIP54. That VacA and VIP54 interact has been independently confirmed by co-immunoprecipitation and direct binding experiments. The result of the interaction of these two proteins has not yet been determined. VIP54 is expressed in many tissues and appears to interact with intermediate filaments within cells.

The pathology of anthrax is due to the toxins secreted by *Bacillus anthracis*. The organism produces two toxins – lethal factor (LF) and oedema factor (EF) – which function within the target cell cytoplasm (see Section 9.4.1.1.3). LF can cause cytolysis, cytokine synthesis and the death of laboratory animals. LF has a classic zinc-binding motif, suggesting that its actions depend on proteolysis. To identify the substrates for this protein, use has been made of two-hybrid screening using an inactive LF mutant as the bait. This has revealed that the mitogen-activated protein (MAP) kinase kinases (see Section 4.2.4.2)

Mek1 and Mek2 interact directly with LF. Such interaction leads to the cleavage of the N-termini of these kinases, thus interfering with the chain of MAP kinase interactions and thus with correct cell signalling. The consequence of this, among other effects, is the death of the intoxicated cells.

The system has also been used to define the interactions between the *Yersinia* secretion proteins (Ysc) (see Sections 2.5.5 and 8.2.1.1.1). Various pairwise combinations of these genes were made in order to determine which were interacting with one another. More recently, the technique has been used to produce a large-scale map of the protein–protein interactions that occur in *Hel. pylori*. It is envisaged that two-hybrid screening will be increasingly used to determine the identity of the many bacterial and host proteins that need to interact during the process of bacterial infection.

3.9.2 Transgenesis and the generation of gene knockouts

3.9.2.1 Description of the technique

Transferring DNA from one cell to another is the basis of molecular genetics and of the biotechnology industry, whether it be the transformation of *E. coli* with DNA from an elephant to produce elephant proteins or the transfection of a eukaryotic gene from one species into another to see whether the gene product has identical effects in both species. The deliberate transfer of DNA from one somatic cell to another has been going on since the pioneering work of Oswald Avery in the early 1940s. In addition to adding genes, it is important to know what would happen if a particular gene were removed from the genome of an organism. The generation of gene knockouts (isogenic mutants) in bacteria using homologous recombination has been described. It is also possible to generate animals in which specific genes have been inactivated by homologous recombination. Obviously, the process is more complex and time consuming with a mouse than it is with a bacterium, and the process will be described briefly and examples of the uses of gene knockouts in bacterial pathogenesis examined.

In mammals, the technology for knocking out a gene (a process also called insertional inactivation of a gene) has been worked out most comprehensively in the mouse. Step 1 is to choose the gene that needs to be inactivated and to clone it into an appropriate plasmid (see Figure 3.16). The plasmid also contains a copy of the gene for thymidine kinase (*tk*). The secret of making knockouts is to insert a copy of a neomycin resistance gene into the middle of the gene under study. Thus transcription and translation of this novel DNA sequence would produce inactive fragments of the protein of interest plus the protein product of the neomycin resistance gene. The *tk* gene and the neomycin resistance gene allow those cells incorporating the insertionally inactivated gene to be selected for. Having produced the

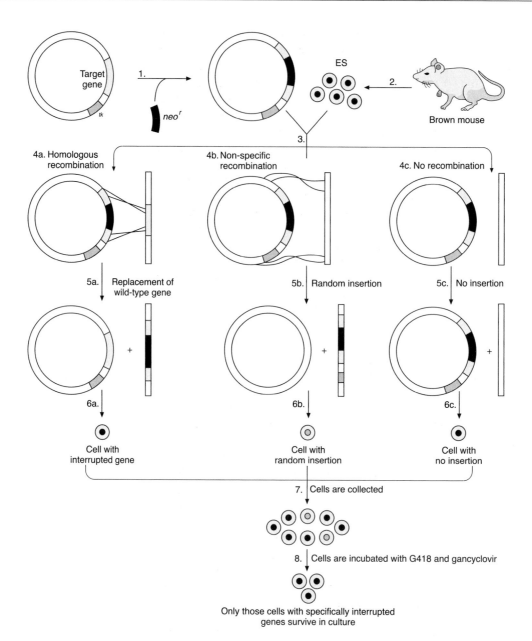

Figure 3.16 How to knockout a mouse gene (part I). A plasmid containing the gene to be knocked out (target gene) and the gene (*tk*) encoding the enzyme thymidine kinase are the basic requirements for making a knockout. The mechanism of 'knocking-out' involves inserting a neomycin-resistance gene (*neo^r*) into the middle of the gene of interest. This technique is also known as insertional inactivation. Once produced, the plasmid is transfected into embryonic stem (ES) cells taken from a brown mouse. In order to select for cells in which the inactivated gene has become incorporated by homologous recombination, use is made of a selective medium containing the antibiotic G418 and the drug gancyclovir. ES cells in which there has been no recombination (the majority) lack the antibiotic resistance gene and are killed by the neomycin analogue G418. ES cells in which recombination has happened randomly also incorporate the thymidine kinase gene and this can activate the gancyclovir that kills the cells. Only those ES cells in which the inactivated gene has become incorporated specifically by homologous recombination will survive. The following figure explains what happens next.

final engineered plasmid, it is transfected into the embryonic stem (ES) cells collected from an Agouti (i.e. brown mouse). These ES cells are totipotent and capable of differentiating into any cell type in the mature animal. The transfection of the plasmid into the ES cells can result in homologous recombination between the interrupted target gene and the homologous wild-type gene, resulting in the replacement of the latter in the cell by the inactive gene complex. Homologous recombination does not result in the insertion of the *tk* gene into the host genome. However, the more likely situation is that there is either non-specific recombination (including the *tk* gene) or no recombination. The problem is now to isolate the cells in which homologous recombination has occurred. This is done by selection in a medium containing the neomycin analogue G418 and the drug gancyclovir. G418 kills all those cells (probably the majority) in which recombination did not occur and which therefore have not incorporated the neomycin resistance gene. Gancyclovir is a guanosine analogue that needs to be phosphorylated by enzymes such as thymidine kinase to be active. Those cells that have incorporated the *tk* gene, because of non-specific recombination, can therefore activate the gancyclovir, and are, in consequence, killed. This only leaves those cells in which there has been homologous recombination.

Having selected the ES cells that have incorporated the interrupted gene, the next problem is to get them back into the whole mouse and generate an animal lacking the chosen gene. To do this, the engineered ES cells are injected into the blastocyst of a mouse destined to have a particular coat colour (either black or white). The blastocyst is then inserted into a surrogate mother, which then produces offspring that are chimeras composed of wild-type and knockout cells (Figure 3.17). This chimera is recognised by the coat colour and this is why the Agouti mouse is chosen. The brown zones of the coat, for example, come from the transplanted engineered cells. Having reached this stage, the rest of the process requires good breeding practice. Chimeric male mice are then mated with black or white mice. In mating with black mice, for example, only brown offspring are selected. These are assessed for the presence of the inactivated gene and appropriate brother–sister matings are continued until mice homozygous for the gene knockout are produced. It is then a question of building up a good colony of knockouts in order to have enough animals to use in biological studies.

3.9.2.2 Knockout mice in bacterial virulence research

There are now many hundreds of knockout mice in which different genes have been inactivated. It is also possible to knock out more than one gene in a mouse strain, although this complicates the whole process. In investigations of infectious diseases, the major knockouts studied have been in genes encoding cytokines. Many of these cytokine knockouts result in mice that are more susceptible to infection. Many hundreds of papers have now been produced on this topic and it would be impossible to cover them in a book of this size. The reader is referred

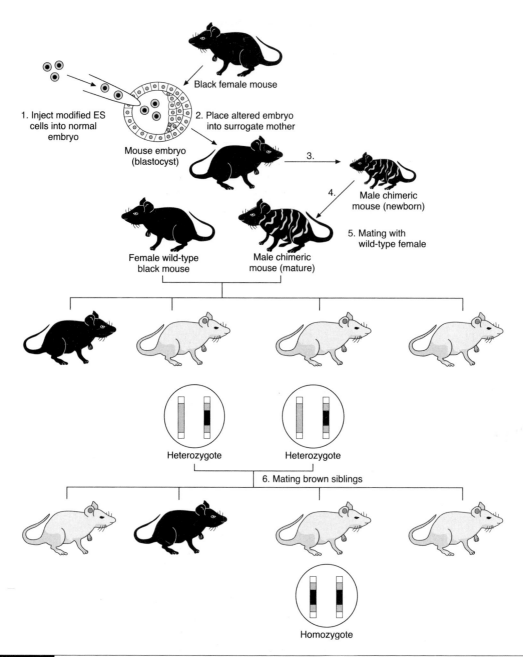

Figure 3.17 How to knock out a mouse gene (part 2). Having prepared the ES cell with an inactivated version of the gene of interest, the next step is to 'make a mouse' from these cells. The first part of this process is to inject the ES cells into a blastocyst-stage embryo taken from a mouse with a different (in this example black) coat colour. Remember, the ES cells are from a brown mouse. The embryo is then transplanted back into the uterus of a surrogate mother. Pups that contain the altered ES cells can be recognised because they are chimeras with recognisable black and brown coat colours. Mature chimeric male mice are then mated with wild-type black females and only the brown offspring are selected. Any black offspring must be derived from the wild-type blastocyst. At this point we have mice that are heterozygous for the inactivated gene. This can be assessed by Southern blotting. To produce mice homozygous at the inactivated gene allele, brother–sister matings between brown mice are continued until it is confirmed that the desired mice have been produced. These homozygous knockout mice can then be bred in large numbers and phenotypic differences investigated.

to Table 3.6, which describes the phenotype of many of these knockouts and to a recent book that is focused entirely on cytokine knockouts (Mak, 1998; see Section 3.13, Books).

3.9.3 Dominant-negative mutants

3.9.3.1 Description of the technique

There is an element of paradox in the concept of a mutant being both dominant and recessive (negative). However, dominant-negative mutants turn out to be common in nature and, because of their activity and specificity, are particularly useful in studying the actions of individual proteins. A dominant-negative mutation gives rise to a protein that lacks its normal functionality but which, because it can still participate in its normal role, can inhibit the action of the non-mutated wild-type protein. A good example of such dominant-negative mutants is the receptor tyrosine kinase (RTK) involved in recognising epidermal growth factor. As will be described in Sections 4.2.4.1 and 4.2.6.6, many cytokine receptors have to dimerise to enable their intracellular domains to function as signal transducers. Mutant genes encoding receptors lacking the intracellular signalling module of the RTK obviously do not signal. However, such proteins, if they dimerise with the normal receptor, can block its signalling action (Figure 3.18).

It is now possible to engineer many genes such that the proteins they produce act as dominant-negative mutants. One classic example is the production of a dominant-negative mutant of fibroblast growth factor receptor (FGFR). When the gene encoding this mutant was engineered to be expressed in frog embryos, the embryos failed to generate a tail, showing the importance of this receptor in the embryology of the animal. Dominant-negative mutations in one of the FGFR genes (FGFR-3) results in the most common form of dwarfism – achondroplasia. Dominant-negative mutations are found to occur naturally in the cell cycle regulatory protein p53. This protein is involved in checking the fidelity of DNA replication. Therefore, the absence of its activity can lead to the accumulation of mutations in daughter cells and the consequence of this can be cancer.

With the knowledge from natural dominant-negative mutants, the generation of specific dominant-negative mutants can be used to probe the role of individual proteins in the function of cells. This involves the generation of the dominant-negative mutant gene and its transfection into the cell of choice. The major problem in this technique is in producing enough of the mutated protein within the cell of interest to block the function of the normal protein. There is increasing interest in the use of this methodology in bacterial virulence and three examples will be given.

Table 3.6. | Phenotypes of cytokine knockout mice

Gene	Phenotype
IL-1β	Resistance to endotoxin shock/LPS-induced pyrogenesis
IL-1α/β	Loss of early phase protective immunity to *Mycobacterium* spp.
IL-1ra	Decreased susceptibility to *Listeria* spp.
IRAK	Reduced response to *Propionibacterium acnes*
ICE	Resistance to endotoxin
IL-2	Develop ulcerative colitis in response to normal gut microflora
IL-4	Deficient TH2 responses; depending on strain, mice either more or less susceptible to *Staphylococcus aureus* sepsis; *Helicobacter pylori* gut inflammation increased
IL-6	Susceptibility to *Listeria* spp.; decrease in early immunity and IFNγ production in response to *Mycobacterium* spp.
IL-8	Dysfunctional neutrophil response to urinary tract infections
IL-10	Develop chronic enterocolitis due to response to gut microflora
IL-12	Increased susceptibility to *Mycobacterium* spp. and *Salmonella* spp.
IL-18	Impaired cytokine and NK response to infection; resistant to endotoxin-induced liver injury but highly susceptible to endotoxin shock
IFNγ	Increased susceptibility to *Mycobacterium* spp. and *Salmonella* spp.
GM-CSF	Susceptible to pulmonary infection with *Streptococcus agalactiae*
G-CSF	Increased susceptibility to *Listeria* spp.
CCSP	Enhanced killing of *Pseudomonas aeruginosa*
CCR2	Increased susceptibility to *Cryptococcus neoformans*
TNFRp55	Increased susceptibility to *Listeria* spp. and *Mycobacterium* spp.
TNFRp70	Plays a minor role in antimycobacterial immunity

IRAK, interleukin-1 receptor-associated kinase; G-CSF, granulocyte colony stimulating factor; CCSP, Clara cell secretory protein; CCR, CC chemokine receptor; for other abbreviations, see list on p. xxii.

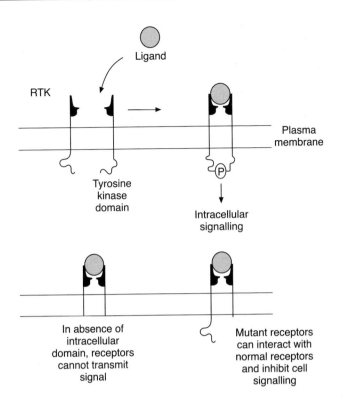

Figure 3.18 An example of a dominant-negative mutant. Receptor tyrosine kinases (RTKs) are a class of cell surface receptor that dimerise when in contact with their ligand and this process activates the tyrosine kinase (TK) domain, switching on the intracellular signal transduction pathways (for more details, see Section 4.2.2.4). Mutant receptors lacking the TK domain are found in nature, and in a cell containing copies of the normal and the mutant receptor the latter can inhibit the function of the former. These mutants can be used experimentally. If a dominant-negative form of RTK is produced and transfected into cells (and produced at significant levels), it can block the action of the RTK.

3.9.3.2 Use of dominant-negative mutants in the study of bacterial virulence

The major *Hel. pylori* exotoxin VacA has already been described briefly. This toxin is membrane-active and interferes with endosomal functioning. A version of the toxin without the hydrophobic N-terminus was constructed and shown to be deficient in its ability to form membrane pores in artificial bilayers. This engineered protein was found to act as a dominant-negative mutant and could inhibit the vacuolating effect of the normal toxin. This protein is, therefore, a very useful tool for probing the function of the wild-type VacA toxin.

In a second study of *Hel. pylori*, the question being asked was of the signalling pathways activated in gastric epithelial cells in contact with this organism. This study utilised dominant-negative mutants of various intracellular kinases, in which the kinase domain was engineered out. Cells were transfected with vectors expressing these mutant kinases, which then acted as specific inhibitors of the chosen kinase. The kinases chosen for inhibition were the I-κB family of kinases (which phosphorylate I-κB and thus activate NF-κB), NF-κB-inducing kinase (NIK) and TRAF-2 and TRAF-6 (for more details, see Section 4.2.6.6.2). These various dominant-negative mutants, when transfected individually into cells, blocked the ability of the gastric epithelial cells to up-regulate NF-κB-induced signalling. These findings could be useful in developing more effective agents for blocking the gastric effects of *Hel. pylori* infection.

The final example of the use of dominant-negative mutants in the study of bacterial pathogenesis focuses on cellular activation by *Clostridium difficile* toxin A. Toxin A stimulates monocytes to produce IL-8 and to undergo cytolysis by an unknown mechanism. This toxin activates the main MAP kinase cascades (for more details, see Section 4.2.4.2) and use was made of selective inhibitors of MAP kinases and of dominant-negative mutants of the p38-activating MAP kinase kinases MKK3 and MKK6 to identify the particular MAP kinase pathways activated by this toxin. The ability to abrogate the activity of specific proteins within cultured cells and also within the whole animal by use of dominant-negative mutants is a powerful tool that will be increasingly used in the study of bacterial pathogenesis.

3.9.4 Oligonucleotide-based gene inactivation

In addition to dominant-negative mutants it is possible to specifically block the 'function' of any chosen protein by use of synthetic oligonucleotides. These can be used to inhibit gene transcription or mRNA translation and thus to selectively block the production of the chosen protein. It is the latter that has been most comprehensively studied and will be described briefly. The blockade of the translation of mRNA uses antisense oligonucleotides and the methodology is called antisense technology or antisense pharmacology.

Antisense reagents are short stretches of modified nucleotides in which the normal phosphodiester bonds are replaced by more stable analogues such as phosphorothionate bonds. These oligonucleotides are designed to hybridise with selected mRNA targets and the product of this interaction either can inhibit the transcription of the selected mRNA or can result in the hydrolysis of the mRNA by RNase H. The end result is the failure of the cell to produce the selected protein.

Antisense technology can be used to target and ablate the synthesis of bacterial proteins or of host proteins. For example, antisense has been used to inhibit the expression of the *Staph. aureus* α-toxin. Bacteria expressing such antisense exhibited lower virulence in *in vivo* models of *Staph. aureus* infection.

Macrophages infected with *Sal. choleraesuis* produced increased levels of tumour necrosis factor (TNF) α and the stress protein hsp70. Inhibition of the TNFα by antibodies increased the number of macrophages dying, showing the host-protective activity of this key cytokine (for more details of the role of TNFα in infection, see Sections 4.2.6 and 10.3). The role of stress proteins is generally protective of cells but it was unclear what role hsp70 was playing in this form of cellular infection. Using antisense to inhibit the translation of hsp70, it was found that the numbers of infected macrophages dying increased significantly, revealing the important role of this stress protein in macrophage infection.

3.10 | Concept check

- New molecular methods are redefining our ideas about bacterial virulence mechanisms.
- As well as expressing virulence properties such as adhesion, invasion and the production of protein exotoxins, bacteria must sense and respond to different environmental conditions in the host.
- Recent refinements of traditional mutagenic approaches to the study of bacterial virulence have resulted in high throughput screening methods for the identification of genes whose *in vivo* expression is essential for bacterial survival (STM, GAMBIT).
- These systems have been augmented by the development of promoter trap systems (IVET, DFI) and mRNA-based systems (differential and subtractive analysis) that allow identification of genes expressed *in vivo* or under defined growth conditions.
- There have been parallel developments in eukaryotic molecular biology techniques that complement the bacterial approaches.
- Bacterial genome sequencing is now fast and relatively inexpensive and is generating vast amounts of data, most of which can be accessed via the Internet. Powerful tools have been developed for searching and analysing these data. Nevertheless, much experimental microbiology and biochemistry is required before we can make full use of genomic sequence information.
- It is now widely accepted that the pattern of mRNA expression may not be an accurate reflection of the proteins that the cell will produce.
- Proteomics, the study of the proteins expressed by cells under given conditions is a relatively new, but very exciting, addition to the molecular microbiologist's toolbox.

3.11 | What's next?

A central theme that will be emphasised throughout this book is that bacteria, long regarded as being very simple creatures, have evolved a variety of ways of undermining the complex signalling pathways of the more evolutionarily advanced eukaryotic cells of their host. By doing so, they can establish themselves in, or on, their host and obtain from it the nutrients and environmental conditions necessary for their growth and proliferation. The next chapter will, therefore, concentrate on the signalling mechanisms employed by host cells and by bacteria – a knowledge of these is essential for understanding many of the succeeding chapters in the book.

3.12 | Questions

1. A number of techniques described in this chapter require fairly advanced molecular genetic systems to be available for the organism under study. What approaches could be used to analyse the virulence mechanisms of an emerging pathogen?

2. What advantages does transposon mutagenesis provide over chemical/ultraviolet light mutagenesis? Describe the additional advantages provided by signature-tagged mutagenesis.

3. What is GAMBIT?

4. IVIAT can be used to identify antigenic proteins expressed during infection. What potentially important surface components might IVIAT miss? You may wish to refer to Chapter 2 for a complete answer.

5. Outline the major approaches for differential and subtractive analysis of mRNA.

6. Describe the three main IVET promoter trap systems.

7. What are the advantages and disadvantages of RIVET over the original IVET systems?

8. What is an orphan gene? Using gridded arrays, how might you infer a functional role for any given orphan?

9. Which mouse genes would you knock out if you wanted to determine the key host responses to infection of a mouse with *E. coli*?

10. Compare and contrast the genome, transcriptome and proteome.

3.13 | Further reading

Books

Accili, D. (2000). *Genetic Manipulation of Receptor Expression and Function.* New York: Wiley-Liss.

Bartel, P. L. & Fields, S. (1997). *The Yeast Two-Hybrid System.* Oxford: Oxford University Press.

Baxevans, A. D. & Ouellette, B. F. F. (1998). *Bioinformatics.* Chichester: Wiley Interscience.

Bishop, M. J. (1999). *Genetics Databases.* London: Academic Press.

Brown, T. A. (1999). *Genomes.* Oxford: Bios.

Glick, B. R. & Pasternack, J. J. (1998). *Molecular Biotechnology.* Washington, DC: ASM Press.

Kreuzer, H. & Massey, A. (2000). *Recombinant DNA and Biotechnology: A Guide for Students.* Washington, DC: ASM Press.

Mak, T. W. (ed.) (1998). *The Gene Knockout FactsBook.* Sidcup: Harcourt Publishers Ltd.

Pennington, S. R. & Dunn, M. J. (2001). *Proteomics: From Protein Sequences to Function.* Oxford: Bios.

Walker, M. R. & Rapley, R. (1997). *Route Maps in Gene Technology.* Oxford: Blackwell Science.

Review articles

Chiang, S. L, Mekalanos, J. J. & Holden, D. W. (1999). *In vivo* genetic analysis of bacterial virulence. *Annu Rev Microbiol* **53**, 129–154.

Clayton, R. A., White, O. & Fraser, C. M. (1998). Findings emerging from complete microbial genome sequences. *Curr Opin Microbiol* **1**, 562–566.

Cole, S. T. (1998). Comparative mycobacterial genomics. *Curr Opin Microbiol* **1**, 567–571.

Cotter, P. A. & DiRita, V. J. (2000). Bacterial virulence gene regulation: an evolutionary perspective. *Annu Rev Microbiol* **54**, 519–565.

Crooke, S. T. (2000). Oligonucleotide-based drugs in the control of cytokine synthesis. In *Novel Cytokine Inhibitors*, ed. by G.A. Higgs & B. Henderson, pp. 83–102. Basel: Birkhauser.

Cummings, C. A. & Relman, D. A. (2000). Using DNA microarrays to study host–microbe interactions. *Emerg Infect Dis* **6**, 513–525.

Diehn, M. & Relman, D.A. (2001). Comparing functional genomic datasets: lessons from DNA microarray analyses of host–pathogen interactions. *Curr Opin Microbiol* **4**, 95–101.

Fraser, C. M. Eisen, J., Fleischmann, R. D., Ketchum, K. A. & Peterson, S. (2000). Comparative genomics and understanding of microbial biology. *Emerg Infect Dis* **6**, 505–512.

Gingeras, T. R. & Rosenow, C. (2000). Studying microbial genomes with high-density oligonucleotide arrays. *ASM News* **66**, 463–469.

Graham, M. R., Smoot, L. M., Lei, B. F. & Musser, J. M. (2001). Toward a genome-scale understanding of group A *Streptococcus* pathogenesis. *Curr Opin Microbiol* **4**, 65–70.

Handfield, M., Brady, L. J., Progulske-Fox, A. & Hillman, J. D. (2000). IVIAT: a novel method to identify microbial genes expressed specifically during human infections. *Trends Microbiol* **8**, 336–339.

Harrington, C. A., Rosenow, C. & Retief, J. (2000). Monitoring gene expression using DNA microarrays. *Curr Opin Microbiol* **3**, 285–291.

Joung, J. K., Ramm, E. I. & Pabo, C. O. (2000). A bacterial two-hybrid selection system for studying protein–DNA and protein–protein interactions. *Proc Natl Acad Sci USA* **97**, 7382–7387.

Judson, N. & Mekalanos, J. J. (2000). Transposon-based approaches to identify essential bacterial genes. *Trends Microbiol* **8**, 521–526.

Lee, S. H. & Camilli, A. (2000). Novel approaches to monitor bacterial gene expression in infected tissue and host. *Curr Opin Microbiol* **3**, 97–101.

Lehoux, D. E., Sanschagrin, F. & Levesque, R. C. (2001). Discovering essential and infection-related genes. *Curr Opin Microbiol* **4**, 515–519.

Lucchini, S., Thompson, A. & Hinton, J. C. D. (2001). Microarrays for microbiologists. *Microbiology* **147**, 1403–1414.

Manger, I. D. & Relman, D. A. (2000). How the host 'sees' pathogens: global gene expression responses to infection. *Curr Opin Immunol* **12**, 215–218.

Shea, J. E. & Holden, D W. (2000). Signature-tagged mutagenesis helps identify virulence genes. *ASM News* **66**, 15–20.

Tang, C. M., Hood, D. W. & Moxon, E. R. (1998). Microbial genome sequencing and pathogenesis. *Curr Opin Microbiol* **1**, 12–16.

Washburn, M. P. & Yates, J. R. (2000). Analysis of the microbial proteome. *Curr Opin Microbiol* **3**, 292–297.

Weinstock, G. M. (2000) Genomics and bacterial pathogenesis. *Emerg Infect Dis* **6**, 496–504.

Papers

Akerley, B. J., Rubin, E. J, Camilli, A., Lampe, D. J., Robertson, H. M. & Mekalanos, J. J. (1998). Systematic identification of essential genes by *in vitro* mutagenesis. *Proc Natl Acad Sci USA* **95**, 8927–8932.

Boyce, J. D. & Adler, B. (2000). The capsule is a virulence determinant in the pathogenesis of *Pasteurella multocida* M1404 (B:2). *Infect Immun* **68**, 3463–3468.

Chakraborty, A., Das, S., Majumdar, S., Mukhopadhyay, K., Roychoudhury, S. & Caudhuri, K. (2000). Use of RNA arbitrarily primed-PCR fingerprinting to identify *Vibrio cholerae* genes differentially expressed in the host following infection. *Infect Immun* **68**, 3878–3897.

Cirillo, D. M., Valdivia, R. H., Monack, D. M. & Falkow, S. (1998). Macrophage-dependent induction of the *Salmonella* pathogenicity island 2 type III secretion system and its role in intracellular survival. *Mol Microbiol* **30**, 175–188.

De Bernard, M., Moschioni, M., Napolitani, G., Rappuoli, R. & Montecucco, C. (2000). The VacA toxin of *Helicobacter pylori* indentifies a new intermediate filament-interacting protein. *EMBO J* **19**, 48–56.

Gygi, S. P., Rist, B., Gerber, S. A., Turecek, F., Gelb, M. H. & Aebersold, R. (1999). Quantitative analysis of complex protein mixtures using isotope-coded affinity tags. *Nature Biotechnol* **17**, 994–999.

Hong, P. C., Tsolis, R. M. & Ficht, T. A. (2000). Identification of genes required for chronic persistence of *Brucella abortus* in mice. *Infect Immun* **68**, 4102–4107.

Houry, W. A., Frishman, D., Eckerskorn, C., Lottspeich, F. & Hartl, F. U. (1999). Identification of *in vivo* substrates of the chaperonin GroEL. *Nature* **402**, 147–154.

Hutchinson, C. A., Peterson, S. N., Gill, S. R., Cline, R. T., White, O., Fraser, C. M., Smith, H. O. & Venter, J. C. (1999). Global transposon mutagenesis and a minimal mycoplasma genome. *Science* **286**, 2165–2169.

Jackson, M. W. & Plano, G. V. (2000). Interactions between type III secretion apparatus components from *Yersinia pestis* detected using the yeast two-hybrid system. *FEMS Microbiol Lett* **186**, 85–90.

Jungblutt, P., Schaible, U., Mollenkopf, H.-J., Zimny-Arndt, U., Raupach, B., Mattow, J., Halada, P., Lamer, S., Hagens, K. & Kaufmann, S. (1999). Comparative proteome analysis of *Mycobacterium tuberculosis* and *Mycobacterium bovis* BCG strains: toward functional genomics of microbial pathogens. *Mol Microbiol* **33**, 1103–1117.

Kilic, A. O., Herzberg, M. C., Meyer, M., Zhao, X. & Tao, L. (1999). Streptococcal reporter gene-fusion vector for identification of *in vivo* expressed genes. *Plasmid* **42**, 67–72.

Larsson, T., Bergstrom, J., Nilsson, C. & Karlsson, K. -A. (2000). Use of an affinity proteomics approach for the identification of low-abundant bacterial adhesins as aplied on the Lewis[b]-binding adhesin of *Helicobacter pylori*. *FEBS Lett* **469**, 155–158.

Link, A. J., Eng, J., Schieltz, D. M., Carmack, E., Mize, G. J., Morris, D. R., Garvik, B. M. & Yates, J. R. (1999). Direct analysis of protein complexes using mass spectrometry. *Nature Biotechnol* **17**, 676–682.

Maeda, S., Yoshida, H., Ogura, K., Mitsuno, Y., Hirata, Y., Yamaji, Y., Akanuma, M., Shiratori, Y. & Omata, M. (2000). *H. pylori* activates NF-kappaB through a signalling pathway involving IkappaB kinase, NF-kappaB-inducing kinase, TRAF2, and TRAF6 in gastric cancer cells. *Gastroenterology* **119**, 97–108.

McGee, D. J., Radcliff, F. J., Mendz, G. L., Ferrero, R. L. & Mobley, H. L. T. (1999). *Helicobacter pylori rocF* is required for arginase activity and acid protection *in vitro* but is not essential for colonization of mice or for urease activity. *J Bacteriol* **181**, 7314–7322.

Rain, J. -C., Selig, L., De Reuse, H., Battaglia, V., Reverdy, C., Simon, S., Lenzen, G., Petel, F., Wojcik, J., Schaechter, V., Chemama, Y., Labigne A. & Legrain, P. (2001). The protein–protein interaction map of *Helicobacter pylori*. *Nature* **409**, 211–215.

Vinion-Dubiel, A. D., McClain, M. S., Czajkowsky, D. M., Iwamoto, H., Ye, D., Cao, P., Schraw, W., Szabo, G., Blanke, S. R., Shao, Z. & Cover, T. L. (1999). A dominant negative mutant of *Helicobacter pylori* vacuolating toxin (VacA) inhibits VacA-induced cell vacuolation. *J Biol Chem* **274**, 37736–37742.

Warmy, A., Keates, A. C., Keates, S., Castagliuolo, I., Zacks, J. K., Aboudola, S., Qamar, A., Pothoulakis, C., LaMont, J. T. & Kelly, C. P. (2000). p38 MAP kinase activation by *Clostridium difficile* toxin A mediates monocyte necrosis, IL-8 production, and enteritis. *J Clin Invest* **105**, 1147–1156.

Wilson, M., DeRisi, J., Kristensen, H. -H., Imboden, P., Rane, S., Brown, P. O. & Schoolnik, G. K. (1999). Exploring drug-induced alterations in gene expression in *Mycobacterium tuberculosis* by microarray hybridization. *Proc Natl Acad Sci USA* **96**, 12833–12838.

3.14 | Internet links

1. http://www.tigr.org
   Website of the Institute for Genome Research.
2. http://www.ncbi.nlm.nih.gov
   Website of the National Centre for Biotechnology Information.
3. http://www.qmw.ac.uk/%7Erhbm001/ketbook/chapter.html
   Internet Resources for Research on Bacterial Pathogenesis.
   Provides a very useful list of links to relevant sites.
4. http://www.tigr.org/tigr-scripts/CMR2/CMRHomePage.spl
   Comprehensive Microbial Resource – information on all of the completed microbial genome sequences. Also has many useful links, including a list of DNA educational sites.
5. http://www.genomechip.com
   A website devoted to DNA microarray technology. Explains the basis of the technique and its applications. It has extensive lists of academic and industrial links related to this technology.

4

Communication
in infection

Chapter outline

Aims

The main aims of this chapter are to introduce the reader to:

- the concept of signalling in biology
- the role of intracellular signalling in normal eukaryotic cell control
- eukaryotic intracellular signalling mechanisms
- cell–cell signalling in eukaryotes
- the role of cytokines as cell–cell signalling controllers
- prokaryotic intracellular signalling
- prokaryotic cell–cell signalling
- prokaryote–eukaryote cell signalling
- the concept that bacteria manipulate eukaryotic signalling
  mechanisms to induce pathology

4.1 | Introduction

Communication – the transfer of information – is part of our everyday
lives and we are all participants in a rapidly expanding planet-wide
communication experiment. It is also obvious that animals communicate – by using smell, sound and visual cues. The dawn chorus is a good

example of animals signalling to each other. However, it may be less apparent that the maintenance of a multicellular organism is absolutely dependent on the correct transmission and reception of signals. To maintain the mammalian body in a healthy state requires an enormous number of signals. There are three major signalling systems in multicellular organisms and these are classified as neural, endocrine and cytokine. These individual systems also interact with each other in ways that are only now becoming apparent.

One way of looking at disease is as a disturbance in the signalling that generates the dynamically stable state known as homeostasis. This term defines the optimal conditions of the multitude of body systems required for normal functioning. Disturbances of endocrine hormone secretion, for example, are well-known causes of disease. Possibly the best-known example of endocrine disturbance is diabetes, caused by a failure to secrete, or respond to, the peptide hormone insulin. The consequences of this one hormone defect can be dramatic and include peripheral neuropathy and blindness.

A new way of looking at bacterial infections is that they are associated with disturbances in the signalling systems of the host, and examples of such signal disturbance are discussed in many of the chapters in this book. This may appear to be an unusual way of viewing infectious diseases. However, it is in line with recent advances in our understanding of bacterial communication. Since the discovery of bacteria in the 19th century, the prevailing viewpoint has been that they are single-celled individuals that have no need to communicate. However, research conducted largely over the past decade has shown that this idea is no longer tenable. Bacteria communicate with each other through the medium of molecularly diverse signals that include peptides, proteins and lactones. Compounds such as the acyl homoserine lactones involved in quorum sensing, which will be described in Section 4.3.2.1, can also interact with host cells. In turn, host cell secretions such as cytokines can interact with bacteria. Thus a new view of infection is of a conversation between pathogen and host cells using a wide variety of signals. It is possible that 's/he who shouts the loudest' has the advantage in this exchange. Cell–cell signalling involves a number of components: (i) intercellular signalling molecules, (ii) the receptors for these molecules, and (iii) a complex system of protein–protein/protein–small molecule interactions (termed 'intracellular signal transduction' or 'intracellular signalling') within cells to induce changes in the target cell. This is a fast-growing, complex and confusing area of science and many readers may shy away from it. However, this would be a mistake as much of our understanding of how bacteria cause disease is emanating from studies of cell signalling and the effects of bacteria on host cell signalling.

This chapter will provide the reader with a brief introduction to cell signalling. This will start with a consideration of signal reception at the eukaryotic cell surface and the intracellular signal transduction pathways involved in amplifying these external signals in such cells. This will be followed by a description of the major intercellular signals in mammals, the cytokines, and their role in infection. The chapter will then turn to a consideration of intracellular and cell–cell signalling in bacteria.

4.2 | Eukaryotic cell signalling

4.2.1 A brief overview of eukaryotic cell–cell signalling

Figure 4.1 Simple schematic diagram showing the two main types of signalling between cells – involving either a secreted molecule or direct contact. In both cases a receptor is required to recognise the signalling ligand. Binding to the receptor sets off a cascade of interactions involving second messengers that may result in (i) changes in enzyme activity, (ii) changes in cell shape, (iii) selective induction/inhibition of gene transcription or (iv) alteration in membrane permeability.

For one cell to communicate with another requires a number of components and the complete package of individual molecules and cellular systems involved is extremely large, and very complex, and is illustrated diagrammatically in Figure 4.1. The focus of this chapter will be on signalling via soluble factors released by one cell and recognised by another. Cell signalling initiated by adhesion to components of the extracellular matrix, or by bacteria mimicking components of the extracellular matrix, will also be discussed. It is possible for cells that are in contact (e.g. epithelial cells in an epithelial layer) to communicate directly, but such signalling will be described only briefly.

For cell–cell signalling to occur (Figure 4.1), cell A must release some signalling molecule that can diffuse from the cell and make contact with the target cell (cell B). In order to be a target, cell B must have some means of recognising the signal molecule. Such recognition molecules

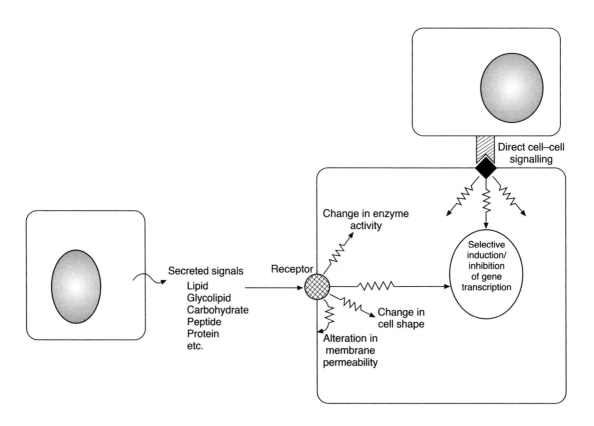

Table 4.1.	Classes of cell surface receptors
Class	Examples
Ion channel receptors	Receptor for acetylcholine, glutamate, GABA, etc.
G protein-coupled receptors	Receptors for chemokines, C5a, PAF, LTB$_4$, glucagon, odours, PTH, etc.
Tyrosine kinase-linked receptors	Many cytokines (e.g. IL-2R, IL-3R, IL-4R, IL-6R, IL-10R, IL-12R)
Receptors with intrinsic enzyme activity	Receptors for atrial natriuretic factor, growth factor, etc.

GABA, γ-aminobutyric acid; LTB, leukotriene B; PTH, parathyroid hormone; for other abbreviations, see list on p. xxii.

are termed receptors. The binding of the signal molecule to its receptor on the host cell sets off a train of intracellular events. These events are extremely complex and result in four possible and overlapping outcomes. These are (i) the stimulation/inhibition of cell metabolism via particular enzymes or proteins, (ii) the stimulation/inhibition of transcription of selected genes, (iii) the modulation of the cell's cytoskeleton, and (iv) alterations in cell membrane permeability. Binding of particular ligands to their cell surface receptors can stimulate all four of these forms of cell behaviour. The reception of signals at the cell surface, together with the mechanisms for passing information about this specific binding to the cytoplasm and nucleus, will be the subject of the next few sections of this chapter. This phenomenon is known as intracellular signalling or signal transduction. In this chapter, intracellular signalling in both eukaryotic and prokaryotic cells will be discussed. It was initially thought that the signal transduction processes in eukaryotes and prokaryotes were different. However, in recent years it has become apparent that both bacteria and eukaryotic cells share many similarities in their signal transduction machinery.

4.2.2 Cell receptors for signal transduction

In order to signal to a cell, the intercellular messenger, for example an endocrine hormone, cytokine or bacterial component, must bind to a receptor on the surface of the target cell. This is the first process involved in cell signalling. Thousands of cell surface receptors have now been identified and this vast number can be segregated into four classes (Table 4.1). These are the: ion channel-linked receptors, G protein-linked receptors, receptors with intrinsic enzymic activity and receptors linked to tyrosine kinases. There is a fifth class of receptor known as the intracellular or steroid/thyroid hormone receptor. This receptor differs from the others in that it is an intracellular protein.

The receptors to be described in this section allow the information contained in the binding of the intercellular message, or ligand, to its specific receptor to be transmitted into the cell and, depending on the receptor and ligand, ultimately to the nucleus.

4.2.2.1 Ion channel receptors

Ion channel receptors are found on all cells, but have been particularly well studied in nerves and muscle. Ion channel receptors, which respond to molecular signals, are more properly known as ligand-gated channels. The acetylcholine receptor is the prototypic ligand-gated ion channel and is composed of five polypeptides, which arrange themselves to form a pore in the membrane. In the absence of acetylcholine, these polypeptides adopt a conformation in which the pore is closed. However, when acetylcholine binds to its receptor there is a conformational change that opens the pore, allowing the passage of positively charged ions, including K^+ and Na^+.

It is interesting that many invertebrate and vertebrate species have evolved many toxins that target sodium, potassium or calcium channels and ligand-gated channels. However, of the many bacterial toxins, only one has so far been found to target ion channel receptors. This is tetrodotoxin, a potent sodium channel blocking toxin, which is produced by a *Vibrio* spp. (see Micro-aside, p. 486). Because of the limited involvement of bacteria in modulating ion channels, there will be no further discussion of these receptors.

4.2.2.2 G protein-coupled receptors (GPCR)

The ligand-gated ion channel receptors simply require binding of the ligand to the receptor to open the channel and cause the appropriate cellular effects. The other classes of cell surface receptors have the binding of the ligand linked to a complex series of intracellular events involving protein–protein interactions. This can lead to the formation of additional signals. These can be in the form of other proteins or they can be small molecules such as: cyclic adenosine monophosphate (c-AMP), lipids (e.g. ceramide) or ions such as Ca^{2+}. These are known as second messengers. This complicated series of events is known as intracellular signal transduction. While complex, much of modern microbiology takes place against a background of bacterial and eukaryotic cell signalling and it is important that these processes are understood in order to be able to grasp the basics of bacterial pathogenesis.

The most numerous receptors on eukaryotic cells are the G protein-coupled receptors (GPCRs), over 1000 receptors having been identified to date. These respond to many ligands including neurotransmitters, neuropeptides, proteins (such as the chemokines), complement components (C5a) and lipid mediators (e.g. leukotrienes). The receptor for bacterial formylated peptides is a GPCR (Table 4.2). They are also vital

| Table 4.2. | Ligands binding to G protein-coupled receptors |

- Neurotransmitters — acetylcholine, adrenaline, glutamate, GABA, dopamine, etc.
- Endocrine hormones — ACTH, glucagon, LH, PTH, TSH, vasopressin, somatostatin, etc.
- Inflammatory mediators — leukotrienes, chemokines, histamine, C5a, FMLP, prostanoids, etc.

GABA, γ-aminobutyric acid; ACTH, adrenocorticotropic hormone; LH, luteinising hormone; PTH, parathyroid hormone; TSH, thyroid-stimulating hormone; FMLP, N-formyl-methionyl-leucyl-phenylalanine.

for our ability to perceive light, odours and tastes. Bacterial toxins target these GPCRs and the, so-called, downstream signalling events induced by binding to these receptors (see Micro-aside, p. 169).

GPCRs have a characteristic structure in which the extracellular N-terminus of the protein is linked to the intracellular C-terminus via a section of the protein that traverses the plasma membrane seven times (Figure 4.2). This has resulted in the receptors being variously called seven membrane span(ning) receptors or serpentine receptors. Binding of an agonist to a GPCR results in a conformational change in the receptor. This conformation change affects the guanine nucleotide-binding protein (or G protein) with which the receptor is associated (Figure 4.3). G proteins are trimers of three different protein subunits (designated α, β and γ) and are often referred to as hetero-trimeric G proteins, which distinguishes them from other cellular guanine nucleotide binding proteins such as the small GTPases, of which Ras, Rho, Rac, etc., discussed in Section 4.2.5.2 are examples. There are many different G proteins and each is able to interact with different receptors, specifically targeting the receptor to distinct intracellular signalling events.

The activation of cells via a GPCR is shown schematically in Figure 4.3. In the non-activated cell, the α-, β- and γ-subunits of the heterotrimeric G proteins form a complex in which the α-subunit is bound to GDP.

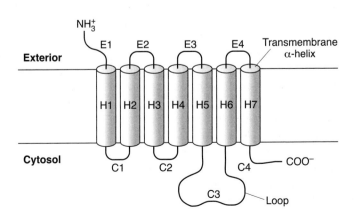

Figure 4.2 Structure of G protein-coupled receptor (GPCR). All receptors of this class are formed of a continuous peptide chain that crosses the plasma membrane seven times — each hydrophobic membrane domain being an α-helix. The peptide loop between helices 5 and 6, and in some cases 3 and 4, is important for interaction with the coupled G protein.

Figure 4.3 Cell signalling via G protein-coupled receptors. For an explanation, see the text.

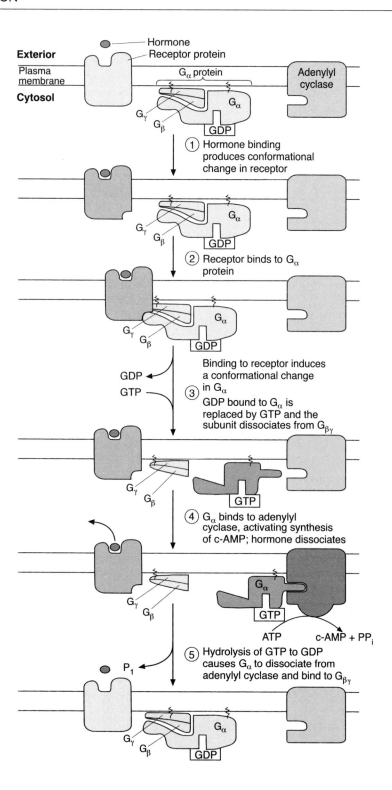

When bound to GDP, the G protein is in an inactive, or off, state. When the receptor is bound by its specific ligand (in this case a hormone), the conformational change in the cytosolic domain of the receptor enables it to interact with the appropriate inactive G protein and to stimulate the release of GDP from the α-subunit and its replacement with GTP. Binding GTP causes the trimeric complex to dissociate into the α-subunit and the βγ-dimer. Both of these dissociated moieties can then interact with cellular targets. The best-known example of a cellular target for G_α is the enzyme adenylyl cyclase, which, when activated, produces c-AMP. Other targets include phospholipase C, which generates inositol lipids involved in releasing calcium from intracellular stores.

Activation of a cell via a particular receptor, and consequent post-receptor signalling, needs to be carefully controlled. In the case of GPCRs, control is exerted at various levels. The α-subunit, with its associated GTP, can be thought of as a time switch as it has intrinsic GTPase activity and hydrolyses the GTP to GDP, thus switching off its ability to activate other proteins such as adenylyl cyclase. Once converted back to its GDP-bound form, the α-subunit can then reassociate with the βγ-dimer to form the original trimeric G protein. The GPCR can also be modulated in various ways. They can be internalised and degraded, or recycled. They can also undergo a process of desensitisation if the stimulus applied to the cell is prolonged.

Bacterial toxins and GPCR

Many bacterial toxins interfere with cell signalling and have been used as probes for determining intracellular signalling mechanisms. This is particularly true with the heterotrimeric G proteins, and cholera toxin and pertussis toxin have been particularly useful in defining aspects of their behaviour. Cholera toxin, the product of *Vibrio cholerae*, is responsible for the massive production of watery diarrhoea that can lead to death by dehydration of patients who have cholera. Indeed, modern treatment is largely directed towards replacing the water and salts lost. The reason for this enormous outflow of fluid from the intestine is because the A subunit (for more details, see Section 9.5.1.1) of the toxin enters intestinal epithelial cells and ADP-ribosylates a stimulatory G_α subunit of a heterotrimeric G protein. This chemical modification results in an active G_α subunit that is unable to hydrolyse the GTP and so continues to activate adenylyl cyclase, resulting in a 100-fold increase in the intracellular levels of c-AMP. In intestinal epithelial cells, the consequence of this is changes in membrane proteins that allow massive release of fluid and the characteristic tissue pathology.

4.2.2.3 The GTPase superfamily

In addition to the trimeric G proteins, eukaryotic cells (and, as has recently been reported, prokaryotes) also contain so-called monomeric G proteins (Table 4.3). These first came to the attention of cell biologists interested in cancer as gene products of cancer-producing retroviruses.

Table 4.3.	Monomeric G protein (small GTPase) families

- Ras
- Rab (at least 30 family members are known)
- Rho (at least 10 family members)
- Arf
- Ran

The prototypic protein is Ras – the product of the *ras* oncogene. As all of these G proteins possess GTPase activity, they have been grouped together as the GTPase superfamily. These proteins, of which there may be as many as 100 grouped into five major families (Table 4.3), are involved in many aspects of cell behaviour, including proliferation, differentiation, vesicular traffic and control of cell shape and movement. Many bacterial toxins, such as the exoenzyme 3 from *Clostridium botulinum*, *Pseudomonas aeruginosa* exotoxin A, cytotoxic necrotising factor (CNF) 1 from *Escherichia coli* and *Cl. difficile* toxin B, inhibit or activate the monomeric G proteins. The actions of these toxins will be described in Chapter 9.

We will take the monomeric G protein Ras as a major example of a protein that couples events at the cell membrane (in this case receptor-linked protein tyrosine kinases) to intracellular signalling events. This protein is found bonded to the inner face of the plasma membrane through a lipid attachment. The protein is a GTPase that cycles between an active GTP-bound state and an inactive GDP-bound state (Figure 4.4). Although these proteins have intrinsic GTPase activity, in order to complete this cycle two other proteins are required. The initial release of GDP is promoted by a protein termed a guanine nucleotide exchange factor (GEF) and the hydrolysis of the GTP is facilitated by a second protein known as a GTPase-activating protein (GAP). As will be described later, the active Ras couples ligand binding to receptors to particular signal transduction pathways, resulting in activation of specific gene transcription.

4.2.2.4 Receptors with inherent enzymic activity

In contrast to the GPCRs, with their seven membrane-span structure, receptors with intrinsic enzyme activity are more heterogeneous, although they share the common features of having an extracellular receptor domain and an intracellular enzymic domain linked by a single membrane-spanning domain. The catalytic domain of the receptor can have kinase, phosphatase or guanylyl cyclase activity. This chapter will concentrate on the receptor kinases. The active site of the kinase can have specificity for the amino acids serine/threonine (called receptor serine/threonine kinases) or for tyrosine (called receptor tyrosine kinases (RTKs)). The first example of a receptor serine/threonine kinase was discovered in the worm *Caenorhabditis elegans* in 1990 and since then it has been discovered that a major group of

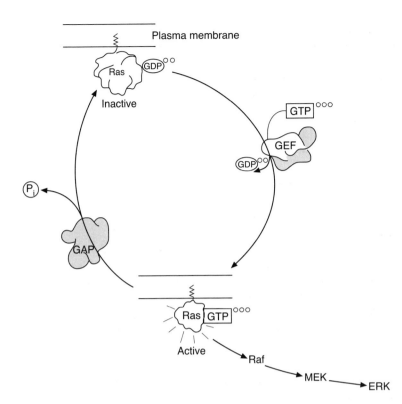

Figure 4.4 Simple schematic diagram showing the cycling of Ras between the active and inactive state. Ras bound to GDP is in an inactive state. It is converted to the active (GTP-bound) state by exchange of GTP for GDP catalysed by the guanine nucleotide exchange factors (GEFs). In the active state, Ras can interact with Raf, a protein serine/threonine kinase, and thus trigger the MAP kinase cascade. Removal of GTP from Ras inactivates the protein. This hydrolysis is catalysed by GTPase-activating proteins (GAPs).

cytokines, the transforming growth factor (TGF) β superfamily, comprising more than 30 proteins, and involved in many aspects of cell and organ physiology, signal through this type of receptor.

It is estimated from the human genome sequencing project that there are around 1100 kinases, of which 150 are protein tyrosine kinases. Currently more than 50 RTKs have been identified in vertebrates (Table 4.4). Binding of ligands to these RTKs results in cellular responses such as proliferation, differentiation, cell motility and change of cell shape as well as selective gene transcription. The binding of ligand to these receptors generally results in the dimerisation of the receptor and this, as will be explained, enables the signal to be transduced intracellularly.

Table 4.4. Examples of receptor tyrosine kinases

- Epidermal growth factor receptor (EGFR)
- Insulin receptor
- Platelet-derived growth factor receptor (PDGFR)
- Fibroblast growth factor receptor (FGFR)
- Nerve growth factor receptor (NGFR)

Table 4.5.	Receptors associated with cytoplasmic protein tyrosine kinases

- Interleukin 2 receptor (IL-2R)
- Interleukin 3 receptor (IL-3R)
- Interleukin 4 receptor (IL-4R)
- Interleukin 5 receptor (IL-5R)
- Interleukin 10 receptor (IL-10R)
- Interferon receptors

4.2.2.5 Receptors linked to a separate cytoplasmic enzyme

This last group of plasma membrane receptors have cell surface, membrane-spanning, and cytoplasmic domains similar to the RTKs. However, the intracellular domains have no inherent enzymic activity but interact with a range of enzymes, and receptor occupation triggers the activation of these enzymes. The majority of the receptors in this 'family' interact with, and activate, protein tyrosine kinases. Many of the receptors for the cytokines that will be described later in this chapter are associated with intracellular protein tyrosine kinases (Table 4.5). Other enzymes with which plasma membrane receptors are associated include protein tyrosine phosphatases, protein serine/ threonine kinases and guanylyl cyclases.

4.2.3 Intracellular signal transduction

Intracellular signal transduction is the name for the process by which the binding of a ligand to its cognate receptor on the cell surface (this is the signal) reaches the cell interior and/or nucleus. This process is exceedingly complex and involves a large number of components including proteins, peptides, lipids, glycolipids, peptidolipids and various ions. This section will try to piece together the complex network of interactions by which signals arriving at the cell surface can activate cell metabolism, alter cell shape and/or reach into the genome and selectively alter patterns of gene transcription.

The starting point is to ask why do we need such a complex process? An agonist, say the cytokine interleukin 1 (IL-1), has to bind to a cell surface receptor to modify cell function. Why has evolution not allowed each cell-activating ligand (e.g. hormone or cytokine) to directly stimulate cellular systems and/or gene transcription? The most likely explanation for the evolved form of cell signalling via receptors and intracellular signal transduction is that it is the most efficient system. If each agonist were able directly to activate cellular enzymes or gene transcription, etc., they would have to be made in very high concentrations in order to bind to cytoplasmic enzymes, or to the appropriate promoter elements on each gene. In addition, with gene activation there would have to be specific promoter sites in each gene for each agonist. This would require that cells produced very large

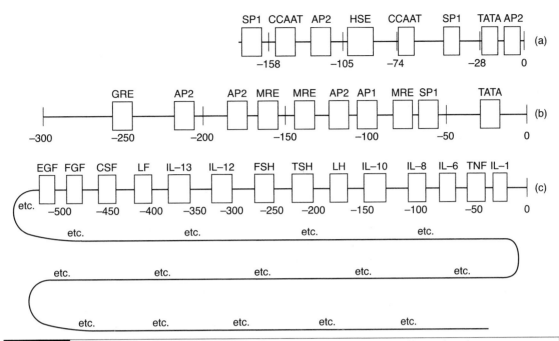

Figure 4.5 What would happen if there was no signal transduction and each agonist was able to induce the transcription of only one particular gene? (a) The transcriptional control sites for the molecular chaperone hsp 70. (b) The transcriptional control elements for metallothionein. The TATA, SP1 and CCAAT boxes bind factors involved in constitutive transcription. The other boxes bind transcription factors, induced by intracellular transduction, to give specific signalling. Note that both genes have at least one similar control element (AP2). These transcriptional control elements utilise a reasonably short span of DNA sequence. Contrast this with (c) the situation in which each agonist able to stimulate the transcription of a gene had to have its own binding site in this region upstream from the transcriptional start site. Very large stretches of DNA would be needed, making the genome very much larger than it is and adding extra complexity to genomic control. In each gene, 0 is the transcriptional start site.

amounts of each signal (remember cells may produce thousands of extracellular signals) and it would require that the non-coding regions of genes be extremely large to accommodate sites for the binding of each agonist (Figure 4.5).

The secret of cell–cell signalling is amplification. This is what allows radio and television companies to beam entertainment into our houses. Our televisions and radios amplify the incoming (weak) signal and turn it into a signal that we perceive as sound and pictures. Likewise, in cell signalling the high affinity receptors on the surfaces of cells act to bind, and thus recognise, the extremely small amounts of individual signals in the external *milieu*. The secret then is to amplify this tiny signal. For example, it is estimated that cells are able to respond to only a few dozen molecules of cytokines such as IL-1. The binding of IL-1, or other agonist, to its receptor then sets in train a series of intracellular events that amplify this signal (Figure 4.6). The amplification is brought about by the enzymic activity of the intracellular signalling systems. One of these enzymes, adenylyl cyclase, has already been mentioned. This enzyme is activated by the action of

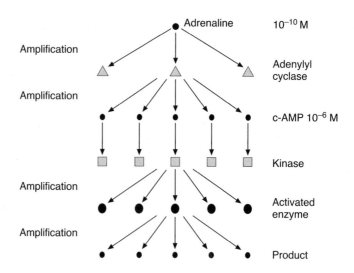

Figure 4.6 Amplification systems in intracellular signal transduction. The binding of very low concentrations of adrenaline to its cell surface receptor activates adenylyl cyclase and induces the first amplification step in the process of signal transduction. This enzyme produces large amounts of c-AMP (10^{-6} M) and each c-AMP can bind to, and activate, a range of kinases, which, in turn, can amplify the response enzymically by producing large amounts of activated enzymes, and so on.

GPCRs and results in the production of c-AMP, a class of molecule known as a second messenger. The binding of one molecule of an agonist to a GPCR could induce the production of thousands of molecules of c-AMP and this constitutes the first level of amplification. In turn, c-AMP interacts with an enzyme, a kinase, in this case a kinase known as c-AMP-dependent kinase or protein kinase A (PKA), and causes this enzyme to become activated. The PKA, in turn, phosphorylates other target proteins to activate them and this is a second level of signal amplification.

4.2.3.1 The basic building blocks of intracellular signal transduction

As the reader will rapidly realise, intracellular signal transduction systems look incredibly complex. A similar feeling of bewilderment is engendered in most people when they look at an electronic circuit. However, it is clear that in such circuit diagrams the complexity is built up of the repetition of a small number of common elements such as resistors, capacitors, diodes and transistors and the associated conducting elements. The same is true of intracellular signal transduction. The addition of phosphate groups onto specific sites in proteins, and their specific removal, is the key element in intracellular signalling. The phosphorylation of proteins is catalysed by enzymes called kinases and, as has been described, the residues in proteins that are subject to phosphorylation in eukaryotic cells are tyrosine (catalysed by protein tyrosine kinases (PTKs)) or serine/threonine (catalysed by protein serine/threonine kinases). In prokaryotic cells, adaptive responses to the environment are largely controlled by so-called two-component regulatory systems, also called histidyl-aspartyl phosphorelay systems (described in Section 4.3.1.2). These systems rely on the activity of a histidine kinase.

Addition of a phosphate group or groups to a protein may activate or inhibit its biological activity. Phosphorylation can alter the activity of enzymes or can confer on proteins the ability to recognise, and bind to, other proteins. The **src homology (SH2) motif** is the best-described example of a protein domain able to recognise phosphorylated residues in proteins. As will be described, this latter activity is very important in building up signalling cascades. As in all control systems, there has to be a mechanism for undoing what has been done. In intracellular signal transduction, the removal of phosphate groups from proteins is catalysed by protein phosphatases. Eukaryotic cells possess both protein tyrosine phosphatases and protein serine/threonine phosphatases (Figure 4.7). Less attention has focused on protein phosphatases, but it is estimated that they must equal in number the protein kinases of cells.

As has been described, phosphorylation/dephosphorylation cycles control the enzymic activity of individual proteins. One of the first examples of the role of kinases in controlling particular aspects of cell function was the regulation of glycogen metabolism in muscle by the adrenal hormone adrenaline (Figure 4.8). This hormone is part of our 'fight-or-flight' response to stress. Binding of adrenaline to its GPCR leads to c-AMP synthesis via the action of the appropriate G protein. The c-AMP in turn results in the activation of PKA. The PKA then phosphorylates two proteins: glycogen synthase (which uses glucose to produce glycogen), and phosphorylase kinase, a kinase distinct from PKA, which phosphorylates and activates glycogen phosphorylase. The phosphorylation of glycogen synthase inhibits its activity and so glycogen synthesis is stopped. Active glycogen phosphorylase

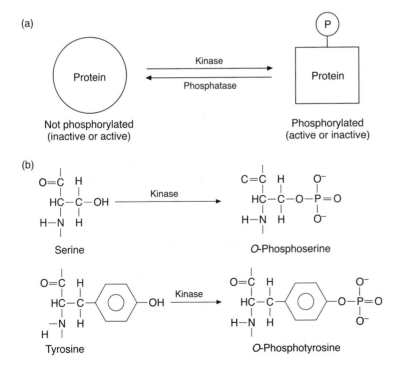

Figure 4.7 Competing activities of kinases and phosphatases. (a) The effect of adding or removing a phosphate group from a protein. (b) The structures of phosphoserine and phosphotyrosine.

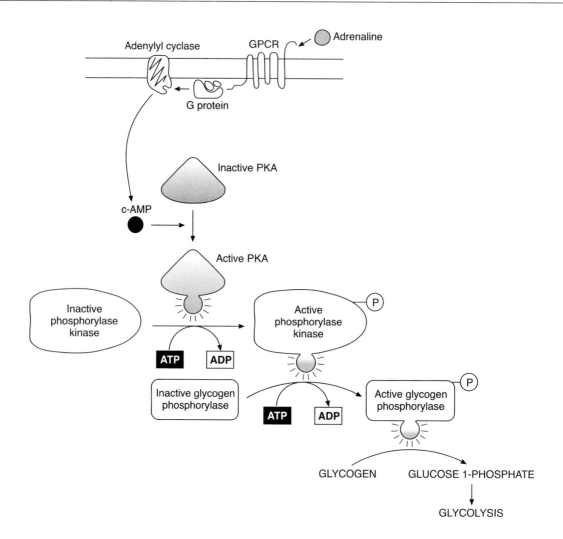

Figure 4.8 Regulation of glycogen metabolism. Binding of adrenaline at the cell surface induces the synthesis of c-AMP, which activates protein kinase A (PKA). In turn, this enzyme phosphorylates, and thus activates, phosphorylase kinase. The now-activated phosphorylase kinase can phosphorylate the inactive glycogen phosphorylase which catalyses the breakdown of glycogen and results in the activation of glycolysis.

hydrolyses glycogen to form glucose 1-phosphate, which then enters the glycolytic pathway to create energy. This exemplifies two characteristics of signal transduction. The first is that kinases, by phosphorylating, can activate (phosphorylase kinase/glycogen phosphorylase) or inhibit (glycogen synthase) other proteins. The second is that much of the intracellular signal transduction process depends upon cascades of kinases – with one kinase acting on a second kinase to activate it, and so on.

In addition to activating enzymes, the induction of c-AMP by hormones such as adrenaline, glucagon, vasopressin and thyroid-stimulating hormone (TSH) can also activate the transcription of specific genes. These genes are controlled by regulatory elements termed the c-AMP response elements (CRE – sequence TGACGTCA), which bind transcriptional factors called **CRE-b**inding (CREB) protein. The CREB is normally bound to the CRE element but does not induce gene transcription. However, when it is phosphorylated by PKA, it stimulates the transcription of these specific genes.

Domain	Interaction	Example of binding protein
SH2 (Src homology region 2)	Phosphotyrosine	Grb2
PTB (phosphotyrosine binding)	Phosphotyrosine	Shc
14-3-3[a]	Phosphoserine	Bcl-2 homologue
SH3 (Src homology region 3)	Proline-rich sequences	SOS (son of sevenless)
PH (Pleckstrin homology)	Lipid and protein	SOS
WW (tryptophan-tryptophan)	Proline-rich sequences	Dystrophin
WD(40) (tryptophan-aspartic acid)	Binds to PH domains	Ankyrin
SPRY (splA and RyR genes)	Proteins	SplA
TRAF (TNF receptor-associated factor)	TNF receptor (TNFR)	Cytoplasmic tail of TNFR
DD (death domain)	TNF receptor	Cytoplasmic tail of TNFR

Table 4.6. | Some protein domains involved in cell signalling

[a] Name is derived from original purification scheme.

4.2.3.1.1 Domain structures and signal transduction

Intracellular signal transduction relies on the generation of specific molecular complexes to carry the signal to the correct site in the cell. The ability to produce these complexes requires the presence, or the generation, of specific binding sites. The role of phosphorylation to introduce a domain recognised by other proteins has been described and the best known of these domains able to recognise phosphoproteins is the SH2 domain. This is a stretch of around 100 amino acid residues in proteins that recognises specific short sequences of peptides on other proteins containing phosphotyrosine. It allows two proteins that would not normally interact to bind to each other for the purpose of transmitting the signal. Other phosphotyrosine-binding domains, such as the PTB domain, exist. Other protein domains recognising phosphoserine have been identified. The 14.3.3 family is a good example of phosphoserine-binding proteins involved in interacting with a range of intracellular proteins and acting in the control of cell signalling. In addition to these, there is a range of other protein domains that recognise specific (non-phosphorylated) sequences in proteins (Table 4.6). The signalling from the IL-1 and tumor necrosis factor (TNF) receptor families, for example, involves a number of such proteins, which have been termed 'adapter' or 'docking' proteins. The literature on the signalling pathways induced by these two receptors is enlivened with tales of 'death domains' and abbreviations such as RIP, TRADD and TRAF, which represent some of these docking proteins. These receptors and their signalling pathways will be described in Section 4.2.6.6.3.

Not all of these domains recognise proteins. For example, the **pleck-strin homology domain** (PHD) recognises specific phospholipids. Examples of these various domains and their interactions will be discussed throughout this chapter.

4.2.3.1.2 Second messengers

Cyclic nucleotides: The first second messenger to be discovered was c-AMP, by Earl Sutherland in the late 1950s. The interaction of this nucleotide with PKA has been described. Large amounts of c-AMP are made by appropriately activated cells and this has to be removed to switch off the signal. The way this is done is by the action of an enzyme, **phosphodiesterase** (PDE), which forms AMP from c-AMP. A number of phosphodiesterase isoforms exist and different cells express different patterns of these isoforms. The development of inhibitors of PDE has been underway for a number of years and the anti-impotence drug sildenafil (Viagra), for example, is an inhibitor of PDE isoenzyme IV. PDE inhibitors may be effective anti-inflammatory drugs with the capacity to block the actions of pro-inflammatory cytokines such as those induced by bacteria.

In the 40+ years since this initial discovery, a number of other low molecular mass second messengers have been discovered. These include another cyclic nucleotide – **cyclic guanosine monophosphate** (c-GMP) – produced by the action of guanylyl cyclases. Fewer activators of guanylyl cyclases are known, but both peptides and the gas nitric oxide (NO) can activate these enzymes. c-GMP is inactivated by a c-GMP phosphodiesterase. Much attention is currently being focused on NO and the inhibition/enhancement of its synthesis. As will be described in Section 6.6.2.3.1, NO is important in phagocyte killing of bacteria and the 1999 Nobel Prize for Medicine and Physiology went to the discoverers of NO, Louis Ignarro, Robert Furchgott and Ferid Murad.

Lipids: Cell membranes are composed of various lipids, mainly phospholipids, which form a lipid bilayer to which proteins are attached in various ways. The phospholipids of cell membranes can form a large collection of lipidic signals including those utilised in signal transduction mechanisms. The formation of such signals is due to the actions of enzymes known as phospholipases, which hydrolyse phospholipids at various sites thus producing a variety of mediators (Table 4.7). Phospholipase A_2 (PLA$_2$) catalyses the formation of arachidonic acid and lysophospholipids. These products have been implicated in signal transduction. They also serve as the precursors of major cell–cell signalling molecules including the prostaglandins, leukotrienes and platelet-activating factor (PAF), which act at GPCRs. There is increasing evidence for a role of such lipid signals in bacterial pathogenesis. PLA$_2$ enzymes are made by both eukaryotes and prokaryotes, and, in the latter case, can function as virulence factors. **Phospholipase C** (PLC) is involved in the production of two key groups of intracellular signalling molecules and these will be discussed in the next paragraph.

Table 4.7.	Cellular phospholipases and signal transduction	
Phospholipase	Products	Actions
PLA$_2$	Arachidonic acid metabolites	External control of inflammation acting at GPCRs
PLC	IP$_3$ and DAG	Multiple effects on intracellular signalling
PLD	Phosphatidic acid and DAG	Effects on intracellular signalling

For abbreviations, see the text.

Phospholipase D (PLD) hydrolyses phospholipids at their terminal phosphodiester bond to produce phosphatidic acid and the release of the polar head group. Phosphatidic acid has been shown to be involved in intracellular signalling and can also be transformed to form an important lipid mediator, **diacylglycerol** (DAG).

The best studied, and arguably the most important form of cellular phospholipase, is PLC, and a large number of isoforms of this enzyme are found in cells. PLC hydrolyses one of the minor phospholipids found on the inner leaflet of the plasma membranes of eukaryotic cells – phosphatidylinositol 4,5-bisphosphate (PIP$_2$). Hydrolysis of this phospholipid produces two signalling molecules: inositol 1,4,5-trisphosphate (IP$_3$) and 1,2-diacylglycerol (DAG) (Figure 4.9). PLC can be activated either by the binding of a ligand to a GPCR via an activated G protein or by interaction, through its SH2 domain, with an activated receptor protein tyrosine kinase (Figure 4.10). IP$_3$ was the first inositol phosphate to be implicated in signal transduction. However, it is now established that eukaryotic cells can produce a wide variety of phosphorylated forms of phosphoinositide by the actions of kinases such as **phosphoinositide 3-kinase** (PI3K) mentioned earlier. Indeed, knockout of PI3K has been shown to result in a greater susceptibility to *Staphylococcus aureus* peritonitis. The complications of IP$_3$ metabolism will not be dealt with in this chapter, but various reviews on this subject are presented in Section 4.9.

What roles do IP$_3$ and DAG play in intracellular signal transduction? If we look at DAG first, it has to be realised that we are dealing here not with one molecule but with a collection of structures dependent on the acyl groups present on the phospholipids hydrolysed by the particular phospholipase. It should also be noted that DAG remains attached to the plasma membrane. The major action of DAG appears to be the activation of a family of serine/threonine kinases known as **protein kinase(s) C** (PKC). This large family of enzymes link phosphoinositide signalling to other major pathways of intracellular signalling such as the mitogen-activated protein (MAP) kinase pathway and the activation,

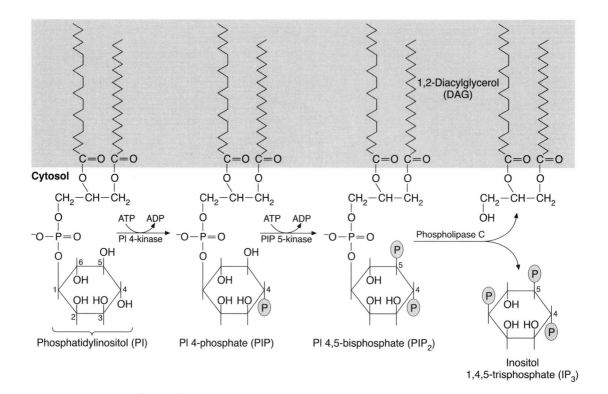

Figure 4.9 In this schematic diagram, the membrane-associated phospholipid, phosphatidylinositol (PI) undergoes phosphorylation on two sites to produce the phosphoinositides PIP and PIP$_2$. The hydrolysis of PIP$_2$ by phospholipase C generates two key second messengers, IP$_3$ and DAG.

in the cytoplasm, of the transcription factor, nuclear factor (NF)-κB. Both of these systems, which are yet to be described in detail, play key roles in the response of the host to bacteria, and the cytokines induced by bacterial infection. Phorbol esters, which are much used in studies of cellular activation, are DAG mimics.

It had been known for many years that certain cellular processes (e.g. fertilisation of amphibian eggs) are associated with rises in intracellular calcium. It is now known that it is lipids such as IP$_3$ that are responsible for these transient rises in intracellular calcium levels. The intracellular calcium concentrations in eukaryotic cells are normally maintained at approximately 10^{-7} M. These levels are maintained by calcium pumps, which can remove calcium from the cell, and pumps that sequester calcium in stores in the endoplasmic reticulum (ER). IP$_3$ causes the release of calcium from these ER stores by binding to receptors that act as ligand-gated calcium channels. The calcium levels rise 10- to 100-fold and can activate a range of target proteins (Table 4.8). An important mediator of the rise in intracellular Ca^{2+} concentrations is the Ca^{2+}-binding protein calmodulin. This dumb-bell-shaped protein has four Ca^{2+}-binding sites and, when they are filled, the protein can interact with, and bind to, a variety of target proteins such as Ca^{2+}/calmodulin-dependent protein kinases.

A second, more recently discovered, lipid-mediator pathway involves sphingolipids. In this pathway the enzyme sphingomyelinase hydrolyses sphingomyelin, releasing the activator ceramide (Figure 4.11).

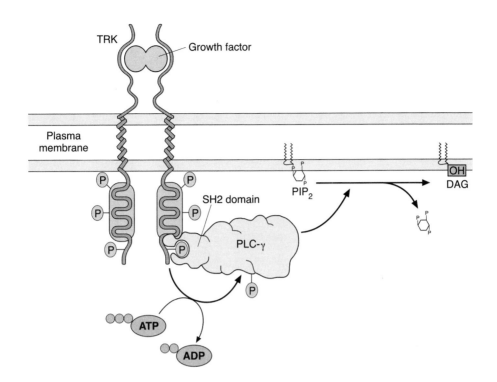

Figure 4.10 The activation of phospholipase C (PLC). In this example, the binding of a growth factor to a tyrosine receptor kinase (TRK) enables the PLC to bind via its SH2 domain to the intracellular portion of the TRK. This enables the PLC to be activated by phosphorylation and to cause cleavage of the membrane-associated PI, as described in Figure 4.9.

Ceramide is released in response to a wide variety of signals including those caused by infection (e.g. bacterial sphingomyelinases) and can interact with a variety of intracellular signalling systems including ceramide-activated protein kinase (CAPK) and ceramide-activated protein phosphatase (CAPP) and a range of other putative targets such as PLC, MAP kinase and Raf.

Nitric oxide: One of the most interesting discoveries of recent years is the role that various gases play in cell signalling. The first example of

Table 4.8. Targets of IP$_3$-gated calcium transients in cells

- Calmodulin
- Troponin
- Ca^{2+}-dependent protein kinase
- Protein kinase C
- Phosphorylase kinase
- Calcium-dependent protein phosphatases
- Phospholipase C
- Nitric oxide synthase
- Adenylyl cyclase
- Pyruvate kinase
- Ca^{2+}/Mg^{2+}-dependent endonuclease
- Ca^{2+}-ATPase

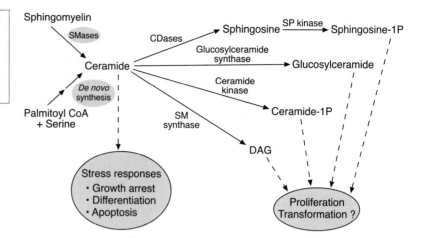

Figure 4.11 Some of the biological actions of ceramide and products derived from ceramide. SM, sphingomyelin; SP, sphingosine; SMases, sphingomyelinases; CDases, ceramidases.

this was the role of ethylene in the ripening of fruit. In 1987 Salvador Moncada and Richard Palmer identified the gas nitric oxide (NO) as a product of vascular endothelial cells that was able to control blood pressure. This gas turns out to be a cell–cell signal, a bactericidal agent and an intracellular signalling molecule, and each of these actions will be covered at some point in this book. In terms of intracellular signalling, NO diffuses into cells from nearby cells and activates the enzyme guanylyl cyclase inducing the formation of c-GMP.

Other gases such as carbon monoxide and the simple chemical hydrogen peroxide have also been implicated as second messengers.

4.2.4 Signal transduction and selective gene transcription

In simplistic terms, intracellular signal transduction subserves three major functions (see Figure 4.1). The first has been described very briefly – namely – the control of cell metabolism – and the example given was the control of glycogen metabolism. The second function of cell signal transduction, which will be discussed in this section, is the control of gene transcription. The third function is the control of the cell cytoskeleton and cell shape. Of course, all three signalling functions are operating in cells all of the time and particular patterns of signalling will occur in cells undergoing major changes in response to the environment, such as cell division, differentiation or apoptosis.

In Section 4.2.1, the problem of signalling into a cell was discussed and the solution – receptors and second messengers – described. A second problem arises when the signalling is designed selectively to activate gene transcription. The process of gene transcription requires the generation of specific transcription factors that must enter the nucleus and interact with the correct DNA sequence upstream from the initiation site of the particular gene to be activated. The mechanism of gene transcription in prokaryotes has been described in Section 2.6.12.

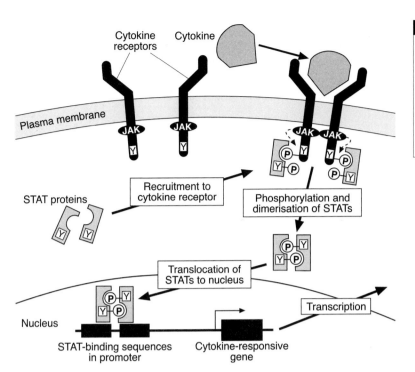

Figure 4.12 JAK/STAT signalling in response to cytokines. The mechanism is detailed in the text. (Reproduced from Abbas, A. K., Lichtman, A. H. & Pober, J. S. (2000). *Cellular and Molecular Immunology*, 4th edition, with copyright permission from W.B. Saunders Company.)

Two examples of signal transduction that result in specific gene activation will be given. Both mechanisms are involved in the host response to infection and may be targeted by bacteria as part of their immune-evasion strategy (see Chapter 10).

4.2.4.1 The JAK/STAT pathway of gene activation

There are five major classes of cytokine receptors (described later in this chapter). The majority of cytokines signal via the class I (haematopoietin) and class II (interferon) receptors, which have common features. The first is that the receptors generally consist of two separate protein subunits and binding of the appropriate cytokine leads to receptor dimerisation. The second is that the cytoplasmic domains of the receptor subunits are associated with distinct protein tyrosine kinases called Janus kinases or JAKs (Figure 4.12). They are so called because, like the Roman god Janus, who had two faces, these proteins have two active sites: one which binds the kinase to the receptor and the second, a catalytic site, which, when activated, has protein tyrosine kinase activity. Thus the binding of a cytokine to a class I/II receptor activates the JAKs to become protein tyrosine kinases and they reciprocally phosphorylate each other. The activated JAKs then phosphorylate various tyrosines in the receptor complex, thereby creating docking sites for inactive cytoplasmic transcription factors termed STATs (signal transducers and activators of transcription). Docking occurs via the SH2 domain on the STATs. The docked STAT is then phosphorylated by the JAK on a key tyrosine residue. Once phosphorylated, the

Table 4.9.	Association of cytokine receptors with JAK/STAT proteins	
Receptor for	JAK	STAT
Interleukin 2 (IL-2)	JAK1 /JAK3	STAT5
Interleukin 3 (IL-3)	JAK2	STAT5
Interleukin 4 (IL-4)	JAK1 /JAK3	STAT6
Interleukin 6 (IL-6)	JAK1	STAT3
Interleukin 10 (IL-10)	JAK1 /Tyk-2	STAT3
Interleukin 12 (IL-12)	JAK2 /Tyk-2	STAT4
Erythropoietin (EPO)	JAK2	STAT5
Interferon γ (IFNγ)	JAK1 /JAK2	STAT1
Interferon α, β (IFNα, β)	JAK1 /Tyk2	STAT2

For abbreviations, see the text.

STATs disassociate from the receptor and associate with each other forming a dimer that undergoes additional phosphorylation on a serine residue to form the active complex, which can then enter the nucleus and cause specific gene transcription. The association of particular cytokine receptors with JAK and STAT gene products is shown in Table 4.9.

The JAK/STAT pathway shows the importance of protein domains, and of specific phosphorylation for the catalysis of protein–protein interactions to generate active complexes. In this example, the protein–protein complexes consist of the receptor–receptor dimer, the JAK–receptor dimer, the STAT–receptor complex and finally the STAT–STAT dimer. In the next example, the MAP kinase pathway, the interactions are even more complex.

4.2.4.2 Gene transcription via the MAP kinase pathways

Eukaryotic cells contain a number of different MAP kinases such as c-Jun N-terminal kinases (JNKs) and p38 kinase (Table 4.10). These proteins are serine/threonine kinases that are activated in the cell cytoplasm via the MAP kinase pathways and are then translocated into the nucleus, where they selectively modulate gene transcription by phosphorylating particular transcription factors. The literature on MAP kinases is confusing because of the use of alternative nomenclature to describe the same protein or family of proteins. For example the MAP kinases are often referred to as ERKs (extracellular signal-regulated kinases – see Table 4.10).

The MAP kinase pathways are literally being discovered as we write and major advances in our understanding of them are being made continually. This section will, therefore, describe only the best-characterised of these signal transduction cascades, which links RTKs

| Table 4.10. | Mammalian MAP kinases |

- ERK1 extracellular signal-regulated kinase (p44 MAP kinase)
- ERK2 (p42 MAP kinase)
- P38 MAP kinase (4 isoforms designated α, β, γ (alternative names ERK6 or SAPK3) and δ
- c-Jun N-terminal kinase/stress-activated protein kinases (JNK/SAPK) – 3 genes and $>$ 10 splice variants
- ERK3 (3 forms encoded by 2 genes)
- ERK4
- ERK5
- ERK7

to MAP kinases (Figure 4.13). This pathway exemplifies many of the foundations of intracellular signalling that have been described earlier in this chapter. It also provides some of the complexities and curious nomenclature that makes understanding cell signalling difficult. The binding of the cognate ligand to the appropriate RTK results in the activation of the small G protein Ras. These small GTPase proteins have been discussed and the cycling between the inactive GDP-bound and active GTP-bound forms described. In this case, the GEF involved in activating the GDP-Ras is known as son of sevenless (SOS) (Figure 4.14). The binding of SOS to the activated (tyrosine phosphorylated) RTK is via a second protein, Grb2, that has an SH2 domain allowing it to dock with the receptor. Exchange of GDP with GTP activates Ras and this active

Figure 4.13 Simplified diagram showing the activation of ERK/MAP kinase via an RTK. Binding of the agonist (in this case, a growth factor) to the receptor activates Ras, which then interacts with Raf, a protein kinase. In turn, Raf activates MEK (MAP kinase/ERK kinase) a threonine/tyrosine kinase that activates ERK. In turn, ERK phosphorylates a variety of cytoplasmic and nuclear proteins including transcription factors.

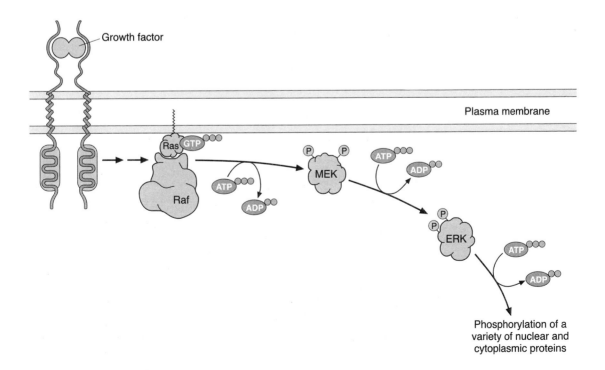

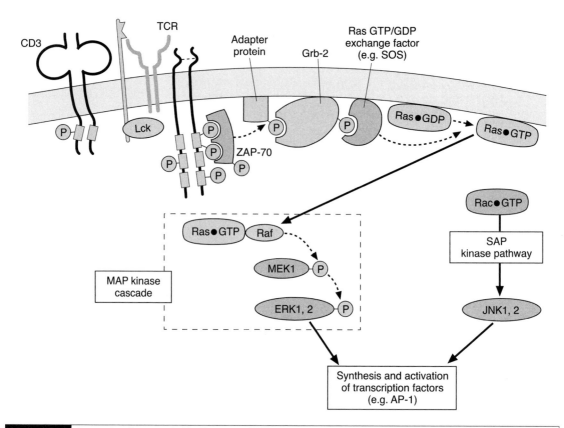

Figure 4.14 Interactions between the T cell receptor (TCR) and MAP kinases in a T lymphocyte. Binding of a peptide–MHC complex to the TCR induces phosphorylation of the ξ-chain resulting in binding of the kinase ZAP-70, which phosphorylates an adapter protein, thus enabling the binding of another adapter protein, Grb-2. This recruits to the membrane a Ras GTP/GDP exchange factor such as son of sevenless (SOS). The SOS then catalyses the exchange of GDP for GTP, thus activating Ras. We then have the sequence Raf/MEK1/ERK1,2 activation as described in Figure 4.13. A second parallel MAP kinase pathway is also activated, involving the small GTPase Rac, which activates JNK1,2. (Reproduced from Abbas, A. K., Lichtman, A.H. & Pober, J.S. (2000). *Cellular and Molecular Immunology*, 4th edition, with copyright permission from W.B. Saunders Company.)

complex then binds to a serine/threonine kinase called Raf. Raf, in its inactive state, is complexed with a phosphoserine-binding protein called 14-3-3. Binding of Ras dissociates this 14-3-3 protein and results in the activation of Raf. In turn, Raf binds to and phosphorylates MEK (MAP kinase/ERK kinase). This is an interesting kinase as it has dual specificity for tyrosine and threonine residues. MEK then phosphorylates MAP kinase, and MAP kinase can then phosphorylate the final group of proteins that mediate the cellular responses to the binding of the hormone. These proteins include AP-1, c-Myc and Elk that act as transcription factors. A simplified diagram showing the response of binding of antigen to the T cell receptor, which encompasses much of what has been said, is shown in Figure 4.14. For details of T cell activation refer to Sections 6.7.2 and 6.7.3. Our current understanding of the control of the various MAP kinases is shown schematically in Figure 4.15, which shows the relationship of these kinases to upstream

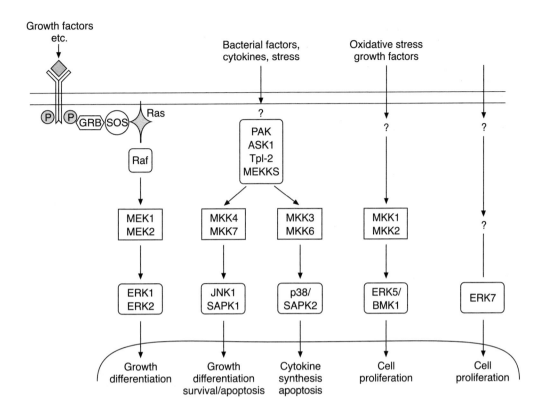

Figure 4.15 The pathways involved in the control of the five major MAP kinase families in mammalian cells. Most of what we know is based upon the study of the binding of cytokines and growth factors to their respective receptors and this has been described in earlier diagrams. In addition, bacterial factors, stress and other effectors of cell function can also activate these MAP kinase pathways. As can be seen in the diagram, there are still areas in which knowledge is limited, denoted by (?). PAK, p21-activated protein kinase; ASK, apoptosis signal-regulating kinase; MEKKs, MEK kinases; BMK, big MAP kinase; SAPK, stress-activated protein kinase; for other abbreviations where available, see list on p. xxii.

controlling kinases. As can be seen, a large amount of information needs to be filled in.

The control of MAP kinase pathways is only now starting to be understood. MAP kinase cascades contain a minimum of three protein kinases and can be said to comprise a module. In the ERK1/2 module the first kinase is one of the Raf kinases. Raf then phosphorylates the MEKs or MAP kinase kinase – the proteins that directly activate the MAP kinases. The MEKs are highly selective, for example MEK1 and MEK2 phosphorylate only ERK1 and ERK2. The dual phosphorylation of the ERKs can increase the enzymic activity by 1000-fold. Other proteins are required for MAP kinase activity. There is growing evidence that the MAP kinase pathways produce large complexes that are stabilised by another protein, Ksr (kinase suppressor of Ras). The concept has emerged that signalling pathways such as the MAP kinase modules are topologically organised within cells, allowing more efficient signalling and preventing, or controlling, cross-talk between signalling pathways. The activity of the MAP kinases appears to be inhibited by a group of proteins called MAP kinase phosphatases, which dephosphorylate on both serine and tyrosine residues.

A variety of bacterial factors can stimulate or inhibit MAP kinases. The best studied, although not the best understood, is lipopolysaccharide (LPS) which activates various MAP kinase pathways, in particular the p38 MAP kinases. In contrast, YopJ/P from *Yersinia* spp. has been found

Table 4.11. | Some bacteria and bacterial factors modulating MAP kinases

Bacterium or factor	MAP kinases affected[a]	Reference[b]
Escherichia coli LPS	p38 MAP kinase (S)	Han *et al.* (1994)
Salmonella and derived LPS	ERK via PI 3-kinase/PLD pathway (S)	Procyk *et al.* (1999)
Clostridium difficile toxin A	p38 sustained/ERK and c-Jun transient (S)	Warny *et al.* (2000)
Helicobacter pylori	ERK1/2 (S)	Meyer-ter-Vehn *et al.* (2000)
Yersinia spp. YopJ/YopP	Inhibits p38 and Jnk MAP kinases	Palmer *et al.* (1998)

[a] (S), stimulates.
[b] See Section 4.9, Papers.

to inhibit MAP kinase activity. A number of other bacteria, or bacterial factors, have been shown to modulate the activity of various MAP kinases (Table 4.11). In the case of YopJ/P, the inhibition of the MAP kinases is due to inhibition of the MAP kinase kinase proteins, preventing phosphorylation and activation of the kinases.

A number of small molecular mass selective inhibitors of MAP kinases have been synthesised and are increasingly being used to determine the role of MAP kinase pathways in the control of inflammation, including that due to bacterial infection.

4.2.5 Signal transduction and the cell cytoskeleton

The shapes of cells, the process of phagocytosis, the ability of cells (such as leukocytes) to cross the walls of blood vessels and move towards an infecting bacterium, the ability of cells to divide and the movement of organelles, such as endosomes, within cells, are all due to the possession of a cytoskeleton. The cell's cytoskeleton consists of three components: intermediate filaments, microtubules and actin filaments (see Section 8.2). The cytoskeleton of the cell is, in turn, connected to, and controlled by, cell–cell contacts and cell-to-extracellular matrix (ECM) contacts. Many cells, such as epithelial cells, are polarised, with one pole of the cell in contact with the basal lamina – an example of an ECM (see Section 5.3.1). The contacts between cells are controlled by cell adhesion proteins called cadherins, and the contact between cells and the ECM is via adhesion receptors such as integrins. The last decade or so of research into the mechanisms of bacterial pathogenesis have shown clearly that bacteria target the cytoskeletal apparatus of the cell and the cell signalling processes that control the cytoskeleton. For example, bacteria can bind to integrin receptors on cells and a growing number of bacterial toxins have been found to target the small G proteins involved in control of the actin cytoskeleton (see Sections 9.5.3.2 and 9.5.4.2). The complexity of the actin cytoskeleton is shown in Figure 4.16.

Spectrin cross-links

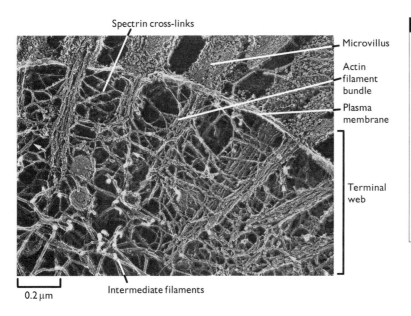

Microvillus

Actin filament bundle

Plasma membrane

Terminal web

Intermediate filaments

0.2 μm

Figure 4.16 Electron micrograph of an intestinal epithelial cell showing the complexity of the cell cytoskeleton. The terminal web is found at the base of the microvilli and is composed mainly of myosin and spectrin. It functions to connect the actin bundles of the microvilli to other parts of the cytoskeleton and gives the microvilli rigidity. (Reproduced from *The Journal of Cell Biology*, 1981, **91**, 399–409, by copyright permission of The Rockefeller University Press.)

4.2.5.1 Integrins and cell signalling

Focal adhesions is the name given to the sites at which the actin cytoskeleton is linked to the ECM through integrin receptors. Most cells are bound to an ECM as described for epithelial cells in Section 5.3.1. Leukocytes have the ability to adhere to the ECM through their integrin receptors. As well as tethering cells, these adhesion complexes can control intracellular signal transduction processes involved in controlling cell migration, proliferation, transformation and apoptosis. Binding of integrins to the ECM triggers a non-receptor protein tyrosine kinase called focal adhesion kinase (FAK). In turn, FAK is linked to a growing number of other signalling systems including the PI 3-kinase and MAP kinases (Figure 4.17).

4.2.5.2 Small GTPases and regulation of the actin cytoskeleton

The small GTPase Ras has already been described in the context of the MAP kinases. A range of other members of the Ras superfamily, the Rho family members (Table 4.12), are involved in the control of the actin cytoskeleton, and are targeted by a range of bacterial toxins that are described in more detail in Sections 9.5.3.2 and 9.5.4.2. Other Ras superfamily members are the Rab proteins, which regulate the fusion of vesicles within cells, and are described in Section 10.4.2.2.2. The role of the small GTPases in controlling the actin cytoskeleton will be described briefly.

All cells contain an actin cytoskeleton. This is built up of the monomeric protein, G (globular) actin, a 43 kDa protein that constitutes some 5–10% of the total protein in all eukaryotic cells. The structure of actin has evolved to allow head-to-tail interactions to occur, thereby enabling this protein to form polymeric filaments. These filaments, in

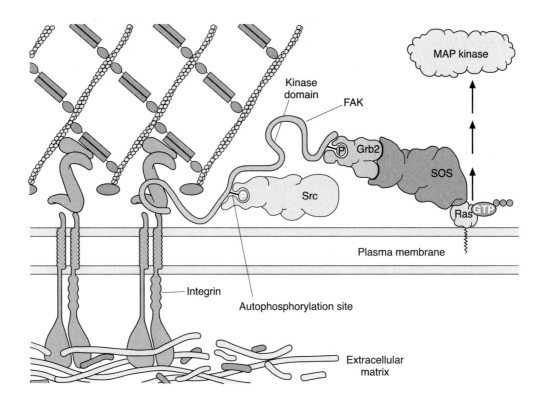

Kinase
domain

FAK

MAP kinase

P Grb2

Src

SOS

Ras GTP

Plasma membrane

Integrin

Autophosphorylation site

Extracellular
matrix

Figure 4.17 Role of focal adhesion kinase (FAK) in linking cell–matrix interactions to intracellular signalling. Binding of the integrins on the cell surface to components of the ECM stimulates FAK activity, leading to autophosphorylation. Members of the Src family of protein tyrosine kinases associate with focal adhesions and can bind to the SH2 domain created by autophosphorylation of FAK. Src and FAK can then signal through Ras and MAP kinases to activate cells. (Reproduced with permission from *The Cell: A Molecular Approach*, G.M. Cooper (1997), ASM Press, Washington, DC.)

conjunction with a very large number of accessory proteins, can form either actin bundles or actin networks that play separate roles in cell function. The actin cytoskeleton allows cells to form various cellular structures such as lamellipodia, filopodia (see Section 8.2) and focal adhesions and is involved in the process of phagocytosis. The formation of these various actin cytoskeletal structures is controlled by different members of the Rho family of small GTPases (Figure 4.18).

In mammals there are 10 members of the Rho family (Table 4.12) and these proteins share more than 50% sequence identity. Like Ras, these

Table 4.12. | Rho family members

- Rho (A, B, C isoforms)
- Rac (1, 2, 3 isoforms)
- Cdc42 (Cdc42Hs, G25K isoforms)
- Rnd1 /Rho6
- Rnd2 /Rho7
- Rnd3 /RhoE
- RhoD
- RhoE
- TC10
- TTF

TTF, translocation three four.

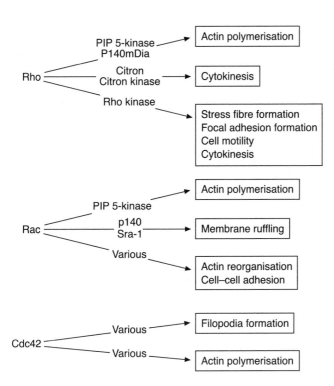

Figure 4.18 The role played by the Rho family of small GTPases in controlling the cellular events modulated by the actin cytoskeleton. For a detailed description, see the text.

proteins exhibit GDP/GTP binding and GTPase activity. The activation/ inhibition of the Rho family depends on a large number of GEFs, on GAPs and on a protein known as a **GDP dissociation inhibitor (GDI)**, which interacts specifically with GDP-Rho and inhibits the dissociation of GDP. This prevents the activation of Rho. A large number of target proteins for Rho have now been identified and the functional significance of certain of these interactions has been defined. Likewise, a number of target proteins have been identified for Rac and Cdc42. To simplify the story about Rho GTPases and the cytoskeleton, it is now established that activated Cdc42 induces the formation of filopodia. These are thin, finger-like extensions of the cell containing actin bundles, whose function is to monitor the cell's extracellular environment. Activated Rac controls the formation of lamellipodia. These are ruffled extensions of the cell membrane that form along the edge of cells. Rho is involved in the formation of stress fibres. These are elongated bundles of actin that traverse the cells and induce attachment to the extracellular matrix through focal adhesions. The consequences of the activation of the Rho GTPases are various changes in the actin cytoskeleton leading to membrane ruffling, filopodia formation and so on. In addition, the Rho GTPases are associated with alterations in cell signalling (Table 4.13). As will be described in other chapters, particularly Chapter 8, bacteria can enter cells and interact with the cytoskeleton and the Rho family GTPases. To show how intimate this bacteria–host interaction can be, it has recently been found that *Sal. typhimurium* encodes a protein, SopE2, which is a GEF for Cdc42 and is

Table 4.13. | Cellular activities controlled by Rho GTPases

Family member	Cytoskeleton	NADPH oxidase	G₁ cell cycle progression	MAPK	NF-κB	Secretion	Cell polarity	Intercellular junctions
Rho	+	−	+	−	+	+	−	+
Rac	+	+	+	+	+	+	−	+
Cdc42	+	−	+	+	+	?	+	+

+, control; −, no control.

required for the entry of bacteria into host cells. There is also growing evidence that pro-inflammatory factors can also modulate the activity of Rho GTPases. Thus bacterial LPS interacts with vascular endothelial cells to cause changes in cell shape and this can be blocked by C3 transferase from *Cl. botulinum* and by a synthetic inhibitor of Rho kinase, Y-27632. The pro-inflammatory cytokines TNFα and IL-1 stimulate Cdc42 in fibroblasts and promote filopodia formation. The various pathways that can stimulate the well-studied members of the Rho GTPases are shown in Figure 4.19.

Bacterial toxins and the Rho GTPases

Our understanding of the role of Rho family members in cytoskeletal organisation has gained substantially from the finding that a variety of bacterial toxins can modify enzyme activity. For example the C3 exoenzyme from *Cl. botulinum*, the C3-like toxins from *Cl. limosum* and *Bacillus cereus* and epidermal cell differentiation inhibitor from *Staph. aureus* are all ADP phosphoribosyltransferases able to transfer the ADP-ribose of NAD to asparagine (Asn) 41 of Rho and, in consequence, inhibit its activity. Another family of Rho-modulating toxins have glycosyltransferase activity. Thus the *Cl. difficile* toxins A and B and the haemorrhagic toxin and lethal toxin of *Cl. sordelli* act to transfer glucose from UDP-glucose to threonine in the effector domain of Rho GTPases.

In contrast, the CNF from *E. coli* and dermonecrotic toxin (DNT) produced by *Bordetella* spp. can activate Rho. These proteins have transglutaminase activity. CNF deamidates glutamine (Gln) 63 in the Rho protein converting it to glutamic acid (Glu). This residue plays a key role in GTP hydrolysis and thus the Rho-Glu-63 remains in a GTP-bound, and thus active, state. DNT also deamidates Rho at the same position, but it has also been shown that this toxin can cross-link this residue with polyamines (e.g. spermidine, spermine, putrescine).

The effects of bacterial toxins on Rho family members will be discussed in various chapters in this book and in more detail in Sections 9.5.3.2 and 9.5.4.2.

4.2.6 Cytokines: key host signalling molecules

While, as discussed in Chapter 9, bacterial exotoxins can directly cause tissue pathology, it is now generally accepted that the inflammatory

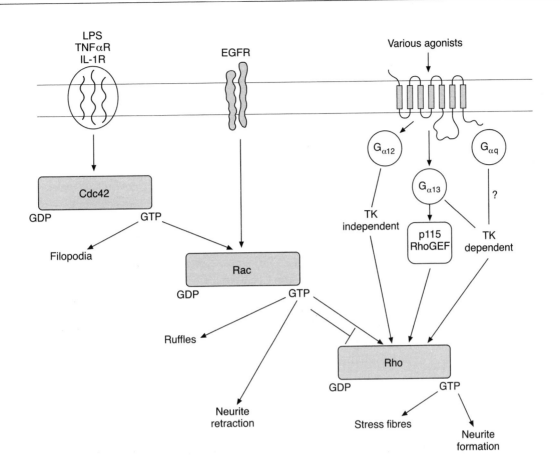

Figure 4.19 An overview of the external ligands controlling the Rho family of small GTPases and a simplified view of the intracellular signalling events causing major changes in cell shape and behaviour. The role of filopodia and ruffles in cellular responses to bacteria is described in Chapter 8. Neurites are projections produced by neurones that act to interconnect these cells. For abbreviations, see list on p. xxii.

and destructive sequelae of bacterial infection are due to the overproduction of key cell–cell signalling proteins known as cytokines. Bacteria produce a bewildering number of molecules that have the ability to induce cytokine synthesis. For example, most bacterial toxins are also potent inducers of cytokine synthesis. Indeed, many bacterial toxins are better inducers of cytokine synthesis than they are intoxicants. The authors have coined the term 'modulin' for bacterial factors that induce cytokine synthesis and have suggested that the modulins are a new class of bacterial virulence factor (for more details, see Section 10.3.1). This section will give a brief introduction to the cytokines involved in bacterial infections and the intracellular signalling mechanisms by which the two major pro-inflammatory cytokines, IL-1 and TNFα, stimulate cells. This will cover signalling systems that are of immense importance in bacterial pathology. Chapters 5, 6 and 10 on immunity to bacteria will also describe the roles of cytokines in controlling immune function.

4.2.6.1 A brief history of cytokines

Cytokines were discovered by scientists and clinicians interested in infection. The first cytokine to be identified was IL-1 in the early

1950s, at which time it was variously termed endogenous pyrogen (EP), as it caused fever, and leukocyte endogenous mediator (LEM), as it stimulated the production of leukocytes. The second cytokine to be discovered was interferon (IFN) in 1957. The 1960s saw the appearance of a whole range of activities, which were given the global title lymphokines, and additional terms such as monokines began to appear. The term cytokine was coined by Stanley Cohen and co-workers in 1974. The 1970s and 1980s brought with them advances in protein purification, protein sequencing, monoclonal antibody production and recombinant DNA technologies and many new cytokines began to be discovered. The 1990s and the new millennium have continued the discovery of new cytokines and the finding of homologous genes in the human EST (expressed signature tag) databases.

It is likely that, if we group together the various classes of proteins known as cytokines, chemokines, IFNs and growth factors under the catch-all title 'cytokines', we are dealing with more than 200 proteins. Increasing numbers of these cytokines are being found to play roles in bacterial infections and there is growing evidence that viruses, bacteria and protozoans have developed means to negate the protective effects of cytokines. It has to be emphasised that cytokines are a major part of the mammal's defences against infectious agents and also the major cause of the pathology resulting from infection.

4.2.6.2 Properties of cytokines

The term 'cytokine' is not particularly exact. It refers to signalling proteins, generally more than 20 kDa in size, which are produced by all cells (with the possible exception of erythrocytes) and act as local hormones by binding to high affinity receptors on cells. Cytokines differ from classical endocrine hormones by not being produced by specialised cells and generally not having one single target cell type. They also differ from classic endocrine hormones by acting over short distances and even interacting with, and activating, the cell that is producing them (Figure 4.20). There is increasing evidence for extensive overlapping of the activity of endocrine hormones and cytokines and the similarities and differences between these two forms of signalling molecules are provided in Table 4.14.

Cytokines have no inherent activity, and their biological actions are dependent on binding to their cognate receptors on the surface of target cells. One of the reasons that cytokines have only been discovered recently is that they are extremely active and, in consequence, are present in biological fluids at low levels (nanomolar (nM, 10^{-9} M) to femtomolar (fM, 10^{-15} M)) and thus bind to receptors that have very high affinities. For many of the cytokines important in bacterial infection, the number of cell surface receptors is of the order of hundreds to thousands per cell. This contrasts with the thousands to hundreds of thousands of receptors for endocrine hormones.

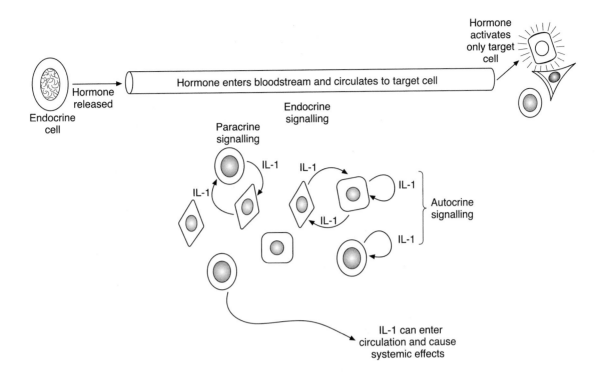

IL-1 can enter circulation and cause systemic effects

4.2.6.3 Cytokine nomenclature

One of the most confusing aspects of cytokine biology is the names given to individual cytokines. This generally reflects the first activity determined for any cytokine. Currently, cytokines are divided into six groups on the basis of their names (Table 4.15) and the reader would reasonably assume that all of the proteins within one such group would have similar functions and properties. Unfortunately, this is not the case. For example, the term interleukin means 'between leuko-cytes' and initially described cytokines produced by one leukocyte and interacting with another. Unfortunately, we now know that a number of the interleukins (e.g. IL-1, IL-6, IL-8) are produced by cells other than leukocytes, such as fibroblasts, chondrocytes or epithelial cells. One interleukin is actually a chemokine. Some growth factors are actually growth inhibitors (e.g. TGFβ). The problem is that cytokines exhibit certain properties that make it difficult clearly to assign them roles. Thus cytokines are pleiotropic; that is, any particular cytokine can exhibit a wide range of functions. In addition, cytokines exhibit redundancy, with many cytokines exhibiting the same activity. This is clearly seen with the three major pro-inflammatory cytokines (IL-1, IL-6 and TNFα) (Table 4.16).

4.2.6.4 Biological actions of cytokines

Cytokines have a very large number of (overlapping) effects on cells. These include cell activation as well as the induction of chemo-taxis, proliferation, apoptosis and differentiation (Figure 4.21). The

Figure 4.20 Diagram comparing signalling properties of cytokines and endocrine hormones. Endocrine hormones are generally made by one specific cell and are released into the bloodstream, where they have to move considerable distances to reach their target cell. This type of signalling is called endocrine signalling. In contrast, many cytokines are made by multiple cells and affect many other cell types – the example given is of IL-1. In general, cytokines act over short distances, although they can enter the circulation and have systemic actions. When cytokines act on nearby cells, the mode of signalling is called paracrine (*para* = nearby). When the cytokine acts on the producing cell, the mode of signalling is known as autocrine.

Table 4.14. | Comparison of the properties of cytokines and endocrine hormones

Biological properties	Cytokines	Endocrine hormones
Role	Maintenance of homeostasis and response to infection and injury	Maintenance of homeostasis
Redundancy	High	Low
Pleiotropy	High	Low
Present in blood	Generally no – with few exceptions (IL-6, EPO)	Yes
Sites of action	Many target cells but generally action is local	Single target cell but hormone may travel large distances
Inducers	Many stimuli including physiological stimuli, infection, injury, etc.	Generally physiological stimuli
Producing cell types	All cells produce some cytokines	Specialised cells
Responding cell types	All cells respond to some cytokines	From one to many, depending on the hormone

EPO, erythropoietin; IL, interleukin.

Table 4.15. | Cytokine nomenclature

Cytokine 'family'	Example	Major biological actions
Interleukins	IL-1 to IL-18	Mainly growth factors for lymphoid cells but have other activities (IL-1, IL-3, IL-8, IL-10)
TNF family	TNFα/β, CD40L, etc.	Many and varied, involved in control of immune function
Interferons	IFNα, β, γ	Antiviral and immunological control
Colony-stimulating factors	IL-3, G-CSF, GM-CSF	Myeloid growth factors
Chemokines	IL-8, MIP-1α, MCP	Chemoattractants for leukocytes
Growth factors	EGF, TGFα	Epithelial and mesenchymal cell growth factors

For abbreviations, see list on p. xxii.

introduction of bacteria into a normally sterile tissue results in the formation of a wide range of cytokines. Initially, the major cytokines to be produced will be those controlling the induction of inflammation, namely IL-1, IL-6, IL-8 and TNFα (Figure 4.22). IL-1 and TNFα function to activate other cells in the area of infection. For example, endothelial cells are activated to switch on the transcription of,

Table 4.16.	Examples of pleiotropy and redundancy among cytokines		
Biological activity	IL-1	IL-6	TNFα
Induces fever	✓	✓	✓
Activates lymphocytes	✓	✓	✓
Induces acute phase protein	✓	✓	✓
Activates cyclo-oxygenase	✓		✓
Stimulates bone resorption	✓	✓	✓

For abbreviations, see the text.

amongst other genes, those controlling the synthesis of the vascular cell adhesion proteins: E-selectin, ICAM (intercellular adhesion molecule) and VCAM (vascular cell adhesion molecule). Other endothelial cell genes are activated to alter the balance of coagulation/fibrinolysis in order to trap bacteria. IL-6, along with IL-1 and TNFα, also has systemic effects, inducing the liver to produce a range of proteins that act to opsonise bacteria. These are the acute phase proteins, molecules such as C-reactive protein (CRP) and serum amyloid A (SAA), which can bind to bacteria and help in their uptake and destruction by phagocytes. The activation of phagocytes by cytokines such as IL-1 and TNFα also helps in the process of bacterial killing. IL-1 and TNFα are also potent inducers of prostaglandin synthesis by other cells, including tissue fibroblasts, macrophages and vascular endothelial cells. These lipid mediators have a wide range of both pro- and anti-inflammatory functions. Thus cytokines can prepare the site of infection to maximise antibacterial function. The induction of vascular cell adhesion proteins

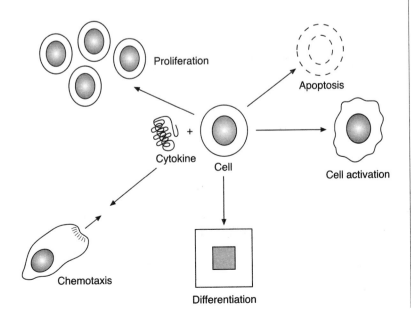

Figure 4.21 The effects of cytokines on cells. The actions of cytokines on cells can be divided simplistically into five categories. Cytokines can induce the activation of cells. A good example is the effect of cytokines on fibroblasts or endothelial cells. These cells produce a wide range of gene products including enzymes that produce lipid mediators (prostaglandins), proteases, tissue factor, adhesion proteins, etc. Such cellular activation is important in the defence mechanisms and pathology of infection-driven inflammation. A large group of small cytokines induce selected leukocyte populations to undergo directed movement (chemotaxis). These chemokines are vital for directing the correct populations of leukocytes into an inflamed site. Certain members of the TNF family can induce cells to apoptose. Other cytokines can cause cells to proliferate and/or to undergo differentiation.

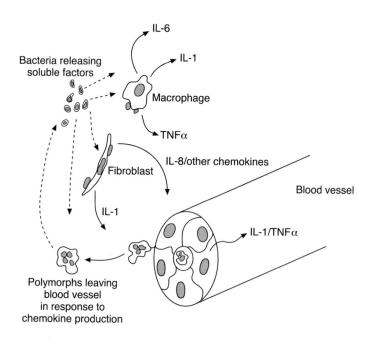

Figure 4.22 Cytokines produced in response to bacteria and their cellular effects. Bacteria, by direct cell contact and release of cytokine-inducing factors, such as LPS, induce the synthesis of the major cell-activating cytokines IL-1 and TNFα. These cytokines can activate vascular endothelial cells, resident macrophages and fibroblasts to generate additional cytokines such as IL-6, IL-8, and other chemokines. These various cytokines, and low molecular mass compounds such as lipids, act to control the influx of leukocytes, the state of activation of these cells, their residency in the site of infection and their eventual apoptosis.

is complemented by the generation of chemokines, proteins that bind to GPCRs and trigger leukocyte recruitment to sites of infection. Inflammation will also induce the formation of colony-stimulating factors, which will enhance the production of appropriate leukocytes in the bone marrow to allow a continuous supply of antibacterial cells to be maintained. In addition to these so-called pro-inflammatory cytokines, the body also produces cytokines that act to inhibit inflammatory events. In an acute bacterial infection, for example, the recruitment of **polymorphonuclear neutrophils (PMNs)** and monocytes to the infected site is normally sufficient to remove the bacteria. Once bacterial signals are on the wane, other cytokines that regulate the potent actions of these pro-inflammatory cytokines can come into play. These include the monocyte-deactivating cytokines such as interleukin-1 receptor antagonist (IL-1ra) and IL-10 and the growth factor TGFβ. If there has been damage to tissues, growth factor production will be switched on as part of the resolution and healing response to infection and injury, and cytokines such as **platelet-derived growth factor (PDGF)**, fibroblast growth factor (FGF) and epidermal growth factor (EGF) will be produced. To give the reader some idea of the range of actions of mammalian cytokines, Table 4.17 gives the biological functions of a range of the cytokines likely to be encountered in bacterial infections.

If bacterial infections are chronic in nature, and those causing tuberculosis, leprosy, periodontal disease and syphilis immediately spring to mind, then the acquired immune system becomes involved in antibacterial defence. This brings in contributions from B and T lymphocytes and cells associated with the activation of lymphocytes such as dendritic cells. Again, cytokines are involved in the regulation

Table 4.17.	Biological actions of selected cytokines involved in bacterial infection		
Cytokine	Producing cell	Target cells	Major activities
IL-1β	Myeloid cells, VEC etc.	Many	Up-regulates many cell functions
IL-1ra	As above	Many	Natural antagonist of IL-1
IL-2	Th1 cells	T cells/NK cells	Growth factor
IL-3	Th, NK, mast cells	Haematopoietic/mast cells	Growth and differentiation
IL-4	Th2, NK, mast cells	B/T cells, macrophages, mast cells	Growth and activating factor
IL-6	Myeloid, fibroblast, Th2	B cells, myeloid cells, etc.	Acute phase proteins/plasma cell formation
IL-8	Macrophages, VEC	Neutrophils	Chemokine
IL-9	Th cells	Some Th cells	Growth factor
IL-10	Myeloid cells, Th2 cells	Myeloid cells	Myeloid cell inactivator
IL-11	Bone marrow stromal cells	B cells, hepatocytes	Acute phase protein synthesis
IL-12	Myeloid cells, B cells	Th0 cells, NK cells	Th1 cell differentiation, IFNγ synthesis
IL-15	T cells, myeloid cells	T cells, NK cells	Growth factor
IL-17	T cells	Macrophages, fibroblasts	Some similarity to IL-1
IL-18	Myeloid cells	T cells, NK cells	Induces IFNγ
IFNα	Leukocytes	Infected cells	Inhibits viral and bacterial growth
IFNβ	Fibroblasts	Infected cells	Inhibits viral and bacterial growth
IFNγ	Th1, NK cells	Infected macrophages	Inhibits intracellular bacteria in macrophages
TNFα	Myeloid, mast cells	Various	Similar actions to IL-1
TNFβ	Th1 and Tc	Macrophages/ neutrophils	Enhances activity
TGFβ	Myeloid, lymphocytes, mast cells	Many	Inflammatory inhibitor
Oncostatin M	Myeloid, T cells	Hepatocytes	Stimulates acute phase proteins
Chemokines	Many cells	Leukocytes	Induces selective leukocyte trafficking

VEC, vascular endothelial cells. For other abbreviations, see list on p. xxii.

of the interactions of these various cell populations and such interactions will be discussed in more detail in Chapter 6.

4.2.6.5 Cytokine networks

The reader is used to the concept that one extracellular signal, for example a hormone, has one mechanism of action on the target cell. Of course that target cell may, at the time it receives this hormonal signal, also be responding to other signals in its external environment. Thus the overall response of the cell will be the summation of all the signals received at any one time. This is obviously complex. However, when we come to deal with cytokines there are the additional complications that a cell responding to one particular cytokine, say IL-1, may, in response, make more IL-1 plus IL-6 and IL-10. These cytokines, in turn, can interact with the producing cells, and with other cells in the vicinity, thereby causing further activation (or inhibition). Most cells can produce many cytokines and can also respond to many cytokines. Thus very rapidly after adding one activating cytokine to a cell, one can find that the number of different cytokines being produced is large. This is an example of network behaviour and it is now clear that the basic unit in cytokine biology is the cytokine network. These networks facilitate the production of antibacterial defences and are also involved in switching off inflammatory responses to infection. One well-studied network is the pattern of cytokines produced by so-called Th1 and Th2 lymphocytes (Figure 4.23). This will be described in more detail in Section 6.7.3.

Another group of cytokines that form complex networks in bacterial infections is the chemokines. These were initially identified as chemotactic cytokines, vital for the correct trafficking of leukocytes to the site of infection. This is a key function of these small cytokines, but there is evidence for them having additional functions in controlling blood vessel formation (angiogenesis), development of Th1/Th2 lymphocytes, wound healing, etc. In addition, a growing number of receptors for chemokines is being revealed and additional functions for these receptors are being discovered. For example, some receptors such as CXCR4 bind the human immunodeficiency virus (HIV) and are part of the mechanism of viral invasion. A list of the known chemokine receptors and their ligands is given in Table 4.18.

Bacterial interleukin 10

IL-10 is a member of a growing group of cytokines that function to control the actions of pro-inflammatory cytokine networks. The importance of this cytokine in infection is shown by the fact that several viruses encode IL-10-like proteins. The role of these IL-10-like proteins is discussed in more detail in Section 10.3.2.1. IL-10 is best known as having the properties of 'deactivating' monocytes and macrophages. These cells, when activated, express MHC class II proteins (see Section 6.7.2.1) and secrete various inflammation-inducing cytokines. IL-10 can block these events and thus 'calm' monocytes and macrophages. The importance of IL-10 in the control of inflammation is clearly seen in mice in which the gene encoding IL-10 has been inactivated. Such animals are known as transgenic knockouts (see Section 3.9.2). The inactivation of the IL-10 gene results in mice that develop inflammation of the colon (colitis). This inflammation is a response to the normal colonic micro-flora, as IL-10 knockouts bred under germ-free conditions do not develop colitis. A fascinating recent experiment has used the commensal bacterium *Lactococcus lactis* engineered to produce the anti-inflammatory cytokine IL-10. When this IL-10-secreting bacterium was delivered into the stomachs of mice with experimental colitis it reduced the severity of inflammation by 50%. More strikingly, in IL-10 knockout mice the administration of this engineered bacterium prevented animals developing colitis. This finding holds enormous promise for the use of our commensal bacteria to control host cytokine networks. Could the enormously prevalent dental diseases such as gingivitis and periodontitis be prevented by colonising our mouths with appropriately engineered bacteria? Could skin diseases be equally treated? Watch this space.

4.2.6.6 Cytokine structure and cytokine receptors

Many reviews and textbooks divide cytokines into two groupings: the cytokines and the growth factors. This convention will be followed in this section. Cytokines bind to a wide variety of receptors. However,

(a)

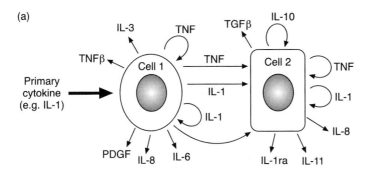

(b)

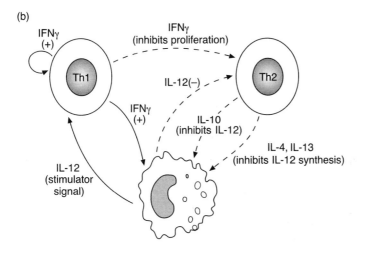

Figure 4.23 The concept of the cytokine network. In (a) a hypothetical network is constructed, although based upon experimental evidence. The input of one key cytokine, in this case IL-1, can give rise to a large number of additional cytokines by autocrine and paracrine signalling. In (b) a simplified version of the cytokine signalling in a well-studied cytokine network is shown. This shows the inter-actions between Th1/Th2 lymphocytes and the macrophage. The cytokines produced by the various cells can either stimulate that cell or inhibit the actions of the other cells. For example, IFNγ activates Th1 cells and macrophages but inhibits the pro-liferation of Th2 cells. IL-12 from the macrophage stimulates the Th1 cell but inhibits the Th2 cell. IL4, IL-10 and IL-13 produced by the Th2 cell can inhibit the macro-phage and thus, in consequence, the Th1 cell. For abbreviations, see the text and list on p. xxii.

Table 4.18.	Human chemokine receptors and their ligands
Chemokine receptor	Ligands
CXCR1	IL-8, GCP-2
CXCR2	IL-8, GCP-2, Groα, Groβ, ENA-78
CXCR3	MIG, IP-10, I-TAC
CXCR4	SDF-1
CXCR5	BLC/BCA-1
CCR1	MIP-1α, MIP-1β, RANTES, HCC-1, 2, 3, 4
CCR2	MCP-1, -2, -3, -4
CCR3	Eotaxin-1, -2, MCP-3
CCR4	TARC, MDC, MIP-1α, RANTES
CCR5	MIP-1α, MIP-1β, RANTES
CCR6	MIP-3α/LARC
CCR7	MIP-3β/ELC
CCR8	I-309
CCR9	TECK
XRC1	Lymphotactin
CX3CR1	Fractalkine/neurotactin

GCP, granulocyte chemotactic protein; Gro, growth-regulated oncogene; ENA, epithelial-derived neutrophil chemoattractant; MIG, monocyte induced by IFNγ; IP, IFN-inducible protein; I-TAC, IFN-inducible T cell alpha chemoattractant; SDF, stromal cell derived factor; BLC, B lymphocyte chemokine; BCA, B cell attracting; RANTES, regulated on activation, normal T expressed and secreted; HCC, human CC chemokine; TARC, thymus and activation-regulated chemokine; MDC, macrophage-derived chemokine; LARC, liver and activation-regulated chemokine; ELC, Epstein–Barr virus-induced molecule 1 ligand chemokine; TECK, thymus-expressed chemokine; for other abbreviations, see list on p. xxii.

over the past decade or so, structural studies have allowed these receptors to be grouped into five families (Table 4.19). The cytokines binding to these receptors are all involved in immune defences against bacteria.

The signalling mechanisms of class I and class II cytokine receptors (via JAK/STAT pathways) and of chemokine receptors (via GPCRs) have already been described in some detail. In this section, the signalling mechanisms of the immunoglobulin superfamily receptors (which includes IL-1 and IL-18 receptors) and of the TNF family receptors will be described. Both of these receptor families are extremely important in host responses to bacterial infections.

Table 4.19. Cytokine receptor families

Immunoglobulin receptor superfamily	Class I receptors	Class II receptors	TNF receptors	Chemokine receptors
IL-1	IL-2	IFNα	TNFα(I/II)	IL-8
M-CSF	IL-3	IFNβ	TNFβ	MIP-1
c-Kit	IL-4	IFNγ	NGF	MCP
IL-18	IL-5	IL-10	CD40	RANTES
	IL-6		Fas	MCAF
	IL-7		HVEM	PF4
	IL-9		CD27	Eotaxin
	IL-11		RANK	IP-10

NGF, nerve growth factor; RANK, receptor activator of NF-κB; HVEM, herpesvirus entry mediator; PF, platelet factor; MCAF, monocyte chemotactic and activating factor; for other abbreviations, see list on p. xxii.

4.2.6.6.1 The IL-1 system and its signalling pathway

One of the most important, and best studied, cytokines in relation to bacterial infections, is IL-1. The actions of this cytokine have been described briefly and will be discussed in other chapters. From the outset, it is important to realise that IL-1 is not simply a cytokine, it is a cytokine system (Table 4.20), and a brief introduction is required before discussing the details of intracellular signalling. The intracellular signal transduction pathways stimulated by IL-1 are very similar to those produced by the other major pro-inflammatory cytokine that is centrally involved in immune responses to bacteria, TNFα. This discussion will deal with the well-studied members of the IL-1 family, but it is important to realise that novel members of the IL-1 fraternity are being described as this chapter is being written and may have additional roles in our response to bacterial infections.

At the present time, five IL-1 proteins are recognised. Discovered in the 1950s, but not cloned until the 1980s, are IL-1α and IL-1β. These proteins are initially translated as 31 kDa pro-forms, which are cleaved to 17 kDa proteins by a protease involved in the process of apoptosis, caspase 1, also known as pro-IL-1β-converting enzyme (ICE). The 31 kDa pro-form of IL-1β is inactive, while the IL-1α pro-form has some biological activity. In addition to these two proteins, a third member of the IL-1 family is the natural antagonist IL-1ra. This protein competes with IL-1α and IL-1β for binding to the receptor. The most recently discovered member of this family is IL-1γ or, as it is more commonly known, IL-18. This is also produced as a 24 kDa pro-form that requires cleavage by caspase 1 to produce the biologically active 18 kDa protein.

There are three receptors for IL-1. **IL-1 receptor type I (IL-1RI)** is an 80 kDa functional receptor, which binds IL-1α, IL-1β and IL-1ra. The

Table 4.20. The IL-1 system. There are also a number of other IL-1-like genes in the human EST database

Component	Function
Pro-IL-1α	Partially active 31 kDa pro-form of IL-1α
Pro-IL-1β	Inactive 31 kDa pro-form of IL-1β
IL-1α	17 kDa pro-inflammatory cytokine
IL-1β	17 kDa pro-inflammatory cytokine
Pro-IL-18	Inactive 24 kDa pro-form of IL-18
IL-18	18 kDa pro-inflammatory cytokine
Caspase 1	Protease that activates pro-form of IL-1β/IL-1α/IL-18
IL-1 receptor antagonist (IL-1ra)	Natural antagonist of IL-1α/β
IL-18 binding protein	IL-18 antagonist
Type I IL-1 receptor (IL-1RI)	Functioning IL-1 receptor
Type II IL-1 receptor (IL-1RII)	Non-functional 'decoy' receptor
IL-1 receptor accessory protein (IL-1RacP)	Accessory receptor required for IL-1RI signalling
Soluble IL-1RacP	Possible antagonist of IL-1
IL-18 receptor	Functional receptor for IL-18
IL-1 receptor-associated kinases (IRAK)	Involved in IL-1 signal transduction

For abbreviations, see list on p. xxii.

second 60 kDa receptor, IL-1 receptor type II (IL-1RII), does not elicit biological responses and has been proposed to be part of the IL-1 control system acting as a 'decoy receptor' to down-regulate the actions of IL-1. The third, and most recently discovered, receptor protein is **IL-1 receptor accessory protein (IL-1RacP)**, which is required for signalling through IL-1RI. IL-18 does not bind to either of these IL-1 receptors but does bind a homologous, so-called orphan, receptor termed IL-1Rrp1 to activate a similar response to that of IL-1.

4.2.6.6.2 Signalling via IL-1 receptors

There was confusion for many years with regard to exactly how IL-1 signalled through its receptor. With increasing understanding of this group of cytokines has come the realisation that they share significant overlap, in terms of intracellular signalling pathways, with the other major pro-inflammatory cytokine TNFα, and with a newly discovered groups of cell surface receptors called the Toll-like receptors (TLRs). These receptors were originally discovered in *Drosophila*, where they play the dual function of being involved in both embryonic development and in the response of the insect to fungi and bacteria. The role of the TLRs in the response of cells to bacterial and fungal virulence

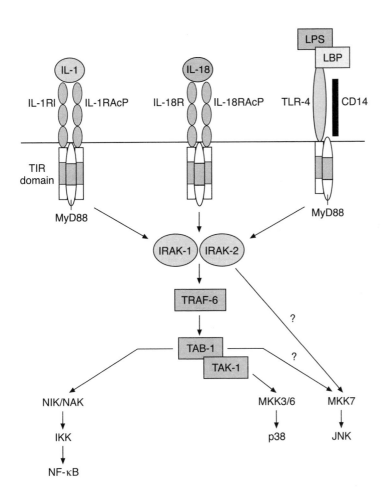

Figure 4.24 The IL-1 signalling pathway. The details of the signalling pathway are described in the text. TAK-1, TGF-β-activating kinase 1; TAB-1, TAK-1-binding protein 1; MKK, MAP kinase kinase. For other abbreviations, see list on p. xxii.

factors will be described in more detail in Chapter 6. As stated earlier, IL-18 also shares the same signalling pathway as IL-1.

From studies in which either the IL-1RI or the IL-1RacP were knocked-out (for a discussion of gene knockouts, see Chapter 3), it is now accepted that IL-1 stimulation of cells requires the participation of both of these receptor proteins for correct intracellular signalling. The intracytoplasmic domain of the IL-1RI receptor is responsible for intracellular signalling and has homology to the intracytoplasmic domains of the TLRs and, in consequence, such domains are known as TIR (Toll/IL-1R) domains. Receptor binding of IL-1α/β results in the formation of a signalling complex that generates the production of a number of activated transcription factors, including the major signalling factor NF-κB (Figure 4.24). The first step in this signalling complex is the recruitment of a protein termed MyD88 (**my**eloid **d**ifferentiation protein), which has a C-terminal TIR domain, to the activated IL-1R. In turn, MyD88 interacts with a family of related serine/threonine kinases called **IL-1R-a**ssociated **k**inases (IRAKs). Phosphorylation of an IRAK leads to its association with another key protein called TNF receptor-associated factor (TRAF) 6. The TRAFs are examples of what are termed 'adapter proteins', which are involved in linking different parts of signalling systems. IRAK-induced activation of TRAF-6 then, it is believed, allows this adapter protein to interact with two other kinases, TAB-1 and TAK-1, which then activate yet other kinases (NF-κB-inducing kinase, NIK; or NF-κB-activating kinase, NAK) leading to NF-κB activation. This is the critical juncture at which IL-1, TLR and TNF signalling converge. Now, a major transcriptional control factor in cells is NF-κB. This is, in fact, a collective term for a family (the Rel or Relish family) of dimeric inducible proteins found in the cytoplasm of cells. Here they exist as a complex with a second protein inhibitory **κB** (IκB) and, in this form, cannot translocate into the nucleus and bind to DNA (as the IκB blocks the nuclear localisation signal). Removal of IκB results in the NF-κB being able to enter the nucleus and induce selective gene transcription. Removal of the IκB requires the phosphorylation of this protein at specific residues by IKK (a kinase containing two catalytic subunits, IKKα and IKKβ, both of which can phosphorylate IκB). Phosphorylation results in the IκB being recognised by a major cellular proteolytic degrading system – the ubiquitin proteolysis system (described in more detail in Chapter 6) – and being degraded. This leaves the NF-κB free to enter the nucleus. Binding to IL-1R/IL-18R/TLR also activates MAP kinase pathways.

A number of bacteria and bacterial products have been shown to be able to influence the activity of the NF-κB system. This will be discussed in more detail in Chapters 6 and 10.

4.2.6.6.3 The TNF receptor family and cell signalling

TNFα is another major pro-inflammatory cytokine intimately involved in host responses to bacterial infection. This cytokine, like IL-1, has been extensively studied because of its obvious involvement in human pathology, particularly the lethal response to bacterial infections known as septic, or toxic, shock.

Again, like IL-1, TNFα is not a single cytokine but is a member of a family of both soluble and cell-bound cytokines that interact with homologous receptors. This family of proteins has a wide range of functions in terms of immune defence, and is involved in induction and control of (i) inflammatory responses, (ii) lymphoid cell development, and (iii) apoptosis. The current family of TNF receptor/ligand proteins is shown in Table 4.21. However, in this chapter only the signalling by TNFα will be described. The other members of the TNF family are introduced only to make the reader aware that the story of TNF in bacterial infections will become more complex with time.

There are two receptors for TNFα. The TNFR1 (also known as CD120a and as the p55 receptor) is the main TNFα receptor involved in innate and acquired immunity and its knockout results in deficient responses to bacteria, particularly intracellular bacteria. The TNFR2 (also known as CD120b or p75) is, in contrast to TNFR1, activated preferentially by cell-bound TNFα and less is known about its role in immune responses generally.

As has already been described, TNFα, like IL-1, IL-18 and Toll ligands, signals mainly through the cytoplasmic transcription factor NF-κB, and the mechanism of action of this protein has been described briefly. This section will explain how the binding of TNFα to its receptor (TNFR1) on the surface of a cell results in the eventual activation of NF-κB and the induction of apoptosis (for details of this process, see Chapter 10). It needs to be pointed out that this is an extremely complex and confusing area of receptor–intracellular signalling and there are more questions than answers at the present time.

The TNF family receptors do not have intrinsic enzyme activity and thus transduction of the signal from the TNFR1 requires adapter or docking proteins such as has been described for the IL-1 receptor, in which MyD88 binds the receptor and the IRAKs. The adapter proteins that connect TNF receptor binding to common intracellular signalling pathways are unique and they play a number of roles, the most important being the activation of NF-κB (and other transcription factors such as AP-1) and the induction of the process of apoptosis.

The ligands for the TNF receptor family are normally trimeric, and binding to the cell surface results in the clustering of receptors and this triggers the signalling events. The two consequences of binding to TNFR1 – NF-κB activation and apoptosis – will be described separately and are shown diagrammatically in Figure 4.25.

NF-κB activation: As stated earlier in this chapter, the cytoplasmic domain of TNFR1 has no enzymic activity. Mutation of this cytoplasmic domain to see whether this would block its signalling activity led to the identification of a functional domain of around 80 amino acid residues that has been termed the 'death domain' (because of its role in

---

Table 4.21. | The TNF family of receptors, receptor-binding ligands and signalling domains

Receptor	Ligand	Death domain	Docking proteins
TNF receptor 1 /CD120a	TNFα/LTα	+	TRADD-FADD, TRAF-2
TNF receptor 2/CD120b	TNFα/LTα	–	TRAF-1, -2
APO-1 /Fas /CD95	APO-1L/FasL/CD95L	+	FADD
Lymphotoxin β receptor	LTα + LTβ dimer	–	TRAF-5
CD27	CD27L/CD70	–	TRAF-2, -5
CD30	CD30L/CD153	–	TRAF-1, -2, -3, -5
CD40	CD40L/CD154	–	TRAF-2, -3, -5, -6
TRAIL receptors 1–4	TRAIL	+	TRADD-FADD
RANK	RANKL	–	?

RANK, receptor activator of NF-κB; for other abbreviations, see list on p. xxii.

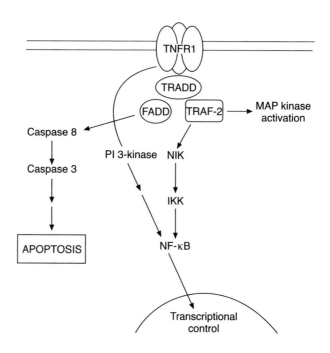

TNF-induced apoptosis). This led on to the identification of a 34 kDa protein with a homologous C-terminal **d**eath **d**omain (DD). This protein has been termed TNF receptor-associated **d**eath **d**omain (TRADD) and is one of the adapter, or docking, proteins that act to pass the signal from the receptor to NF-κB. TRADD, in turn, associates with a second protein, one of a group of proteins, the TRAFs. In the case of the TNFR1, the TRADD complexes with TRAF-2. The TRAFs share a common C-terminus and two other domains – one called a zinc finger and the other a RING finger. It is speculated that these latter two domains are involved in the activation of downstream kinases. TRAF-2 interacts with, and activates, NIK, which is a member of the MAP kinase group of proteins. In turn, it is believed that NIK activates IκB kinase (IKK) leading to phosphorylation and release of IκB from the NF-κB/IκB complex, as described earlier.

Apoptosis: This is a normal cell process important in many aspects of homeostasis and will be described in detail in Chapter 10. Bacteria have evolved to modulate the processes of apoptosis, which is why the process will also be described at various points in this book. The TNF receptor superfamily plays a major role in controlling apoptosis and a brief description of the role of TNFR1 in controlling this process is described.

The two main TNF receptors for inducing apoptosis are TNFR1 and CD95 (normally known as Fas and, previous to this, APO-1). These bind TNFα and CD95 ligand (CD95L – previously Fas ligand, FasL), respectively. Ligand binding with TNFR1 results in interaction with TRADD. In turn, TRADD complexes with another death domain-containing protein called Fas-associated DD (FADD). This is an example of what is termed homotypic binding of proteins – two similar domains recognising and binding to each other. FADD acts as a 'bridge' between the TNFR and a group of proteases (termed caspases) that catalyse the cellular breakdown constituting apoptosis. The initial event is the binding of FADD to the protease caspase 8 (Figure 4.25). The caspases are a family of cysteine proteases that cleave proteins after an aspartic acid residue. We have already dealt with one caspase in this chapter, the protease involved in activating pro-IL-1α, pro-IL-1β and IL-18. This protease was originally known as pro-IL-1β-converting enzyme or caspase 1.

The caspases are produced as inactive zymogens that have to be activated by proteolytic cleavage and produce a cascade of proteases whose function is to activate the breakdown of the various constituents of the cell. The process of apoptosis is complex and involves a large number of separate proteins as well as a cellular organelle, the mitochondrion. The process will be described in detail in Chapter 10.

Another example of the close interaction between bacteria and their hosts has come from the expression of the death domain (DD) of FADD in *E. coli*. In such recombinants, cells showed an elongated morphology, nicked chromosomal DNA and an increased rate of mutation. The addition of oxygen-derived free radical scavengers blocked these effects of the DD, suggesting that its effects were due to generation of free radicals. It is unclear whether the actions of the DD on bacteria have physiological relevance.

4.3 | Prokaryotic cell signalling

The preceding part of this chapter has attempted to give a brief overview of the main intracellular signalling pathways in eukaryotic cells and how they interact with the major intercellular signalling molecules involved in bacterial infection, the cytokines. The discussion will now move from the eukaryotic cell to the prokaryotic cell and the signalling systems used by bacteria will be reviewed.

4.3.1 Intracellular signalling in prokaryotes

4.3.1.1 Introduction

In the chapters that follow, we will describe how bacteria, in order to successfully initiate and maintain an infectious process, must possess a number of virulence factors. It is important now to address the question of how the production of these virulence factors is regulated. Obviously, it would not be cost effective, in terms of its energy or materials balance, for a bacterium to synthesise all of these factors all of the time. For example, this would be particularly wasteful during the periods in which it is being transmitted between hosts. Furthermore, particular virulence factors are needed only at particular stages of the infectious process. Consider, for example, the pathogen

Figure 4.26 Diagram showing the various environments that *Salmonella typhi* will encounter and in which it will have to survive and, in some cases, proliferate.

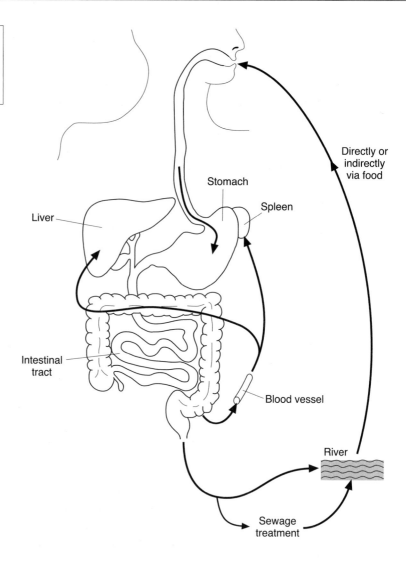

Salmonella typhi. This organism causes a systemic disease (typhoid fever) in humans and is transmitted predominantly by the faecal–oral route, i.e. it is expelled from its host in diarrhoea, gains access to water supplies and is then ingested by a new host via contaminated water. Following passage through the stomach and into the intestines, it crosses the intestinal epithelium, accesses the blood circulation via endothelial cells (or else within macrophages) and is carried to, and colonises, a number of organs. The organism therefore encounters a large number of different environments – water/sewage, stomach, the intestinal lumen, epithelial cells, blood, phagocytes, liver and spleen (Figure 4.26). In each of these situations it will have to: (i) cope with different physicochemical conditions (pH, osmotic pressure, oxygen content, redox potential); (ii) use a different range of molecules as nutrients; (iii) produce different adhesins to enable colonisation of the different surfaces it encounters; and (iv) adopt different anti-host strategies depending on where it is within the host. In order to grow (or, at least, survive) in each of these environments, the organism must be able to sense the prevailing external conditions and adjust its structure, physiology and behaviour (i.e. its phenotype) to suit that particular environment. It does this by up- or down-regulating appropriate genes, expressing only those needed to survive under the specific environmental conditions to which it is being exposed. Evidence that bacteria produce a particular virulence factor only when it is required, demonstrating environmental control over virulence gene expression, came initially from studies comparing freshly isolated and laboratory-grown strains of an organism. For example, freshly isolated strains of *Neisseria gonorrhoeae*, but not laboratory-grown strains, are resistant to complement-mediated killing. During the early studies of toxin production by *B. anthracis*, scientists found it very difficult to get the organism to produce an exotoxin in laboratory media. More recent studies, using modern molecular techniques, have confirmed that the expression of particular virulence factors is tightly regulated by bacteria, and numerous examples will be given of this in subsequent chapters.

A particular environmental signal, such as a change in pH or temperature, will often alter the expression of a number of genes encoding those virulence factors needed at a particular stage in the infectious process. These genes will often be organised in a regulon, i.e. although they are located on different parts of the chromosome they are under the control of the same regulatory protein (see Chapter 2). For example, *Yersinia* spp. have a temperature-dependent regulon that regulates nearly all of the virulence factors of the organism, including adhesins, invasins, enterotoxin and outer membrane proteins. A group of regulons that are affected by the same environmental stimulus but respond in different ways are said to comprise a stimulon. Environmental parameters particularly important in controlling virulence gene expression in many organisms include temperature, pH, osmolarity, the composition of the gaseous phase and the concentrations of iron, phosphate and calcium.

4.3.1.2 Two-component signal transduction

Bacteria have evolved sophisticated systems for eliciting adaptive responses to their environment. It would appear that the response of many bacterial species to a variety of environmental stimuli occurs by a similar mechanism, which involves a two-component signal transduction system (TCSTS). One component of this system (the sensor, located in the cytoplasmic membrane) monitors some environmental parameter and signals the other component (the response regulator, a cytoplasmic protein), which initiates the response (Figure 4.27). While the response usually involves some change in gene expression, this is not always the case. For example, in motile organisms, the response may ultimately involve an alteration in flagellar motion so that the organism moves towards a source of nutrients or away from toxic chemicals.

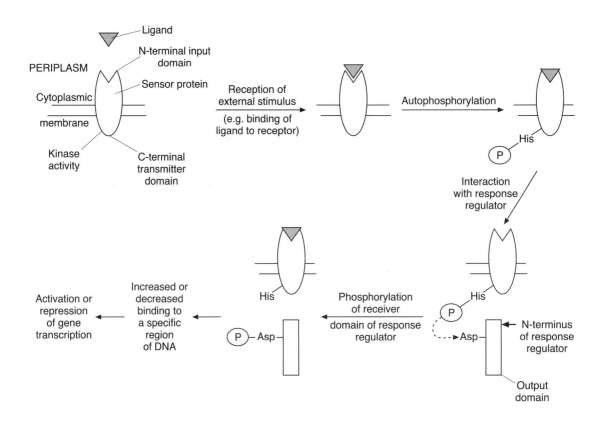

Figure 4.27 General features of two-component signal transduction systems (TCSTSs) of bacteria. A sensor protein (which spans the cytoplasmic membrane) receives a signal from the external environment and responds by undergoing autophosphorylation at its C-terminal transmitter domain (located inside the cytoplasm). The phosphorylated sensor protein then interacts with the N-terminus (the receiver domain) of a specific response regulator and phosphorylates the latter. Phosphorylation of the response regulator alters the ability of its output domain to interact with a particular region of the chromosome, thereby altering (activating or repressing) gene transcription.

Let us now look at the structure and function of the two-component regulatory system in more detail. Essentially, signal transduction involves signal recognition and its conversion to gene activation/suppression or to some other response. Typically, the sensor will have an N-terminal input domain located in the periplasmic space, which can respond to some specific stimulus it receives from the environment. This stimulus may be the presence of a particular molecule (a nutrient, toxic chemical, waste product, etc.) or physical state (pH, osmotic pressure, redox potential, etc.). On detecting one of these stimuli, it must convey this 'information' to the cell interior. It does this by using its C-terminal region (located in the cytoplasm) as a 'transmitter' to the response regulator. Before we explain how this occurs, the first thing to consider is how the external stimulus results in activation of the transmitter region of the molecule. The means by which the sensor detects a particular stimulus is, in some cases, by ligand binding, for example a molecule such as glucose binding to its complementary receptor on the sensor protein. This probably induces a conformational change in the sensor protein, thereby activating the cytoplasmic transmitter domain. All sensor proteins have kinase activity (and are often referred to as 'sensor kinases') and this activity is located in their C-terminal region. The conformational change induced in the molecule when it receives an environmental stimulus activates this kinase activity, resulting in autophosphorylation of a specific histidine residue on the molecule. Phosphorylation of the sensor kinase enables it to inter-act with its specific response regulator, thereby transmitting the signal one more step along its pathway. The response regulator usually has an N-terminal receiver module that detects the incoming signal from the sensor and then alters the activity of its C-terminal output domain and thereby initiates a response. Signal transmission to the response regulator involves the transfer of the phosphate group from the histi-dine residue of the receptor kinase to an aspartate residue on the N-terminal receiver domain of the response regulator. The latter accounts for the alternative name of the TCSTS – the histidyl-aspartyl phospho-relay. The latter term, although less widely used, is preferred, as it is now evident that the signalling pathway may involve far more than two proteins. The receiver domain of the response regulator is usually connected to its output domain by flexible linkers. The output domain generally has DNA-binding activity, and phosphorylation of the receiver alters this activity, thereby increasing or decreasing its ability to bind to DNA. Hence, the transcription of one or more target genes is affected. Control of the whole system is achieved by phospha-tases (particularly histidine and aspartate phosphatases) acting at various points in the signal transduction pathway. The signal may be terminated, for example, by dephosphorylation of the activated recep-tor kinase.

The functional domains of each of the proteins comprising the system are highly conserved. It is possible, therefore, to identify the presence of such domains in bacteria either by searching genome sequence databases or by using appropriate DNA probes. Hence, although such approaches have revealed that TCSTSs are widely dis-tributed among bacteria, few have actually been subjected to detailed experimental analysis. A single organism may contain many TCSTSs. For example, it has been shown that in E. coli more than 50 such systems operate (Table 4.22). Examples of such systems known to control viru-lence gene expression are given in Table 4.23.

Two-component signal transduction systems in *Helicobacter pylori*

TCSTSs are widespread signal transduction devices that enable bacteria to regulate their cellular functions in response to changing environmental conditions and, as already pointed out, large numbers of such systems appear to be present in many organisms. However, there was a surprise in store for those who searched the genome sequence of *Hel. pylori* for genes encoding such systems. Only four open reading frames (ORFs) with homology to TCSTS sensor kinase ORFs were found, along with six genes encoding response regulators. Of the four histidine kinase response regulator pairs detected, one is thought to be involved in chemotaxis, while another is involved in regulating the synthesis of flagellar components. The possession of such a small number of TCSTSs is unusual and possibly reflects the adaptation of this organism to the extreme environment provided by the human stomach. It also implies that the organism would have difficulty in adapting to any en-vironment other than that of the human stomach.

Table 4.22. Examples of two-component signal transduction systems in *Escherichia coli*

Sensor	Response regulator	Signal	Response
NtrB	NtrC	Concentration of nitrogen-containing compounds in the environment	Transcription of genes involved in uptake and metabolism of nitrogenous compounds such as glutamine, arginine, histidine, nitrate and nitrite
PhoR	PhoB	Concentration of phosphate in the environment	Phosphate regulation
EnvZ	OmpR	Osmolarity of environment	Osmoregulation, regulation of porin expression
EvgS	EvgA	Low temperature, Mg^{2+}, nicotinic acid	Regulation of OmpC expression
PhoQ	PhoP	Concentration of Mg^{2+} in environment	Transcription of loci essential for growth at low concentrations of Mg^{2+}

Table 4.23. Examples of two-component regulatory systems used in the control of virulence gene expression

Organism	Signal	Sensor	Response regulator	Response
Salmonella typhimurium	Mg^{2+}	PhoQ	PhoP	Ability to survive low pH and withstand antibacterial peptides
Salmonella typhimurium	Low pH within macrophage phagosome	EnvZ	OmpR	Activates a type III secretion system
Bordetella pertussis	Temperature, Mg^{2+}, nicotinic acid, SO_4^{2-}	BvgS	BvgA	Synthesis of filamentous haemagglutinin, pertussis toxin and adenylyl cyclase toxin
Enterococcus faecalis	Vancomycin	VanS	VanR	Enzymes required for vancomycin resistance
Shigella flexneri	Osmolarity	EnvZ	OmpR	Porin expression
Pseudomonas aeruginosa	Osmolarity	AlgR2	AlgR1	Alginate synthesis
Klebsiella pneumoniae	N_2	NtrC	NtrA	Urease production
Vibrio cholerae	pH, osmolarity, temperature	ToxS	ToxR	Synthesis of toxin and pili

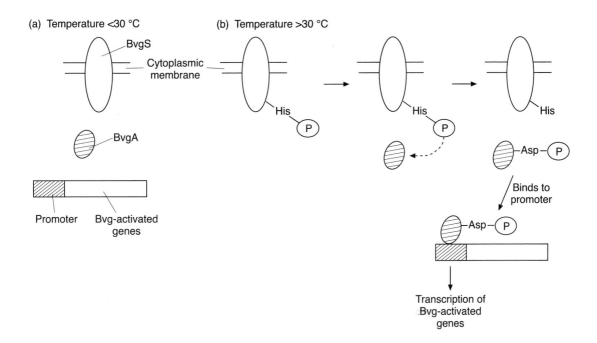

(a) Temperature <30 °C

(b) Temperature >30 °C

BvgS

Cytoplasmic membrane

His

P

BvgA

His

P

His

Asp—P

Binds to promoter

Promoter   Bvg-activated genes

Asp—P

Transcription of Bvg-activated genes

One example will be described of the way in which a histidyl-aspartyl phosphorelay system is involved in the regulation of virulence genes in a human pathogen – *Bordetella pertussis*. Important virulence determinants of this organism include a number of adhesins (see Chapter 7) and exotoxins (see Chapter 9). The production of three of these toxins – the pertussis toxin, adenylyl cyclase toxin and the DNT – is regulated by the Bvg (*Bordetella* virulence gene) TCSTS, which consists of the sensor kinase BvgS (a 135 kDa transmembrane protein), and the response regulator BvgA (a 23 kDa cytoplasmic protein). BvgA is a transcriptional activator that has a C-terminus with a helix-turn-helix motif – a characteristic of a DNA-binding protein. The Bvg system is affected by a number of environmental factors including temperature, sulphate ions and nicotinic acid. Temperatures below 30 °C, or a high concentration of sulphate ions or nicotinic acid, are known to reduce expression of the three toxins mentioned above. At temperatures greater than 30 °C (or in the absence of sulphate ions or nicotinic acid), autophosphorylation of BvgS occurs and this is followed by transfer of the phosphate moiety to BvgA. Phosphorylation of BvgA confers on it an increased affinity for promoter regions of the Bvg-activated genes to which it binds and thereby activates transcription of these genes (Figure 4.28). Hence, at low temperatures (such as those encountered outside its human host) no toxins are produced. However, when the organism colonises the respiratory epithelium (and hence is exposed to a temperature greater than 30 °C) toxin production is induced.

Figure 4.28 The Bvg two-component signal transduction system (TCSTS) of *Bordetella pertussis*. At temperatures below 30 °C (a) no toxins are produced whereas at temperatures above this (b) toxins are synthesised and released.

4.3.1.3 Other signal transduction pathways in bacteria

Until fairly recently, the histidyl-aspartyl phosphorelay system was thought to be the only signal transduction pathway in bacteria. However, it is now known that other signalling pathways, similar to those found in eukaryotic organisms, operate in certain bacterial species. Hence serine/threonine protein kinases (see Section 4.2.2.4) have been detected in *Mycobacterium tuberculosis*, *Myxococcus xanthus* and *Streptomyces coelicolor* and have been shown to be involved in processes such as spore development and antibiotic production. Furthermore, protein tyrosine kinases have also been identified in *Streptomyces griseus*. Whether control of virulence gene expression in human pathogens involves such 'eukaryotic-like' signalling molecules remains to be established.

4.3.1.4 Monitoring of the internal environment of the bacterium

Complementary to the signalling pathways that monitor changes in the environment external to the bacterium are those sensing the intercellular environment. These depend on sensor molecules whose sensory domains (known as PAS domains) are located in the cytoplasm. The term PAS is an acronym based on those proteins in which the characteristic structure of the PAS domain was first recognised – the *Drosophila* period clock protein, the vertebrate aryl hydrocarbon receptor nuclear translocator and the *Drosophila* single-minded protein. PAS domains monitor changes in redox potential, oxygen, light intensity, small ligands and the overall energy level of a cell. Most PAS domains are located on histidine kinase sensor proteins and are associated with transducing and regulatory proteins. While the exact role of PAS-mediated signal transduction in bacterial virulence is not known, it is certain to be crucial to the survival of pathogens in the wide range of environments encountered within the host.

4.3.2 Cell–cell signalling in bacteria

Until fairly recently, individual bacterial cells were considered to lead very lonely lives. Each cell appeared to exist as an independent unit, concerned only with feeding, growing, excreting and undergoing binary fission, without any sense of community. However, due to some fascinating studies involving deep-sea fish and the phenomenon of bioluminescence, this view of the lifestyle of bacteria has gone forever and we must now think of bacteria in terms of communities in which the individual members are able to communicate by a process known as quorum sensing.

4.3.2.1 Quorum sensing

4.3.2.1.1 Introduction

In the 1960s it was discovered that sea water from almost anywhere on earth contained bioluminescent bacteria. Light generation in these organisms is due to the activity of the enzyme luciferase, which is able to convert chemical energy (stored in ATP) to light energy (photons). However, the light emitted was of such a low intensity that it was difficult to imagine that it could serve any useful physiological function. Almost a decade later, it was observed that when these organisms (e.g. *Vibrio fischeri*) were grown in the laboratory, the quantity of light emitted increased dramatically when the cultures reached the mid-logarithmic phase of growth. It was concluded that the bacteria were producing a chemical (an 'auto-inducer') that activated the transcription of the luciferase-encoding genes but this occurred only when a certain concentration of the auto-inducer was reached. At low cell concentrations, insufficient auto-inducer is produced to initiate luciferase production but, as the cells grow, the auto-inducer accumulates and reaches the critical concentration required to initiate transcription of the luciferase genes. The ability of bacteria to initiate the transcription of certain genes only when a certain population density is reached is known as 'quorum sensing'. But what about a functional role for this phenomenon? Although *V. fischeri* are found as free-living planktonic cells in the sea, they are also found concentrated in enormous numbers in the light organs of certain aquatic animals such as the pony fish (*Leiognathus fasciatus*) and the bobtail squid (*Euprymna scolopes*). The symbiosis between this bacterium and the bobtail squid has been extensively investigated and makes a fascinating story that is described in greater detail in Section 11.1.6. Suffice to say at this stage that the light emitted by the bacteria is important in camouflaging the squid, while the bacteria are provided with a conducive environment for their survival and growth.

4.3.2.1.2 Quorum sensing in Gram-negative bacteria

The molecular basis of quorum sensing in Gram-negative bacteria has been investigated extensively. In the case of *V. fischeri*, the auto-inducer is a small molecule (*N*-3-oxohexanoyl-L-homoserine lactone, OHHL) belonging to the acyl-homoserine lactone (AHL) family that is continually synthesised and secreted by the organism. Under certain circumstances, a threshold concentration of OHHL is reached in the environment (approximately 10 nM) and at this point a signal transduction cascade is initiated that results in the production of luciferase and light generation (Figure 4.29). Luciferase, and the proteins required for the synthesis of the substrate for this enzyme (tetradecanal), are encoded by the operon *luxCDABEG*. *luxA* and *luxB* encode the α- and β-subunits of luciferase, while the other genes in the operon encode polypeptides involved in the synthesis of tetradecanal. The OHHL auto-inducer is synthesised by the product of the *luxI* gene (OHHL synthase) and can bind to the N-terminal region of the transcriptional activator protein (LuxR). The C-terminal region of LuxR, when activated by binding to OHHL, can bind to an operator region of DNA upstream from *luxCDABEG*, resulting in gene transcription and hence luciferase production. The whole process, of course, is a type of signal

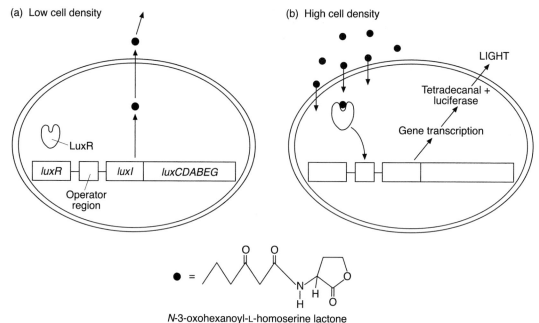

N-3-oxohexanoyl-L-homoserine lactone

Figure 4.29 Quorum sensing in *Vibrio fischeri*. (a) At low cell densities the auto-inducer (an acyl homoserine lactone) diffuses away from the cell and so the transcriptional activator (LuxR) cannot induce transcription of the *luxCDABEG* operon. (b) At high cell densities, the auto-inducer binds to LuxR, which is then able to bind to DNA, so inducing transcription of the genes encoding luciferase and the proteins required to synthesise its substrate, thereby generating light.

transduction system in which the organism responds to an external signal (in this case the 'population density' of the organism) by transcribing certain genes.

The signalling system outlined above has been shown to form the basis of density-dependent gene expression in over 30 different species of Gram-negative bacteria including some important pathogens of humans, other animals and plants (Table 4.24).

What is obvious from the table is that a range of AHLs can function as auto-inducers (also termed pheromones or microbial hormones) and that a wide variety of genes, or processes, are controlled by this signalling system. All of the systems investigated to date consist basically of a AHL synthase (homologous to LuxI) and a transcriptional activator (homologous to LuxR). In the LuxR-type proteins, the C-terminal third of the molecule is homologous to the DNA-binding domain of a group of transcriptional regulators (sometimes termed the LuxR superfamily) and contains a 19 amino acid residue sequence with a helix-turn-helix motif (characteristic of a DNA-binding region) that can bind to the target promoter. The N-terminal region of the molecule prevents such binding taking place but this inhibition is reversed by the auto-inducer (presumably by binding to some, as yet, unspecified region) so that the C-terminal can then bind to the target promoter and activate gene transcription. Each LuxR homologue is activated by a particular auto-inducer, although high concentrations of other members of the AHL family can also result in activation of gene transcription.

Inhibition of quorum sensing – a new antimicrobial approach?

Quorum sensing appears to be involved in regulating gene expression in a wide range of pathogens of humans, other animals and plants. It is reasonable, therefore, to expect that higher organisms could have evolved some means of interfering with quorum sensing in order to prevent virulence gene expression by colonising/invading bacteria – such a system has been found in the red marine alga *Delisea pulchra*. This alga is able to resist colonisation by a number of marine bacteria and does so by secreting molecules such as halogenated enones and furanones. Some of these molecules have structures similar to those of acylhomoserine lactones and so it was suggested that they could act as inhibitors of AHL-mediated gene expression. This hypothesis was tested by examining the ability of these compounds to inhibit two AHL-induced phenomena – swarming motility in *Serratia liquefaciens* and bioluminescence in *V. fischeri* and *V. harveyi*. Two of the furanones secreted by the alga were able to inhibit both processes and were effective at concentrations likely to be encountered in the natural environment. Furthermore, it has been shown that these algal products exert their inhibitory activity by displacing AHL from its binding site on the regulatory protein, LuxR. Such compounds have great potential for controlling plant and animal pathogens in which virulence gene expression is regulated by AHLs.

Table 4.24. | Density-dependent gene expression in Gram-negative bacteria

Organism	LuxI/LuxR homologues	Signalling molecule	Genes (or processes) regulated
Aeromonas salmonicida	AsaI/AsaR	BHL	Exoproteases
Pseudomonas aeruginosa	LasI/LasR	OHHL	Elastase, alkaline protease, LasA protease, exotoxin A, Type II secretion apparatus
	RhlI/RhlR	BHL	Rhamnolipids, elastase, lipase
	LasI/LasR	ODL	Biofilm formation
Burkholderia cepacia	CepI/CepR	OHL	Protease secretion
Erwinia carotovora	ExpI/ExpR	OHHL	Exoenzymes
Chromobacterium violaceum	?	HHL	Exoenzymes
Serratia liquefaciens	?	BHL	Secretion of a lipopeptide involved in swarming
Yersinia pseudotuberculosis	YpsI/YpsR	OHHL	Motility
Yersinia enterocolitica	YenI/YenR	OHHL	Not determined

OHL, *N*-octanoylhomoserine lactone; BHL, *N*-butanoylhomoserine lactone; OHHL, *N*-3-oxanoylhomoserine lactone; HHL, *N*-hexanoylhomoserine lactone; ODL, *N*-3-oxododecanoyl homoserine lactone.

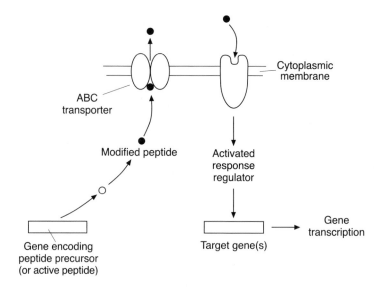

Figure 4.30 Quorum sensing in Gram-positive bacteria. The oligopeptide signalling molecule is exported across the cytoplasmic membrane via an ABC transporter. The molecule then binds to the sensor kinase of a specific two-component signal transduction system, thereby activating its response regulator, which induces transcription of the target gene(s).

4.3.2.1.3 Quorum sensing in Gram-positive bacteria

Gram-positive bacteria do not use AHLs as signalling molecules and do not employ a LuxI/LuxR type of signalling system. Instead, they use oligopeptides as signalling molecules and, in many cases, these are secreted via ATP-binding cassette (ABC) transporter proteins (see Section 2.5.4), in contrast to the AHLs, which simply diffuse out through the cytoplasmic membrane (Figure 4.30). The signalling peptide is usually (but not always) synthesised as an inactive precursor that undergoes some form of modification before being exported. The secreted peptide then interacts with a specific TCSTS, as described in section 4.3.1.2. The genes encoding the signalling peptide precursor, the ABC exporter and the TCSTS that recognises it are usually transcriptionally linked. Examples of this type of system are given in Table 4.25. A range of processes appear to be regulated in this way, particularly competence, conjugation, bacteriocin production, sporulation, and the production of virulence factors.

Multiple signalling pathways in *Bacillus*

The induction of competence and sporulation in *B. subtilis* involves a complex regulatory network of peptide signals. The ComX competence pheromone is a short, modified peptide whose concentration in the extracellular medium is monitored by ComP, the sensor kinase of a TCSTS. Once the ComX pheromone reaches a critical concentration, ComP autophosphorylates and subsequently phosphorylates ComA, the response regulator of this TCSTS, which leads to activation of late competence genes required for DNA uptake and homologous recombination with the bacterial chromosome. Several studies, however, also demonstrated that cells with mutations in the oligopeptide permease operon, *spo0K*, demonstrated delayed competence phenotype. The *spo0K* operon encodes an import ABC transporter that mediates the ATP-dependent uptake of short oligopeptides. Thus it appears that full competence and sporulation in *B. subtilis* requires two signals: one that is secreted and detected extracellularly by a TCSTS, and a second signal that is internalised via an oligopeptide permease and functions intracellularly in the target cell. The second signal, competence-stimulating factor, was identified as a peptide with five amino acid residues that was cleaved from a larger precursor, encoded by *phrC*. The *phrC* gene is co-transcribed with *rapC*, encoding an intracellular phosphatase. It has been proposed that the internalised competence-stimulating factor interacts with RapC phosphatase to inhibit its activity. Consequently, the RapC phosphatase is unable to dephosphorylate active (phosphorylated) ComA and so ComA remains phosphorylated and thus able to activate the gene transcription required for the development of competence. A similarly integrated system is thought to exist for the initiation of sporulation in *B. subtilis*.

4.3.2.1.4 Other quorum sensing systems

Vibrio harveyi appears to have two quorum sensing systems and these have some of the features of the systems found in both Gram-negative and Gram-positive organisms. One system depends on an AHL auto-inducer (designated AI-1) but, rather than simply diffusing back into the bacterium (or into a neighbouring cell) through the cytoplasmic membrane, this transmits its signal via an unusual TCSTS. LuxN, required for response to the AI-1 molecule, contains regions of sequence with homology to *both* sensor kinase and response regulator elements

Table 4.25.	Quorum sensing in Gram-positive bacteria		
Organism	Signalling molecule	Signalling pathway	Genes (or processes) regulated
Staphylococcus aureus	Cyclic octapeptides	Two-component system	Production of exotoxins, exoenzymes, coagulase, protein A
Streptococcus pneumoniae	Heptadecapeptide (competence-stimulating peptide)	Two-component system	Competence
Streptococcus sanguis	Peptide	Two-component system	Competence
Lactobacillus plantarum	Plantaricin A (a 26 AA residue peptide)	Two-component system	Bacteriocin production
Enterococcus faecalis	Oligopeptides	Internalisation of molecule	Plasmid transfer
Bacillus subtilis	ComX (a 10 AA residue peptide)	Two-component system	Competence

AA, amino acid.

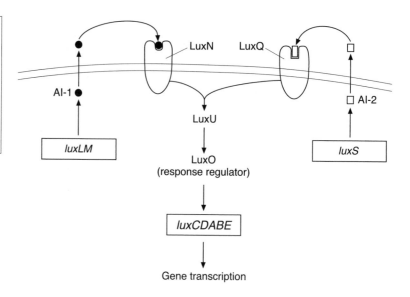

Figure 4.31 The two quorum sensing systems operating in *Vibrio harveyi*. When threshold concentrations of the auto-inducers (AI-I and AI-2) are reached, a signalling cascade is initiated via different sensor kinases that utilise a common response regulator (LuxO).

of normal TCSTSs. These components are usually on separate protein molecules. The identity of the second auto-inducer (AI-2, synthesised by the product of the *luxS* gene) is not known, but, like AI-1, it is recognised by an unusual TCSTS (LuxQ) that consists of a single protein with both sensor kinase and response regulator functions. LuxU and LuxO comprise a phosphorelay system that serves to integrate sensory input from signalling systems 1 and 2. LuxU is a shared integrator protein that can be phosphorylated either by LuxN or LuxQ. LuxU subsequently conveys the signal to the response regulator protein LuxO. LuxO is a negative regulator of the *luxCDABE* operon. Phosphorylation of LuxO by LuxU inactivates its repressor activity, allowing transcription of the *luxCDABE* operon and, consequently, light production (Figure 4.31). This is likely to be an important quorum sensing system, as homologues of the *luxS* gene have been found in a wide range of Gram-positive and Gram-negative organisms including *Sal. typhimurium*, *Haemophilus influenzae*, *Strep. pneumoniae*, *Staph. aureus*, *Cl. difficile*, *E. coli*, *Hel. pylori* and *N. gonorrhoeae*. The *luxS* genes do not show any homology with any other gene involved in auto-inducer production. The genes controlled by this system remain to be established in most cases. However, it has recently been shown that siderophore production by *V. harveyi* is controlled in this way.

Bacterial cross-talking

So far, we have described quorum sensing in terms of an auto-inducer produced by one particular bacterium regulating the genes expressed by that bacterium, i.e. self-signalling or autocrine signalling. However, it has recently been proposed that an auto-inducer produced by one organism may regulate gene expression in another (albeit closely related) organism. The *luxS* gene has been detected in both enterohaemorrhagic (EHEC) and enteropathogenic (EPEC) strains of *E. coli* and, as explained in Section 4.6.2.2, the AI-2 produced by this gene can regulate the expression of the genes required to produce the attaching and effacing lesion induced by these organisms in intestinal epithelial cells. However, as the infectious dose of EHEC can be as few as 50 cells it is difficult to understand how density-dependent expression of virulence genes could occur in such circumstances. It has been suggested that density-dependent induction of virulence gene expression results from the accumulation of auto-inducers synthesised by non-pathogenic strains of *E. coli* normally resident in the large intestine.

A reporter strain of *V. harveyi* in which the sensor for AI-1 had been inactivated has been used to screen the culture supernatants of a diverse range of bacteria for the production of AI-2 molecules. Culture supernatants from *E. coli*, *Salmonella* spp., *Hel. pylori* and *Porphyromonas gingivalis* were all found to be capable of inducing light production by the reporter strain. This finding raises the exciting possibility that this signalling system could be the basis of a means of interspecies and intergeneric communication.

4.3.2.1.5 Benefits of quorum sensing

We have seen that quorum sensing is a widely distributed ability among bacteria and that a number of different systems have evolved to enable bacteria to sense their population density and thereby control gene expression. Furthermore, in organisms such as *Ps. aeruginosa*, a remarkably large number of genes and processes have been found to be regulated by quorum sensing (Table 4.26). The advantages that such a system confers on the organism remain the subject of much speculation. In their natural environments bacteria must generally

Table 4.26. Genes and processes in *Pseudomonas aeruginosa* that are known to be regulated by quorum sensing

Gene (or process) regulated	Gene product (or process affected)	Quorum sensing system(s) involved
lasA	LasA elastase	*las*
aprA	Alkaline protease	*las, rhl*
toxA	Exotoxin A	*las*
xcpP	Type II secretion system	*las /rhl*
rhlAB	Rhamnolipid	*rhl /las*
lasR	Transcriptional activator	*las*
lasI	Auto-inducer synthase	*las*
rhlR	Transcriptional activator	*las*
?	Pyoverdine production	*las*
Export or assembly of pilin	Twitching motility	*rhl*
Biofilm formation	?	*las*

?, not currently known.

derive their nutrients from complex polymers, and the effective degradation of these requires the concerted secretion of enzymes from large numbers of cells. An individual cell, or a population in which only some of the members were secreting appropriate enzymes, would not constitute an effective means of utilising the available nutrient resources. This may apply to the quorum-dependent secretion of proteases by organisms such as *Ps. aeruginosa*. The advantage of competence and conjugation being regulated by a density-dependent process is more obvious – DNA transfer is not possible in the absence of other cells! The ability to limit virulence factor secretion until a large bacterial population has been achieved may also serve as a protective measure against host defence systems. Hence, if only a few bacteria were to start secreting a particular virulence factor (small concentrations of which would be unlikely to cause serious damage) this could alert the host, which might then be able to dispose of this threat effectively – something it is less likely to achieve if a large population is present. Recently it has been reported that mutants of *Ps. aeruginosa* lacking the two quorum sensing systems of the organism are almost avirulent in neonatal mice.

4.3.2.2 Bacterial cytokines

As described in Section 4.2.6, cytokines are important intercellular signalling molecules in mammalian hosts and are involved in the integration of cell and tissue behaviour in multicellular organisms. One of the most important ways in which cytokines carry out these functions is by controlling the growth and differentiation of a range of host cell types. Molecules able to affect cellular growth and/or differentiation have now been found in other types of organism including plants and even unicellular eukaryotes. There is now evidence that cytokine-like molecules are also produced by bacteria. For example, it has been known for many years that the supernatant from a log-phase culture of a bacterium will often shorten the lag phase when it is added to another batch culture of the same organism. This implies that the organism produces a substance capable of stimulating its own growth. Such substances have been called pheromones, auto-inducers, bacterial hormones, bacterial cytokines and bacteriokines. The identity of most of these bacterial cytokines, however, has not been determined. In the case of *B. subtilis*, the cytokine has been shown to be a siderophore while in *Micrococcus luteus* it is a 17 kDa protein. The cytokine from *Mic. luteus* has been investigated the most extensively. Picomolar concentrations of this protein (also known as resuscitation promoting factor, Rpf) are able to reduce the lag phase of cultures of the organism and to promote the resuscitation of dormant cells, i.e. those which have been in the stationary phase of growth for prolonged periods of time. The gene encoding Rpf has been sequenced and genes similar to *rpf* have been detected in many Gram-positive bacteria that have a high G + C content, for example *Mycobacterium* spp., *Corynebacterium* spp. and *Streptomyces* spp. Interestingly, Rpf is also able to stimulate the growth of a number of *Mycobacterium* spp., including *M. tuberculosis*, *M. leprae* and *M. avium*. This suggests that these organisms (which do have the *rpf* gene) also produce a similar (or even identical) cytokine. However, the receptor for Rpf (on *Mic. luteus* or *Mycobacterium* spp.) has not been identified and nothing is known of the signal transduction pathways that it induces.

The study of bacterial cytokines is in its infancy but is receiving increasing attention due, partly, to the possibility that such substances would constitute targets for new classes of antibiotics.

4.4 | Bacterial reception of host signals

We have seen that a bacterium is well equipped to respond to a range of environmental stimuli, as well as to signals emanating from other bacteria. But what about signals emanating from host cells? Can bacteria respond to these? Death, of course, is a response and bacteria can certainly be killed by host-generated substances such as antimicrobial peptides and a range of antibacterial substances produced by phagocytes. Host cells will also be continually excreting end-products of metabolism and some of these (e.g. organic acids) can be utilised by bacteria as nutrients – the response in this case is to grow and reproduce. As will be described in Chapter 10, bacteria can bind to, and even degrade, host cell signalling molecules such as cytokines and thereby disrupt cytokine networks. Furthermore, the growth of some bacteria (e.g. *E. coli*) has been shown to be stimulated by certain cytokines such as IL-2 and granulocyte-macrophage colony-stimulating factor (GM-CSF). There is also evidence that bacteria have receptors for mammalian hormones and that hormones such as noradrenaline can stimulate the growth of *E. coli*. Other examples of host signalling molecules influencing bacteria include (i) stimulation of the growth of Gram-negative bacteria by catecholamines, (ii) stimulation of the growth of *Enterococcus faecalis* by serotonin and (iii) inhibition of the growth of *Ps. pseudomallei* by insulin.

A more intimate interaction between a bacterium and a host cell may be expected to result in a mutual exchange of signals. As we will describe in far greater detail in Chapters 7 and 8, adhesion to, and invasion of, host cells by bacteria results in changes in both types of cell as a result of 'cross-talk' between them. During both the adhesion and the invasion processes, it is known that bacterial genes are up- and down-regulated. However, it is generally not known whether these changes result from an altered environment or are due to specific signals emanating from the host cell. A fuller discussion of examples of bacterial responses to host cells, and the activation of bacterial gene transcription by host cells, is given in Chapters 7 and 8. As will be made clear in these chapters, bacteria are certainly able to interfere with host signal transduction and use this for their own benefit. It is worth pre-empting these chapters to some extent by briefly mentioning some important examples of the means by which bacteria accomplish this.

Table 4.27. | Bacterial interference with host cell signal transduction. Phosphatases and kinases delivered to host cells by type III secretion systems

Organism	Enzyme	Activity	Host cell targets	Effect on host cell
Yersinia pseudotuberculosis	YopH	Tyrosine phosphatase	p130^{cas}, paxillin and focal adhesion kinase	Cytoskeletal rearrangements
Yersinia pestis	YpkA	Serine/ threonine kinase	Not known	Possibly cytoskeletal rearrangements
Yersinia enterocolitica	YopO	Serine/ threonine kinase	Not known	Possibly cytoskeletal rearrangements
Salmonella typhimurium	SptP	Tyrosine phosphatase and GTPase-activating protein activity	Rac, Cdc42	Actin rearrangements
Salmonella dublin	SopB	Inositol phosphatase	Hydrolyses several inositol and phosphatidylinositol phosphates	Induction of Cl$^-$ flux
Salmonella typhimurium	SigD	Inositol phosphatase	As for SopB	As for SopB
Shigella spp.	IpgD	Inositol phosphatase	Not known	Not known

As was described in Section 2.5.5, some bacteria possess a type III secretion system, which is used to inject effector molecules directly into the cytoplasm of the host cell. Some of these effector molecules are, in fact, phosphatases or kinases and, furthermore, they have been found to have significant homology to the corresponding eukaryotic enzymes (Table 4.27). As described earlier in this chapter (Section 4.2.3.1) phosphorylation/dephosphorylation reactions are the cornerstones of host cell signalling pathways. Injection into the host cell of bacterial phosphatases and kinases, therefore, enables the organism to interfere with host cell signal transduction. The consequences of this are discussed in greater detail in Chapters 8 and 10.

The various signalling and communication pathways used by bacteria are summarised in Table 4.28.

Table 4.28. Communication and signalling systems used by bacteria

Type of communication system	Signalling mechanism	Principle function	Distribution among bacterial species
Intracellular	Two-component signal transduction systems	To monitor external environment and respond accordingly	Universal?
Intracellular	Phosphorelay system	To monitor intracellular environment and respond accordingly	Universal?
Intercellular (involving only bacteria)	Quorum sensing involving homoserine lactones	To monitor population density and respond accordingly	Present in species from many Gram-negative genera including human pathogens
Intercellular (involving only bacteria)	Quorum sensing involving other signalling molecules (e.g. synthesised by LuxS)	To monitor population density and respond accordingly. Possibly enables interspecies and intergenera communication	Present in species from many Gram-negative genera and some Gram-positive genera including human pathogens
Intercellular (involving only bacteria)	Quorum sensing involving other signalling molecules (mainly peptides)	To monitor population density and respond accordingly	Present in species from some Gram-positive genera including human pathogens
Intercellular (involving only bacteria)	Involves bacterial cytokines – mechanism unknown	Promotes bacterial growth and resuscitation of dormant cells	Present in species from some Gram-positive genera including human pathogens
Intercellular (involving bacteria and host cells)	Involves bacterial manipulation of host cell signal transduction pathways (see Chapters 7, 8 and 9)	To enable: (i) bacterial invasion, (ii) transfer of bacterial products into host cell, (iii) inhibition of phagocytosis	Carried out by many human pathogens
Intercellular (involving bacteria and host cells)	Involves bacterial utilisation of host cell signalling molecules – mechanism unknown	Stimulates bacterial growth and disrupts intercellular signalling pathways in host	Not widely investigated but detected in a few human pathogens

4.5 | Concept check

- Eukaryotic cells perceive their environment by using cell surface receptors that bind selected ligands (hormones, cytokines, secreted molecules from pathogens) with high affinity.
- Ligand binding to specific receptors induces particular patterns of intracellular signalling – an amplification mechanism for ensuring that the small amounts of biological signals present can be recognised and acted upon.
- Binding to ligands induces various changes in cells and, in particular, changes in cell metabolism (e.g. by activating selected enzymes), cell shape or patterns of gene transcription.
- Intracellular signalling involves a very large number of proteins with kinase or phosphatase activity or with specific domain binding characteristics.
- Intracellular signalling also involves the actions of small molecular mass compounds called second messengers.
- Specific patterns of protein–protein interactions are involved in transmission of messages from the cell surface to the cell's interior, including the nucleus, for example the MAP kinase pathway.
- A major class of extracellular signals perceived by cells and involved in controlling responses to infectious agents is the cytokines.
- Bacteria are able to sense, and respond to, the external environment using two-component signal transduction systems (TCSTSs).
- An organism will usually have many TCSTSs, each recognising a particular environmental signal such as pH, oxygen concentration, etc.
- The TCSTS generally responds to a particular environmental signal by altering the transcription of a particular gene or operon.
- Bacteria are able to sense their own population density – this is known as quorum sensing.
- Quorum sensing enables bacteria to co-ordinate, within a population, many physiological processes and also the expression of virulence factors.
- The ability of an organism to express virulence factors only when a critical population density is reached is likely to be important in helping it to survive within its host.
- Bacteria produce cytokine-like molecules (bacteriokines) that are involved in regulating growth and differentiation.
- Bacterial interference with, and manipulation of, host signalling pathways is a key feature of bacteria–host cell interactions both in health and in disease.

Table 4.29. | Examples of proteins involved in the regulation of virulence factors of *Streptococcus pyogenes*. These proteins are similar to the response regulator elements of two-component signal transduction systems but the corresponding sensor components have not yet been identified

Response regulator	Environmental factors influencing response	Virulence factors affected
Mga	Temperature, pH, CO_2, bacterial growth phase	M protein, C5a-peptidase, serum opacity factor, inhibitor of complement cascade
RofA	Oxygen	Fibronectin-binding protein F
Nra	Growth phase	A collagen-binding protein, and a fibronectin-binding protein PrtF2

Mga, multiple gene regulator of group A streptococci; RofA, regulator of protein F – a fibronectin-binding adhesin protein; Nra, negative regulator of group A streptococci.

4.6 | Signalling in the paradigm organisms

4.6.1 *Streptococcus pyogenes*

While the existence of intracellular signalling mechanisms (of the TCSTS type) has been established in this organism, there is, as yet, little evidence that it has any intercellular signalling systems.

4.6.1.1 Intracellular signalling

The most extensively investigated TCSTS in this organism is CsrR-CsrS (capsule synthesis regulator). This has been shown to regulate the expression of a number of important virulence factors including the hyaluronic acid capsular polysaccharide, streptokinase, streptolysin S and mitogenic factor. However, the environmental factors responsible for inducing a signal from the sensor protein CsrS have not yet been identified. Other TCSTSs, as yet incompletely characterised, which have been identified in *Strep. pyogenes* are shown in Table 4.29.

4.6.1.2 Intercellular signalling

The only evidence that this organism is capable of intercellular signalling comes from the finding that disruption of the chromosomal *opp* (oligopeptide permease) operon inhibits the production of one of its exotoxins – SpeB. Opp systems are responsible for the recognition and transport of oligopeptides across the bacterial cytoplasmic membrane and so are involved in intercellular communication (which is often mediated by such molecules) in Gram-positive bacteria. The above finding implies that SpeB production may be regulated by a peptide-based signalling system, although there is no other evidence in support of this.

4.6.1.3 Host–bacteria signalling

The normal habitat of *Strep. pyogenes* is the human pharynx, and the means by which it colonises this region is still the subject of debate. However, it has recently been found that *Strep. pyogenes* contains what appears to be part of the glycolytic pathway on its outer surface. The glycolytic pathway is involved in the oxidation of glucose to pyruvate or lactate and the production of ATP. One of these surface-associated glycolytic proteins has been termed streptococcal surface dehydrogenase (SDH) and is a 36 kDa protein with structural and functional similarity to the glycolytic enzyme glyceraldehyde-3-phosphate dehydrogenase (GAPDH). SDH also has ADP-ribosylating activity similar to toxins such as cholera toxin, described in Section 9.5.1.1. Incubation of cultured human pharyngeal cells with purified SDH has revealed that it binds selectively to a 30 to 32 kDa membrane protein and results in the activation of both serine/threonine and tyrosine kinases. A number of specific proteins are phosphorylated on serine/threonine and tyrosine as a result of SDH binding. Additional proteins are also phosphorylated when pharyngeal cells are incubated with whole *Strep. pyogenes*, suggesting that proteins, in addition to SDH, are involved in bacteria–host cell signalling. The importance of this interaction between SDH and related proteins, host cell kinase activation and bacterial infection lies in the finding that inhibitors of protein kinases such as the tyrosine kinase inhibitor genistein or the protein kinase C inhibitor staurosporine inhibit the invasion of pharyngeal cells by *Strep. pyogenes*.

4.6.2 *Escherichia coli*

4.6.2.1 Intracellular signalling

As mentioned previously in this chapter, more than 50 TCSTSs have been detected in *E. coli* and some of these are listed in Table 4.22. A typical example of one of these is the EnvZ/OmpR system, which senses, and regulates the response of the organism to, changes in the osmotic pressure of the environment. One of the main responses of the organism to such changes is to alter the relative proportions of two outer membrane proteins, OmpF and OmpC. When the osmolarity is low, OmpF predominates whereas at high osmolarity OmpC becomes dominant. EnvZ is the histidine kinase that senses changes in the osmotic pressure of the environment and undergoes autophosphorylation at His-243. The phosphorylated EnvZ then phosphorylates the response regulator, OmpR, at Asp-55. The phosphorylated form of OmpR (OmpR-phosphate) has an increased DNA-binding ability and binds to the upstream sites of the *ompF* and *ompC* genes, thereby regulating their transcription. As well as being able to phosphorylate OmpR, EnvZ can also induce the dephosphorylation of OmpR-phosphate and so can control its intracellular concentration. A low level of OmpR phosphate activates transcription of the *ompF* gene whereas a high level represses this gene but activates trancription of *ompC* (Figure 4.32).

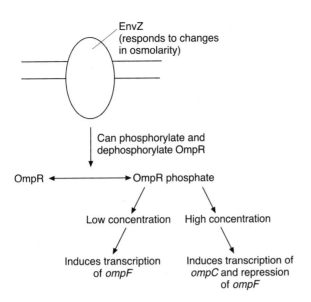

Figure 4.32 Regulation of the relative proportions of the outer membrane proteins (OmpC and OmpF) of *Escherichia coli* by the EnvZ/OmpR two-component signal transduction system (TCSTS).

4.6.2.2 Intercellular signalling

Although *E. coli* does not possess quorum sensing systems of the LuxI/ LuxR type or involving AI-1, several strains of the organism have been shown to produce an auto-inducer with an activity similar to that of AI-2. However, unlike other quorum sensing systems, the AI-2 signal is degraded when the bacteria enter stationary phase. The production of the auto-inducer has been shown to be affected by a number of environmental factors. Hence, production is increased by rapid logarithmic growth, certain carbon sources, low pH and high osmolarity. In contrast, low levels of the auto-inducer are found when (i) the organism enters the stationary phase (ii) certain carbon sources are used (iii) the pH is neutral and (iv) the osmolarity is low. Although the chemical identity of this auto-inducer has not been established, a gene (*luxS*) essential for its synthesis has been detected in *E. coli*, and in many other organisms. Recently it has been shown that the production of key virulence factors of enteropathogenic *E. coli* responsible for the formation of the attaching and effacing lesion (described in Chapter 7) are controlled by a quorum sensing system that utilises AI-2. Specifically, AI-2 controls expression of those operons encoding the type III secretion system, Tir and intimin (see Chapter 7).

4.6.2.3 Host–bacteria signalling

Enteropathogenic *E. coli* is a major cause of diarrhoea in the developing world and the means by which it does this are described in detail in Chapters 1 and 7. This brief section will deal only with the signalling aspect of the pathogenesis of this organism. The ability of this strain to cause pathology is encoded by a pathogenicity island termed the locus of enterocyte effacement (LEE), which encodes a type III secretion system (see Chapters 2 and 7). The pathology induced by EPEC is dependent on its ability to bind to intestinal epithelial cells. This is done initially by the bacteria attaching to the microvilli via the bundle-forming pilus. Once bound, the bacterium uses its type III secretion system to inject proteins into the host cell. Included among these proteins is one termed the translocated intimin receptor (Tir). This is a receptor for the bacterium that binds to a bacterial surface protein known as intimin. Tir is also tyrosine phosphorylated and this is necessary for its biological activity. The changes that then occur in the intestinal epithelial cells are due to the ability of the bacterium to control aspects of the intracellular signal transduction machinery of the enterocyte. The exact nature of the intracellular signalling programme that is induced by EPEC is not fully understood. From the available data it appears that the interaction of EPEC with intestinal epithelial cells results in activation of PLCγ by tyrosine phosphorylation. It is not clear whether Tir is involved in PLCγ activation. Along with the activation of PLCγ, EPEC-stimulated cells have an activated PKC. In addition there have been reports that EPEC-stimulation of cells leads to the synthesis of IP_3 and the release of calcium from intracellular stores. However, the role of this calcium flux is under debate. It is therefore clear that the binding of EPEC to host intestinal epithelial cells results in complex changes in intracellular signalling that must eventually result in changes in ion metabolism and the release of water from the intestinal cells to cause the diarrhoea.

4.7 | What's next?

Because of the persistent presence of microbes (both as part of our normal microflora and in our environment) we have evolved very effective means of protecting ourselves against infectious diseases. At any given point in time, therefore, most of the human population of this planet do not suffer from a bacterial infection. The next two chapters will describe the (sometimes very complex) defence mechanisms that enable us to live in harmony (for most of the time) with our normal microflora and to keep exogenous pathogens at bay. We will start by describing, in Chapter 5, the role played by mucosal surfaces – the first line of defence – in preventing infection.

4.8 | Questions

1. Explain, with examples, the methods used by eukaryotic cells to carry messages from the cell surface receptor to the nucleus.
2. Explain what advantage bacteria gain from disrupting host cell signalling.
3. What roles do cytokines play in controlling host responses to bacterial infection?
4. Compare and contrast IL-1-induced and TNF-induced cell signalling mechanisms.

5. Give examples of host cell signalling influenced by bacteria or their products.

6. Explain the essential differences between the quorum sensing mechanisms used by Gram-positive and those used by Gram-negative bacteria.

7. What advantages does quorum sensing confer on those bacteria that possess this ability?

8. Describe the means by which bacteria sense, and respond to, changes in their environment.

9. Why are two-component signal transduction systems also termed 'histidyl-aspartyl phospho-relay systems'?

10. Compare and contrast intracellular signal transduction systems in eukaryotic and prokaryotic cells.

4.9 | Further reading

Books

Dunny, G. M. & Winans, S. C. (1999). *Cell–Cell Signalling in Bacteria.* Washington, DC: ASM Press.

England, R., Hobbs, G., Bainton, N. & Roberts, D. M. (1999). *Microbial Signalling and Communication.* Cambridge: Cambridge University Press.

Hancock, J. T. (1997). *Cell Signalling.* Reading, MA: Addison-Wesley.

Helmreich, E. J. M. (2001). *The Biochemistry of Cell Signalling.* Oxford: Oxford University Press.

Henderson, B., Poole, S. & Wilson, M. (1998). *Bacteria–Cytokine Interactions in Health and Disease.* London: Portland Press.

Higgs, G. A. & Henderson, B. (eds.) (2000). *Novel Cytokine Inhibitors.* Basel: Birkhauser.

Hoch, J. A. (1995). *Two-Component Signal Transduction.* Washington, DC: ASM Press.

Mantovani, A., Dinarello, C. A. & Ghezzi, P. (2000). *Pharmacology of Cytokines.* Oxford: Oxford University Press.

Review articles

Bakal, C. J. & Davies, J. E. (2000). No longer an exclusive club: eukaryotic signalling domains in bacteria. *Trends Cell Biol* **10**, 32–38.

Bassler, B. L. (1999). How bacteria talk to each other: regulation of gene expression by quorum sensing. *Curr Opin Microbiol* **2**, 582–587.

Bassler, B. (2001). Signals and cross-signaling. *Annu Rev Microbiol* **55**, 165–199.

Chatterjee-Kishore, M., van den Akker, F. & Stark, G. R. (2000). Association of STATs with relatives and friends. *Trends Cell Biol* **10**, 106–111.

de Kievit, T. R. & Iglewski, B. H. (2000). Bacterial quorum sensing in pathogenic relationships. *Infect Immun* **68**, 4839–4849.

DeVinney, R., Steele-Mortimer, O. & Finlay, B. B. (2000). Phosphatases and kinases delivered to the host cell by bacterial pathogens. *Trends Microbiol* **8**, 29–33.

DiRita, V. J., Engleberg, N. C., Heath, A., Miller, A., Crawford, J. A. & Yu, R. (2000). Virulence gene regulation inside and outside. *Philos Trans R Soc Lond B Biol Sci* **355**, 657–665.

Dunny, G. M. & Leonard, B. A. B. (1997). Cell–cell communication in Gram-positive bacteria. *Annu Rev Microbiol* **51**, 527–564.

Eberl, L. (1999). *N*-Acyl homoserine lactone-mediated gene regulation in Gram-negative bacteria. *Syst Appl Microbiol* **22**, 493–506.

English, J., Pearson, G., Wilsbacher, J., Swantek, J., Karandikar, M., Xu, S. & Cobb, M. H. (1999). New insights into the control of MAP kinase pathways. *Exp Cell Res* **253**, 255–270.

Fabret, C., Feher, V. A. & Hoch, J. A. (1999). Two-component signal transduction in *Bacillus subtilis*: how one organism sees its world. *J Bacteriol* **181**, 1975–1983.

Foussard, M., Cabantous, S., Pédelacq, J.-D., Guillet, V., Tranier, S., Mourey, L., Birck, C. & Samama, J.-P. (2001). The molecular puzzle of two-component signaling cascades. *Microbes and Infection* **3**, 417–424.

Fruman, D. A., Rameh, L. E. & Cantley, L. C. (1999) Phosphoinositide binding domains: embracing 3-phosphate. *Cell* **97**, 817–820.

Fu, H., Subramanian, R. R. & Masters, S. C. (2000). 14-3-3-proteins: structure, function and regulation. *Annu Rev Pharmacol Toxicol* **40**, 617–647.

Groisman, E. A. (2001). The pleiotropic two-component regulatory system PhoP-PhoQ. *J Bacteriol* **183**, 1835–1842.

Hannum, Y. A. & Luberto, C. (2000). Ceramide in the eukaryotic stress response. *Trends Cell Biol* **10**, 73–80.

Hoch, J. A. (2000). Two-component and phosphorelay signal transduction. *Current Opin Microbiology* **3**, 165–170.

Hunter, T. (2000). Signalling – 2000 and beyond. *Cell* **100**, 113–127.

Inoue, J.-I., Ishida, T., Tsukumoto, N., Kobayashi, N., Naito, A., Azuma, S. & Yamamoto, T. (2000). Tumor necrosis receptor-associated factor (TRAF) family: adapter proteins that mediate cytokine signalling. *Exp Cell Res* **254**, 14–24.

Kjoller, L. & Hall, A. (1999). Signalling to Rho GTPases. *Exp Cell Res* **253**, 166–179.

Konkel, M. E. & Tilly, K. (2000). Temperature-regulated expression of bacterial virulence genes. *Microbes and Infection* **2**, 157–166.

Miller, M. B. & Bassler, B. L. (2001). Quorem sensing in bacteria. *Annu Rev Microbiol* **55**, 165–199.

Morrison, D. A. & Lee, M. S. (2000). Regulation of competence for genetic transformation in *Streptococcus pneumoniae*: a link between quorum sensing and DNA processing genes. *Res Microbiol* **151**, 445–451.

Olson, M. F. & Marais, R. (2000). Ras protein signalling. *Semin Immunol* **12**, 63–73.

Parsek, M. R. & Greenberg, E. P. (2000). Acyl-homoserine lactone quorum sensing in Gram-negative bacteria: a signaling mechanism involved in associations with higher organisms. *Proc Natl Acad Sci USA* **97**, 8789–8793.

Perraud, A. L., Weiss, V. & Gross, R. (1999). Signalling pathways in two-component phosphorelay systems. *Trends Microbiol* **7**, 115–120.

Pirrung, M. C. (1999). Histidine kinases and two-component signal transduction systems. *Chem Biol* **6**, R167–R175.

Rodrigue, A., Quentin, Y., Lazdunski, A., Mejean, V. & Foglino, M. (2000). Two-component systems in *Pseudomonas aeruginosa*: why so many? *Trends Microbiol* **8**, 498–508.

Rumbaugh, K. P., Griswold, J. A. & Hamood, A. N. (2000). The role of quorum sensing in the *in vivo* virulence of *Pseudomonas aeruginosa*. *Microbes Infect* **2**, 1721–1731.

Schindler, C. (1999). Cytokines and JAK-STAT signalling. *Exp Cell Res* **253**, 7–14.

Spiik, A. K., Meijer, L. K., Ridderstad, A. & Pettersson, S. (1999). Interference of eukaryotic signalling pathways by the bacteria *Yersinia* outer protein YopJ. *Immunol Lett* **68**, 199–203.

Stock, A. M., Robinson, V. L. & Goudreau, P. N. (2000). Two-component signal transduction. *Annu Rev Biochem* **69**, 183–215.

Strauss, E. (1999). A symphony of bacterial voices. *Science* **284**, 1302–1304.

Whitehead, N. A., Barnard, A. M. L., Slater, H., Simpson, N. J. L. & Salmond, G. P. C. (2001). Quorem-sensing in Gram-negative bacteria. *FEMS Microbiol Rev* **25**, 365–404.

Whiteley, M., Parsek, M. R. & Greenberg, E. P. (2000). Regulation of quorum sensing by RpoS in *Pseudomonas aeruginosa*. *J Bacteriol* **182**, 4356–4360.

Williams, P., Camara, M., Hardman, A., Swift, S., Milton, D., Hope, V. J, Winzer, K., Middleton, B., Pritchard, D. I. & Bycroft, B. W. (2000). Quorum sensing and the population-dependent control of virulence. *Phil Trans R Soc Lond B* **355**, 667–680.

Wisdom, R. (1999). AP-1: one switch for many signals. *Exp Cell Res* **253**, 180–185.

Withers, H., Swift, S. & Williams, P. (2001). Quorum sensing as an integral component of gene regulatory networks in Gram-negative bacteria. *Curr Opin Microbiol* **4**, 186–193.

Papers

Bernish, B. & van de Rijn, I. (1999). Characterization of a two-component system in *Streptococcus pyogenes* which is involved in regulation of hyaluronic acid production. *J Biol Chem* **274**, 4786–4793.

Bowie, A. & O'Neill, L. A. (2000). The interleukin-1 receptor/Toll-like receptor superfamily: signal generators for pro-inflammatory and microbial products. *J Leukoc Biol* **67**, 508–514.

Han, J., Lee, J.-D., Bibbs, L. & Ulevich, R. J. (1994). A MAP kinase targeted by endotoxin and hyperosmolarity in mammalian cells. *Science* **265**, 808–811.

Hassett, D. J., Ma, J. F., Elkins, J. G., McDermott, T. R., Ochsner, U. A., West, S. E., Huang, C. T., Fredericks, J., Burnett. S., Stewart, P. S, McFeters, G., Passador, L. & Iglewski, B. H. (1999). Quorum sensing in *Pseudomonas aeruginosa* controls expression of catalase and superoxide dismutase genes and mediates biofilm susceptibility to hydrogen peroxide. *Mol Microbiol* **34**, 1082–1093.

Heath, A., DiRita, V. J, Barg, N. L. & Engleberg, N. C. (1999). A two-component regulatory system, CsrR-CsrS, represses expression of three *Streptococcus pyogenes* virulence factors, hyaluronic acid capsule, streptolysin S, and pyrogenic exotoxin B. *Infect Immun* **67**, 5298–5305.

Hirsch, E., Katanaev, V. L., Garlanda, C. Azzolino, O., Pirola, L., Silengo, L., Sozzani, S., Mantovani, A., Altruda, F. & Wymann, M. P. (2000). Central role for G-protein-coupled phosphoinositide 3-kinase γ in inflammation. *Science* **287**, 1049–1053.

Kumar, S., McDonnell, P., Lehr, R. Tierney, L., Tzimas, M. N., Griswold, D. E., Capper, E. A., Tal-Singer, R., Wells, G. I., Doyle, M. L. & Young, P. R. (2000). Identification and initial characterisation of four novel members of the interleukin-1 family. *J Biol Chem* **275**, 10308–10314.

Lee, S. W., Ko, Y.-G., Bang, S., Kim, K. S. & Kim, S. (2000). Death effector domain of a mammalian apoptosis mediator, FADD, induces bacterial cell death. *Mol Microbiol* **35**, 1540–1549.

Meyer-ter-Vehn, T., Covacci, A., Kist, M. & Pahl, H. L. (2000). *Helicobacter pylori* activates mitogen-activated protein kinase cascades and induces expression of the proto-oncogenes *c-fos* and *c-jun. J Biol Chem* **275**, 16064–16072.

Orth, K., Palmer, L. E., Bao, Z. Q., Stewart, S., Rudolph, A. E., Bliska, J. B. & Dixon, J. E. (1999). Inhibition of the mitogen-activated protein kinase superfamily by a *Yersinia* effector. *Science* **285**, 1920–1923.

Palmer, L. E., Hobbie, S., Galan, J. E. & Bliska, J. P. (1998). YopJ of *Yersinia pseudotuberculosis* is required for the inhibition of macrophage TNFα production and downregulation of the MAP kinases p38 and JNK. *Mol Microbiol* **27**, 953–965.

Pancholi, V. & Fischetti, V. A. (1997). Cell-to-cell signalling between group A streptococci and pharyngeal cells. Role of streptococcal surface dehydrogenase (SDH). *Adv Exp Med Biol* **418**, 499–504.

Pearson, J. P., Feldman, M., Iglewski, B. H. & Prince, A. (2000). *Pseudomonas aeruginosa* cell-to-cell signaling is required for virulence in a model of acute pulmonary infection. *Infect Immun* **68**, 4331–4334.

Procyk, K. J, Kovarik, P., von Gabain, A. & Baccarini, M. (1999). *Salmonella typhimurium* and lipopolysaccharide stimulate extracellularly regulated kinase activation in macrophages by a mechanism involving phosphatidylinositol 3-kinase and phospholipase D as novel intermediates. *Infect Immun* **67**, 1011–1017.

Robinson, V. L., Buckler, D. R. & Stock, A. M. (2000). A tale of two components: a novel kinase and a regulatory switch. *Nat Struct Biol* **7**, 626–633.

Rumbaugh, K. P., Griswold, J. A., Iglewski, B. H. & Hamood, A. N. (1999). Contribution of quorum sensing to the virulence of *Pseudomonas aeruginosa* in burn wound infections. *Infect Immun* **67**, 5854–5862.

Stender, S., Friebel, A., Linder, S., Rohde, M., Mirold, S. & Hardt, W.D. (2000). Identification of SopE2 from *Salmonella typhimurium*, a conserved guanine nucleotide exchange factor for Cdc42 of the host cell. *Mol Microbiol* **36**, 1206–1221.

Throup, J. P., Koretke, K. K., Bryant, A. P., Ingraham, K. A., Chalker, A. F., Ge, Y., Marra, A., Wallis, N. G., Brown, J. R., Holmes, D. J., Rosenberg, M. & Burnham, M. K. (2000). A genomic analysis of two-component signal transduction in *Streptococcus pneumoniae. Mol Microbiol* **35**, 566–576.

Visick, K. L. & McFall-Ngai, M. J. (2000). An exclusive contract: specificity in the *Vibrio fischeri–Euprymna scolopes* partnership. *J Bacteriol* **182**, 1779–1787.

Warny, M., Keates, A. C., Keates, S., Castagliuolo, I., Zacks, J. K., Aboudola, S., Qamar, A., Pothoulakis, C., LaMont, J. T. & Kelly, C.P. (2000). p38 MAP kinase activation by *Clostridium difficile* toxin A mediates monocyte necrosis, IL-8 production, and enteritis. *J Clin Invest* **105**, 1147–1156.

4.10 | Internet links

1. http://vl.bwh.harvard.edu/signal_transduction.shtml
   The signal transduction section of the WWW Virtual Library of Cell Biology. Consists of many links to every conceivable aspect of signal transduction in eukaryotic cells.

2. http://www-isu.indstate.edu/thcme/mwking/signal-transduction.html
   On-line lectures on signal transduction. Part of the on-line course in Medical Biochemistry at the Terre Haute Center for Medical Education.

3. http://copewithcytokines.de
   Everything you wanted to know about cytokines courtesy of Horst Ibelgaufts.

4. http://www.bmb.psu.edu/nixon/micro401/overview.htm
   This site provides an introduction to two-component signal transduction. It contains a comprehensive summary of the history of two-component systems, their distribution amongst organisms, and their target functions and pathways.

5. http://www.nottingham.ac.uk/quorum/what.htm
   This website is based at the University of Nottingham, which has a team of scientists recognised as being among the world leaders in the field of quorum sensing. It contains essential information about this signalling mechanism and stresses its importance in bacterial virulence. It describes the work of the group based at the university and there are also links to some other groups working on quorum sensing.

5

The mucosal surface: the front line of antibacterial defence

Chapter outline

Aims

To introduce the reader to:

- the concept of antibacterial defence
- the mucosal surfaces of the body
- the role of epithelial cells in antibacterial defence
- antibacterial biomolecules including antibacterial peptides
- the concept of the normal microflora as a defence mechanism
- the mucosal epithelium as an antibacterial watchdog for the whole body

5.1 | Introduction

The planet we live on is alive with unimaginable numbers, and diverse types, of bacteria. It is only possible to give the roughest estimate of how many such species may exist and the figures that are derived are in the range of millions. As described in Chapter 1, our own bodies are not, as we imagine, single organisms, but are colonies containing between 1000 and 3000 (or more) different species of bacteria. Indeed, in the average human body there are approximately 10 bacteria for each human cell. In addition to bacteria, we are also host to a range of viruses, fungi, protozoa and multicellular creatures such as the mites that live on our skin and hair.

Our general view of bacteria as disease-causing organisms is incorrect. This is obvious, given that we normally share our bodies with huge numbers of diverse bacteria. What prevents these astronomical numbers of bacteria from causing disease is not understood. However, it is likely to involve the very complex systems of immunity that mammals have co-evolved with bacteria to enable both types of organism to live in harmony. We are generally aware of these systems only when something goes wrong and our own defence mechanisms produce some form of pathology. This chapter is the first of three (the other two being Chapters 6 and 10) concerned with our defences against bacterial infection. It will deal with our first, but least understood, line of defence – the mucosal/epithelial surface.

5.2 | An overview of antibacterial defence

Recent advances in our understanding of the defences mounted against bacteria are showing both how ancient such defences are, and how large a proportion of our genes are devoted to defence against pathogens. Preventing disease-causing bacteria from infecting our bodies is a 24 hour per day, 7 day per week, 365 day per year occupation. Any failure in these systems results in infection. What happens when one facet of our defences is deficient is most clearly seen in those infected with human immunodeficiency virus (HIV), the cause of AIDS. The AIDS virus depresses the numbers of circulating CD4 T lymphocytes (for details on lymphocytes, see Section 6.3.4). AIDS patients suffer from common infections (e.g. pneumonia and septicaemia), have a higher incidence of tuberculosis and can become infected with organisms, such as *Mycoplasma avium* and *Cryptococcus neoformans*, which rarely infect humans.

The antibacterial defences of the human body are numerous and all-encompassing (Figure 5.1) and include (i) physical systems such as the hairs in the nose and the flow of saliva, (ii) chemical defences such as the low pH in the stomach and the salt and lipids produced by the skin, (iii) barriers, such as the epithelial layers of the body, preventing entry

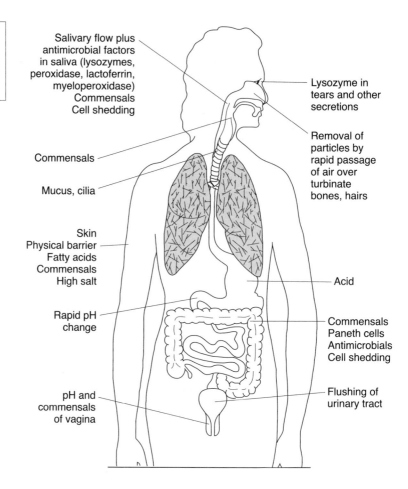

Figure 5.1 Schematic diagram showing the various levels of antibacterial defence mechanisms at mucosal surfaces. For details, see the text.

Salivary flow plus antimicrobial factors in saliva (lysozymes, peroxidase, lactoferrin, myeloperoxidase)
Commensals
Cell shedding

Lysozyme in tears and other secretions

Removal of particles by rapid passage of air over turbinate bones, hairs

Commensals

Mucus, cilia

Skin
Physical barrier
Fatty acids
Commensals
High salt

Acid

Rapid pH change

Commensals
Paneth cells
Antimicrobials
Cell shedding

pH and commensals of vagina

Flushing of urinary tract

of bacteria, (iv) cellular replacement and detachment – (a system for removing bacteria from the external body surfaces), (v) the normal microflora, (vi) antibacterial proteins produced by epithelial cells on our external surfaces, (vii) antibacterial peptides produced by most cells in the body, (viii) the acute phase response and (ix) the panoply of cells and mediators known as acquired immunity. A key element of antibacterial defence, and one that will be emphasised in this book, is the ability of our defence systems to recognise that we are being infected. The systems used by the host to recognise infectious organisms appear to be ancient and highly efficient, and are only now, as this text is being written, becoming recognised. These include a growing number of receptors initially discovered in insect immunity called the Toll-like receptors (TLRs), which will be dealt with in detail in this and later chapters.

Thus our defences against bacteria involve two major systems: (i) physical and chemical/biochemical systems that are constantly working and, as far as can be determined, do not need to recognise infectious agents; and (ii) cell-based systems that generally need to recognise the infecting agent to be set in motion. These two systems are often described as innate and acquired immunity, respectively. Innate systems are essentially pre-formed and ready to operate as soon as infection begins. Acquired systems (antibodies and T cells) require the build-up of specific lymphocytes and may take days to weeks to become effective. However, this is probably an oversimplification of the real situation and in this book we will divide our description of antibacterial defences into those functioning at epithelial or mucosal surfaces and those that function when the bacterium has penetrated past the mucosal surface and has invaded the host. These former systems have generally been overlooked in textbooks on Microbiology and Immunology, but are now being recognised for the important contributions they make to our overall defences against bacteria. It is important to realise from the beginning that all of the systems of immunity that will be described in this book work together in a seamless and highly interactive fashion. Unfortunately, we only vaguely understand the nature of such interactions.

In Chapter 6, the systems of innate and acquired immunity will be compared and contrasted. It is important to realise that invertebrates, which make up some 95% of the multicellular organisms on our planet, have only innate immune mechanisms to cope with bacterial infections. This emphasises the importance of the apparently simple systems of defence provided by epithelia, epithelial cell products and phagocytic cells – the major components of innate or natural immunity. Moreover, as will be described in Chapter 6, these more ancient innate immune systems play a major role in controlling and focusing our innate immune responses.

5.3 | The mucosal surface

Huge sums of money are spent each year in packaging an enormous range of products. Imagine the results if soap powders or chocolates were not properly packaged. In a properly ordered world it would make great ecological sense, but in our real world the powder and the chocolate would soon spoil. If packaging of these simple commodities is important, think how much more important it is that the human body is properly 'packaged'. A major packaging for the human body is the skin. It is often assumed that the skin is the only part of our body exposed to the environment. But a moment's consideration will reveal that large parts of our body are exposed to the outside environment. As seen in Figure 5.2, the human body is actually open to the outside from the mouth to the anus to allow feeding, digestion and excretion. In addition, the respiratory and urogenital tracts are also open to the outside to fulfil their functions. All these areas of the body are exposed to bacteria, and they need to be covered with a protective coating. The major cellular element of this protective coating is a specialised cell known as an epithelial cell and the name given to the covering exposed to the outside environment is either epithelial (when referring to the skin) or mucosal when the epithelial layers are covered

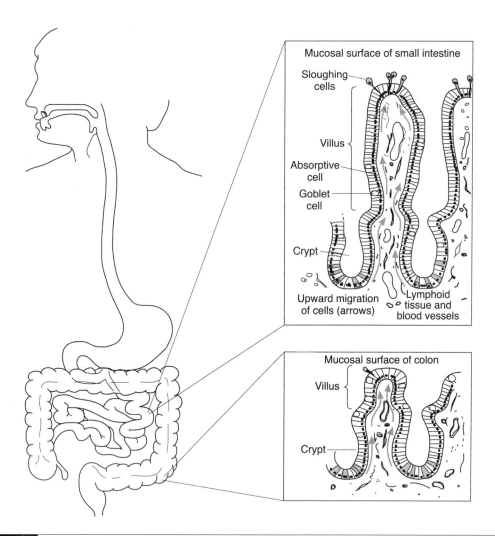

Figure 5.2 This diagram shows that, in addition to the skin, the human body is open to the environment from the mouth to the anus. The lungs are generally sterile and are not included in the diagram. The largest area of interface between the normal microflora and the host is the intestinal tract, largely because of the convoluted nature of its internal surface, comprised as it is of villi. The latter are finger-like projections and the diagram shows their structure and cellular composition. (Reproduced with permission from *Bacterial Pathogenesis: A Molecular Approach*, by A.A. Salyers & D.D. Whitt (1994), ASM Press: Washington, DC.)

with a mucin or mucus layer. In simpler terms, we have 'dry' (skin) and 'wet' epithelia. Rub your finger over the skin on the back of your hand then rub it over the inside of your cheek to feel the difference between these two epithelial surfaces. The mucosal surfaces of the small intestine and colon are shown schematically in Figure 5.2.

5.3.1 Epithelial cells and epithelia

Our bodies are composed of about 200 different cell types that produce the various tissues that make up our normal body plan: epithelia,

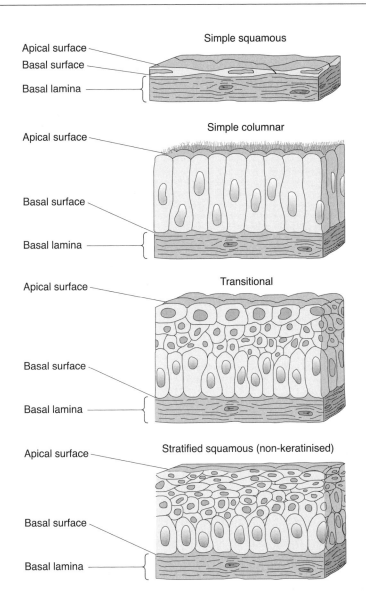

Figure 5.3 The different types of epithelial surfaces in the human body. Note that cells have two distinct faces, the apical surface, which faces the cavity or lumen of the particular tissue or organ, and the basolateral surface, which abuts the basal lamina and underlying supportive connective tissue (see also Figure 5.6). Simple squamous epithelia line the various body cavities. The simple columnar epithelia are found in the stomach, intestine and cervical tract. When the epithelium consists of several layers of cells with different shapes it is said to be a transitional epithelium. This is found in cavities subject to expansion and contraction such as the urinary bladder. If the cells in the epithelium show a progressive flattening then it is termed a stratified squamous (non-keratinised) epithelium. This type of epithelium is found in the mouth and vagina and is designed not for absorption but for mechanical resistance to forces.

connective tissue, nervous tissue, muscle, blood and germ cells. The majority of cells in a multicellular organism are, in fact, epithelial cells and these can be thought of as the jacks-of-all-trades of the body. One characteristic of epithelial cells is that they join together with other epithelial cells, through specialised junctions, to form multicellular sheets (called epithelia). These sheets may be only one cell thick and, depending on the shape of the cell, may be defined as squamous, cuboidal or columnar (Figure 5.3). Simple squamous epithelia, for example, line many of the body cavities. Simple columnar epithelial cells line the stomach, the cervical tract and the intestine. If more than one layer of cells is present then epithelia are said to be stratified (Figure 5.3). If the cells have different shapes, the epithelia are called transitional and are found in tissues that are subject to change

Figure 5.4 Skin – a stratified epithelium. A longitudinal section through a hair follicle showing the presence of bacteria in close association with the epithelium. (Kindly supplied by Dr Richard Bojar, Skin Research Centre, Division of Microbiology, University of Leeds.)

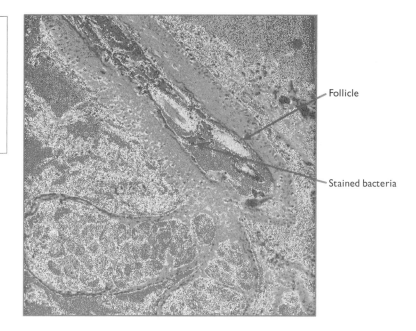

Follicle

Stained bacteria

of shape, such as the bladder. If the layers of cells show a progressive change in shape they are called stratified squamous epithelia (Figure 5.3) and such epithelia line the oral cavity and the vagina. Stratified squamous epithelia can either be non-keratinised (mouth and vagina) or keratinised. Skin is a good example of a stratified keratinised epithelial surface, and Figure 5.4 shows the intimate relationship our bounding layer has with bacteria. Epithelia are the subject of much scientific investigation because 90% of cancers occur in the epithelia of the body.

Epithelial layers often contain additional cell populations. The best known accompanying cell is the goblet cell, so called because of its resemblance to a goblet. These cells produce the constituents, such as glycoproteins and proteoglycans, of the mucus that covers 'wet' epithelia. Other cell populations that reside within the epithelial cell layers of the body appear to be part of our antibacterial defence mechanisms. The M (microfold) cells of the intestine and lung are involved in controlling passage of 'antigens' to underlying lymphoid tissue. This seems to allow the immune system to sample bacterial antigens in the external environment. Paneth cells of the crypts of the small intestine produce antibacterial proteins and peptides. The intestine also has a large population of so-called intra-epithelial lymphocytes (IELs), which play a role, yet to be fully elucidated, in our defence against exogenous bacteria (Figure 5.5). These various cell populations will be described in more detail later in Sections 5.7.3 and 6.3.4.

Epithelial cells have a wide range of functions and can be secretory or absorptive. For example, the epithelial cells lining the stomach are involved in secreting hydrochloric acid into the lumen of the stomach whereas the epithelial cells lining the alveoli of the lungs are involved in absorbing oxygen from inhaled air. All epithelia are joined together

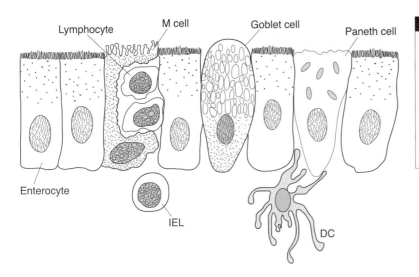

Lymphocyte | M cell | Goblet cell | Paneth cell

Enterocyte

IEL

DC

Figure 5.5 Schematic diagram showing the various cell populations found at mucosal surfaces in the intestine. These cells include: the mucin-producing goblet cell; the Paneth cell, which produces antibiotic peptides and phospholipase A_2; intra-epithelial lymphocytes (IEL); M cells; and dendritic cells (DC).

by various types of cell junctions formed from specialised regions of the plasma membrane. These include tight junctions, which 'seal' epithelial cells together so that epithelial sheets are impermeable to all but the smallest molecules. Gap junctions between epithelial cells allow the passage of ions between cells and help to integrate the functions of epithelia. Other junctions in epithelial cells such as hemidesmosomes attach the plasma membranes of cells onto the extracellular matrix of the basal lamina. Thus all epithelial cells have two 'faces'. That part of the cell which is free and exposed to the external environment is termed the apical surface. The opposite face (termed basolateral) is attached to the basal lamina through hemidesmosomes. Hence epithelial cells are generally polarised (Figure 5.6). The key function of epithelial cells is to provide a barrier between the internal and external spaces of an organism. In this respect they can be considered analogous to the plasma membranes of eukaryotic cells.

Now, before describing the antibacterial functions of the mucosal epithelia of the body, it is worth pointing out that the surface area of

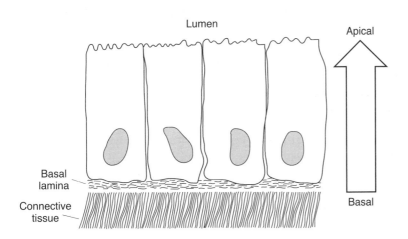

Lumen

Apical

Basal lamina

Connective tissue

Basal

Figure 5.6 Polarisation of epithelia showing the basal surface of the cell attached to the underlying connective tissue and the apical surface facing the lumen.

the mucosae is around $400\,\text{m}^2$ – equivalent to the area of a tennis court. It is this surface that has to be exposed to the environment if we are to survive. It is this surface that is, therefore, at risk of encountering, and allowing to enter, a range of bacterial pathogens. In this chapter, and to some extent Chapter 6, the mechanisms by which our mucosal surfaces protect us from bacteria will be described.

5.4 | Mucosal antibacterial defences

For ease of discussion, the defence mechanisms of the skin and the mucous membranes will be divided into those that are physical, chemical or biological. These divisions, like most in biology, are not hard and fast and overlaps occur.

5.4.1 The mucosal surface as a physical barrier to bacteria

The inside of the body is sterile. That this is so is due largely to the fact that the external-facing epithelial surfaces of the body form, by and large, an unbroken barrier. As described, the epithelial cells are glued together by tight junctions, which prevent all but the smallest molecules penetrating and certainly can keep out bacteria. The skin is the one epithelial barrier that we can all see. This thin, and potentially fragile, covering appears to be resistant to bacterial penetration and no bacteria have been identified that can broach our horny outsides. In addition to preventing the entry of cells, as would be the case for any epithelial/mucous barrier, skin is dry, which inhibits bacterial growth. The principal sources of nutrients for skin bacteria are sweat and the product of the sebaceous gland, sebum. Metabolism of the lipids present in sebum leads to the accumulation of fatty acids, which are not only toxic to many bacterial species but also lower the pH to a value unsuitable for many of these organisms. The continuous evaporation of sweat results in the accumulation of high concentrations of sodium chloride on the skin surface, which, again, many species cannot tolerate. Bacteria in pores, hair follicles and sweat glands will be confronted by additional antibacterial factors including lysozyme (an enzyme that breaks down peptidoglycan, one of the main constituents of the bacterial cell wall – see Section 2.2.4), antibodies produced by skin-associated lymphoid tissue and antibiotic peptides (described in Section 5.4.3.11). Another factor tending to reduce bacterial colonisation is the shedding of the outer layer of the skin – the stratum corneum. This consists of dead cells filled with keratin (so providing a tough protective layer), which are continually peeling off at such a rate that the surface of the body is completely renewed, on average, every two days. The cells tend to peel off the surface in small clumps or sheets (known as squames) carrying with them any attached organisms.

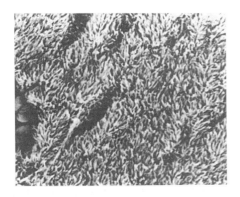

Figure 5.7 Scanning electron micrograph showing ciliated respiratory epithelium. Magnification 1440×. (Reprinted from *Small Ruminant Research*, **37**, Kahwa, C. K., Balemba, O. B. & Assey, R. J. The pattern of ciliation and the development of the epithelial lining of the respiratory tract in the neonatal kid: a scanning electron microscopic study 27–34. Copyright 2000, with permission from Elsevier Science.)

In addition to being a purely physical barrier, epithelia are a backdrop for physical mechanisms that limit bacterial colonisation. The removal of bacteria from epithelial surfaces by fluid flow is responsible, in part, for preventing colonisation of the outer surfaces of the eyes (by use of tears) and fluid flow (in this case, saliva) over the various mucosal surfaces in the mouth (tongue, cheeks, gums, palate, etc.) washes bacteria down into the stomach. Fluid flow also plays a role in removing bacteria in the gut. In the case of the mouth and eye, fluid flow is largely passive, although aided by chewing and blinking, respectively.

In contrast, in the case of the airway epithelia of the nasopharynx and lungs, fluid movement is active. The reader will accept that breathing is an important process and one that brings air from a bacterially contaminated environment into a very delicate and complicated mechanism – the tissues of the lung. It is vital that bacteria are excluded from the lungs (or are killed if they enter) and the passage of air to the lungs has evolved to be as difficult as possible for bacteria to 'maintain the course'. Air entering the nose first encounters the nasal hairs, often thought of as unsightly and the source of some humour. However, these hairs function as a coarse filter for particulates (including bacteria). Air is then directed over the large mucus-covered surfaces of the inner bone of the nose (the turbinates) to trap bacteria. In addition to trapping bacteria on the mucus-covered epithelia, the epithelia of the upper and lower respiratory tracts contain ciliated cells (Figure 5.7). In the lower respiratory tract, these cells act to move bacteria, trapped in the layer of mucus, away from the lungs towards the oesophagus. A similar mechanism exists in the nasal cavity, carrying bacteria to the back of the throat, where they are swallowed. This has been called the mucociliary escalator. Interference with this system of removing bacteria, the best-known example being cigarette smoking, can lead to lung infection. Indeed, it is surprising that lung infections are not more widespread.

The final example of the use of physical means to remove bacteria is the urogenital tract. Here the flow of urine from the bladder to the outside of the body acts to wash away all but the most tenacious organisms.

5.4.2 Antibacterial chemicals produced by the mucosae

All life forms have a preferred pH optimum, and this is near neutral (pH 7) for most organisms. Lowering or raising the local pH can therefore inhibit cell growth. An increased production of protons is used by the skin, vagina and stomach to limit bacterial growth. The pH of healthy skin is acidic (pH 5), as is that of the vagina. In the former, the lowered pH is due to lactic acid produced by sweat and to the accumulation of acidic end-products of bacterial metabolism, while in the adult vagina the low pH is attributable primarily to the acids produced by lactic acid bacteria. These are examples of the normal microflora acting to limit bacterial colonisation. The pH of the stomach is approximately 2, and this kills most organisms, with a few notable exceptions such as *Helicobacter pylori*. In addition to lowering pH locally, the skin also produces lipids that have antibacterial activity. The lower intestine is relatively alkaline and this can also discourage the growth of particular organisms. Thus it is clear that, by maintaining a pH lower or higher than that conducive for the growth of many bacterial species, the body, either on its own or with help from its normal microflora, can, by this simple expedient, limit bacterial colonisation.

5.4.3 Antibacterial biomolecules produced by the mucosae

In addition to the mechanisms described above for limiting bacterial colonisation and host infection, a growing list of antibacterial proteins and peptides have been discovered to be involved in defending the host against exogenous bacteria (Table 5.1). These antibacterial molecules are both produced constitutively and are inducible and, as described in Section 11.3.2, some are being replicated by the pharmaceutical industry as novel therapeutics. These various bioactive molecules are described below. Of interest is the recent finding that the concentration of antibacterial polypeptides in nasal secretions varies significantly between donors, suggesting that polymorphic genes are encoding these defensive molecules.

It is interesting how many of the secreted peptides/proteins of the mucosal surface recognise lipopolysaccharide (LPS). This is one of the major bacterial signals recognised by the host and is responsible for the lethal condition known as septic shock. It would appear that the recognition and inhibition of the activity of LPS on mucosal surfaces is required for antibacterial defence. As will be seen later in this chapter (Section 5.4.3.6 and Figure 5.15) and in Section 10.3.1, mucosal epithelial cells respond to LPS by producing antibacterial peptides and cytokines to both protect against, and warn the body of, the invasion of bacteria. While the mucosal epithelial cells are responsive to bacterial factors and produce a range of antibacterial components, it is not clear how the normal mucosal surface still maintains its specific normal microflora. These bacteria may have become insensitive to

Table 5.1.	Antibacterial proteins and peptides produced by mucosal surfaces
Antibacterial molecule	Function of molecule
Mucins	Prevention of bacterial contact/adhesion with the mucosa?
Lysozyme	Breakdown of peptidoglycan/non-enzymic antibacterial action
Lactoferrin	Iron-chelating protein inhibits bacterial growth
Lactoperoxidase	Produces damaging superoxide radical
Secretory PLA$_2$	Membrane-damaging enzyme
SLPI	Inhibits bioactivity of LPS
Acute phase proteins	Opsonise bacteria
Collectins	Opsonic proteins
Secretory IgA	Inhibition of adhesion of bacteria to mucosa
Trefoil peptides	Involved in wound healing
Antibacterial peptides	Direct killing of bacteria by membrane damage

PLA$_2$, phospholipase A$_2$; SLPI, secretory leukocyte protease inhibitor; LPS, lipopolysaccharide.

the production of such antibacterial components by some, as yet unknown, manner.

5.4.3.1 Mucin

The mucus layer that overlays epithelial surfaces is composed of a variety of factors but its main properties are attributable to the presence of mucins, which are produced mainly by goblet cells in the mucosa (Figure 5.8). Mucins are complex linear polymorphic glycoproteins – the products of at least eight genes in humans. The secreted mucin monomer can form dimers and these dimers can then form oligomers, producing macromolecular assemblies that can be up to several micrometres in length (Figure 5.9). Mucus has a number of perceived functions, including keeping epithelia moist, acting as a lubricant, trapping bacteria and keeping bacteria from contact with the epithelia. Mucus also contains a number of substances secreted by mucosal epithelial cells that have antibacterial activity. Surprisingly little work has been done on the mucins of the mucosal surfaces. There is now evidence that the production of mucins can be controlled by exogenous bacterial signals and that signalling pathways involving mitogen-activated protein (MAP) kinase and nuclear factor (NF) κB are involved (for details of signalling pathways, see Chapter 4).

5.4.3.2 Lysozyme

Bacterial cell walls contain a complex macromolecule, peptidoglycan, also known as murein (Figure 5.10). As described in Section 2.2.4, peptidoglycan is the main structural component of the cell walls of

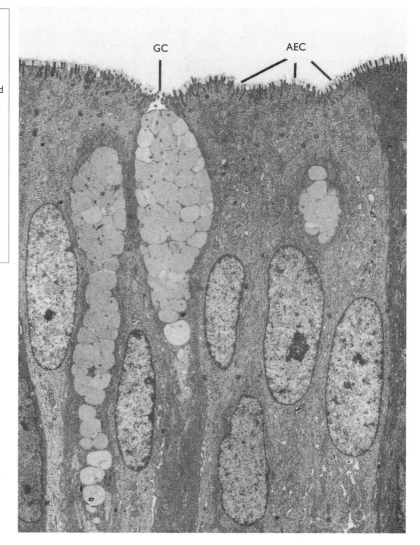

Figure 5.8 Transmission electron micrograph of intestinal epithelium showing a goblet cell (GC), and absorptive epithelial cells (AEC) with associated villi. Magnification 4675×. (Reproduced with permission from Mayhew, T. M., Elbrond, V. S., Dantzer, V. Skaghauge, E. & Moller, O. Structural and enzymatic studies on the plasma membrane domains and sodium pump enzymes of absorptive epithelial cells in the avian lower intestine. *Cell and Tissue Research*, 1992, **270**, 577–585. Copyright 1992 Springer-Verlag.)

eubacteria. Alexander Fleming, the discoverer of penicillin, was also responsible for the discovery of lysozyme, an enzyme that cleaves the β1–4 glycosidic bond between *N*-acetylglucosamine and *N*-acetylmuramic acid in peptidoglycan (Figure 5.10). This enzyme is present at a concentration of approximately 1 mg/ml in tears. Gram-positive cells, which contain large amounts of peptidoglycan, are more susceptible to lysozyme than are Gram-negative cells in which the peptidoglycan is overlaid by the outer membrane. Surprisingly, it has been reported in recent years that the antibacterial activity of lysozyme can be independent of its enzymic activity. As for many other antibacterial proteins, it has also been found that lysozyme can interact and modulate the activity of LPS. This is a major component of the outer membrane of Gram-negative bacteria and is one of the bacterial constituents recognised by host systems to warn of infection. Another finding that appears to be generic is that lysozyme can interact with other secreted

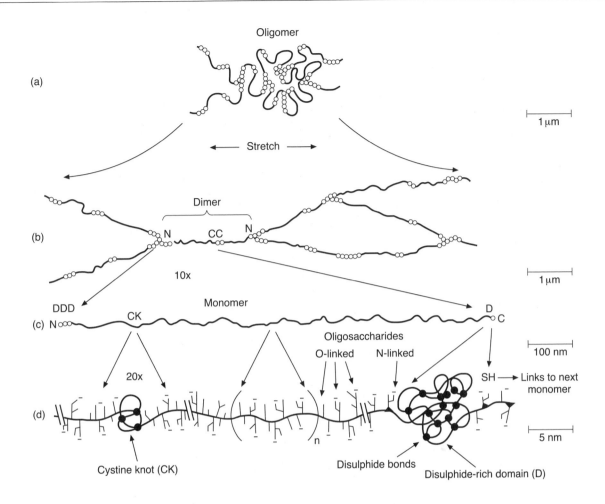

Figure 5.9 Monomeric and oligomeric structures of mucin. In (a) a number of mucin monomers (denoted by lines) are linked together (linkages denoted by circles) in an oligomeric gel. In (b) the gel has been stretched to show more clearly the linkages between the individual monomers. N and C denote the N- and C-termini of the individual mucin monomers. In (c) an individual monomer is schematically denoted, showing the presence of the D domains, which are involved in forming disulphide bonds between monomers. In (d) the structure of the monomer is shown in more detail. It should be noted that the monomer contains many O- and N-linked oligosaccharides.

products of epithelial cells to increase the overall antibacterial activity. This type of interaction is known as a synergistic (Greek synergia = co-operation) interaction. Lysozyme is found in sweat, nasal secretions and tears and is produced by mucosal surfaces generally.

5.4.3.3 Lactoferrin

The metal iron is essential for all life. Therefore one strategy that has co-evolved between bacteria and their hosts is competition for iron. Lactoferrin is a glycoprotein produced by epithelial cells and **polymorphonuclear neutrophils** (PMNs) with a high binding affinity for iron. The presence of this protein at mucosal surfaces results in the levels of free iron being of the order of 10^{-18} M. In addition to preventing bacterial growth by removing iron, lactoferrin also has antibacterial actions. For example, lactoferrin can bind to LPS, via the lipid A moiety, and this is important for this protein's antibacterial activity against Gram-negative organisms. It can also kill Gram-positive bacteria by interacting with outer cell wall components. Proteolytic cleavage of lactoferrin produces an antimicrobial peptide, lactoferricin.

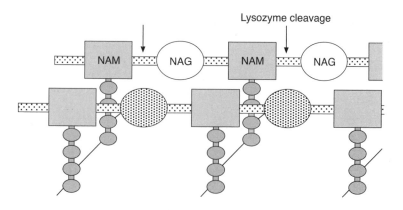

Figure 5.10 The structure of peptidoglycan showing the β(1–4) linkage between alternating N-acetylglucosamine (NAG) and N-acetylmuramic acid (NAM). These polysaccharide chains are linked via peptide side chains (see Section 2.2.4). The sites of cleavage of the molecule by lysozyme are denoted by the arrows.

Lactoferrin and lysozyme have been found to act in a co-operative manner to kill bacteria. In addition to the effects described, lactoferrin has also been reported to block the binding of several bacterial species to host epithelial cells.

Bacterial defence against iron or 'you can have too much of a good thing'

All living organisms require iron to grow and the fight between the host and the bacterium for this precious metal has been referred to in this and other chapters. Now, a number of the bacterial pathogens of humankind also spend some of their time in the external environment, where iron concentrations can be high, even toxic. It has recently been discovered that *Salmonella* spp. protect themselves against toxic concentrations of iron, specifically the ferric (Fe^{3+}) ion. It turns out that the known two-component system, PmrA/PmrB, which controls resistance to the natural cyclic peptide antibiotic polymyxin B, is also responsive to Fe^{3+} concentration. In *Salmonella* spp. with a defective PmrB sensor protein, the bacteria do not respond to iron via PmrA and, in consequence, are sensitive to raised iron concentrations. The relationship between iron sensitivity and polymyxin B resistance appears to be due to the fact that *Salmonella* in soil would be in an environment both high in Fe^{3+} and replete with those bacteria that produce the antibiotic polymyxin B. Responding to iron switches on protection to the antibiotic and ensures cell survival.

5.4.3.4 Lactoperoxidase

As Chapter 6 will describe, PMNs and macrophages can kill ingested bacteria by producing activated oxygen compounds such as superoxide. This radical can interact with various biomolecules and cause damage to them. The enzyme lactoperoxidase is an 80 kDa glycoprotein produced by mucosal cells that generates superoxide and is believed to have a function in controlling bacteria at mucosal surfaces. It has also been shown that lactoperoxidase can catalyse the peroxidation of halides to produce additional antibacterial products.

5.4.3.5 Secretory phospholipase A$_2$

Phospholipases remove fatty acids from phospholipids and thus target, and damage, cell membranes. Tears contain a phospholipase A$_2$ (PLA$_2$) that is very active in killing Gram-positive organisms, such as *Staphylococcus aureus* and *Listeria monocytogenes*. However, this enzyme lacks bactericidal activity against Gram-negative organisms. Paneth cells of the small intestine produce, among other antimicrobial peptides and proteins, a PLA$_2$ that has been shown to kill *E. coli* and *Lis. monocytogenes*.

5.4.3.6 Secretory leukocyte protease inhibitor

It is now established, and will be discussed in several chapters of this book, that both vertebrates and invertebrates have complex systems for recognising and mounting defence against LPS. These include bacterial permeability-increasing protein (BPI) produced by PMNs and secretory leukocyte protease inhibitor (SLPI) produced by both PMNs and epithelial cells. SLPI, although it acts as a protease inhibitor, also interacts with, and inhibits, the activity of LPS. It has also been established that this protein has antibacterial activity. The synthesis of SLPI is also increased in response to LPS and to cytokines such as interleukin (IL)-6 and IL-10, though not to IL-1 or tumour necrosis factor (TNF). Thus SLPI is part of the responsive epithelial system that recognises, and reacts to, bacteria and bacterial components in the external environment.

5.4.3.7 Trefoil peptides

This family of peptides is so-called because of their peptide structure, which has three loops. These peptides are synthesised and secreted mainly by mucin-producing epithelial cells of the gastrointestinal tract and they have a close association with mucins. The functions of these very stable peptides is not completely defined. They promote mucosal repair by stimulating the movement of epithelial cells. They have not been shown to have a direct antibacterial effect but it is envisaged that any factor that promotes mucosal re-epithelialisation would have, as a matter of course, anti-infective activity. Much more needs to be learned about the role of these factors in mucosal immunity.

Trefoil peptides, cystic fibrosis and bacteria

The trefoil factor family (TFF) domain peptides 1 to 3 are small, disulphide-bonded peptides that associate with mucin molecules on mucosal surfaces, particularly in the gut. These peptides have been proposed to be involved in defending epithelial surfaces against damaging agents and also in wound healing. For example, knockout of TFF1 results in mice with an atrophic gastric mucosa, and knockout of TFF3 produces mice with impaired healing of the intestinal mucosa.

It has now been observed, using a range of techniques such as immunohistochemistry, immunofluorescence, western blotting and reverse transcriptase polymerase chain reaction (RT-PCR), that these peptides are also produced by the lung epithelium. The obvious question with regard to the TFFs is: do they play any role in controlling bacteria? In specimens of sputum from patients with cystic fibrosis it was found that TFF1 and TFF3 associated with bacteria such as *Staph. aureus* and *Pseudomonas aeruginosa*. This raises the possibility that the trefoil peptides may play a role in antibacterial defence.

5.4.3.8 Acute phase proteins

Section 6.4.1 will deal in more detail with acute phase proteins whose levels in serum can rise 1000-fold during infection. It was always believed that these proteins were produced by the liver. However, data from the culture of epithelial cells, and from *in situ* hybridisation studies, are indicating that mucosal epithelial cells can produce acute phase proteins such as serum amyloid A (SAA). These proteins generally act to opsonise bacteria, although they have other roles. It is not absolutely clear what function they play at epithelial surfaces.

5.4.3.9 Collectins

Clinical syndromes often offer clues to disease mechanisms. For example, certain children have been described who suffer from recurring upper respiratory tract bacterial infections. These children were subsequently found to have decreased concentrations of the serum protein mannose-binding lectin (MBL). MBL, along with lung surfactant proteins A and D (SP-A and D) are known as collectins. These are oligomeric proteins composed of C-type lectin domains (a lectin being a protein that recognises and binds to specific carbohydrates) connected to a domain with homology to collagen. The collectins have multiple carbohydrate-binding domains and can, therefore, act to cross-link carbohydrate-containing components. The function of these proteins is the subject of intensive review and their full range of functions is not yet known. Their role in defence against bacteria has been strengthened by the finding that mice in which SP-A has been knocked out have increased susceptibility to infection with *Streptococcus agalactiae*.

The collectins are produced by most mucosal surfaces. Even the gut epithelia produce SP-A and SP-D. They bind to bacteria through surface carbohydrates and can cause aggregation of said bacteria. They act as opsonins as SP-A and D increase the uptake of both Gram-negative and Gram-positive bacteria by human neutrophils. SP-A has also been reported to inhibit LPS-induced cytokine and NO production.

As will be described in Section 6.6.1.1, the collectins are believed to be one of the pattern recognition receptors important in the ability of innate immunity to recognise that we have been infected with bacteria (or viruses).

5.4.3.10 Secretory IgA

As will be emphasised throughout this book, all organisms are composed of interacting systems. Thus far, this chapter has dealt only with relatively simple mechanisms of defence at mucosal surfaces. However, as will be explained later in this and subsequent chapters, mucosal epithelial cells have actions over and above simply being a barrier and a producer of secreted components. They act as watchdogs and guardians of the internal milieu of the body. In this context, they are associated with systems of lymphoid cells that are responsible for the activation of B lymphocytes to produce one form of immunoglobulin (Ig), in this case IgA. The external mucosal surfaces of the body, with the exception of the genital tract, which contains more IgG than IgA, contain large amounts of so-called secretory IgA. Indeed IgA is the immunoglobulin produced in greatest amount, with an average human producing 5–15 g of IgA per day into his/her external secretions. This is done by actually transferring the IgA from the plasma cells producing it in the submucosa, through the cytosol of the epithelial cells, and on to the surface of these cells. The action of these secreted antibodies is believed to be the inhibition of the adhesion of bacteria to the mucosal surfaces. However, it should be noted that deficiency of IgA production in humans is relatively common and the effects are variable, with some individuals showing minimal abnormalities. This suggests that mucosal IgA has a limited role in antibacterial defences.

5.4.3.11 Antibacterial peptides

The study of natural (or innate) immunity has been a relative backwater of research for many years. The feeling has been that innate immunity is an ancient and relatively unimportant facet of our immune defences, and that our real defence system lies in the clonal system of immunity, with its T and B lymphocytes. The flaw in this argument is that invertebrates such as insects have no T and B lymphocytes and yet appear to survive quite happily in a world that swarms with bacteria and fungi. For example, how do dung beetles survive a lifetime's exposure to animal excreta? Invertebrates constitute 95% of the multicellular species on our planet. Thus it must be the case that innate immunity is an efficient and effective system of antibacterial defence.

A number of antibacterial defence mechanisms have been described. The question this raised was: were they enough? If only animals had evolved antibiotics. This would solve, or at least partially solve, the problem of bacterial infections. Having discovered that bacteria and fungi produced antibiotics it was perhaps short-sighted not to ask the question: do all living creatures produce such agents? The eventual discovery that all organisms produce some form of antibiotic, or more correctly antibacterial, peptide, was made by two groups of researchers. One was the group of workers, largely in the USA, who were studying the antibacterial activity of PMNs. These circulating leukocytes produce a number of proteins, such as the cationic antimicrobial proteins (CAPs), and BPI. These cells also produce antibacterial peptides such as the defensins. The other group was led by Hans Boman of the Karolinska Institute in Stockholm, who was studying anti-infective

Table 5.2.	Antibiotic peptides

Classification

1. Amphiphilic α-helical peptides without cysteine, such as the magainins and insect and pig cecropins
2. Cysteine-disulphide ring peptides with or without amphiphilic tails, such as bovine bactenicin, and frog ranalexin and brevinins
3. Cysteine-rich, amphiphilic β-sheet peptides such as the α- and β-defensins found in mammals, the tachyplesins found in the crab and the porcine protegrins
4. Linear peptides with one or two predominant residues such as porcine PR-39, a proline-rich peptide.

mechanisms in insects. The fact that many insects survive in environments rich in bacteria (faeces or rotting matter) suggests that they have very effective antibacterial defences.

After 10 years of research, Boman reported on the first insect antibiotic peptides, termed cecropins, in the 1980s (Table 5.2). The first mammalian antibacterial peptides were isolated from the small intestine of the pig in 1989. Only a little over a decade later, at least 400 different antibacterial peptides have been discovered. It turns out that every living creature on our planet (as far as can be ascertained) produces such peptides. Bacteria make a range of such peptides including the LPS-inactivating cyclic peptide polymyxin B. As Table 5.2 briefly illustrates, the ancient languages have been pillaged to provide the apparently bizarre names used to describe these peptides, such as cecropins, magainins, defensins, etc.

As stated, most cells in vertebrates and invertebrates can produce some antibacterial peptides but there are only a few cell populations able to produce large amounts of these peptides. The circulating phagocytes produce a range of both antibacterial proteins and peptides, and their functions will be discussed in Section 6.6.2. Epithelial cells of skin and mucosal linings are active producers of antibiotic peptides such as the human β-defensins. In the epithelium of the small intestine resides a specialised cell population, the Paneth cell (Figure 5.11), which produces a number of antibacterial compounds including lysozyme, phospholipase A_2 and antibacterial peptides such as α- and β-defensins. Interestingly, the mouse produces many more intestinal defensins than does the human.

5.4.3.11.1 Structure of antibacterial peptides

Antibacterial peptides of multicellular organisms are generally between 10 and 40 amino acid residues in length and fall into four general chemical groups (Table 5.2). The first are linear peptides that do not contain the amino acid cysteine and therefore cannot have disulphide bonds. These peptides can be further subdivided into those that

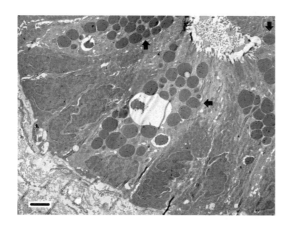

Figure 5.11 Human small intestine immunostained for α-defensin (human defensin-5). Defensin-rich granules (arrows) were found exclusively in Paneth cells. Bar = 2 mm. (Reproduced with permission from Porter, E. M., Liu, L. Oren, A., Anton, P. A. & Ganz, T. Localization of human intestinal defensin 5 in Paneth cell granules. *Infection and Immunity*, 1997, **65**, 2389–2395.)

form amphiphilic α-helices and those that are unable to adopt this structure. The cecropins, temporins, magainins and the dermaseptins all fall into this first category. The second and third types are cysteine-containing peptides where disulphide bonds between the cysteines stabilise the peptide. Examples include the tachyplesins (found in horseshoe crabs (*Limulus polyphemus*)) and the α- and β-defensins (Figures 5.12 and 5.13). The fourth group of antibacterial peptides have high proportions of specific amino acids. These are often proline (Pro or P in single amino acid code), arginine (Arg, R) or, in one example, tryptophan (Trp, W). The apidaecins, produced by the honeybee, were the first example of this group of peptides to be discovered. One of the major families of such peptides is the cathelicidins and these are all initially translated as pre-proteins which have to be proteolytically processed to form the active peptide. Humans are now known to express six different α-defensins and two β-defensins.

The role of human β-defensin-1 in cystic fibrosis

Cystic fibrosis (CF) is an inherited disease caused by a defect in a phosphorylation-regulated chloride channel called the cystic fibrosis transmembrane conductance regulator (CFTR) located in the apical membranes of involved epithelia, for example the lung. Patients with CF suffer from severe lung disease that is due largely to recurrent infections with organisms such as *Ps. aeruginosa*. It is the sequelae of these recurrent infections that causes progressive lung damage and eventual respiratory failure. It has been shown that incubation of common CF pathogens with normal lung airway epithelia results in the rapid killing of the bacteria. In contrast, the same cells from patients with CF are unable to kill bacteria. However, if the raised NaCl concentration produced by the CF epithelial cells (due to defects in the CFTR) is corrected then the cells do kill bacteria. The explanation for these findings is that the major defence of lung airway epithelia against bacteria is the antibacterial peptide human β-defensin-1. This peptide is not active at the high salt concentration in the extracellular fluids of the lungs of CF patients and thus they are unable to kill common respiratory bacteria.

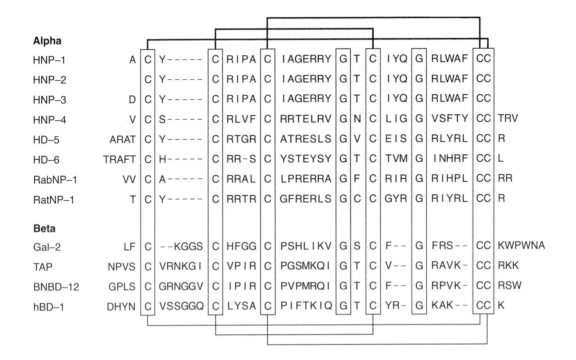

Alpha

HNP–1	A	C	Y-----	C	RIPA	C	IAGERRY	G	T	C	IYQ	G	RLWAF	CC	
HNP–2		C	Y-----	C	RIPA	C	IAGERRY	G	T	C	IYQ	G	RLWAF	CC	
HNP–3	D	C	Y-----	C	RIPA	C	IAGERRY	G	T	C	IYQ	G	RLWAF	CC	
HNP–4	V	C	S-----	C	RLVF	C	RRTELRV	G	N	C	LIG	G	VSFTY	CC	TRV
HD–5	ARAT	C	Y-----	C	RTGR	C	ATRESLS	G	V	C	EIS	G	RLYRL	CC	R
HD–6	TRAFT	C	H-----	C	RR-S	C	YSTEYSY	G	T	C	TVM	G	INHRF	CC	L
RabNP–1	VV	C	A-----	C	RRAL	C	LPRERRA	G	F	C	RIR	G	RIHPL	CC	RR
RatNP–1	T	C	Y-----	C	RRTR	C	GFRERLS	G	C	C	GYR	G	RIYRL	CC	R

Beta

Gal–2	LF	C	--KGGS	C	HFGG	C	PSHLIKV	G	S	C	F--	G	FRS--	CC	KWPWNA
TAP	NPVS	C	VRNKGI	C	VPIR	C	PGSMKQI	G	T	C	V--	G	RAVK-	CC	RKK
BNBD–12	GPLS	C	GRNGGV	C	IPIR	C	PVPMRQI	G	T	C	F--	G	RPVK-	CC	RSW
hBD–1	DHYN	C	VSSGGQ	C	LYSA	C	PIFTKIQ	G	T	C	YR-	G	KAK--	CC	K

Figure 5.12 The primary sequences of a number of α- and β-defensins, showing the conserved cysteines and the intra-chain disulphide bonds.

5.4.3.11.2 Biological actions of antibacterial peptides

As a group, antibiotics have a wide range of actions. Penicillin, for example, prevents the synthesis of peptidoglycan, while a very large number of antibiotics (e.g. tetracyclines, chloramphenicol and strepto-mycin) inhibit protein synthesis by targeting the bacterial ribosome. Nalidixic acid and novobiocin inhibit DNA gyrases and block DNA synthesis. In contrast, the antibacterial peptides seem to target the bacterial cell membrane (Figure 5.14). For example, early studies of the interaction of the cecropins with artificial membranes revealed that they formed voltage-dependent anion channels. This can be due to the peptide spanning the membrane in multimeric complexes. Some peptides are too short to span the membrane and must produce stacked aggregates to form a pore. Pore formation causes a collapse of the membrane voltage gradient, which disrupts the normal functioning of the bacterium. Studies of artificial membrane interactions with antibiotic peptides have also suggested that membrane damage, without pore formation, can also occur. It is believed that the reason that these antibacterial peptides have little, or no, effect on mammalian cells is that the cytoplasmic membranes of these cells, in contrast to those of bacteria, contain cholesterol.

As will be described in Section 10.2.2, bacteria have, not surprisingly, evolved resistance mechanisms against these host antibacterial pep-tides.

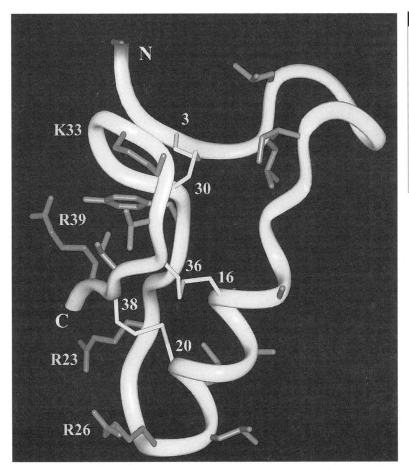

Figure 5.13 Crystal structure of defensin-1. (Reprinted with permission from Yang, Y. S., Mitta, G., Chavanieu, A., Calas, B., Sanchez, J. F., Roch, P. & Aumelas, A. Solution, structure and activity of the synthetic four-disulfide bond Mediterranean mussel defensin (MGD-1). *Biochemistry*, 2000, **39**, 14436–14447. Copyright 2000, American Chemical Society.)

5.4.3.11.3 Control of the synthesis of antibacterial peptides

A major feature of our current understanding of host–pathogen interactions is the commonality of mechanisms used in host defences. For example, shortly after the discovery of antibacterial peptides it was found that control of transcription of these genes in *Cecropia* was through the transcription factor Cif which is part of the Relish (NF-κB) system described in Section 4.2.6.6.2 and which will be discussed in more detail in Sections 6.6.1.2 and 6.7.2.2.2. It turns out that in *Drosophila*, the control of antifungal and antibacterial peptides is via two cell surface receptors, 18-wheeler and Toll, respectively, that are linked to activation of the NF-κB transcription control system (Figure 5.15). Homologues of these two receptors are found in humans where they have been named the Toll-like receptors (TLRs). The TLRs and the receptor complex for IL-1 are very similar and the mechanism by which IL-1 stimulates intracellular signalling is detailed in Section 4.2.6.6. The details of the 18-wheeler and Toll intracellular transduction systems are described in Sections 4.2.6.6.2 and 6.6.1.2.

Studies of the human intestine have established that the production of the two major epithelial cell antibiotic peptides, human β-defensin 1

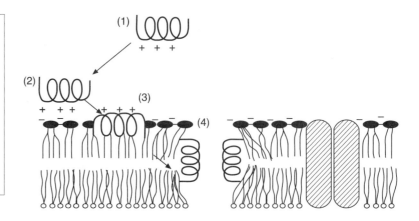

Figure 5.14 Proposed mechanism of action of an antibacterial peptide (1). The key event is the binding (2) and insertion (3) of the peptide into the bacterial cytoplasmic membrane to produce a pore (4) through which cytosolic contents can leak. This is not the only mechanism possible, but it is believed that most antibacterial peptides in some way increase membrane permeability.

and 2 is under the control of the NF-κB transcription system and that, in addition to bacteria and bacterial factors, the pro-inflammatory cytokine IL-1 can stimulate antibacterial peptide production.

5.4.4 Overview of the synthesis of antibacterials by mucosal epithelial cells

Mucosal immunity is the Cinderella of immunology and the innate mechanisms involved in antibacterial defence received little attention until very recent years. In spite of this, it is clear that mucosal epithelial cells are actively producing and secreting a whole range of proteins and peptides that have antibacterial functions (Figure 5.16). The concept also appears to be emerging that these various non-specific antibacterial proteins can interact and synergise to kill or inhibit the growth of bacteria.

5.5 | The mucosal epithelium as a shedding surface

Many cells in the body are regularly replaced by the process of apoptosis and cell division. Other cells, neurones being the classic example, are never replaced. The epithelia of the body consist of one group of cells that are constantly being replaced. The best-known example of this is the skin. The fact that the skin is being constantly replaced is obvious to all of us. The dust that we have to remove on a regular basis from our rooms consists largely of skin particles. Epithelia constantly shed the outer layers of epithelial cells and replace them with fresh cells from the underlying surface. This means that the bacteria adherent to, or in contact with, the individual epithelial cell will be removed when the cell is replaced (Figure 5.17). The new epithelial cell will then have to be repopulated with bacteria.

It is envisaged that epithelial cell replacement plays some role in controlling the numbers of commensal bacteria supported on an epithelial surface. An idea of the efficacy of the mechanisms employed by mucosal epithelia, including cell replacement, in limiting bacterial

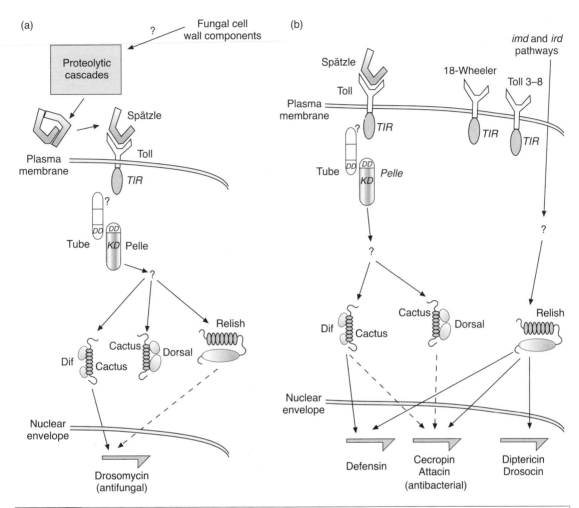

Figure 5.15 Schematic showing the control of transcription of antibacterial and antifungal peptides in the fat body of *Drosophila*. The activation of the transcription of these anti-pathogen peptides depends on the binding of components of the pathogen to a family of receptors, the prototype of which is Toll. (a) Fungal cell wall components (such as mannans) bind to Toll, a protein involved in dorsal/ventral patterning in *Drosophila* embryos, triggering signalling via an NF-κB signalling pathway (see Section 4.2.6.6.2) in which Cactus/Relish (equivalent to IκB in mammalian cells) is displaced by phosphorylation allowing the NF-κB analogues Dif or Dorsal to enter the nucleus and induce transcription of the antifungal peptide drosomycin. In (b), bacterial components (such as LPS) can interact via a number of potential Toll-like receptors such as 18-wheeler and induce the transcription of antibacterial peptides such as defensin, cecropin, diptericin.

numbers is obtained by asking what happens to bacterial numbers on a surface that does not shed? The main non-shedding surface in our bodies is provided by our teeth. We all know how rapidly our teeth become 'furry' when we stop brushing them. In Figure 7.2 electron micrographs of dental plaque show the huge numbers of bacteria that can build up on a non-shedding surface. Would such numbers be able to build up on epithelia that failed to shed?

Is the rate of cell shedding controlled by the bacteria that live on epithelial surfaces? The answer to this question is not yet known.

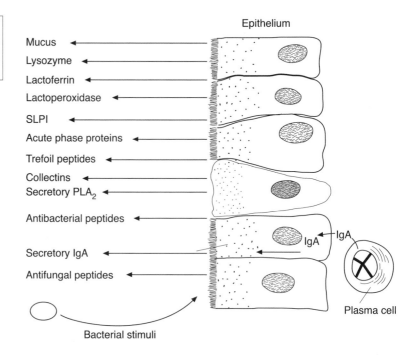

Figure 5.16 Summary of the secretion of antibacterial components at mucosal surfaces. For abbreviations, see list on p. xxii.

However, there is growing evidence that bacteria produce factors that are able to control the cell cycle kinetics of eukaryotic cells. The reader should turn to Section 10.6 for more information on these novel bacterial virulence factors that control the host cell cycle and the process of apoptosis.

Synergy, synergy, synergy

This chapter has described a large number of mucosal factors believed to be involved in the protection of the mucosal surface from exogenous pathogens. There is some evidence that these various factors may act in synergy. However, few studies have deliberately examined the possible synergistic interactions between the various antibacterial substances. A recent study has examined the combined action of six antimicrobial peptides found in the fluid that bathes our airways. This study was carried out to examine the factors that might be responsible for bacterial growth in the lungs of patients with CF. It was found that the paired combinations of lysozyme–lactoferrin, lysozyme–secretory leukocyte protease inhibitor (SLPI), and lactoferrin–SLPI were synergistic. In studies in which the combination of three factors – lysozyme, lactoferrin and SLPI – was examined there was even greater synergy. Other combinations involving the human β-defensins, LL-37 and tobramycin (an antibiotic that is often administered to CF patients by inhalation) were additive. It is known that the salt concentration of the airway surface liquid may be elevated in CF patients and this may inhibit the activity of the defensins. The effect of salt on the synergistic combinations was therefore examined. As the ionic strength increased, synergistic interactions were lost. These findings suggest that the antibacterial potency of the mucosal fluid in the lungs may be significantly increased by synergistic and additive interactions between antimicrobial factors.

These results also suggest that increased salt concentrations that can exist in CF could inhibit airway defences by diminishing these synergistic interactions.

5.6 | The normal microflora as an antibacterial agent

For the non-microbiological reader, and in fact for most microbiology students, the finding that the human body contains 10 bacteria for every human cell comes as a huge surprise. Moreover, it is not simply the number of bacteria that causes wonderment, it is their sheer diversity. There are certainly 1000, and there may be as many as 10 000, different bacterial species living on our mucosal surfaces (for a more detailed description of the normal microflora, see Chapter 1). Compare this with the 50 or so bacterial species known to cause disease. Now, while we are beginning to understand much of the interactions going on between pathogenic bacteria and their hosts, we know very little about the interactions that preserve and control the diversity of our commensal bacteria.

The role of the commensal microflora in generating pathology has been dealt with in Section 1.3.2. In Section 11.6 the effects of this microflora on the shape and functions of the host organism will be described. In this section the antibacterial actions of the commensal microflora will be discussed briefly.

This ability of the commensal microflora to inhibit the colonisation of exogenous pathogens was first discovered in the 1950s by individuals working on *Salmonella* spp. or on *Vibrio cholerae*. Since these pioneering

Figure 5.17 Epithelial cell replacement and removal of bacteria. Two examples are given. In (a) the replacement of enterocytes in the crypts of the small intestine acts to remove attached bacteria, while in (b) the differentiation of stratified epithelium in the mouth/vagina results in the removal of attached bacteria.

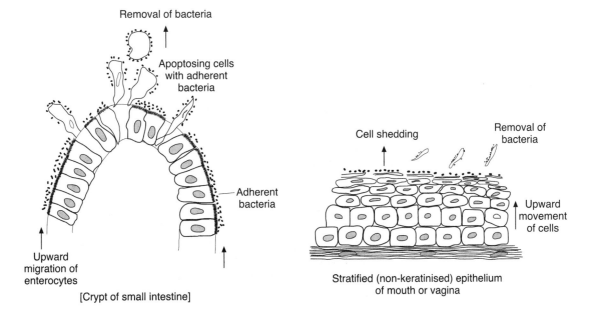

Removal of bacteria

Apoptosing cells with adherent bacteria

Adherent bacteria

Upward migration of enterocytes

[Crypt of small intestine]

Cell shedding

Removal of bacteria

Upward movement of cells

Stratified (non-keratinised) epithelium of mouth or vagina

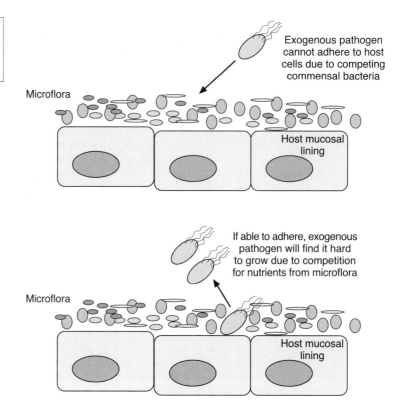

Figure 5.18 Effects of the commensal microflora on colonisation by exogenous bacterial pathogens.

experiments, a range of other pathogens have been shown to be excluded by the normal microflora. This phenomenon has been given various names: bacterial antagonism, bacterial interference or, most commonly, colonisation resistance. The possible mechanisms underlying this phenomenon are represented schematically in Figure 5.18.

Studying bacterial antagonism, which has focused largely on the gut, is obviously complicated by the diversity of the microflora. However, a number of studies have shown that for *Shigella flexneri* and *Clostridium difficile*, the most effective species excluding these pathogens are nonpathogenic strains of the same organism. This implies that competition for adhesion sites is important.

Bacterial interference – an alternative to antibiotics?

It is widely recognised that our ability to combat infectious diseases is seriously undermined by the growing prevalence of bacteria that are resistant to many antibiotics. This, together with the limited number of new antibiotics currently under development (but see Section 11.3), has prompted many clinicians and scientists to suggest that we may be entering a new 'dark age' in which many infectious diseases will be untreatable. There is, therefore, a great need to develop new approaches to killing pathogenic organisms that could be used therapeutically. Investigations of the normal microflora, and of the interactions that occur between members of these complex communities, have suggested one new approach. As there is intense competition between members of the microflora of a particular site for adhesion sites and nutrients, some bacteria have evolved means of maintaining a competitive edge over their neighbours. Hence some species can secrete chemicals that are antagonistic to others or else they can alter the local environment to one favouring their own growth but inhibiting that of others. This phenomenon is known as bacterial interference and could be used to prevent or treat infectious diseases. This would be achieved by implanting an organism of low virulence (usually a member of the normal microflora of the particular site) that is antagonistic towards a particular pathogen. For example, certain α-haemolytic streptococci (members of the normal microflora of the oropharynx) are known to produce proteins (known as bacteriocins) that are inhibitory to major pathogens including *Strep. pneumoniae* and *Strep. pyogenes*. They also compete with these pathogens for similar nutrients. In a number of clinical trials, it has been shown that application of certain strains of α-haemolytic streptococci to the mouths of children with recurrent tonsillitis (a disease caused by *Strep. pyogenes*) either abolishes, or considerably reduces, recurrence of the infection. Many other studies of upper respiratory tract infections (e.g. otitis media, sinusitis) have shown that the absence in individuals of members of the normal microflora (usually α-haemolytic streptococci) with the ability to inhibit respiratory pathogens (such as *Haemophilus influenzae*, *Strep. pneumoniae*) results in a much greater risk of infection.

The results of a recent study in a population of sex workers in Kenya have shown that vaginal colonisation with lactobacilli results in a decreased chance of acquiring both HIV type I infection and gonorrhoeae.

Consider, for a moment, what the requirements are for a pathogenic bacterium that finds itself near the mucosal surface of its host organism. For it to infect, it must adhere to the mucosal epithelium and be able to obtain sufficient nutrients to grow. On an appropriate epithelial surface, say in the intestine, if there were no bacteria already present, both requirements would be met. Fortunately for the host, the pathogen finds itself more in a situation of a customer in a college bar at closing time on a Friday night – you simply can't get near the bar to get a drink! Likewise, as mentioned above, the pathogen cannot get near the epithelial barrier to adhere. Moreover, even if it were able to adhere, it would be in competition with the established normal microflora for nutrients. Many studies, using largely continuous flow culture of intestinal or oral bacteria, have shown that competition for nutrients plays a major role in excluding exogenous pathogens. There is obviously an interplay between nutrition, bacterial numbers and competition for adhesion sites. An additional barrier to colonisation is the production of antibacterial compounds by the microflora. Such compounds include simple gases such as hydrogen sulphide (H_2S), short-chain volatile fatty acids and antibacterial compounds produced by bacteria called bacteriocins.

With the growing realisation of the importance of our normal microflora, there is increasing interest in trying to maintain, or modify, its composition to promote good health. In the case of the gut, this can be done by using probiotics (such as live yoghurt or proprietary products), which contain live bacteria. The alternative is to use prebiotics, which

Table 5.3.	Specialised cell populations in mucosal surfaces
Cell population	Proposed function
M (microfold) cells	Allow the passage of bacteria and bacterial antigens from the external environment to meet with antigen-presenting cells in the submucosa. Allow the immune system to recognise external antigens to aid in immune responses to the microflora
Paneth cells	Active producers of antibacterial molecules, including lysozyme, secretory PLA_2 and human defensins
Intra-epithelial lymphocytes	Surveillance of intestines for microbial pathogens; removal of damaged epithelial cells; production of trophic factors to maintain epithelial integrity and regulation of local immune and inflammatory responses
Goblet cells	Produce mucin
Enteroendocrine cells	Release hormones in response to environmental cues
Tuft cells	Cells associated with absorptive epithelium but with elongated microvilli; may function as chemical sensory cells
Cup cells	Cells have short microvilli with a cup-like concavity. Their function is unknown

are carbohydrate-containing materials thought to be able to change the bacterial speciation of the gut microflora. This is one area of Microbiology that deserves much greater study and is likely to be a focus of major research interest in the coming decades.

5.7 | Other cell populations in mucosal epithelia

The mucosal surfaces of the body are composed largely of epithelial cells with a smaller proportion of non-epithelial cells. The role of goblet cells has been described briefly. A range of other cell populations exist within mucosal surfaces of the body, mainly in the intestines, including M cells, Paneth cells, IELs, enteroendocrine cells, tuft cells and cup cells (Table 5.3). Only the first three cell populations will be described briefly, as they are known to be involved in antibacterial defence. It is important to realise that the gut contains a large proportion of the lymphocytes that populate the human body.

5.7.1 M cells

All external epithelia are 'connected' to adjacent immunological systems, which are known generically as the **mucosa-associated lymphoid tissues** (MALT). This system in the gut is called the **gut-ALT**

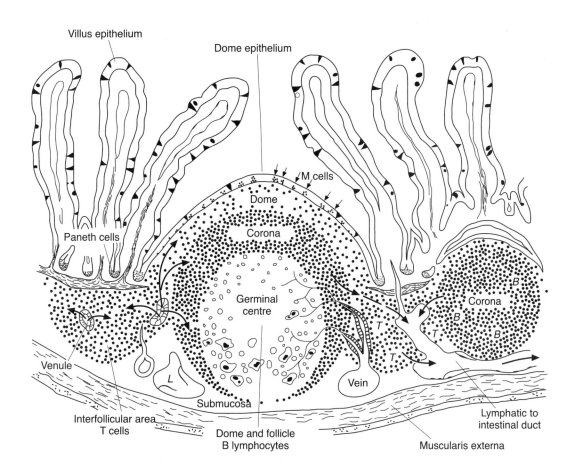

(GALT) and in the skin, quite appropriately, the skin-ALT (SALT). The structure of the GALT is shown in Figure 5.19. Another name for this system is the follicle-associated epithelium (FAE). In specialised areas of the mucosal epithelium, particularly in the gut overlying the GALT or, as it is more commonly known Peyer's patches, there is a population of specialised cells called M or microfold/membranous cells.

The origin of M cells is still the subject of controversy, with two major hypotheses being proposed to account for their existence: (i) they are a pre-programmed cell lineage or (ii) they derive from differentiated enterocytes (intestinal epithelial cells) by phenotypic conversion. M cells are present in small numbers in the intestine and no markers defining these cells have yet been found. It is therefore not possible to isolate and study these cells and so what little we know of them is derived largely from histomorphological studies. Fortunately, M cells have a characteristic morphology when examined by transmission or scanning electron microscopy and this can be seen in Figure 5.20. A characteristic finding is that M cells form a pocket in which lymphocytes and antigen-presenting cells (e.g. macrophages) associate (see Figure 5.5). The height of the M cell is less than that of

Figure 5.19   Schematic diagram showing the organisation of an ileal Peyer's patch in the mouse. These Peyer's patch lymphoid aggregates are composed of three anatomically distinct areas: (i) follicular area, containing the germinal centre, corona and dome; (ii) an interfollicular area containing T cells (T) and (iii) the dome epithelium containing M cells. L, lymph vessel; B, B cell.

Figure 5.20 (a) Normal mouse follicular-associated epithelium showing enterocytes and M cells (depressed cells). (b) Changes induced in M cell surface morphology by *Salmonella typhimurium*. M cells in the centre exhibit membrane ruffles. Bacteria (arrowed) can be seen associated with the lower, left-hand ruffle. Bars = 10 μm. (Reprinted from *Trends in Microbiology*, **6**, Jepson, M. A. & Clark, M. A. Studying M cells and their role in infection, 359–365. Copyright 1998, with permission from Elsevier Science.)

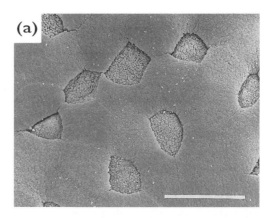

(a)

(b)

its neighbouring enterocytes and the role this plays in cell functioning will become apparent.

What do M cells do? Their function appears to be to take up material from the lumen of the intestine and to transport it in endocytotic vacuoles to the basolateral surface of the cell where it can contact the underlying lymphoid elements of the Peyer's patch. It is assumed that this function allows the immune system to sample the particulate antigens, particularly the bacterial antigens, present in the intestinal lumen and to stimulate an immune response. One of the reasons that there has been increasing interest in M cells in recent years is the finding that a number of bacteria (*Salmonella typhimurium*, *Yersinia* spp., enteropathogenic *Escherichia coli*, *Shig. flexneri*, *Lis. monocytogenes* and even *Strep. pneumoniae*) are able to enter the body and cause infection by passing through M cells. The role of this portal in bacterial infection is discussed in more detail in Section 8.2.1.

5.7.2 Paneth cells

These cells are a subpopulation of the epithelial cells of the intestine found in the crypts of Lieberkühn in the small intestine. They are pyramidal in shape and have a granular appearance. They have come to the notice of microbiologists and immunologists in recent years, with the finding that they produce a variety of antibacterial molecules including antibacterial peptides (the α-defensins, also known as cryptidins), lysozyme and PLA_2.

5.7.3 Intra-epithelial lymphocytes

The gastrointestinal tract contains many of the lymphocytes that populate the body. These are found in the submucosa (or lamina propria), Peyer's patches and scattered within the epithelial layer. It is to the latter population that the term intra-epithelial lymphocytes (IELs) is applied. T lymphocytes are classically divided into CD4 and CD8 cells (see Section 6.3.4) with the majority of these cells bearing a T cell receptor (TCR) composed of $\alpha\beta$ heterodimers. A smaller population of T cells exist with TCRs consisting of two related proteins termed $\gamma\delta$ and these cells are therefore termed $\gamma\delta$ T lymphocytes. The IELs have a number of characteristics that make them stand out from the T lymphocytes in blood and thymus. They express a wider variety of TCRs including the unique expression of lymphocytes with a TCR composed of two α-subunits (CD8$\alpha\alpha$ homodimer). It is believed that these T lymphocytes, in contrast to the normal situation, develop outside the thymus. The majority of T cells in the circulation have the $\alpha\beta$ TCR, with perhaps only a small percentage being $\gamma\delta$ T cells. In the IELs, the majority of the cells are CD8 positive and approximately one third may express the $\gamma\delta$ TCR. The majority of these cells have a restricted repertoire of antigen receptors, suggesting that they have evolved to recognise a limited number of antigens commonly encountered in the gut. One of the antigens recognised by these cells is the heat shock protein chaperonin 60. Of interest is the finding that the replenishment of the IELs is dependent on thyroid hormone interactions with the gut. It has also been proposed that the IELs can produce growth factors for epithelial cells when they are stimulated by antigens. In germ-free animals the numbers of IELs is significantly reduced. The functions of these IELs are described in Table 5.3.

5.8 | The mucosal epithelium as a watchdog

Mucosal epithelial cells clearly play a major defence role in the very early stages of infection by acting as a barrier and producing a wide range of antibacterial molecules. From Section 5.7 it is clear that the M cells in the intestinal mucosal epithelium communicate with submucosal cells and, in particular, the lymphoid elements. There is increasing evidence that the epithelial cells of the mucosal surfaces of the body are also programmed to signal to the innate and acquired systems of immunity (which will be discussed in Chapter 6) to induce the chemoattraction of leukocytes. This has been termed the 'watch-dog' function of the mucosal epithelium.

As has been briefly described (see Figures 5.2 and 5.8) the lumen of the intestine is lined by the villi which in turn are lined by a single layer of polarised epithelial cells. The face of the epithelial cell lining the lumen is termed the apical surface. The other end of the cell, which is attached to the basal lamina is termed the basolateral surface. This basolateral face of the enterocyte is in close contact with the lymphoid and myeloid cells of the submucosa and the myriad blood vessels supplying the gut with nutrients.

It is now established that epithelial cells, such as those of the gut, can produce a wide range of cytokines (see Section 4.2.6) including some of the chemotactic cytokines known as chemokines. Epithelial cells also have receptors for a range of cytokines. The scenario that is now believed to occur when an intestinal epithelial cell makes contact with a bacterial pathogen, for example *Sal. typhimurium*, is that the epithelial cell becomes activated by a mechanism controlled largely and integrated by the cytoplasmic transcription factor NF-κB. The biological function of NF-κB is described in more detail in Sections 4.2.6.6.2, 4.2.6.6.3 and 6.7.2.2.2. Bacterial stimulation results in the polarised release of a number of chemokines. Thus IL-8 is released from the basolateral surface of epithelial cells. This so-called CXC chemokine, acts to recruit neutrophils to sites of inflammation. However, attracting neutrophils to the basolateral surface of the epithelium would be of little use when the bacteria that are the targets of the neutrophils are in the lumen. It is therefore proposed that, in the case of intestinal infection, a second chemoattractant is released from the apical surface of the cell to further attract neutrophils into the lumen. This second chemoattractant which has been termed pathogen-elicited epithelial chemoattractant (PEEC) is a small (1 to 3 kDa) molecule that has not been identified at the time of writing. It is likely to be a peptide. Studies using gridded DNA arrays to identify genes switched on in epithelial cells exposed to *Sal. dublin*, have shown that only a small number of chemokine genes, those encoding IL-8 and MIP-2α, are substantially stimulated, suggesting that these are the major chemokines involved in cell trafficking to the infected site.

One of the major unanswered questions in the inflammatory pathology of the infectious diseases is: what switches off inflammation? This is not as obvious a question as it seems. We know that many people suffer from chronic idiopathic (unknown cause) inflammation in conditions such as psoriasis and arthritis. Understanding how inflammation is switched off in infections is therefore important. One possible off-switch is the lipid mediator lipoxin A_4. This is the product of the enzymic action of lipoxygenases on arachidonic acid. This compound has been shown to be able to prevent *Salmonella*-induced migration of neutrophils in a model of intestinal polarised epithelial cell monolayers (Figure 5.21). This is a simplified model of the situation at mucosal surfaces, and other cytokines are likely to be involved. However, it shows the importance of the interaction of

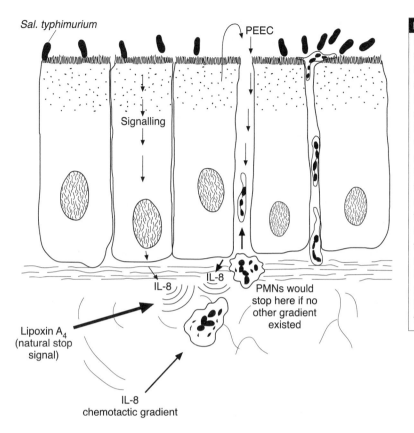

Figure 5.21 Signalling between bacteria, mucosal epithelial cells and the underlying immune systems. This diagram shows one proposed mechanism involving two chemokines (IL-8 and PEEC) for chemoattracting PMNs to the lumen of the gut, which has been infected with *Salmonella typhimurium*. The IL-8 released by epithelial cells can get the PMNs to the basolateral surface of the cell. It is proposed that a second chemokine (PEEC) released from the apical surface of the enterocyte is necessary to attract the PMNs onto the lumenal surface. It is also proposed that the lipid mediator lipoxin A_4 can act to switch off the inflammatory response.

specific cytokines to create a functional network that, in this case, is designed to get a specific cell population to a specific site in the body.

One of the major areas of ignorance is how the epithelial surfaces signal to the underlying immune cells both to prevent and to induce trafficking of functional leukocytes to the defence of the epithelium. This will be one of the major growth areas of the new science of cellular microbiology and will be a major challenge for scientists.

5.9 | Concept check

- The human body has an enormous surface area exposed to the external environment.
- Our only protection from this environment is our epithelium.
- Our internal surfaces that are exposed to the environment have a moist epithelium termed the mucosa.
- The mucosa has a very large surface area – approximately $400 \, m^2$ – most of which is heavily colonised by bacteria.
- The mucosa acts as the first line of immune defence.
- The mucosa acts as a mechanical barrier.
- The mucosa produces a number of antibacterial chemicals.

- The mucosa produces a wide range of antibacterial biomolecules including antibacterial peptides.
- The microflora of the mucosa plays a key role in defence against exogenous pathogens.
- The mucosa is a watchdog able to alert the underlying immune tissues to the presence of bacteria.

5.10 | The paradigm organisms and mucosal surfaces

5.10.1 *Streptococcus pyogenes*

The key message in this book is that bacteria and their hosts are both responsive to each other. Two examples of this interactivity will be discussed. The first concerns the role of mucin in bacteria–host interactions. To make contact with host epithelia, bacteria must traverse a mucin layer. What is the nature of the interaction between bacteria, specifically *Strep. pyogenes* and mucins? It has now been demonstrated that *Strep. pyogenes* binds to salivary/tracheobronchial-type mucin via both the M protein and a 39 kDa cell-wall-associated protein. In culture, this binding to mucin increases the adhesion of bacteria to human pharyngeal cells. This suggests that bacteria have evolved mechanisms of migrating through mucin, perhaps by releasing mucolytic enzymes. Supernatants from bacteria such as *Strep. pyogenes*, when added to epithelial cell cultures, have been found to enhance the expression of mucin genes such as those for MUC2 or MUC5AC. Such bacterially induced expression of mucins was found to be due to tyrosine phosphorylation events, suggesting the mode of cell activation. The nature of the activating signals from bacteria has not been identified. Is this increase in production of mucin a protective response? This seems the obvious answer. However, again we can take nothing for granted in science. This is exemplified by the recent findings that certain Gram-positive bacteria encode serine/threonine-rich proteins that are also likely to function as mucins. It is too early to suggest what role such bacterial mucins play but it looks like we may be seeing a 'battle of the mucins' appearing in host–bacterial interactions.

Different bacteria respond differently to the same set of environmental conditions due to the presence or absence of specific genes. A good case in point is *Strep. pyogenes*, which, because of its inability to synthesise haem, lacks the enzymes such as catalase and cytochromes required for survival in oxygen. Of course, this organism must survive in oxygenated environments to infect the human host. An answer to the question of how *Strep. pyogenes* survives in the presence of oxygen has come from the results of knocking out the gene for superoxide dismutase (SOD). This enzyme is responsible for the breakdown of superoxide, a damaging oxygen radical. To determine the importance of SOD to the survival of *Strep. pyogenes*, the *sod* gene was insertionally inactivated. The SOD-negative mutant was unable to grow in aerobic conditions on solid media. In wild-type *Strep. pyogenes* the levels of SOD protein increase in response to oxygenation. The binding of *Strep. pyogenes* to cells or extracellular matrix is facilitated by the adhesin protein F. It has been shown that protein F expression is increased by the presence of oxygen or superoxide. In the SOD knockout, the gene encoding protein F, *prtF*, showed enhanced transcriptional activation under oxidatively stressful conditions. This work suggests that *Strep. pyogenes* has an environmentally sensitive signal transduction system that senses superoxide and can co-ordinately control the expression of both *prtF* and *sod*. This system may function at epithelial surfaces and be switched on by the actions of lactoperoxidase.

5.10.2 *Escherichia coli*

As described in Section 1.6.2, as many as eight different pathogenic strains of *E. coli* have been recognised. Each of these strains has a specific virulence factor or set of virulence factors allowing it to cause disease. However, in order to cause disease, the bacterium must survive on the mucosal surface of the gut. In order to survive, it must gain nutrients including the vital metal iron. How does *E. coli* compete with its host for the very limited supply of iron present on the epithelial surface of the gut? The importance of iron for infection can be seen from experimental infection of animals with *E. coli*, in which one group of animals was also injected with iron-containing compounds. In the absence of exogenous iron, 10^8 bacteria were required to kill the guinea pigs. However, when iron compounds were also administered, only 10^3 bacteria were required to kill the animals. This is a 100 000-fold reduction and shows the enormous importance of iron both for the bacterium and for ourselves.

Free iron in the host is kept at an absolute minimum by the presence of high affinity iron-binding proteins. We have already described the production of lactoferrin by the mucosal epithelium and this protein, as its name suggests, is also found in milk. The low levels of extracellular iron in blood and extracellular fluid is due to binding of iron by another protein, transferrin. These proteins are responsible for maintaining the free iron levels in the body at around 10^{-18} M. How does *E. coli* cope with this tiny amount of free iron, which is far too low to support growth?

Four main mechanisms are employed by pathogenic bacteria to capture iron from the host. The first is to use proteases to cleave the iron-binding proteins and obtain released iron. The second is to reduce the Fe^{3+} in the protein complex to Fe^{2+}, which causes the iron to be released. The third method has been described earlier for *Strep. pneumoniae*, namely the use of receptors for iron-binding proteins. The final method, involving siderophores, will be described in a little more detail.

Siderophores are low molecular mass organic compounds produced by a variety of bacteria, including *E. coli*, which have the ability to bind iron with high affinity. Siderophores are produced by bacteria only under conditions of iron limitation and thus represent a key aspect of

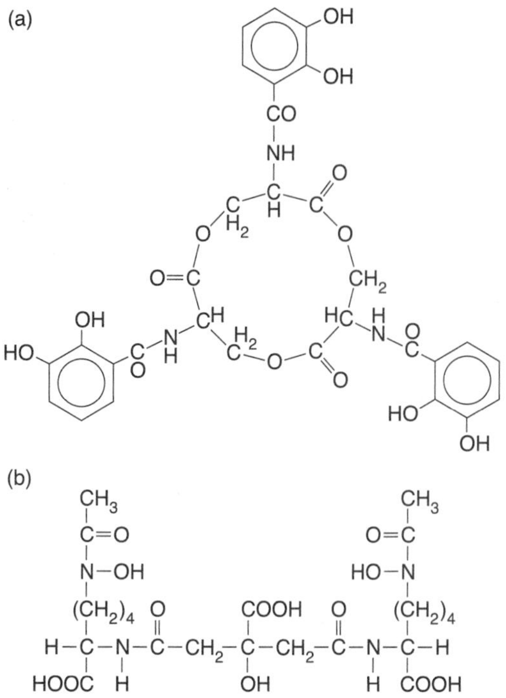

Figure 5.22 Siderophores produced by *Escherichia coli*. (a) The structure of enterobactin; (b) the structure of aerobactin.

bacterial behaviour – their ability to respond to their environment. *Escherichia coli* (as well as *Sal. typhimurium* and *Klebsiella pneumoniae*) produces two major siderophores – enterobactin and aerobactin (Figure 5.22). Enterobactin is a cyclic triester of 2,3-dihydroxy-*N*-benzoyl-L-serine and is able to remove iron from iron-binding proteins and deliver it into the cell cytoplasm. The iron is bound by the three catecholamine groups and the affinity of enterobactin for ferric iron is enormous – the formation constant at neutral pH being 10^{52}. Thus this simple chemical can compete very effectively with iron-binding proteins for their bound iron.

The synthesis of enterobactin by *E. coli* requires the concerted action of seven genes located in the enterobactin gene cluster on the bacterial chromosome. Each enterobactin molecule is used only once, as the iron-containing molecular complex has to be cleaved by a specific protease to release the bound iron. This is obviously one of the problems of evolving such a high affinity iron-binding compound: how do you get the iron back? The uptake of enterobactin into *E. coli* is dependent on the presence of enterobactin receptors, which are synthesised in concert with the siderophores when the bacterium finds itself in conditions of iron limitation.

The competition for iron between bacteria and their environment is an ongoing one of evolutionary push and shove. Bacteria on mucosal surfaces have to obtain enough iron to survive and the development of siderophores is one example of evolution coming up with two molecularly distinct, but functionally identical, means of sequestering an essential nutrient.

5.11 | What's next?

Chapter 6 will continue the story of our antibacterial defences. Attention will turn from the epithelial surface to the submucosal defence systems, which are termed innate and acquired immunity. The recognition that we have become infected and the generation of the various means of combating bacteria, through the role of phagocytes and the participation of complement, antibodies and T lymphocytes, will be described.

5.12 | Questions

1. Compare and contrast the physical, chemical and biological mechanisms of bacterial control on mucosal surfaces.
2. What attributes make epithelial cells such good guardians of the body against bacteria?
3. Epithelial cells are watchdogs at mucosal surfaces: explain.
4. Everybody makes antibiotics: explain.
5. Discuss the role of the normal microflora as a component of the antibacterial defence system of mucosal surfaces.
6. Describe the role played by each of the following cell types in defending mucosal surfaces against bacteria: (a) M cells, (b) Paneth cells, (c) intra-epithelial lymphocytes.
7. Discuss the role of secreted proteins in the defence of mucosal surfaces against bacteria.
8. Your commensal microflora takes a holiday: what should you do?
9. Explain the importance of metal ions for life and for death.
10. Explain the difference between probiotics and prebiotics.

5.13 | Further reading

Books

Ogra, P. L, Mestecky, J., Lamm, M. E. & McGee, J. E. (1999). *Mucosal Immunology*, 2nd edn. London: Academic Press.

Sussman, M. (1997). *Escherichia coli: Mechanisms of Virulence*. Cambridge: Cambridge University Press.

Tannock, G. W. (1999). *Medical Importance of the Normal Microflora*. Dordrecht: Kluwer Academic Publishers.

Review articles

Anderson, K. V. (2000). Toll signalling pathways in the innate immune response. *Curr Opin Immunol* **12**, 13–19.

Belley, A., Keller, K., Gottke, M. & Chadee, K. (1999). Intestinal mucins in colonisation and host defence against pathogens. *Am J Trop Med Hyg* **60**, (suppl), 10–15.

Boman, H. G. (1998). Gene-encoded peptide antibiotics and the concept of innate immunity: an update review. *Scand J Immunol* **48**, 15–25.

Brook, I. (1999). Bacterial interference. *Crit Rev Microbiol* **25**, 155–72.

Collins, M. D. & Gibson, G. R. (1999). Probiotics, prebiotics, and synbiotics: approaches for modulating the microbial ecology of the gut. *Am J Clin Nutr* **69**, 1052S–1057S.

Crouch, E., Harthorn, K. & Ofek, I. (2000). Collectins and pulmonary innate immunity. *Immunol Rev* **173**, 52–65.

Feizi, T. (2000). Carbohydrate-mediated recognition systems in innate immunity. *Immunol Rev* **173**, 79–88.

Forstner, G. (1995). Signal transduction, packaging and secretion of mucins. *Annu Rev Physiol* **57**, 585–605.

Hancock, R. E. W. & Diamond, G. (2000). The role of cationic antimicrobial peptides in innate host defences. *Trends Microbiol* **8**, 402–410.

Hancock, R. E. W. & Lehrer, R. (1998). Cationic peptides: a new source of antibiotics. *TibTech* **16**, 82–88.

Jepson, M. A. & Clark, M. A. (1998). Studying M cells and their role in infection. *Trends Microbiol* **6**, 359–365.

LeVine, A. M. & Whitsett, J. A. (2001). Pulmonary collectins and innate host defense of the lung. *Microbes Infect* **3**, 161–166.

Macfarlane, G. T. & Cummings, J. H. (1999). Probiotics and prebiotics: can regulating the activities of intestinal bacteria benefit health? *BMJ* **318**, 999–1003.

Niedergang, F. & Kraehenbuhl, J.-P. (2000). Much ado about M cells. *Trends Cell Biol* **10**, 137–141.

Pancholi, V. (2000). Streptococci-mediated host cell signalling. In *Gram-Positive Pathogens*, ed. V. A. Fischetti, R. P. Novick, J. J. Ferretti, D. A. Portnoy & J. I. Rood, pp. 87–95. Washington, DC: ASM Press.

Reid, G., Howard, J. & Siang Gan, B. (2001). Can bacterial interference prevent infection? *Trends Microbiol* **9**, 424–428.

Svanborg, C., Godaly, G. & Hedlund, M. (1999). Cytokine responses during mucosal infections: role in disease pathogenesis and host defence. *Curr Opin Microbiol* **2**, 99–105.

Travis, S. M., Singh, P. K. & Welsh, M. J. (2001). Antimicrobial peptides and proteins in the innate defense of the airway surface. *Curr Opin Immunol* **13**, 89–95.

Wong, W. M., Poulsom, R. & Wright, N. A. (1999). Review: trefoil peptides. *Gut* **44**, 890–895.

Zhang, P., Summer, W. R., Bagby, G. J. & Nelson, S. (2000). Innate immunity and pulmonary host defense. *Immunol Rev* **173**, 39–51.

Papers

Boris, S. & Barbés, C. (2000). Role played by lactobacilli in controlling the population of vaginal pathogens. *Microbes Infect* **2**, 543–546.

Brook, I. & Gober, A. E. (1999). Bacterial interference in the nasopharynx and nasal cavity of sinusitis prone and non-sinusitis prone children. *Acta Otolaryngol* **119**, 832–836.

Cole, A. M., Dewan, P. & Ganz, T. (1999). Innate antimicrobial activity of nasal secretions. *Infect Immun* **67**, 3267–3275.

Dohrman, A., Miyata, S., Gallup, M., Li, J. D., Chapelin, C., Coste, A., Escudier, E., Nadel, J. & Basbaum, C. (1998). Mucin gene (MUC2 and

MUC5AC) upregulation by Gram-positive and Gram-negative bacteria. *Biochim Biophys Acta* **1406**, 251–259.

Gibson, C. M. & Caparon, M. G. (1996). Insertional inactivation of *Streptococcus pyogenes sod* suggests that *prtF* is regulated in response to a superoxide signal. *J Bacteriol* **178**, 4688–4695.

Hammerschmidt, S., Bethe, G., Remane, P. H. & Chhatwal. G. S (1999). Identification of pneumococcal surface protein A as a lactoferrin-binding protein of *Streptococcus pneumoniae*. *Infect Immun* **67**, 1683–1687.

Lemaitre, B., Reichhart, J.-M. & Hoffmann, J. A. (1997). Drosophila host defence: differential induction of antimicrobial peptide genes after infection by various classes of microorganisms. *Proc Natl Acad Sci USA* **94**, 14614–14619.

Okada, N., Tatsuno, I., Hanski, E., Caparon, M. & Sasakawa, C. (1998). *Streptococcus pyogenes* protein F promotes invasion of HeLa cells. *Microbiology* **144**, 3079–3086.

Roos, K., Håkansson, E. G. & Holm, S. (2001). Effect of recolonisation with 'interfering' streptococci on recurrences of acute and secretory otitis media in children: randomised placebo controlled trial. *BMJ* **322**, 210–213.

Ryan, P. R., Pancholi, V. & Fischetti, V. A. (1998). Binding of group A streptococci to mucin and its implications in the colonisation process. In *Proceedings of the 98th meeting of the American Society for Microbiology*, Abstract B-4.

Urieli-Shoval, S., Cohen, P., Eisenberg, S. & Matzner, Y. (1998). Widespread expression of serum amyloid A in histologically normal tissues. Predominant localization to the epithelium. *J Histochem Cytochem* **46**, 1377–1384.

5.14 | Internet links

1. http://www.mercury.faseb.org/aai/default.asp
   American Association of Immunologists website.
2. http://www.immunology.org
   British Society for Immunology website.
3. http://www.cat.cc.md.us/courses/bio141/lecguide/unit3/index.html
   A set of lecture notes covering a range of immunological topics which includes a section on mucosal defences.
4. http://www.biology.arizona.edu/immunology/immunology.html
   A tutorial from the University of Arizona covering basic aspects of innate and acquired immunity.
5. http://www-micro.msb.le.ac.uk/312/BS312.html
   Lecture notes (from the University of Leicester) on various host defence systems.
6. http://www.meddean.luc.edu/lumen/MedEd/Histo/frames/Histo18.html
   A set of slides illustrating the various cells and structures of the gastrointestinal mucosa.
7. http://www.med.sc.edu:85/book/immunol-sta.htm
   A comprehensive on-line course in immunology with nice sets of images. Includes sections on innate immunity and complement.
8. http://www.slic2.wsu.edu:82/hurlbert/micro101/pages/Chap12.html
   Introductory notes on non-specific defence mechanisms.

6

Immune defences against bacteria

Chapter outline

Aims

To introduce the reader to:

- the concept of immunity against bacterial infections
- the cell populations involved
- the soluble mediators involved
- the organisation of the immune system
- innate immunity and recognition of pathogens
- mechanisms of phagocytosis and bacterial killing: linking innate and acquired immunity
- acquired immunity, generation of diversity and antigen presentation
- how these systems work together to kill bacteria

6.1 | Introduction

Many of our defence mechanisms against infection function at mucosal surfaces, are continually active and are generally invisible to us. These various defence systems have been described in Chapter 5 and the linkage between epithelial 'immunity' and what would classically be called *immunity* has been made. This chapter will consider the antibacterial defence mechanisms that can be thought of as being 'post-mucosal' and will describe the systems of cells and mediators that are normally thought of when the term *'immunity'* is used. The reader has to understand that, for the purposes of teaching, the various systems that defend us against bacterial infection are parcelled up into distinct categories ('innate' versus 'acquired', or 'humoral' versus 'cellular'). However, in reality, they work as one seamless mechanism. Unfortunately, we do not yet understand all the links that exist between these systems and the challenge is to identify the connections.

Before entering the fascinating realm of antibacterial immunity, the reader must be warned that modern immunology, which has developed largely since the 1960s, was created at a period when the world thought itself safe from bacterial infections. Because of this, immunologists have viewed 'immunology' more as a mechanism that goes wrong – producing autoimmunity and transplant rejection – than as a working system designed to protect us. If there has been a focus on the immunology of infection it has been concentrated on antiviral immunity. This deficit in our understanding of immunology is beginning to be corrected with the realisation that bacterial diseases have not been defeated by antibiotics. However, it is only within the last decade or so that immunologists have turned their attention to the immunology of bacterial infections. This revisiting of immunology is rapidly changing the way that the immune system is viewed and our understanding of its basic mechanisms. There is now a growing feeling amongst immunologists that immunity is a protective and useful system. However, most immunology textbooks, even those written after 2000, still pay no more than lip service to antibacterial immunity. This has to be realised when the reader uses current immunology texts. However, this is changing and the textbooks referred to under Further reading (Section 6.13) do pay some attention to antibacterial immunity. In this text, the immune system is viewed primarily as a means of defence against exogenous infectious agents.

Having made this statement, it is surprising to realise that Immunology was created in the crucible of 19th century Bacteriology, and our current understanding owes much to the study of viral infections. However, major advances in our understanding of the role of the two major systems of immunity – innate and adaptive – in antibacterial defences are being made as this chapter is being written. We will focus attention here on these new advances, which are radically altering the conventional paradigm of modern Cellular Immunology and its role in antibacterial immunity. This chapter will describe the two main divisions of Immunology. The first is the immediately available, germ-line encoded, innate immune system consisting of such components as the myeloid cells, complement and acute phase proteins. The second system is termed 'acquired' (or specific) immunity and consists of clonal effector cells (T and B lymphocytes) whose antigen receptors are not encoded by the germline and whose participation in bacterial infections may take days to weeks to 'kick-in'. We now realise that these two systems are intimately interlinked. However, before going into the details it is pertinent to pose the following question to the reader.

6.2 | How would you design an immune system to defend yourself against bacteria?

The problem facing you is to construct a system that will defend your multicellular body against single-celled bacteria.

The first problem that has to be addressed is: where will you find the bacteria? The answer is: they can be anywhere in your body. The system you design must therefore constantly patrol all areas of your body. Of course, most bacteria will be on mucosal surfaces and it would be sensible to concentrate resources in this vicinity, and this is what appears to happen (as described in Chapter 5).

If we refine the answer to this question, then bacteria are to be found in four distinct sites within our bodies. The first is on the epithelial surfaces. As long as they are not attaching, or signalling, to the mucosal cells, which act as our outer boundary and as guardian cells, then we need not worry about them. If bacteria invade the body and bypass our mucosal surfaces, we can find them within three types of location. The first is free-living, either in the blood or extracellular fluid, as plank-tonic organisms or as biofilms on the many surfaces in our bodies. The second site that bacteria may end up in is within the cytoplasm of cells. The third location, in which we may find bacteria, is within the vacuo-lar apparatus of macrophages and polymorphonuclear neutrophils (PMNs or neutrophils). This seems a very unlikely site for colonisation but is popular with many bacteria. For further information on bacteria entering cells the reader should refer to Chapter 8, where bacterial invasion of host cells is described in detail, and Chapter 10, where the ability of bacteria to enter cells and evade host immune (phago-cytic) defences is discussed.

Thus, to combat bacteria, the system we develop must be equipped to deal with: (i) organisms, either singly or in biofilms, living outside cells; (ii) bacteria within the cytoplasm (cytosol) of cells; or (iii) bacteria within the vacuolar system of phagocytes (Figure 6.1). The obvious minimal requirements of a system of defence are as follows:

1. The ability to recognise that bacteria have invaded.
2. The ability to find the bacteria and ensure that the appropriate

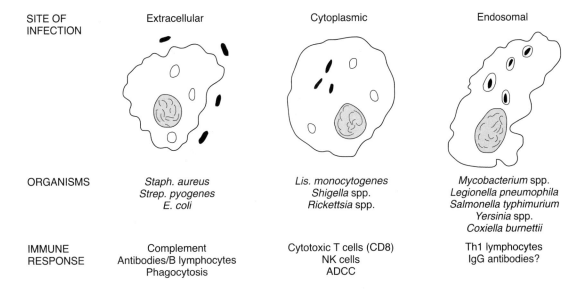

SITE OF INFECTION	Extracellular	Cytoplasmic	Endosomal
ORGANISMS	*Staph. aureus* *Strep. pyogenes* *E. coli*	*Lis. monocytogenes* *Shigella* spp. *Rickettsia* spp.	*Mycobacterium* spp. *Legionella pneumophila* *Salmonella typhimurium* *Yersinia* spp. *Coxiella burnettii*
IMMUNE RESPONSE	Complement Antibodies/B lymphocytes Phagocytosis	Cytotoxic T cells (CD8) NK cells ADCC	Th1 lymphocytes IgG antibodies?

Figure 6.1 Where would you expect to find infecting bacteria in the body? This schematic diagram shows the three sites within the body (excluding bacteria on mucosal surfaces) at which bacteria can be found. The first is free-living in the body fluids or attached to various extracellular matrices. Many common pathogens have this lifestyle (see Chapters 1 and 7). The major defences against them are complement- and antibody-mediated opsonisation and phagocytic killing. To evade these mechanisms, it is now appreciated that a growing number of bacteria can enter into various cell types and take up residence in the cytosol of the cell (see Section 8.4.1). To kill these pathogens, the immune system generally has to kill the host cell and this is done by CD8 or cytotoxic T cells and natural killer (NK) cells or by the process of antigen-dependent cell-mediated cytotoxicity (ADCC also involving NK cells). A most unexpected site to find bacteria is within the endosomal compartment of macrophages. However, many bacteria have evolved to defeat the killing power of the macrophage (see Section 10.4.2). To kill these bacteria, the macrophage has to be activated by antigen-specific Th1 CD4 T lymphocytes. There is also the suggestion that antibodies play some role in killing endosomal pathogens.

measures can be brought to this site (e.g. leukocytes, complement components, etc.).

3. A mechanism for killing the bacteria, or the cells containing the bacteria.
4. A mechanism for remembering the bacteria so that, if they invade again, immune systems can be rapidly mobilised.

Of course, killing bacteria may not be enough to solve the problem of infection. Some bacteria are still lethal after they die. The best example of this is botulism caused by the pre-formed toxin produced by *Clostridium botulinum* (see Chapter 9, which discusses toxins). Indeed, the bacterium does not even need to be present to cause disease – as long as the potent toxin is present. *Streptococcus pneumoniae* releases a very potent pro-inflammatory toxin, pneumolysin, but only after the bacterium has died. These, and many other exotoxins and endotoxins, need to be 'defused' to protect the host. The mechanism that has evolved in vertebrates, although, as stated earlier, missing in invertebrates, is the antibody molecule.

The final 'icing on the cake' is memory. Think how much more efficient an antibacterial defence system would be if it did not have to start from scratch each time it was exposed to a particular bacterial pathogen.

What has been described is a simplistic model of our immune system, which, in reality, comprises elements of innate and acquired immunity. In this chapter, the cells and mediator molecules that constitute the immune system will be described and their role in combating bacterial infections discussed. As it is envisaged that the reader will not have had much experience of immunology, the chapter will attempt to explain from first principles how the immune system works, but from the viewpoint that its major function is to combat infecting pathogens.

6.3 | The cell populations involved in immunity to bacteria

The cells that are classically involved in immune and inflammatory responses to bacteria are the circulating leukocytes (white blood cells), the so-called fixed tissue leukocytes and the vascular endothelial cells that make up the blood vessels that permeate all tissues (Table 6.1). Other cells that play roles in antibacterial defence are the mucosal epithelial cells (whose actions have been described in Chapter 5), mast cells and fibroblasts. In reality, it would be fair to say that most host-derived cells could probably be implicated in some form of antibacterial defence. For example, most produce antibacterial peptides. However, in this chapter we will concentrate on those cells that are classically implicated in immune responsiveness.

All of our circulating blood cells are produced in the bone marrow (Figure 6.2). These include erythrocytes, platelets, monocytes, lymphocytes, PMNs, etc. As can be seen in Figure 6.2, all of these cells arise from a cell type called a pluripotent stem cell. This stem cell population is self-renewing and populates our bone marrow throughout life. These pluripotent stem cells produce progenitor stem cells that can differentiate along two distinct pathways. The lymphoid stem cell gives rise to all of the distinct lymphoid cell populations, including the natural killer (NK cell). The T cell populations require further so-called 'education' in the thymus before they can become functional. The myeloid stem cell produces a large range of cells including the monocyte, dendritic cell, osteoclast, PMN, eosinophil, basophil, mast cell, platelet and erythrocyte. The key to the differentiation of these distinct cell populations is the capacity to produce, and respond to, different patterns of cytokines. These proteins exert powerful influences on cells and lie at the heart of the control of inflammation and immunity. Bacteria contain powerful inducers of cytokine synthesis and much of the pathology of bacterial infections is due to the overproduction of certain pro-inflammatory cytokines. The biology of the cytokines, in

Table 6.1.	Cells involved in immune defence against bacteria
Cell type	**Function**
Neutrophil (PMN)	Phagocytic cell with oxidative and non-oxidative killing mechanisms
Monocyte	Circulating immature macrophage – limited killing power but can produce cytokines
Macrophage	Phagocytic cell that can kill bacteria and present their antigens to T cells
Dendritic cell	Phagocytic cell that is the most active cell in presenting bacterial antigens to T cells
Mast cell	Releases signalling molecules, e.g. TNF to aid in antibacterial defence
Basophil	Actions similar to those of mast cells
NK cell	Kills cells infected with cytoplasmic bacteria
B lymphocyte	When activated produces antibodies
CD8 T lymphocyte	Cytotoxic T cell – kills cells containing cytosolic bacteria
CD4 T lymphocyte	Acts as a helper cell for B lymphocytes and activates macrophages with vesicular pathogens
γδ T lymphocyte	Controls inflammatory response and produces trophic factors
Endothelial cell	When activated, allows passage of leukocytes to site of bacterial infection
Fibroblast	Regulates inflammation and acts in repair after infection
Platelet	Produces clotting, which acts as a barrier to bacterial entry

relation to bacterial infections, will be discussed in more detail later in Section 6.4.4 and also in Sections 4.2.6 and 10.3.

6.3.1 Monocytes, macrophages, dendritic cells

Approximately 2–5% of the circulating leukocytes in humans are monocytes (Figure 6.3). These cells are released from the bone marrow and spend less than a day in the circulation before entering the tissues of the body, where they differentiate into fixed tissue macrophages (from the Greek *makros* = large, *phagein* = to eat) and dendritic cells. The names given to these various macrophage populations in different tissues are shown in Table 6.2. In tissues, macrophages and related dendritic cells play major roles in recognising invading bacteria and presenting bacterial antigens to lymphocytes to generate a humoral or cell-mediated immune response. The macrophage also has a direct role in killing pathogenic bacteria by phagocytosis and oxidative/non-oxidative damage. The term phagocyte is from the Greek (*phagein* = to eat, *kytos* = a vessel) and macrophages and PMNs are the major phagocytes

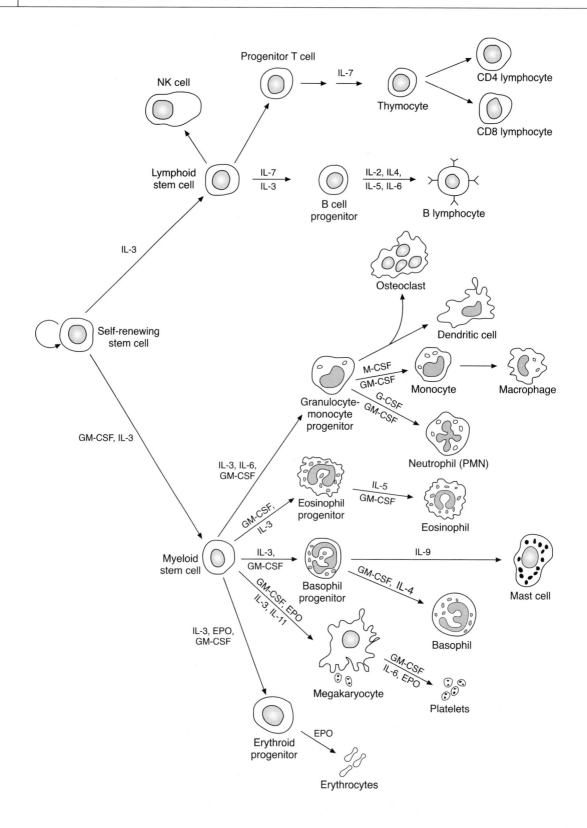

Figure 6.2 Regulation of leukocyte differentiation in the bone marrow. The bone marrow contains pluripotent stem cells that give rise to the distinct lymphoid and myeloid stem cell populations. The lymphoid stem cell gives rise to the B cell and the two populations of T cells (CD4 and CD8). Note that these cells need to be 'educated' in the thymus before they enter the circulating pool of T lymphocytes. The myeloid stem cell gives rise to the monocyte, which can further differentiate into macrophages, dendritic cells, osteoclasts (all shown), epithelioid cells and giant cells (not shown). In addition, the myeloid stem cell differentiates into the neutrophil, eosinophil, mast cell, basophil, platelet and erythrocyte. Some of the cytokines involved in controlling the differentiation of these diverse cell populations are shown. For abbreviations, see list on p. xxii.

of the body. Dendritic cells (DCs) are less active in killing bacteria (although immature DCs can actively pinocytose and phagocytose), and are more active in presenting bacterial antigens to lymphocytes. These fixed tissue myeloid cells are also major producers of cytokines, complement proteins and enzymes. The cytokines and complement proteins are involved in activating cells such as those of the vascular endothelium and in controlling the process by which additional bone marrow-derived cells are attracted to the site of infection. A partial list of the molecules (proteinaceous and non-proteinaceous) produced by activated macrophages may be found in Table 6.3.

6.3.2 Granulocytes

A number of circulating cells contain stainable granules. The most common of these cells (constituting some 50–70% of the total population of circulating leukocytes) are the polymorphonuclear leukocytes known variously as neutrophils, PMNs or simply 'polys' (Figure 6.4). These short-lived cells have a characteristically shaped multilobed nucleus, which makes their identification easy. Unlike monocytes, PMNs do not normally enter tissues, and when they do

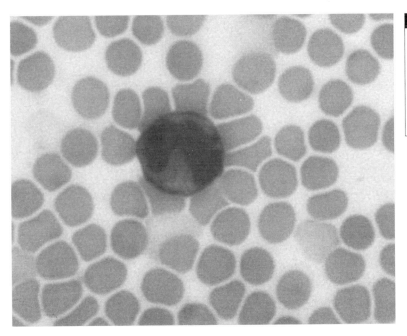

Figure 6.3 Light photomicrograph of a circulating monocyte. The non-nucleated cells are erythrocytes. (Image kindly supplied by Dr Tim Arnett, Department of Anatomy and Developmental Biology, University College London.)

Table 6.2.	Tissue macrophages
Tissue	Cell description
Blood	Monocyte
Inflammatory site	Macrophage, epithelioid cell, giant cell
Lung	Alveolar macrophage
Liver	Kupffer cell
Kidney	Mesangial cell
Brain	Microglia
Joint	Synovial type A cell
Bone	Osteoclast

so it is in response to chemotactic signals emanating from the process of infection (or from other causes such as tissue injury). PMNs are active phagocytes, containing a large number of bactericidal proteins within their primary and secondary granules, and they utilise both oxygen-dependent and oxygen-independent

Table 6.3.	Proteins produced by active macrophages at sites of infection
Protein	Function
Cytokines	To stimulate/inhibit inflammation and promote tissue repair
Growth factors	Promote cell growth and repair
Cyclo-oxygenase II	Production of prostanoids – controllers of inflammation
Lipoxygenase enzymes	Production of a variety of lipid products with pro- and anti-inflammatory actions
Inducible nitric oxide synthase	Antibacterial by producing nitric oxide
Enzymes of PAF metabolism	Similar to cyclo-oxygenase and lipoxygenase enzymes
Complement components	Opsonisation/inflammogens
NADPH oxidase	Killing of bacteria
Antibacterial peptides/proteins	Killing of bacteria
Proteases	Modulation of bacterial function/tissue remodelling

PAF, platelet-activating factor.

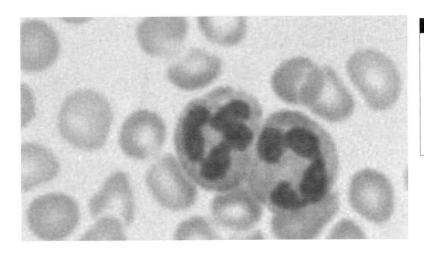

Figure 6.4 Light photomicrograph of two circulating neutrophils. The non-nucleated cells are erythrocytes. (Image kindly supplied by Dr Tim Arnett, Department of Anatomy and Developmental Biology, University College London.)

cell-killing mechanisms. The latter include the antibacterial proteins and peptides discussed in Section 5.4.3.11.

The other granulocytes in blood include eosinophils and basophils. These cells are present in smaller numbers and have been thought to be more involved in defences against parasitic worms. However, recent evidence is suggesting that mast cells, and possibly also basophils, are involved in antibacterial defences.

6.3.3 Mast cells and basophils

Mast cells are highly visible in tissues because of their large, stainable granules, and were identified more than 100 years ago. Indeed, the term mast cell comes from the German *mastzellen*, meaning well-fed cells, because their cytoplasm is stuffed with prominent granules. They are closely associated with allergic diseases and it is only within the past five years that evidence has accrued to support the hypothesis that mast cells, and also basophils, are important in antibacterial responses.

Mast cells – the forgotten cells of antibacterial immunity

Over the past decades mast cells have gained a poor press as the causative cells of allergic diseases. The cytoplasm of these cells is full of granules containing a range of molecules including (i) vasoactive amines such as histamine and serotonin, (ii) proteoglycans such as heparin and chondroitin sulphate, and (iii) proteases (chymase and tryptase). In addition, mast cells have the capacity to make a range of mediators involved in the pathology of allergy, including various lipid products such as the leukotrienes. These cells have therefore been classified as troublemakers, with a normal role in defending us from multicellular parasites. However, over the past five years evidence has begun to accrue to support the hypothesis that mast cells have antibacterial roles. Mast cells can phagocytose, and kill, bacteria. They can also process bacterial antigens and present them to T cells and they contain TNFα, a key cytokine in antibacterial defence. Although the evidence is limited, transgenic mice have been used to ascertain the importance of mast cells in bacterial infections. These studies have shown that mast cells seem to be important in the clearance of *Klebsiella pneumoniae* and in limiting the pathology of experimental peritonitis. Much more work is needed to determine the role of mast cells in innate and acquired immune responses to bacteria. However, the current evidence means that we have to welcome the mast cell as another one of the group of defence cells protecting us from exogenous bacteria.

6.3.4 Lymphocytes

These constitute some 20–40% of the circulating leukocytes in normal human blood and 99% of the cells in the lymph (Figure 6.5). These cells are long lived and have the ability to circulate throughout the tissues of the body, entering into tissues through specialised areas of the vasculature called **high endothelial venules** (HEV) and exiting into the lymphatics. Lymphocytes can be divided into three major populations, based upon (i) the presence of particular proteins on the outer surface of the plasma membranes of these cells and (ii) their function. The surface proteins of lymphocytes are known by their CD nomenclature (Table 6.4). Cells are now identified on the basis of specific CD proteins or on patterns of CD proteins. This will be described in more detail as the chapter unfolds.

CD nomenclature

One of the most confusing aspects of modern immunology is the CD nomenclature. CD stands for 'cluster of differentiation'. For the derivation of this term we have to go back to the 1970s, when it first became possible to make monoclonal antibodies. This is a process by which one immunises animals with a particular immunogen or population of immunogens, takes the B cells and fuses them with a cell that makes the individual plasma cells (the antibody-producing cells) immortal. Using a process called limiting-dilution analysis, it is possible to dilute the cells until one has a single cell in a well of a 96 well plate. This single cell can be induced to proliferate to produce a clone of immortalised B cells all with the same immunoglobulin (Ig) receptor. This B cell makes one, and only one, antibody, which is why they are termed monoclonal antibodies (mAbs). Using this technique, it is possible to take complex mixtures of antigens, to immunise mice with them and to tease out single clones of B cells making an antibody of a single specificity. Thus, if one takes a complex mixture of antigens, such as the population of proteins on the surface of a leukocyte, and uses this as an immunogen, then the individual mAbs that are produced will recognise, it is to be hoped, one of the cell surface proteins. Individual scientists began doing this worldwide and many, many mAbs were produced that reacted against proteins on the surfaces of leukocytes. This gave rise to an enormous number of mAbs and a huge number of potential cell surface antigens. People then began getting together to compare what their mAbs recognised and patterns of antibody binding began to emerge. These cell surface molecules were said to be recognised by clusters (collections) of mAbs, which identified the stage of differentiation of the leukocyte. Thus the term 'cluster of differentiation' arose. At the present time there are around 200 proteins defined on the basis of this CD nomenclature. The CD numbers that appear time and time again in immunology, and will be used extensively in this book, include CD1, CD3, CD4, CD8, CD11, CD14, CD16, CD18, CD40, CD154, to name but a few. The proteins that these CD numbers define are described in Table 6.4.

Circulating lymphocytes are divided into three basic populations: B cells, T cells and null cells. The null cell population fails to express surface proteins that characterise the other two lymphocyte populations. Moreover, the cells are large and granular. These cells are considered to be something of a halfway-house between the innate and acquired systems of immunity and have been termed NK cells. There is evidence that these cells are involved in the killing of bacteria. In contrast, the circulating T and B cells are small, non-motile cells that do not phagocytose. Each T and B cell contains a surface receptor that will recognise one particular molecular shape or, in immunological parlance, antigen. When the cell meets this antigen (a rare event) it becomes activated. In general terms, B and T cells can exist in four states (Figure 6.6). The first state is the cell that has not encountered antigen and which has, therefore, never been activated. These are known as naïve (unprimed or virgin) lymphocytes. Cells that recognise and bind their cognate antigen (e.g. some bacterial component) become activated and undergo the central process of acquired immunity – clonal expansion – to form what are known as effector cells. The effector B cell, for example, is the antibody-forming cell known as the plasma cell. The third state that lymphocytes can attain occurs after activation, when a proportion of the clonally activated cells become memory cells. It is this ability to 'memorise' antigens that allows secondary immune responses to be mounted much more rapidly than primary responses. A fourth state lymphocytes can assume – a state where they cannot be activated and in which they are called anergic – will be described in more detail in Section 10.5.2.

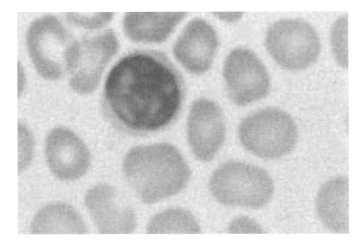

Figure 6.5 Light photomicrograph of a circulating lymphocyte. The non-nucleated cells are erythrocytes. (Image kindly supplied by Dr Tim Arnett, Department of Anatomy and Developmental Biology, University College London.)

Table 6.4. Some CD antigens involved in immunity

CD number	Synonym	Molecular definition	Main cells expressing	Functions
CD1a,b,c,d	R4 (for CD1a)	MHC class I-like	APCs, T cells	Antigen presentation
CD2	LFA 2	Adhesion molecule	T cell, NK cell	T cell activation
CD3	Leu-4	TCR complex	T cells	T cell activation
CD4	Leu-3	Ig superfamily	T cells	CD4 T cell co-receptor
CD5	Ly-1	Scavenger receptor family	T/B cells	Signalling – binds CD72
CD8	Leu-2	34 kDa dimer	CD8 T cells	CD8 T cell co-receptor
CD10	CALLA	100 kDa	Many	Metalloproteinase
CD11a	LFA-1	180 kDa	Many	Cell-cell adhesion
CD13	Aminopeptidase N	150 kDa	Monocyte, PMN	Aminopeptidase
CD14	LPS receptor	53 kDa	Mono/Mac, PMN	LPS co-receptor
CD16	Fc receptor	50–70 kDa	PMN	Low affinity Fc receptor
CD19	B4	95 kDa Ig superfamily	B cells	B cell activation
CD21	CR2	145 kDa RCA family	B/dendritic cell	C3d receptor
CD25	IL-2Rα chain	55 kDa RCA family	B/T cell/mac	IL-2 receptor subunit
CD28	TP44	44 kDa homodimer	CD4 T cells	Co-stimulatory molecule
CD35	CR1	190–285 RCA family	Many	Binds C3b/C4b
CD39	Apyrase	78 kDa	B/T/plasma cells	Ectoenzyme with ATPase activity
CD40		Homodimer	Many	Binds CD40L (CD154)
CD44	Hermes	80–100 kDa	Leukocytes	Binds hyaluronan/adhesin
CD46	MCP	52–58 kDa RCA family	Leukocytes	Regulator of complement activation
CD50	ICAM-3	110–140 kDa Ig superfamily	Leukocytes	Adhesion receptor
CD54	ICAM-1	75–114 kDa Ig superfamily	Many	Binds β$_2$-integrins
CD58	LFA-3	40–70 kDa GPI-linked	Many	Leukocyte adhesion
CD62E	E-selectin	115 kDa selectin family	Endothelium	Leukocyte adhesion
CD71	transferrin receptor	95 kDa chain homodimer	Many	Iron metabolism/cell growth
CD163	M130	130 kDa scavenger receptor	Monocyte/mac	Unknown

CALLA, common acute lymphoblastic leukaemia antigen; Mac, macrophage; for other abbreviations, see list on p. xxii.

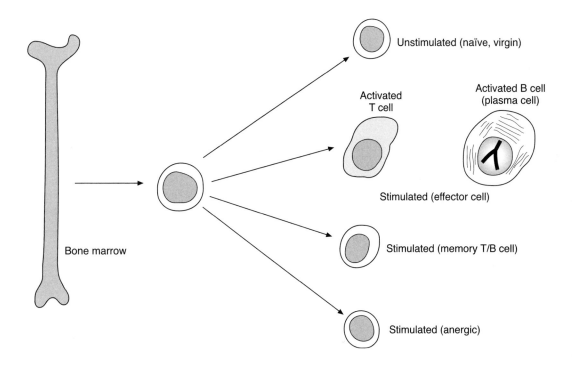

Unstimulated (naïve, virgin)

Activated B cell (plasma cell)

Activated T cell

Stimulated (effector cell)

Bone marrow

Stimulated (memory T/B cell)

Stimulated (anergic)

There are two types of B cell. The major population are known as B-2 B cells and recognise primarily protein antigens. The members of the smaller subpopulation, B-1 B cells, have two key differences. Firstly, they can replicate outside of the bone marrow and, secondly, they have a preference for recognising the polysaccharides or lipids of bacteria. These B-1 B lymphocytes are generally found at mucosal surfaces and surfaces lining body cavities (serosal surfaces); it is increasingly believed that they are involved in antibacterial responses, acting as sentinel cells. The antibodies that are produced by B-1 B cells are known as natural antibodies.

Circulating T cells come in two varieties. Cells that express the CD4 protein on their surfaces are known as helper T cells. Cells that express the protein CD8 are called cytotoxic T cells. The function of the latter cell is obvious. When activated, they can kill cells bearing the cognate antigen. Their major role is to kill cells containing intracellular parasites. Thus they constitute the major defence against viruses. They can also kill cells containing intracellular bacteria (e.g. *Salmonella typhimurium*, *Listeria monocytogenes* and *Mycobacterium tuberculosis*) and their importance in this respect is only now being appreciated. Immunology is a complex subject and, to complicate matters further, the CD4 T cell population is subdivided into two functional effector cells called Th1 and Th2 cells (Figure 6.7). The Th1 cell is very important in antibacterial immunity. Firstly, it is a helper T cell involved in promoting B cell synthesis of complement-fixing antibodies. Secondly, the Th1 cell has evolved to activate macrophages infected with intracellular bacteria. The Th2 cell is involved in immune defences against multicellular

Figure 6.6 The four states that lymphocytes can attain. Cells leaving the bone marrow, and before they encounter antigen, are classified as naïve or virgin cells. When lymphocytes are exposed to antigen they are activated and become what is known as effector cells. Thus the CD4 Th1 T lymphocyte, or the plasma cell, are examples of antigen-activated effector cells. A proportion of the activated T and B lymphocytes are long-lived memory cells, which enables the immune response to the second (and subsequent) exposure to a particular bacterial pathogen to happen much more rapidly than the primary response. Finally, if lymphocytes are inappropriately triggered they can become unresponsive to further stimulation. This state is known as immunological anergy.

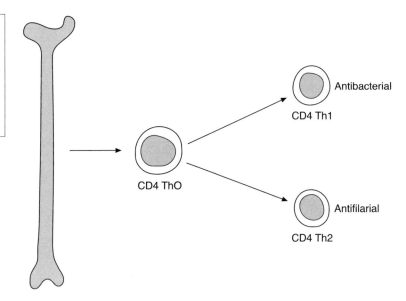

Figure 6.7 CD4 T lymphocytes can, depending on the signals they receive, be stimulated to become either CD4 Th1 cells with antibacterial activity or CD4 Th2 cells with anti-multicellular parasite (e.g. antifilarial) activity.

parasites such as worms. There is a reciprocal relationship between Th1 and Th2 cells that will be discussed later. There is increasing evidence that the balance between Th1 and Th2 cells is important in antibacterial responses and that it can also modulate allergic responses as well. The increasing incidence of the allergic condition asthma may be due to a Th1 : Th2 imbalance (for more details, see Section 11.5.3.3).

A T cell is defined by its T cell receptor (TCR), which is a cell surface protein composed of two different subunits (i.e. it is heterodimeric). The majority of T cells in the human circulation have TCRs composed of α- and β-subunits. These were the first T lymphocytes to be discovered. A second population of T lymphocytes exists, that has a different T cell receptor comprising γ- and δ-subunits. These cells tend to be found in the mucosa-associated lymphoid tissues where they are called intra-epithelial lymphocytes (IELs), although not all IELs are γδ T cells. This subclass of T cell has a limited T cell repertoire or, in other words, they recognise fewer antigens than do the conventional αβ T cells. They may be important in responses to specific components of bacterial pathogens and it is suggested that they play a role in down-regulating the actions of inflammatory macrophages.

6.3.5 M cells

These flattened cells are found in the mucosae overlying Peyer's patches in the gut. They are important in allowing the immune system to sample gut antigens and they have been described in detail in Section 5.7.1.

6.3.6 Vascular endothelial cells

A major problem in dealing with bacterial infections is that the leukocytes required to kill the invading bacteria are essentially on the

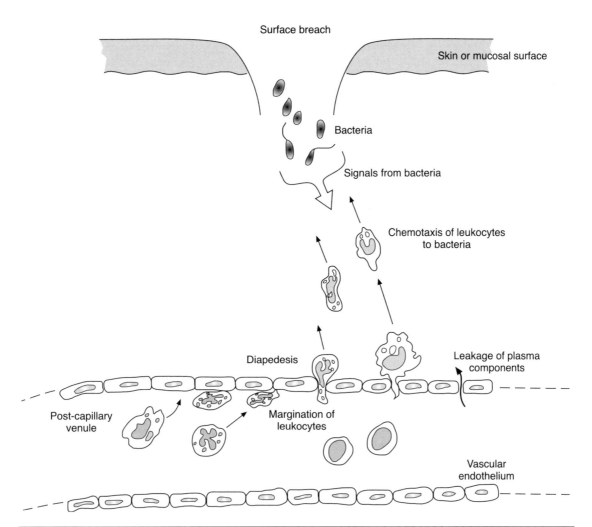

Figure 6.8 Bacterial infection necessitates the recruitment of leukocytes to the site of infection to kill the invading organisms. Here a breach in the integrity of an epithelial surface has allowed bacteria to enter past the epithelial (mucosal) barrier and into the body. Unfortunately, the leukocytes that can kill these bacteria are in the blood vessels. The inflammatory response involves (i) the recognition of signals emanating from the invading bacteria, (ii) the activation of the local vasculature, (iii) the trapping of leukocytes on the activated vasculature and their passage through the blood vessel, and (iv) the chemotaxis of the leukocytes to the bacteria. It should be noted that the passage of leukocytes through the endothelium can be accompanied by leakage of plasma, which brings with it many protein mediators of complement, clotting and fibrinolysis.

outside of the infected tissue, in the vasculature. These cells have to cross the vascular cell wall (Figure 6.8) in order to gain access to the bacteria. In this process, the vascular endothelial cells, which form the lumen of the blood vessels, play, not a passive role, but a very active one and are an integral part of the immune response. They also form the central element of the process known as inflammation, the universal response to bacterial infection. It is an inherent part of the immune response and the term largely describes the vascular events. The basic responses in inflammation were described by the Roman encyclo-paedist Celsus, almost 2000 years ago. He termed these responses the

Table 6.5. | The cardinal signs of inflammation

Sign	Causation
Rubor (redness)	Increased blood flow to site of inflammation causes redness
Tumour (swelling)	Increased blood flow and vascular leakage causes swelling
Calor (heat)	Increased blood flow to site of swelling increases temperature
Dolor (pain)	Mediators produced at site of inflammation sensitise nerve endings
Functio laesa (loss of function)	Pain and other changes make inflamed site (e.g. joint) unable to function normally

Table 6.6. | The secreted effector molecules of immunity

Effector molecule	Function
Acute phase proteins	Opsonisation of bacteria and other undefined actions
Complement	Opsonisation of bacteria, inflammogens
Immunoglobulins	Neutralisation of toxins, opsonisation of bacteria
Cytokines	Integrators of immune and inflammatory responses

'*cardinal signs*' of inflammation (Table 6.5). As the reader will know very well, inflamed sites are red, swollen, hot and painful. The explanation for these visible signs relate to changes in the local vasculature at the site of inflammation.

To recap, this section has introduced the reader, albeit very briefly, to the cell populations involved in immune and inflammatory responses. How these cells interact to defend us against bacterial infections will be described in Sections 6.6, 6.7 and 6.8. Section 6.4 will describe briefly the soluble effector molecules that play major roles in our antibacterial defences.

6.4 | The soluble effector molecules of inflammation and immunity

In addition to the various cell populations described, our protective immune system requires four groups of secreted proteins to function properly (Table 6.6).

6.4.1 Acute phase proteins

The first is a group of proteins, termed the 'acute phase proteins', that are produced by the liver, macrophages and mucosal epithelium. The

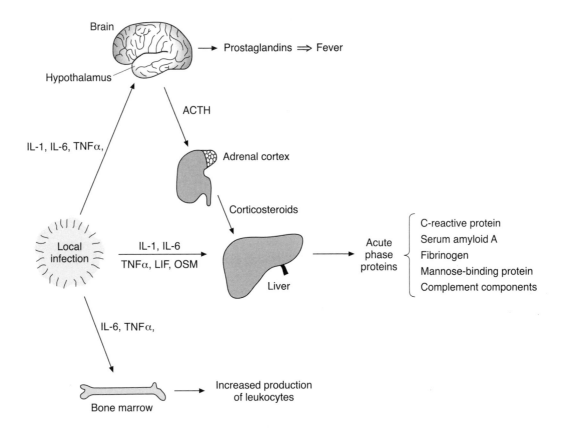

name derives from the initial set of host inflammatory responses to infection – the 'acute phase response', in which the levels of these proteins in serum rise enormously, up to 1000-fold (Figure 6.9). The induction of the synthesis of these proteins is due to the actions of a small number of pro-inflammatory cytokines such as: interleukin (IL) 1, IL-6, tumour necrosis factor (TNF) α, leukaemia-inhibitory factor (LIF) and oncostatin M (OSM), which are induced, in turn, by the action of bacterial components (lipopolysaccharide (LPS), peptidoglycan, etc.). The acute phase proteins include molecules such as C-reactive protein (CRP) and serum amyloid A (SAA). These proteins were initially thought to act as opsonins to promote bacterial phagocytosis. However, they have other functions, including activation of complement and modulation of host cell activity.

6.4.2 Complement system

The second group of proteins that are present constitutively in the blood and that can be up-regulated during infection is the complement system (Figure 6.10). There are now recognised to be three pathways of complement activation and these pathways involve the participation of approximately 30 soluble and membrane-bound proteins. The complement system is designed to recognise either immune complexes (i.e. the combination of an antibody and its antigen) or the cell surfaces of

Figure 6.9 The acute phase response and acute phase proteins. Inflammation due to bacterial infection releases key pro-inflammatory cytokines such as IL-1, IL-6 (and related cytokines LIF and OSM) and TNFα. These enter the circulation and interact with the brain, liver and bone marrow. Signals from the inflammatory site, in concert with those from the brain and adrenal cortex, stimulate the liver to synthesise the acute phase proteins. LIF, leukaemia inhibitory factor; OSM, oncostatin M; ACTH, adrenocorticotrophic hormone.

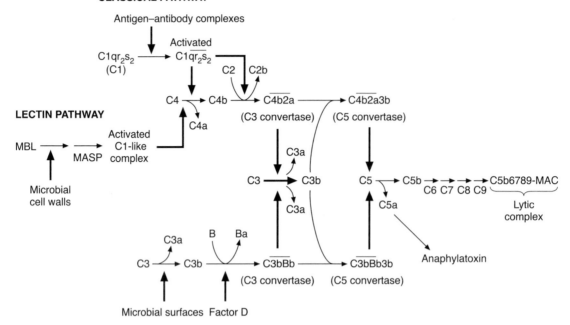

CLASSICAL PATHWAY

LECTIN PATHWAY

ALTERNATIVE PATHWAY

Figure 6.10 Simple diagram illustrating the three pathways of complement activation: the classical pathway, which is stimulated by immune complexes; the alternative pathway, activated by bacterial surfaces; and the lectin pathway, which is stimulated by bacterial cell walls. For details of the pathways, see the text. MBL, mannose-binding lectin; MASP, mannose-binding protein-associated serine protease; MAC, membrane attack complex.

bacteria. Once recognised, the complement pathway is activated, producing three major products. The first are proteins (e.g. C3b, C4b) that bind covalently to the bacterial cell surface and act as opsonins. Phagocytic cells have receptors for C3b and can bind to these C3b-coated bacteria. This receptor interaction makes it much easier for the phagocyte to ingest the bacterium (Figure 6.11). In addition, proteolytic processing of C3b to C3d allows complement to enhance B lymphocyte responses to the opsonised bacterium. The second set of products of complement activation are the proteins (C3a, C4a and C5a) that have been called anaphylatoxins. C5a is a potent chemoattractant involved in attracting more leukocytes into the site of infection. All three anaphylatoxins are able to bind to, and activate, mast cells and basophils. The third product of complement activation is the terminal complex or membrane attack complex (MAC) – a macromolecular assembly that forms pores in cell membranes. This complex is very effective in punching holes through sheep erythrocytes and most immunology textbooks have this image. However, it appears to be less effective in causing bacterial membrane damage due to effective counter-measures evolved by bacteria (as described in Section 10.4.1).

6.4.3 Antibodies

The product of the activated B cell (plasma cell) is the antibody molecule or immunoglobulin (Ig) (Figure 6.12). This is a molecular machine with functional sites at each end of the protein. One end,

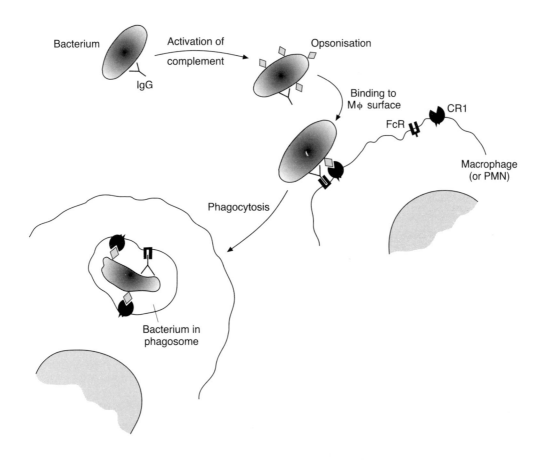

called the variable region, binds to the cognate antigen, for example a bacterial cell wall protein or a bacterial toxin. The other, the constant region (abbreviated Fc), is a ligand for so-called Fc receptors on phagocytes and for the C1 component of complement. There are five classes of antibody molecules in human sera: IgG, IgM, IgA, IgE and IgD (Table 6.7). The first three classes are involved in antibacterial immunity. There are four subclasses of IgG antibodies in humans, denoted IgG1 to 4. These differ largely in their ability to activate complement when they form immune complexes with antigen. This combination changes the conformation of the Fc region of the Ig molecule, enabling it to bind to the first component of the complement pathway, C1q. IgG3 is the best complement-activating, or in immunological jargon, 'fixing', member of the IgG class of immunoglobulins. The importance of antibodies in antibacterial defences is that they enable neutralisation of exotoxins and opsonisation of bacteria, allowing them to be phagocytosed by Fc-receptor-bearing phagocytes (Figure 6.11).

Figure 6.11 The route by which bacteria are opsonised by either antibody or the complement component C3B and phagocytosed by phagocytes bearing the receptors for these opsonins. CR1, C3B receptor; FcR, Fc receptor; Mφ, macrophage.

6.4.4 Cytokines

The final class of proteins involved in immune responses are the cytokines, to which the reader has already been introduced in Section 4.2.6.

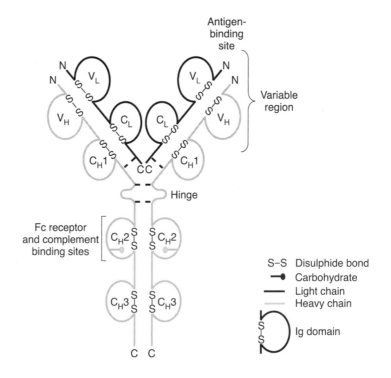

Figure 6.12 Schematic structure of the antibody molecule showing variable regions (V$_H$/V$_L$) that constitute the antigen-combining site and the constant domains C$_H$1 to 3, which contain the Fc region and bind complement components and Fc receptors.

These are a heterogeneous and very large group of peptides, proteins and glycoproteins that act as intercellular signals to integrate and control those cells involved in immune responses. There is a bewildering variety of these molecules – probably several hundred. A list of some of the major cytokines involved in combating infection, and in producing the pathology of infection, is provided in Table 6.8. Cytokines have a wide variety of roles in controlling myeloid and lymphoid cell function (Figure 6.13). The role of cytokines in cell–cell signalling has been described in more detail in Section 4.2.6.

6.5 | Organisation of the immune system

The reader has now been provided with a brief description of the cells and mediators involved in innate and acquired immunity. This is analogous to being provided with all the pieces of a car and told to guess how it works. Two additional pieces of the puzzle need to be provided. The first is the anatomical organisation of the immune system. The second is the role of the many cell surface receptors whose occupation by a ligand activates, or inhibits, leukocyte function and controls the overall operation of immunity. In this section, a brief description will be given of the structure of the all-pervading immune system.

For the proper working of antibacterial acquired immunity, T and B cells possessing the correct antigen receptors must come into contact with the invading bacterial pathogen. As the activation of these

Table 6.7.	Some facts about human antibodies				
Antibody	Subtypes	Serum concentration (mg/ml)	Serum halflife (days)	Molecular mass (kDa)	Function
IgA	IgA1	3	6	150	Mucosal immunity
	IgA2	0.5	6	150	Mucosal immunity
IgD	None	Trace	3	180	Antigen receptor on naive B cells
IgE	None	0.05	2	190	Anti-helminth responses
IgG	IgG1	9	23	150	Opsonisation,
	IgG2	3	23	150	neutralisation,
	IgG3	1	23	150	complement
	IgG4	0.5	23	150	activation, etc.
IgM	None	1.5	5	960	Complement activation

Ig, immunoglobulin.

lymphocytes requires the participation of other leukocytes (mainly macrophages and dendritic cells) in the process known as antigen presentation, an absolute requirement is that these cell populations must come into contact. The immune system has therefore evolved as a whole body organisation in which specialised tissues allow antigen-presenting cells (APCs – macrophages, dendritic cells, B lymphocytes, etc.) to meet with their cognate lymphocytes in specialised tissues (Figure 6.14). As Figure 6.14 reveals, our immune system pervades our body from head to toe. The system can be divided into three discrete sets of tissues. The first are the tissues that generate the lymphocytes (and, as described earlier, all other leukocytes). These are termed the primary lymphoid organs. The bone marrow is the ultimate source of all lymphocytes. However, the T cells generated in the bone marrow need to enter the thymus and undergo a process of maturation (often termed 'thymic education') in order to become the functional circulating CD4 and CD8 lymphocytes. The thymus is thought to remove T cells that have specificity for self-antigens and, because of this, could cause autoimmunity. Thus the bone marrow and thymus are the primary lymphoid organs. The lymph nodes and spleen are classically termed the secondary or peripheral lymphoid tissues. The lymph nodes are situated along the lymphatic channels and pervade our body (Figure 6.14). All tissues have a vascular supply, which provides oxygen and nutrients. The smallest blood vessels, the capillaries, are leaky and produce what is termed the 'interstitial fluid'. This fluid is collected, along with circulating lymphocytes, in the lymphatics where it is returned to the blood. The lymph nodes sample this fluid and any bacteria or bacterial antigens it contains. These are generally

Table 6.8.	Some representative cytokines important in bacterial infection
Cytokine	Roles in bacterial infection
Innate immunity	
IL-1	Activation of leukocytes/endothelial cells, induction of fever and acute phase response
IL-1 receptor antagonist	Inhibitor of IL-1, switch-off signal for inflammation
IL-6	Induction of acute phase response/B cell growth factor
IL-8	Neutrophil chemoattractant
Chemokines	Chemoattractants for leukocyte populations to sites of inflammation
IL-10	Macrophage deactivator
IL-12	Inducer of IFNγ
IL-13	Macrophage deactivator
IL-15	NK and T cell activator
IL-18	IL-1-like cytokine inducing IFNγ
TNFα	Similar range of actions to IL-1
IFNγ	Macrophage activator
Acquired immunity	
IL-2	Lymphocyte growth factor
IL-4	Th2 differentiation factor
IL-5	B cell growth factor
IL-12	Th1 differentiation factor
IFNγ	Macrophage activation factor
TGFβ	T/B cell inhibitory factor

For abbreviations, see text and list on p. xxii.

bound to APCs. The spleen does a similar job in sampling the blood and identifying bacterial antigens. It is within the lymph nodes and spleen that the coming together of the appropriate lymphocyte receptor with its cognate antigen occurs. This process is known as antigen presentation and leads to the clonal proliferation of lymphocytes and the production of effector lymphocytes.

Although generally classed as part of the secondary lymphoid tissues, the mucosal epithelia lining the external environment of the body, as described in Section 5.3, represent a major fraction of the body's lymphoid tissues. If one estimates the surface area of the mucosal epithelium exposed to the environment (and to the bacteria in the environment) it comes to about $400 \, m^2$. This is an enormous

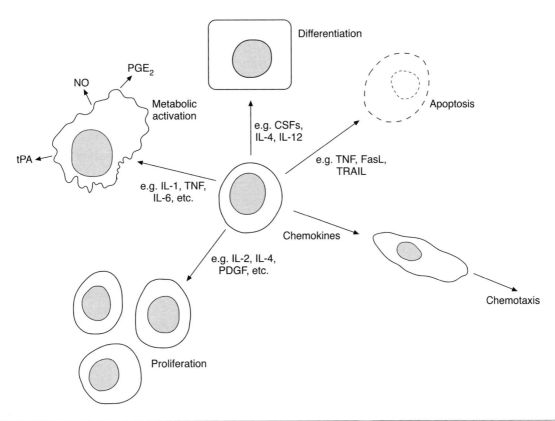

Figure 6.13 Schematic diagram outlining the major cellular functions of the cytokines. Binding of a cytokine to a target cell can result in one, or more than one, of the fates depicted. Many cytokines are able to activate cells to produce a range of mediators such as cytokines, lipid mediators, and proteases. The pro-inflammatory cytokines IL-1 and TNFα are good examples of activating cytokines. A growing number of cytokines, known as chemokines, stimulate the directed movement of specific leukocytes and thus act as traffic controllers in the inflammatory process. Some cytokines have the ability to induce cells to apoptose. Others act as mitogens and stimulate cell proliferation. Yet other cytokines can cause cells to differentiate. Note that many cytokines have more than one of the activities indicated. PGE_2, prostaglandin E_2; tPA, tissue plasminogen activator; for other abbreviations, see list on p. xxii.

surface area and it is not surprising that it has evolved an associated immune system. This anatomically heterogeneous immune system is known as the mucosal-associated lymphoid tissue (MALT) and ranges from loosely organised collections of immune cells to well-known, and sometimes troublesome, organs such as the tonsils and appendix. Another organised structure comprises the Peyer's patches found in the submucosa of the intestines (Figure 6.15). The MALT is present in all externally facing mucosal linings and is a major force in our immune system. For example, the number of plasma cells in the MALT far exceeds that of the other lymphoid organs – bone marrow, spleen and lymph nodes – combined.

Having given a brief overview and introduced the reader to the cells, the mediators and the anatomical organisation of the human immune system, the discussion will now switch to the working of immunity. The starting point for this discussion will be the paradoxical nature of the

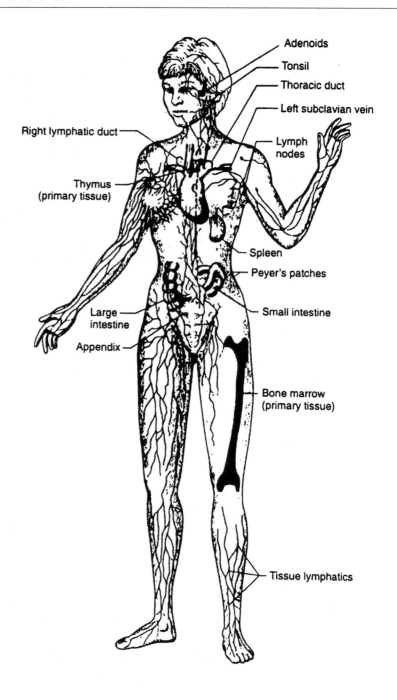

Figure 6.14 Anatomical organisation of the human immune system. This diagram shows the anatomical sites of the primary (bone marrow/thymus) and secondary (spleen/lymph nodes/Peyer's patches, etc.) lymphoid organs and their inter-connections via the lymphatics. The key point that this diagram illustrates is that the immune system is a whole body mechanism as it has to be able to cope with parasites that can invade at any site.

immune response. We know that the antigen receptors on T and B lymphocytes can recognise billions of molecular shapes, in other words antigenic epitopes. However, to be activated, a lymphocyte must receive its cognate antigen, plus other requisite signals, from an APC (macrophage, dendritic cell, B cell). This implies that the non-clonal innate immune system must recognise that it has been invaded by pathogenic bacteria and thus control the activity of the acquired immune system. This realisation is revolutionising our understanding

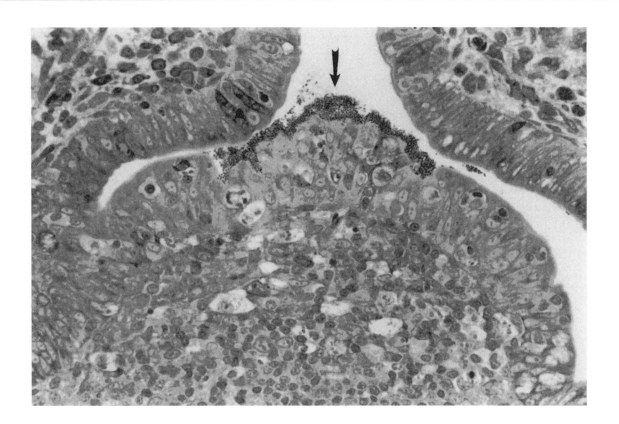

of innate immunity and emphasising its importance for the working of acquired immunity.

6.6 | Innate immunity

When we think of immunity to bacteria we generally think of the acquired immune responses that are produced by vaccination. For example, almost all of the readers of this book will have been immunised against *M. tuberculosis* (which causes tuberculosis) and *Bordetella pertussis* (the causative agent of whooping cough). Antibacterial acquired immunity, as the name suggests, requires prior contact with the causative organism or its immunogenic components. The result is the activation and generation of clones of reactive B and T lymphocytes.

In contrast, innate immunity does not require this prior contact phase that results in clonal proliferation but, instead, relies on preformed cells and mediators. We have already dealt with one aspect of innate immunity, epithelial innate defence, in Chapter 5. In this section, the focus will be on the major role that phagocytes and complement play in killing bacteria. It was the Russian zoologist Élie Metchnikoff, in the 19th century, who discovered the role that certain motile cells played in innate immunity. He coined the term 'phagocytosis' for the cellular process by which these motile cells can engulf invading

Figure 6.15 Light micrograph of rabbit Peyer's patch lymphoid follicle 24 hours post inoculation with *Escherichia coli*. Arrow depicts adherence of bacteria to Peyer's patch lymphoid follicle epithelium. (Reproduced with permission from Von Moll, L. K. & Cantey, J. R. Peyer's patch adherence of enteropathogenic *Escherichia coli* strains in rabbits. *Infection and Immunity*, 1997, **65**, 3788–3793.)

pathogens and defined the existence of the two major phagocytic leukocytes in the blood. The discovery of the phagocyte gave rise to the concept that these cells could be stimulated to provide therapeutic benefit. This idea is put into the mouth of Sir Ralph Bloomfield Boningon in George Bernard Shaw's play *The Doctor's Dilemma*, written in 1906: 'Stimulate the phagocytes. Drugs are a delusion'. This is a sentiment that could still find support today, although few modern playrights are likely to be so *au fait* with current biology.

6.6.1 How do you recognise a pathogen?

While acquired immunity, with its mechanism for generating antigen receptor diversity, the process known as generation of diversity (GOD, to be discussed in Section 6.7.1), was being dissected into smaller and smaller pieces in the period from the 1960s to 1990s, little attention was being paid to what was considered to be an ancient, simple and relatively unimportant aspect of our immune defences – innate immunity. However, in the past decade there has been a major reappraisal of innate immunity and the realisation has dawned that this system of cells and mediators is extremely important. Indeed, it is now recognised that innate immunity acts to focus and control acquired immunity. As will be described, the acquired immune system on first encountering a new bacterium will require 7 to 10 days or more to produce sufficient numbers of so-called effector B and T lymphocytes to deal with the infection (Figure 6.16). What happens in the meantime? The answer is that the non-clonal innate immune system must cope. Indeed, as has been said on a number of occasions, in this chapter and Chapter 5, for 95% of the multicellular creatures on this planet (the invertebrates), innate immunity is all they have for defence against infection. Indeed, a paradox at the centre of our immune defence system is the fact that the clonal system of immunity requires that cells of the innate system – the APCs – must present the bacterial antigens in a pre-packaged form to allow clonal proliferation and effector cell production to occur. So how do these agents of innate immunity recognise that we are being infected and, of equal importance, recognise what we are being infected with?

6.6.1.1 Pattern recognition receptors

The beginnings of the answer to this question can be traced back to a number of seminal papers written by the immunologist Charles Janeway in the late 1980s and early 1990s. In attempting to understand how the innate immune system could recognise the entry of infectious bacteria, Janeway proposed that cells of the innate immune system (macrophages, dendritic cells, neutrophils, basophils) had receptors, to which he gave the name pattern recognition receptors (PRRs). These receptors, he hypothesised, recognised bacterial molecules that were so important for bacterial function that they could not change over evolutionary time. These bacterial molecules were termed

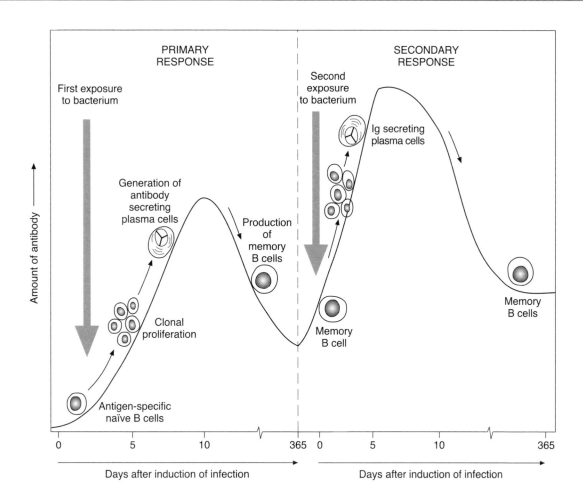

Figure 6.16 The kinetics of effector lymphocyte formation in primary and secondary responses to bacterial infection. When first exposed to bacteria, the immune system takes some considerable time to produce sufficient quantities of effector B cells and antibody (and also T cells, not shown) to deal with the infection. It may take up to 10 days to reach a peak of antibody levels in the circulation. When the host encounters the same bacterium again, the memory cells created at the first exposure ensure that the secondary response is much more rapid.

pathogen-associated molecular patterns (PAMPs) (Figure 6.17). Examples of the PAMPs include LPS, peptidoglycan, the lipoarabinomannan of mycobacteria, and bacterial CpG-rich DNA, etc. (Table 6.9). Of course, as these PRRs can also recognise members of the normal microflora, PAMPs may more correctly be termed microbe-associated molecular patterns or MAMPS.

The first PRR to be identified was a cell surface molecule found largely on monocytes and known as CD14. This is an example of a membrane protein that is tethered to the membrane by a glycosylphosphatidylinositol (GPI) molecule – often called a 'GPI anchor'. Binding of LPS to the CD14 receptor was promoted by another serum protein known as LPS-binding protein (LBP). It was also found that CD14 was present in serum and could bind to LPS and activate cells that did not express surface-bound CD14. The fact that CD14 had no intracellular domain, and therefore could not signal to the cell, suggested that this membrane protein must interact with some additional plasma membrane partner protein to activate cells. The nature of this partner protein (a Toll-like receptor, see Section 6.6.1.2) was discovered in the late 1990s and the answer, to say the least, was unexpected.

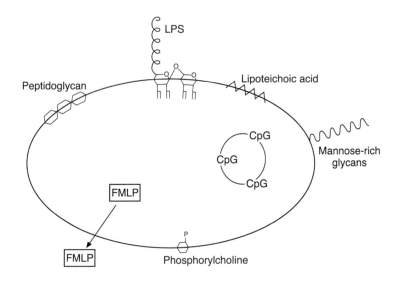

Figure 6.17 Schematic diagram showing the various pathogen-associated molecular pattern (PAMPs) molecules that are believed to be recognised by innate immunity via pattern-recognition receptors (PRRs). CpG, unmethylated dinucleotide sequence; FMLP, *N*-formyl-methionyl-leucyl-phenylalanine.

Before describing the mysterious LPS receptor, the role of CD14 in recognising LPS and in recognising apoptotic cells, will be described briefly to bring home the concept of context in cell signalling. If human macrophages are exposed to tiny quantities (picograms) of *Escherichia coli* LPS they will generate large amounts of inflammatory cytokines such as IL-1, IL-6, IL-8, IL-12 and TNFα. This is why Gram-negative septic shock is such a serious condition; the human is responsive to very low levels of LPS. In 1998 it was discovered that CD14 is also a receptor on macrophages involved in the recognition of apoptotic cells. It must be emphasised that CD14 is only one of a number of

Table 6.9. Bacterial pathogen-associated molecular patterns (PAMP) and pattern recognition receptors

PAMP	PRR	Consequence of binding
LPS	CD14/TLR4	Macrophage activation
Peptidoglycan	CD14/TLR2/TLR4	Macrophage activation
Lipoarabinomannan	CD14/TLR2 plus others	Macrophage activation
Mannose-rich glycans	Macrophage mannose receptor Serum mannose binding protein	Opsonisation/macrophage activation
Phosphorylcholine and related molecules	Serum C-reactive protein	Opsonisation, complement activation
Unmethylated CpG nucleotides	TLR9	Macrophage activation
FMLP	FMLP receptor	Neutrophil and macrophage activation
Chaperonin 60	CD14/TLR2/TLR4	Macrophage activation

For abbreviations, see the text.

receptors involved in such recognition. However, it is interesting that many of these receptors, such as scavenger receptors, CD36 etc., also recognise bacteria or opsonised bacteria. What is striking is that when CD14 is bound by the ligands on apoptotic cells it does not trigger cytokine production. This is presumably because additional signalling creates a signalling network that fails to activate cytokine transcription. It is not known whether pathogens can utilise this apoptotic cell recognition pathway to evade immune responses.

6.6.1.2 *Drosophila*, Toll, Toll-like receptors and the *lps* locus

Insects have three systems of immune defence. Firstly, there is a protease-dependent process in the haemolymph, resulting in coagulation, free-radical formation and microbial opsonisation. Secondly, blood cells can phagocytose microorganisms. The third line of defence is the induction of very high levels of antibacterial and antifungal peptides. These antimicrobial peptides (see Section 5.4.3.11) are believed to be very important in insect antimicrobial defences.

Our current understanding of how innate immunity recognises microbial pathogens has come from three seemingly distinct lines of research – two in the insect *Drosophila* and one in the study of insensitivity to LPS in the mouse. In *Drosophila*, the study of dorsoventral (back front) patterning in the development of the embryo had identified the protein Toll, a cell surface receptor that bound to a protein fragment termed Spätzle. This triggered an intracellular signalling pathway described in Section 4.2.6.6.2, which resulted in the entry of the nuclear factor (NF) κB homologue, dorsal, into the nucleus. In the absence of Toll, embryos do not develop ventral structures. Those studying the transcriptional control of antimicrobial peptides in insects also found that the NF-κB system of transcriptional regulation was involved. The experiments that tied these two independent areas of study together involved the generation of mutants of this Toll signalling pathway that could be switched on in the adult. In such mutant *Drosophila*, the flies succumbed to fungal infections because they failed to produce the antifungal peptide drosomycin. This revealed that, in adults, the Toll signalling pathway stimulated the synthesis of antimicrobial peptides. More importantly, these studies demonstrated that *Drosophila* could discriminate between bacterial and fungal infections. This was the first evidence that innate immunity had a discriminatory ability.

The C3H/HeJ mouse strain is insensitive to LPS and this insensitivity has been mapped to a chromosomal locus called *lps*. Using a technique known as positional cloning, a procedure that uses information about the chromosomal localisation of a gene to obtain a clone of the gene, a gene was identified with homology to the *Drosophila* Toll gene. When cloned and sequenced, this murine gene, which encodes the protein known as Toll-like receptor (TLR) 4, was found to have a point mutation leading to the substitution of a histidine residue for an invariant proline within the cytoplasmic domain of the receptor. A comparison of the receptor/ligand/signalling systems used by *Drosophila* and vertebrates involving Toll and the TLRs is shown in Figure 6.18.

Since 1998, when the first mammalian TLRs were identified, at least 10 TLR genes have been identified (Table 6.10). We now recognise that TLR4 is the LPS receptor and that TLR2 recognises, among other components, peptidoglycan. The ligands for the other receptors are currently being defined. The cytoplasmic domains of the TLRs are homologous to the cytoplasmic domains of the receptors for IL-1 and IL-18 and all these receptors share similar signalling pathways (Figure 6.19; and see Section 4.2.6.6.2).

The discovery of Toll has identified a receptor-based system for the recognition and discrimination of microbial pathogens. The TLRs can form both homo- and hetero-dimers and, with a growing number of TLR genes being discovered, it is likely that this receptor system provides the specificity for the innate system of immunity, allowing it to recognise whether the host is being infected by Gram-negative or Gram-positive bacteria or fungi. The exact nature of the discriminatory power of this TLR system is, however, still the subject of active research.

How the *Toll* gene got its name

The ability to generate genetic mutants and ascertain their phenotype is much used in all aspects of biology and this book has dealt in detail with the deliberate knockout of genes in bacteria to define the function of the protein products of known or unknown genes. This technique has been used in *Drosophila* since the early 20th century. Christiane Nusslein-Volhard and Eric Wieschaus were workers studying the genetics of embryonic patterning. When one particular mutant was shown by Wieschaus to Nusslein-Volhard, the latter exclaimed 'Toll', a German word that translates as 'crazy' or 'far out' — and this is how the *Toll* gene got its name.

6.6.1.3 Complement: a recognition and effector system of antibacterial defence

A brief description has been given of the complement system. Because it is an important defence mechanism required both for the recognition of bacteria and for their killing (directly and by enhancing phagocytosis), and because bacteria have developed a number of mechanisms to evade complement-mediated killing, a more detailed description of this system will now be provided.

The complement system is a series of interrelated amplificatory cascades, kept in check by a range of control proteins, and able to generate products that (i) enhance microbial phagocytosis (the process of opsonisation), (ii) attract leukocytes to the site of infection, and (iii) directly kill bacteria (Figure 6.20). It is now known that there are three pathways of complement activation (Figure 6.10). The key components of these pathways are detailed in Table 6.11. The first to be described, the

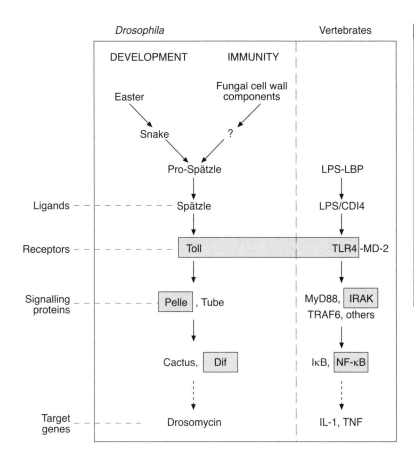

Figure 6.18 Comparison of the Toll and TLR4 pathways of cell bactivation in *Drosophila* and in vertebrates. In *Drosophila*, Toll can be activated as part of a developmental process or by fungal components. In the latter case, the Toll receptor induces the transcription of the gene encoding drosomycin, an anti-fungal peptide. In vertebrates, the TLRs are able to bind components of microorganisms. In the case shown, LPS/CD14 signals via TLR4 to induce the transcription of pro-inflammatory cytokines. Homologous proteins are shaded. For abbreviations, see list on p. xxii.

Table 6.10.	Toll-like receptors (TLR) and their specificities	
Receptor	Leukocytes expressing	Specificity
TLR1	All	ND
TLR2[a]	Myeloid cells only	Peptidoglycan, components of *Lis. monocytogenes*, lipopeptide
TLR3	Dendritic cells	ND
TLR4	Myeloid cells only	LPS
TLR5	Myeloid cells only	Flagella
TLR6[a]	Myeloid cells	Peptidoglycan
TLR7	ND	ND
TLR8	ND	ND
TLR9	ND	Bacterial DNA

ND, no data available.

[a] TLR2 and TLR6 can interact to recognise bacteria.

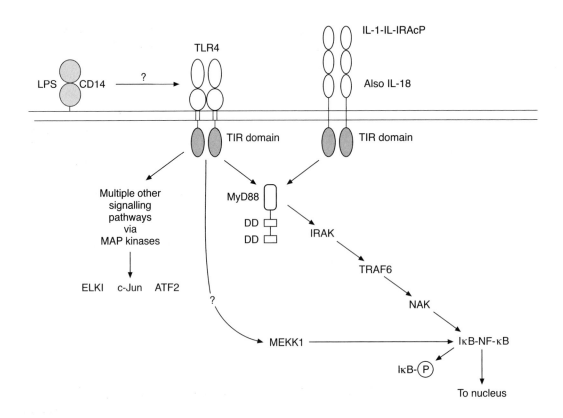

Figure 6.19 Similarity of the signalling pathways induced by binding of LPS to TLR4 and ligands to the IL-1 or IL-18 receptors. Currently the exact mechanism of interaction between LPS and TLR4 is not clear but does involve prior binding of the LPS to CD14. Binding of the TRAF-6/NAK/NF-κB ligands to these receptors triggers signalling via the MyD88/IRAK/TRAF-6/NAK/NF-κB pathway and via MAP kinases. For more details of these signalling pathways, see Chapter 4. MEKK1, MEK kinase 1; for most other abbreviations, see list on p. xxii.

classical pathway, is triggered by the binding of the protein C1q (which resembles the collectin proteins, see Section 5.4.3.9) to the Fc domain of immunoglobulins, forming immune complexes on the surfaces of bacteria. The most active immunoglobulins in stimulating complement activation are IgM, IgG1 and IgG3. Such binding triggers a cascade of protein–protein interactions in which the complement protein C4 binds to C1q. C4 is proteolytically cleaved by C1s, which forms part of the C1 complex (Table 6.11), to form C4b and C4a. C4b can bind co-valently to nearby bacterial surfaces via a short-lived reaction that generates a thioester bond. C2 then associates with C4b and is also cleaved by C1s to form C2a, which associates with C4b and C2b. This complex of C4b and C2a is a key component of the complement cascade. It is able proteolytically to cleave C3, the most abundant of the complement proteins, to form C3b and C3a. This is the major amplificatory part of the three complement pathways and the protein complexes that catalyse it are called C3 convertases. The products of these proteases are the opsonin C3b and the anaphylatoxin C3a (see Figure 6.10). C3b, like C4b, can form a thioester bond with nearby surfaces and the complex of C3b/C2a/C4b forms the proteolytic specificity to cleave the next component in the complement cascade, C5. This is the first of the so-called terminal reactions of the complement cascade and will be dealt with later.

To understand the alternative pathway of complement activation, it has to be realised that there is a continuous low rate of cleavage of C3

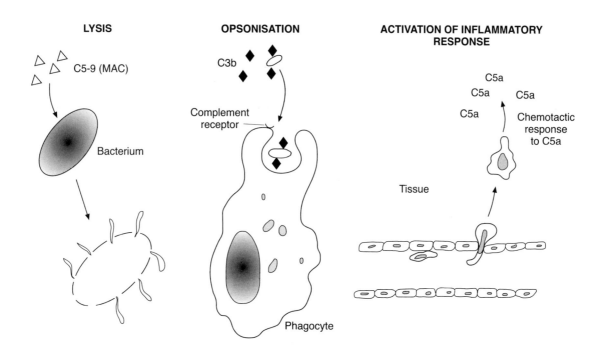

LYSIS

△ △
△ △ C5-9 (MAC)
△

Bacterium

OPSONISATION

C3b ◆ ◆ ◯

Complement
receptor

◆

◆

0

◯◯
◯
0

Phagocyte

ACTIVATION OF INFLAMMATORY RESPONSE

C5a
C5a C5a
C5a

Chemotactic
response
to C5a

Tissue

into C3a and C3b. This so-called 'C3 tickover', can lead to complement activation only if the C3b formed can bind to nearby membranes. If this does not happen, the thioester bond that forms and covalently links C3b hydrolyses and the C3b becomes inactive. C3b binding to host cell membranes is prevented by several regulatory proteins (see Table 6.11). In the alternative pathway of complement activation, the small amount of spontaneously formed C3b produced can be stabilised by binding to the surfaces of bacteria. These do not contain host regulatory proteins and, in consequence, can trigger complement activation. Bound C3b on the bacterial surface then binds to a plasma protein called factor B. A plasma serine protease – factor D can then cleave factor B to produce Bb, which remains bound to C3b and a soluble component Ba. This C3bBb complex is the C3 convertase of the alternative pathway.

In the most recently discovered pathway of complement activation, the lectin pathway, the binding of the acute phase protein mannose binding protein (MBP) to mannose residues on bacterial cell surface polysaccharides allows the binding and activation of MASP (MBP-associated serine protease). This complex can then activate C4 and then C2 to form the same C3 convertase that initiates the classical pathway.

The three complement pathways differ only in the mechanisms by which they generate the C3/C5 convertase. The terminal reactions of all three pathways are identical. Thus C5 convertase cleaves C5 into the anaphylatoxin C5a and the membrane-bound fragment C5b. This is the last protease-dependent reaction in the complement cascade. The remaining reactions involve the sequential binding of four structurally similar proteins to form a membrane pore. It is now clear that the pore

Figure 6.20 Schematic diagram showing how the complement pathways generate three pro-inflammatory/antibacterial activities: opsonins, anaphylatoxins and the membrane attack complex (MAC). Proteolysis of C3 generates the opsonin C3b, which binds covalently to bacteria and enables them to be phagocytosed by neutrophils and macrophages. Proteolysis of C5 generates the anaphylatoxin C5a, which is a potent chemoattractant for phagocytes. The terminal components of the complement pathway generate the MAC, which can lyse certain bacteria.

Table 6.11. The proteins involved in the complement pathways

Protein	Structure	Function
Classical pathway		
C1 complex (C1qr₂s₂)	750 kDa complex	Initiates classical pathway
C1q	Hexameric complex	Binds to Fc in immune complexes
C1r	85 kDa dimer	Serine protease – cleaves C1s to activate
C1s	85 kDa dimer	Serine protease – cleaves C4 and C2
C4	210 kDa trimer	C4b binds to microbial surface and to C2 for cleavage by C1s. C4a is an anaphylatoxin
C2	102 kDa monomer	C2a is a serine protease and cleaves C3 and C5. C2b generates C2 kinin (vasoactive peptide)
C3	185 kDa heterodimer	C3b – opsonin; C3a – anaphylatoxin
Alternative pathway		
C3 as above		
Factor B	93 kDa monomer	Bb, a serine protease that acts as C3/C5 convertase
Factor D	25 kDa monomer	Serine protease cleaves factor B when bound to C3b
Properdin	two to four 56 kDa monomers	Stabilises C3 convertase (C3bBb) on bacterial surface
Lectin pathway		
MBP	18 chain 540 kDa complex	Binds bacterial surface components
MASP	94 kDa monomer	C3/C3 convertase
Terminal complement components		
C5	190 kDa heterodimer	C5b initiates assembly of membrane attack complex C5a – anaphylatoxin
C6	110 kDa monomer	Binds C5b and to C7

C7	100 kDa monomer	Binds to C5b and C6 and inserts into bacterial membrane
C8	155 kDa heterotrimer	Binds C5b, C6, C7 and C9 – initiates polymerisation of C9
C9	79 kDa monomer	Binds other components and polymerises to form membrane pore

Complement control proteins (regulators of complement activation – RCA proteins)

C1 inhibitor	104 kDa monomer	Binds C1r/C1s and dissociates from C1q
C4-binding protein	570 kDa	Binds C4b and displaces C2
Factor H	150 kDa	Binds C3b and displaces Bb
Factor H-like protein	42 kDa	Similar to factor H
Factor I	88 kDa dimer	Serine protease – cleaves C3b and C4b
CR1 (CD35)	100–250 kDa	Dissociates C3 convertases
CR2 (CD21)	145 kDa	Role in control still under study
CD46 (MCP)	45–70 kDa	Cofactor for factor I-mediated C3b/C4b cleavage
CD55 (DAF)	70 kDa	Displaces C2b from C4B and Bb from C3b

formed by the terminal components of the complement pathway is similar to the pores formed by perforin, the cytolytic granule protein described in Section 6.7.3.

The complement pathways thus generate three major products (Figure 6.10). The cleavage of C3 generates covalently bound C3b on the bacterial surface. This is a major opsonin recognised by the CR1 (complement receptor 1) receptor on phagocytes (described in Section 6.6.2). The small molecular mass products of the cleavage of C3 and C5 (C3a and C5a) are anaphylatoxins important in stimulating inflammatory responses. C5a is a potent chemoattractant for phagocytic leukocytes. The final major product is the MAC, which can form pores in bacterial membranes. In addition, in recent years it has been found that C3d, a further breakdown product of C3b, can enhance antibody formation.

Genetic deficiencies of complement activation are now well documented and enable a perspective to be gained of the importance of individual complement proteins in antibacterial defences. Deficiencies in the early components C1/C2/C4 are not normally associated with increased susceptibility to infection. However, deficiency in C3 is associated with frequent and serious pyogenic bacterial infection. Deficiencies in alternative pathway components lead to similar defects in antibacterial defences but deficiency in formation of the terminal pore results only in increased susceptibility to infection with *Neisseria* spp.

The ability of bacteria to evade complement-mediated attack is described in Section 10.4.1.

6.6.2 Phagocytes and phagocytosis of bacteria

The ability of immune effector cells such as macrophages and neutrophils to recognise a bacterium is vital and, as will be described, is involved in priming the acquired immune system in the process known as antigen presentation. Another key function that these cells perform is the killing of invading pathogens. This is done through the process of phagocytosis. Bacteria taken into the vacuolar apparatus of the phagocyte are then killed by a variety of mechanisms, divided into what are termed oxidative and non-oxidative killing. The various steps in this process will be described briefly. They involve (i) recognition of the pathogen via cell surface receptors, (ii) uptake into a phagosome, with concurrent alterations in cell signalling, (iii) vacuolar trafficking and fusion with lysosomes, and (iv) bacterial killing (Figure 6.21). As mentioned earlier, the reader should be aware that the process of phagocytosis is also important for the removal of senescent cells from the body.

6.6.2.1 Receptors involved in phagocytosis

Phagocytes have two major classes of receptor for recognising bacteria (or senescent/apoptotic cells). The first recognises components of the bacterial cell wall or cell membrane. The second recognises components

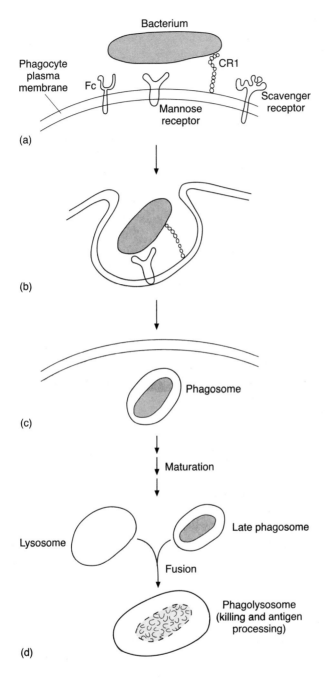

(a)

(b)

(c)

(d)

Figure 6.21 Simple diagram showing the various steps in bacterial phagocytosis from recognition of the bacterium to its killing. (a) The bacterium can be recognised by a number of receptors on the phagocyte plasma membrane, including Fc, CR1, mannose and scavenger receptors. (b) Once triggered, the plasma membrane extends a cup-like extrusion of the membrane around the bacterium and then, when the membranes meet, they are zippered together to form a membrane vacuole called a phagosome. (c) The phagosome undergoes a process of maturation and (d) fuses with the lysosome to produce a phagolysosome, in which killing takes place.

of the immune response (complement components or immunoglobulins) that have bound to the bacterial surface in the process of opsonisation (Table 6.12). Scavenger receptors (SRs) are membrane glycoproteins that were discovered initially by scientists studying atherosclerosis, because of the SRs' ability to bind and internalise modified serum lipoproteins. It is believed that it is this activity that forms the modified macrophage known as the foam cell, which contributes to the formation of the atherosclerotic plaque and the consequent conditions of

Table 6.12. | Phagocyte receptors involved in phagocytosis

Receptor	Type	Ligand
SR-AI	Scavenger	Fucoidin, polyguanylic/polyinosinic acid
SR-AII	Scavenger	As above
MARCO	Scavenger	Unidentified bacterial components
CD36	Scavenger	Phosphatidylserine
SR-BI	Scavenger	PS
SR-CI	Scavenger	Polyguanylic/polyinosinic acid, β-glucan
Lox-I	Scavenger	Polyguanylic/polyinosinic acid, phosphatidylserine
Mannose receptor	Mannose receptor	Glycosylated ligands
Endo180	Mannose receptor	Glycosylated ligands
$\alpha_5\beta_1$	Integrin receptor	Fibronectin
$\alpha_v\beta_3$	Integrin receptor	Fibronectin
$\alpha_M\beta_2$	Integrin receptor	iC3b
$FC_\gamma RI$ (CD64)	Fc receptor	IgG-opsonised bacteria
$FC_\gamma RIIA$ (CD32)	Fc receptor	IgG-opsonised bacteria
$FC_\gamma RIIA$ (CD16)	Fc receptor	IgG-opsonised bacteria
$FC_\alpha R$ (CD89)	Fc receptor	IgA-opsonised bacteria
CRI(CD35)	Complement receptor	C3B (lower affinity for C4B/iC3b)

MARCO, macrophage receptor with collagenous structure; PS, polysaccharide; for other abbreviations, see list on p. xxii.

coronary heart disease and stroke. A large number of these receptors have now been found and a surprising facet of their activity is the breadth of molecules that they can bind to with high affinity. The term 'molecular flypaper' has been coined to describe their binding 'specificity'. It is now recognised that these scavenger receptors can recognise bacterial cell surface molecules such as LPS and lipoteichoic acid. Binding of bacteria to these scavenger receptors stimulates tyrosine kinase, protein kinase C and the actin cytoskeleton, processes described in Sections 4.2.4 and 4.2.5. Phagocytes of transgenic mice lacking SR-AI ingest significantly fewer organisms than do wild-type animals and are more susceptible to bacterial infections, showing the importance of these receptors in bacterial recognition.

The mannose receptor is a 175 kDa transmembrane receptor expressed on resident macrophages and dendritic cells and is part of a family of proteins comprising the phospholipase A_2 (PLA_2) receptor, the DEC-205/MR6 receptor and endo180. The mannose receptor recognises complex carbohydrates such as those found on the surface of bacteria.

Evidence that this receptor was involved in phagocytosis came from transfection studies in which it was shown that transfection of non-phagocytic cells with the cDNA for this receptor enabled them to phagocytose bacteria.

The integrins are a superfamily of heterodimeric (α/β-chain) cell adhesion molecules that can be divided into several subfamilies on the basis of the usage of the β-chain. The β_1 or VLA (very late antigen) family of integrins recognise a range of extracellular matrix components such as collagen, fibronectin, vitronectin and laminin. The β_2 or leukocyte-cell adhesion molecule (Leu-Cam) family of integrins are important for leukocyte trafficking to inflammatory sites and recognise adhesion receptors on vascular endothelial cells (the so-called intercellular adhesion molecule, ICAM-1/2/3 receptors). In addition, as described, $\alpha_M\beta_2$ recognises iC3b. The β_3 or cytoadhesin family of integrins recognise a range of extracellular matrix (ECM) and serum proteins including fibronectin, vitronectin, Von Willebrand factor and fibrinogen. In simplistic terms, the β_2-integrins can be thought of as the receptors enabling leukocytes to adhere strongly to activated vascular endothelial cells and thus helping them to exit the vasculature and enter the site of infection. The β_1-integrins enable lymphocytes to adhere to the ECM and thus tether these cells and prevent them from escaping from inflammatory sites.

Fibronectin is a multivalent protein that can bind to many macromolecules, including those found on the surfaces of bacteria (see Section 7.3.4). A number of the integrins recognise fibronectin and are thus involved in the phagocytosis of fibronectin-coated components. The iC3b-recognising $\alpha_M\beta_2$-integrin is involved in the internalisation of several bacterial pathogens.

A variety of receptors recognising the Fc region of immunoglobulin molecules form immune complexes. The Fc$_\gamma$RI (CD64) receptor, found on macrophages and neutrophils has a very high affinity (10^{-9} M) for IgG1 and IgG3. The other Fc receptors (Fc$_\gamma$RIIA (CD32) and Fc$_\gamma$RIIIA/B (CD16)) have substantially lower (10^{-6} to 10^{-7} M) affinities. These receptors bind to IgG coated on the surfaces of bacteria. The FcRIIIA gene product is expressed primarily on NK cells.

While each of these phagocyte receptors has been described separately, it has to be remembered that they could all be involved in the binding and phagocytosis of particular bacteria. The non-opsonic receptors (and the receptors for complement) would be responsible for recognising bacteria that the body had previously not encountered. Bacteria to which an acquired immune response had been raised would also ligate Fc receptors. Chapter 10 will describe how bacteria can defeat the host's phagocytic processes by either preventing the bacterium binding to phagocyte receptors or, if taken up into the phagocyte, by inhibiting the formation of the killing power of the phagolysosome.

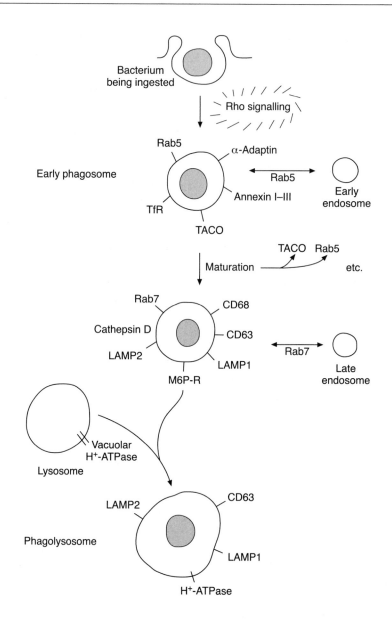

Figure 6.22 Pathway of phagolysosome formation in the macrophage. Ingestion of the bacterium into the early phagosome involves specific signalling via the Rho GTPase family and then a series of maturation steps occurs involving the gain or loss of a range of marker proteins that have distinct functions during the lifetime of the endosomal compartment containing the bacterium. For details of this process, see the text. TfR, transferrin receptor; M6P-R, mannose 6-phosphate receptor; TACO, tryptophan aspartate-containing coat protein; for other abbreviations, see the text.

6.6.2.2 Phagosomes, vesicular transport and generation of the phagolysosome

Having recognised and bound to a bacterium through its surface receptors, the phagocyte has then to ingest the prokaryotic cell, move it through its cytosol to fuse with a lysosome, and then kill it (Figure 6.22). Macrophages and immature dendritic cells also use the lysosomal apparatus to present bacterial antigens to lymphocytes. To understand the process of bacterial uptake into cells, it must be appreciated that all cells have systems for producing and moving vacuoles (which contain many different cellular constituents) through the cell from one compartment to another. Such vacuolar movement requires the participation of the cytoskeleton and is controlled by specific intracellular signalling pathways involving primarily Rho GTPases. Most work on the vacuolar trafficking in phagocytes has been carried out on macrophages.

Receptor binding to bacteria and/or opsonins at the phagocyte surface results in intracellular signalling via Rho GTPases and G protein-linked phospholipases (see Sections 4.2.3.1.2 and 4.2.5.2), which induces cytoskeletal reorganisation and the formation of pseudopodia, which engulf the bacterium. The intracellular vacuole, which forms from the pinching-off of the plasma membrane is called an early phagosome. This early phagosome can be recognised by the fact that it contains the transferrin receptor and other proteins such as α-adaptin and TACO (**t**ryptophan **a**spartate-containing **co**at protein). The phagosome undergoes a process of maturation that requires the participation of the small GTPase Rab5. The maturation process can be followed by changes in specific proteins such as LAMP-1 (lysosome-associated membrane protein), vacuolar proton pump ATPase and hydrolytic enzymes. The presence of the proton pump lowers the intravacuolar pH. The maturation of early phagosomes to become late phagosomes requires the participation of another small GTPase, Rab7. The fusion of the late phagosome with a lysosome to form a phagolysosome is the final process and requires additional GTPases. The fusion with the lysosome turns the phagosome into a vacuole with low pH and high degradative activity.

Various bacteria have evolved mechanisms to block different parts of this phagosome maturation process and so allow the organism to survive within this normally lethal environment. A brief survey of these bacterial anti-phagocytic mechanisms will be given in Section 10.4.2.2.2.

6.6.2.3 Bacterial killing in the phagolysosome

The phagolysosomes of macrophages and neutrophils have a large array of mechanisms for killing bacteria (Figure 6.23). These can be most conveniently divided into oxidative and non-oxidative mechanisms.

6.6.2.3.1 Oxidative killing

First described in 1933 as the 'extra respiration of phagocytosis' it is now known that the phagocytosis of bacteria activates a pentameric protein complex in the cell membrane of the phagosome known as NADPH oxidase. This enzyme reduces oxygen to the superoxide free radical and in the process oxidises reduced NADPH. The superoxide free radical is a very reactive species that can chemically modify all classes of biomolecules and disrupt the functioning of intact bacteria. The importance of this enzyme complex is seen by the fact that, in the absence of oxygen, neutrophils are deficient in their ability to kill bacteria. In the natural human disease, chronic granulomatous disease (CGD), this respiratory burst associated with bacterial phagocytosis is absent and sufferers have a severe predisposition to recurrent

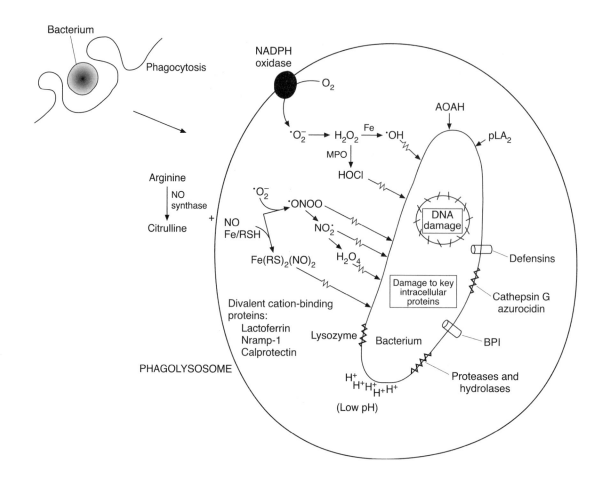

Figure 6.23 Schematic diagram illustrating the many oxidative and non-oxidative mechanisms that phagocytes use to kill bacteria. The details of the various killing mechanisms are given in the text. AOAH, acyloxyacyl hydrolase; for other abbreviations, see list on p. xxii.

infections with bacteria and fungi. Knockout (in mice) of the gp91[phox] subunit of the NADPH oxidase complex, which is the molecular basis of CGD, leads to a disease in mice that is similar to CGD. As will be described in Chapter 10, bacteria have evolved ways to deal with these reactive oxygen species or intermediates (ROS/ROI).

Alternative oxidative species used for killing intracellular pathogens are reactive nitrogen species/intermediates (RNS/RNI). The enzyme responsible for the production of the major reactive nitrogen compound, nitric oxide, is an inducible enzyme called inducible nitric oxide synthase (iNOS). The discovery of the importance of iNOS in macrophage killing of pathogens was derived initially by using selective enzyme inhibitors. Mice in which iNOS has been knocked out have increased susceptibility to infections with *M. tuberculosis* and *Lis. monocytogenes*.

6.6.2.3.2 Non-oxidative killing

Two pieces of information about the killing of bacteria by phagocytes are informative. Firstly, patients with CGD are not constantly infected and, indeed, some are not identified until adulthood. Secondly, mice in which both the NADPH oxidase and iNOS have been knocked out can still deal effectively with certain bacterial infections. Thus there are additional non-oxidative mechanisms used by phagocytes to kill bacteria. These include the acidification of the phagolysosome, the sequestration of iron, the antibacterial actions of peptides, cationic proteins and various enzymes.

For many years, the acidification of the phagolysosome was considered to play a major role in the killing of ingested bacteria. This was supported by the finding that certain bacteria can survive in macrophages in which acidification is inhibited. Unfortunately, pharmacological inhibition of phagolysosomal acidification does not inhibit the killing of ingested bacteria and may even enhance such killing. It may be that the acidic conditions act in synergy with other antimicrobial mechanisms to promote bacterial killing.

Competition between bacteria and the host at mucosal surfaces for iron was described in Section 5.4.3.3. Divalent metal ions such as iron, zinc and manganese are required for both host and bacterial growth and sequestration of these ions by the host would limit bacterial growth. Over the last 10 to 15 years it has emerged that innate resistance or susceptibility to infection with several intracellular bacteria in the mouse is under the control of one genetic locus, *Bcg*, and the gene responsible has been identified. The product of this gene is termed **natural resistance-associated macrophage protein 1** (Nramp1) and is found in the vacuolar apparatus of macrophages. On the basis of homology to other proteins that are divalent cation transporters, it is proposed that the function of Nramp1 in the phagosome is to deplete this vacuole of divalent ions such as iron and manganese and thus limit the growth or survival of bacteria. As will be described in Section 10.4.2.2.2, bacteria also encode homologous proteins.

The mechanisms described above are present in the macrophage and, to a lesser extent, in the neutrophil. The next group of non-oxidative components is found largely within neutrophil granules and were discovered during the 1990s (Table 6.13). These include a number of antibacterial peptides and proteins (defensins, bacterial permeability-inducing protein (BPI), cathelicidins) with actions similar to those discussed in Section 5.4.3.11. Lysozyme and other hydrolytic enzymes are found in the neutrophil granules. Of interest is the finding of enzymes able to target bacterial outer membrane lipids such as lipid A and phospholipids. The presence of lactoferrin and calprotectin shows the importance of sequestering metal ions in the killing of bacteria.

6.7 | Acquired immunity

As has been commented on earlier, there is a certain degree of artificiality in dividing immunity into innate and acquired systems. In Section 6.6, the requirement for antibodies to opsonise bacteria to enhance phagocytosis and the role of complement components in enhancing B cell antibody formation clearly show the overlap in these

Table 6.13. Non-oxidative microbial killing mechanisms of neutrophils

Localisation	Protein	Function
Primary granules	Defensins BPI Lysozyme Hydrolases	Bacterial membrane permeabiliser LPS-binding/permeabilisation Peptidoglycan cleavage Cleavage of bacterial macromolecules
Secondary granules	Cathelicidins Lactoferrin Lysozyme	Multiple antibacterial effects Iron-binding Peptidoglycan cleavage
Unspecified	Phospholipase A_2 Acyloxyacyl hydrolase	Bacterial membrane destruction Deacylates lipid A
Cytosol	Calprotectin	Zinc and calcium chelator

BPI, bactericidal permeability-inducing protein.

systems. The reason for these divisions is purely to aid the understanding of the enormous complexities of our immune systems. In this section, the mechanisms responsible for enabling B and T lymphocytes to recognise, and respond to, bacteria will be described and the consequences of inducing lymphocyte effector cells reviewed. Before describing how the innate and acquired systems of immunity link up to generate effector cells, it is essential to explain, albeit very briefly, how the immune system generates the enormous diversity of lymphocyte antigen-binding specificities (antibodies and TCRs). This is known as GOD.

Immunology and the 'Danger Model'

Immunology is among the most philosophical of the biological sciences and the paradigm that has developed over the past 30 years to explain how the cellular system known as immunity recognises antigens is the so-called 'self-non-self model'. Thus the immune system 'looks' for foreignness and only recognises this by relating this back to 'self'. This is quite a complex hypothesis to explain. Fortunately, in the last decade another model for the immune system has been proposed. This has been championed by Polly Matzinger, who has termed her hypothesis the 'Danger Model'. In this hypothesis the immune system is not seeking to identify foreignness but to recognise the consequences of infection or injury. In infection, cells are often killed. In homeostasis, cell death occurs by the process known as apoptosis or programmed cell death. This process leaves no cellular 'residues' that would activate nearby cells. However, in infection or injury, cells are killed (e.g. by bacterial toxins) by a process known as cytolysis. In this process, many activating cellular constituents can be released. It is these released cellular constituents that act as the 'danger signals' in Matzinger's model and it is these signals, it is proposed, that activate the immune response. One of the major danger signals is suggested to be the stress proteins produced by all cells – proteins such as heat shock protein 60 secreted by bacteria or by damaged host tissues.

There is currently a heated debate about the Danger Model, with many individuals taking entrenched positions. This is part of the fabric of science and is extremely healthy. Debates of this kind stretch the mind and produce new ways of thinking about the complex systems that we, as biologists, have to deal with.

6.7.1 Generation of diversity (GOD)

At the time of writing, with 90% of the human genome sequenced, it is estimated that the number of genes in our cells is approximately 32 000. Estimates of the numbers of different antibody molecules and TCRs that can be produced by our immune systems are from 10^{10} upwards. Remember, each lymphocyte produces only one specific antibody-combining site or TCR. How is it possible to generate such a diverse repertoire of receptors from approximately 10^4 genes? Solving this problem has been one of the most satisfying aspects of modern molecular immunology and won Susumu Tonegawa the 1987 Nobel Prize for Medicine and Physiology. Both the antibody molecule and the TCR contain a variable and a constant region (Figure 6.12). The variable region is the recognition site for the antigen or peptide/major histocompatibility complex (MHC) complex, and it is in this region that the diversity is maintained. In this section, the mechanisms of generating diversity of the antibody molecule will be described. These mechanisms are very similar to those used to generate TCR diversity.

The antibody binding site is composed of the variable domains of the immunoglobulin light and heavy chain (Figure 6.12). The variable domain proteins are not the product of single genes but are the products of gene fusions. The variable domain of the immunoglobulin light chain is the product of two genes termed V_L and J (for joining). There are, in fact, two distinct light chain gene families called κ and λ. Each immunoglobulin heavy chain is the product of three fused genes – V_H, J and D (for diversity). It is estimated that there are 51 V_H, 40 V_κ, 30 V_λ, 27 V_D, 6 heavy chain V_j genes and 9 light chain V_j genes (Table 6.14). During B cell development in the bone marrow, each B cell (and T cell) chooses a particular set of the available V, D and J genes to produce its immune receptor protein. The combination of these genes is catalysed by recombinases encoded by recombination-activating genes (RAG-1 and 2). The statistics of the shuffling together of these various genes to encode an antibody-combining site composed of V_H and V_L chains provides 2×10^6 potential antibody specificities (or TCRs). Further mechanisms contributing to antibody diversity, and thereby creating the estimated $> 10^{10}$ antibody-combining sites or $> 10^{10}$ TCRs, include several mechanisms for ensuring that variation is introduced into the gene when the individual segments are fused. In addition, for the antibody molecule there is an additional process called somatic hypermutation. This process is not fully understood, but basically the rate of

Table 6.14. | Antibody diversity due to combinatorial gene fusion in humans

Germline genes	Heavy chain genes	Light chain genes	
		κ	λ
V	51	40	30
D	27	0	0
J	6	5	4
Combinations	$51 \times 27 \times 6 = 8262$	$40 \times 5 = 200$	$30 \times 4 = 120$

Possible $V_L \times V_H$ combinations $= 8262 \times (200 + 120) = 2.6 \times 10^6$.

mutation of the genes encoding antibodies in B lymphocytes is 100 000 times greater than the normal rate.

The genetics of antibody and TCR formation is fascinating and the reader is referred to the various immunology textbooks in Section 6.13 for more information.

6.7.2 Major histocompatibility complex proteins and antigen presentation

In this section, the activation of specific clones of T and B cells in response to cognate antigen will be described. The starting point for this is the APC, which is generally either a macrophage, a dendritic cell or a B lymphocyte, and its role in presenting bacterial antigens to T cells. Before describing the cell biology of this process, the reader will have to be briefly introduced to the MHC. This is because T lymphocytes do not recognise protein antigens but recognise only the combination of fragments of immunogenic proteins linked to MHC proteins.

6.7.2.1 MHC proteins

Located on human chromosome 6 is the human leukocyte antigen (HLA)/MHC complex, which contains two groups of polymorphic genes encoding cell surface proteins. The MHC class I genes encode MHC class I proteins, which are found on all nucleated cells and consist of one polypeptide chain encoded by the MHC complex and a non-covalently linked protein called β_2-microglobulin (Table 6.15). The MHC class II locus is also polymorphic and encodes MHC class II proteins found on a restricted group of so-called professional APCs; these include macrophages, dendritic cells and B lymphocytes. The structure of both classes of MHC proteins is similar. The key feature is the presence, in the extracellular domain, of a peptide-binding groove linked to a pair of immunoglobulin domains (Figure 6.24). This groove contains the polymorphic amino acid residues that recognise particular peptides.

T cells can recognise an enormous number of foreign antigens, and

Table 6.15.	Characteristics of MHC class I and class II proteins	
Feature	Class I	Class II
Composition	44–47 kDa α-chain	32–34 kDa α-chain
	12 kDa β$_2$-microglobulin	29–32 kDa β-chain
Cellular location	All nucleated cells	Macrophage/dendritic cell/B cell
Size of peptide accommodated	8–11 AA residues	10–30 AA residues
Binding site for T cell co-receptor	CD8 binds to α3 region	CD4 binds to β2 region
Human nomenclature	HLA-A, HLA-B, HLA-C	HLA-DR, HLA-DP, HLA-DQ
Mouse nomenclature	H-2K, H-2D, H-2L	I-A, I-E

AA, amino acid.

how they do this will be explained briefly in Section 6.7.2.2. However, it turns out that they cannot recognise whole protein antigens. They can recognise only peptide fragments. Not only that, they can recognise such fragments only when they are bound to MHC proteins. Thus T cells recognise, and respond to, the complex of a peptide with the MHC class I or class II protein. The MHC proteins can bind to a range of peptides. Binding is of low affinity, but the peptide–MHC complexes are stable. This means that the APCs limit the ability of T lymphocytes to recognise antigens.

The two distinct classes of MHC proteins stimulate different T lymphocyte populations. The MHC class I proteins present antigens to CD8 lymphocytes. The MHC class II proteins present antigens to CD4 lymphocytes. These different patterns of protein–protein interactions relate to the nature of the antigens being presented and the presence in cells of pathogens in different cell compartments.

As has been touched upon, the MHC class I and II genes are extremely polymorphic and contain an extremely large number of alleles. In human class I molecules there are 60 A alleles, 110 B alleles and 40 C alleles. This polymorphism creates an enormous diversity of peptide-binding structures. This has consequences for the susceptibility of populations to disease, and during the past 30 years the role of the MHC in such human susceptibility has been studied in great detail. Most of this information has been derived from the study of autoimmune diseases and much less is known about the role of the MHC in susceptibility/protection with regard to infection, particularly bacterial infection (see Micro-aside, p. 27).

MHC class I and II proteins bind to and 'present' peptide antigens, thus inducing CD4 or CD8 T cell immune responses. In recent years a third locus, not on chromosome 6 and containing genes homologous to the MHC locus, has been found to encode CD1 proteins and these have been termed 'non-classical class I molecules'. In the human there are five CD1 genes (CD1a–e) that encode heavy chains that associate

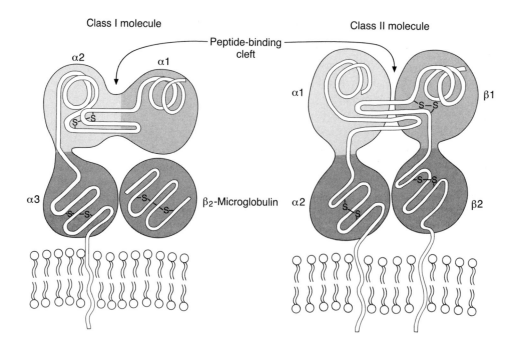

Class I molecule   Class II molecule

Peptide-binding cleft

α2   α1

α1   β1

α3   β₂-Microglobulin   α2   β2

Figure 6.24 Schematic diagram of the MHC class I and II proteins. The major difference is that MHC class I proteins, comprise two separate proteins one of which is β₂-microglobulin. In MHC class II proteins, two proteins of similar size are combined to form the final protein. In both class I and class II proteins, the molecules have a peptide-binding cleft for binding peptides for presentation to the T lymphocyte.

with β₂-microglobulin. CD1 expression is normally restricted to haematopoietic cells, particularly macrophages and dendritic cells. The importance of these proteins is that they present non-protein antigens to T cells. The importance of the CD1 system in antimicrobial immunity will be described in Section 6.7.2.2.4.

6.7.2.2 An introduction to antigen presentation

Antigen presentation can be regarded as the central process in acquired immunity and the major point of contact between the innate immune and acquired immune systems. The purpose of this process is to find the appropriate antigen-reactive lymphocytes and to provide them with the signals that will cause them to enter into the cell cycle and produce large numbers of effector cells. This process is known as clonal proliferation and is the great strength of acquired immunity. Thus our bodies do not have to maintain large numbers of cells able to respond to the potentially enormous numbers of antigens derived from pathogens that can infect us. All we need is one cell responding appropriately to the antigen and within a few days to a week large numbers of active lymphocytes are generated. The main process of antigen presentation requires the interaction of APCs with T lymphocytes. This process can generate activated CD4, CD8 and B lymphocytes. However, some antigens can directly activate B lymphocytes in a T cell-independent manner and this process will be described briefly.

6.7.2.2.1 B lymphocyte activation by thymus-independent antigens

B lymphocytes respond directly, in a T cell-independent manner, to certain antigens known as thymus-independent (TI) antigens. In turn, these TI antigens are further subdivided into type 1 (TI-1) and type 2 (TI-2) antigens, which work by different mechanisms. TI-1 antigens – the Gram-negative molecule LPS is the prototypic example – are so-called polyclonal B cell activators, or mitogens, and are able to activate B cells irrespective of their antigenic specificity. Such TI-1 antigens can stimulate proliferation of large proportions of the B cells of mammals. The reader should compare this activity with that described for bacterial superantigens (which stimulate T lymphocytes) in Section 10.5.2. It is not clear how these TI-1 molecules act, but it is established that the response of B cells to them is completely independent of T lymphocytes.

TI-2 antigens are generally molecules with repetitive structures (i.e. polyvalent), such as polysaccharides, glycolipids or nucleic acids. They cause B cell activation by cross-linking immunoglobulin on antigen-specific B cells. There is evidence that the complement component C3d, which can be produced by complement activation induced by such polyvalent molecules, can provide a second signal for B cell activation and thus enhance B cell responses to these (normally bacterial) components. TI-2 antigens do require the presence of macrophages and/or T cells and it is believed that these cells provide the appropriate cytokine milieu for activation.

The predominant antibodies produced by TI antigens are IgM, and class switching (a process in which the Fc domain of an antibody is changed for another, normally IgG for IgM) does not occur. The so-called natural antibodies that recognise bacterial carbohydrates are produced by this TI mechanism. Individuals with immunoglobulin deficiencies do not make such antibodies and are susceptible to infections with encapsulated bacteria such as *Strep. pneumoniae*, *Neisseria meningitidis* and *Haemophilus* spp.

6.7.2.2.2 Antigen presentation to T lymphocytes: formation of the MHC–peptide complex

The reader has already been introduced to MHC class I and class II proteins and has been told that CD4 and CD8 T cells respond to the combination of small peptides derived from intact antigenic proteins, and the MHC proteins themselves. These peptides sit in the groove of the MHC molecules. The question is: how does the cell produce the combination of antigenic peptide and MHC protein recognised by the T lymphocyte? Again, the reader should note that, classically, CD8 lymphocytes respond to cytoplasmic protein antigens (viruses, cytoplasmic bacteria) produced within the cell and linked to class I MHC proteins. CD4 T lymphocytes respond to exogenous parasitic proteins endocytosed by the APC and presented as peptides on MHC class II proteins.

All cells can present endogenous antigens via MHC class I proteins to CD8 lymphocytes. Presentation of exogenous antigen to CD4 lymphocytes is more the province of the professional APC, although a number of cells have been shown to possess the ability to present antigens to these cells. These include dendritic cells, macrophages, B lymphocytes, mast cells, endothelial cells, etc. However, the dendritic cell and macrophage are, by far and away, the most active in this respect.

Presentation to CD4 lymphocytes: For an APC to present exogenous antigens to CD4 T lymphocytes the process involves the uptake of the antigen into an endosome/phagosome (by the process of endocytosis or phagocytosis), the maturation of these endosomes/phagosomes (see Section 6.6.2) with a decrease in endosome pH and the proteolytic processing of the antigen(s) (Figure 6.25). The nature of the proteases involved is not fully resolved. However, it is known that knockout mice lacking the lysosomal protease cathepsin S have defects in class II-associated antigen presentation. The endosomes containing proteolysed bacterial proteins then fuse with exocytotic vesicles transporting class II proteins to the cell surface. These exocytotic vesicles are generated in the endoplasmic reticulum (ER) and contain MHC class II proteins. To prevent these class II proteins binding to self-peptides, they are generated as a complex with an additional protein called the invariant chain. This blocks the MHC groove and prevents peptide attachment. The invariant chain also helps in the folding of the MHC complex. Once fused, the invariant chain is proteolytically removed, leaving a 24 residue peptide called the class II-associated invariant chain peptide (CLIP). This has to be removed to allow antigenic peptides to bind. Removal is facilitated by a non-polymorphic MHC class II-like protein called HLA-DM. This is the first example of a lysosomal protein that acts as a molecular chaperone – i.e. a protein that 'catalyses' the folding of other proteins. Knockout of HLA-DM results in mice that are deficient in presenting extracellular proteins. Once the CLIP is removed, the peptides generated by proteolysis of endocytosed bacteria can bind to the class II MHC groove. The binding of such peptides also stablises the class II MHC proteins. The exocytotic vesicle/endosomal vesicle product moves to the cell surface, where the class II MHC–peptide complexes can be expressed and activate antigen-specific CD4 T lymphocytes.

Presentation to CD8 lymphocytes: The presentation of antigens to CD8 T lymphocytes involves a distinct pathway of proteolysis (Figure 6.26). Again, the distinction between MHC class I and class II antigen presentation is that the former uses peptides derived from endogenous proteins found in the cell cytoplasm as opposed to exogenous antigens. In the context of this book, the key proteins are those derived from bacteria present in the cell cytoplasm. However, the major source of pathogen protein will be derived from viruses. There is also evidence that proteins derived from phagocytosed material can be presented on class I proteins.

Proteolysis of ingested bacterial proteins takes place in the endo-

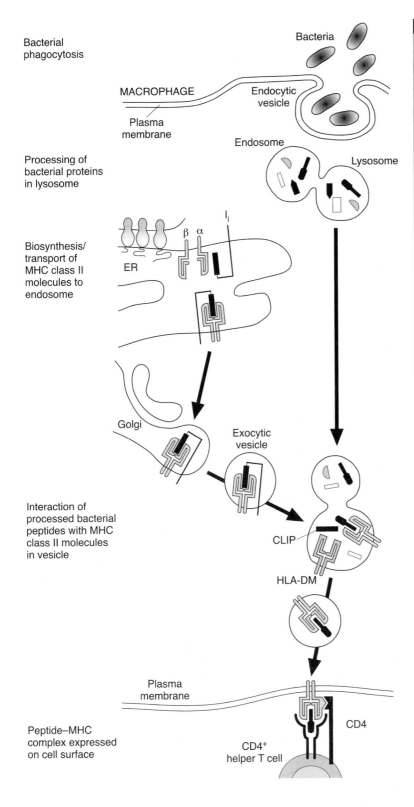

Bacterial
phagocytosis

MACROPHAGE

Plasma
membrane

Endocytic
vesicle

Bacteria

Endosome

Lysosome

Processing of
bacterial proteins
in lysosome

Biosynthesis/
transport of
MHC class II
molecules to
endosome

ER

β   α   I$_i$

Golgi

Exocytic
vesicle

Interaction of
processed bacterial
peptides with MHC
class II molecules
in vesicle

CLIP

HLA-DM

Plasma
membrane

CD4

Peptide–MHC
complex expressed
on cell surface

CD4$^+$
helper T cell

Figure 6.25 Presentation of exogenous bacterial antigens to CD4 T lymphocytes. Uptake of bacteria into the endosome/ phagosome results in the breakdown of the proteins (and other components) into molecules small enough to be incorporated into, in this case, MHC class II proteins. These are produced in the endoplasmic reticulum (ER) and are exported into the phago-somal vacuole where they can bind to bacterial peptides and be presented on the cell surface of the APC. The details of this process are described in the text. CLIP, class II-associated invariant chain peptide. (Reproduced from Abbas, A. K., Lichtman, A. H. & Pober, J. S. (2000). *Cellular and Molecular Immunology*, 4th edition, with copyright permission from W.B. Saunders Company.)

Figure 6.26 Presentation of endogenous antigens to CD8 T lymphocytes. The major difference between class I and class II presentation is that, with a class I antigen, the peptides that bind to it are produced in the cell's cytoplasm in a process driven by the large proteolytic 'machine' known as a proteasome. This will proteolyse all denatured proteins within the cytoplasm. Peptides so produced, including those from cytosolic bacteria, will enter into the endoplasmic reticulum (ER) where they can bind to appropriate MHC class I proteins. The complex of class I and peptide is then transported to the cell surface where it can recognise cognate CD8 T lymphocytes. TAP, transporter associated with antigen processing; β_2m, β_2-microglobulin. (Reproduced from Abbas, A. K., Lichtman, A. H. & Pober, J. S. (2000). *Cellular and Molecular Immunology*, 4th edition, with copyright permission from W.B. Saunders Company.)

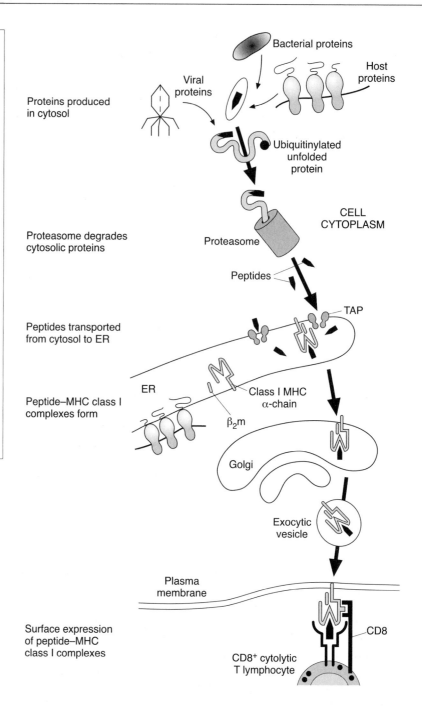

somal compartment. In contrast, the 'protease' that digests cytoplasmic proteins is a unique multiprotein enzyme complex, almost a nanomachine, called the proteasome. Two forms of the proteasome are found in cells. There is a 700 kDa complex that has the crystal structure of a cylinder composed of two inner and two outer rings stacked on top of each other. Each ring is composed of seven subunits and three of these contain the catalytic sites for proteolysis. A second, 1500 kDa, proteasome is composed of the basic 700 kDa proteasome structure but contains additional subunits that regulate protease activity. Many of these larger proteasomes contain catalytic subunits, termed low molecular mass polypeptide (LMP) 2 and LMP-7, that are encoded by genes within the MHC locus and appear to be involved in the generation of peptides (6 to 30 residues long) destined for binding to MHC class I proteins. LMP-2 and LMP-7 are induced by interferon (IFN) γ and this is one of the mechanisms by which this key cytokine enhances antigen presentation.

The proteasome is involved in a number of cellular activities. It catalyses the turnover of cytoplasmic proteins, removing unfolded proteins that are targeted to the proteasome by the addition of a small polypeptide, ubiquitin. This process is, for example, important in controlling the response of cells to a range of activators that stimulate the NF-κB gene transcription system. As described in Section 4.2.6.6.2, the activation of NF-κB requires the removal of an inhibitory binding protein, IκB. This is done by phosphorylating the latter protein, making it a target for ubiquitinylation and proteolysis by the proteasome. Indeed, it is possible to inhibit the activation of cells via NF-κB by adding proteasome inhibitors.

The peptides generated by the proteasome – being within the cytoplasm – need to come into contact with class I proteins. However, as described earlier, the MHC proteins are produced in the ER. The transmembrane passage of these peptides is catalysed by two proteins termed transporter associated with antigen processing (TAP1 and 2). These proteins are homologous to ATP-dependent membrane transport proteins that catalyse the transport of peptides into the lumen of the ER. The specificity of these transporter proteins is for peptides of 6–30 residues with appropriate structure/chemistry for MHC class I binding. Knockout of the TAP proteins in mice results in animals deficient in class I expression and presentation. Interestingly, the few rare cases of human TAP deficiency show increased susceptibility to infections with some bacteria.

Once peptides have been transported into the ER they can attach to the MHC class I proteins. These proteins are brought together and stabilised by a number of molecular chaperones. The class I α-chain binds to a membrane-bound molecular chaperone called calnexin. This complex then binds the β_2-microglobulin, releasing the calnexin and promoting binding of two additional proteins calreticulin and tapasin (TAP-associated protein). It is this latter complex that promotes binding of the antigenic peptides to class I, and not class II, MHC proteins. The other reason that TAP-imported peptides do not bind to class II proteins is the presence of the invariant chain in the class II groove. Once the MHC class I proteins have been loaded with their antigenic peptides, they are then packaged into exocytotic vesicles and trafficked to the plasma membrane, where they can be recognised by CD8-positive T lymphocytes.

6.7.2.2.3 New thoughts on bacterial antigen presentation

The dogma that has been presented to the reader is that exogenous antigens are presented on MHC class II proteins to CD4 T lymphocytes and that endogenous (cytoplasmic) antigens are presented on MHC class I proteins to CD8-positive T cells. However, over the past decade cracks have begun to appear in this dogma. In Section 6.7.2.2.4, the role of the non-polymorphic MHC class I-like proteins known as CD1 and their ability to present non-peptide antigens to T cells will be described. The potential role of these CD1 proteins in modulating T cell responses to bacteria is now being actively investigated. Major weaknesses in the dogma have also been introduced by detailed studies of the T cell responses to specific bacterial infections. It is now rapidly becoming established that both Gram-negative and Gram-positive bacteria can elicit CD4 and CD8 responses. This may not be surprising if the bacteria or bacterial proteins are present in the cytosol. For example, bacterial proteins injected into the cytoplasm of cells by type III secretion systems may be expected to be proteolysed by the proteasome and presented on MHC class I proteins. This is true also for the cytosolic pathogen *Lis. monocytogenes*. However, bacteria such as *M. tuberculosis* that are taken up into endosomes in APCs would not be expected to induce CD8-positive responses. In spite of this, it is now clear that CD8 lymphocytes play an important role in immunity to *M. tuberculosis*.

It now appears that the classical model of CD8/MHC I and CD4/MHC II interactions is, in fact, a simplification of the real situation. Alternative pathways of antigen presentation exist. The processing of exogenous antigens for presentation on MHC class I proteins has now been described for a variety of cell types including macrophages, dendritic cells, B cells and even mast cells. Similarly, MHC class II proteins have now been demonstrated to present endogenous antigens to CD4 T lymphocytes. Most attention has focused on the loading of exogenous bacterially derived antigens on to class I proteins, and a number of distinct mechanisms have been proposed to account for the binding of endosomal bacterial antigens on to ER MHC class I proteins. The exact mechanisms of this process have not been established and they will not be discussed further, although the possibility exists that different patterns of proteases (from the classic class I/II systems) may be required. The reader is referred to the reviews by Wick and Ljunggren (1999), Brodsky *et al.* (1999) and Maksymowych & Kane (2000) (detailed in Section 6.13) for further reading on this fascinating topic. However, we will return to this topic in Chapter 10, in which the ability of bacteria to evade the antigen-processing mechanisms of APCs will be discussed in the context of bacterial evasion of immunity.

6.7.2.2.4 Antigen presentation to T lymphocytes: role of CD1

One of the major intellectual foundations of modern cellular and molecular immunology was the hypothesis that APCs presented peptide antigens, and only peptide antigens, to T cells and that this presentation required polymorphic MHC proteins encoded by the MHC locus on chromosome 6. A second MHC-like locus was discovered and the protein product was one of the first to have monoclonal antibodies made against it. As described in Table 6.4, the CD nomenclature defines cell surface proteins to which antibodies have been raised. This protein was thus termed CD1. CD1 proteins, it now turns out, can present non-peptide antigens to T lymphocytes.

It is now known that in the human there are a family of five non-polymorphic, closely linked, CD1 genes on chromosome 1. Four of these genes encode the proteins CD1a, b, c and d. The fifth gene has no, as yet discovered, protein product. These proteins are subdivided, on the basis of nucleotide and amino acid homology, into two groups. Group I CD1 includes the products of the CD1A, B and C genes in the human and their homologues in other species. Group II CD1 defines the products of the CD1D gene and homologues. Of interest is the finding that mice, commonly used as human surrogates in infectious diseases research, lack group I CD1 genes and have only two very similar group II CD1 genes.

The CD1 gene products are polypeptides with a molecular mass of approximately 33 kDa. These proteins are expressed on the surface of a range of myeloid cells as type I transmembrane proteins associated non-covalently with β_2-microglobulin. This is similar in structure to the MHC class I proteins and the crystal structure of CD1 shows a three-dimensional structure very similar to that of class I proteins.

What do CD1 proteins recognise? It is still early days in the study of CD1 specificity, but the conclusion to date is that these surface proteins bind to and present bacterial glycolipids such as mycolic acid, phosphatidylinositol mannoside and α-glycosylacylated phytosphingosines. Most of the work on CD1 has been in the context of diseases caused by mycobacteria and these glycolipid antigens are components of such bacteria (Figure 6.27).

The pathways of class I and class II peptide loading have been described. How does the CD1 pathway of antigen loading compare? It appears that both exogenous (endosomal) and endogenous (cytoplasmic) glycolipids can associate with CD1 proteins in CD1-positive cells. The presentation of the complex LPS-like molecule of mycobacteria, lipoarabinomannan (LAM), requires that this molecule be taken up into endosomes via the macrophage mannose receptor. Uptake can be blocked by soluble mannans or antibodies to that receptor. Presentation of glycolipids can be blocked by preventing endosomal acidification (see Section 6.6.2), suggesting that CD1 proteins are present in the same lysosomal compartments as MHC class II proteins. There is also an endogenous pathway for loading CD1 proteins with glycolipids, although less is known about the mechanisms involved.

The strongest evidence for a role for CD1 proteins in human bacterial infection has come from the study of leprosy. This is a chronic inflammatory disease caused by *M. leprae*. Now, to deal effectively with infection with this mycobacterium, the individual needs to mount a strong cellular immune response. As will be described, this cellular immune

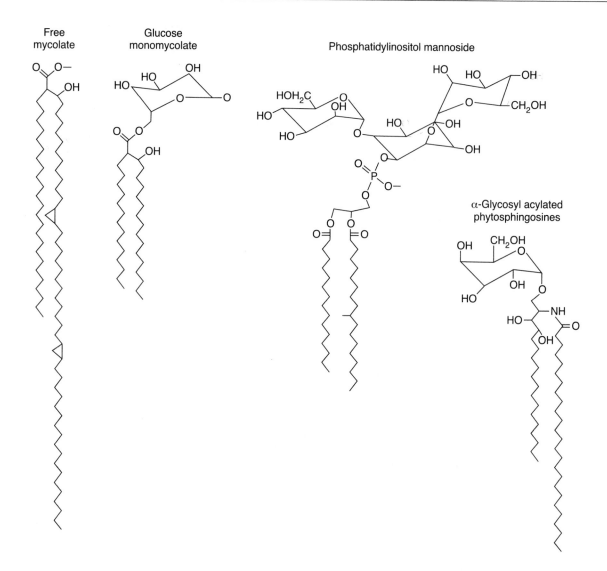

Free mycolate

Glucose monomycolate

Phosphatidylinositol mannoside

α-Glycosyl acylated phytosphingosines

Figure 6.27 Not all lymphocytes recognise peptides. This figure shows some of the non-peptidic antigens recognised by CD1 and able to be presented to lymphocytes.

response is called a Th1 response. However, if the infected individual were to make not a cellular immune response but an antibody-mediated response, then this would be relatively ineffective. Clinically, patients with leprosy exhibit a distribution of antimycobacterial responses ranging from those that have tuberculoid leprosy and are able to cope with the infection, owing to their strong cell-mediated immune responses, to those with lepromatous leprosy. In this latter group of patients, the disease is poorly controlled because the cellular immune response is limited. In a study of skin biopsies from patients with both forms of leprosy, the frequency of CD1-positive dendritic cells in patients with tuberculoid leprosy was 10-fold higher than in those patients with unresponsive lepromatous leprosy.

Although most work has been done with mycobacteria, there is increasing evidence that CD1-bearing APCs can present non-peptidic

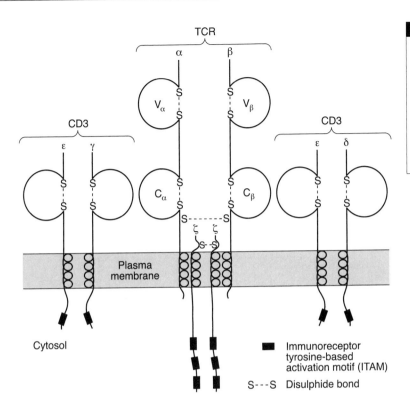

antigens from other bacteria. In this context, it has been demonstrated that H2-M3, the so-called non-classical MHC class I molecule in the mouse, binds to bacterial antigens such as the formylated peptides released by bacteria as part of the process of protein synthesis. This MHC protein is believed to be important in presenting antigens from *Lis. monocytogenes*.

6.7.2.2.5 Antigen presentation to T lymphocytes: APC-lymphocyte surface receptor interactions

Having packaged peptides derived from a cytoplasmic pathogen on to MHC class II or class I proteins and exported them to the surface of the cell, the next step in the process is the recognition of the MHC–peptide complex by lymphocytes. The activation of CD4 T lymphocytes will be discussed first, followed by a brief description of the activation of CD8 cells. The actions of the effector cells induced by the process of antigen presentation in antibacterial defences will be discussed in Section 6.7.3.

As described, the TCR is the complex protein that recognises the specific peptide linked to the MHC class II protein (Figure 6.28). The TCR is a multiprotein complex comprising the two-chain TCR, with its V_α and V_β domains, linked to two CD3 protein molecules and two ξ-chains. As will be discussed in later chapters, the V_β domain is recognised by bacterial superantigens. The binding of this complex to the MHC class II protein–peptide complex provides the specificity for T

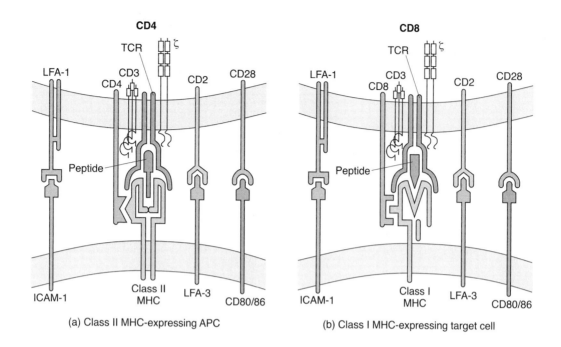

(a) Class II MHC-expressing APC (b) Class I MHC-expressing target cell

Figure 6.29 Accessory molecules required for the activation of T lymphocytes. In order to stimulate CD4 or CD8 lymphocytes, protein–protein interactions in addition to the TCR-antigen/ MHC binding event are required. These include the binding of CD4 or CD8 to the respective MHC class molecule. CD2 recognises LFA-3, and LFA-1 recognises ICAM-1. In addition it is important that the pairing of CD28 and CD80/86 occurs. This is the very important co-stimulatory signal that lymphocytes need to receive to enter into productive clonal proliferation. LFA, lymphocyte function-associated antigen. (Reproduced from Abbas, A. K., Lichtman, A. H. & Pober, J. S. (2000). *Cellular and Molecular Immunology*, 4th edition, with copyright permission from W.B. Saunders Company.)

cell proliferation. However, this single binding event does not provide all of the signals required for the activation of the cognate T cell (Figure 6.29). It turns out that the interaction of a CD4 T lymphocyte and its antigen-bearing APC is a more complex interactional process. Included in the cognate binding events required to stimulate CD4 T cells is the binding of (i) CD4 to the MHC class II molecule (ii) LFA-1 on the T cell to ICAM-1 on the APC, (iii) CD2 on the T cell to LFA-3 on the APC and (iv) CD28 on the T cell to CD80/CD86 (previously called B7-1/7-2) on the APC. This last coupling of CD80/86 to CD28 is a key binding event in the triggering of CD4 lymphocyte activation. B and T lymphocytes require two signals to stimulate them. This is a failsafe mechanism built into the immune system, presumably to prevent inappropriate lymphocyte activation. The first signal is provided by antigen binding to the appropriate antigen receptor and provides the specificity in the system. The second signal (CD80/86-CD28) is provided by so-called co-stimulatory molecules. In the context of bacterial infections, CD80/86 is not present on inactivated macrophages, although it is expressed by immature dendritic cells. To up-regulate the levels of these proteins on macrophages, the cells have to be exposed to bacterial factors (LPS, lipoproteins, CpG DNA, etc.). These factors interact through the CD14–TLR system to induce these co-stimulatory molecules. This then allows the cognate CD4 T lymphocytes to become activated in response to the APC. In the absence of co-stimulatory molecules, CD4 T cells recognising MHC class II-peptide complexes would undergo a series of cellular alterations that would induce an unresponsive state termed 'T cell anergy'. This requirement for co-stimulatory signals explains why it is essential, if one wishes to raise an antibody response to a protein, to provide the protein along with a bacterial adjuvant. The most powerful adjuvants are bacteria in various forms (whole cells, lysates, sonicates). *Mycobacterium tuberculosis* is a particularly powerful adjuvant. We now understand that the bacteria switch on the co-stimulatory signals, allowing the acquired immune response to focus on the immunising protein and produce T and B cell responses. Cytokines induced by microbial products are also involved in this process. This shows, yet again, the close connection between innate and acquired immune mechanisms. The result of these various ligand–receptor interactions is the stimulation of appropriate intracellular signal transduction pathways. These pathways will not be described in any detail.

CD8 T lymphocytes are involved largely in the killing of virally infected cells, although there is recent evidence that CD8 lymphocytes are important in protection against intracellular bacteria such as *Lis. monocytogenes*, a cytosolic pathogen, and even *M. tuberculosis*, the prototypic CD4-stimulating bacterial pathogen. The reader is referred to the review by Stenger & Modlin (1998) (see Section 6.13) for more details. The stimulation of CD8 cells requires a similar panoply of surface receptor–ligand interactions (Figure 6.29).

Once appropriately stimulated, T cells undergo an activation phase where they produce cytokine growth factors and then enter a state called clonal proliferation to produce large numbers of mature effector cells (Figure 6.30).

6.7.3 Functions of effector T cell populations

T cells responding to cognate antigen begin to divide within 18 hours and then have an estimated doubling time of 6 hours. Thus, within a few days to a week, there are very large numbers of T lymphocytes available to deal with invading microbes. As seen in Figure 6.30, there are a number of differentiated effector T cells. In general terms, the functions of CD4 T lymphocytes are to activate phagocytes and B lymphocytes by generating specific networks of cytokines. The function of CD8 T lymphocytes is to kill antigen-bearing target cells, mainly cells infected with cytoplasmic pathogens – usually viruses. In addition, clonal populations of T lymphocytes also produce memory cells required for secondary immune responses.

T cells are involved in providing help for B cells in inducing antibody production. Cell–cell interactions provide appropriate cytokines for proliferation and clonal expansion and for class (or isotype) switching – a process where antibodies of one class change their Fc domains (e.g. IgM switching to IgG). The roles of antibodies in neutralising bacterial exotoxins, and other virulence molecules, and in opsonisation have been described earlier in this chapter and in other chapters in this book and will not be discussed further.

CD8, or cytolytic, T cells interact with antigen-specific target cells and kill them by two potentially interdependent mechanisms. The first involves the secretion of a pore-forming protein called perforin and a

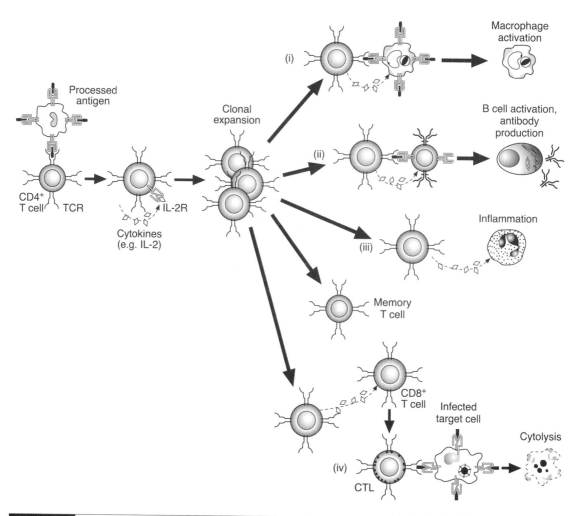

Figure 6.30 The three phases of the T cell response. In the recognition phase, an antigen-specific T cell contacts an appropriately primed APC and signalling events caused by the interactions of both cells induce the activation of the T cell and the synthesis and release of cytokine growth factors, principally IL-2. The release of this growth factor enables it to act in an autocrine fashion and induce lymphocytes to undergo clonal proliferation or expansion. These cells are called effector cells and can (i) activate macrophages, (ii) help B lymphocytes make antibody, (iii) induce delayed-type inflammatory responses or (iv) kill cells infected with cytoplasmic pathogens. In addition, antigen-activated cells can give rise to memory cells. CTL, cytotoxic T lymphocyte.

group of serine proteases called granzymes (Figure 6.31). The perforin forms pores in the plasma membrane of the target cell and the granzymes can enter through such pores. The granzymes can activate pro-caspases and induce apoptosis. The second mechanism utilised by CD8 lymphocytes is **Fas l**igand (FasL). As explained in Section 10.6.2.1.1, FasL binding to Fas (CD95) on target cells induces apoptosis through the activation of caspase 3 via death domain interactions. There is now evidence to support the hypothesis that CD8 cells play a role in defence against *Lis. monocytogenes*, *Chlamydia* spp. and *M. tuberculosis*.

The γδ T lymphocyte

There is a second type of T lymphocyte with a TCR composed, not of α- and β-chains, but of γ- and δ-chains – thus these cells are called γδ T lymphocytes. These cells are present in low numbers in the blood and are localised mainly in mucosal tissues. In the gut, they form a proportion of the so-called IELs (see Section 5.7.3). Most γδ T lymphocytes do not express CD4 or CD8 on their plasma membranes. Nor are these cells MHC-restricted. So what do they do?

The answer to this question is: we are not sure. However, there is growing evidence that they are important in monitoring bacterial infections and that they may play important roles in dealing with such infections, especially in the young host. Examination of what certain clones of γδ T lymphocytes recognise has found that they can be activated by small phosphorylated organic molecules and alkyl amines. Such molecules are very abundant among bacteria. There is also evidence that γδ T lymphocytes recognise host molecules such as heat shock proteins (molecular chaperones). This has suggested that γδ T lymphocytes may monitor cells in their microenvironment (such as epithelial cells at mucosal surfaces) and, if they recognise that these cells are being stressed, they may be able to provide appropriate signals including trophic factors.

The role of γδ T lymphocytes in bacterial infections is unclear. In experimental animals, infection of the peritoneal cavity with *Lis. monocytogenes* results in a 10-fold increase in the numbers of these cells. However, the peak of this expansion occurred two to three days after resolution of the infection. When strains of *Lis. monocytogenes* lacking the toxin listeriolysin were used (see Section 9.4.1.3.3), these increases in γδ T lymphocyte numbers did not occur. One interpretation is that the γδ T lymphocytes were not responding to the bacteria *per se* but to the stress induced in the host cells by the actions of the bacterial toxin.

A recent hypothesis has proposed that γδ T lymphocytes, which are dipropor-tionately abundant in young animals, may play a key role in protection against infection in the early days of life. This hypothesis is currently undergoing experi-mental testing. Another concept that is emerging as this book is being written is that γδ T lymphocytes act to control the activity of inflammatory macrophages and may be employed to switch off inflammatory responses to microbial infections (see Egan & Carding, 2000 (Section 6.13, Papers). How immune and inflammatory responses are down-regulated is a poorly understood area of infectious diseases research and these special lymphocytes may play a role in this activity.

Thus the γδ T lymphocyte population is still mysterious and its precise functions are not fully defined. The next decade should see the resolution of these questions about function and allow the placing of these cells within the framework of innate and acquired defences against bacteria.

The other functions of CD4 T lymphocytes relate to their abilities to deal with microbial pathogens that exist in the vacuolar apparatus of macrophages or large multicellular pathogens. There is a further division of labour in the CD4 T lymphocyte population into Th1 and Th2 lymphocytes. This paradigm has caught the popular imagination and will be described below.

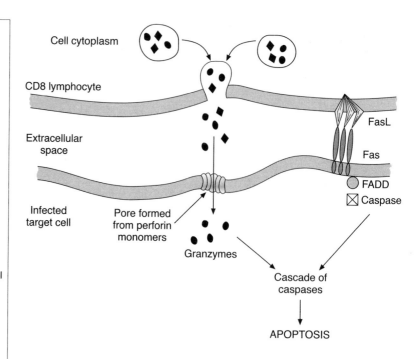

Figure 6.31 Interaction of CD8 lymphocytes (cytotoxic T lymphocyte (CTL)) with a target cell showing the two mechanisms of induction of cell killing: the granule exocytosis pathway and the Fas pathway. The interaction of the CD8 cytotoxic T cell with an infected target cell produces a rise in intracellular calcium that triggers exocytosis of granules containing two components: perforin monomers (represented by diamonds) and granzymes (represented by ovals). Within the intercellular environment the perforin monomers undergo a calcium-dependent conformational change that allows them to insert into the target cell membrane and polymerise to form pores. The granzymes can enter via these pores and induce a cascade of caspase activation leading to apoptosis (for more details, see Section 10.6.2). In addition, the binding of the CTL to the target cell can result in binding of Fas ligand (FasL) with Fas. This causes association of FADD (Fas-associated death domain) and the activation of caspase cascade and programmed cell death. For more details about Fas/FasL/FADD, see Sections 4.2.6.3.3 and 10.6.2.1.1.

6.7.3.1 CD4 effector mechanisms: Th1 and Th2 lymphocytes

To recap, bacteria found outside of cells are dealt with by (i) opsonisation by complement and antibodies and (ii) uptake and killing by phagocytes. Bacterial toxins are neutralised by antibodies. If the phagocytes are macrophages and dendritic cells, phagocytosis is also part of the system of antigen presentation to lymphocytes. If the bacterium is found in the cytoplasm of cells then it is likely to be the target of cytolytic (CD8) T lymphocytes. For those bacteria that inhabit the vacuolar apparatus of macrophages, then the requirement is for a Th1 CD4 T lymphocyte.

6.7.3.2 The Th1/Th2 paradigm

The study of cloned murine CD4 T lymphocytes in the 1980s resulted in the discovery that different clones could produce one of two patterns of cytokine production. Thus some clones produced IL-2 and IFNγ, while others produced IL-4 and IL-5, and IL-13. The lymphocytes producing the former pattern of cytokines were termed **T helper 1** (Th1 or T_h1) CD4 lymphocytes, those producing the latter, Th2. The differences in the cytokine/cytokine receptor expression patterns, and behaviour, of these two cell populations is shown in more detail in Table 6.16. These two CD4 T cell populations now seem to be reciprocally modulated, with the Th1 population being important for cell-mediated immune responses to intracellular pathogens of macrophages, and the Th2 lymphocytes playing a role in immunity to multicellular parasites such as parasitic worms. There is increasing evidence that the

Table 6.16.	Properties of CD4 Th1 and Th2 lymphocytes		
Property		Th1	Th2
Cytokine production			
IL-2, IFNγ, TNFα		Yes	No
IL-4, IL-5, IL-13		No	Yes
IL-10		+	+++
Chemokine receptor expression			
CXCR3		++	+/−
CCR4		+/−	++
CCR3		+/−	++
Cytokine receptor expression			
IL-12β chain		++	−
IFNγβ chain		−	++
Biological functions			
Activates macrophages		+++	−
Delayed type hypersensitivity		+++	−
Eosinophil/mast cell induction		−	+++
Antibody isotypes induced		IgG2a	IgE, IgG1
Ligands for E/P selectins		++	+/−

For abbreviations, see list on p. xxii. Plus and minus signs indicate level of production/expression of property.

inappropriate generation of these two CD4 subsets can cause disease, or limit immune responses to particular pathogens. It is also possible that pathogens can utilise this redirection of CD4 responses as an immune evasion mechanism (see Sections 10.3.2.1 and 10.3.3). The Th1/Th2 paradigm has caught the scientific imagination and has been applied widely to try to explain many aspects of immune functioning in bacterial, protozoal and viral infections. The reader should be aware that this paradigm is a simplification of the real-life situation.

6.7.3.3 Generation of Th1 and Th2 lymphocytes

Now, it is important to realise that the Th1 and Th2 subsets of lymphocytes are not genetically defined but are determined by environmental factors (Figure 6.32). Both cell populations arise from undifferentiated (or, as immunologists call them, 'naïve') CD4 T lymphocytes (also referred to as Th0 cells). The signals that cause their differentiation into the appropriate cells are mainly cytokines. Thus IL-12 is the major inducer of Th1 cells and IL-4 of Th2 lymphocytes. It is proposed that

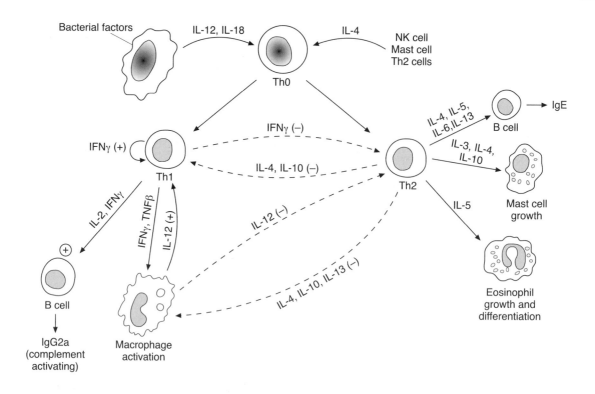

Figure 6.32 The generation of Th1 and Th2 lymphocytes and their interactions. To generate these two CD4 subpopulations, the activated Th0 cell has to be exposed to specific cytokines. To produce Th1 cells, the major signal is IL-12. This can come from macrophages stimulated with bacterial products. To generate Th2 lymphocytes, the main signal required is IL-4 from natural killer (NK) and other cell populations. Once induced, the Th1 cells produce cytokines able to activate macrophages. This can be directly by signals such as IFNγ, or indirectly by inducing B cells to produce complement-fixing antibodies. Th2 cells produce a panoply of cytokines that generate mainly antiparasite responses. There is significant negative cross-talk between the Th1 and Th2 lymphocytes, as shown in this diagram. For abbreviations, see the list on p. xxii.

bacteria, and factors secreted by bacteria (e.g. LPS, LAM, toxins, etc.), induce myeloid cells to secrete IL-12. Endocytosis of bacteria by macrophages is also a stimulus for IL-12 synthesis. The IL-12 binds to receptors on antigen-stimulated CD4 lymphocytes to promote their differentiation into Th1 cells. Likewise, the infestation of the host with helminths stimulates the production of IL-4, causing antigen-sensitised CD4 T cells to differentiate into Th2 lymphocytes.

The cytokine networks generated by these two different CD4 T cell populations have reciprocal effects on the other population. For example IFNγ secreted by Th1 cells preferentially inhibits the proliferation of the Th2 subset. Likewise, IL-4 and IL-10 secreted by Th2 cells inhibit secretion of the key Th1-inducing cytokine IL-12. These cytokines also have differential effects on the target cells of Th1/Th2 cells. For example, IFNγ secreted by Th1 T lymphocytes promotes complement-fixing IgG2a production by B cells, but inhibits the production of antihelminth IgG1 and IgE antibodies. The opposite is true of IL-4.

6.7.3.4 Effector functions of Th1 lymphocytes

As this book is focused on antibacterial immune responses, the actions of only the Th1 cell population will be described. Many of the readers of this book will be familiar with the working of the Th1 cell. If they can remember back to their BCG test for tuberculosis, the red, swollen and potentially painful lesion which occurred at the test site was due to the actions of these Th1 cells. This skin response is described by immunologists as a **delayed type hypersensitivity** (DTH) response. In individuals

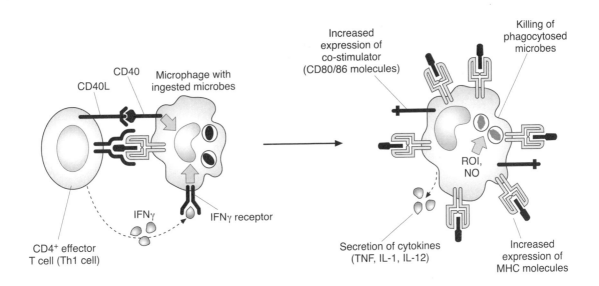

who have bacteria infecting their macrophages (such as *M. leprae* or *M. tuberculosis*) the requirement is to activate these macrophages to enable them to kill the resident organisms. The processes of antigen presentation to T lymphocytes and their differentiation into Th1 cells have already been described. What happens when an antigen-specific Th1 CD4 T lymphocyte makes contact with a macrophage acting as a host to intravacuolar bacteria? The interactions between these two cell populations is described schematically in Figure 6.33. Recognition of the macrophage by the Th1 T lymphocyte results in two signals being transmitted to the macrophage. The first is the macrophage-activating cytokine IFNγ. Knockout of IFNγ or the IFNγR leads to increased susceptibility to intracellular bacteria. The second signal produced by the Th1 T lymphocyte is the cell surface ligand, and member of the TNF family, CD40 ligand (CD40L or CD154) – a type II membrane protein. This binds to CD40 on the surface of the macrophage. Binding to CD40 activates the macrophage in a manner similar to that produced by TNFα. The importance of this CD40L–CD40 interaction is shown from the finding that, in humans, inherited disorders of CD40L are associated with severe deficiency in immune responses to intracellular infections. These two signals from the activated Th1 T lymphocytes activate the antibacterial mechanisms of the macrophage, including the production of reactive oxygen and nitrogen intermediates, the secretion of cytokines and the up-regulation of co-stimulatory molecules (CD80/86). Such activation allows the macrophage to kill intracellular bacteria and also promotes the antigen-presenting activity of these cells.

Figure 6.33 Interaction of a CD4 Th1 cell with a bacterially infected macrophage, showing the changes that occur in the activated macrophage as a result of signals from the lymphocyte. For abbreviations, see the list on p. xxii. (Reproduced from Abbas, A. K., Lichtman, A. H. & Pober, J. S. (2000). *Cellular and Molecular Immunology*, 4th edition, with copyright permission from W.B. Saunders Company.)

6.8 | A brief overview of the immune response to bacterial infections

The reader has been rapidly introduced in this chapter to many of the cellular and humoral components of innate and acquired immunity as it pertains to defence against bacterial infection. In this section, the temporal sequence of events occurring in response to a 'general' bacterial infection will be described very briefly.

Infection: Bacteria that evade mucosal antibacterial responses and invade into the tissues will begin to grow and produce signals in the form of surface components, such as LPS or peptidoglycan.

Local cell activation: The components released from bacteria can be recognised by cells bearing members of the TLR family of receptors. These are part of the pattern recognition receptor family. The cells bearing TLRs will include macrophages and DCs. Binding to these receptors will activate these cells and cause them to release pro-inflammatory cytokines (IL-1, IL-6, IL-8 and related chemokines, IL-12, IL-18, TNFα). In addition to the induction of cytokines, the APCs (macrophages, dendritic cells) will up-regulate their production of cell surface receptors such as CD80 and CD86 (co-stimulatory signals) TLRs and other key receptors, and can also begin to produce a range of other pro-inflammatory proteins such as complement components.

Actions of dendritic cells: DCs are specialised APCs. There are two basic types of DC – one arising from myeloid precursors and the other from lymphoid precursor cells. Immature DCs, either within the infected tissue or attracted by appropriate chemokines induced by the bacteria, can phagocytose whole bacteria or endocytose fragments derived from bacterial breakdown by macrophages. These cells can then detach from the site of infection and carry bacterial antigens to the local lymph nodes to find cognate T cells in order to stimulate T cell activation and associated cell-mediated and humoral immune responses (Figure 6.34).

Activation of endothelium: A key requirement of the inflammatory response to bacterial infection is to increase the numbers of leukocytes able to deal with the infecting bacteria. This is important, as bacteria can have doubling times as short as 20 minutes. Thus, in, say, a 6 hour period, one bacterium infecting the host could multiply to form approximately 132 000 'offspring'. The only problem is that the antibacterial leukocytes required for bacterial killing are in the blood and there is a thick blood vessel wall preventing them from entering the site of infection. The solution to this problem is to activate the blood vessels (specifically the post-capillary venules, PCVs) at the site of infection so that they bind to circulating leukocytes and allow leukocyte extravasation (Figure 6.35). This involves signals from the bacteria, and

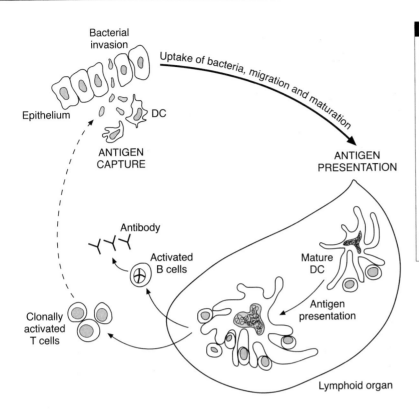

Figure 6.34 Role of DCs in antibacterial immunity. Immature DCs in tissues interact with infecting bacteria and ingest whole organisms and bacterial antigens. They then migrate to local lymph nodes (and in the process begin to mature), where they can present antigens to cognate lymphocytes, resulting in B and T cell activation. Activated lymphocytes so produced can then migrate and reach the infected tissue, where they can promote antigen-specific responses.

from the cytokines induced by the bacteria, activating these PCVs so that they express two distinct adhesive proteins – selectins (P and E) and ICAM/VCAM (vascular cell adhesion molecule). The selectins bind with low affinity to specific oligosaccharides on the surfaces of leukocytes. This allows the leukocytes to roll along the surface of the blood vessels and in so doing they become activated by chemokines such as IL-8. This activation process results in conformational changes in the β_2-integrins (e.g. CR3) on the leukocyte surface (as a result of phosphorylation). This enables these integrins to bind strongly to ICAM or VCAM and enter into the site of inflammation by a process known as diapedesis (Figure 6.35). Once leukocytes enter the site, they can chemotact to the bacteria and phagocytose them. Leukocytes also express β_1-integrins, which bind to the extracellular matrix and trap these cells at the site of infection until the infecting organisms, and their associated activating signals, are removed. Complement activation and production of C5a will also act as a further chemotactic signal for the entry of leukocytes.

Blood vessel leakage: The activation of the endothelium by bacterial/host cytokine signals may also make the local blood vessels leaky. This will bring in additional factors including complement components, preformed antibodies (natural or from previous infections) and other factors such as fibrinogen (which can give rise to fibrin and act to seal off the site of infection) and acute phase proteins. The antibodies

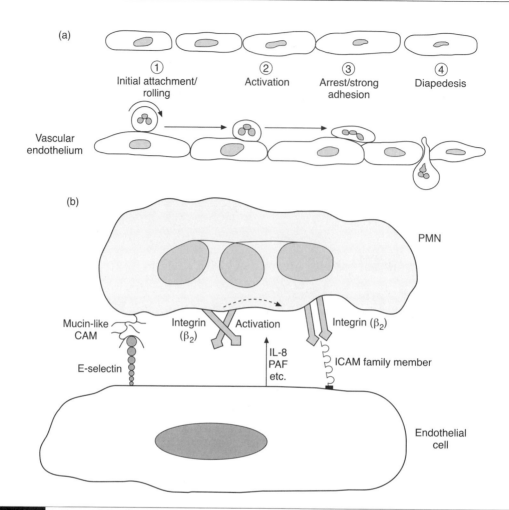

Figure 6.35 In Figure 6.8 the effect of bacterial invasion on leukocytes was described. In this figure, additional details of the process of leukocyte recruitment are provided. In (a) the process of capture of leukocytes is shown. The initial event is the binding of E- and P-selectin on the endothelial surface (induced by bacterial factors/cytokines) to oligosaccharides on the leukocyte plasma membrane. This initial low affinity event starts slowing down the leukocytes; they roll along the endothelial surface and begin to be activated by signals coming from the inflamed site (e.g. IL-4, PAF, leukotriene B$_4$ (LTB$_4$)). Such signals activate the β$_2$-integrins, which bind to ICAM induced on the endothelial cell surface. This is the signal for the leukocytes to leave the blood vessels and enter into the tissues. The various interactions described are shown in (b). Here the mutual binding partners can be seen. The initial slowing and rolling is due to low affinity interactions between E- (or P-) selectin and oligosaccharides. The adhesion events that stop leukocytes and prepare for diapedesis involve binding of ICAM family members to the β$_2$-integrins. For abbreviations, see list on p. xxii.

and the acute phase proteins, for example, can be used to opsonise the infecting organisms.

Immune cell function: Antigen presentation by macrophages and DCs, and by B lymphocytes, will continue as long as the infection persists. In short-lived bacterial infections, antigen presentation (in local lymph nodes or in other lymph organs if the infection spreads) will occur only for a limited period. If such responses are not utilised to any great extent in primary infections, they will become of use if the individual is reinfected with the same organism. In this case, B and T cell responses occur very much more quickly – a few days instead of a few weeks. In infections lasting longer than a few days, antigen-specific B cells will begin to produce antibodies and release them into the circulation, from where they can enter the site of infection. In addition, antigen-specific T cells will target to the site of infection and perform their functions as cytolytic or Th1 effector cells.

6.9 | Concept check

- Our immune response to bacteria can be divided into innate and acquired systems.
- These two systems are closely interrelated.
- Both immune systems utilise humoral factors and the actions of cells.
- The innate immune system recognises bacteria via pattern recognition receptors such as TLR4.
- The immune system has distinct methods for dealing with free-living, cytosolic and intravacuolar bacteria.
- Antibodies and complement are used to opsonise bacteria and antibodies can also neutralise bacterial toxins.
- Macrophages and neutrophils phagocytose and kill free-living bacteria.
- CD8 T lymphocytes kill cells with cytosolic bacteria.
- CD4 lymphocytes help B cells make antibody and activate macrophages infected with bacteria.

6.10 | Immune defences and the paradigm organisms

6.10.1 *Streptococcus pyogenes*

Streptococcus pyogenes causes a remarkable range of diseases, from localised infections such as pharyngitis and pyoderma to highly invasive diseases such as sepsis, necrotising fasciitis and toxic shock-like syndrome. What is worrying is the recent increase in the rate of infection with it and the increase in the severity of its pathology. There is obviously a need for a vaccine against it. As will be described in Section 10.8.1, *Strep. pyogenes* has a number of useful tricks for evading host immune defences. Being a free-living bacterium its major foe is the phagocyte. It has therefore evolved a number of means of evading phagocytosis, including a protease that cleaves the major anaphylatoxin, C5a, and a cell surface protein (M protein) that acts as an adhesin and inhibits complement activation. M protein binds to a complement regulatory protein called C4-binding protein and, by covering the bacterium in this protein, prevents complement activation. However, M protein is also highly immunogenic and was believed to be a suitable target for vaccine development. The initial enthusiasm for this protein as a target for immunisation foundered on the realisation that the N-terminus of the protein is hypervariable. More than 100 M types (serotypes) of *Strep. pyogenes* have now been discovered. Unfortunately, protective immunity is serotype specific. In addition, structural similarities between M protein and several host proteins could trigger immunological cross-reactivity, with the possibility of the induction of autoimmunity.

The search is still on for a candidate vaccine. In a recent paper, Guzman *et al.* (1999) (see Section 6.13, Papers) have looked to the fibronectin-binding protein SfbI as a vaccine candidate. Fibronectin is a large glycoprotein found in serum/plasma, the extracellular matrix and on the surfaces of cells. Thus bacterial fibronectin-binding proteins such as SfbI can enable bacteria to attach to extracellular matrices or to the surface of cells. Guzman and colleagues have immunised mice with SfbI and have found that it protects them against experimental lung infections due to *Strep. pyogenes*. Thus the results of this experiment would suggest that SfbI is a potential protective antigen that could be useful for vaccination in humans. Now, the mouse is patently not a human, and much more work is now required to show that immunisation of humans with SfbI would be both protective and not likely to give rise to autoimmune side effects.

6.10.2 *Escherichia coli*

There is a pressing need to discover how to improve immunity to this organism, as it is an important cause of a range of intestinal infections as well as being the major causative organism of infections of the urinary tract. The well-described antigens of *E. coli* include LPS (the so-called O antigens), capsular components (K antigens), adhesive fimbriae, outer membrane proteins, lipoprotein and toxins. There is also a common antigen called the enterobacterial common antigen. We are colonised by *E. coli* in the first few days of life and there is good evidence that breast milk, with its constituent antibodies, protects infants against diarrhoea. Breast milk contains antibodies that react with a variety of O and K antigens and various adhesins. It takes children a few months to establish secretory and serum antibody titres of IgA, IgG and IgM antibodies and up to nine months before serum IgG reaches the same avidity as that of the IgG obtained from the mother.

The presence and titres of antibodies to *E. coli* in the blood of normal individuals varies considerably. Naturally occurring antibodies to *E. coli* are predominantly IgM, and the major specificity is for the O-specific side chains of LPS. Patients with acute *E. coli* infection of the kidney and urinary tract demonstrate a significant rise in antibacterial antibody titre and this is generally an IgM response directed against the O antigen. Unfortunately, such antibodies do not seem to be protective.

Low titres of anti-fimbrial antibodies occur in the majority of patients with acute pyelonephritis. Again, the protective role of these antibodies is uncertain.

In *E. coli* enteric infections, exotoxins can play a major role in pathology. Four main exotoxins have been identified: heat-labile toxin, Shiga-like toxins, cytotoxic necrotising factor and cytolethal distending toxin (see Section 10.6.1.2). What role do antibodies to these toxins play in protection? Antibodies to the heat-labile and Shiga-like toxins are found in the appropriate infections and are believed to be protective, although the evidence for this thesis is not enormous. Blocking of both of these toxins is a major aim of experimental vaccines. Little is known about the role of antibody responses to the other two toxins.

As will be explained in Section 10.8.2, both the heat-labile and cytolethal distending toxins are immunomodulatory and this may be one of the reasons that *E. coli* producing these toxins are pathogenic.

Given the fact that *E. coli* is the best-studied cell on our planet, there is an enormous amount of information yet to be determined about the host's immune response to this organism, the role of such a response in protection and the ability of the bacterium to nullify protective immunity.

6.11 | What's next?

For a bacterium to infect its host it must first of all bind to that host. In Chapter 7 the mechanisms that bacteria use to bind to host cells or extracellular tissues will be described and the consequences of such binding reviewed.

6.12 | Questions

1. Innate and acquired immunity are inextricably linked – discuss.
2. How does the immune system cope with bacteria in the different sites in the body where they can be found?
3. Explain the importance of opsonisation in bacterial killing.
4. Describe the process by which antigens are presented to CD4 and CD8 T lymphocytes.
5. Explain the importance of Th1 and Th2 lymphocytes in bacterial infection.
6. Describe how phagocytes sense, engulf and kill bacteria.
7. Describe the basic structure of each of the main classes of immunoglobulin molecules and explain the functions of the various parts of these molecules.
8. List, and describe the functions of the various antigen-presenting cells involved in antibody formation.
9. Explain the differences between innate and acquired immunity.

10. Describe and explain the clonal selection theory of antibody production.
11. Compare and contrast cell-mediated and antibody-mediated immunity.
12. Describe the role played by cytokines in cell-mediated and antibody-mediated immunity.
13. Describe the complement cascade and explain its role as an anti-bacterial defence system.

6.13 | Further reading

Books

Abbas, A. K., Lichtman, A. H. & Pober, J. S. (2000). *Cellular and Molecular Immunology*, 4th edn. Philadelphia: W.B. Saunders.

Cunningham, M. & Fujinami, R. S. (2000). *Molecular Mimicry, Microbes, and Autoimmunity*. Washington, DC: ASM Press.

Goldsby, R. A., Kindt, T. J. & Osborne, B. A. (2000). *Kuby Immunology*, 4th edn. New York: Freeman.

Oelschlaeger, T. A. & Hacker, J. (2000). *Bacterial Invasion into Eukaryotic Cells*. New York: Kluwer Academic/Plenum Publishers.

Review articles

Abbas, A. K. & Janeway, C. A. (2000). Immunology: improving on nature in the twenty-first century. *Cell* **100**, 129–138.

Abraham, S. N. & Arock, M. (1998). Mast cells and basophils in innate immunity. *Semin Immunol* **10**, 373–381.

Aderem, A. & Hume, D. A. (2000). How do you see CG? *Cell* **103**, 993–996.

Anderson, K. V. (2000). Toll signaling pathways in the innate immune response. *Curr Opin Immunol* **12**, 13–19.

Braud, V. M., Allan, D. S. J. & McMichael, A. J. (1999). Functions of nonclassical MHC and non-MHC-encoded class I molecules. *Curr Opin Immunol* **11**, 100–108.

Brodsky, F. M., Lem, L., Solache, A. & Bennett, E. M. (1999). Human pathogen subversion of antigen presentation. *Immunol Rev* **168**, 199–215.

Galli, S. J., Maurer, M. & Lantz, C. S. (1999). Mast cells as sentinels of innate immunity. *Curr Opin Immunol* **11**, 53–59.

Gregory, C. D. (2000). CD14-independent clearance of apoptotic cells: relevance to the immune system. *Curr Opin Immunol* **12**, 27–34.

Hayday, A. C. (2000). γδ cells: a right time and a right place for a conserved third way of protection. *Annu Rev Immunol* **18**, 975–1026.

Janeway, C. A. (1989). Approaching the asymptote? Evolution and revolution in immunology. *Cold Spring Harb Symp Quant Biol* **54**, 1–13.

Krause, K. H. (2000). Professional phagocytes: predators and prey of microorganisms. *Scheiz Med Woschenschr* **130**, 97–100.

Krutzik, S. R., Sieling, P. A. & Modlin, R. L. (2001). The role of Toll-like receptors in host defense against microbial infection. *Curr Opin Immunol* **13**, 104–108.

Lambris, J. D., Reid, K. B. M. & Volanakis, J. E. (1999). The evolution, structure, biology and pathophysiology of complement. *Immunol Today* **20**, 207–211.

Lineham, S. A., Martin-Pomares, L. & Gordon, S. (2000). Macrophage lectins in host defence. *Microbes Infect* **2**, 279–288.

Maksymowych, W. P. & Kane, K. P. (2000). Bacterial modulation of antigen processing and presentation. *Microbes Infect* **2**, 199–211.

McCracken, V. J. & Lorenz, R. G. (2001). The gastrointestinal ecosystem: a precarious alliance among epithelium, immunity and microbiota. *Cell Microbiol* **3**, 1–11.

O'Garra, A. & Arai, N. (2000). The molecular basis of T helper 1 and T helper 2 cell differentiation. *Trends Cell Biol* **10**, 542–550.

Porcelli, S. A. & Modlin, R. L. (1999). The CD1 system: antigen-presenting molecules for T cell recognition of lipids and glycolipids. *Annu Rev Immunol* **17**, 297–329.

Stenger, S. & Modlin, R. L. (1998). Cytotoxic T cell responses to intracellular pathogens. *Curr Opin Immunol* **10**, 471–477.

Théry, C. & Amigorena, S. (2001). The cell biology of antigen presentation in dendritic cells. *Curr Opin Immunol* **13**, 45–51.

Wick, M. J. & Ljunggren, H.-G. (1999). Processing of bacterial antigens for peptide presentation on MHC class I molecules. *Immunol Rev* **172**, 153–162.

Papers

Campbell, D. J., Serwold, T. & Shastri, N. (2000). Bacterial proteins can be processed by macrophages in a transporter associated with antigen processing-independent, cysteine protease-dependent manner for presentation by MHC class I molecules. *J Immunol* **164**, 168–175.

Egan, P. J. & Carding, S. R. (2000). Down-modulation of the inflammatory response to bacterial infection by γδ T cells cytotoxic for activated macrophages. *J Exp Med* **191**, 2145–2158.

Guzman, C. A., Talay, S. R., Molinari, G., Medina, E. & Chhatwal, G. S. (1999). Protective immune response against *Streptococcus pyogenes* in mice after intranasal vaccination with the fibronectin-binding protein SfbI. *J Infect Dis* **179**, 901–906.

Peiser, L., Gough, P. J., Kodama, T. & Gordon, S. (2000). Macrophage class A scavenger receptor-mediated phagocytosis of *Escherichia coli*: role of cell heterogeneity, microbial strain, and culture conditions *in vitro*. *Infect Immun* **68**, 1953–1963.

Sieling, P. A., Ochoa, M.-T., Jullien, D., Leslie, D. S., Sabet, S., Rosat, J. P., Burdick, A. E., Rea, T. H., Brenner, M. B., Porcelli, S. A. & Modlin, R. L. (2000). Evidence for human CD4$^+$ T cells in the CD1-restricted repertoire: derivation of mycobacteria-reactive T cells from leprosy lesions. *J Immunol* **164**, 4790–4796.

6.14 | Internet links

1. http://www.cehs.siu.edu/fix/medmicro/genimm.htm
   Extensive set of notes on all aspects of immunology relevant to bacterial infections.

2. http://dunn1.path.ox.ac.uk/~cholt/sgInflam.pdf
   A set of lecture notes on inflammation.
3. http://copewithcytokines.de/
   Everything you wanted to know about cytokines, courtesy of
   Horst Ibelgaufts.
4. http://www-micro.msb.le.ac.uk/MBChB/Merralls/Merralls.html
   Tutorial on complement and how bacteria interact with the
   system, from the Department of Microbiology and Immunology,
   University of Leicester.
5. http://www.cellsalive.com/antibody.htm
   A simple animation explaining how lymphocytes produce
   antibodies against bacteria.
6. http://www.antibodyresource.com/educational.html
   Antibody resource page with links to more than 50 sites on
   antibodies.
7. http://www.kcom.edu/faculty/chamberlain/msimm.htm
   An extensive set of lecture notes on all aspects of immunology.
8. http://www.med.sc.edu:85/book/immunol-sta.htm
   A comprehensive on-line course in immunology (from the
   Department of Microbiology and Immunology, University of
   South Carolina) with nice sets of images. Includes sections on the
   structure, formation and function of antibodies.
9. http://www.path.cam.ac.uk/~mrc7/mikeimages.html
   Basic information on immunoglobulin structure and function with
   some animated images.

Bacterial adhesion as a virulence mechanism

Chapter outline

Aims

The principal aims of this chapter are to describe:

- the molecular basis of bacterial adhesion
- the range of adhesins produced by bacteria
- the types of bacterial structures involved in adhesion
- the nature of the host receptors for bacterial adhesins
- tissue tropism
- the consequences of adhesion for the bacterium and for the host cell
- the possibility of anti-adhesion approaches as prophylactic or therapeutic measures for bacterial infections

7.1 | Introduction

The ability of a bacterium to adhere to its host is an often-overlooked aspect of virulence and in this chapter we will consider this crucial first stage in the induction of an infectious disease. It is difficult to understand why adhesion should be given such short shrift by many microbiologists as it is an essential preliminary to any interaction (including one of the most intimate interactions possible, disease) between a bacterium and its host. We would also like to stress that adhesion should not be regarded simply as a somewhat random, serendipitous 'sticking' of a bacterium to its host. Rather, the adhesion process is accomplished by highly specific molecular interactions and is accompanied by changes in the phenotype of the bacterium and, when the substratum is cellular, changes in the behaviour of the host cell.

Before we go any further, it is important to consider how we should go about investigating the interactions between bacteria and host cells. Practical bacteriology is traditionally based on the use of a 'pure culture' (axenic) of an organism. Obviously, if we are to investigate the interactions between a particular bacterium and its host, we can no longer deal with axenic cultures; there must be at least two types of cell – the bacterium and the host cell. Interactions between bacteria and host cells can be investigated at many different levels of complexity, particularly with respect to the latter, which may be present in an intact host, in an isolated organ, in an individual tissue or in the form of a pure culture of a single cell type. While examination of biopsies from patients has contributed to our knowledge of bacterial adhesion to (and invasion of) host tissues, most of what is known about these two phenomena has been derived from studies that have employed animals and laboratory cultures of eukaryotic cells. It is important, therefore, to point out some of the drawbacks of these approaches and to caution against extrapolating too far from these *in vivo* and *in vitro* models.

Animals have been used in the study of infectious diseases of humans since the original studies of Pasteur in the 1890s. The most commonly used species are mice, rats, rabbits and guinea pigs. The reasons why these particular mammals are used include their ease of handling, their rapid rates of breeding, the availability of inbred lines (to reduce genetic variability) and comparatively low maintenance costs. However, as well as ethical considerations, there are a number of important scientific limitations regarding the use of such animals, including that their physiologies differ from that of humans, that the pathophysiology of the disease process may differ from that occurring in humans and that they may not be susceptible to the same strain of bacterium responsible for the particular human disease. Some of these problems may be overcome as a result of advances in biotechnology that will enable the production of animals with some human characteristics. For example, the humanised severe combined immunodeficiency (SCID-hu) mouse has a human-like immune system.

During the past decade, however, because of ethical considerations and the above-mentioned drawbacks of animal experimentation, the majority of investigations of bacterial adhesion and invasion have employed tissue culture models. Although these are easier to work with and facilitate studies of adhesion and invasion at the cellular and molecular level, they also have their drawbacks in that (i) many studies utilise immortalised cell lines and these differ from primary (i.e. freshly isolated) cell cultures, (ii) the cells are in an undifferentiated state, (iii) the cells are in a monoculture, which is artificial, (iv) the cells are often used in a non-polarised state so that there is no differentiation into apical and basolateral surfaces (see Section 5.3.1) and (v) environmental conditions under which the cells are grown differ considerably from those that would exist *in vivo* and this can have a dramatic effect on gene expression. All of these factors should make us exercise care in extrapolating too far the results obtained using these models.

7.2 | To what do bacteria adhere?

7.2.1 Adhesion to external body surfaces

In Section 1.1 we referred to the fact that, at any point in time, the average human is carrying around approximately 10^{14} microbes, most of which are bacteria. With the exception of those organisms that are present in faeces and in saliva, all of these bacteria maintain themselves within their host by adhering to some surface. The vast majority of these adherent bacteria will do no harm, but for those that are destined to initiate disease it must be recognised that adhesion represents the first, essential, step in the chain of events that will result in pathology. However, it must be remembered that, while an organism must adhere to a particular surface to maintain itself within its host, in order for it to survive and grow there must also exist appropriate environmental conditions at that adhesion site. The body of a normal, healthy individual provides three main types of surface (termed substrata) to which a bacterium may adhere – mucosae, skin and teeth (Table 7.1). However, none of these is a uniform structure and each exists in a variety of forms that provide different types of site for adhesion and/or different environments in which the adherent bacterium must survive. The various mucosal surfaces, for example, include the oral, respiratory, gastric, intestinal and genitourinary mucosae. Furthermore, subdivisions of each of these can be recognised. For example, the oral mucosa (see Figure 1.4) includes cheek, gingival (gum), palatal and lingual (tongue) surfaces, while the nature of the respiratory mucosa varies at different anatomical sites, for example, the pharynx, bronchi and alveoli (see Figure 1.5). Likewise, the skin, while exhibiting less structural variation at different anatomical sites, provides a range of environments that dictate the type of organism to be found at a particular site. For example, the axilla and groin provide a warm, moist environment that supports a large microbial population, while the back is a cooler and drier region with a more sparse microflora. The surfaces mentioned so far are not, of course, passive bystanders in this process and, as discussed in previous chapters, employ a variety of mechanisms to prevent adhesion of disease-inducing organisms and their subsequent colonisation. One of the

Table 7.1. | Substrata for bacterial adhesion available in humans

Substratum	Major types	Subdivisions	Examples of adherent organisms
Mucosae	Oral	Cheek	Streptococci, *Actinomyces* spp., *Haemophilus* spp.
		Gingivae	Streptococci, *Actinomyces* spp.
		Tongue	Streptococci, *Haemophilus* spp., *Veillonella* spp.
		Palate	Streptococci, *Actinomyces* spp.
	Respiratory	Anterior nares	Staphylococci, *Corynebacterium* spp.
		Nasopharynx	*Neisseria* spp., *Haemophilus* spp.
		Oropharynx	Streptococci, *Mycoplasma* spp.
		Trachea	Normally sterile
		Bronchus	Normally sterile
		Alveolus	Normally sterile
	Gastrointestinal	Oesophagus	Normally sterile
		Stomach	Streptococci, lactobacilli, *Helicobacter pylori*
		Duodenum	Streptococci, lactobacilli
		Jejunum	Streptococci, lactobacilli
		Ileum	*Escherichia coli*, *Bacteroides* spp., *Clostridium* spp.
		Colon	*Bacteroides* spp., *Eubacterium* spp., *Bacillus* spp., *Peptostreptococcus* spp.
	Genitourinary	Urethra	*Escherichia coli*, lactobacilli, staphylococci
		Vagina	Lactobacilli, staphylococci, streptococci
Skin	Forehead		Staphylococci, *Propionibacterium acnes*
	Arm pits		Staphylococci, *Propionibacterium avidum*
	Between toes		*Brevibacterium epidermidis*
Teeth	Fissures on tooth crown		Streptococci, *Actinomyces* spp.
	Approximal regions (i.e. between teeth)		As above but more anaerobes (e.g. *Veillonella* spp.)
	Gingival region		As above but a greater proportion of anaerobic Gram-negative rods (e.g. *Fusobacterium nucleatum*, *Bacteroides* spp., *Prevotella* spp., *Porphyromonas* spp.)

most important characteristics of these surfaces from this point of view is that they are shedding, i.e. these are multilayered tissues, the outer layer of which is being continuously shed, taking with it any adherent organisms (for a diagram showing the structure of a typical epithelium, see Figure 5.3). It is difficult, therefore, for huge numbers of bacteria to build up on such surfaces. It has been estimated that a typical epithelial cell of the oropharynx will have, on average, between 5 and 25 bacteria adhering to it (Figure 7.1). In contrast to mucosae and skin, the teeth

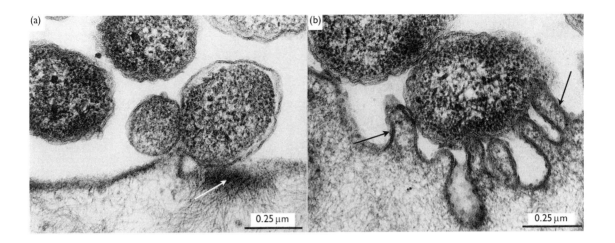

are non-shedding and, consequently, huge numbers of bacteria can build up on the tooth surface, where they form complex communities known as biofilms (Figure 7.2; see also Section 2.7). These biofilms (known as dental plaques), if allowed to accumulate, can induce diseases such as dental caries, gingivitis and periodontitis. These oral diseases are unique in that they are the only infectious diseases requiring daily prophylactic measures, i.e. brushing and flossing to remove the biofilms. Although the teeth are the only naturally occurring non-shedding surfaces, modern medical and surgical practice make great use of prosthetic devices such as catheters, artificial joints and heart valves, hence creating additional non-shedding surfaces for bacterial adhesion and, inevitably, biofilm formation (Figure 7.3). Biofilm formation on such devices and subsequent bacteraemia is one of the major problems associated with the use of such devices, as biofilms display a marked resistance to antibiotics (detailed in Section 2.7) and, in most cases, the only recourse is to remove the contaminated device and replace it. This, of course, is a great inconvenience to the patient and a drain on health service resources.

Figure 7.1 Transmission electron micrographs showing adhesion of *Haemophilus influenzae* to human oropharyngeal cells. In (a) an electron-dense region (arrowed) in the epithelial cell beneath the point of attachment of the bacterium can be seen. This region contains a dense assembly of cytoskeletal filaments. Some of the epithelial cells have produced microvillus-like protrusions (arrowed) that can be seen in (b). (Reproduced with permission from Holmes & Bakaletz, 1997. See Section 7.11.)

7.2.2 Adhesion to internal surfaces

In some diseases, adhesion of a bacterium to a tissue is followed by invasion of the tissue by that bacterium. The bacterium may then grow and reproduce within the tissue, possibly exiting from it, as in the case of some *Shigella* spp., or it may be disseminated to other sites within the host. In the latter case, there will be a need for the organism to adhere to cells of other tissues, such as endothelial cells and the many specialised cells constituting the various organs of the body, or to any of a number of polymers comprising the extracellular matrix (ECM) (proteoglycans, collagen, etc.) of connective tissues (Figure 7.4). Adhesion to, and invasion of, endothelial cells of blood vessels by bacteria are essential for their dissemination throughout the body via the bloodstream and occurs in diseases such as meningitis,

Figure 7.2 (a) Scanning electron micrograph of dental plaque, magnification 4260×. (b) Transmission electron micrograph of a cross-section through dental plaque, magnification 2900×. Although both electron micrographs show large numbers of tightly packed bacteria, the processing involved in electron microscopy destroys the structure of the biofilm so that the extracellular matrix and other features characteristic of biofilms cannot be seen. (Courtesy of Ms N. Mordan, Electron Microscopy Unit, Department of Biomaterials, Eastman Dental institute, University College London.)

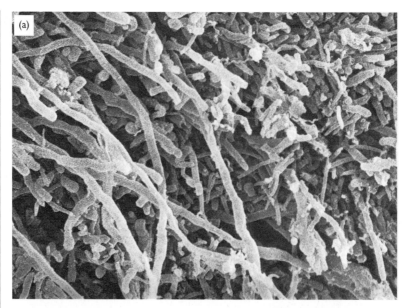

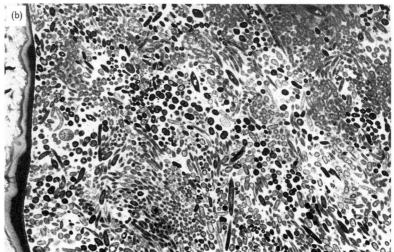

typhoid fever, brucellosis and endocarditis. In some cases, bacteria (e.g. *Yersinia pestis*, *Mycobacterium tuberculosis*, *Rickettsia typhi*) can reach other parts of the body via the lymphatic system, in which case adhesion to, and invasion of, endothelial cells of lymphatic vessels takes place.

7.2.3 Adhesion under the microscope – to what do bacteria really adhere?

Although Section 7.2.2 has described the range of host surfaces, both natural and artificial, on which bacteria can be found, we need to take a closer look at the interaction between the bacterial cell and the substratum to really understand what is going on. First of all, do bacteria

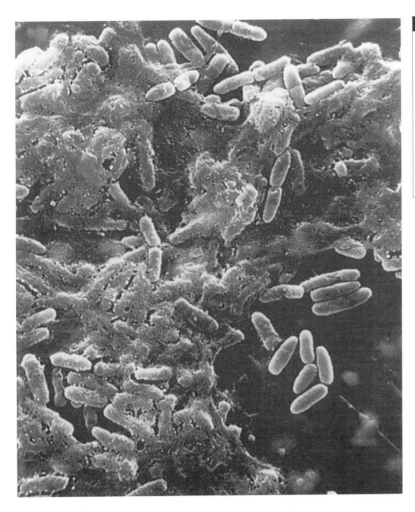

Figure 7.3 Scanning electron micrograph of a biofilm on an intravascular catheter, magnification 5660×. Micro-colonies of *Pseudomonas aeruginosa* are present, some of which are covered by extracellular matrix material. Processing of the specimen has destroyed the structure of the biofilm. Reproduced with permission from Yassien *et al.*, 1995. (See Section 7.11.)

really adhere to cells of the tissues mentioned above? The response, as is the case with many other questions in biology, has to be yes and no. While it is certainly true that bacteria do adhere directly to cells (by mechanisms to be described below), in many cases the external layers of the body are, in fact, coated by host secretions. All of the mucosal surfaces (as implied by their name) are, of course, bathed in mucus, i.e. an aqueous solution containing glycoproteins and various salts. Bacteria arriving at a mucosal surface, therefore, cannot initially interact with the epithelial cells of the mucosal surface because the first thing with which they come into contact is the mucous secretion (Figure 7.5). Entrapment of bacteria in this secretion, followed by expulsion of the fluid (by the mucociliary escalator in the case of the respiratory tract) is a major host defence mechanism (see Sections 1.2.3 and 5.4.1). Organisms that can maintain themselves on mucosal surfaces must adhere quickly to the epithelial cells themselves before they are expelled (Figure 7.6).

As for non-shedding surfaces, bacteria will rarely, if ever, come into direct contact with either enamel (in the case of teeth) or the material

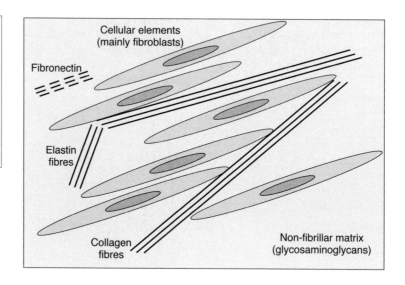

Figure 7.4 Diagram showing the main constituents of connective tissue. Cells are scattered throughout a gel-like extracellular matrix. A variety of fibres may be present (depending on the type of connective tissue) as well as blood vessels. The main functions of such tissues are protection and support.

Cellular elements (mainly fibroblasts)

Fibronectin

Elastin fibres

Collagen fibres

Non-fibrillar matrix (glycosaminoglycans)

from which a prosthetic device is fabricated. This is because in both cases the surface is always covered in a layer of adsorbed molecules (known as a 'conditioning film') whose nature will depend on the anatomical location (Figure 7.6). Teeth and oral prostheses (e.g. dentures) are covered in glycoproteins derived from saliva, while prosthetic devices in contact with blood (e.g. prosthetic heart valves) have a layer of adsorbed plasma proteins. When we talk about bacterial adhesion to these surfaces, therefore, what is really involved is bacterial adhesion to the conditioning film. Furthermore, as these surfaces are non-shedding, bacteria can accumulate in large numbers and can provide new surfaces to which other bacteria (of the same, or different, species) can adhere. In this way, a dense biofilm can be produced containing species that may well have not been able to adhere to the original surface. A similar phenomenon, i.e. adhesion of an organism to an already adherent organism, may also, of course, take place on a shedding surface but the dense bacterial aggregates characteristic of biofilms rarely accumulate because the epithelial cells are shed. The exception to this generalisation is the vaginal mucosa, which does, in fact, support an appreciable biofilm composed mainly of lactobacilli.

When it comes to internal body surfaces, as well as adhering directly to cells of various tissues, bacteria can also adhere to those polymers that comprise the ECM. These include collagen, fibronectin, fibrinogen, glycosaminoglycans, heparins, etc.

7.3 | Mechanisms involved in bacterial adhesion

7.3.1 Pre-adhesion events

Although a wide variety of surfaces can function as a substratum for bacterial adhesion, the molecular basis of the adhesion process is

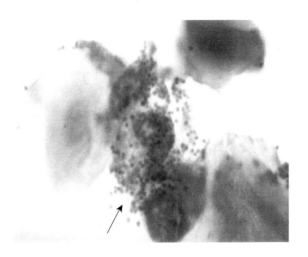

Figure 7.5 Adhesion of *Staphylococcus aureus* to mucin. Epithelial cells were scraped from the nasal cavities of volunteers, incubated with *Staph. aureus* and then stained with mucicarmine. The mucicarmine stains mucin but not epithelial cells. The bacteria can be seen to adhere to the mucin (arrowed) but not to the epithelial cells. Magnification 27 350×. (Reproduced with permission from Shuter *et al.*, 1996. See Section 7.11.)

similar in all cases. Let us consider a bacterium (remember that a 'typical' bacterium has a diameter of approximately 1 μm) approaching the surface of a substratum – which may be a host cell, another bacterium or an inanimate material (Figure 7.7). When the distance between them is of the order of tens of nanometres the two objects are affected by van der Waals and electrostatic forces. At a distance of >50 nm, only van der Waals forces operate: these are the result of the induction of dipoles in the two objects causing their mutual attraction. As the distance between the objects decreases (10–20 nm), electrostatic forces become significant and, as most bacteria and surfaces have a negative charge, the net effect is repulsion. However, these repulsive forces decrease with increasing ionic strength and in most natural environments the ionic strength is sufficient to reduce or overcome this repulsion. As the bacterium approaches the surface more closely, intervening water molecules act as a barrier to attachment. However, hydrophobic molecules on the surface of the bacterium, the substratum or both can exclude these water molecules. Hydrophobic

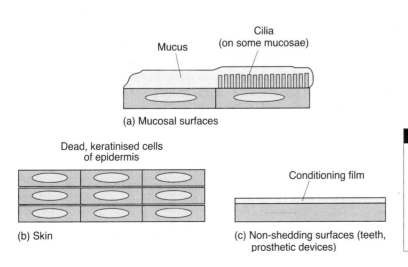

(a) Mucosal surfaces

(b) Skin

(c) Non-shedding surfaces (teeth, prosthetic devices)

Figure 7.6 The three major types of surfaces with which bacteria frequently come into contact: (a) moist, shedding mucosal surfaces (b) the dry, shedding surfaces of the skin; and (c) non-shedding surfaces such as teeth and prosthetic devices.

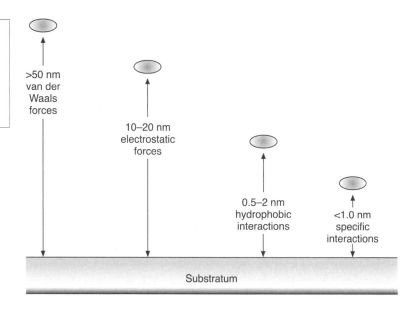

Figure 7.7 Physicochemical forces involved in bacterial adhesion. As indicated in the diagram, these operate at different distances between the bacterium and the substratum to which it adheres.

interactions between the bacterium and the substratum can then result in adhesion or can enable a close enough approach (< 1.0 nm) for other adhesive interactions to occur. These include hydrogen bonding, cation bridging and receptor–ligand interactions, i.e. the specific binding of a molecule (ligand) on the bacterial surface to a complementary substrate molecule (receptor) on the other surface. Bacterial adhesion to host cells is thought to be mediated primarily by hydrophobic interactions, cation bridging and receptor–ligand binding (Figure 7.8). While hydrophobic interactions are recognised as being important in the adhesion of bacteria to host cells and to inanimate substrata, more is known about the role of receptor–ligand interactions. The actual molecules on the bacterial surface responsible for adhesion are known as adhesins.

7.3.2 Adhesive interactions

As mentioned above, adhesion between bacteria and host cells is mediated primarily by hydrophobic, ion-bridging and receptor–ligand interactions. These will now be described in more detail. However, it must be borne in mind that, in many cases, the nature of the adhesive interaction has not been determined, even though it may have been established that a particular bacterial adhesin is involved.

7.3.2.1 Hydrophobic interactions

When non-polar molecules on the bacterium and the substratum are in close proximity, the intervening ordered layers of water are displaced and the consequent increase in entropy results in adhesion being energetically favourable. In contrast to receptor–ligand interactions (see Section 7.3.2.3), hydrophobic bonding is usually considered to be non-specific, as there are no apparent stereospecific interactions between

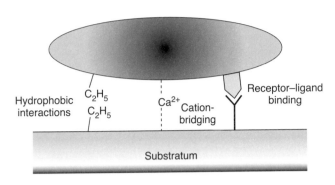

the molecules involved. However, this may, of course, simply reflect our lack of knowledge of the identity of the molecules involved and the nature of their physicochemical interactions. The surface components responsible for the interactions are known as hydrophobins and include hydrocarbons, aromatic amino acids, fatty acids and mycolic acids. It is worth noting that approximately 55% of the accessible surface of an average protein is non-polar. Even in carbohydrates, that are generally considered to be highly polar, hydrophobic regions do exist. For example, galactose contains six adjacent carbon atoms that form a continuous hydrophobic region, and when polysaccharides containing this sugar interact with proteins the hydrophobic regions are invariably packed against aromatic side chains in the protein.

7.3.2.2 Cation-bridging

The surfaces of bacteria, host cells and inanimate materials invariably carry a net negative charge in an aqueous environment. The mutual repulsion between these negatively charged surfaces can be counteracted by divalent metal ions such as calcium ions, which thereby act as a bridge between the two. Such interactions are thought to be important in adhesion of oral bacteria to tooth surfaces, in co-aggregation between similar and different bacterial species and, possibly, between bacteria and negatively charged molecules on the surfaces of host cells.

7.3.2.3 Receptor–ligand binding

A molecule (receptor) on the surface of a host cell (or on a molecule attached to an inanimate surface) can 'recognise' a molecule (ligand) on the bacterial surface with a complementary structure and form a strong (but non-covalent) bond. The process is similar to the 'lock-and-key' relationship between the catalytic site of an enzyme and its substrate. The binding usually involves only a portion of each of the molecules involved and the molecular structures responsible are known as epitopes (or adhesiotope in the case of the epitope on the receptor molecule). A variety of molecules can function as receptors or ligands, including proteins, polysaccharides, glycoproteins and glycolipids. When recognition involves the interaction between a protein and a carbohydrate epitope (which may be on a glycoprotein, glycolipid, etc.) the interaction is known as 'lectin binding' (Figure 7.9).

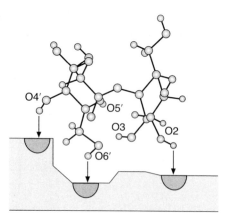

Figure 7.9 Interaction between the 18 kDa adhesin of *Streptococcus suis* and its Galα(1–4)Gal receptor – an example of lectin binding. The hydrogen bonds formed between the disaccharide and the adhesin on the bacterial surface that are responsible for maintaining adhesion are designated by the arrows.

Although the interaction between a receptor and its complementary ligand is highly specific, it must be remembered that other, often unrelated, molecules may also have the same epitope and are, therefore, able to block adhesion. This could form the basis of a new approach to preventing infectious diseases, as described in Section 7.6 (see also Section 11.3.1).

7.3.3 Bacterial structures involved in adhesion

Although it is apparent that a number, indeed most, of the surface structures of bacteria described in Chapter 2 play a role in adhesion, less is known about the molecular basis of the interaction between the structure involved and the target surface. Most attention has been directed at trying to define important receptor–ligand interactions between the bacterium and the surface to which it adheres and in identifying the bacterial adhesin and its complementary receptor responsible for adhesion.

7.3.3.1 What bacterial structures are involved in adhesion?

It can be said with confidence that every known structure identified on the bacterial surface has been shown to function in the adhesion of some bacterial species to some surface. Many of these structures, however, are multifunctional and are involved in processes other than adhesion. Examples are given in Table 7.2. Nevertheless, certain structures produced by bacteria appear to be designed solely for the purpose of mediating adhesion to a surface, whether this be cellular or acellular (e.g. a tooth or prosthetic device); these are known as fimbriae or pili (see Section 2.2.5). Another specialised type of adhesive organelle found in a more limited range of organisms (*Escherichia coli* and *Salmonella* spp.) are the curli – coiled surface structures that mediate attachment to epithelial cells and to certain host proteins such as fibronectin and plasminogen (Figure 7.10). In addition, some organisms produce specialised structures involved in adhesion (and may have other functions) only under certain circumstances. For example, within 15

Table 7.2. Surface structures of bacteria involved in adhesion

Structure	Bacterium	Surface/molecule to which the structure adheres
S layer	*Aeromonas salmonicida*	Fibronectin, laminin, vitronectin
	Lactobacillus acidophilus	Intestinal epithelial cells
Capsule	*Klebsiella pneumoniae*	Mucus
	Staphylococcus epidermidis	Various biomaterials
	Streptococcus pyogenes	Keratinocytes
	Bacteroides fragilis	Intestinal epithelial cells
Fimbriae	*Escherichia coli*	Intestinal and urinary epithelial cells
	Salmonella typhimurium	Intestinal epithelial cells
	Vibrio cholerae	Intestinal epithelial cells
	Bordetella pertussis	Laryngeal epithelial cells
	Actinomyces spp.	Bacteria (e.g. *Streptococcus* spp.)
	Pseudomonas aeruginosa	Epithelial cells of respiratory tract
Flagella	*Pseudomonas aeruginosa*	Epithelial cells
Cell wall	*Streptococcus pyogenes*	Epithelial cells
	Staphylococcus aureus	Epithelial cells
Fibrils	*Haemophilus influenzae*	Epithelial cells
	Streptococcus sanguis	Blood platelets, salivary proteins
Curli	*Escherichia coli*	Fibronectin, laminin, plasminogen, intestinal epithelial cells
	Salmonella spp.	Fibronectin, laminin, plasminogen, intestinal epithelial cells

minutes of contacting an epithelial cell, *Salmonella typhimurium* produces surface appendages (known as invasomes) that are very different from flagella or pili, being almost three times thicker than the former and only one tenth of their length (Figure 7.11). These invasomes are thought to be involved in maintaining adhesion of the organism to the host cell and are also an essential prerequisite for cellular invasion; they disappear once internalisation has commenced. Bacteria that infect mucosal surfaces often form discrete microcolonies and this is particularly evident in the case of enteropathogenic strains of *E. coli*. This phenomenon is known as localised adherence and is due to the production of bundles of filaments (bundle-forming pili, bfp) by the bacteria once they have adhered to the epithelial cells (Figure 7.12). The bfp form a network that envelopes neighbouring bacteria and this results in the formation of a microcolony. The formation of microcolonies could be a bacterial defence mechanism, as these structures are known to be less susceptible to host defence mechanisms and to antibiotics.

Figure 7.10 Transmission electron micrograph of *Escherichia coli* showing curli (arrowed). The bacteria were probed with an antibody against the main protein subunit of the curli (CsgA) and then labelled with protein A–gold particles (10 nm in diameter). The particles bind (via protein A) to the anti-CsgA antibodies, which are themselves bound to the CsgA protein of the curli. Magnification 490 000×. (Reproduced with permission from Bian *et al.*, 2000. Copyright 2000 University of Chicago Press.) See Section 7.11.

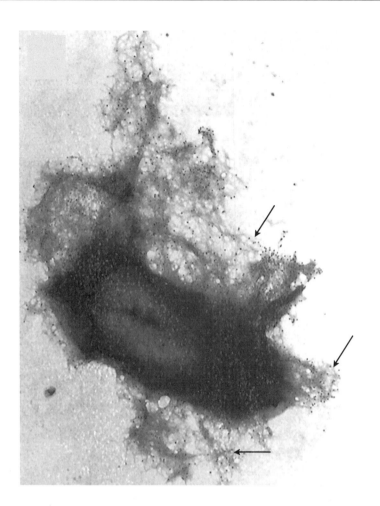

Sticky fingers – the mycoplasma adherence organelle

Mycoplasma spp. are frequent inhabitants of many mucosal surfaces and some members of the genus are responsible for major diseases, e.g. *Myc. pneumoniae* (pneumonia), *Myc. hominis* (genitourinary infections) and *Myc genitalium* (urethritis). One of the most distinguishing features of these bacteria is that they do not possess a cell wall and so are devoid of the usual adhesive structures used by most bacteria. Their outermost layer is, like host cells, the cytoplasmic membrane and it has recently been shown that this can be used to produce a distinct adhesion organelle. This consists of a characteristically tapered extension of the membrane that has a button-shaped terminus and appears to consist of tightly packed cytoskeletal proteins (Figure 7.13). Although little is known about these proteins, at least one of them (high molecular weight protein-2) has regions exhibiting homology to the filamentous coiled-coil portion of the myosin type II heavy chain found in muscle microfilaments! Two of the surface proteins found in the organelle, P1 and P30, are thought to function as adhesins.

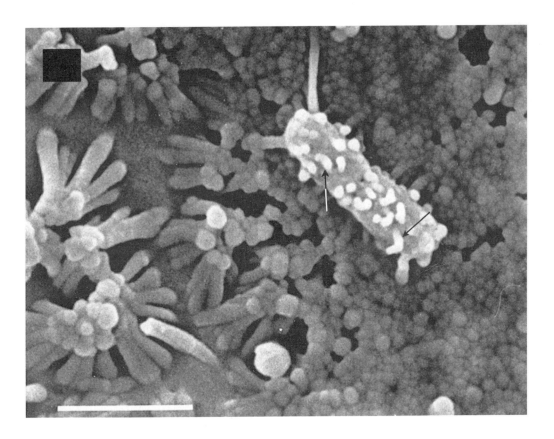

Figure 7.11 Scanning electron micrograph showing short appendages (invasomes, arrowed) on the surface of *Salmonella typhimurium* adhering to the microvilli of an epithelial cell. Bar = 2 μm. (Reproduced with permission from Reed *et al.*, 1998. See Section 7.11.)

7.3.3.2 What is the nature of the adhesins on the structures involved in adhesion?

A wide variety of molecules present on adherent structures of bacteria can function as adhesins (Table 7.3). The molecular identity of many of these adhesins, as well as the corresponding receptors for them, have been identified (see below). In many cases, the adhesins are located at the end of structures (fimbriae, fibrils, flagella) that protrude relatively long distances from the bacterial cell wall. This would be advantageous to such organisms in that adhesion could be achieved at a separation distance greater than that at which repulsion, due to negative charge, takes effect.

A recurring theme in bacteriology is the multifunctionality of bacterial macromolecules and this is also evident in the case of bacterial adhesins. One striking example is that of the enzyme glyceraldehyde-3-phosphate dehydrogenase (GAPDH). This is a key enzyme of the glycolytic pathway, where it catalyses the phosphorylation of glyceraldehyde 3-phosphate to 1,3-diphosphoglycerate and, until recently, this was thought to be its main function. However, GAPDH is also found on the surface of a number of bacteria including *Staphylococcus aureus*, *Staph. epidermidis* and *Streptococcus pyogenes*. This surface-located GAPDH

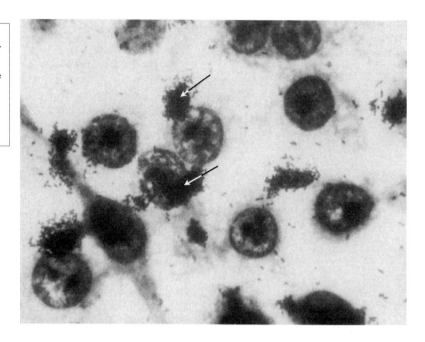

Figure 7.12 Transmission electron micrograph of an entero-pathogenic strain of *Escherichia coli* displaying localised adherence (arrows) to human intestinal epithelial cells. (Reproduced with permission from Rodrigues *et al.*, 1996. See Section 7.11.)

not only functions as an adhesin (being able to bind to a range of host extracellular matrix molecules) but is also involved in the acquisition of transferrin-bound iron and in bacteria–host signalling (see Section 4.6.1.3).

From Tables 7.2 and 7.3 it can be seen that a particular organism often posesses a number of adhesins, although these are unlikely to be expressed at the same time. This is particularly the case with invasive species (e.g. *Neisseria meningitidis*, *Sal. typhi*, *Strep. pyogenes*) as such organisms will encounter a variety of surfaces at different points during the initiation and development of the disease process. Taking *N. meningitidis* as an example, this organism will have to interact with the following substrata during the course of a typical episode of meningitis: mucus, nasopharyngeal epithelial cells, the ECM and endothelial cells. An organism that causes osteomyelitis (e.g. *Staph. aureus*) will, in addition, also have to interact with the mineralised tissue, bone. Expression of the genes encoding these various adhesins must be under tight control (necessitating the organism to have a detailed 'knowledge' of its environment) so that a particular adhesin is produced only at the appropriate point during the journey of the organism from its initial site of colonisation to its ultimate target organ.

The fact that bacteria have to express a particular adhesin on their surface in order to adhere to a host receptor renders them somewhat vulnerable to host defence mechanisms. Hence, the host will eventually produce secretory IgA antibodies that can bind to the adhesin and so prevent adhesion. Some organisms have evolved ways of combating this defence mechanism by undergoing antigenic or phase variation (see Sections 2.6.2 and 10.5.1.2) of their adhesive structures, so enabling them to remain one step ahead of the host for much of the time.

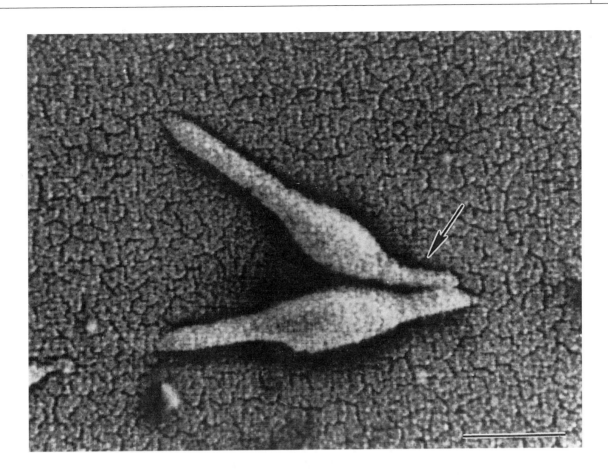

A taste of honey – the glycosylation of pili of *Neisseria* spp.

Until comparatively recently, bacteria, in contrast to eukaryotic organisms, were thought to be unable to glycosylate proteins. Increasing reports of glycoproteins in bacteria during the last few years have, however, shown this view to be untenable. For example, the pilin proteins comprising the pili of the important human pathogen *N. meningitidis* have been shown to be glycosylated, the carbohydrate substituent being the trisaccharide Gal(β1–4)Gal(α1–3)2,4-diacetimido-2,4,6-trideoxyhexose. This organism, which is responsible for a life-threatening meningitis, can be found in the nasopharynx of up to 10% of the normal population and up to 20% during a meningitis epidemic. Under certain circumstances, it can invade the nasopharynx and be carried via the bloodstream to the meninges. The pili of the organism have long been known to be important in effecting adhesion to both epithelial and endothelial cells and the adhesins responsible are the pilin glycoproteins of these structures. Recent evidence suggests that the epithelial cell receptor for these adhesins is membrane cofactor protein (CD46), a member of the regulators of complement activation (RCA) family of proteins (described in more detail in Section 10.4.1.3). The pilin proteins of a closely related human pathogen, *N. gonorrhoeae*, the causative agent of gonorrhoea, are also known to be glycosylated. In this case the carbohydrate is the disaccharide galactose α(1–3)*N*-acetylglucosamine.

Figure 7.13 Scanning electron micrograph of *Mycoplasma pneumoniae* attached to a glass cover slip. The attachment organelle (arrowed) is long and tapering. Bar = 0.5 μm. (Reprinted from *Trends in Microbiology* **6**, Krause, D. C., *Mycoplasma pneumoniae* cytadherence: organization and assembly of the attachment organelle, 15–18, 1998, with permission from Elsevier Science.)

Table 7.3. | Molecular nature of bacterial adhesins

Adhesin	Location	Organism	Substratum
Lipoteichoic acid	Cell wall	*Staphylococcus aureus*	Epithelial cells
		Streptococcus pyogenes	Buccal epithelial cells
		Lactobacillus spp.	Intestinal epithelial cells
		Streptococcus pneumoniae	Erythrocytes
Protein	Cell wall	*Streptococcus pyogenes*	Epithelial cells
Protein	Fimbriae	*Streptococcus parasanguis*	Salivary proteins
		Salmonella typhimurium	Intestinal epithelial cells
		Actinomyces spp.	Streptococci
		Escherichia coli	Epithelial cells
		Pseudomonas aeruginosa	Respiratory epithelial cells
Protein	Fibrils	*Streptococcus gordonii*	Fibronectin, other bacteria
Protein	Outer membrane	*Salmonella typhimurium*	Macrophages
		Neisseria spp.	Epithelial and endothelial cells
		Campylobacter jejuni	Epithelial cells
		Vibrio cholerae	Epithelial cells
		Haemophilus influenzae	Mucin
		Escherichia coli	Intestinal epithelial cells
Polysaccharides	Capsule	*Klebsiella pneumoniae*	Mucin
		Pseudomonas aeruginosa	Respiratory epithelial cells
		Escherichia coli	Biomaterials
		Staphylococcus epidermidis	Biomaterials
		Bacteroides fragilis	Intestinal epithelial cells
		Streptococcus pyogenes	Epidermal keratinocytes
			Pharyngeal epithelium
		Klebsiella pneumoniae	Uroepithelial cells
Lipopolysaccharide	Cell wall	*Helicobacter pylori*	Gastric epithelial cells
		Salmonella typhimurium	Macrophages
		Escherichia coli	Epithelial cells
		Aeromonas hydrophila	Epithelial cells
		Pseudomonas aeruginosa	Corneal epithelial cells
Enzymes	Cell wall	*Porphyromonas gingivalis*	Epithelial cells, bacteria
		Haemophilus influenzae	Epithelial cells
		Streptococcus mutans	Salivary proteins, bacteria
		Streptococcus pyogenes	Extracellular matrix proteins

7.3.4 Host molecules functioning as receptors

As mentioned previously, humans provide a number of different types
of surface for bacterial attachment. However, regardless of whether the
surface is cellular, acellular (e.g. ECM), mineralised (e.g. teeth or bone)
or inanimate (e.g. a prosthetic device), adhesion is mediated, at least in

part, by receptor–ligand binding. A considerable amount of research has been undertaken into determining the nature of those receptor(s) on host cells that are involved in bacterial adhesion. Figure 7.14 shows, in a very simplified fashion, the major components of the surface (i.e. the cytoplasmic membrane) of a typical host cell. This has the classic lipid bilayer structure in which proteins are embedded. The major lipids present are phosphatidylcholine, phosphatidylserine, phosphatidylinositol, phosphatidylethanolamine, sphingomyelin, cholesterol and various glycolipids. Several of these contain molecular structures (i.e. epitopes) that function as binding sites for bacterial adhesins. Although outnumbered by lipid molecules, proteins are responsible for most membrane functions including: (i) transport; (ii) recognition and binding of hormones, cytokines and ECM molecules; and (iii) signal transduction. Certain amino acid sequences of proteins, and the carbohydrates of glycoproteins, function as important receptors for bacterial adhesins. Mammalian cells also have receptors for a range of signalling, defensive and structural molecules such as hormones, immunoglobulins, cytokines and ECM molecules. Certain bacteria have evolved to take advantage of the host cell's requirement for such receptors and have produced adhesins that can bind to these receptors, thereby usurping their natural function. Host cells, of course, rarely exist in isolation and are often embedded in a complex mixture of polymers (e.g. fibronectin, fibrin, collagen, proteoglycans) that constitute the ECM. Not only does the ECM have a structural function but, as its components are linked to cell surface receptors (e.g. integrins see Section 4.2.5), it also affects a number of cellular activities including migration, proliferation and differentiation. Bacteria can adhere to the ECM and also to the host cell receptors for these ECM molecules (Table 7.4). For bacterial pathogens that adopt an extracellular lifestyle within the host (e.g. *Staph. aureus*), adhesion to ECM molecules is an important feature of their existence. The adhesins responsible are termed **microbial surface components recognising adhesive matrix molecules** (MSCRAMMs). Binding to integrins is used by a wide range of microbes (e.g. *Yersinia* spp., *Bordetella pertussis*, *Borrelia burgdorferi*, *Leishmania mexicana*, *Histoplasma capsulatum*, echovirus 1 and adenovirus) to enable colonisation, and sometimes invasion, of host cells. Another class of host cell receptors to which bacteria can adhere are the **carcinoembryonic-antigen-related cellular adhesion molecule** (CEACAM) receptors. These molecules are found on a wide range of host cells, including epithelial and endothelial cells, and are involved in cell–cell adhesion and cellular differentiation. They appear also to function as receptors for an important group of adhesins (the Opa proteins) of *N. gonorrhoeae* and *N. meningitis*.

In summary, therefore, bacteria can adhere to host cells in three ways: (i) directly to the lipid bilayer, (ii) directly to cell surface receptors for host molecules, or (iii) indirectly to host molecules already bound to the host cell surface. Examples of the receptors for some bacterial adhesins are shown in Table 7.5.

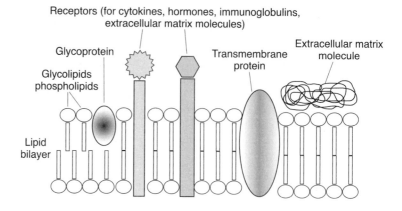

Figure 7.14 Diagram showing the major types of molecule to be found on the surface of a typical mammalian cell. Most of these can function as receptors for bacterial adhesins.

One very important, and fascinating, exception to the concept that adhesion is mediated by a bacterial adhesin binding to a receptor molecule provided by the host is enteropathogenic *E. coli* (EPEC). This organism actually carries around its own receptor, which it injects into its target cell! Adhesion of EPEC to host cells will be described in greater detail in Section 7.5.2.1.2.1.

Table 7.4. | Adhesion of bacteria to extracellular matrix molecules

Extracellular matrix molecule	Organism
Collagen	*Staphylococcus aureus, Streptococcus pyogenes*
Elastin	*Staphylococcus aureus,* CNS
Laminin	*Staphylococcus aureus, Streptococcus pyogenes, Streptococcus sanguis, Pseudomonas aeruginosa, Helicobacter pylori*
Fibrinogen	*Staphylococcus aureus, Streptococcus pyogenes,* other streptococci
Fibronectin	*Staphylococcus aureus,* CNS, *Streptococcus pyogenes, Streptococcus pneumoniae, Streptococcus sanguis, Escherichia coli, Treponema pallidum*
Vitronectin	*Staphylococcus aureus,* CNS, *Streptococcus pyogenes, Escherichia coli, Helicobacter pylori*
Thrombospondin	*Staphylococcus aureus, Streptococcus pyogenes*
Heparan sulphate	*Staphylococcus aureus, Listeria monocytogenes, Helicobacter pylori*
Heparin	*Staphylococcus aureus, Mycobacterium tuberculosis, Helicobacter pylori, Listeria monocytogenes, Neisseria gonorrhoeae*
Hyaluronan	*Treponema denticola*

CNS, coagulase-negative staphylococci.

Wherever I leave my hat, that's my home – the intimin receptor of *E. coli*

Although it has been known for a number of years that an outer membrane protein of EPEC, known as intimin, functioned as an important adhesin in the interaction of this organism with intestinal epithelial cells, the identity of the complementary receptor remained uncertain. Initially, evidence suggested that the receptor was a 90 kDa host protein on the epithelial cell surface. However, recent work has revealed the astonishing fact that the receptor is actually a bacterial protein (translocated intimin receptor, Tir) that is injected into the host cell following the activation of a type III protein secretion system once the bacterium has adhered. Tir is secreted by EPEC as a 78 kDa protein which then undergoes phosphorylation on one or more of its tyrosine residues within the host cell. The resulting protein (with an apparent molecular mass of 90 kDa) then becomes incorporated within the host cell membrane, where it interacts with intimin, so enabling adhesion of the bacterium to the epithelial cell. Binding of intimin to Tir then apparently triggers a whole series of events resulting in cytoskeletal rearrangements that eventually lead to pedestal formation.

If EPEC 'carries around' its own receptor why then cannot it adhere to any cell? Why does it exhibit tropism for only certain tissues of the bowel? It is possible that the type III secretion system responsible for injecting Tir into the host cell is activated only by adhesion to certain types of host cell. Also, it may be that other, tissue-specific, adhesins are involved in ensuring initial adhesion of EPEC to the target cell prior to activation of the type III secretory machinery.

7.4 | Tissue tropism

It has long been recognised that a particular bacterial species will tend to adhere to, colonise, and possibly induce pathology in, only one type of tissue. This phenomenon is referred to as tissue tropism and has been invaluable in helping clinicians to identify the causative agent of an infectious disease. Hence, the most frequent bacterial cause of tonsillitis, for example, is *Strep. pyogenes* – less commonly other bacteria may be involved, including *Corynebacterium* spp. and spirochaetes. Otitis media, an ear infection very common in young children, is invariably due to *Strep. pneumoniae*, *Haemophilus influenzae* or *Strep. pyogenes*. The ability of an organism to colonise a particular type of tissue stems from the operation of two major selective forces: adhesion and environment. We have seen that specific ligand–receptor interactions are the basis of bacterial adhesion, so it is easy to appreciate that the absence in a tissue of an appropriate receptor to which a bacterium can bind will result in the failure of that organism to colonise the tissue. However, although ligand–receptor interactions are undoubtedly specific, some receptor epitopes are present in a variety of tissues, hence raising the possibility that bacteria could colonise a number of tissues. This, of

Table 7.5. Bacterial adhesins and their receptors

Adhesin	Organism	Receptor
Opc – an outer membrane protein	Neisseria meningitidis	$\alpha_v\beta_3$ integrin on endothelial cells, extracellular matrix proteins
Opa proteins (outer membrane proteins)	Neisseria meningitidis, Neisseria gonorrhoeae	CEACAM receptors on epithelial and endothelial cells
Invasin – an outer membrane protein	Yersinia enterocolitica	$\alpha_5\beta_1$-integrins
25 kDa protein	Helicobacter pylori	N-acetylneuraminyl-α(2–3)lactose residues in laminin
16 kDa protein	Streptococcus mutans	Collagen
P2 – an outer membrane protein	Haemophilus influenzae	Sialic acid-containing oligosaccharides of mucin
92 kDa protein	Staphylococcus aureus	Fibrinogen
Glyceraldehyde-3-phosphate dehydrogenase	Streptococcus pyogenes	Fibronectin, myosin, actin
Hyaluronic acid	Streptococcus pyogenes	CD44 (a hyaluronic acid-binding protein) on epithelial cells
Arginine-specific gingipain	Porphyromonas gingivalis	Fibrinogen
Filamentous haemagglutinin	Bordetella pertussis	Galactose N-acetyl glucosamine residues in glycolipids of host cells
Protein I/IIf	Streptococcus mutans	N-acetylneuraminic acid residues of host cells
LPS	Pseudomonas aeruginosa	Galectin-3 protein in corneal epithelial cells
FimHS – a 35 kDa fimbrial protein	Salmonella typhimurium	Mannosylated proteins on epithelial cells
BabA protein	Helicobacter pylori	Lewis b blood group antigen
Fimbrial protein	Escherichia coli	Galactose disaccharide of glycolipids
PAK protein	Pseudomonas aeruginosa	β-D-GalNAc(1–4)β-D-Gal residues of gangliosides of epithelial cells
35 kDa fimbrial protein	Enterobacter cloacae	Man$_9$(GlcNAc)$_2$-tyrosinamide residues in epithelial cells
46 kDa protein	Haemophilus influenzae	Phosphatidylethanolamine in epithelial cells
Pilus protein	Burkholderia cepacia	Cytokeratin-13 of buccal epithelial cells

Table 7.6.	Examples of tissue tropism
Organism	Tissue
Neisseria meningitidis	Nasopharyngeal epithelium
Neisseria gonorrhoeae	Urethral epithelium
Vibrio cholerae	Intestinal epithelium
Bordetella pertussis	Respiratory epithelium
Salmonella typhimurium	Intestinal epithelium
Helicobacter pylori	Gastric mucosa
Streptococcus pyogenes	Pharyngeal epithelium
Campylobacter jejuni	Intestinal epithelium
Mycoplasma pneumoniae	Respiratory epithelium

course, certainly happens to some extent. However, environmental factors then come into play to counteract the colonisation of a wide range of different tissues by a large number of different species. The environment of the skin (high salt content due to evaporation of sweat, quite dry, low pH due to fatty acids), for example, differs considerably from that of the nasopharynx (low salt content, very moist, neutral pH) enabling the survival of only those organisms adapted to such environments regardless of the presence of appropriate receptors for bacterial adhesins. Examples of tissue tropism are given in Table 7.6.

7.5 | Consequences of bacterial adhesion

So far we have considered the means by which bacteria adhere to different substrata provided by the host. Is that then the end of the story? As alluded to at the beginning of the chapter, the answer is no: there is far more to bacterial adhesion than the 'sticking' aspect. It is becoming increasingly obvious that the act of adhering to a substratum may induce profound changes in the adherent organism, regardless of the nature of the substratum. Such changes have been more extensively investigated (because it is technically easier) when the substratum is inanimate. Hence, it has been known for many years that adhesion of an organism to such a substratum profoundly affects its rate of growth, carbohydrate utilisation, protein synthesis and energy generation as well as its cell wall composition and production of exopolymers. Evidence is accumulating to show that adhesion of a bacterium to host cells and to other substrata encountered in the host also affects bacteria in a number of ways. In comparison to what is known of the effects of adhesion on bacteria, an enormous amount of information is available concerning the effects on host cells subsequent to bacteria adhering to them. This is not surprising given the human's egocentric attitude — we have always been more interested in the consequences for 'us' than in the consequences for 'them'. It is apparent, nevertheless, that the host cell and its adherent bacterium exchange signals (i.e. they engage in molecular 'cross-talk'; see Section 4.4) which brings about changes in both cell types.

7.5.1 Effects on the bacterium

There are a number of reasons why we might expect a bacterium to adjust its phenotype once it has adhered to a substratum on or within its host. Firstly, prior to any interaction with its host, the bacterium is likely to have been exposed to environments (e.g. air, water, soil, vegetation, clothing) that may not have been conducive to its growth. Having reached an epithelial surface (which may not necessarily be its ultimate destination) the organism will have to adapt to living in this new environment (even though this may be only for a short time) and this will require the up-regulation and/or suppression of a number of gene products. Such changes will, undoubtedly, be induced in response to changes in a number of environmental factors (e.g. pH, osmolarity, nutrient concentration), but some are known to be triggered by the adhesion process itself (Table 7.7). Secondly, attachment to a host cell may be only the first step in a sequence of events that could involve invasion of the cell followed by dissemination of the organism to other sites within the host. Adhesion of the bacterium might be expected, therefore, to be used as a signal for the organism to begin the synthesis of those molecules essential to initiate invasion of the host cell. As shown in Table 7.7, there are many reports suggesting that bacteria do respond in this way to contact with the host cell.

Until recently, one dramatic change thought to be induced in bacteria solely by contact with host cells was the activation of a type III secretion system (see Section 2.5.5). A range of Gram-negative pathogens (e.g. *Salmonella* spp., *Yersinia* spp., *Shigella* spp., *E. coli*, *Pseudomonas* spp.) were thought to behave in this way and the secretion system has even been termed 'contact-dependent secretion'. However, while contact with host cells is certainly known to activate the system, it is now apparent that activation can be achieved without the necessity for such contact. Hence, in the case of *Salmonella* spp., the secretion system can be induced by a change in the pH of the culture medium from acidic to mildly alkaline, while type III secretion can occur in EPEC when the organism is transferred to tissue culture medium. Activation of type IV secretion systems (see Section 2.5.3.3), however, appears to be dependent on contact with host cells.

Come on baby light my fire – adhesion-induced gene expression in *Yersinia* spp.

A visually exciting demonstration of adhesion-induced changes in a bacterium is provided by an experiment that involved the secretion of YopE (a cytotoxin that depolymerises actin microfilaments in host cells) by *Y. pseudotuberculosis*. The promotor for the *yopE* gene was fused to a *luxAB* operon, the products of which catalyse the emission of photons at a wavelength of 490 nm in the presence of *n*-decanal. Activation of the *yopE* gene therefore results in light emission. *Yersinia pseudotuberculosis* containing this reporter system was added to epithelial cells and the bacteria were seen to light up on attachment to the epithelial cells. Un-attached bacteria failed to emit light. This experiment elegantly demonstrates the attachment-induced production of a key bacterial virulence factor.

7.5.2 Effects on the host

As mentioned previously, far more is known about the changes in host cells that are induced following attachment of bacteria. At a micro-scopic level, adhesion may induce no apparent change in the host cell (although this may conceal up- and/or down-regulation of many gene products) or the cell may undergo dramatic changes in morphol-ogy resulting in invasion by the bacteria, cell death or both (Figure 7.15). The outcome of this interaction between a bacterium and its cellular target depends very much on the nature of both the bacterium and the host cell. We will now give examples of the changes that can be induced by the adhesion of bacteria to four types of host cell with which bac-teria often come into contact – epithelial cells, fibroblasts, endothelial cells and phagocytic cells.

7.5.2.1 Epithelial cells

With only a few exceptions (the lower respiratory tract, bladder and the urethra) all of the epithelial surfaces of a healthy individual are colon-ised by bacteria. Given that all of these organisms contain molecules that are potentially harmful to mammalian cells (e.g. lipopolysacchar-ide (LPS), peptidoglycan, lipoteichoic acid (LTA)) it is remarkable that most adherent organisms have, in general, such minimal effects on epithelial cells. The normal residents of these mucosal surfaces have obviously co-evolved with humans to establish a commensal, or poss-ibly mutualistic, association. However, as summarised in Table 7.8, the situation can be very different when a disease-inducing organism adheres to an epithelial cell.

7.5.2.1.1 Adhesion without any apparent effect

Most interactions between bacteria and epithelial cells involve members of the normal microflora. Adhesion of such organisms to epithelial cells does not appear to induce any dramatic changes in

Table 7.7. Examples of bacterial responses following their adhesion to host cells

Organism	Effect
Neisseria gonorrhoeae	Stimulation of growth following attachment to epithelial cells
Escherichia coli	Stimulation of growth following attachment to epithelial cells
Escherichia coli	Expression of intimin and EspA is decreased following attachment to epithelial cells and formation of the A/E lesion
Escherichia coli	Inhibition of growth after adhering to uroepithelial cells
Campylobacter jejuni	Adhesion to epithelial cells induces the synthesis of 14 proteins required for cellular invasion
Escherichia coli	Attachment to epithelial cells induces the production of bundle-forming pili, which results in the formation of microcolonies. Contact also activates a type III secretion system (see Section 2.5.5)
Salmonella typhimurium	Attachment to epithelial cells induces the synthesis of structures (invasomes) required for invasion and induces a type III secretion system
Shigella flexneri	Induction of type III secretion system (see Section 2.5.5) following adhesion to epithelial cells or extracellular matrix polymers
Yersinia spp.	Induction of type III secretion system (see Section 2.5.5) following adhesion to epithelial cells or macrophages
Yersinia pseudotuberculosis	Increased rate of transcription of virulence genes
Escherichia coli	Up-regulation of siderophore synthesis on attachment to epithelial cells
Porphyromonas gingivalis	Adhesion to epithelial cells induces the secretion of five proteins and down-regulates the production of extracellular proteases
Helicobacter pylori	Adhesion to host cells activates a type IV secretion system resulting in the translocation of a protein (CagA) into the host cell (Section 2.5.3.3)
Neisseria meningitidis	Adhesion to host cells results in up-regulation of contact-regulated gene A (crgA). CrgA is a transcriptional regulator involved in the control of capsule and pilus expression

A/E, attaching and effacing.

the behaviour of either member of the association, which, apparently, continue their existence in a state of mutual indifference. However, this is a poorly studied area as researchers have found it far more exciting to investigate the dramatic changes induced by the adhesion of pathogenic bacteria to epithelial cells. This is unfortunate, as the ability of the normal microflora to adhere to, and live in harmony with, epithelial cells raises some very important points. Firstly, epithelial cells are known to secrete a range of antibacterial peptides (see Section 5.4.3.11), so how do members of the normal microflora avoid being killed by these? It may be that epithelial cells have 'designed' the

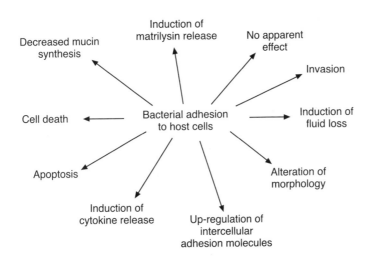

Figure 7.15 Bacterial adhesion to a host cell can induce one or more of a number of changes in the host cell, the most common of which are shown in the diagram.

antibiotics so that only pathogenic organisms are killed or perhaps members of the normal microflora can 'persuade' the host cells to down-regulate the production of these hostile chemicals. Secondly, members of the normal microflora contain a range of molecules (e.g. LPS, LTA, etc.) that can induce the release of pro-inflammatory

Table 7.8. Effects of bacterial adhesion on epithelial cells

Effect on epithelial cell	Examples
No apparent effect	Most members of the normal microflora, *Bordetella pertussis*, *Vibrio cholerae*
Altered morphology	Enteropathogenic *Escherichia coli*, *Treponema denticola*, *Streptococcus pyogenes*, *Helicobacter pylori*
Induction of cytokine release	Oral streptococci, uropathogenic *Escherichia coli*, *Helicobacter pylori*, *Streptococcus pyogenes*, *Haemophilus influenzae*
Expression of intercellular adhesion molecules	Uropathogenic *Escherichia coli*, *Helicobacter pylori*, *Haemophilus influenzae*
Inhibition of mucin secretion	*Helicobacter pylori*
Stimulation of matrilysin secretion	*Escherichia coli*, *Bacteroides thetaiotamicron*
Release of pathogen-elicited epithelial chemoattractant	*Salmonella typhimurium*
Cell death	*Mycoplasma* spp.
Apoptosis	*Yersinia enterocolitica*, *Helicobacter pylori*
Invasion	Many examples including: *Salmonella* spp., *Shigella* spp., *Neisseria* spp., *Haemophilus* spp., *Yersinia* spp., *Listeria* spp.

cytokines from epithelial cells (see Sections 4.2.6 and 10.3.1). Why then do these bacteria not induce a state of chronic inflammation? Do these bacteria also produce molecules that are capable of inhibiting the release of such cytokines or perhaps neutralise their activity? Clearly, we need to know more about the interactions between host cells and members of the normal microflora.

A number of bacterial species capable of inducing pathology in humans also adhere to epithelial cells without inducing any apparent change in these cells. However, although the adhesion process itself may not affect the structure or function of the host cell, the bacterium may also produce toxins, enzymes, etc. that may ultimately damage these cells. *Bordetella pertussis* is the causative agent of whooping cough and, although the organism is non-invasive and remains in the respiratory tract, it releases toxins (pertussis toxin and an adenylyl cyclase toxin) that are responsible for the characteristic symptoms of the disease (see Sections 9.4.1.4.3 and 9.5.2.1). The organism colonises the respiratory mucosa by adhering to ciliated epithelial cells, which it ultimately kills due to the actions of its tracheal cytotoxin. Adhesion of *Bord. pertussis per se* does not appear to affect the behaviour of ciliated cells but the ultimate outcome of the interaction is cell death owing to the toxins it produces.

Vibrio cholerae is responsible for cholera, one of the most serious gastrointestinal infections of humankind. The disease is contracted by drinking water contaminated with the organism, which then colonises the intestinal tract without any apparent effect on the epithelial cells. It secretes an exotoxin that induces massive water loss and electrolyte imbalance resulting, in many cases, in death. Enterotoxigenic strains of *E. coli* cause a gastroenteritis with symptoms similar to those of cholera. As in the case of cholera, the bacteria do not alter the morphology of epithelial cells, nor is there much inflammation in the tissues.

With a little help from my friends – the toxin co-regulated pilus of *Vibrio cholerae*

Several studies have shown that the toxin co-regulated pilus (TCP) of *V. cholerae* is essential for colonisation by this organism of the intestinal tract of humans and other animals. TCP belongs to the type IV group of pili, which are common in other Gram-negative pathogens such as EPEC and enterotoxigenic strains of *E. coli* (ETEC). It consists of a polymer of the pilin TcpA. TCP appear as long bundles of laterally associated fibres similar to the bfp produced by EPEC. Like the bfp of EPEC, the TCP enable autoagglutination of the organism which results in the formation of microcolonies – this may help to protect the organism against host defence systems. The exact role of TCP in colonisation, however, has not yet been established and no receptor for the pilin has been detected on epithelial cells. Surprisingly, TCP also functions as a receptor for the bacteriophage (CTXφ) that carries the gene encoding the cholera toxin – a key virulence factor for this organism (see Section 9.5.1.1). The genes encoding the production of TCP are located on a large pathogenicity island (see Section 2.6.4) known as *V. cholerae* pathogenicity island (VPI). Interestingly, it has now been shown that the VPI is itself the genome of another bacteriophage (VPIφ) and that the TcpA pilin is, in fact, a coat protein of this phage.

To recap, here we have a situation in which one phage (VPIφ) infects a bacterium (V. cholerae) and its genome is incorporated into that of its host as a pathogenicity island. This pathogenicity island then directs the synthesis and export of the viral coat protein in the form of a pilus, which not only functions in adhesion of the organism to the bacterium's host but also acts as a receptor for another phage (CTXφ) that carries the genes encoding a toxin (cholera toxin) essential for the virulence of the bacterium!

7.5.2.1.2 Induction of morphological alterations

7.5.2.1.2.1 Enteropathogenic *Escherichia coli*

EPEC strains are the major cause of diarrhoea in children worldwide. Infection with these organisms has a high mortality rate in developing countries and it has been estimated that more than one million children die each year as a result of infection with EPEC. While the organism does not produce either enterotoxins or cytotoxins, and is not usually invasive, it induces a characteristic and striking alteration in the intestinal epithelium known as an **attaching and effacing** (A/E) lesion. First of all, the microvilli of the epithelial cells disappear and then a region of the membrane of each affected cell balloons outwards to form a pedestal (which can extend for up to 10 μm) on which the bacterium resides (Figure 7.16). What is truly remarkable is that the organism can induce such profound changes in an epithelial cell and yet remains extracellular. All of the bacterial genes involved in the induction of the A/E lesion are located on a 35 kb region of the chromosome known as the locus of enterocyte effacement (LEE), that is an example of a pathogenicity island (see Section 2.6.4). Transfer of the LEE to an *E. coli* strain which normally cannot induce an A/E lesion will confer this ability on the recipient.

Several stages can be recognised in the interaction between EPEC and epithelial cells: initial adhesion, protein translocation, intimate attachment and pedestal formation (Figure 7.17). The first stage in the process involves adhesion of the organism to the epithelium in a patchy arrangement of scattered microcolonies. This phenomenon is known as localised adhesion and is a consequence of the formation of specialised structures known as bfp (which are type IV pili), which are synthesised in response to the organism contacting an epithelial cell. The expression of bfp involves the transcription of 14 genes and these are located on a 90 kb plasmid known as the EPEC adherence factor plasmid. Bfp may also be involved in inter-bacterial adhesion and so may be responsible for the formation of the characteristic EPEC microcolonies (Figure 7.18). As well as inducing bfp production, adhesion to the epithelial cell also activates the assembly of a type III secretion system (described in Section 2.5.5).

Figure 7.16 Scanning electron micrograph showing characteristic attaching and effacing lesions, in human epithelial cells, resulting from infection with enteropathogenic *Escherichia coli*. The arrow indicates a filamentous organelle involved in the translocation of proteins into the epithelial cell. Bar= 0.25 μm. (Reproduced by permission of Oxford University Press from Knutton, S., Rosenshine, I., Pallen, M. J., Nisan, B. C., Bain, C., Wolff, C., Dougan, G. & Frankel, G. A novel EspA-associated surface organelle of enteropathogenic *Escherichia coli* involved in protein translocation into epithelial cells. *EMBO Journal*, 1998, **17**, 2166–2176.)

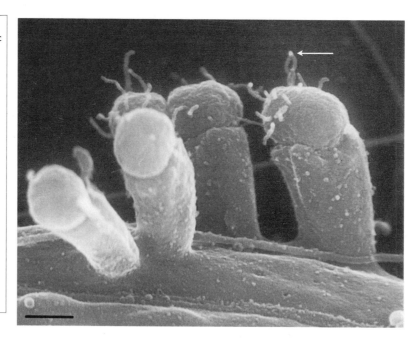

One of the first proteins to be secreted by this system is EspA (25 kDa), which has been found in filamentous structures on the bacterial surface (Figure 7.16). It has been suggested that this functions as a tubular channel (connecting the bacterium to the epithelial cell) through which bacterial proteins are injected into the host cell (Figures 7.16 and 7.19). Proteins that are currently known to be injected into the host cell are Tir, EspB (EPEC secreted protein B), EspD and EspF. While the function of EspF is unknown, EspB and EspD are inserted into the host cell membrane, where they are thought to be involved in pore formation and thereby may form the 'anchoring point' for the protein-transporting channel. Tir (with a molecular mass of 78 kDa), undergoes phosphorylation of one or more of its tyrosine residues in the host cell (increasing its apparent molecular mass to 90 kDa) and becomes inserted into the membrane of the epithelial cell, where it functions as a receptor for the EPEC adhesin intimin – a 94 kDa outer membrane protein (Figure 7.19). Binding of intimin to its receptor, Tir, results in intimate attachment of EPEC to the host cell. This intimate attachment induces the repolymerisation of actin monomers and the rearrangement of other cytoskeletal components into a configuration necessary for pedestal formation – a process accompanied by loss of microvilli. It may well be that the latter is due to depolymerisation of actin microfilaments in the microvilli, the resulting actin then being repolymerised to form the pedestal. Actin accumulates at the tip, and along the length, of the pedestal whereas α-actinin, ezrin, talin, plastin and villin are found only along its length. Interestingly, the pedestals are motile so that attached EPEC are moved along the surface of the epithelial cell at a speed of 0.07 μm/s. Pedestal motility is inhibited by cytochalasin D, indicating that it is dependent on the polymerisation

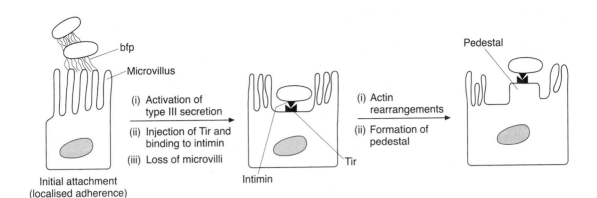

Figure 7.17 Stages involved in the formation of an attaching/effacing lesion by enteropathogenic *Escherichia coli*. The bacterium produces bundle-forming pili (bfp) following contact with the epithelial cell and these mediate initial attachment to the host cell. Activation of a type III secretion system results (among other things) in the injection of the bacterial translocated intimin receptor, Tir, into the host cell. After its phosphorylation and insertion into the membrane of the epithelial cell, Tir acts as the receptor for the *E. coli* adhesin, intimin. At the same time, cytoskeletal rearrangements take place, resulting in loss of microvilli and the formation of a characteristic pedestal.

of actin. Although microscopic observations have revealed some details of the cytoskeletal rearrangements involved in pedestal formation, little is known about the signalling events or mechanisms involved in the process. Intimate attachment is known to activate signal transduction pathways involving protein kinase C, phospholipase Cγ and inositol phosphate. It is possible that EspB, which is injected into the host cell as well as forming a pore in its membrane, is responsible for activating these pathways. Recently, evidence has accumulated to suggest a role in pedestal formation for the Wiskott–Aldrich syndrome protein (WASP), a molecule well known for its involvement in cytoskeletal rearrangements.

The other characteristic feature of EPEC infections is, of course, diarrhoea. In the majority of gastrointestinal infections, diarrhoea is the result of active ion secretion from host cells and, as efflux of chloride ions from epithelial cells in response to EPEC infection has been detected, this is true also for EPEC. Chloride secretion from the host cells is a consequence of the activation of the protein kinase C signalling pathway. The loss of microvilli during the formation of the A/E lesion would decrease the surface area available for water absorption and this would also contribute to diarrhoea.

Several other organisms, including *Hafnia alvei* (another gastrointestinal pathogen of humans), *Citrobacter rodentium* (a gastrointestinal pathogen of mice) and enterohaemorrhagic *E. coli*, also characteristically induce an A/E lesion, but these have been less extensively studied.

Keep on running – diarrhoea due to *Escherichia coli*

Until a few years ago, *E. coli* was regarded as being a rather inocuous member of the normal microflora of the intestinal tract, which caused disease only when it gained entry to other anatomical sites such as the urinary tract. However, it is now recognised that strains of this organism are a major cause of diarrhoeal diseases worldwide. As many as eight main types of diarrhoea-inducing *E. coli* have been recognised and, of these, only one involves invasion of the intestinal epithelium. The main distinguishing characteristics of these organisms are summarised in Table 7.9.

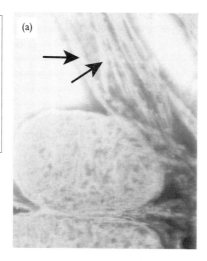

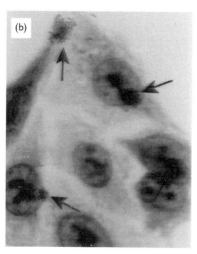

7.5.2.1.2.2 *Helicobacter pylori*

There has been a marked increase in interest in this organism during the last few years as it has been shown to be the major cause of chronic gastritis in humans, is associated with peptic ulcer formation and is a risk factor for the development of gastric carcinoma. Indeed, the World Health Organization has designated the organism as being a type 1 carcinogen. It has been estimated that in the USA more than 80% of adults are infected with the bacterium. Surface molecules such as flagellin, lipopolysaccharide, urease, various outer membrane proteins,

Figure 7.19 Attachment of enteropathogenic *Escherichia coli* to an epithelial cell activates a type III secretion system, which results in the injection of a number of effector proteins into the host cell. This triggers several signal transduction pathways (resulting in cytokine release, etc.) and cytoskeletal rearrangements (resulting in pedestal formation).

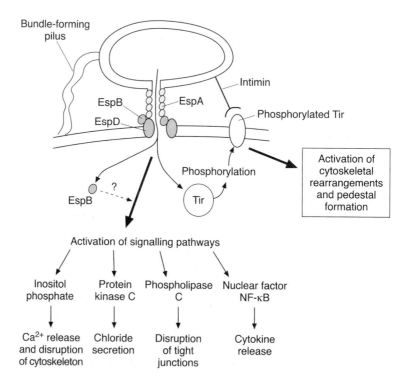

Table 7.9. Classification of enteropathogenic strains of *Escherichia coli*

Organism	Main features of disease pathogenase	Main susceptible populations
Enteropathogenic (EPEC)	Forms an A/E lesion	Infants in developed and developing countries
Enterotoxigenic (ETEC)	Produce a heat-labile and/or heat-stable enterotoxin	Infants and travellers in developing countries
Shiga toxin-producing (STEC)	Produce verotoxin	Sporadic outbreaks mainly in developed countries
Enteroaggregative (EAEC)	Adhere to epithelial cells as characteristic aggregates	Persistent diarrhoea in children in developing countries
Diffusely adherent (DAEC)	Do not adhere as localised aggregates	Children in developed and developing countries
Enteroinvasive (EIEC)	Invade intestinal epithelium	Infants and travellers in developing countries
Cell-detaching (CDEC)	Produce an α-haemolysin	Children younger than 18 months, mainly in developing countries
Cytolethal distending toxin-producing (CLDTEC)	Produce CDT	?

A/E, attaching and effacing; CDT, cytolethal distending toxin.

chaperonin 60 and a haemagglutinin have all been proposed to function as adhesins in *Hel. pylori*. Putative receptor molecules on gastric epithelial cells for these adhesins include phosphatidylethanolamine, lysophosphatidylethanolamine and blood group antigens. Although the exact means by which *Hel. pylori* adheres to the gastric mucosa remains to be elucidated, it is thought that the BabA adhesin (which binds to Lewis X and Y epitopes on gastric cells) is important in this process. Examination of the gastric mucosa of patients infected with the organism has revealed that it induces effacement of the microvilli of the gastric epithelial cells and adheres to cellular projections termed 'adherence pedestals'. Such ultrastructural changes are similar to those induced in intestinal epithelial cells by EPEC. *In vitro* studies have shown that loss of microvilli and pedestal formation occurs within 3 hours of attachment of *Hel. pylori* to gastric epithelial cells. The pedestals are formed from the bases of the damaged microvilli and consist mainly of microfilaments.

Recent studies *in vitro* using gastric epithelial cell lines have shown that contact with an epithelial cell results in the injection of a protein, CagA (cytotoxin-associated gene A product), into the host cell via a type IV secretion system (see Section 2.5.3.3). This protein then undergoes tyrosine phosphorylation, leading to cytoskeletal rearrangements that

Figure 7.20 Phase contrast micrograph showing characteristic 'humming bird' appearance of human gastric epithelial cells infected with *Helicobacter pylori*. Such cells are characterized by spreading and elongated growth, the presence of lamellipodia (thin actin sheets at the edge of the cell), and filopodia (finger-like protrusions containing a tight bundle of actin filaments). (Reproduced with permission from Segal, E. D., Cha, J., Lo, J., Falkow, S. & Tompkins, L. S. Altered states: Involvement of phosphorylated CagA in the induction of host cellular growth changes by *Helicobacter pylori*. *Proceedings of the National Academy of Sciences*, USA, 1999, **96**, 14559–14564. Copyright 1999 National Academy of Sciences, USA.)

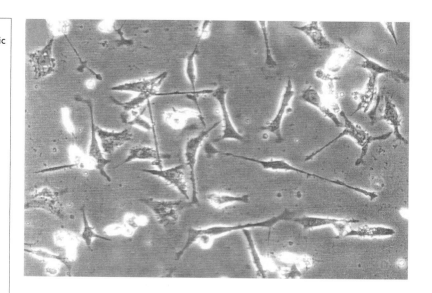

give rise to a characteristic 'humming bird' appearance owing to the production of filapodia and lamellipodia (Figure 7.20).

7.5.2.1.3 Induction of functional changes

Streptococci constitute one of the numerically dominant genera of bacteria in the oral cavity. A wide variety of species is usually present and most of these rarely cause disease, the exceptions being *Strep. mutans* and *Strep. sobrinus* (both of which are responsible for dental caries) as well as *Strep. sanguis*, which may cause endocarditis in patients with cardiac abnormalities or cardiac prostheses when it gains entry to the bloodstream. Among the many adhesins expressed by streptococci are a group of rhamnose/glucose polymers and a group of antigenic proteins (antigen I/II) with molecular masses of 180 to 200 kDa. The latter appear to be present on the surfaces of many oral streptococci and have been shown to mediate their binding to human epithelial cells. Adhesion to epithelial cells has been shown to stimulate the release of the pro-inflammatory and chemotactic cytokine interleukin 8 (IL-8). IL-8 is also released from the cells following binding of purified adhesins (rhamnose-glucose polymer and protein I/IIf) from *Strep. mutans*. IL-8 is a pro-inflammatory cytokine with a number of biological activities, the most important of which is the ability to stimulate the migration and degranulation of polymorphonuclear neutrophils (PMNs) to deal with any infecting microbes. The release of IL-8 by epithelial cells following adhesion of oral streptococci would be expected, therefore, to induce an inflammatory response and this would be accompanied by an influx of PMNs. As this clearly does not happen (otherwise the epithelia of our mouths would be continuously inflamed) other factors must be operating *in vivo* to counteract this response.

One of the main defence mechanisms of the urinary tract is the flushing action of urine and this is an important deterrent to colonisation by infecting organisms. The ability to adhere to urinary epithelial cells is, therefore, a prime requirement for a uropathogenic organism. The most frequent cause of urinary tract infections (UTIs) is *E. coli*, and adhesion of this organism to urinary epithelial cells has been studied extensively. *Escherichia coli* posesses a variety of adhesins that can be classified broadly as fimbrial, fibrillar and non-fimbrial. Those considered to be important in adhesion to the urinary epithelium are the fimbriated type. Binding of *E. coli* to urinary epithelial cells induces the release of the pro-inflammatory cytokines IL-1α, IL-1β, IL-6 and IL-8. Interestingly, it has been shown that urinary IL-6 levels increase in patients with a UTI and in volunteers infected with *E. coli*. IL-6 is a pro-inflammatory cytokine that can stimulate the release of acute phase proteins, activate B and T cells, stimulate immunoglobulin synthesis and stimulate the proliferation of fibroblasts (see Section 6.4.4, Figure 6.13 and Table 6.8). A characteristic feature of any bacterial infection is the migration of PMNs towards the site colonised by the infecting organism and, in the case of UTIs, this results in the presence of PMNs in urine. In response to a chemotactic gradient from the site of the infection, PMNs adhere to, and then traverse, the vascular endothelium. This process involves the interaction between E-selectin, intercellular adhesion molecule (ICAM) 1 and ICAM-2 on the surface of the endothelial cell and sialyl Lewis X and β_2-integrins (CD11a/CD18 and CD11b/CD18) on the PMN surface (Figure 7.21). Initially, E-selectin binds to sialyl Lewis X, which causes the PMNs to roll along the vessel wall, then interactions between the β_2-integrins and ICAM-1 and ICAM-2 result in adhesion to, and migration across, the endothelium. A great deal is known about the mechanisms involved in these processes (see Section 6.8) but very little is known about how PMNs then cross the epithelial lining into the mucosal space, an event that occurs in infections of the urinary, intestinal and respiratory tracts. As described above, adhesion of *E. coli* to epithelial cells can trigger the release of IL-8, which is a potent chemoattractant for PMNs and will therefore stimulate the migration of PMNs to the urinary epithelium. At the same time, the adherent *E. coli* up-regulates the expression of ICAM on the surface of the cells, which binds to the CD11b/CD18 integrin on the surface of the arriving PMNs. Details of the mechanism by which the PMNs then traverse the epithelium remain to be established.

A similar series of events leading to the activation of host defence mechanisms has recently been described when *H. influenzae*, a member of the normal flora of the nasopharynx, adheres to respiratory epithelial cells. Bacterial adhesion stimulates the release of the chemokine IL-8 (but not IL-1 or tumour necrosis factor (TNF)) and ICAM-1, which results in increased adhesion of neutrophils to the epithelial cells. *Helicobacter pylori* is another organism that, on adhering to epithelial cells, induces expression of ICAM-1. PMN recruitment has also been shown to result from the adhesion of *Sal. typhimurium* to intestinal epithelial cells. In addition to the IL-8 released following bacterial adhesion, the epithelial cells were also shown to release a novel factor, pathogen-elicited epithelial chemoattractant (PEEC). PEEC is secreted

Figure 7.21 Mechanisms involved in the bacteria-induced migration of PMNs from a capillary and across an epithelial surface. IL-8 released by epithelial cells (following bacterial adhesion) functions as a potent chemoattractant of PMNs, which migrate across the endothelium. Integrins on the PMN bind to ICAM-1 (the expression of which has been up-regulated following bacterial adhesion) and the PMNs cross the epithelium by a mechanism that is, as yet, largely unknown. However, in intestinal infections due to *Salmonella* spp., trans-epithelial migration appears to be driven by pathogen-elicited epithelial chemoattractant – a chemokine secreted from the apical surfaces of epithelial cells (see Section 5.8). For abbreviations, see the text.

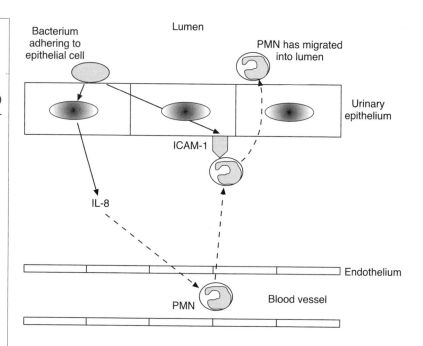

apically from the cells and appears to direct the final step of PMN migration across the epithelial monolayer to the intestinal lumen (see Section 5.8).

Adhesion of bacteria to epithelial cells has been shown to activate another aspect of the host defence system – matrilysin secretion. Matrilysin is a matrix metalloprotease that can degrade a number of matrix proteins including elastin and fibronectin. As well as having a major role in the repair of epithelia, matrilysin also functions as an activator of antimicrobial peptides (e.g. defensins) that are secreted by epithelial cells in an inactive form (as pro-defensins). These antimicrobial peptides are a key component of the innate defence system (see Section 5.4.3.11). Recently it has been shown that adhesion of *E. coli* to a range of epithelial cell types results in the up-regulation of matrilysin production. Furthermore, while intestinal epithelial cells of germ-free mice failed to secrete matrilysin, infection of these animals with *Bacteroides thetaiotamicron* (a member of the normal intestinal microflora) stimulated the secretion of matrilysin within 10 days.

In order to protect itself against acid, pepsin and other potentially damaging agents, the gastric mucosa produces a mucus layer that consists mainly of a mucin-based, water-insoluble gel (see Section 5.4.3.1). The gastric pathogen *Hel. pylori* not only is able to degrade this mucus layer by means of extracellular enzymes but can also decrease mucin synthesis by gastric epithelial cells following adhesion to them *in vitro*.

7.5.2.1.4 Invasion

Adhesion of a pathogenic bacterium to an epithelial cell is often followed by invasion of that cell. As adhesion itself induces a series of changes in the host cell that eventually leads to uptake of the organism, it would be artificial to discuss only adhesion in isolation from subsequent cellular events. Therefore those organisms that invade epithelial cells, and the adhesion mechanisms involved, will be discussed in Chapter 8.

7.5.2.1.5 Cell death

Growth and replication of bacteria within host cells often leads to the death of the colonised cell due to nutrient depletion, degradation of key intercellular constituents and the accumulation of toxic end-products of bacterial metabolism: this is termed 'necrotic cell death'. However, it has become apparent in recent years that bacteria can also induce 'programmed cell death' (known as 'apoptosis' and described in detail in Section 10.6.2.1) in the invaded cell. Apoptosis is a natural process by which the host controls cell numbers and it is also important in embryogenesis. As it is accompanied by characteristic changes in the cell (membrane blebbing, chromatin condensation and fragmentation and the presence of apoptotic bodies), it can be distinguished from necrotic cell death (Figure 7.22). In contrast to the latter, apoptosis is a highly regulated process that involves the active participation of the cell and is controlled by complex signal transduction pathways activating a group of cysteine aspartyl proteases (caspases) that initiate the apoptotic process. However, some bacteria have evolved ways of triggering this process, thereby bringing about the premature death of the cell. In some cases this can be achieved 'at a distance' by the secretion of apoptotic toxins (e.g. *Bord. pertussis*, *Corynebacterium diphtheriae* and *Pseudomonas aeruginosa*; see Chapter 9) but in this section we will consider only apoptosis induced by adhesion of bacteria.

Death of intestinal cells due to EPEC has been detected in biopsies from infected human individuals but only recently has it been suggested that death may be due to apoptosis. Indeed, epithelial cell lines infected with EPEC have been reported to display some of the characteristic features of apoptosis. Killing was found to be contact dependent and to require live bacteria. Apoptosis has also been demonstrated in gastric epithelial cells infected with *Hel. pylori* both *in vivo* and *in vitro* (Figure 7.22).

7.5.2.2 Fibroblasts

Fibroblasts are a major component of connective tissue and are involved in the production of ECM material and in maintaining the integrity of the connective tissue. They can also secrete cytokines and other inflammatory mediators and so are important in host defence. Interference with their function can, therefore, have profound

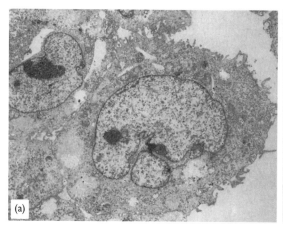 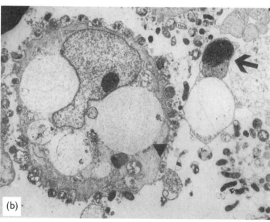

(a) (b)

Figure 7.22 Transmission electron photomicrographs of human gastric epithelial cells before (a) and after (b) infection with *Helicobacter pylori*. The uninfected cells exhibit a normal cellular morphology whereas the infected cells have the characteristic features of programmed cell death: cytoplasmic vacuolation (arrowhead) and apoptotic body formation (arrow). Magnification 32 480×. (Reproduced with permission, from Jones *et al.*, 1999. See Section 7.11.)

consequences for immunobiology and tissue integrity. Bacteria can gain access to, and interact with, fibroblasts either by invading mucosal surfaces or via breaches in the epithelium due to wounds, abrasions or burns. A number of organisms can adhere to, and alter the behaviour of, fibroblasts.

Attachment of *Treponema denticola* to human gingival fibroblasts (HGFs), for example, induces a number of alterations in these cells including: (i) retraction of pseudopods followed by rounding of the cells and the formation of membranous blebs; (ii) detachment of the cells from their substratum, possibly as a result of the degradation of fibronectin by surface-associated proteases; and (iii) cell death. Death of the cells would, obviously, have adverse consequences for tissue integrity if it occurred *in vivo*. However, the cytoskeletal changes induced by the organism would also be detrimental, as they would inevitably interfere with the ability of fibroblasts to phagocytose collagen fibres, which is important in tissue remodelling.

Haemophilus ducreyi is responsible for the sexually transmitted disease chancroid. The organism gains access to the dermis via microbreaks in the epidermis and then induces the formation of ulcers. *In vitro* studies using human foreskin fibroblast (HFF) monolayers have shown that the organism forms microcolonies on the surface of these cells and often becomes embedded in deep invaginations of the cell membrane. The bacteria can also penetrate between cells but do not invade them. Attachment to the HFFs results in a distinctive cytopathic effect – disruption of cell monolayers, changes in cell morphology and cell death (Figure 7.23). The organism produces a cell-surface-associated haemolysin that may be responsible for the adherence-induced cytopathic effects.

7.5.2.3 Endothelial cells

Endothelial cells form a continuous monolayer lining the blood vessel walls and function not only as a barrier between the blood and the vessel walls but also in the regulation of blood vessel tone and permeability, blood coagulation, leukocyte and platelet reactivity, and angiogenesis. Vascular endothelial cells are the source of vascular mediators including NO, prostacyclin and endothelin. In order to colonise internal tissues, blood-borne bacteria have to adhere to, and traverse, the endothelium. *Staphylococcus aureus*, *N. meningitidis*, *H. influenzae*, *Strep. pyogenes*, *Bor. burgdorferi*, *Listeria monocytogenes*, *E. coli*, *Ps. aeruginosa* and *Porphyromonas gingivalis*, for example, have all been shown to adhere to, and invade, endothelial cells. As invasion is the inevitable consequence of bacterial adhesion to such cells, this topic will be discussed in Chapter 8.

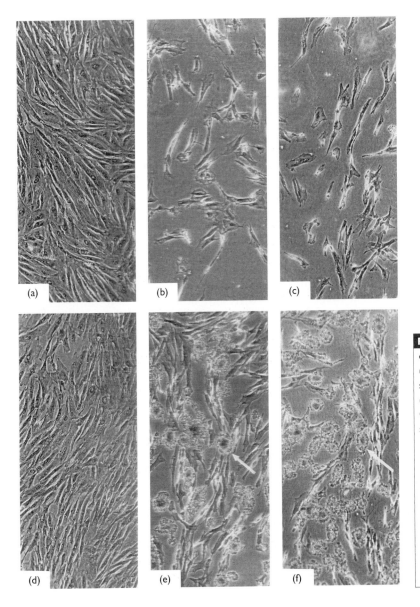

(a) (b) (c) (d) (e) (f)

Figure 7.23 Photomicrographs of human foreskin fibroblasts (HFFs) infected with various strains of *Haemophilus ducreyi* (b, c, e, f) or *H. influenzae* (d). Control, uninfected HFFs are shown in (a). A cytopathic effect (disruption of the cell monolayers and changes in cell morphology) are evident in those HFFs infected with *H. ducreyi* but not in control cultures or those infected with *H. influenzae*. Microcolonies formed by *H. ducreyi* are indicated by the white arrows in (e) and (f). (Reproduced with permission, from Alfa *et al.*, 1995. See Section 7.11.)

7.5.2.4 Phagocytic cells

The interaction of some bacteria with phagocytic cells (PMNs and macrophages) constitutes a fascinating aspect of the host–parasite relationship and illustrates that, with bacteria, all things are possible. On the one hand, adhesion of bacteria to phagocytic cells is important in host defence, as it is the primary means by which bacteria are ingested and then disposed of. However, some bacteria (e.g. *M. tuberculosis*, *Bord. pertussis*, *Y. enterocolitica* and *Lis. monocytogenes*) are able to survive within such cells. Furthermore, other organisms (e.g. *Legionella pneumophila*, *Salmonella* spp., *Mycobacterium* spp. and *Shigella* spp.) actually invade their arch enemies and can live quite happily inside them – a very effective way of dealing with a major host defence mechanism! Adhesion of bacteria to phagocytes may occur either directly (i.e. involving the interaction between adhesins and receptors on the phagocyte) or indirectly by interacting with host components (e.g. immunoglobulins or complement components) that then bind to phagocyte receptors. Whichever mechanism is involved, adhesion induces endocytotic uptake of the bacterium into a phagosome, which then fuses with a lysosome, resulting, usually, in the death of the organism. Both invasion and phagocytosis involve profound adhesion-induced changes in the phagocyte and are discussed in Sections 6.6.2 and 10.4.2. However, it has also been shown that adhesion of some invasive organisms (including *Leg. pneumophila* and *Sal. typhimurium*) induces cytokine release by the host cell prior to internalisation of the bacteria. Attachment of *Leg. pneumophila* to murine macrophages in the presence of cytochalasin (which inhibits uptake of the bacteria) has been shown to induce increased synthesis of mRNA for the cytokines IL-1β, IL-6 and granulocyte-macrophage colony-stimulating factor (GM-CSF) and the chemokines macrophage inflammatory protein (MIP) 1β, MIP-2 and KC (the murine form of the human chemokine MGSA (melanoma growth-stimulatory activator)). Interestingly, α-methyl-D-mannoside inhibited the induction of increased levels of cytokine mRNA (but not chemokine mRNA) implying that adhesion-induced cytokine synthesis is mediated by the interaction between a mannose-sensitive receptor and its complementary ligand. This suggests that different receptor–ligand systems are responsible for adhesion-mediated up-regulation of cytokine and chemokine synthesis in these host cells (Figure 7.24).

Another possible consequence of bacterial adhesion to phagocytic cells that does not involve bacterial internalisation is apoptosis. This occurs, for example, when the enteric pathogen *Y. enterocolitica* binds to macrophages. Once this organism has invaded the intestinal mucosa it resists phagocytosis and maintains an extracellular lifestyle. However, it adheres to macrophages and this induces death of these phagocytes, which display all of the features characteristic of apoptosis – shrinkage of the cytoplasm, condensation of nuclear chromatin and fragmentation of DNA. The ability of *Y. enterocolitica* to induce the suicide of one of

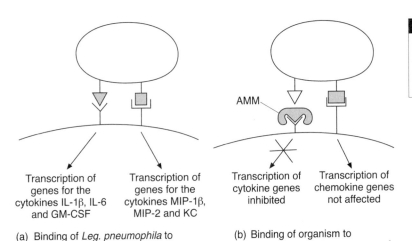

AMM

Transcription of genes for the cytokines IL-1β, IL-6 and GM-CSF

Transcription of genes for the cytokines MIP-1β, MIP-2 and KC

Transcription of cytokine genes inhibited

Transcription of chemokine genes not affected

(a) Binding of *Leg. pneumophila* to macrophage in absence of α-methyl-D-mannoside (AMM)

(b) Binding of organism to macrophage in the presence of AMM

Figure 7.24 Cytokine and chemokine synthesis in macrophages induced by adhesion of *Legionella pneumophila*. For further details, see the text.

the key effector cells of the host defence system is likely to constitute an important survival strategy for this extracellular pathogen.

7.6 | Prevention of bacterial adhesion

In this chapter we have emphasised the importance of adhesion as being the first step in the initiation of an infectious disease. It follows, therefore, that blocking the adhesion of an organism to its target cell/tissue should prevent disease. If the specific molecules (or epitopes) involved in the ligand–receptor interaction that underpins the adhesion process are known, then it should be possible to design molecules that could be used to block their interaction and so prevent adhesion. A number of approaches to this are possible, as outlined in Figure 7.25. Firstly, the use of antibodies to the adhesin is an approach already used by the host, and the immune response to many infectious diseases involves the production of secretory IgA antibodies capable of preventing bacterial adhesion to mucosal surfaces. There is currently great interest in the possibility of providing protection against a number of diseases (e.g. those due to *Strep. pyogenes*, *Bord. pertussis*, *Staph. aureus*, *E. coli* and *Strep. pneumoniae*) by immunisation with bacterial adhesins. Another approach would be to use molecules that can bind to the bacterial adhesin – such molecules may be those that actually constitute the epitope of the receptor or may be analogues of these. Hence, it has been shown that molecules such as mannose or α-methyl-mannoside can inhibit mucosal infections (in experimental animals) by certain *E. coli*, *Klebsiella pneumoniae* and *Shigella flexneri* strains. Other molecules proving successful in animal experiments include N-acetylglucosamine (for preventing *Strep. pneumoniae* infections) and Galα4GalβOMe (for *E. coli* infections). A third possibility is to use adhesin analogues that would bind to the host cell receptor,

Figure 7.25 Strategies for inhibiting adhesion of bacteria to host cells. One or more of these could form the basis of new approaches for preventing and/or treating bacterial infections.

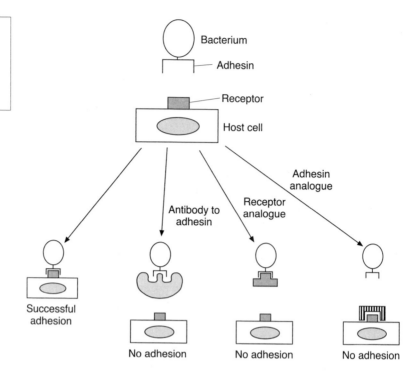

thereby preventing access by the bacterial adhesin. This approach has recently been shown to be capable of preventing colonisation of teeth by *Strep. mutans*. Anti-adhesins would represent a new class of drugs based on an understanding of the infectious process at the molecular level. Such drugs are desperately needed to augment (and possibly replace) the dwindling supply of classical antibiotics to which pathogens are developing resistance at an alarming rate.

Move over darling – prevention of bacterial adhesion

Streptococcus mutans, the major causative organism of dental caries, binds to certain salivary glycoproteins (particularly one with a molecular mass of >200 kDa) that cover the tooth surface. One of the major adhesins responsible for this is a protein known as streptococcal antigen I/II. Researchers have identified the adhesive epitope involved and have shown it to be located between amino acid residues 1025 and 1044 of the molecule. Having established this, a peptide with the same amino acid sequence as this region was synthesised. This peptide should bind to the salivary receptors for the streptococcal adhesin thereby 'blocking' binding of *Strep. mutans*. Laboratory studies have shown that the peptide could prevent adhesion of *Strep. mutans* to salivary glycoproteins. Furthermore, when it was applied to the teeth of human volunteers, it was able to prevent colonisation of the tooth surfaces by the organism. This anti-adhesive approach may well be applicable to the prevention, or treatment, of other infectious diseases.

7.7 | Concept check

- Adhesion involves long-range (non-specific) and short-range (specific) forces.
- The range of surfaces available for bacterial adhesion is wide and includes both cellular and acellular substrata.
- Bacteria utilise a wide range of molecules (adhesins) to effect adhesion to the host.
- Adhesins may be located on a variety of bacterial structures.
- Adhesins bind to specific receptors present either on the host cell or on macromolecules coating cellular or acellular surfaces.
- Lectin-carbohydrate binding is a common mechanism involved in adhesion.
- Adhesion induces phenotypic changes in the bacterium.
- When a host cell is the substratum for the bacterium, adhesion affects the behaviour of the host cell in a number of ways.
- A frequent consequence of the adhesion of a bacterium to a host cell is invasion of that cell.
- Adhesion of a bacterium to a phagocyte is an essential part of the host defence system.
- Prevention of adhesion may form the basis of new approaches to treat or prevent infectious diseases.

7.8 | Adhesion of the paradigm organisms

7.8.1 *Streptococcus pyogenes*

The main habitat of this organism is the oropharynx, where it can be recovered from up to 10% of healthy adults and up to 20% of children. It is also found occasionally on the skin. We are, therefore, interested primarily in the means by which the organism adheres to the substratum offered by the oropharynx. This substratum consists of stratified, non-ciliated epithelial cells (the majority of which are non-keratinised) and these are bathed in mucus. We would expect, therefore, *Strep. pyogenes* to have adhesins that would enable it to bind to both mucus constituents as well as to nasopharyngeal epithelial cells (Figure 7.26). Furthermore, as this organism is capable of invading tissues and causing diseases elsewhere (e.g. myositis, impetigo, cellulitis, lymphangitis, bone infections) it is likely to be able to produce adhesins specific for receptors in other tissues. In fact, *Strep. pyogenes* does display a wide variety of adhesins and some of these are listed in Table 7.10.

Which of these adhesins function at different stages in an invasive infection due to *Strep. pyogenes* remains to be established. However, LTA supported on a 'scaffold' of M protein is thought to be one of the principal adhesins involved in the attachment of the organism to pharyngeal epithelial cells (Figure 7.27). The M protein is abundant

Figure 7.26 Scanning electron micrographs showing adhesion of *Streptococcus pyogenes* to (a) human respiratory epithelial cells (3970×) and (b) human tonsillar epithelial cells (8000×) *in vitro*. (Reproduced with permission from: (a) Bennett-Wood, V. R. Carapetis, J. R. & Robins-Browne, R. M. Ability of clinical isolates of group A streptococci to adhere to and invade HEp-2 epithelial cells. *Journal of Medical Microbiology*, 1998, **47**, 899–906; (b) Caparon, M. G., Stephens, D. S., Olsen, A. & Scott, J. R. (1991). Role of M protein in adherence of group A streptococci. *Infection and Immunity*, **59**, 1811–1817.)

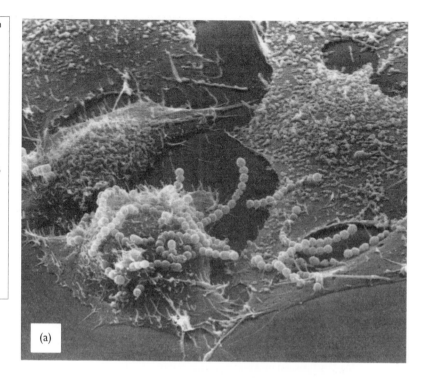

(a)

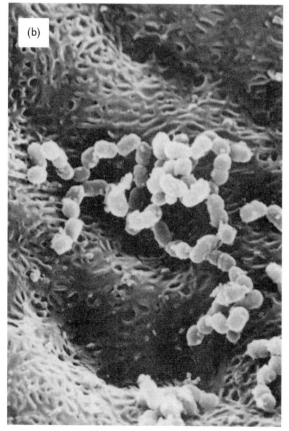

(b)

| Table 7.10. | Adhesins of *Streptococcus pyogenes* | |

Adhesin	Receptor	Target cell/substratum
Lipoteichoic acid	Fibronectin	Buccal epithelial cells, fibroblasts, pharyngeal cells
Fibronectin-binding proteins (Sfbl, FBP54, serum opacity factor)	Fibronectin	Fibronectin, buccal cells, epidermal Langerhans cells
M protein	(i) Galactose, fucose, fibrinogen (ii) β_1-Integrin (iii) Fibronectin	(i) Pharyngeal cells, keratinocytes (ii) Epithelial cells (iii) Various cell types
Vitronectin-binding proteins	Vitronectin	Pharyngeal cells
Glyceraldehyde-3-phosphate dehydrogenase	Fibronectin, lysozyme, myosin, actin, 32 kDa membrane protein	Pharyngeal cells
Laminin-binding proteins	Laminin	Epithelial cells, basement membrane
Collagen-binding proteins	Collagen	Extracellular matrix
70 kDa galactose-binding protein	Galactose	Pharyngeal cells
Hyaluronic acid	CD44 – a cell adhesion molecule with an affinity for hyaluronic acid	Skin keratinocytes, pharyngeal epithelial cells

on the surface of the organism, where it appears (by electron microscopy) as a dense mat of hair-like projections (Figure 7.28). This fibrillar protein is coated with LTA molecules, possibly by ionic interaction of the negatively charged polyglycerol phosphate backbone with positively charged regions of the M protein. The lipid portion of the LTA molecule could then protrude outwards from the complex enabling it to bind (by hydrophobic interactions) to specific receptors (fatty acid binding sites) on the host cell surface. Fibronectin, a glycoprotein found in host cell membranes, has the necessary fatty acid binding sites to function as a receptor for LTA. This would result in relatively weak binding to the host cell and would not account for the tissue tropism displayed by the organism, as fibronectin is found in many types of epithelial cells as well as those of the nasopharynx. It is thought that this initial adhesion is followed by the binding of the M protein (or another adhesin) to a specific receptor (one or more of those listed in Table 7.10) on the nasopharyngeal epithelial cells. Recently it has been shown that CD44 (a cell adhesion molecule expressed by many types of host cell that mediates various cell–cell and cell–ECM interactions) can function as a receptor for the hyaluronic acid capsule of *Strep.*

Figure 7.27 Interaction of lipoteichoic acid with M protein to form an adhesive structure for *Streptococcus pyogenes*.

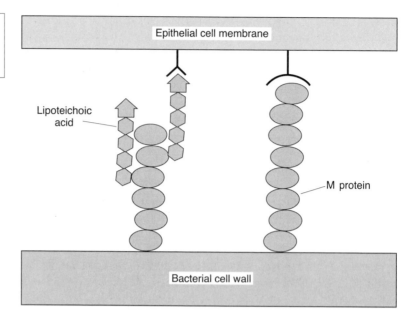

pyogenes. An anti-CD44 antibody was able to significantly inhibit binding of *Strep. pyogenes* to mouse keratinocytes, while pre-treatment of mice with hyaluronic acid significantly reduced colonisation by the organism.

A number of morphological changes in host cells have been observed as a consequence of adhesion of the organism. Hence, attachment of *Strep. pyogenes* to human pharyngeal, but not buccal, epithelial cells induces ruffling of the host cell membrane; this is not seen in the case of viridans streptococci. Furthermore, tonsillar epithelial cells produce protrusions that appear to 'grip' the organism from several directions. There are fewer reports of any functional changes in the

Figure 7.28 Electron micrograph of *Streptococcus pyogenes*, showing the dense layer of M protein on its surface. Magnification 35 850×. (Reproduced with permission from V. A. Fischetti, Laboratory of Bacterial Pathogenesis and Immunology, the Rockefeller University, New York, USA. Website: http://www.rockefeller.edu/vaf)

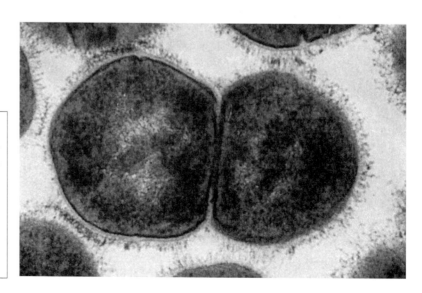

host cell following adhesion. However, adhesion of the organism to human keratinocytes has been shown to induce the synthesis of a number of pro-inflammatory cytokines (IL-1α, IL-1β, IL-6 and IL-8) and prostaglandin E_2, while adhesion to epithelial cells leads to the release of IL-6. Treatment of a human pharyngeal cell line with strains of the organism that express the adhesin GAPDH has been shown to induce the phosphorylation of a number of host cell proteins.

7.8.2 *Escherichia coli*

The normal habitat of *E. coli* is the gastrointestinal tract, where different strains of the organism are responsible for various forms of diarrhoeal disease (see Chapter 1). Some (uropathogenic) strains are also able to cause UTIs. Occasionally, the organism is responsible for meningitis in neonates. The main host substrata for attachment of the organisms are, therefore, intestinal and urethral epithelial cells. A large number of potential adhesins have been identified in *E. coli* (Table 7.11), although not all have been shown to function *in vivo* and not all have been detected in every strain of the organism.

In uropathogenic strains of the organism, the most important structure involved in adhesion to urinary tract epithelium is the type P fimbria – also known as the **p**yelonephritis-**a**ssociated **p**ilus (PAP). The adhesin on PAP is a protein known as PapG and this binds to a galactose α(1–4)-galactose residue present in certain glycolipids of epithelial cells.

Many enteropathogenic strains of the organism adhere to the intestinal epithelium by means of colonisation factor antigens (CFAs). CFAs (of which there are four main classes – CFA I, II, III, IV) are proteinaceous adhesins located on fimbriae and they bind to certain sialoglycoproteins on the surface of epithelial cells.

One of the most fascinating aspects of the adhesion of *E. coli* to host cells is the fact that EPEC strains carry around their own receptor (Tir) for one of their adhesins (intimin), which is inserted into the target host cell; this has been described in detail earlier in the chapter.

7.9 | What's next?

This chapter has been concerned with the crucial first stage in the interaction between a bacterium and its human host, regardless of whether or not that interaction will result ultimately in disease. In most cases, the interaction results in a state of mutual indifference but, as has been described in the chapter, adhesion of some bacteria is the prelude to an infectious process often mediated by toxins secreted by the organism. In yet other cases, adhesion is merely a prequel to invasion of the host and this is the subject of Chapter 8.

Table 7.11. | Adhesins of *Escherichia coli*. In many cases neither the molecular nature of the adhesin nor that of the receptor has been identified

Adhesin	Bacterial location	Receptor (or host cell targeted)
Intimin (94 kDa protein)	Outer membrane	Tir (translocated intimin receptor)
		β_1-Integrins on M cells of Peyer patches
?	Bundle-forming pilus	Epithelial cells and other cells of *E. coli*
?	Aggregative adherence fimbriae I	Intestinal epithelial cells
?	Aggregative adherence fimbriae II	Intestinal epithelial cells
Colonisation factor antigens (CFA) I–IV	Fimbrae	Sialoglycoproteins on intestinal epithelial cells
Fim H	Type I fimbriae	Mannose-containing receptors on uroepithelial cells
PapG	P fimbriae	Galactose disaccharide of glycolipids in urinary epithelial cells
Curlin (a 15 kDa protein)	Curli	Fibronectin, laminin, intestinal cells

?, unidentified.

7.10 | Questions

1. List the major organisms colonising the nasopharynx and describe the bacterial, host and environmental factors that account for the presence of these organisms at that particular site.
2. Discuss the importance of bacterial biofilms in human diseases.
3. Detailed knowledge of the mechanisms involved in bacterial adhesion to host cells could lead to the development of new measures of preventing infectious disease: discuss.
4. Human disease-causing bacteria can adhere to host cells, mineralised tissue, inanimate materials and other bacteria. Give examples of each type of organism and discuss the common themes underlying bacterial adhesion to these substrata.
5. Bacteria maintain themselves on their human host by the interaction of a bacterial adhesin with a complementary receptor produced by a host cell: discuss.
6. Host cells remain largely unaffected by the adhesion of bacteria to them: discuss.
7. Why are members of a bacterial species usually found at a particular anatomical site on a human being?
8. Describe what is known about the effects adhesion has on the functions of a bacterium.

9. Compare and contrast the events occurring subsequent to the adhesion of *Escherichia coli* and *Streptococcus pyogenes* to an epithelial cell.

10. List those organisms likely to encounter endothelial cells during the course of the diseases they induce. What is known about the adhesion of these organisms to endothelial cells?

7.11 | Further reading

Books

An, Y. H. & Friedman, R. J. (2000). *Handbook of Bacterial Adhesion: Principles, Methods and Applications.* Totowa, NJ: Humana Press.

Henderson, B., Wilson, M., McNab, R. & Lax, A. (1999). *Cellular Microbiology: Bacteria-Host Interactions in Health and Disease.* Chichester: John Wiley Ltd.

Kahane, I. & Ofek, I. (1996). Toward Anti-adhesion Therapy for Microbial Diseases. *Advances in Experimental Medicine and Biology,* vol. 408. New York: Plenum Press.

Ofek, I. & Doyle, R. J. (1994). *Bacterial Adhesion to Cells and Tissues.* New York: Chapman & Hall.

Review articles

Anderson, D. M. & Schneewind, O. (1999). Type III machines of Gram-negative pathogens: injecting virulence factors into host cells and more. *Curr Opin Microbiol* **2**, 18–24.

Celli, J., Deng, W. & Finlay, B. B. (2000). Enteropathogenic *Escherichia coli* (EPEC) attachment to epithelial cells: exploiting the host cell cytoskeleton from the outside. *Cell Microbiol* **2**, 1–9.

Censini, S., Stein, M. & Covacci, A. (2001). Cellular responses induced after contact with *Helicobacter pylori. Curr Opin Microbiol* **4**, 41–46.

DeVinney, R., Knoechel, D. G. & Finlay, B. B. (1999). Enteropathogenic *Escherichia coli:* cellular harassment. *Curr Opin Microbiol* **2**, 83–88.

Doyle, R. J. (2000). Contribution of the hydrophobic effect to microbial infection. *Microbes Infect* **2**, 391–400.

Ernst, P. B. & Gold, B. D. (2000). The disease spectrum of *Helicobacter pylori:* the immunopathogenesis of gastroduodenal ulcer and gastric cancer. *Annu Rev Microbiol* **54**, 615–640.

Finlay, B. B. (2000). Fluorescence and confocal microscopy reveal host-microbe interactions. *ASM News* **66**, 601–608.

Foster, T. J. & Hook, M. (1998). Surface protein adhesins of *Staphylococcus aureus. Trends Microbiol* **6**, 484–488.

Goosney, D. L., Gruenheid, S. & Finlay, B. B. (2000). Gut feelings: enteropathogenic *E. coli* (EPEC) interactions with the host. *Annu Rev Cell Dev Biol* **16**, 173–189.

Hedlund, M., Duan, R.D., Nilsson, A., Svensson, M., Karpman, D. & Svanborg, C. (2001). Fimbriae, transmembrane signaling, and cell activation. *J Infect Dis* **183** (Suppl 1), S47–S51.

Hudson, M. C., Ramp, W. K. & Frankenburg, K. P. (1999). *Staphylococcus aureus* adhesion to bone matrix and bone-associated biomaterials. *FEMS Microbiol Lett* **173**, 279–284.

Joh, D., Wann, E. R., Kreikemeyer, B., Speziale, P. & Hook, M. (1999). Role of fibronectin-binding MSCRAMMs in bacterial adherence and entry into mammalian cells. *Matrix Biol* **18**, 211–223.

Kaper, J. B. (1998). EPEC delivers the goods. *Trends Microbiol* **6**, 169–172.

Knight, S.D., Berglund, J. & Choudhury, D. (2000). Bacterial adhesins: structural studies reveal chaperone function and pilus biogenesis. *Curr Opin Chem Biol* **4**, 653–660.

Krause, D. C. (1998). *Mycoplasma pneumoniae* cytadherence: organization and assembly of the attachment organelle. *Trends Microbiol* **6**,15–18.

Law, D. (2000). Virulence factors of *Escherichia coli* O157 and other Shiga toxin-producing *E. coli*. *J Appl Microbiol* **88**, 729–745.

Rosenshine, I., Knutton, S. & Frankel, G. (2000). Interaction of enteropathogenic *Escherichia coli* with host cells. *Subcell Biochem* **33**, 21–45.

Russell, J. C. (2000). Bacteria, biofilms, and devices: the possible protective role of phosphorylcholine materials. *J Endourol* **14**, 39–42.

Sauer, F. G., Mulvey, M. A., Schilling, J. D, Martinez, J. J. & Hultgren, S. J. (2000). Bacterial pili: molecular mechanisms of pathogenesis. *Curr Opin Microbiol* **3**, 65–72.

Schilling, J.D., Mulvey, M.A. & Hultgren, S.J. (2001). Structure and function of *Escherichia coli* type 1 pili: new insight into the pathogenesis of urinary tract infections. *J Infect Dis* **183** (Suppl 1), S36–S40.

Soto, G. E. & Hultgren, S. J. (1999). Bacterial adhesins: common themes and variations in architecture and assembly. *J Bacteriol* **181**, 1059–1071.

Vallance, B. A. & Finlay, B. B. (2000). Exploitation of host cells by enteropathogenic *Escherichia coli*. *Proc Natl Acad Sci USA* **97**, 8799–8806.

Watnick, P. & Kolter, R. (2000). Biofilm, city of microbes *J Bacteriol* **182**, 2675–2679.

Wizemann, T. M., Adamou, J. E. & Langermann, S. (1999). Adhesins as targets for vaccine development. *Emerg Infect Dis* **5**, 395–403.

Zarrilli, R., Ricci, V. & Romano, M. (1999). Molecular response of gastric epithelial cells to *Helicobacter pylori*-induced cell damage. *Cell Microbiol* **1**, 93–99.

Papers

Alfa, M. J., Stevens, M. K., DeGagne, P., Klesney-Tait, J., Radolf, J. D. & Hansen, E. J. (1995). Use of tissue culture and animal models to identify virulence-associated traits of *Haemophilus ducreyi*. *Infect Immun* **63**, 1754–1761.

Bennett-Wood, V. R., Carapetis, J. R. & Robins-Browne, R. M. (1998). Ability of clinical isolates of group A streptococci to adhere to and invade HEp-2 epithelial cells. *J Med Microbiol* **47**, 899–906.

Bian, Z., Brauner, A., Li, Y. & Normark, S. (2000). Expression of and cytokine activation by *Escherichia coli* curli fibers in human sepsis. *J Infect Dis* **181**, 602–612.

Caparon, M. G., Stephens, D. S., Olsen, A. & Scott, J. R. (1991). Role of M protein in adherence of group A streptococci. *Infect Immun* **59**, 1811–1817.

Crane, J. K., Majumdar, S. & Pickhardt, D. F. (1999). Host cell death due to enteropathogenic *Escherichia coli* has features of apoptosis. *Infect Immun* **67**, 2575–2584.

Ginocchio, C. C., Olmsted, S. B., Wells, C. L. & Galan, J. E. (1994). Contact with epithelial cells induces the formation of surface appendages on *Salmonella typhimurium*. *Cell* **76**, 717–724.

Holmes, K. A. & Bakaletz, L. O. (1997). Adherence of non-typeable *Haemophilus influenzae* promotes reorganization of the actin cytoskeleton in human or chinchilla epithelial cells *in vitro*. *Microb Pathogen* **23**, 157–166.

Jones, N. L., Day, A. S., Jennings, H. A. & Sherman, P. M. (1999). *Helicobacter pylori* induces gastric epithelial cell apoptosis in association with increased Fas receptor expression. *Infect Immun* **67**, 4237–4242.

Knutton, S., Rosenshine, I., Pallen, M. J., Nisan, I., Neves, B. C., Bain, C., Wolff, C., Dougan, G. & Frankel, G. (1998). A novel EspA-associated surface organelle of enteropathogenic *Escherichia coli* involved in protein translocation into epithelial cells. *EMBO J* **17**, 2166–2176.

Luo, Y., Frey, E. A., Pfuetzner, R. A., Creagh, A. L., Knoechel, D. G., Haynes, C. A., Finlay, B. B. & Strynadka, N. C. (2000). Crystal structure of enteropathogenic *Escherichia coli* intimin–receptor complex. *Nature* **405**, 1073–1077.

McCormick, B. A., Parkos, C. A., Colgan, S. P., Carnes, D. K. & Madara, J. L. (1998). Apical secretion of a pathogen-elicited epithelial chemoattractant activity in response to surface colonization of intestinal epithelia by *Salmonella typhimurium*. *J Immunol* **160**, 455–466.

Park, Y. & Lamont, R. (1998). Contact-dependent protein secretion in *Porphyromonas gingivalis*. *Infect Immun* **66**, 4777–4782.

Phillips, A. D., Giron, J., Hicks, S., Dougan, G. & Frankel, G. (2000). Intimin from enteropathogenic *Escherichia coli* mediates remodelling of the eukaryotic cell surface. *Microbiology* **146**, 1333–1344.

Ramer, S. W., Bieber, D. & Schoolnik, G. K. (1996). BfpB, an outer membrane lipoprotein required for the biogenesis of bundle-forming pili in enteropathogenic *Escherichia coli*. *J Bacteriol* **178**, 6555–6563.

Reed, K. A., Clark, M. A., Booth, T. A., Hueck, C. J., Miller, S. I., Hirst, B. H. & Jepson, M. A. (1998). Cell-contact-stimulated formation of filamentous appendages by *Salmonella typhimurium* does not depend on the type III secretion system encoded by *Salmonella* pathogenicity island 1. *Infect Immun* **66**, 2007–2017.

Rodrigues, J., Scaletsky, I. C., Campos, L. C., Gomes, T. A., Whittam, T. S. & Trabulsi, L. R. (1996). Clonal structure and virulence factors in strains of *Escherichia coli* of the classic serogroup O55. *Infect Immun* **64**, 2680–2686.

Rosqvist, R., Magnusson, K. E. & Wolf-Watz, H. (1994). Target cell contact triggers expression and polarized transfer of *Yersinia* YopE cytotoxin into mammalian cells. *EMBO J* **13**, 964–972.

Segal, E. D., Cha, J., Lo, J., Falkow, S. & Tompkins, L. S. (1999). Altered states: involvement of phosphorylated CagA in the induction of host cellular growth changes by *Helicobacter pylori*. *Proc Natl Acad Sci USA* **96**, 14559–14564.

Shuter, J., Hatcher, V. B. & Lowy, F. D. (1996). *Staphylococcus aureus* binding to human nasal mucin. *Infect Immun* **64**, 310–318.

Tesh, V. L. (1998). Virulence of enterohemorrhagic *Escherichia coli*: role of molecular cross talk. *Trends Microbiol* **6**, 228–233.

van Schilfgaarde, M., van Ulsen, P., Eijk, P., Brand, M., Stam, M., Kouame, J., van Alphen, L. & Dankert, J. (2000). Characterization of adherence of nontypeable *Haemophilus influenzae* to human epithelial cells. *Infect Immun* **68**, 4658–4665.

Virji, M., Evans, D., Griffith, J., Hill, D., Serino, L., Hadfield, A. & Watt, S.M. (2000). Carcinoembryonic antigens are targeted by diverse strains of typable and non-typable *Haemophilus influenzae*. *Mol Microbiol* **36**, 784–795.

Yassien, M., Khardori, N., Ahmedy, A. & Toama M. (1995). Modulation of biofilms of *Pseudomonas aeruginosa* by quinolones. *Antimicrob Agents Chemother* **39**, 2262-2268.

7.12 | Internet links

1. http://www.med.monash.edu.au/microbiology/teaching/science/ MIC3032/LectureNotes.html
   A set of lecture notes (as Powerpoint presentations) by Dr L. Hartland on bacterial adhesins and the adhesion process.
2. http://www.biotech.ubc.ca/faculty/finlay/homepage.htm
   The homepage of Dr Brett Finlay, one of the leading researchers in the field of bacteria–host cell interactions. Information, and a beautiful collection of images, on the interactions between host cells and *E. coli* and *Salmonella* spp.
3. http://www.helico.com
   The website of the Helicobacter Foundation (founded by Dr Barry Marshall). Everything you ever wanted to know about *Helicobacter pylori*.
4. http://www.bact.wisc.edu/microtextbook/disease/colinize.html
   A set of lecture notes on bacterial adhesion from the undergraduate microbiology course at the University of Wisconsin-Madison, USA.

Bacterial invasion as a virulence mechanism

Chapter outline

Aims

The principal aims of this chapter are:

- to explain what is meant by the term 'invasion'
- to describe the various mechanisms underlying bacterial invasion of host cells
- to emphasise that invasion involves bacterial manipulation of the signalling pathways and cytoskeleton of the host cell
- to provide examples of invasive organisms and the diseases associated with them
- to indicate the consequences of invasion for both the bacterium and the host cell
- to describe the different ways in which bacteria can survive within the host cell

8.1 | Introduction

We have seen in Chapters 1 and 7 that the ability of a bacterium to adhere to a body surface (the mucosa or skin) or structure (e.g. a tooth) is an attribute of many organisms and is not limited to those species

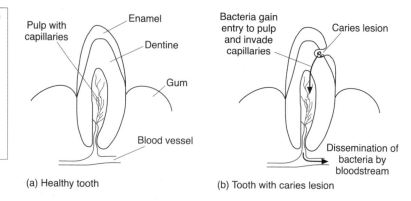

Figure 8.1 Destruction of the enamel layer of the tooth enables bacteria to gain access first to the underlying dentine and then into the pulp with its capillary supply. Invasion of these capillaries then provides a means by which the bacteria can be carried by the bloodstream to all parts of the body.

(a) Healthy tooth

(b) Tooth with caries lesion

that induce disease. Hence, colonisation of epithelial surfaces by members of the normal microflora is, ultimately, of benefit to the host as it can protect against the adhesion of pathogenic species. The adhesion of such benign organisms rarely results in a pathological process, unless the host defences are defective in some way, for example in an immunocompromised individual. However, for some organisms, adhesion is only the first step in a pathological process that involves invasion of the host. Invasion of the epithelium may result in penetration of the organism to only the superficial layers of tissue, as in the case of cholera and gonorrhoea or it may, ultimately, involve deeper penetration, resulting in dissemination of the organism throughout the body. The latter may be achieved via the bloodstream or lymphatic system or, in some cases, may be due to the ability of the organism to survive within phagocytes, which then carry it to other parts of the body. Dissemination of the organism to tissues other than those associated with its portal of entry is characteristic of many infections such as typhoid, meningitis and tuberculosis. The causative organisms of these diseases have evolved an array of virulence factors to protect themselves against the panoply of host defence systems that will come into play once the epithelial barrier has been breached (see Chapters 6 and 10). Such host defence mechanisms are generally very effective as, for most of our lives, the tissues underlying the epithelium are sterile. While we generally associate invasion with the breaching of an epithelial barrier by a bacterium, the term may also be applied to two other types of infectious disease that involve dissemination of the bacterium throughout the body. Firstly, some members of the oral microflora (in particular *Streptococcus mutans*, lactobacilli and *Actinomyces* spp.) can erode the outer surface of the tooth (the enamel) and invade the underlying tissue (dentine) and, in some cases, gain access to the bloodstream (Figure 8.1).

Dental caries

This is an infection that is characterised by the destruction of tooth tissue (enamel and dentine) and is perhaps the most common infectious disease of humans as it affects most of the population of industrialised nations. The surface of the tooth is colonised by bacteria present in saliva (containing approximately 10^8 bacteria/ml), and these then grow and reproduce to form a biofilm. Many factors influence the exact bacterial composition of the resulting biofilm and these include the host individual's genetic make-up, age, diet, oral hygiene practice, etc. Nevertheless, the predominant organisms tend to be viridans streptococci (particularly *Strep. sanguis* and *Strep. oralis*), *Actinomyces* spp. and *Haemophilus* spp. When the individual's diet is rich in fermentable carbohydrate (e.g. sucrose), the acid produced as an end-product of metabolism induces a shift in the bacterial composition of the biofilm, resulting in the selection of organisms (e.g. *Strep. mutans* and lactobacilli) that can tolerate the low pH levels produced. These organisms produce lactic acid as a metabolic end-product and when the pH falls below 5.5, demineralisation of the enamel layer (which consists mainly of hydroxyapatite – a calcium phosphate mineral – and is the hardest structure in the body) of the tooth begins. If unchecked, acid-tolerant bacteria can continue to demineralise the enamel layer and, eventually, the underlying dentine, which consists of collagen fibres and 75% hydroxyapatite. During the progression of the lesion, the microflora often changes. In general, *Strep. mutans* is often the most important organism in the initiation of the carious lesion, while lactobacilli tend to predominate when the lesion has invaded the dentine. Once in the dentine, organisms can then readily gain access to the pulp tissues and hence the bloodstream.

This pathological process (known as caries) is, perhaps, one of the most common invasive diseases of humans. Also, some bacteria, while being incapable themselves of breaching the epithelium, can be 'injected' into the underlying tissues by an arthropod, for example an insect or spider. One could also extend the list of 'invasive' infections by including those arising as a result of damage to the epithelium by trauma (e.g. burns, wounds, animal bites) or tooth extraction, as both of these can result in dissemination, throughout the body, of organisms that are not normally capable of breaching the epithelium themselves (Table 8.1).

In this chapter we will concentrate mainly on those organisms capable of breaching the epithelium of their own volition and those that are transmitted to humans via arthropods. Both groups of organisms can be considered to be 'professional' inhabitants of human internal tissues as they are well adapted to life in this bacterially hostile environment. It is pertinent at this point to ask why bacteria should want to invade a human. As mentioned on several occasions, our tissues are superbly equipped to deal with hostile intruders. The epithelium itself has a range of specific and non-specific antibacterial mechanisms and, once this barrier is overcome, the invader, in order to survive, must be able to protect itself against phagocytic cells (polymorphonuclear neutrophils (PMNs), macrophages), antibody-mediated defences, the complement system, etc. (see Chapters 5 and 6). So why go to all this trouble? The short answer is that any organism that has evolved means of dealing with these elaborate defences will have at its disposal an ideal environment (in terms of its physical and

Table 8.1. | Main categories of infections that may result in the dissemination of bacteria throughout the host

Organism is capable of breaching the epithelium	Organism erodes the tooth structure	Organism is injected into the host by an arthropod	Organism gains access to the underlying tissues following trauma to the epithelium
Salmonella typhi (typhoid fever)	*Streptococcus mutans* (caries)	*Borrelia burgdorferi* (Lyme disease, following tick bite)	Some oral streptococci (infective endocarditis, following tooth extraction)
Neisseria meningitidis (meningitis)	*Streptococcus sobrinus* (caries)	*Yersinia pestis* (plague, following flea bite)	*Pseudomonas* spp. (infections of burns and wounds)
Shigella dysenteriae (dysentery)	*Lactobacillus* spp. (caries)	*Francisella tularaemia* (tularaemia, following tick bite)	*Pasteurella multocida* (pasteurellosis, following cat or dog bite)
Treponema pallidum (syphillis)	*Actinomyces* spp. (caries)	*Rickettsia typhi* (typhus, following flea bite)	*Streptobacillus moniliformis* (rat bite fever, following rat bite)
Mycobacterium tuberculosis (tuberculosis)			*Bacteroides* spp. (peritonitis, following appendicitis or abdominal surgery)
Listeria monocytogenes (listeriosis)			

nutritional needs) for growth and proliferation and there will be virtually no bacterial competitors for this relatively boundless resource.

So far we have implied that 'invasion' involves only the penetration of the epithelium, followed by the survival of bacteria in the underlying tissues. However, the subsequent dissemination of pathogens throughout the body requires the penetration of other cellular structures. In order to gain access to other internal sites, the pathogen may take direct advantage of the circulatory system and simply be carried there by fluid flow. This, of course, will require penetration of the endothelium of the blood or lymphatic vessels. Alternatively, some organisms have adopted a 'Trojan horse' tactic and have evolved means of surviving within phagocytic cells (macrophages or PMNs) and so can be transported within these to other tissues. Some pathogens are capable of doing both. Examples of these different strategies are provided in Table 8.2. Yet another, rather crude, method of reaching other internal sites is for the bacterium to simply 'digest' its way through tissues by enzymic degradation of extracellular matrix molecules. This strategy is sometimes adopted by organisms such as *Clostridium* spp., which produce an enzyme, collagenase, that can degrade muscular tissue. The lifestyle adopted by an organism once it has breached the epithelium is often used as the basis of a classification system for invasive species. Hence, bacteria that generally live within host cells are termed 'obligately intracellular'; those that tend to avoid

| Table 8.2. | Examples of organisms that have been shown to be able either to invade endothelial cells or to survive within phagocytic cells. Some of these organisms (e.g. *Listeria monocytogenes*, *Salmonella typhimurium*) are capable of doing both |

Organisms capable of invading endothelial cells	Organisms capable of surviving in phagocytic cells
Neisseria meningitidis (meningitis)	*Mycobacterium tuberculosis* (tuberculosis)
Streptococcus pneumoniae (meningitis)	*Listeria monocytogenes* (listeriosis)
Escherichia coli (neonatal meningitis)	*Legionella pneumophila* (Legionnaires' disease)
Borrelia burgdorferi (Lyme disease)	*Coxiella burnetii* (Q fever)
β-Haemolytic streptococci (septicaemia, pneumonia)	*Brucella* spp. (brucellosis)
Haemophilus influenzae (meningitis)	*Chlamydia trachomatis* (trachoma, genital infections)
Bartonella henselae (cat-scratch fever)	*Francisella tularensis* (tularaemia)
	Salmonella typhi (typhoid fever)
	Salmonella typhimurium (gastroenteritis)

living in host cells are described as being 'extracellular'. A third category, the 'facultative intracellular bacteria' are those that appear to have no preference but are capable of doing both. Examples of organisms adopting these different lifestyles are provided in Table 8.3.

8.2 | Invasion mechanisms

In Section 7.1 we mentioned the various experimental models that can be used to investigate bacteria–host interactions. Although all of these have been used to study bacterial invasion, most of our knowledge of this process has come from experiments involving host cells in tissue culture.

The main types of host cell invaded by bacteria are epithelial and endothelial cells. It has been mentioned, also, that certain bacteria can also survive within phagocytes, i.e. PMNs and macrophages. However, it is debatable as to whether this constitutes 'invasion' of these professional 'bacteria-eaters'. Essentially, such organisms are phagocytosed and have evolved means of surviving within phagocytes. This particular lifestyle, and the means by which organisms achieve it, is best viewed as a circumvention of the normal host defence mechanisms and, as such, will be described in Section 10.4.2.2.

Bacteria appear to have adopted one of two general strategies for gaining entry to host cells – either they induce changes in the cell's cytoskeleton so that the bacterium is engulfed by the cell or, far less commonly, they forcibly enter the cell without any involvement of the

Table 8.3. | Classification of invasive bacteria based on their preference for living within or outside host cells. The classification is not rigid, as it depends to some extent on host factors. Furthermore, both intracellular and extracellular bacteria must, of necessity, spend at least some of their time respectively outside and within host cells

Obligate intracellular bacteria	Extracellular bacteria	Facultative intracellular bacteria
Chlamydia spp. (pneumonia and genital infections)	*Escherichia coli* (gastroenteritis, meningitis)	*Salmonella* spp. (typhoid, gastroenteritis)
Rickettsia spp. (typhus, Rocky mountain spotted fever)	*Haemophilus influenzae* (meningitis, respiratory tract infections)	*Legionella pneumophila* (Legionnaires' disease)
Coxiella burnetii (Q fever)	*Mycoplasma* spp. (pneumonia, genital infections)	*Brucella* spp. (brucellosis)
Mycobacterium spp. (tuberculosis, leprosy)	*Streptococcus pneumoniae* (pneumonia, meningitis)	*Francisella tularensis* (tularaemia)
Ehrlichia spp. (ehrlichiosis)		*Listeria monocytogenes* (listeriosis) *Shigella* spp. (dysentery) *Yersinia* spp. (plague, gastroenteritis)

cytoskeleton. The cytoskeletal components involved in the former mechanism are either the microfilaments or the microtubules. As some features of the invasion process also depend on the type of host cell being invaded, the mechanisms underlying the invasion of each of these cell types will be described separately. However, prior to any discussion of the mechanisms underlying invasion of host cells, it is essential for the reader to know something about the cytoskeleton of the host cell. The cytoskeleton, unlike the skeleton of vertebrates, is a dynamic structure that is constantly being reorganised to meet the ever-changing needs of the cell. Hence, the cytoskeleton is involved in cellular movement, the trafficking of vesicles within the cell, mitosis and the formation of a range of protrusions. It comprises three major types of protein filaments (Figure 8.2): actin filaments (diameter 6 nm), microtubules (diameter 23 nm) and intermediate filaments (diameter 10 nm). These may be linked together, joined to cell organelles or attached to the cytoplasmic membrane – such linkages often involve the participation of a number of accessory proteins.

An actin filament (also often termed a microfilament) consists of a polymer of a 43 kDa globular protein, actin, arranged to form a helical structure. In the cell, these filaments are organised in the form of linear bundles, two-dimensional networks or three-dimensional gels. The formation of such arrangements is controlled by a large number of accessory proteins; these are listed in Table 8.4. Actin filaments are concentrated at the cell periphery and form a three-dimensional network beneath the cytoplasmic membrane, often known as the 'cell cortex'. This network determines cell shape and movement. It is

ACTIN FILAMENTS

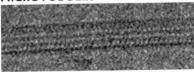

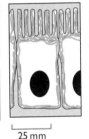

25 nm

25 mm

Actin filaments (also known as *microfilaments*) are two-stranded helical polymers of the protein actin. They appear as flexible structures, with a diameter of 5–9 nm, that are organised into a variety of linear bundles, two-dimensional networks, and three-dimensional gels. Although actin filaments are dispersed throughout the cell, they are most highly concentrated in the cortex, just beneath the plasma membranes.

MICROTUBULES

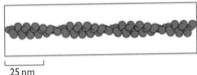

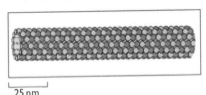

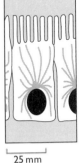

25 nm

25 mm

Microtubules are long, hollow cylinders made of the protein tubulin. With an outer diameter of 25 nm, they are much more rigid than actin filaments. Microtubules are long and straight and typically have one end attached to a simple microtubule-organising center called a centrosome as shown here.

also responsible for maintaining structures such as the microvilli found on many epithelial cells, and for producing transient cell protrusions such as lamellipodia and filopodia. Lamellipodia (or ruffles) are broad, sheet-like extensions of the cell whereas filopodia are long, thin extensions (Figure 8.3). As will be seen later in this chapter, these two structures are invariably produced during bacterial invasion of host cells.

Microtubules are polymers of the globular protein tubulin, which consists of two closely-related 55 kDa proteins – α-tubulin and β-tubulin. These dimers are arranged to form a long, stiff, hollow tube. The microtubules in a cell are attached to a structure known as the centrosome. The major role of these filaments (which can be rapidly polymerised and depolymerised, depending on the needs of the cell) is in the intracellular movement and positioning of cell components such as chromosomes, membrane vesicles and organelles. They act rather like rails along which the above-mentioned structures travel, the motive power being provided by large ATP-hydrolysing proteins (kinesins and dyneins) known as 'motor proteins'. Microtubules are also used to build cilia and flagella.

Intermediate filaments are more varied than the other cytoskeletal components described and can be classified into a number of types depending on the nature of the constituent protein. These groups include keratins, vimentin-related proteins and nuclear lamins. They all have great tensile strength and form a network extending throughout the cell. They enable cells to withstand the stresses associated with

Figure 8.2 Two of the three types of filament that comprise the cytoskeleton of eukaryotic cells. Actin filaments form a network beneath the cytoplasmic membrane and are involved in the formation of microvilli and temporary protrusions (lamellipodia and filopodia). Microtubules emanate from the centrosome and are responsible for the transport and positioning of cell organelles. (Copyright (1994) from *Molecular Biology of the Cell*, 3rd edn, by B. Alberts, D. Bray & J. Lewis. Reproduced by permission of Routledge, Inc., part of the Taylor & Francis Group.)

Table 8.4.	Accessory proteins associated with actin microfilaments
Protein	Function
Tropomyosin	Strengthens actin microfilaments
Thymosin	Binds actin monomers and prevents polymerisation
Cofilin	Regulates actin polymerisation
Fimbrin	An actin-bundling protein that forms closely packed bundles
α-Actinin	An actin-bundling protein that forms more loose structures than fimbrin
Filamin	Cross-links filaments into a gel
Villin	Actin-bundling protein in microvilli
Myosin-II	Moves actin filaments
Ezrin	Links actin filaments to the cytoplasmic membrane

stretching. They do not appear to be involved in bacterial invasion and so will not be discussed further.

8.2.1 Invasion of epithelial cells

The epithelium constitutes the interface between the constantly changing external environment and the controlled, stable conditions that mammalian cells require for their existence. There are, of course, several types of epithelium (keratinised, stratified, ciliated, etc.) depending on their principal functions; the latter include protection, gaseous exchange and nutrient uptake. Regardless of their anatomical location, epithelial surfaces are colonised by the normal microflora and will come into contact with microbes from the external environment that may be highly pathogenic. The epithelium, therefore,

(a) (b)

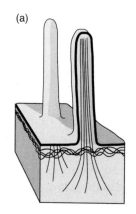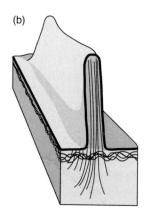

Figure 8.3 The drawings show the temporary protrusions that can be formed by a cell due to rearrangements of actin filaments. Filopodia (a) are long, thin extensions whereas lamellipodia (b) are sheet-like extensions. Either or both of these types of protrusion may be involved in the invasion of host cells by bacteria.

constitutes the first physical barrier preventing access of bacteria to deeper tissues of the body, and invasion of the cells comprising this, physically rather fragile, barrier is the first step in the initiation of a systemic infection. Not surprisingly, in view of their intimate association with the normal microflora, epithelial cells are not generally capable of engulfing particles as large as bacteria. Of the epithelial surfaces exposed to bacteria, those that have evolved to enable uptake of molecules (e.g. oxygen and nutrients) from the external environment (i.e. mucosal epithelia of the lungs and large intestine) constitute the weak links in the host's ability to exclude bacteria from internal tissues as the latter are separated from the external environment by only a single layer of cells.

8.2.1.1 Invasion involving actin rearrangements

The majority of invasive bacteria gain entry into epithelial cells by inducing the rearrangement of the microfilaments of the cytoskeleton. These microfilaments (also known as 'actin filaments') form a network throughout the cell and are particularly dense just beneath the cytoplasmic membrane. They are involved in maintaining the shape of the cell and play a key role in enabling cellular movement and division. Remarkably, however, some bacteria can direct the reorganisation of these filaments within the cell to ensure bacterial uptake. Bacterial invasion of epithelial cells, therefore, is generally not a crude attack on the cell but, rather, a sophisticated manipulation of the cell's own internal structural components. The most extensively studied bacteria utilising this means of entry include *Yersinia* spp., *Salmonella* spp., *Shigella* spp. and *Listeria monocytogenes*. While all of these species have in common the utilisation of microfilament rearrangements as their means of gaining entry to the epithelial cell, there are important differences in the actual invasion process and the subsequent fate of the invading organism. These differences are summarised in Table 8.5.

8.2.1.1.1 *Yersinia* spp.

Three species of the genus *Yersinia* are able to cause disease in humans: *Y. pestis* (plague), *Y. enterocolitica* (gastroenteritis) and *Y. pseudotuberculosis* (gastroenteritis). The latter, however, is primarily a pathogen of animals and is only occasionally responsible for infections in humans. The invasive potential of the three species differs considerably. *Yersinia pestis* invariably spreads systemically whereas the other two species usually invade only the submucosal tissues. A considerable amount is known about epithelial cell invasion by *Y. enterocolitica* and so this will be described in detail below. When the bacterium adheres to an epithelial cell, close contact is made at many points on the bacterial surface – this process is described as 'zippering'. It induces the uptake of the organism into an endocytic vacuole and the bacterium appears to sink into the membrane of the epithelial cell (Figure 8.4). Within a few minutes of bacterial entry, the host cell regains its normal appearance.

Table 8.5. Characteristic features of epithelial cell invasion by bacteria that induce rearrangements of host cell microfilaments

Feature	Yersinia spp.	Salmonella spp.	Shigella spp.	Listeria monocytogenes
Attachment to host cell	Multiple contact	Localised contact	Localised contact	Multiple contact
Morphological alterations in host cell	Pseudopod formation Transient effect	Pseudopod formation Membrane ruffling	Pseudopod formation Membrane ruffling	Pseudopod formation Transient effect
Fate of bacteria	Reside in vacuole	Reside in vacuole	Escape from vacuole	Escape from vacuole
	Little/no replication	Replicate in vacuole	Multiply in cytoplasm	Multiply in cytoplasm
			Move through cytoplasm	Move through cytoplasm
Cell–cell invasion	No	No	Yes	Yes

The first stage in the invasion process involves binding of the bacterium to the surface of the epithelial cell. A number of adhesins of *Y. enterocolitica* have been identified as being important in mediating adhesion to host cells: invasin (encoded by the chromosomal *inv* gene), the Ail protein (encoded by the chromosomal *ail* (attachment–invasion locus) gene) and YadA (encoded by the *yadA* (*Yersinia* adherence) gene located on a 75 kb virulence plasmid). Invasin is an outer membrane protein that binds to those integrins on the surface of the host cell that contain the β_1-subunit. The main function of integrins is to anchor the cell to extracellular matrix (ECM) molecules (e.g. fibronectin and collagen) and to neighbouring cells. Recently, examination of the crystal structure of the invasin of *Y. pseudotuberculosis* has shown that it is very similar to that of fibronectin, even though the amino acid sequences of these two proteins are very different. Interestingly, the invasin has a 100-fold greater affinity for β_1-containing integrins than their natural ligand, fibronectin. Invasin shows homology to the host cell-binding proteins (intimins) of enteropathogenic and enterohaemorrhagic strains of *E. coli* (see Section 7.5.2.1.2.1). The N-terminal regions of these proteins are homologous and near the C-terminus they have two cysteine residues, which, in the case of invasin, form a disulphide bond essential to the proper functioning of the protein. The YadA adhesin which forms a fibrillar matrix on the bacterial surface also binds to β_1-containing integrins as well as to fibronectin, collagen and laminin. The YadA molecules associate on the surface of the bacterium to produce lollipop-shaped complexes. Each complex is approximately 23 nm in length and consists of a rod (which can bind

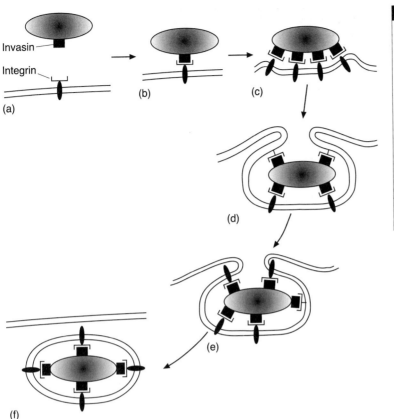

Invasin

Integrin

(a)

(b)

(c)

(d)

(e)

(f)

Figure 8.4 Invasion of an epithelial cell by *Yersinia* spp. Bacterial adhesion (a, b) involves interaction between invasin and integrins (containing β_1-chains) and this induces cytoskeletal rearrangements that result in only modest changes in host cell morphology (c, d). The host cell membrane then 'flows' (d, e) over the surface of the bacterium, resulting in internalisation of the latter. The bacterium ends up within the host cell cytoplasm enclosed in a vacuole (f).

to molecules of the ECM) with a terminal, round-shaped 'head' that is able to mediate adhesion of the organism to host cells. The receptor for Ail (which is an outer membrane protein) remains to be identified. Invasin appears to be the main adhesin involved in mediating invasion of epithelial cells and this process will now be described in detail.

Invasin molecules bind to integrins at a number of points along the bacterium/host cell interface, resulting in very close contact between the two cells. The bacterium then sinks into the cytoplasmic membrane of the host cell (Figure 8.4). To understand the mechanisms underlying this process, it is important to review the normal functions of the integrins. As stated above, these molecules mediate adhesion of the cell to other cells and to the ECM. When cells come into contact with, for example, a surface coated with ECM molecules, the integrin molecules bind to constituents of the matrix and this induces the cell to spread over the surface. This spreading phenomenon leads to tighter binding of the cell to the surface and results from the clustering and immobilisation of the integrins at the site of attachment. This integrin clustering induces the actin polymerisation necessary for cell spreading (Figure 8.5) by a complex signalling process involving protein tyrosine kinases (PTK), guanine nucleotide exchange factors (GEFs)

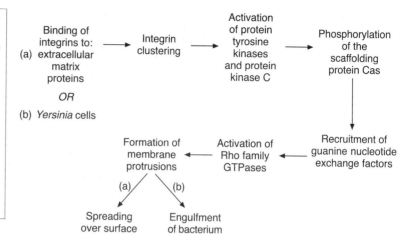

Figure 8.5 Invasion of a host cell by *Yersinia* spp. involves usurping the normal consequences of integrin binding. (a) Usually, integrin binding to host extracellular matrix proteins induces the host cell to 'spread' over the protein, thereby inducing tight binding to the matrix itself or to matrix-coated cells. (b) Binding of a bacterium to the integrin induces 'spreading' but this results in engulfment of the bacterium.

and guanine triphosphatases (Rac, Cdc42 and Rho) also known as monomeric or small G proteins (for a discussion of cell signalling, see Chapter 4). In the case of bacterial invasion, a very similar process occurs. Binding of the bacteria to integrins induces clustering of these molecules and this activates a signalling pathway involving PTK (invasion is blocked by inhibitors of PTK such as genistein) resulting, ultimately, in actin polymerisation (invasion is blocked by inhibitors of this process, e.g. cytochalasin). The net result is that the cell is 'fooled' into spreading over a surface. In this case, however, the 'surface' is a bacterium and this results in engulfment of the latter. Hence the bacterium has manipulated a normal host cell process (spreading) to induce its uptake. The internalised bacterium will, of course, be enclosed within a vacuole that is surrounded by polymerised actin. Within this vacuole, the bacterium can survive, although it appears to be unable to reproduce.

Although *Y. enterocolitica* is able to invade intestinal epithelial cells *in vitro* it is doubtful whether this occurs *in vivo* in the manner described above. This is because integrins are not found on the apical surfaces (i.e. facing the intestinal lumen) of epithelial cells but only on the opposite (basolateral) surfaces. This neatly illustrates the danger of extrapolating too far from studies carried out using inadequate laboratory models. Whenever possible, *in vitro* studies need to be supplemented by *in vivo* investigations, either in humans or in other animal species that suffer a similar pathology. Studies of infections due to *Yersinia* spp. in animals have shown that the bacteria attach to and invade the microfold cells (M cells) of Peyer's patches (see Sections 5.7.1 and 6.3.5). M cells are specialised epithelial cells capable of actively transporting antigenic material (both soluble and particulate) from the intestinal lumen across the epithelium. The material is exocytosed from the basolateral membrane, which is usually invaginated to form a pocket. This pocket contains lymphocytes (as well as macrophages and other cells) that therefore come into direct contact with the transcytosed antigenic material. M cells are, thus, important in

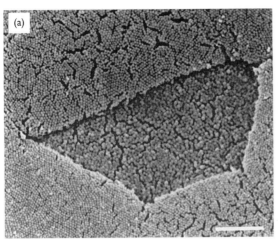
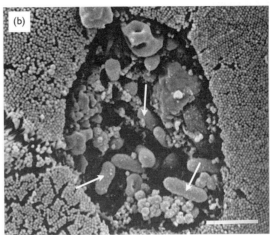

initiating mucosal immune reactions. The apical membrane of M cells contains integrins to which invasin binds, thereby inducing bacterial uptake as described above. Evidence in support of this route of invasion comes from studies involving mice. Yersiniosis in mice closely resembles human *Yersinia* infections and it has been shown that *Y. enterocolitica* selectively invades the Peyer's patches of mice via M cells – bacteria can be seen inside M cells within 1 hour of infection (Figure 8.6). As *Yersinia* spp. can evade phagocytosis (see Section 10.4.2.1), the bacteria are able to multiply in the underlying tissue. Destruction of the follicle-associated epithelium is observed several days after infection. It would appear, therefore, that invasion of the intestinal epithelium by *Y. enterocolitica in vivo* occurs via the M cells of Peyer's patches. Once the organism has traversed the epithelium, it may, of course, invade epithelial cells from the basolateral surface and this could take place in a manner similar to that for invasion of epithelial cells *in vitro* above (Figure 8.7). It appears that M cells can be used as a portal of entry to underlying tissues by a number of organisms including *Salmonella typhi*, *Sal. typhimurium*, *Shigella flexneri*, and *Campylobacter jejuni*. In this chapter we have described how *Yersinia* spp. invade host cells. However, an essential feature of the pathogenesis of diseases due to these organisms is their ability to resist uptake by phagocytic cells (a process mediated by the products of a type III secretion system; see Section 2.5.5) and this will be described in Section 10.4.2.1.

Figure 8.6 Scanning electron micrograph images of mouse Peyer's patch follicle-associated epithelium incubated for 60 minutes with *Yersinia pseudo-tuberculosis*. Both images show a central M cell surrounded by enterocytes. The M cell in (a) does not have any adherent bacteria and has a normal surface morphology. In (b) several bacteria (arrowed) have adhered to, and invaded, the M cell thereby disrupting its surface. Bars = 2 μm. (Reproduced with permission from Clark *et al.*, 1998. See Section 8.9.)

8.2.1.1.2 *Listeria monocytogenes*

The Gram-positive organism *Lis. monocytogenes* is the cause of an uncommon, but serious, form of food poisoning, with a mortality rate as high as 70%. It is one of the few bacteria capable of crossing the placenta and this can result in pre-term labour, stillbirth, neonatal meningitis or sepsis.

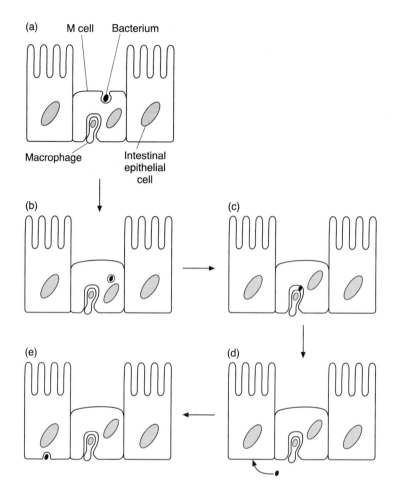

Figure 8.7 Invasion of intestinal epithelial cells. *Yersinia* spp. can cross the intestinal epithelium by invading an M cell (a, b) and, as it can resist phagocytosis by associated macrophages (c, d), it can then invade an epithelial cell via its basolateral surface (d, e).

Invasion of epithelial cells by the organism takes place by a zippering mechanism similar to that described for *Yersinia* spp. (Figure 8.8). In contrast to invasion by *Salmonella* and *Shigella* (see Table 8.5), no membrane ruffling occurs and the cytoskeletal rearrangements involved are highly localised. The bacteria are able to escape from the vacuole and then multiply and move within the cytoplasm by a mechanism involving the formation of actin tails (described in Section 8.4.1.2.1).

Very little is known about the means by which the organism binds to host cells, although it has been suggested that the surface protein ActA (which is involved in intercellular movement of the bacterium, see below), may be an important adhesin – the receptor(s) for this being heparin and/or heparan sulphate. Pre-treatment of bacteria with either of these molecules inhibits invasion of epithelial cells, and heparinase treatment of host cells, which degrades the receptors on these cells, results in a substantial reduction in bacterial adhesion and invasion. Other potential adhesins include a 104 kDa surface protein (p104) and Ami, a surface-exposed autolysin.

Following adhesion to an epithelial cell, invasion is mediated by an

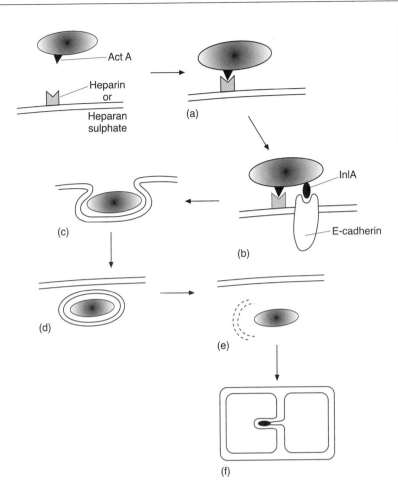

Figure 8.8 Invasion of epithelial cells by *Listeria monocytogenes*. Initial adhesion (a) is achieved by binding of the adhesin ActA to its receptor (heparin or heparan sulphate). This is followed by binding of InlA to its receptor, E-cadherin (b), and this triggers actin rearrangements similar to those accompanying invasion by *Yersinia* spp. The organism is taken up in a vacuole (c, d) but escapes from this (e) and moves around within the cell. It can then invade a neighbouring cell via a membrane protrusion (f).

80 kDa surface-associated protein known as **internalin A** (InlA). Mutations in the *inlA* gene render the organism incapable of invading intestinal cells, while coating beads with InlA ensures their uptake by such cells (Figure 8.9). Near its N-terminus, InlA contains 15 successive copies of a 22 amino acid residue leucine-rich motif ('leucine-rich repeat') and its C-terminus is similar to that of the M protein of *Strep. pyogenes*. The leucine-rich repeats are characteristic of a protein involved in strong protein–protein interactions. The receptor for this molecule is E-cadherin, a transmembrane glycoprotein that plays a central role in cell–cell adhesion. The molecule consists of three main regions: a cytoplasmic domain that interacts with cytoskeletal components (via catenins, which also have signalling functions), a transmembrane region, and an extracellular domain that mediates cell–cell adhesion. The organism also has another important adhesin, internalin B (InlB). This is a 67 kDa protein that also has leucine-rich repeats but the host cell receptor for this molecule has not yet been identified. It would appear that InlA and InlB mediate invasion in different cell types. InlA is crucial for the invasion of human intestinal

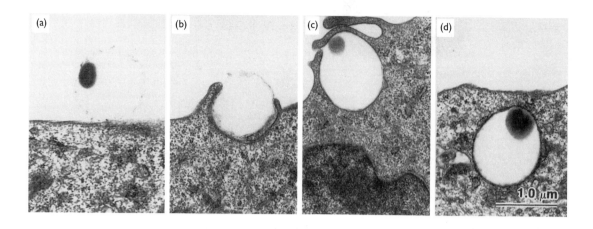

(a) (b) (c) (d)

Figure 8.9 Internalisation of InlA-coated latex beads by human epithelial cells. The sequence of electron micrographs illustrates the 'zipper' type of internalisation mechanism. (Reproduced with permission, from Lecuit *et al.*, 1997. See Section 8.9.)

cell lines (e.g. Caco-2 cells) whereas InlB functions as an invasin for HeLa cells, Vero cells, endothelial cells and hepatocytes.

Little is known of the mechanisms underlying invasion of epithelial cells by *Lis. monocytogenes*. Invasion can be inhibited by cytochalasin D, implicating actin rearrangements in the process, and by tyrosine kinase inhibitors (e.g. genistein), suggesting the involvement of these signalling molecules. Recent evidence points to an early step in the process that may involve the activation of phosphoinositide 3-kinase (see Section 4.2.3.1.2), which is a signalling protein involved in actin polymerisation. Activation of this lipid kinase requires the participation of a tyrosine kinase. Internalisation of the organism results in entrapment of the bacterium within a vacuole from which it subsequently escapes, as described in Section 8.4.1.2.1 .

Listeriosis

Infections due to the food-borne pathogen *Lis. monocytogenes* are now recognised as being a major public health problem in many countries. The organism has unique invasive abilities and is one of the few capable of crossing three of the main barriers against infection – the intestinal epithelium, the blood–brain barrier and the placental barrier – hence causing a wide variety of diseases including gastroenteritis and meningitis. Its ability to cross the placental barrier and cause infections *in utero* is very unusual and the consequences can be tragic – spontaneous miscarriage, still-birth, premature delivery and neonatal sepsis (often with meningitis). Unfortunately, the recognition of an infection *in utero* is very difficult, and it is usually too late to save the fetus, as the symptoms in the mother are often minimal. Little is known of the means by which the organism crosses the placenta. Another distinguishing characteristic of *Lis. monocytogenes* is its very wide temperature range for growth – from 4 °C to 44 °C. Hence, unlike most food-borne pathogens, it is able to grow in household refrigerators. This means that even though a food may initially be minimally contaminated with the organism, growth to disease-inducing levels may occur during storage in a refrigerator. Preventing the contamination of food, therefore, is a very important means of controlling listeriosis. Unfortunately, this is rendered difficult by yet another disturbing characteristic of the organism – it is ubiquitous in the environment and is present in the gastrointestinal tracts of a wide range of animals, including humans. It has been estimated that up to 15% of all foods are contaminated with *Lis. monocytogenes*.

8.2.1.1.3 *Shigella* spp.

This genus consists of four species, *Shig. dysenteriae*, *Shig. flexneri*, *Shig. boydii* and *Shig. sonnei*, all of which can cause gastrointestinal infections in humans. The severity of the infection depends on which species is involved, but all induce a diarrhoea that may contain mucus and/or blood. In severe cases, destruction of the colonic mucosa occurs. The disease is contracted by ingestion of contaminated water or food and is generally self-limiting in adults. In children, however, it can be fatal and it is a major killer of infants in some developing countries, where it is believed to be responsible for the death of more than 500 000 children per year.

Shigella spp. are not able to invade intestinal epithelial cells via the apical surfaces but can do so via their basolateral surfaces and, like *Yersinia* spp., they cross the intestinal epithelium via M cells. The organism is then usually ingested by the macrophages present in the basolateral pocket of the M cell but, as will be described in Section 8.4.1.2.2, can escape from these cells after inducing apoptosis in them. They can then invade epithelial cells via their basolateral membranes. Invasion is, visually, a more spectacular event than that occurring with *Yersinia* spp. and *Listeria* spp. and is described as occurring by a 'trigger' mechanism rather than by 'zippering'. Within minutes of the bacterium contacting an epithelial cell, dramatic cytoskeletal rearrangements occur, resulting in the formation of large pseudopods (filopodia), that can extend from the cell surface for tens of micrometres, and membrane ruffles (lamellipodia) – the overall effect resembles, visually, a stone being dropped into water (Figure 8.10). These pseudopods extend beyond the bacterium, fold around it, then fuse together, so entrapping the organism in a membrane-bound vacuole, the whole process taking between 5 and 10 minutes. The appearance of the cell's surface then returns rapidly to normal. The overall process resembles macropinocytosis. The ingested bacterium soon escapes from the vacuole, multiplies, and moves through the cytoplasm until it meets the cell membrane. A protrusion is then formed that penetrates the adjacent cell and this is followed by lysis of the membranes, so enabling the organism to gain access to the other cell (Figure 8.11). The bacteria do not normally spread systemically but are confined to the subepithelial tissues.

Little is known of the adhesive structure or the adhesin responsible for mediating attachment of *Shigella* spp. to colonic epithelial cells, although lipopolysaccharide (LPS) has been implicated in this respect. Thirty-two bacterial genes are associated with epithelial cell invasion and these are arranged in two regions of a 220 kb virulence plasmid. One of these contains the *ipa* (invasion plasmid antigen) genes, which

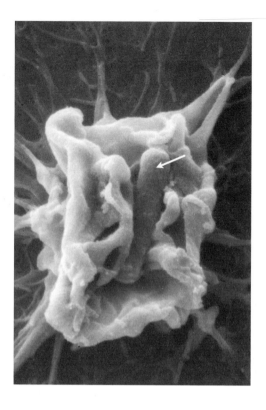

Figure 8.10 Scanning electron micrograph showing the membrane ruffling induced by *Shigella flexneri* (arrowed) on contacting an epithelial cell. (Courtesy of Ariel Blocker, Sir William Dunn School of Pathology, University of Oxford.)

encode a series of proteins (IpaA–D) crucial to the entry process. The other consists of the *mxi-spa* (**m**embrane e**x**pression of **I**pas-**s**urface **p**resentation of **a**ntigens) locus, which is a type III secretion system responsible for ensuring the export of the Ipa proteins. Not much is known of the organisation or control of this secretory system in *Shigella* spp., although the secretory apparatus (also termed the 'secreton' or 'injectisome') has a structure similar to that found in *Salmonella* spp. (Figure 8.12). As the components of such systems are highly conserved at the protein sequence level, comparison of *Shigella* proteins with those of known function from other organisms gives us some idea of their likely function (Table 8.6). Within 5 minutes of contact with the epithelial cell, approximately 15 proteins are secreted via the secretory channel. The most abundant of these are the four Ipa proteins (A, B, C, D) and Ipg (**i**nvasion **p**lasmid **g**ene) and, of these, IpaB, IpaC and IpaD are essential for the invasion of HeLa cells and for virulence *in vivo*. The IpaB and IpaC proteins form a soluble complex, known as the Ipa complex, while IpaD (in association with IpaB) functions as a 'plug' blocking the secreton. On release, the IpaB/C complex binds to the fibronectin receptor ($\alpha_5\beta_1$-integrin) present on the basolateral surface of epithelial cells and can also bind to the hyaluron receptor, CD44. Both $\alpha_5\beta_1$-integrin and CD44 are linked to the underlying cytoskeleton and so they could, following their interaction with the Ipa complex, trigger the cytoskeletal changes associated with invasion. However, it is thought that binding of the IpaB/C complex to these molecules is a

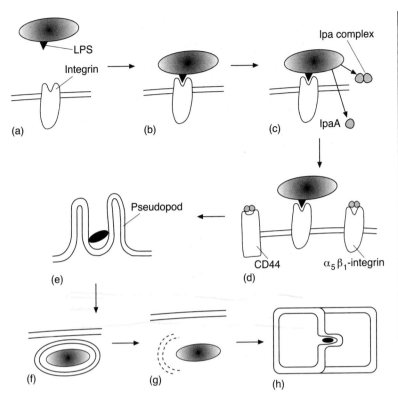

Figure 8.11 Invasion of epithelial cells by *Shigella* spp. Initial adhesion is, possibly, achieved by binding of lipopolysaccharide (LPS) to an integrin (a, b) and this induces the injection of IpaA into the host cell (via a type III secretion system) and the secretion of the Ipa complex (c). The latter is able to bind to both the $\alpha_5\beta_1$-integrin and to the hyaluron receptor (CD44) and this results in extensive cytoskeletal rearrangements and pseudopod formation (d, e). The bacterium is taken up in a vacuole (f) from which, like *Listeria monocytogenes*, it escapes (g) and can move within the cell. The bacterium is able to invade a neighbouring cell via a cytoplasmic protrusion (h).

transient event and is followed by insertion of the complex into the host cell membrane to form a pore through which other secreted bacterial proteins (particularly IpaA, a 78 kDa protein) can enter the host cell. The pore-forming IpaC protein and the injected IpaA protein trigger signalling pathways that induce a complex series of cytoskeletal rearrangements resulting in the uptake of the bacterium. IpaC induces the activation of small GTPases of the Rho family – Cdc42, Rac and Rho – which are involved in the formation of filopodia and lamellipodia (see Section 4.2.5.2). These structures result from actin reorganisation, and inhibitors of F-actin, such as cytochalasin, prevent invasion by *Shigella* spp. F-actin polymerisation occurs beneath the point of contact with the bacterium and in the extensions that form around it. The IpaA injected into the host cell binds directly to vinculin, which induces the accumulation of α-actinin and talin to the membrane of what is destined to become the bacterium-enclosing vacuole (Figure 8.13). The protein tyrosine kinase pp60^{c-src}, which is involved in actin cytoskeletal rearrangements, also accumulates at the site of bacterial entry and is thought to be responsible for tyrosine phosphorylation of cortactin. This protein co-localises with newly formed actin filaments and is probably involved in the cytoskeletal remodelling essential for bacterial uptake.

Following internalisation, the bacteria soon escape from the vesicle into the cytoplasm, where they are able to grow and divide (see Section 8.4.1.2.2).

Figure 8.12 The secretory apparatus of the Mxi-Spa type III secretion system of *Shigella* spp. The structures (50 to 100/cell) can be seen dispersed over the cell surface (a) with a needle-like structure protruding into the external environment (insert). Arrows indicate secretons at the margin of the bacterium. Bar = 200 nm. Image (b) shows that each structure is composed of three parts: (i) a needle 11 nm in diameter, (ii) a neck 10 nm high and 21 nm wide, and (iii) a bulb 44 nm wide and 27 nm high. (Reproduced from *The Journal of Cell Biology*, 1999, **147**, 683–693, by copyright permission of The Rockefeller University Press.)

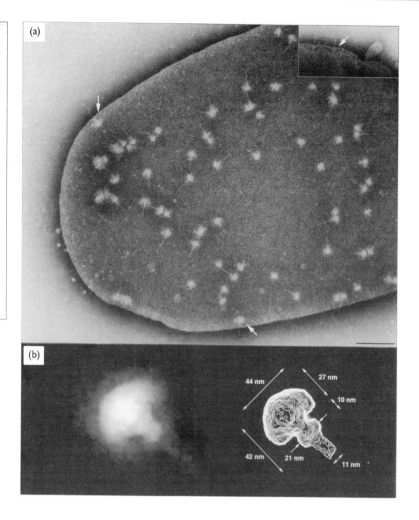

8.2.1.1.4 *Salmonella* spp.

The genus *Salmonella* includes a number of human pathogens. Examples include *Sal. typhi*, the causative agent of typhoid fever, a systemic infection contracted from contaminated food or water, and *Sal. typhimurium*, which rarely penetrates beyond the submucosal tissues and causes gastroenteritis following its ingestion in food, milk or water. Most studies of invasion by *Salmonella* spp. have been carried out using *Sal. typhimurium*, which, in mice, causes a systemic disease similar to typhoid fever in humans.

Invasion of intestinal epithelial cells by *Salmonella* spp. occurs by a trigger mechanism following adhesion to the microvilli. Like *Shigella* spp., therefore, invasion by these organisms also involves dramatic changes in the morphology (pronounced membrane ruffling in a 'splash' arrangement) of the epithelial cell. Unlike *Yersina* spp. and *Shigella* spp., *Salmonella* spp. can invade the apical surfaces of the intestinal epithelium. However, *Salmonella* spp. can also cross the

| Table 8.6. | Components of the type III secretion system displaying homology among intestinal pathogens |

Salmonella spp.	Shigella spp.	EPEC	Yersinia spp.	Possible location	Possible function in one or more species
InvA	MxiA	EscV	LcrD	Cytoplasmic membrane	Forms channel in cytoplasmic membrane
SicA	IpgC	–	SycD	Cytoplasm	Molecular chaperone
InvC	Spa47	EscN	YscN	Cytoplasm / membrane	ATPase – energiser of secretion system?
InvG	MxiD	EscC	YscC	Outer membrane	Forms channel in outer membrane
InvE	MxiC	–	YopN	Outer membrane?	Regulation of secretion
PrgK	MxiJ	EscJ	YscJ	Outer membrane	?
InvL	Spa24	EscR	YscR	Cytoplasmic membrane	Translocase
SipB	IpaB	–	YopB	Outer membrane or secreted	Pore formation in host cell membrane

EPEC, enteropathogenic *Escherichia coli*; Inv, invasion; Mxi, membrane expression of Ipa; Esc, EPEC secretion; Lcr, low calcium response; Sic, *Salmonella* invasion chaperone; Ipg, invasion plasmid gene; Syc, specific Yop chaperone; Spa, surface presentation of antigen; Ysc, Yop secretion; Yop, *Yersinia* outer membrane protein; Prg, *phoP*-repressed gene; Ipa, invasion plasmid antigen; Sip, *Salmonella* invasion protein.

intestinal epithelium via M cells and this may well be the main portal of entry *in vivo*.

At least four different types of fimbriae have been identified on *Salmonella* spp.: (i) type 1, (ii) plasmid-encoded, (iii) long polar and (iv) thin, aggregative (known as curli). While each of these has been shown to mediate adhesion to one or more cell types, their role (if any) in invasion remains to be established. It is possible that the rather loose bacterium/host cell association achieved by fimbriae may act as a stimulus for the synthesis of, as yet unidentified, adhesins that mediate the close contact necessary to induce invasion of the host cell. Fimbria-mediated contact with the host cell has been shown to induce the appearance of bacterial surface appendages (termed invasomes), comprising unidentified proteins, that disappear immediately prior to invasion (Figure 8.14). Mutants unable to produce invasomes are able to adhere to, but not invade, epithelial cells. The receptor(s) for the adhesin(s) responsible for inducing bacterial uptake has not yet been identified.

Following adhesion of the organism to an epithelial cell, the microvilli disappear and membrane ruffling occurs, resulting in the uptake

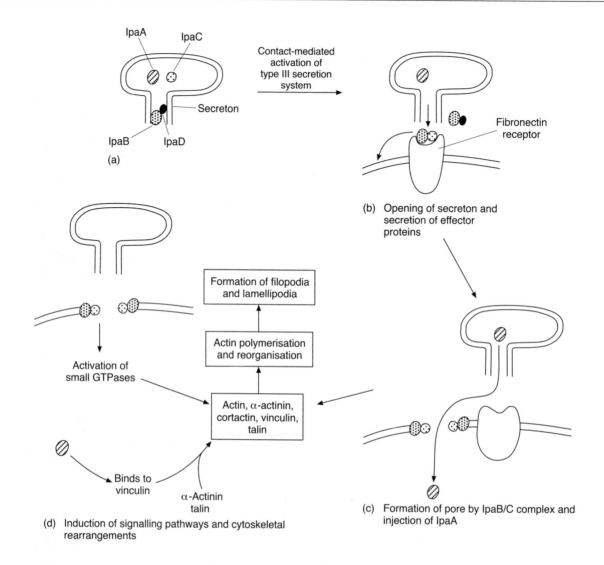

IpaA
IpaC

Contact-mediated
activation of
type III secretion
system

Secreton
Fibronectin
receptor

IpaB   IpaD

(a)

(b) Opening of secreton and
secretion of effector
proteins

Formation of filopodia
and lamellipodia

Actin polymerisation
and reorganisation

Activation of
small GTPases

Actin, α-actinin,
cortactin, vinculin,
talin

Binds to
vinculin
α-Actinin
talin

(c) Formation of pore by IpaB/C complex and
injection of IpaA

(d) Induction of signalling pathways and cytoskeletal
rearrangements

Figure 8.13 Sequence of events occurring prior to, and following, adhesion of *Shigella* spp. to epithelial cells. Adhesion (a) activates a type III secretion system (b) resulting in the secretion of effector molecules, some of which are injected into the host cell (c). This results in induction of signalling pathways and cytoskeletal rearrangements (d).

(macropinocytosis) of the bacterium into a large fluid-filled vacuole. The whole process takes only a few minutes (Figure 8.15).

It has been demonstrated that invasion is dependent on the presence of at least 28 genes located on a pathogenicity island known as *Salmonella* pathogenicity island 1 (SPI-1). SPI-1 has genes encoding a type III secretion system as well as secreted proteins and regulatory components, as summarised in Table 8.7. Recently, the secretory apparatus (secreton) has been detected by electron microscopy and has been shown to consist of a syringe-shaped structure (of which there are between 10 and 100 per cell) that spans the inner and outer membranes, with the narrow 'needle' portion projecting out of the cell (Figure 8.16). The components of the secretion apparatus include some of the Inv, Spa and Prg (*phoP*-repressed gene) proteins as well as OrgA (oxygen-related gene A). The precise location and role of each of these many proteins have not yet been determined but, based on

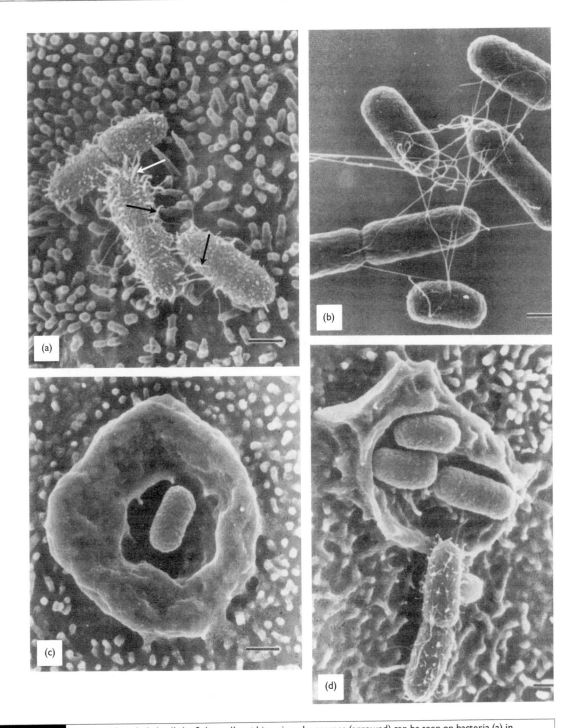

Figure 8.14 Invasion of epithelial cells by *Salmonella typhimurium*. Invasomes (arrowed) can be seen on bacteria (a) in contact with epithelial cells but are absent from those that are not in contact (b). Part (c) shows the formation of a characteristic membrane ruffle around a bacterium that is now devoid of invasomes. Bacteria that have not yet induced membrane ruffling still have their invasomes (d). Bars = 0.5 μm. (Reprinted from *Cell*, **76**, Ginocchio, C. C., Olmsted, S. B., Wells, C. L. & Galan, J. E., Contact with epithelial cells induces the formation of surface appendages on *Salmonella typhimurium*, 717–724, 1994, with permission from Elsevier Science.)

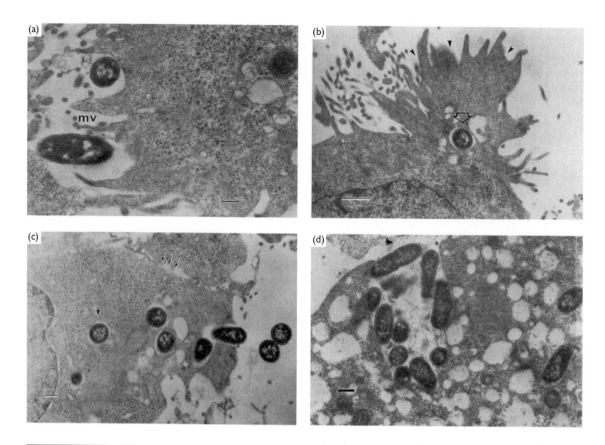

Figure 8.15 Invasion of human intestinal epithelial cells by *Salmonella typhi*. After 10 minutes (a), the bacteria are seen to interact with shortened microvilli (mv). After 20 minutes (b), a bacterium can be seen to be engulfed (open arrow) by the cell and a characteristic membrane 'splash' (arrowheads) can be seen on the host cell. (c) After 30 minutes, many bacteria can be seen to have been internalised at a site adjacent to the junctional space (arrows). After 2 hours (d), the entire monolayer shows the presence of numerous vacuoles and clusters of internalised bacteria. (a), (c), and (d), bar=0.5 μm; (b), bar =1 μm. (Reproduced with permission, from Huang et al., 1998. See Section 8.9.)

homology with the better-characterised secretons of *Shigella* spp. and *Yersinia* spp., certain predictions have been made. Hence, the base (i.e. that portion embedded in the cytoplasmic membrane) is likely to be composed of InvA, InvC, SpaP, SpaQ, SpaR, SpaS and PrgH, while InvG and InvH are though to comprise the ring embedded in the outer membrane. This secretion system enables the injection of a number of effector molecules that induce the cytoskeletal rearrangements involved in the uptake of the bacterium by the cell. While some of these effector molecules are encoded by SPI-1 (SipA, SptP (secreted protein tyrosine phosphatase P), AvrA (avirulence protein A)), others (e.g. Sop (**S**almonella **o**uter **p**rotein) B and SopE) are not.

The means by which the uptake of the bacterium into the host cell is induced is a very active area of research and the roles of individual effector molecules in this process remain to be elucidated (Figure 8.17). There is even some doubt regarding which molecules are the actual 'effectors', as some of the proteins (e.g. SipB, SipC) involved in translocation of recognised effectors may also modulate host cell function. Currently, the role of only a few effector molecules has been investigated in any detail (Table 8.8) and two of these (SopE and SptP) appear to induce cytoskeletal rearrangements by regulating the activity of the Rho family (Rac, Rho and Cdc42) of GTPases. Activation of any of the latter results in actin cytoskeletal rearrangements. SopE is capable of activating Cdc42 and Rac1 and, when injected into host cells, induces

Table 8.7. | Proteins involved in invasion which are encoded by genes on *Salmonella* pathogenicity island 1 (SP1-1)

Function of protein	Examples
Component of secretion apparatus	InvH, InvG, InvA SpaO
Involved in regulation of the secretion process	InvF, HilA
Cytoplasmic chaperones	SicA, InvI
Involved in translocation of effector molecules into host cell	SipB, SipC, SipD
Effector molecules (demonstrated or putative)	InvJ, SptP, SopE

Inv, invasion; Hil, hyperinvasive locus; Sic, salmonella invasion chaperone; Sip, *Salmonella* invasion protein; SptP, secreted protein tyrosine phosphatase; Sop, *Salmonella* outer protein.

membrane ruffling. SptP is a protein tyrosine phosphatase that requires SipB, SipC and SipD for translocation into the host cell cytoplasm. The N-terminal region is homologous to the YopE of *Yersinia* spp. (which is involved in disruption of the host cell cytoskeleton), while the C-terminal domain is similar to the catalytic domain of eukaryotic protein tyrosine phosphatases and of the tyrosine phosphatase of *Yersinia* spp., YopH. SptP can also inactivate Cdc42 and Rac1 and so counteracts the action of SopE. It may, therefore, be involved in restoring normal cell architecture once bacterial invasion has taken place. SipA, once injected into the host cell, accumulates at the site of membrane ruffles, where it forms a complex with T-plastin and F-actin. The complex stimulates actin polymerisation (and inhibits its depolymerisation) and helps to extend the membrane ruffles to form the pseudopodia that eventually engulf the bacterium. SipC, which is involved primarily in protein translocation, also appears to function as an effector molecule. After delivery into the host cell, it inserts itself into the cytoplasmic membrane where it induces actin polymerisation

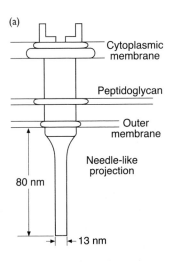

(a)

Cytoplasmic membrane

Peptidoglycan

Outer membrane

Needle-like projection

80 nm

13 nm

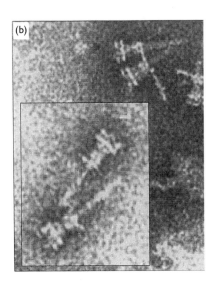

(b)

Figure 8.16 (a) Diagram showing the basic structure of the type III secretion apparatus (secreton) of *Salmonella* spp. (b) Electron micrograph of the secreton of *Sal. typhimurium*. Bar = 50 nm. (Reproduced with permission from Kimbrough, T. G. & Miller, S. I. Contribution of *Salmonella typhimurium* type III secretion components to needle complex formation. *Proceedings of the National Academy of Sciences, USA*, 2000, **97**, 11008–11013. Copyright 2000, National Academy of Sciences, USA.)

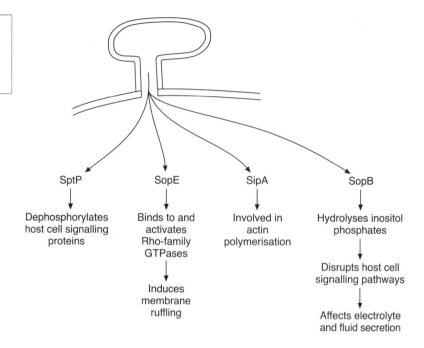

Figure 8.17 Signalling pathways and cytoskeletal rearrangements involved in the uptake of *Salmonella* spp. by epithelial cells.

and bundles together actin filaments. It may, therefore, be involved in the initiation of membrane ruffling. SopB of *Sal. dublin* (which is 98% homologous to SigD of *Sal. typhimurium*) is encoded by another pathogenicity island (SPI-2) but is secreted via the SPI-1 secretion system. It hydrolyses a range of phosphoinositides and therefore can interfere with a number of host cell signalling pathways.

Once complete, the invasion process results in the bacterium residing in a large, fluid-filled vacuole known as a 'spacious phagosome' that is surrounded by polymerised actin and other proteins such as talin and α-actinin (Figure 8.18). The subsequent development of these vacuoles is described later (Section 8.4.1.1.2). As the vacuoles do not become acidified and do not contain lysosomal hydrolytic enzymes, the bacteria survive. Four to six hours after invasion, proliferation

Table 8.8. | Effector proteins secreted by the SPI-1 secretion system

Effector protein	Activity	Function in invasion
SopE	Activates Cdc42 and Rac1	Initiation of membrane ruffling
SptP	Inactivates Cdc42 and Rac1	Restoration of normal cytoskeleton
SipA	Stimulates actin polymerisation	Formation of pseudopodia
SipC	Polymerises actin and bundles actin filaments	Initiation of membrane ruffling
SopB	Inositol phosphate phosphatase	Interferes with signalling pathways

Sop, *Salmonella* outer protein; SptP, secreted protein tyrosine phosphatase; Sip, *Salmonella* invasion protein.

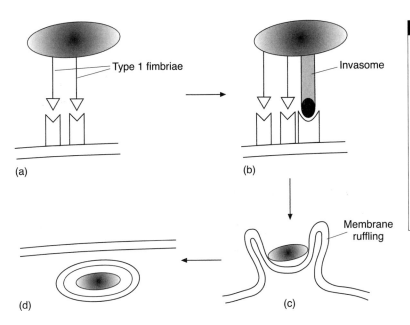

Type 1 fimbriae

Invasome

(a)

(b)

Membrane
ruffling

(d)

(c)

Figure 8.18 Invasion of epithelial cells by *Salmonella* spp. Initial adhesion (a) is mediated by fimbriae, which then induce the formation of invasomes (b). Proteins are injected into the host cell via the secretons and these interfere with host cell signalling pathways and induce the cytoskeletal rearrangements essential for engulfment of the bacterium (c). The bacterium is taken up by the cell and resides in a vacuole (d).

takes place and this is accompanied by the formation of lysosomal glycoprotein-containing fibrillar structures (*Salmonella*-induced filaments, Sifs) attached to the phagosome. The formation of Sifs, which are thought to be involved in the delivery of nutrients to the bacteria, is a microtubule-dependent process.

It has been known for a long time that invasion is affected by a number of environmental factors including oxygen levels, osmolarity and bacterial growth phase. Hence invasion is enhanced under anaerobic conditions, when the bacteria are in the stationary phase of growth and when osmolarity is high. More recently, it has been established that expression of SPI-1 genes is maximal when the temperature is 37 °C, oxygen is limiting, the pH is neutral, osmolarity is high and bacteria are in the late-log phase of growth. However, little is known with regard to how these environmental factors control the expression of genes associated with invasion, although it is thought to involve a two-component regulatory system and the products of the *hilA* (hyperinvasive locus) and *sirA* (*Salmonella* invasion regulator) genes.

Invasion of mast cells – a cellular refuge from host defence systems

Mast cells are an important component of the immune response to bacterial invasion (see Section 6.3.3). These cells have IgE bound to their surface and, when bacterial antigens bind to these cell-associated antibodies, the cells are activated. Activation results in the release of vasoactive amines (histamine and serotonin) that increase blood flow to the area, as well as vascular permeability, thereby causing an accumulation of fluid and PMNs in the surrounding tissue. Activation also induces the release of a number of cytokines that help to orchestrate an effective immune response to the infecting bacteria. It has been reported recently that caveolae (recesses on the surface that are involved in endocytosis and signalling) on mouse bone marrow-derived mast cells contain the receptor (CD48) for strains of *Escherichia coli* with the FimH adhesin. Binding of bacteria to these receptors was found to induce their uptake by the cell. The resulting vacuoles did not fuse with lysosomes, so enabling the bacteria to survive. Bacteria can, therefore, 'hide' within mast cells, thereby avoiding not only cells of the immune system but also many antibiotics.

8.2.1.2 Invasion involving microtubules

Microtubules are an important cytoskeletal component and are dynamic structures that are constantly undergoing polymerisation and depolymerisation. Their involvement in bacterial invasion can be demonstrated by using specific inhibitors of these processes. Colchicine, for example, induces depolymerisation of microtubules, while taxol stabilises them and therefore inhibits depolymerisation. Inhibition (either total or partial) of bacterial invasion by one of these agents therefore implicates microtubules in the invasion process. While it is apparent that microfilaments are involved in the invasion of host cells by many bacterial pathogens, in some cases rearrangement of microtubules is crucial to the invasion process (Table 8.9). However, bacteria rarely invade epithelial cells by inducing rearrangements of only the microtubules – generally, microfilaments are also involved. One of the best-studied examples of the former type of organism is *Camp. jejuni*. This organism is often present on poultry and, when ingested by humans, may cause diarrhoea; it is, in fact, one of the most common

Table 8.9. | Organisms that induce rearrangement of microtubules during the invasion process

Bacterium	Cells invaded	Involvement also of microfilaments?
Campylobacter jejuni	Human intestinal epithelium	Depends on strain used; in some cases there is no involvement
Citrobacter freundii	Bladder epithelium	Depends on strain used; in some cases there is no involvement
Uropathogenic *Escherichia coli*	HeLa cells	Depends on strain used; in some cases there is no involvement
Porphyromonas gingivalis	Oral epithelium	Depends on strain used; in some cases there is no involvement
Klebsiella pneumoniae	Human bladder epithelium	Yes
	Human intestinal epithelium	Yes
Proteus mirabilis	Human bladder epithelium	Yes
Vibrio hollisae	HeLa cells	Yes

causative agents of human diarrhoeal disease. The clinical symptoms vary from a watery diarrhoea to a febrile, blood-containing diarrhoea. In some cases, bacteraemia can occur and the organism has been detected within intestinal biopsies from patients, suggesting that bacterial invasion has taken place. Invasion of a number of intestinal epithelial cell lines by various strains of *Camp. jejuni* has been demonstrated and, in the case of some bacterial strain/epithelial cell combinations, invasion was found to be unaffected by cytochalasin D – implying that actin microfilaments are not crucial to the invasion process. In contrast, colchicine, a microtubule depolymeriser, caused a marked reduction in invasion in these cases. It is important to note that the involvement of microfilaments and microtubules in the invasion process is very much dependent on the particular bacterial strain and epithelial cell line used; with some strain/cell line combinations, microfilament formation is crucial to the invasive process. Interestingly, orthovanadate, an inhibitor of dynein, prevented invasion. Dynein is a 'molecular motor' responsible for propelling vesicles and organelles along microtubules. Microscopic studies of the invasion process have shown that the organism binds to the tip of a finger-like protrusion of the cell that contains a small number of microtubules. It is possible that the protrusion is induced by some diffusible bacterial product. Approximately 1 hour after adding the bacteria to the cells, bacteria are seen to be associated with microtubules in the projections. Within 4 hours, the bacteria are localised mainly around the cell nucleus – probably after being transported along microtubules in dynein-propelled vesicles (Figure 8.19).

Some organisms, while not dependent on microtubule rearrangements for invasion of epithelial cells, appear to use microtubules for other purposes once they are within the cell. Thus uptake of *Actinobacillus actinomycetemcomitans* by oral epithelial cells involves actin rearrangements. However, once the organism has invaded, it uses microtubules to move within the cell, to exit from the cell and to invade neighbouring epithelial cells.

8.2.1.3 Paracytosis

In order to function as an effective physical barrier, epithelial cells must adhere tightly to one another and this is achieved with the help of molecules such as E-cadherins and occludin. Some bacteria, however, can penetrate these junctions and so traverse an epithelial layer by passing between cells rather than through them – a process termed paracytosis. Strictly speaking, therefore, it is an example of tissue invasion rather than cellular invasion. A good example of an organism with this ability is *Haemophilus influenzae*. This resides in the human upper respiratory tract and is responsible for a wide range of diseases including localised respiratory tract infections and systemic infections such as meningitis, arthritis and cellulitis. Examination of biopsies of the respiratory mucosa of patients with bronchitis due to this organism has revealed the presence of bacteria in the subepithelial

Figure 8.19 Invasion of human intestinal cells by *Campylobacter jejuni*. (a) A confocal fluorescence microscopic image of cells 1 hour after infection. Bar = 10 µm. The filamentous structures are the microtubules inside the host cell (which is itself not visible), while the bright rods (arrowed) are the bacteria. (b) Lower power image of cells after 4 hours incubation with bacteria. Bacteria (arrowed) can be seen near the nuclei. Bar = 2 µm. (Reproduced with permission from Hu & Kopecko, 1999. See Section 8.9.)

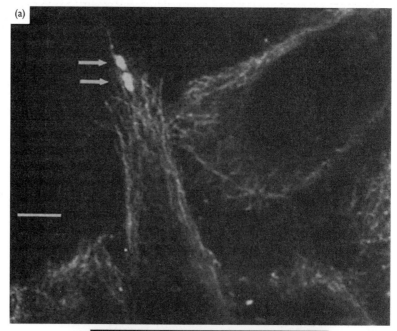

(a)

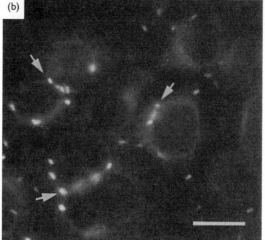

(b)

layers. As the epithelium itself was undamaged, this implies that paracytosis may have taken place. An *in vitro* study using a human lung epithelial cell line has shown that the organism can pass through layers of these cells (up to three cells thick) without affecting their viability or the integrity of the layer. Furthermore, bacteria could be seen between the cells. Penetration is thought to involve disclosure of the intercellular junctions and is prevented by inhibitors of bacterial protein synthesis. Spirochaetes are also widely reported as having the ability to cross epithelial layers by paracytosis; this is considered to be of importance in the pathogenesis of periodontal diseases, as oral *Treponema* spp. use this process to gain access to the soft tissues surrounding the tooth.

Table 8.10.	Main features of invasion of vascular endothelial cells

Characteristic features	Examples
Invasion followed by intracellular persistence without multiplication	*Staphylococcus aureus, Escherichia coli, Streptococcus pneumoniae, Pseudomonas aeruginosa, Streptococcus agalactiae*
Invasion followed by intracellular replication	*Rickettsia rickettsii, Chlamydia pneumoniae, Listeria monocytogenes, Citrobacter freundii, Bartonella henselae*
Traversal without cell disruption	Spirochaetes
Invasion within phagocytes	*Listeria monocytogenes, Chlamydia pneumoniae*

Other examples of paracytosis include the passage of *Campylobacter* spp. across intestinal epithelia and the traversal of endothelial cell monolayers by *Treponema pallidum* and *Borrelia burgdorferi*.

8.2.2 Invasion of vascular endothelial cells

Once an organism has breached the epithelium it may not be capable of penetrating any further and so ends up residing in, and causing damage to, subepithelial tissues. However, many organisms are able to spread throughout the body, causing pathology at sites distant from the site of entry, for example *Neisseria meningitidis, Strep. pneumoniae, H. influenzae, Shig. dysenteriae, Sal. typhi*. In order to do so, they must enter the bloodstream and this entails crossing another cellular barrier – the endothelium. This layer of cells lines the entire vascular system and controls the passage of materials, including blood cells, into and out of the bloodstream. Although invasion of endothelial cells by bacteria is not as well documented as invasion of epithelial cells it appears to follow one of four main courses, as shown in Table 8.10.

8.2.2.1 *Neisseria meningitidis*

Meningitis is a life-threatening infection of the central nervous system characterised by inflammation of the meninges and invasion of the subarachnoid space by the causative organism. One of the most common agents of the disease is *N. meningitidis*, which may be present in the nasopharynx of as many as 10% of the healthy population. The organism possesses a bewildering array of adhesins, any of which may also function as invasins and induce the uptake of the organism by host cells. The expression of the various adhesins is affected by the host environment as different adhesins/invasins will be needed at different stages of the disease process: the organism must adhere to respiratory mucosa, cross this epithelium and then adhere to, and cross, possibly several endothelia. Which of the many putative adhesins/invasins mediates these processes is a controversial issue, but currently it is believed that the major invasin responsible for inducing uptake by endothelial cells is the 28 kDa outer membrane protein Opc. However, this does not interact directly with a receptor on the host cell surface but first binds to serum proteins containing Arg-Gly-Asp (RGD) sequences (e.g. vitronectin) which then interact with the $\alpha_V\beta_3$-vitronectin receptor on the surface of the endothelial cell. The ability of Opc to mediate invasion, however, is affected by other bacterial surface molecules. Hence, sialylation of lipopolysaccharide (LPS) (which may occur during the course of infection with the organism) and the presence of a capsule have been shown to inhibit significantly the ability of Opc to mediate adhesion to endothelial cells. In contrast, pili significantly increase invasion. The most invasive phenotype, therefore, is an Opc-expressing, piliated, non-capsulated strain that does not have sialylated LPS. Variation in the expression of surface structures, including capsule, pili and sialylation of LPS, has been shown to occur during the course of infection with *N. meningitidis*. Invasion has been shown to involve actin microfilaments. Subsequent to invasion, the bacteria can emerge from the cells and thereby can successfully traverse monolayers of **h**uman **u**mbilical **v**ascular **e**ndothelial **c**ells (HUVEC).

Mother was wrong – cleaning your teeth can be really bad for you (if you have gum disease)

Cardiovascular disease is the leading cause of death in the Western world. Although this condition is associated with a number of well-known risk factors (smoking, obesity, high blood pressure), as many as 25% of fatal cases do not score highly with respect to these factors. It has been suggested, therefore, that such cases may have an infectious aetiology. Recent epidemiological data imply that periodontitis (gum disease) is a strong risk factor for coronary heart disease. Hence, in a study involving almost 10 000 subjects, those with periodontitis were found to have a 25% greater risk of coronary heart disease than those without the disease. Although the biological basis for this association has not been established, it may be that organisms associated with periodontitis (e.g. *Porphyromonas gingivalis*) gain access to the bloodstream during tooth-brushing, flossing, chewing, etc. and then damage endothelial cells, resulting in an inflammatory response that leads ultimately to atherosclerosis. Evidence supporting this mechanism comes from the finding that periodontal patients have transient bacteraemias, organisms responsible for periodontitis can be detected in atheromatous tissues and that these organisms can invade endothelial cells. In a recent study, it was found that two organisms associated with periodontitis (*Por. gingivalis* and *Prevotella intermedia*) were able to invade human coronary artery endothelial cells and coronary artery smooth muscle cells *in vitro*. Invasion was inhibited by cytochalasin, implying the involvement of actin polymerisation in the process.

8.2.2.2 *Listeria monocytogenes*

Listeria monocytogenes invades endothelial cells in a manner similar to that which it employs to invade epithelial cells (see above). Internalin mediates binding of the organism to the endothelial cell and this induces endocytosis of the bacterium (Figure 8.20). Recently it has been shown that InlB may also function as an invasin for HUVEC. The organism then escapes from the resulting vacuole, replicates within the cytoplasm and moves through the cytoplasm and into neighbouring cells. It has also been shown that monocytes containing the organism can enter endothelial cells, after which the bacteria escape from the monocytes into the cytoplasm of the endothelial cell.

8.2.2.3 *Citrobacter freundii*

Citrobacter freundii is an opportunistic pathogen that can cause neonatal meningitis with a high fatality rate (25–50%). Studies of invasion of human brain microvascular endothelial cells (HBMEC) by this organism have revealed that the process involves both microfilaments and microtubules. Once taken up by the cells, the bacteria can replicate within the resulting vacuole (Figure 8.21).

Investigations using monolayers of HBMEC have shown that the organism can traverse such layers and that this involves transcytosis rather than paracytosis.

8.2.2.4 *Bartonella henselae*

Bartonella henselae is the causative agent of cat-scratch disease, which is contracted from cats and is characterised by a pronounced lymphadenopathy, often accompanied by fever, fatigue and headaches. *In vitro* studies have shown that, following their adhesion to a HUVEC, bacteria are transported by the leading lamella of the cell to a site just ahead of the nucleus, where they form an aggregate. Membrane protrusions (formed as a result of actin rearrangements) then surround and engulf the bacterial aggregate, resulting in a globular structure (5 to 15 µm diameter) known as an invasome. The overall morphology of the cells is also affected and they become spindle shaped. The whole process is relatively slow, requiring approximately 24 hours for completion (Figure 8.22).

8.2.2.5 β-Haemolytic streptococci

Streptococcus agalactiae is a β-haemolytic streptococcus that is a frequent cause of septicaemia, meningitis and pneumonia in neonates. It is able to invade human BMEC *in vitro* by an endocytotic mechanism that involves both microfilaments and microtubules. The bacteria remain within the resulting vacuoles and do not appear to multiply. However, they can survive for at least 20 hours and damage the host cells, possibly due to the action of their haemolysin.

8.2.2.6 *Streptococcus pneumoniae*

The way in which this organism adheres to, and invades, endothelial cells is particularly interesting, as it makes use of the consequences of cytokine–host cell interactions to increase both its adhesive and

Figure 8.20 Invasion of human brain microvascular endothelial cells by *Listeria monocytogenes*. The images are scanning electron micrographs taken 35 minutes post infection. The bacteria can be seen on the cell surface (a), or in contact with microvilli (b–d). In (e) a bacterium can be seen in the process of invading a cell while in (f) invasion has already taken place (indicated by the arrowhead). Bars 1 μm. (Reproduced with permission, from Greiffenberg *et al.*, 2000. See Section 8.9.)

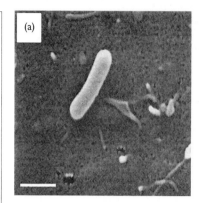

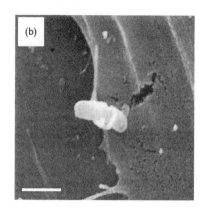

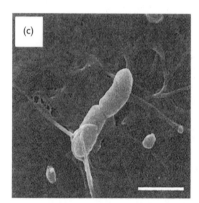

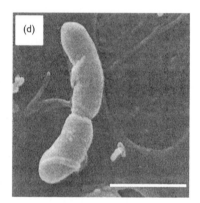

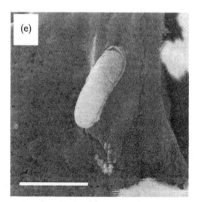

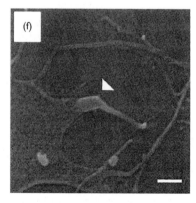

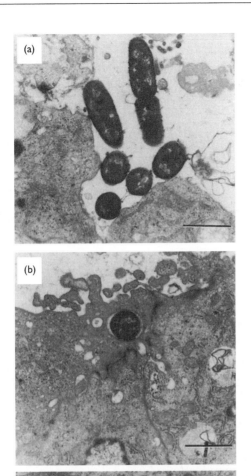

Figure 8.21 Invasion of human brain microvascular endothelial cells by *Citrobacter freundii*. In (a) bacteria can be seen attaching to the host cells, while in (b) a bacterium is present within a membrane-bound vacuole. Bacterial replication is evident within the vacuole shown in (c). Magnification 14000×. (Reproduced with permission, from Badger *et al.*, 1999. See Section 8.9.)

invasive potential. *Streptococcus pneumoniae* can bind to either of two receptors on HUVEC that have epitopes consisting of GalNAcβ(1–3)Gal and GalNAcβ(1–4)Gal. Treatment of HUVEC with either of the pro-inflammatory cytokines tumour necrosis factor (TNF) α or interleukin (IL) 1α has been shown to increase adhesion of the organism to these cells by almost 200% and 100%, respectively. This increased adhesion is accompanied by the appearance of a new receptor (platelet-activating

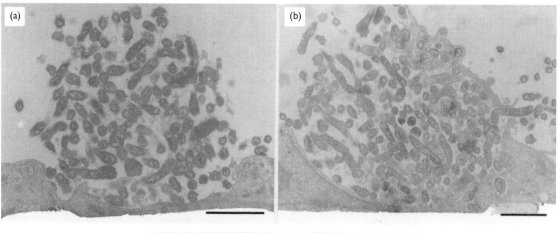

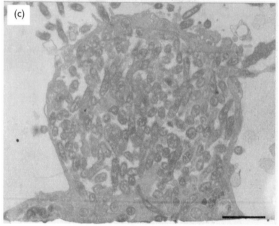

Figure 8.22 Invasion of human brain microvascular endothelial cells by *Bartonella henselae*. In (a) a bacterial aggregate has formed and is then gradually engulfed by the host cell (b and c). Bars = 2 μm. (Reproduced with permission of the Company of Biologists Ltd from Dehio *et al.*, 1997. See Section 8.9.)

factor (PAF) receptor) for the organism. The bacterial adhesin for the PAF receptor has been shown to be the phosphorylcholine residue on its teichoic acid. Invasion of HUVEC that had not been treated with cytokines was shown to be minimal, but significant invasion (3.7% of adherent bacteria) was observed after they have been exposed to TNFα. Invasion of HBMEC by pneumococci is also increased by prior exposure to inflammatory cytokines. However, it has been shown in this case that initial adhesion of the bacteria is mediated by choline-binding protein A, the receptor for which may be the PAF receptor on the HBMEC. Invasion involves both microfilaments and microtubules and results in bacterial uptake in a vacuole within which the cells do not replicate. Ultimately, the bacteria exocytose via the basolateral surface of the cell (Figure 8.23).

Lipid rafts – another portal of entry for bacteria

Lipid rafts are regions of the cytoplasmic membrane of eukaryotic cells that are characterised by detergent insolubility, light density and high concentrations of cholesterol, glycosphingolipids and glycosylphosphatidylinositol-linked proteins.

Some of these lipid rafts form a flask-like indentation in the membrane that is rich in caveolin (a cholesterol-binding protein) and these are known as caveolae. Lipid rafts are important in signal transduction and have high concentrations of signalling molecules such as receptor tyrosine kinases, mitogen-activated protein kinases, adenylyl cyclase and lipid signalling intermediates. They are also able to form vacuoles, so enabling internalisation of material from the cell's environment. Recently it has been shown that certain bacteria (*E. coli*, *Camp. jejuni* and *Mycobacterium bovis*) are able to invade cells via caveolae. In the case of *E. coli*, macrophages are invaded following binding of the organism's FimH adhesin to its CD48 receptor located in caveolae. The resulting vacuole is rich in lipid rafts and differs from the vacuole formed as a consequence of antibody-mediated phagocytosis. The internalised bacteria are able to survive within these vacuoles, although the means by which they do so have not yet been established.

8.3 | Consequences of invasion

Invasion of a host cell by a bacterium has consequences for both cells. These may be quite difficult to discern or may be very dramatic, resulting, for example, in the death of either or both of the participants.

8.3.1 Effect on host cells

Invasion of a host cell by a bacterium can affect the former in a number of ways ranging from the barely perceptible to the ultimate effect – death. In many cases, however, the effects are intermediate between these extremes and may be transient or long lived, depending on the type of host cell involved, the nature of the invading organism and whether the host cell becomes permanently colonised by the

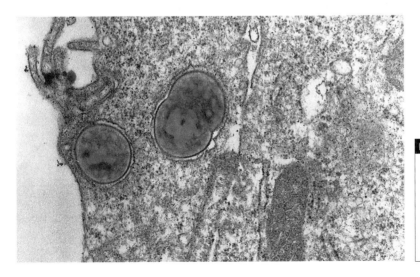

Figure 8.23 Invasion of rat brain microvascular endothelial cells by *Streptococcus pneumoniae*, magnification 46 800×. Streptococci can be seen clearly within vacuoles. (Reproduced with permission from Ring *et al.*, 1998. See Section 8.9.)

bacterium. The adoption of modern molecular approaches to investigations in this area will undoubtedly lead to a greater understanding of the response, at the molecular level, of host cells to invasion. For example, high density cDNA arrays have recently been used to investigate changes in gene expression occurring in intestinal epithelial cells following their invasion by *Sal. dublin*. This enabled the determination of the mRNA expression profile of more than 4300 host cell genes after bacterial invasion. Interestingly, it was found that invasion induced the up-regulation of mRNA expression by only a relatively small number of genes. These included genes encoding cytokines (e.g. granulocyte colony-stimulating factor (GCSF), IL-8, macrophage inflammatory protein (MIP) 2α), kinases, transcription factors (e.g. interferon regulatory factor-1) and HLA class 1.

The main effects of invasion on host cell function that have been documented are described below.

8.3.1.1 Cytokine release

Many types of host cell respond to bacterial invasion by secreting cytokines involved in the activation of host defence systems. However, their overproduction or sustained release may have adverse consequences for the host. Human colon epithelial cells, for example, have been shown to secrete four pro-inflammatory cytokines (IL-8, monocyte chemotactic protein 1 (MCP-1) granulocyte-macrophage colony-stimulating factor (GMCSF) and TNFα) in response to invasion by a range of enteropathogens including *Sal. dublin*, *Shig. dysenteriae*, *Y. enterocolitica*, *Lis. monocytogenes* and enteroinvasive *E. coli*. This is a specific, rather than a generalised, response as the synthesis of a wide range of other cytokines is not affected by bacterial invasion. As IL-8, MCP-1, GM-CSF and TNFα are all involved, directly or indirectly, in chemotaxis and activation of inflammatory cells, the release of these cytokines by epithelial cells activates the mucosal inflammatory response to deal with the invading organisms. What is particularly interesting is that, despite the fact that the bacteria listed above employ very different strategies to induce their uptake into epithelial cells, they nevertheless induce the up-regulation of the same, limited set of cytokine-encoding genes. Recently it has been shown that this is due to activation of the transcriptional regulator nuclear factor (NF) κB, via an invasion-activated signalling pathway involving TNF receptor-associated factor-2 and inhibitory (I) κB kinases.

Other examples of cytokine release by host cells induced by bacterial invasion are shown in Table 8.11. The chemokine IL-8 is a potent chemoattractant and activator of PMNs and the induction of its release by epithelial cells owing to bacterial invasion serves as an early warning for the immune system. However, certain bacteria can reduce the impact of this warning. For example, it has been shown that virulent strains of *Y. enterocolitica* stimulate the release of significantly lower levels of IL-8 than do non-virulent strains. This would be of great benefit to the invading organism, especially during the early stages

Table 8.11.	Cytokines released by host cells in response to bacterial invasion	
Bacterium	Host cell	Cytokines released (or cytokine gene transcription up-regulated)
Staphylococcus aureus	Human umbilical vein endothelial cells	IL-6 and IL-1β
	Bovine mammary epithelial cells	IL-1β and TNFα
Neisseria gonorrhoeae	Human mucosal epithelial cells	TNFα
Campylobacter jejuni	Human intestinal epithelial cells	IL-8
Chlamydia pneumoniae	Human monocytes	TNFα, IL-1β, MCP-1, IL-8 and IL-6
Mycobacterium avium	Laryngeal epithelial cells	IL-8

For abbreviations, see the text.

of infection. Another interesting example of this phenomenon is *Por. gingivalis*. When it invades oral epithelial cells, not only does this organism fail to induce the secretion of IL-8 but it also prevents infected cells from secreting this cytokine in response to other bacteria that would normally induce such a response. This phenomenon has been termed 'local chemokine paralysis'.

8.3.1.2 Prostaglandin release

Prostaglandins are another class of inflammatory mediators released by host cells in response to bacterial invasion. Indeed, the production of diarrhoea in response to enteropathogenic species is mediated, in part, by prostaglandins (particularly prostaglandin E_2, PGE_2), which are important regulators of gastrointestinal fluid secretion. Prostaglandins are derived from PGH, which is produced by the action of the enzymes (termed cyclo-oxygenases, Cox) on arachidonic acid, which is derived from membrane phospholipids. The cyclo-oxygenase activity is the product of two distinct genes: *cox-1* and *cox-2*. The former encodes a constitutive enzyme, while the latter is inducible by pro-inflammatory cytokines. Infection of intestinal epithelial cells with the invasive organisms *Sal. dublin*, *Y. enterocolitica*, *Shig. dysenteriae* and enteroinvasive *E. coli* both *in vitro* and *in vivo* has been shown to rapidly up-regulate expression of *cox-2* and the secretion of PGE_2 and PGF_2. In contrast, non-invasive organisms have little effect on *cox-2* expression.

Invasion of lung endothelial cells by *Strep. agalactiae* has also been shown to induce the release of PGE_2 and prostacyclin.

8.3.1.3 Effects on the expression of adhesion molecules and neutrophil adhesion

The intestinal epithelium, as well as constituting a physical barrier, also functions as part of the host's immune system. For example, it can process and present antigens as well as produce components of the complement system. Furthermore, these cells secrete pro-inflammatory cytokines in response to bacterial invasion, so signalling the presence of bacteria and inducing the recruitment of neutrophils and mononuclear phagocytes to the site of invasion (see Section 5.8). Invasion of human colon epithelial cells by a number of organisms (including *Sal. dublin*, *Y. enterocolitica*, *Lis. monocytogenes* and enteroinvasive *E. coli*) has been shown to up-regulate the expression of intercellular adhesion molecule (ICAM) 1 (the receptor for the β_2-integrins, lymphocyte function-associated antigen (LFA) 1 and Mac-1, on neutrophils), thereby increasing the adhesion of neutrophils to epithelial cells. The net effect of this is to keep neutrophils at the epithelial surface, so helping to combat the invading bacteria. Similarly, the invasion of human cervical or endometrial epithelial cells by *N. gonorrhoeae* has been shown to increase ICAM-1 expression by these cells.

In contrast, invasion of oral epithelial cells by *Por. gingivalis* was found to decrease ICAM-1 expression. When PMNs were added to the system, they failed to transmigrate across the cell monolayer and transmigration induced by stimuli such as *E. coli* was also inhibited once the cells had been invaded by *Por. gingivalis*. This ability to block neutrophil migration across the oral epithelium would impair the ability of the host to deal with bacteria accumulating on the tooth surface and so may contribute to the initiation of periodontal diseases.

Bacterial invasion of endothelial cells can, likewise, affect the expression of adhesion molecules. For example, invasion of HUVEC by *Lis. monocytogenes* up-regulates the expression not only of ICAM-1 but also of E-selectin, and vascular cell adhesion molecule (VCAM) 1. This has been shown to result in increased adhesion of PMNs to infected monolayers. Similarly, invasion of HUVEC by *Chlamydia pneumoniae* up-regulates the expression of the same three adhesion molecules.

8.3.1.4 Cell death

A wide range of organisms can induce apoptosis of epithelial and endothelial cells (Figure 8.24) subsequent to invasion (Table 8.12). Although little has been discerned regarding the mechanisms involved, in the case of *Staphylococcus aureus*, invasion of bovine mammary epithelial cells has been shown to activate caspase 3 and caspase 8, which are considered to be key components of the proteolytic cascade leading to apoptosis (described in Section 10.6.2).

Apoptosis of macrophages following uptake of a wide range of bacteria has also been observed. As this is an effective means of circumventing an important host defence system, it will be described in more detail in Section 10.6.2.

8.3.1.5 Synthesis of tissue factor

Tissue factor is a 45 kDa transmembrane protein produced by endothelial cells in response to a variety of stimuli including pro-inflammatory cytokines and LPS. Its expression on the surface of endothelial cells triggers the activation of the clotting system and results in the

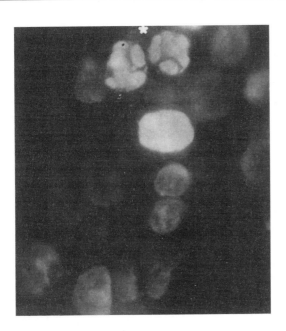

Figure 8.24 Apoptosis in human colon epithelial cells following infection with *Salmonella dublin*. In the upper margin of the figure (*), two cells are visible displaying the nuclear fragmentation typical of cells that have undergone apoptosis. (Reproduced with permission from Kim *et al.*, 1998. See Section 8.9.)

production of fibrin, which coats the cells (thereby increasing bacterial adhesion) and any bacteria present (protecting them from host defence systems). The net effect is the accumulation of bacteria on the walls of blood vessels; when this occurs on heart valves (known as an endocardial vegetation) this can result in the life-threatening condition, bacterial endocarditis. Invasion of endothelial cells by *Staph. aureus*,

Table 8.12. | Bacteria able to induce apoptosis following invasion

Organism	Susceptible cell
Streptococcus pyogenes	Human alveolar epithelial cells
	Human epidermal epithelial cells
Salmonella dublin	Human colon epithelial cells
Enteroinvasive *Escherichia coli*	Human colon epithelial cells
Legionella pneumophila	Human alveolar epithelial cells
Staphylococcus aureus	Bovine mammary epithelial cells
	Human umbilical vein endothelial cells
	Human keratinocytes
Pseudomonas aeruginosa	Human epithelial cells
Salmonella typhi	Human keratinocytes
Chlamydia psittaci	Human epithelial cells

Table 8.13. | Organisms in which genes essential for virulence, or survival in the host, have been identified. Details of the techniques used to identify these genes are given in Chapter 3

Organism	Experimental animal	Technique
Salmonella typhimurium	Mouse	IVET
Pseudomonas aeruginosa	Mouse	IVET
Actinobacillus pleuropneumoniae	Pig	IVET
Listeria monocytogenes	Mouse	IVET
Streptococcus gordonii	Rabbit	IVET
Staphylococcus aureus	Mouse	IVET
Mycobacterium tuberculosis	Mouse	STM
Salmonella enterica	Mouse	STM
Legionella pneumophila	Guinea pig	STM
Yersinia enterocolitica	Mouse	STM

IVET, *in vivo* expression technology; STM, signature-tagged mutagenesis.

Chl. pneumoniae and *Chl. trachomatis*, all of which can cause endocarditis, has been shown to induce the synthesis of tissue factor. The production of this molecule by the endothelial cells is enhanced by the presence of IL-1.

8.3.2 Effects on bacteria

Once a bacterium has entered a host cell, it will find itself in an environment very different from that in which it existed prior to invasion. The invading organism will have to adapt to these new conditions and this is achieved by the expression of gene products needed to survive in the new habitat and by the down-regulation of those that are no longer required. This is accomplished by means of regulatory systems described in Sections 2.6 and 4.3. These systems enable the organism to respond to changes in a variety of environmental factors that it may encounter, including temperature, pH, osmolarity and the concentration of oxygen, carbon dioxide, macro- and micro-nutrients. Even within the invaded cell, a number of very different habitats will be encountered. Hence, the conditions (e.g. pH) within the vacuole itself alter with time (see Sections 6.6.2 and 10.4.2.2) and, for some bacterial species, escape into the cytosol (with, again, a different environment) is possible. Until relatively recently, little was known concerning which bacterial genes were up- or down-regulated *in vivo*. However, new molecular techniques such as IVET and STM (see Sections 3.2.3 and 3.5) are beginning to add greatly to our knowledge in this area (Table 8.13).

These techniques have also been used to identify which bacterial genes are affected after the invasion of host cells – mainly macro-

Table 8.14.	The potential fate of a bacterium once it has invaded a host cell

Behaviour of bacterial invader	Examples
Remains within vacuole	*Salmonella typhimurium*, *Chlamydia* spp., *Legionella pneumophila*, *Coxiella burnettii*
Exits vacuole and colonises cytosol	*Shigella* spp., *Listeria monocytogenes*, *Rickettsia* spp., *Actinobacillus actinomycetemcomitans*
Exits vacuole and cell and then remains extracellular	*Yersinia* spp.

phages. However, some investigators have applied them to determining which genes are expressed following the invasion of other cell types. Hence, IVET has been used to demonstrate that expression of the *sitABCD* operon (involved in iron transport) of *Sal. typhimurium* is induced when the organism invades murine hepatocytes. Other techniques have shown that invasion of human epithelial cells by this organism is accompanied by the up-regulation of at least four genes (*ssaR*, *sopB/sigD*, *pipB* and *iicA*) and that 34 bacterial proteins are uniquely synthesised when it invades human intestinal epithelial cells.

8.4 | Bacterial survival and growth subsequent to invasion

Invasion of a host cell by a bacterium results in the latter being enclosed within a vacuole. A number of options are then available with regard to where the bacterium subsequently grows and replicates: it may continue to inhabit the vacuole, it may exit from the vacuole and live in the cytoplasm or it may exit the cell and maintain an extracellular existence (Table 8.14).

However, before we discuss these different possibilities, it is appropriate to describe here what would normally be the fate of a vacuole produced by the host cell. Vacuole (or 'vesicle') formation by endocytosis is a continually occurring process in eukaryotic cells as it provides the cell with a means of sampling its external environment and it has been estimated that a surface equal to that of the entire plasma membrane is internalised every hour by a mucosal cell. The plasma membrane has, of course, a range of receptors for binding a variety of solutes (e.g. transferrin, low density lipoprotein) and these become part of the internal surface of the vacuole. The resulting vacuole encloses both soluble matter (either bound to receptors or in the fluid phase) and particles, which are then normally delivered to structures known as early endosomes. These pleiomorphic structures are located near the plasma membrane and their contents are slightly acidic (pH 6.0 to 6.2). The incoming vacuole fuses with the early endosome and receptors are stripped of their ligands and some (e.g. the transferrin and low density lipoprotein receptors) are recycled back to the cell surface. Portions of the early endosome then detach to form endosomal carrier vesicles (ECVs) containing material destined for degradation. The ECVs are transported to structures known as late endosomes (usually located near the nucleus), with which they fuse. Late endosomes are lamellar structures that characteristically have mannose 6-phosphate receptors (MP-Rs) and lysosomal glycoproteins in their membranes. The late endosome then 'matures' into a lysosome; although it is not clear exactly how this occurs, the process does involve fusion with enzyme-containing vesicles from the Golgi network. It has also been suggested that the late endosome fuses with an existing lysosome. The resulting lysosome has certain characteristics – low pH (approximately 5.0) and a variety of hydrolytic enzymes (proteases, glycosidases, lipases) – which make it an effective organelle for the degradation of a range of molecules. The lysosomal membrane contains two characteristic, highly glycosylated proteins (lysosome-associated membrane protein 1 (LAMP-1) and LAMP-2), but, unlike late endosomes, does not have mannose 6-phosphate receptors.

As will be described in Section 8.4.1, bacteria that have been taken up by endocytosis have developed a number of strategies for interfering with the process described above, thereby avoiding what should be their fate – killing and degradation by the contents of the lysosome.

8.4.1 Intracellular lifestyle

It is appropriate to remind the reader at this point that we are dealing with the survival of bacteria that have invaded non-phagocytic cells. The means by which bacteria survive following their phagocytosis by PMNs or macrophages will be described in Section 10.4.2.2, which focuses on the ways in which bacteria can circumvent host defence mechanisms. Surprisingly, whereas a considerable amount is known regarding the behaviour and survival of bacteria within professional phagocytes, relatively little is known of the means by which bacteria survive within non-phagocytic cells.

8.4.1.1 Survival within vacuoles

8.4.1.1.1 *Chlamydia* spp.

Chlamydia spp. are responsible for a range of infections in humans and other animals, including trachoma (*Chl. trachomatis*), genital infections (*Chl. trachomatis*), respiratory tract infections (*Chl. pneumoniae*) and psittacosis (*Chl. psittaci*). They are very unusual bacteria in that they do not contain peptidoglycan in their cell wall and are thought to be incapable of synthesising ATP (but see Micro-aside, p. 450).

As they depend on their hosts to satisfy their energy requirements, they are obligate intracellular parasites. Relatively little is known about chlamydia–host cell interactions as the organisms are very difficult to work with: *Chl. trachomatis* has been the most intensively

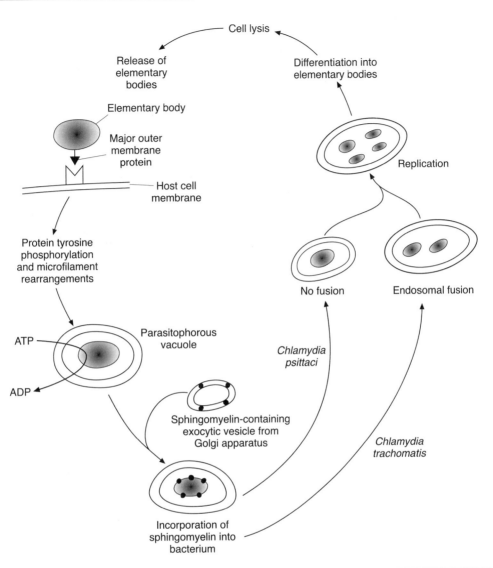

Cell lysis

Release of elementary bodies

Differentiation into elementary bodies

Elementary body

Major outer membrane protein

Replication

Host cell membrane

Protein tyrosine phosphorylation and microfilament rearrangements

No fusion

Endosomal fusion

ATP

Parasitophorous vacuole

Chlamydia psittaci

ADP

Sphingomyelin-containing exocytic vesicle from Golgi apparatus

Chlamydia trachomatis

Incorporation of sphingomyelin into bacterium

Figure 8.25 Diagram showing the intracellular lifestyle of *Chlamydia* spp. For explanation, see the text.

investigated (Figure 8.25). This organism exists in two forms; an elementary **body** (EB) and a **reticulate body** (RB), both of which are bounded by an outer membrane similar to that found in Gram-negative bacteria. The EB is a rigid structure (approximately 0.3 µm in diameter) that is metabolically inactive and constitutes the infectious phase of the organism whereas the RB is an intracellular, non-infectious form resembling a Gram-negative coccus, with a diameter of approximately 1.0 µm. The adhesin(s) involved in the attachment of the organism to epithelial cells has not been determined although a 40 kDa **m**ajor **o**uter **m**embrane **p**rotein (MOMP) has been shown to bind to heparan sulphate on epithelial cells. The EB is endocytosed by a process involving protein tyrosine phosphorylation and microfilament rearrangements. Internalisation is associated with an increase in respiration and glucose catabolism by the host cell. Within 6 to 8 hours of being taken up by the host cell, an EB will have differentiated into an RB. There is considerable evidence that the vacuole in which the organism resides is not processed along the normal route described previously. Hence, the MP-R (a late endosomal–pre-lysosomal marker) cannot be detected in the vacuole, acid phosphatase and cathepsin D (characteristic lysosomal enzymes) are absent and LAMP-1 and LAMP-2 are not found. The vacuole (known as a 'parasitophorous' vacuole or 'inclusion body') is not acidified but is transported to the Golgi apparatus, where it fuses with sphingomyelin-containing vesicles. The sphingomyelin is incorporated into the outer membrane of the bacterium. The subsequent fate of the bacteria-containing endosomes is dependent on the particular species of *Chlamydia*. In the case of *Chl. trachomatis*, the endosomes fuse to form a large inclusion whereas, for *Chl. psittaci*, the vacuoles remain as single units. The RBs then replicate within the vacuole and these daughter cells give rise to EBs (within 1 to 3 days post infection), which are released on lysis of the host cell.

Chlamydia trachomatis – an energy parasite or not?

This organism belongs to one of the few bacterial genera that are obligate intracellular parasites. It is responsible for a sexually transmitted disease which, in the USA, is the most frequently reported infectious disease. In 1997 the number of cases in the USA was estimated to be 3 million. Females are more commonly infected than males – the infection rate being approximately 350 per 100 000 in the former and a fifth of this in the latter. Up to 40% of infected females develop pelvic inflammatory disease if untreated and, of these, 20% will become infertile and 9% will have a life-threatening tubal pregnancy. In males, the most common manifestation of infection is urethritis. The organism is also responsible for one of the major causes of preventable blindness – trachoma. This affects approximately 300 million people, mainly in the developing countries.

For many years the organism has been thought to be incapable of synthesising its own ATP and so has been considered an 'energy parasite', dependent on uptake of ATP from the host cell it inhabits. This view was based on a failure to detect glycolytic enzymes or components of the electron transport chain in the organism and was supported by the presence of an ATP–ADP translocase that enables the transport into the cell of extracellular ATP. However, this concept of 'energy parasitism' is now being called into question following the recent sequencing of the organism's genome. The latter has shown that *Chl. trachomatis* has genes encoding several energy-generating enzymes. Furthermore, the genes encoding the pyruvate kinase and phosphoglycerate kinase in this organism were cloned and expressed and found to be capable of producing ATP by substrate level phosphorylation. Additional work has shown that the organism has complete and functional Embden–Meyerhof and pentose phosphate pathways as well as portions of the tricarboxylic acid cycle. These findings suggest that *Chl. trachomatis* has the functional capacity to produce its own ATP, although whether it does so within the host cell remains to be determined.

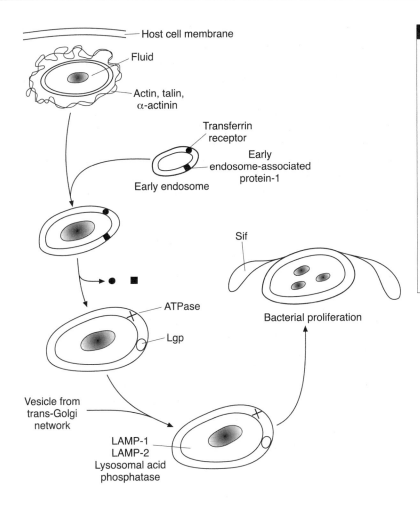

- Host cell membrane
- Fluid
- Actin, talin, α-actinin
- Transferrin receptor
- Early endosome-associated protein-1
- Early endosome
- Sif
- ATPase
- Lgp
- Bacterial proliferation
- Vesicle from trans-Golgi network
- LAMP-1 LAMP-2 Lysosomal acid phosphatase

Figure 8.26 Fate of *Salmonella* spp. within an epithelial cell. Following uptake of the bacterium, the resulting vacuole fuses with an early endosome and then acquires some late endosomal markers (ATPase and lysosomal glycoproteins (Lgps)). The vacuole then fuses with a vesicle from the trans-Golgi network but does not become acidified. The bacterium then proliferates and this is accompanied by the formation of *Salmonella*-induced filaments (Sifs), which are thought to serve some nutritional function. For other abbreviations, see the text.

8.4.1.1.2 *Salmonella typhimurium*

As described previously, uptake of *Sal. typhimurium* by epithelial cells occurs by macropinocytosis. As this involves the intake of relatively large quantities of extracellular fluid, the bacterium comes to reside in a large, fluid-filled vacuole that is surrounded by polymerised actin and other proteins such as talin and α-actinin (Figure 8.26). The organism remains in this vacuole and begins to replicate after several hours. However, during this time, the vacuole undergoes a number of changes. At first, it appears to fuse with an early endosome as it acquires two of the characteristic markers of this structure, the transferrin receptor and early endosome-associated protein 1. The resulting vacuole then rapidly loses these markers but gains some of the characteristics of late endosomes – ATPase and some lysosomal glycoproteins (Lgps) – although there does not appear to be a direct fusion of the vacuole with late endosomes. The vacuole then acquires some lysosomal molecules (LAMP-1, LAMP-2 and lysosomal acid phosphatase) but not others such as MC-Rs and soluble lysosomal enzymes. Incorporation of the lysosomal proteins is the result of fusion of the vacuole with vesicles from the trans-Golgi network. The vacuoles do not become acidified and do not contain lysosomal hydrolytic enzymes. Four to six hours after invasion, bacterial proliferation takes place and this is accompanied by the formation of lysosomal glycoprotein-containing fibrillar structures (*Salmonella*-induced filaments, Sifs) attached to the vacuole. The formation of these structures, which are thought to be involved in the delivery of nutrients to the bacteria, is a microtubule-dependent process. Recently it has been shown that at least one of the effectors (SpiC) produced by the type III secretion system of SPI-2, together with another protein, SifA, are involved in maintaining the special properties of the vacuole that enable *Salmonella* to survive within host cells.

8.4.1.2 Survival in the cytoplasm of the host cell

Following invasion of a host cell, a number of bacteria escape from the resulting vacuole and colonise the cytoplasm. Examples of species exhibiting this behaviour pattern include *Lis. monocytogenes*, *Shigella* spp., *Rickettsia* spp., *A. actinomycetemcomitans* and *H. ducreyi*.

8.4.1.2.1 *Listeria monocytogenes*

Listeria monocytogenes soon escapes from its vacuole (Figure 8.8) by a process involving three molecules: listeriolysin, a phosphatidylinositol-specific phospholipase C, and a broad range phospholipase C. Listeriolysin O is a pore-forming haemolysin that displays optimum activity at a pH (5.5 to 6.0) similar to that found in the vacuole. Pore formation in the vacuole results in a rise in vacuolar pH that prevents further maturation of the vacuole and enables the two phospholipases to degrade the membrane. The bacterium is thereby liberated into the cytosol, where it begins to replicate with a mean generation time of approximately 50 minutes. Another consequence of its release from the vacuole is that the bacterium becomes coated in host actin filaments and actin-binding proteins. After a few replications, the bacterium begins to move through the cytoplasm at a rate of between 6 and 60 μm per minute, propelled by the formation of an actin tail. Movement results from the continuous assembly of microfilaments, followed by their release and cross-linking behind the bacterium – the greater the speed of the bacterium the longer the tail (Figure 8.27).

The bacterial protein responsible for actin polymerisation is ActA, which consists of 610 amino acid residues and accumulates at the poles of the organism, with the highest concentration at the pole adjacent to the actin tail. The C-terminal domain of ActA anchors the protein to the bacterial surface whereas the N-terminal region has high homology to the actin-binding region of human vinculin – a cytoskeletal protein – and is responsible for actin polymerisation. The central domain is able to bind to VASP (vasodilator-stimulated phosphoprotein), which in turn binds to profilin. The actin tails contain a variety of proteins including α-actinin, tropomyosin, fimbrin, profilin, vinculin, ezrin, talin and

(a)

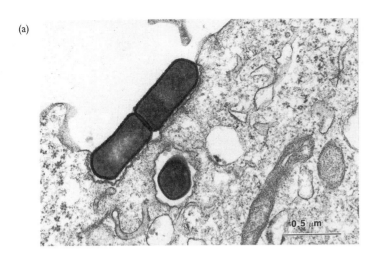

(b)

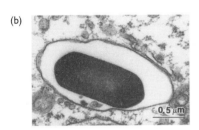

(c)

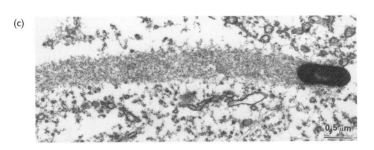

(d)

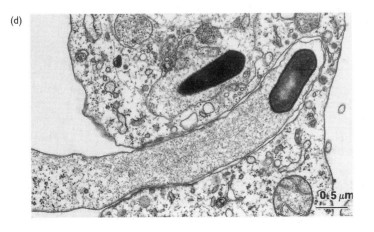

Figure 8.27 Intracellular lifestyle of *Listeria monocytogenes*. (a, b) Bacteria outside and within vacuoles of an epithelial cell. After escaping from the vacuole the organism displays actin-based motility (c) and can force its way into a neighbouring cell (d). (Reproduced by permission of Oxford University Press from Cossart, P. & Bierne, H. The use of host cell machinery in the pathogenesis of *Listeria monocytogenes*. *Current Opinion in Immunology*, 2001, **13**, 96–103.)

Figure 8.28 Formation of actin 'tails' by motile cells of *Shigella flexneri*. The arrows indicate the location of the bacteria, each of which has a long (white) actin 'tail'. (Reproduced with permission from Charles *et al.*, 1999. See Section 8.9.)

(a)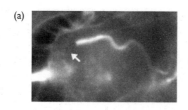

(b)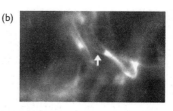

VASP. However, the role of many of these proteins in inducing bacterial movement has not yet been established.

On contacting the cell membrane, the bacterium continues to move forwards, producing a protrusion into the adjacent cell. The intervening membranes are then lysed – a process dependent on the two bacterial phospholipases and a metalloprotease, Mpl – and so the bacterium gains access to the neighbouring host cell.

8.4.1.2.2 *Shigella* spp.

Details of how *Shigella* spp. escape from their vacuole are lacking but the process is probably dependent on the lytic activity of the secreted protein IpaB. This activity is stimulated by low pH and so the natural acidification of the vacuole that follows its formation could act as a trigger for lysis of the vesicular membrane and release of the bacteria. The bacteria then become coated in actin and, by polymerising it at one end to form a tail, move through the cytoplasm at speeds of up to 24 µm per minute (Figure 8.28).

The formation of the actin tail is dependent on a 120 kDa outer membrane protein, IcsA, which accumulates at the point of tail formation. Surprisingly, IcsA has no sequence homology with ActA, the protein responsible for motility in *Listeria* spp. The N-terminal 706 amino acid residues (the α-domain) of the molecule protrudes into the cytoplasm, while the rest of the molecule (the β-domain) is embedded in the outer membrane. An N-terminal 95 kDa fragment is slowly released from the molecule owing to the action of IcsP, a *Shigella* protease. The 95 kDa fragment has ATPase activity and is thought to induce actin condensation, hence propelling the bacterium through the cytoplasm ahead of the actin tail. On contacting the cytoplasmic membrane of the host cell, the bacteria are pushed into the adjacent cell, forming a protrusion. The membranes separating the two cells are then breached, possibly by the *Shigella* protein IcsB, and the bacteria invade the adjacent cell.

8.4.1.2.3 *Rickettsia* spp.

Rickettsia spp. comprise a group of obligate intracellular parasites responsible for a number of important human infections including typhus (*R. prowazekii*) and Rocky Mountain spotted fever (*R. rickettsii*). The organisms are transmitted to humans by arthropods. The target cells of the host are the endothelial cells of small blood vessels; however, many studies of invasion by these organisms have been

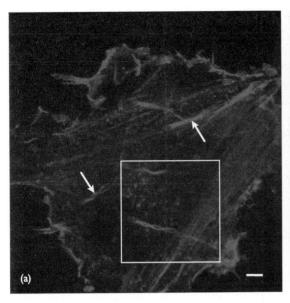
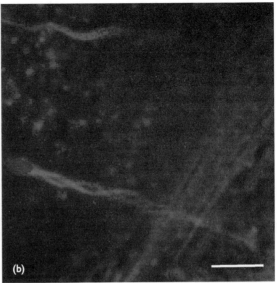

carried out using fibroblasts. Within approximately 15 minutes of invasion, *R. rickettsii* escapes from the endocytic vacuole and becomes surrounded by an actin cloud. A polar actin 'comet tail' is then usually formed within 30 minutes and propels the cell forward by a mechanism similar to that demonstrated in *Shigella* spp. and *Listeria* spp. These tails usually grow considerably longer (>30 μm) than those formed by *Listeria* spp. and *Shigella* spp. Furthermore, they differ from the actin tails of the latter organisms in that they tend to consist of separate 'streamers' of actin filaments rather than compact tails (Figure 8.29). When the bacterium reaches the cell membrane, a short pseudopodium (~5 μm long) is formed that protrudes into the adjacent cell. Presumably, dissolution of the membrane allows access to the neighbouring cell. In contrast, *R. prowazeckii*, while colonising and reproducing within the cytoplasm of the host cell, does not display actin-based motility. This organism simply reproduces within the cytoplasm, producing huge numbers of progeny, and ultimately induces cell lysis, permitting liberation of the bacteria.

Figure 8.29 Confocal microscopic images of motile *Rickettsia rickettsii* within Vero cells. In (a) it can be seen that the long actin tails (arrowed) often comprise multiple, twisting, distinct F-actin bundles. (b) High magnification of an *R. rickettsii* actin tail in panel (a) comprising two F-actin bundles. Bars = 5 μm. (Reproduced with permission from Van Kirk *et al.*, 2000. See Section 8.9.)

8.4.2 Extracellular lifestyle

Some invasive bacteria, having breached the protective epithelium and gained access to the internal tissues of the host, avoid uptake by host cells and maintain an extracellular existence. Examples of such organisms are given in Table 1.10. Among the problems confronting these bacteria are the constant threat they face from phagocytic cells and other host defence mechanisms. Evasion of the antibacterial systems of the host is a major evolutionary adaptation of these species, and the strategies they employ to achieve this are described in detail in Chapter 10.

8.5 | Concept check

- Invasion of non-phagocytic cells can provide the bacterium with a conducive environment that is free from host defence systems.
- Invasion involves a manipulation by the bacterium of the signalling pathways of the host cell.
- The invading bacterium 'coaxes' the host cell to rearrange its cytoskeleton in such a way that internalisation of the organism is achieved.
- Host cell actin microfilaments are involved in the invasion of most bacteria but a few species utilise microtubules – other species use a combination of both.
- The invading bacterium may continue to reside in the vacuole produced during its uptake into the host cell.
- Some invading organisms escape from the vacuole and others move around within the host cell.
- Invasion is accompanied by dramatic changes in gene transcription by both bacteria and host cells.
- Host cells respond to invasion in a variety of ways including the secretion of cytokines, the production of prostaglandins and the expression of adhesion molecules.
- Invasion may result in either necrotic cell death or apoptosis in the host cell.

8.6 | Invasion of host cells by the paradigm organisms

8.6.1 *Streptococcus pyogenes*

Although the commonest infection caused by *Strep. pyogenes* is pharyngitis, a disease not usually accompanied by tissue invasion, the organism is also responsible for a number of diseases in which invasion of underlying tissues, and dissemination to other sites within the host, does occur. Diseases involving tissue invasion, but not necessarily dissemination to other sites, include necrotising fasciitis, myositis, toxic shock syndrome and cellulitis. Puerperal fever and meningitis are examples of invasive diseases in which *Strep. pyogenes* does gain access to other body sites. *In vitro* studies employing a range of mammalian cell lines have shown that the organism can invade a variety of cell types, including human lung epithelial cells, BMEC, human pharyngeal squamous cells, human cervical epithelial cells, human laryngeal epithelial cells, HUVEC and murine macrophage-like cells (Figure 8.30).

What is known about the mechanisms involved in the invasive process will now be described. In Section 7.8.1, a number of different adhesins of *Strep. pyogenes* were mentioned. Of these, it would appear that two major groups (M proteins and fibronectin-binding proteins) are

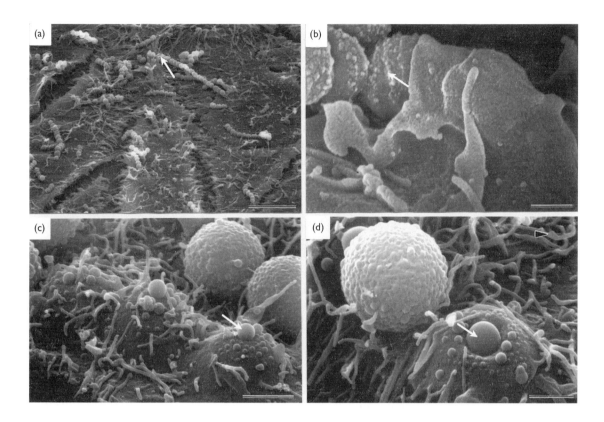

also capable of functioning as invasins. The two main fibronectin-binding proteins that mediate invasion of host cells by *Strep. pyogenes* are Sfb1 (**s**treptococcal **f**ibronectin-**b**inding protein 1) and F1 (also known as Prt1). These bacterial surface molecules bind to fibronectin, which then attaches to (or is already attached to) one of the host cell fibronectin-binding proteins. The latter consist of various integrins such as $\alpha_V\beta_3$. The fibronectin molecule therefore acts as a bridge between the bacterium and the host cell prior to invasion of the latter. Latex beads coated with Sfb1 have been shown to be rapidly internalised by epithelial cells, suggesting that the expression of this protein alone by *Strep. pyogenes* is sufficient to enable host cell invasion. However, *Strep. pyogenes* must possess other invasins, as strains lacking Sfb1 and PrtF1 are also capable of invading epithelial cells. The invasion of one such strain was found to be dependent on the expression of one of the M proteins (M1). Inactivation of the gene encoding this protein resulted in a dramatic decrease in the strain's invasive ability. Interestingly, the M1 protein, like Sfb1 and PrtF1, appears also to use fibronectin as a bridging molecule to the host cell. In this case, the host cell receptor for the fibronectin molecule is the $\alpha_5\beta_1$-integrin. Other members of the M protein family of surface molecules, M3 and M18, have also been shown to be capable of mediating invasion of host cells. Laminin is another host extracellular matrix molecule that can function as a bridge between *Strep. pyogenes* and a host cell and mediates

Figure 8.30 Scanning electron micrographs showing the interaction between human laryngeal epithelial cells and either *Streptococcus pyogenes* or latex beads coated with streptococcal fibronectin-binding protein 1 (Sfb1). In (a) chains of streptococci can be seen adhering to the epithelial cell monolayer (arrow) whereas in (b) invasion is taking place (arrow). In the other two micrographs, latex beads coated with Sfb1 (c) adhere to (arrow) and (d) invade (arrow) the cells. Bars: (a) 5 μm, (b) 0.25 μm, (c) 2 μm and (d) 1 μm. (Reprinted from *Current Opinion in Microbiology* **2**, Molinari, G. & Chhatwal, G. S., Streptococcal invasion, 56–61, 1999, with permission from Elsevier Science.)

invasion of the latter. The bacterial protein involved has not been identified but the receptors for the laminin molecule, like fibronectin, are various host cell integrins including $\alpha_1\beta_1$, $\alpha_2\beta_1$, $\alpha_3\beta_1$ and $\alpha_6\beta_1$. Once binding of the bacterium to the host cell has taken place, subsequent steps in the invasive process appear to be similar to those occurring when Y. pseudotuberculosis invades host cells, i.e. a zippering mechanism is involved. Electron microscopy shows that the bacteria are in close contact with the host cell microvilli, which often extend across the bacterial surface to entrap the organism. Invasion is blocked by cytochalasin D, which suggests that actin polymerisation is involved in the uptake process. The resulting bacteria-containing vacuoles have been shown to be surrounded by polymerised actin. There is great uncertainty with regard to what happens next. The bacteria appear to be incapable of multiplication within the vacuoles but they may survive for several hours or even days. Some vacuoles containing live bacteria move to the surface of the host cell, where they fuse with the cytoplasmic membrane, hence expelling the bacteria, which could then invade other cells.

8.6.2 *Escherichia coli*

Invasion is a feature of a number of diseases caused by E. coli. Hence, meningitis due to E. coli in neonates involves invasion of both epithelial and endothelial cells, while enteroinvasive strains of the organism are responsible for outbreaks of diarrhoea.

A number of adhesins have been shown to mediate attachment of E. coli to BMEC: these include a 14 kDa adhesin (**S f**imbria **a**dhesin **S**, SfaS) located at the tip of the S fimbriae of the organism and a 35 kDa outer membrane protein (OmpA). However, recent reports have shown that invasion of BMEC can be blocked by an antibody to OmpA and that an OmpA$^-$ mutant has a considerably diminished invasive ability, suggesting that OmpA can function as an invasin. The OmpA protein binds to BMEC receptors containing the GlcNacβ(1–4)GlcNac epitope. Invasion takes place by a zipper-type mechanism similar to that which occurs when Yersinia spp. and Listeria spp. enter epithelial cells. However, unlike the latter two organisms, the bacterium remains enclosed within the resulting vacuole and, presumably, transcytoses within this structure (Figure 8.31).

Uptake of the bacterium is dependent on actin microfilaments, involves protein tyrosine kinases and is accompanied by the localised accumulation of polymerised actin (Figure 8.32). However, as microtubule-depolymerising drugs also partially inhibited invasion, it would appear that these components of the cytoskeleton are also involved in the process.

The initial steps in invasion of BMEC by E. coli may, therefore, be summarised as follows:

(1) Fimbrial-mediated binding of the organism to a 65 kDa BMEC protein containing NeuAcα(2–3)-galactose residues.

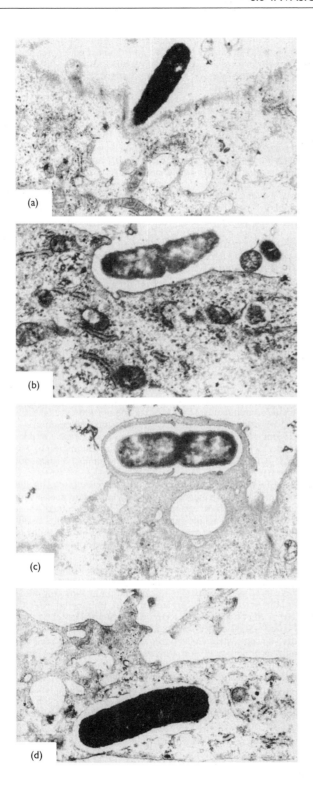

Figure 8.31 Transmission electron micrographs showing invasion of bovine brain microvascular endothelial cells by *Escherichia coli*. After 15 minutes incubation, bacteria can be seen (a) adhering to and (b, c) being taken up by the endothelial cells. After 30 minutes, bacteria can be observed within vacuoles (c, d). Magnifications (a) 282 960× and (b, c, d) 284 530×. (Reproduced with permission from Prasadarao *et al.*, 1999. See Section 8.9.)

Figure 8.32 Light micrographs showing the accumulation of F-actin (labelled with a specific fluorescent stain) in bovine microvascular endothelial cells infected with *Escherichia coli*. (a, c) The position of the bacteria (using ordinary light microscopy); (b, d) fluorescence micrographs that show the fluorescent F-actin (white), with arrows indicating the positions of the bacteria (which do not fluoresce and so cannot be seen). F-actin was found in the host cells directly beneath groups of adherent bacteria. (Reproduced with permission from Prasadarao *et al*., 1999. See Section 8.9.)

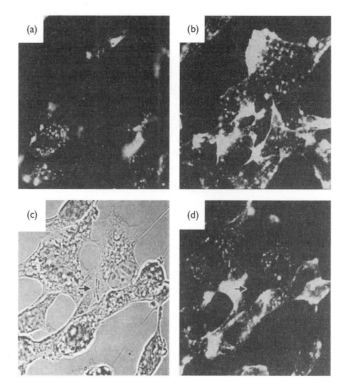

(2) Attachment of OmpA to GlcNAcβ(1–4)GlcNAc epitopes of BMEC glycoproteins.

(3) Activation of signalling pathways involving protein tyrosine kinases.

(4) Actin polymerisation and microtubule assembly.

Invasion is also a characteristic feature of infections due to enteroinvasive *E. coli* (EIEC). Strains of this organism are very closely related to *Shigella* spp. and are responsible for food- and water-borne outbreaks of diarrhoea. The diarrhoea produced is usually watery, with very few patients producing blood and mucus. Although the pathogenic mechanisms of EIEC have not been as extensively investigated as those of *Shigella* spp., the work that has been carried out suggests that they are virtually identical with those of the latter organism. Both organisms invade the colonic epithelium and both secrete enterotoxins. Five stages can be recognised in the invasion sequence: (i) uptake by the epithelial cell, (ii) lysis of the endocytic vacuole, (iii) bacterial multiplication, (iv) movement through the cytoplasm and (v) invasion of adjacent cells.

The genes necessary for invasion are located on a 140 MDa plasmid (pInv) and these include the *mxi* and *spa* loci, which encode a type III secretion system. The latter is responsible for the secretion of four effector proteins (IpaA–D). In *Shigella* spp., IpaC is involved in the uptake of the bacterium by the epithelial cell, while IpaB is thought to be responsible for lysis of the vacuole.

8.7 | What's next?

In this and in the previous chapter we have described how bacteria adhere to, and invade, host cells and tissues. We have seen that, in many cases, these activities result in damage to the host either directly (e.g. by impairing cell function or by killing the cell) or indirectly by inducing an inappropriate (or prolonged) host response, for example, the overproduction of pro-inflammatory cytokines. This type of damage can be viewed as largely 'unintentional' or simply as a byproduct of the bacterium going about its normal 'business', i.e. survival, growth and replication. In Chapter 9 we will consider those bacterial products that have evolved to harm or to kill host cells – the exotoxins.

8.8 | Questions

1. Describe the experiments you would perform to determine whether microfilaments or microtubules are involved in the invasion of a particular type of host cell by a bacterium.
2. Bacteria do not 'invade' host cells but 'persuade' the host cell to ingest them: discuss.
3. Describe the various environments encountered during the passage of *H. influenzae* from its normal habitat to the meninges. List the various substrata and cell types to which it will have to adhere and/or invade.
4. List the ideal attributes a successful invasive pathogen should have and explain why each of these is necessary.
5. The exact functions of many of the secreted virulence factors of gastrointestinal pathogens have not been determined experimentally but have been deduced from their homology to proteins of known function. Do you consider this to be a valid approach? Give reasons for your answer.
6. Compare and contrast the mechanisms involved in the invasion of epithelial cells by *Yersinia* spp. and *Shigella* spp.
7. Describe, giving suitable examples, the different possible fates of a bacterium subsequent to it having invaded an epithelial cell.
8. Describe the possible effects bacterial invasion has on host cell function, emphasising the implications of these for the disease process.
9. Discuss the role of type III secretion systems in bacterial invasion of epithelial cells.
10. Compare and contrast the life cycles of two obligate intracellular parasites of humans.

8.9 | Further reading

Books

Brogden, K., Roth, J., Stanton, T., Bolin, C., Minion, C. & Wannemuehler, M. (2000). *Virulence Mechanisms of Bacterial Pathogens*. Washington, DC: ASM Press.

Cossart, P., Boquet, P., Normark, S. & Rappouli. R. (2000). *Cellular Microbiology*. Washington, DC: ASM Press.

Henderson, B., Wilson, M., McNab, R. & Lax, A. (1999). *Cellular Microbiology: Bacteria–Host Interactions in Health and Disease*. Chichester: John Wiley Ltd.

McCrae, M. A., Saunders, J. R., Smyth, C. J. & Stow, N. D. (1997). *Molecular Aspects of Host–Pathogen Interactions*. Cambridge: Cambridge University Press.

Oelschlaeger, T. A. & Hacker, J. (2000). *Bacterial Invasion into Eukaryotic Cells*. New York: Kluwer Academic/Plenum Publishers.

Review articles

Bavoil, P. M., Hsia, R. & Ojcius, D. M. (2000). Closing in on *Chlamydia* and its intracellular bag of tricks. *Microbiology* **146**, 2723–2731.

Boland, A. & Cornelis, G. R. (2000). Interaction of *Yersinia* with host cells. *Subcell Biochem* **33**, 343–382.

Bourdet-Sicard, R., Egile, C., Sansonetti, P. J. & van Nhieu, G. T. (2000). Diversion of cytoskeletal processes by *Shigella* during invasion of epithelial cells. *Microbes Infect* **2**, 813–819.

Braun, L. & Cossart, P. (2000). Interactions between *Listeria monocytogenes* and host mammalian cells. *Microbes Infect* **2**, 803–811.

Cheng, L. W. & Schneewind, O. (2000). Type III machines of Gram-negative bacteria: delivering the goods. *Trends Microbiol* **8**, 214–220.

Cleary, P. P. & Cue, D. (2000) High frequency invasion of mammalian cells by beta hemolytic streptococci. *Subcell Biochem* **33**, 137–166.

Cornelis, G. & Van Gijsegem, F. (2000). Assembly and function of type III secretory systems. *Annu Rev Microbiol* **54**, 735–774.

Cossart, P. & Bierne, H. (2001). The use of host cell machinery in the pathogenesis of *Listeria monocytogenes*. *Curr Opin Immunol* **13**, 96–103.

Darwin, K. H. & Miller, V. L. (1999). Molecular basis of the interaction of *Salmonella* with the intestinal mucosa. *Clin Microbiol Rev* **12**, 405–428.

Donnenberg, M.S. (2000). Pathogenic strategies of enteric bacteria. *Nature* **406**, 768–774.

Finlay, B. B. & Brumell, J. H. (2000). *Salmonella* interactions with host cells: *in vitro* to *in vivo*. *Phil Trans R Soc Lond B* **355**, 623–631.

Frischknecht, F. & Way, M. (2001). Surfing pathogens and the lessons learned for actin polymerisation. *Trends Cell Biol* **11**, 30–38.

Galan, J. E. & Collmer, A. (1999). Type III secretion machines: bacterial devices for protein delivery into host cells. *Science* **284**, 1322–1328.

Gilbert, D. N., Ewald, P. W., Cochran, G. M., Grayston, T., Saikku, P., Leinonen, M., Siscovick, D. S., Schwartz, S. M., Caps, M. *et al.* (2000). The potential etiologic role of *Chlamydia pneumoniae* in atherosclerosis. *J Infect Dis* **181** (Suppl 3).

Goebe, l. W. & Kuhn, M. (2000). Bacterial replication in the host cell cytosol. *Curr Opin Microbiol* **3**, 49–53.

Goosney, D. L., Knoechel, D. G. & Finlay, B. B. (1999). Enteropathogenic *E. coli*, *Salmonella*, and *Shigella*: masters of host cell cytoskeletal exploitation. *Emerg Infect Dis* **5**, 216–223.

Heithoff, D. M., Sinsheimer, R. L., Low, D. A. & Mahan, M. J. (2000). *In vivo* gene expression and the adaptive response: from pathogenesis to vaccines and antimicrobials. *Phil Trans R Soc Lond B* **355**, 633–642.

Ireton, K. & Cossart, P. (1998). Interaction of invasive bacteria with host signalling pathways. *Curr Opin Cell Biol* **10**, 276–283.

Isberg, R. R. & Barnes, P. (2001). Subversion of integrins by entero-pathogenic *Yersinia*. *J Cell Sci* **114**, 21–28.

Isberg, R. R., Hamburger, Z. & Dersch, P. (2000). Signaling and invasin-promoted uptake via integrin receptors. *Microbes Infect* **2**, 793–801.

Kwaik, Y. A. (2000). Invasion of mammalian and protozoan cells by *Legionella pneumophila*. *Subcell Biochem* **33**, 383–410.

Lee, V. T. & Schneewind, O. (1999). Type III secretion machines and the pathogenesis of enteric infections caused by *Yersinia* and *Salmonella* spp. *Immunol Revs* **168**, 241–255.

Molinari, G. & Chhatwal, G. S. (1999). Streptococcal invasion. *Curr Opin Microbiol* **2**, 56–61.

Monack, D. & Falkow, S. (2000). Apoptosis as a common bacterial virulence strategy. *Int J Med Microbiol* **290**, 7–13.

Niebuhr, K. & Sansonetti, P. J. (2000). Invasion of epithelial cells by bacterial pathogens. The paradigm of *Shigella*. *Subcell Biochem* **33**, 251–287.

Oelschlaeger, T. A. & Kopecko, D. J. (2000). Microtubule dependent invasion pathways of bacteria. *Subcell Biochem* **33**, 3–19.

Ohl, M. E. & Miller, S. I. (2001). *Salmonella*: a model for bacterial pathogenesis. *Annu Rev Medicine* **52**, 259–274.

Philpott, D. J., Edgeworth, J. D. & Sansonetti, P. J. (2000). The pathogenesis of *Shigella flexneri* infection: lessons from *in vitro* and *in vivo* studies. *Phil Trans Roy Soc Lond B* **355**, 575–586.

Pizarro-Cerdá, J., Moreno, E. & Gorvel, J.-P. (2000). Invasion and intracellular trafficking of *Brucella abortus* in nonphagocytic cells. *Microbes Infect* **2**, 829–835.

Rosenberger, C. M., Brumell, J. H. & Finlay, B. B. (2000). Microbial pathogenesis: lipid rafts as pathogen portals. *Curr Biol* **10**, R823–R825.

Ruckdeschel, K. (2000). *Yersinia* species disrupt immune responses to subdue the host. *ASM News* **66**, 470–478.

Sansonetti, P. J. (2001). Rupture, invasion and inflammatory destruction of the intestinal barrier by *Shigella*, making sense of prokaryote–eukaryote cross-talks. *FEMS Microbiol Rev* **25**, 3–14.

Sansonetti, P. J. & Phalipon, A. (1999). M cells as ports of entry for enteroinvasive pathogens: mechanisms of interaction, consequences for the disease process. *Sem Immunol* **11**, 193–203.

Schechter, L. M. & Lee, C. A. (2000). *Salmonella* invasion of non-phagocytic cells. *Subcell Biochem* **33**, 289–320.

Svanborg, C., Godaly, G. & Hedlund, M. (1999). Cytokine responses during mucosal infections: role in disease pathogenesis and host defence. *Curr Opin Microbiol* **2**, 99–105.

Swanson, M. S. & Hammer B. K. (2000). *Legionella pneumophila* pathogenesis: a fateful journey from amoebae to macrophages. *Annu Rev Microbiol* **54**, 567–613.

Vazquez-Torres, A. & Fang, F. C. (2000). Cellular routes of invasion by enteropathogens. *Curr Opin Microbiol* **3**, 54–59.

Wallis, T. S. & Galyov, E. E. (2000). Molecular basis of *Salmonella*-induced enteritis. *Mol Microbiol* **36**, 997–1005.

Weinrauch, Y. & Zychlinsky, A. (1999). The induction of apoptosis by bacterial pathogens. *Annu Rev Microbiol* **53**, 155–187.

Papers

Badger, J. L., Stins, M. F. & Kim, K. S. (1999). *Citrobacter freundii* invades and replicates in human brain microvascular endothelial cells. *Infect Immun* **67**, 4208–4215.

Blocker, A., Gounon, P., Larquet, E., Niebuhr, K., Cabiaux, V., Parsot, C. & Sansonetti, P. J. (1999). The tripartite type III secreton of *Shigella flexneri* inserts IpaB and IpaC into host membranes. *J Cell Biol* **147**, 683–693.

Charles, M., Magdalena, J., Theriot, J. A. & Goldberg, M. B. (1999). Functional analysis of a rickettsial OmpA homology domain of *Shigella flexneri* IcsA . *J Bacteriol* **181**, 869–878.

Clark, M. A., Hirst, B. H. & Jepson, M. A. (1998). M-cell surface β1 integrin expression and invasin-mediated targeting of *Yersinia pseudotuberculosis* to mouse Peyer's patch M cells. *Infect Immun* **66**, 1237–1243.

Cossart, P. & Lecuit, M. (1998). Interactions of *Listeria monocytogenes* with mammalian cells during entry and actin-based movement: bacterial factors, cellular ligands and signaling. *EMBO J* **17**, 3797–3806.

Dehio, C., Meyer, M., Berger, J., Schwarz, H. & Lanz, C. (1997). Interaction of *Bartonella henselae* with endothelial cells results in bacterial aggregation on the cell surface and the subsequent engulfment and internalisation of the bacterial aggregate by a unique structure, the invasome. *J Cell Sci* **110**, 2141–2154.

Ginocchio, C. C., Olmsted, S. B., Wells, C. L. & Galan, J. E. (1994). Contact with epithelial cells induces the formation of surface appendages on *Salmonella typhimurium*. *Cell* **76**, 717–724.

Greiffenberg, L., Goebel, W., Kim, K. S., Daniels, J. & Kuhn, M. (2000). Interaction of *Listeria monocytogenes* with human brain microvascular endothelial cells: an electron microscopic study. *Infect Immun* **68**, 3275–3279.

Hoiczyk, E., Roggenkamp, A., Reichenbecher, M., Lupas A. & Heesemann, J. (2000). Structure and sequence analysis of *Yersinia* YadA and *Moraxella* UspAs reveal a novel class of adhesins. *EMBO J* **19**, 5989–5999.

Hu, L. & Kopecko, D. J. (1999). *Campylobacter jejuni* 81-176 associates with microtubules and dynein during invasion of human intestinal cells. *Infect Immun* **67**, 4171–4182.

Huang, X., Tall, B., Schwan, W. R. & Kopecko, D. J. (1998). Physical limitations on *Salmonella typhi* entry into cultured human intestinal epithelial cells. *Infect Immun* **66**, 2928–2937.

Kim, J. M., Eckmann, L., Savidge, T. C., Lowe, D. C., Witthöft, T. & Kagnoff, M. F. (1998). Apoptosis of human intestinal epithelial cells after bacterial invasion. *J Clin Invest* **102**, 1815–1823.

Kimbrough, T. G. & Miller, S. I. (2000). Contribution of *Salmonella typhimurium* type III secretion components to needle complex formation. *Proc Natl Acad Sci USA* **97**, 11008–11013.

Lecuit, M., Ohayon, H., Braun, L., Mengaud, J. & Cossart, P. (1997). Internalin of *Listeria monocytogenes* with an intact leucine-rich repeat region is sufficient to promote internalization. *Infect Immun* **65**, 5309–5319.

Prasadarao, N. V., Wass, C. A., Stins, M. F., Shimada, H. & Kim, K. S. (1999). Outer membrane protein A-promoted actin condensation of brain microvascular endothelial cells is required for *Escherichia coli* invasion. *Infect Immun* **67**, 5775–5783.

Ring, A., Weiser, J. N. & Tuomanen, E. I. (1998). Pneumococcal trafficking across the blood-brain barrier: molecular analysis of a novel bidirectional pathway. *J Clin Invest* **102**, 347–360.

Van Kirk, L. S., Hayes, S. F. & Heinzen, R. A. (2000). Ultrastructure of *Rickettsia rickettsii* actin tails and localization of cytoskeletal proteins. *Infect Immun* **68**, 4706–4713.

8.10 | Internet links

1. http://www.cdc.gov/ncidod/dvbid/index.htm
   The website of the Division of Vector-borne Infectious Diseases of the Centers for Disease Control and Prevention, USA. It has detailed information on all aspects of Lyme disease (*Bor. burgdorferi*) and plague (*Y. pestis*).

2. http://www.salmonella.org/links.html
   Site devoted to *Salmonella* spp. with many links to other related sites.

3. http://www.hhmi.org/grants/lectures/biointeractive/web_video/salmonella.htm
   Howard Hughes Medical Institute. Has an animated sequence showing invasion of an epithelial cell by *Salmonella*. Also has a sequence showing adhesion of *E. coli* to an epithelial cell.

4. http://cmgm.stanford.edu/theriot/movies.htm The Theriot Lab movie collection from the Department of Biochemistry, Beckman Center, Stanford University Medical School. Excellent series of movies showing invasion of (and, in some cases, movement within) host cells by a number of organisms including *Lis. monocytogenes*, *Sal. typhimurium* and *Shig. flexneri*.

Bacterial exotoxins

Chapter outline

Aims

The principal aims of this chapter are:

- to describe the various categories of bacterial exotoxins and their cellular effects
- to provide examples of toxins from each category and their role in pathogenesis
- to emphasise that exotoxins may have subtle effects at low concentrations that do not cause target cell death
- to describe the role that exotoxins have played in the dissection of eukaryotic signalling pathways
- to highlight some therapeutic uses of exotoxins

9.1 | Introduction

Bacterial exotoxins are a diverse collection of proteins responsible for many of the symptoms caused by pathogens during infection. Exotoxins are actively exported by bacteria or else are released on cell lysis. Some toxins can kill cells outright by a number of different means. Others interfere with fundamental cellular activities to the ultimate benefit of the bacterium.

A number of bacteria produce a barrage of toxic activities, some of which may appear at first glance to be 'blunt instruments'. These include toxins that degrade host cell membrane components, and the channel-forming toxins that punch holes in target cell membranes. Closer inspection of some of these toxins, however, reveals an underlying degree of sophistication in function that may result, for example, in interference with host immune defence mechanisms. Other toxins are perhaps more analogous to the stiletto dagger, being exquisitely specific in their target selection. This is perhaps best exemplified by the clostridial neurotoxins.

Whatever their means of activity, toxins play an essential role in the disease process. Relatively harmless commensal organisms can become virulent through the acquisition of DNA encoding one or more toxins. Additionally, toxin gene inactivation experiments have demonstrated the necessity of some toxins for bacterial survival in the host and for causing disease symptoms. This chapter will describe bacterial exotoxins on the basis of their structure, function and their effects on target cells.

9.2 | Classification of toxins by their activity

Once released by bacteria, toxins bind to specific receptors on the target cell surface. Bound toxins can exert their effects directly on the target membrane, as is the case for type I (membrane-acting) and type II (membrane-damaging) toxins (Figure 9.1). In contrast, type III intracellular toxins are taken up by receptor-mediated endocytosis, whereupon the toxin gains access to the cytoplasm to exert its, often very specific, activity. Some intracellular toxins are 'injected' directly from the bacterial cytoplasm into the cytoplasm of the target cell by a specialised protein secretion apparatus (type III or type IV secretion; see Sections 2.5.3.3 and 2.5.5). Rigid classification of toxins into one or other of these categories, however, is often not advisable, nor indeed possible, as we shall see.

However, a number of themes do appear throughout the broad spectrum of exotoxin biology. The means of accessing the target cell cytoplasm is one example. Also, certain eukaryotic proteins appear again and again as targets, although the bacterium's chosen method of interfering with the target may differ. Thus the small GTPases, Rho and Ras (see Section 4.2.5.2), can be ADP-ribosylated, glucosylated or even deamidated, often with very different results. Similarly, a particular toxin enzymic activity (such as ADP-ribosylase) may be delivered to the target cell in a number of different toxin formats, often with quite different target specificities. Nevertheless, classification into types I to III provides a useful framework for this chapter's foray into the bacterial toxin arsenal.

Figure 9.1 Classification of toxins by their activity. Type I toxins are membrane-acting whereas type II toxins damage membranes. Type III toxins act inside the target cell. Many type III toxins are taken up by receptor-mediated endocytosis and then gain access to the cytoplasm directly or else via the endoplasmic reticulum (ER) (for details, see the text). Other intra-cellularly acting toxins are directly 'injected' into target cells by a type III secretion system (see Section 2.5.5). N, nucleus; for other abbreviations, see list on p. xxii.

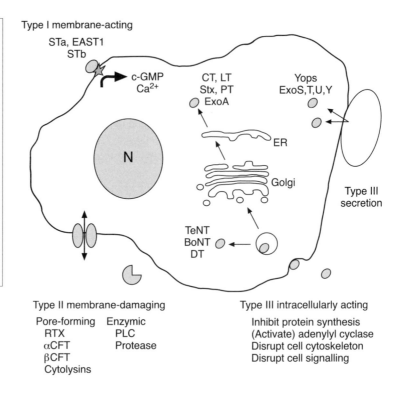

9.3 | Type I (membrane-acting) toxins

Type I toxins act from outside the target cell, although their effects are felt within. These toxins interfere with host cell signal transduction by inappropriate activation of cellular receptors. As we will see in Section 9.5, which deals with type III (intracellular-acting) toxins, many more bacterial toxins can interfere with signal transduction events from within the target cell, leading to either activation or suppression of various signalling pathways.

Toxins that disrupt host cell signalling – tools for cellular biology

Cell death is a common outcome of intoxication. For example, some channel-forming toxins and those toxins that inhibit protein synthesis can result in cell death. Other toxins exert more subtle effects on cell function by interfering with signal transduction pathways (Table 9.1). For example, many toxins give rise to elevated concentrations of the secondary messenger molecule cyclic AMP (c-AMP). This leads to stimulation of protein kinase A (PKA) and, in general, results in down-regulation of the Raf kinase/mitogen-activated protein (MAP) kinase pathway (see Section 4.2.4.2). Elevated c-AMP levels are achieved by different means depending on the toxin: some toxins are adenylyl cyclases while others stimulate endogenous adenylyl cyclase or else block normal inhibition of endogenous adenylyl cyclase activity. However, elevation of c-AMP levels may have different outcomes at the whole animal level, reflecting different sites of activity. Thus cholera toxin (CT) causes fluid accumulation and diarrhoea by acting on intestinal epithelial cells, whereas the repeat in toxin (RTX) adenylyl cyclase activity of *Bordetella pertussis*, CyaA, impairs the function of macrophages.

A second group of toxins affects the cytoskeleton, the integrity of which is essential for most cell processes. The cytoskeleton is also increasingly recognised as being an integral part of the cell's intracellular communication system (see Section 4.2.5). Some toxins, such as C2 of *Clostridum botulinum*, block actin polymerisation. Others interfere with the function of small GTP-binding proteins (G proteins) of the Ras superfamily that are involved in the regulation of the cytoskeleton (Table 9.1).

These toxins, and others, have been invaluable in the investigation of eukaryotic cellular processes and in the dissection of eukaryotic signal transduction pathways. For example, tetanus toxin has been used as a tool for studying the mechanisms of neuronal cell death in the mammalian brain and for the study of exocytosis. The receptor-binding and translocation domains of diphtheria toxin (DT) have been used as a carrier to deliver other peptides or proteins into cells in order to investigate signal transduction pathways activated by that factor. The range of Rho protein-activating and -inhibiting toxins have played a central role in dissecting the regulatory networks that mediate numerous cellular responses, including the control of the actin cytoskeleton and the regulation of cell cycle progression.

9.3.1 Stable toxin (ST) family

Many *Escherichia coli* strains produce enterotoxins that can be divided broadly into two groups on the basis of their heat stability. Thus heating destroys the activity of heat-labile enterotoxins (LTs). LTs are type III (intracellularly acting) AB_5 toxins that are functionally and immunologically very similar to CT, as is described in Section 9.5.1.1. In contrast, heat-stable enterotoxins (STs) retain toxic activity after a 30 minute treatment at 100 °C. All STs are small peptides (up to approximately 60 amino acid residues long) whose structure is stabilised by multiple intrachain disulphide bonds. STa and STb, expressed by enterotoxigenic *E. coli* (ETEC) strains, have different membrane receptors and modes of action.

Enteroaggregative *E. coli* (EAEC) produces a third type of heat stable enterotoxins – EAST1 (EAEC heat-stable enterotoxin 1). EAST1 is expressed also by other *E. coli* isolates including enterohaemorrhagic (EHEC), enterotoxigenic (ETEC) and enteropathogenic (EPEC) strains, and acts in a manner similar to that of STa.

The family of STa enterotoxins (18 or 19 amino acid residues in length) is characterised by a conserved 13 amino acid residue sequence at the C-terminal end of the peptide that is essential for activity. Within this sequence are six cysteine residues that form three disulphide bonds. STa binds to, and activates, cytoplasmic membrane-associated guanylyl cyclase (Figure 9.2). This results in an increase in intracellular cyclic GMP (c-GMP), which in turn activates protein kinase G and thus

Table 9.1. Toxins that interfere with host cell signalling events

Organism/toxin	Target	Cellular and physiological effects
Escherichia coli STa, EAST	Guanylyl cyclase	Stimulation of guanylyl cyclase
		Increased c-GMP levels, activation of protein kinase G
		Diarrhoea
Vibrio cholerae CT	α-Subunit Gs Activation	Stimulation of adenylyl cyclase
		Increased c-AMP levels, activation of protein kinase A
		Diarrhoea
Bordetella pertussis PT	α-Subunit Gi Inhibition	As above
		Blocks phagocyte function
Bacillus anthracis EF	ATP	Action of adenylyl cyclase results in increased c-AMP
		Oedema
Clostridium difficile TcdA, B	Rho, Rac, Cdc42	Glucosylation of Rho proteins blocks their interaction with effectors.
		Diarrhoea, cell death
Bordetella pertussis DNT	Rho, Rac, Cdc42	Deamidation of Rho proteins inhibits GTPase activity
		Activated Rho stimulates actin reorganisation and phosphorylation of FAK
Escherichia coli CNF1, 2		Stimulates DNA synthesis but blocks cell division
Pasteurella multocida PMT	(i) α-subunit Gq	(i) Mitogenic. IP$_3$ release, stimulation of phospholipase C and protein kinase C
	(ii) ?	(ii) Rho-dependent tyrosine phosphorylation of p125^{FAK}, cytoskeletal rearrangements

Yersinia spp.		
YopE	Rho, Rac, Cdc42	YopE exibits GTPase-activating protein activity leading to inhibition of Rho function. Blocks phagocytosis
YopP/YopJ	MAP kinase kinases (MAPKKs)	YopJ binds to MAPKKs blocking phosphorylation and subsequent activation. Prevents cytokine synthesis and promotes apoptosis
YopH	PI25^{FAK}	Tyrosine phosphatase activity of YopH interferes with early signals during phagocytosis
Pseudomonas aeruginosa		
ExoS	Ras	ADP ribosylation of Ras uncouples Ras-mediated signal transduction. Cell death

For abbreviations, see list on p. xxii.

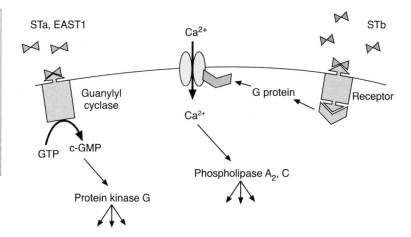

modulates several signalling pathways, leading to secretion of water and electrolytes from the cell. STa-like enterotoxins are expressed also by a number of other enteropathogens including *Yersinia enterocolitica* and *Vibrio cholerae* strains. EAST1 (38 amino acid residues in length) is immunologically distinct from STa and contains only four cysteine residues but nevertheless stimulates guanylyl cyclase through the same receptor-binding region as does STa.

STb is associated mainly with pig isolates of ETEC, though it is found in some human strains. STb is secreted as a 58 amino acid residue peptide whose tertiary structure is stabilised by two disulphide bridges. In contrast to STa and EAST1, STb does not exert its toxic effects by altering c-GMP levels. Instead, STb binds to an as yet unidentified receptor, resulting in G protein-mediated opening of a Ca^{2+} channel (Figure 9.2). The resulting increase in intracellular Ca^{2+} activates phospholipase, A_2 and C, which in turn elevate levels of prostaglandin E_2, leading to fluid and electrolyte secretion.

The genes encoding heat-stable enterotoxins are often found on transposable elements that are carried in turn by a variety of plasmids.

9.3.2 Bacterial superantigens

Bacterial superantigens are a family of functionally related proteins that play a pivotal role in the pathogenesis of a number of human diseases. Superantigens interact with immune cells in a nonconventional manner and stimulate potent immune responses directly by acting as mitogens for subsets of T cells (see Section 10.5.2). As a result, stimulated cells release massive amounts of pro-inflammatory cytokines, including interleukins IL-1, IL-6 and IL-8 and tumour necrosis factor (TNF), which results in a severe shock response.

Superantigens bind directly to the T cell receptor (TCR) and to MHC class II protein on antigen-presenting cells (APCs), thereby enhancing intracellular interactions (see Section 6.7.2.2). Binding of superantigens to these receptors occurs outside the groove where conventional processed antigenic peptides bind, thereby bypassing the need for processing by the APCs. Consequently, they activate a whole subset or subsets of T cells of a particular vβ TCR phenotype. This can translate to activation of up to 20% of the T cell population instead of the 0.0001–0.01% expected with a conventional antigen.

Streptococcus pyogenes and *Staphylococcus aureus* each secrete multiple related superantigens, amongst which are the toxic shock syndrome toxin (TSST) 1 of *Staph. aureus*, and streptococcal pyrogenic exotoxins (SPE-A, C, F–H and J) and mitogenic factor.

9.4 | Type II (membrane-damaging) toxins

9.4.1 Channel-forming toxins

These toxins insert into the target cell membrane to generate a channel. In the simplest interpretation, the channel allows the influx and efflux of small molecules and ions and the disruption of the membrane electrical potential (for example *Staph. aureus* alpha toxin and streptolysin-O from *Strep. pyogenes*). Cell death can occur by osmotic lysis or by apoptosis (for apoptosis, see Section 10.6.2.1) depending on the toxin, the size of the channel formed and the target cell. Additionally, some toxin molecules are known to have other activities and effects upon cells that may not be related to channel formation, for example the induction of cytokine release. Other channel-forming toxins (CFTs) may generate a pore for the translocation of a toxic enzymic activity to the target cell cytoplasm. Examples of this include *Bord. pertussis* adenylyl cyclase and some members of the AB toxin family (see Sections 9.4.1.4.3 and 9.5.1).

The challenge faced by CFTs is to be able to convert from a soluble extracellular protein to one that is inserted into, and spans, the target cell membrane. The solution adopted is to sequester hydrophobic domains from the aqueous environment when in the soluble form and then to expose these during conformational changes and/or oligomerisation when the toxin is associated with the target membrane. Some toxins also employ proteolytic nicking to initiate the channel-forming events. Two different structural approaches to channel formation have been well defined. These employ the protein secondary structure motifs of either β-sheet or α-helix, as is described in Sections 9.4.1.1 and 9.4.1.2. For other CFTs, however, such as the RTX family of toxins (see Section 9.4.1.4) or the so-called thiol-activated cytolysins, less molecular detail is known about the mechanism of channel formation.

9.4.1.1 Channel formation involving β-sheet-containing toxins

9.4.1.1.1 α-Toxin of *Staphylococcus aureus*

The α-toxin (or α-haemolysin) of *Staph. aureus* is considered to be a paradigm for the family of channel-forming toxins that utilise a

Table 9.2. | Examples of some β-barrel structure channel-forming toxins. Toxins from heptameric transmembrane structures with a central hydrophilic channel of approximately 1 to 2 nm

Organism	Toxin	Comments
Staphylococcus aureus	α-Toxin (haemolysin) γ-Toxin (haemolysin) Leukocidin	Related in primary amino acid sequence and three-dimensional structure
Aeromonas hydrophila	Aerolysin	Proteolytically activated
Clostridium septicum	α-Toxin	Proteolytically activated
Bacillus anthracis	Anthrax protective antigen	AB toxin component. Translocated active subunits disrupt eukaryotic cell signalling

β-barrel structure (βCFTs) for channel formation (Table 9.2). The toxin is expressed by most pathogenic *Staph. aureus* strains and is active against erythrocytes (hence the alternative name α-haemolysin) as well as a range of other cells including peripheral blood monocytes and endothelial cells. Secreted α-toxin protomer is a hydrophilic protein of approximately 33 kDa. Once associated with the target cell membrane, seven protomers assemble into a pre-pore complex that subsequently inserts into the membrane to form the channel. The final heptamer structure resembles a mushroom with the cap and rim domains on the plasma membrane surface and the stem forming the transmembrane channel. Structurally, the channel is formed by a β-barrel constructed from 14 anti-parallel β-strands, two being donated from each protomer (Figure 9.3).

The stages in channel formation can be outlined therefore as follows: (i) soluble monomer, (ii) membrane-associated monomer, (iii) nascent oligomer/pre-pore complex and (iv) membrane insertion and pore formation. The soluble monomer assumes a conformation that masks protein–protein and protein–membrane interfaces prior to cell binding and oligomer assembly. Receptor-mediated binding of toxin protomers to membranes is a high affinity interaction, though as yet the receptor(s) have not been identified. However, at higher concentrations, the protomer will bind non-specifically to lipid bilayers and this accounts for its channel-forming activity on human erythrocytes and planar lipid membranes, both of which lack the receptor. Insertion of the pre-pore complex, once assembled on the cell surface, into the lipid bilayer requires energy, but this is not provided by ATP hydrolysis. Instead, the soluble protomer undergoes considerable conformational changes on membrane binding and oligomerisation to expose the protein and membrane interfaces, the interactions of which provide the energy to insert the channel-forming domain into the lipid bilayer.

The size of the channel formed by α-toxin depends on the target

(a)

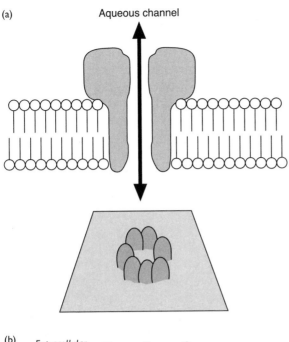

Aqueous channel

Figure 9.3 Alpha-toxin of *Staphylococcus aureus* forms trans-membrane channels. (a) Schematic diagram of the channel formed by insertion of seven α-toxin subunits into a target membrane. (b) Ribbon diagram showing the characteristic 14-strand β-barrel. (Reproduced with permission from Gouax 1998. See Section 9.11.)

(b)

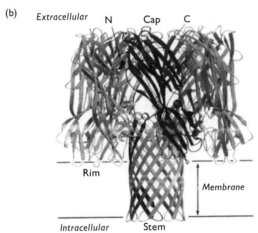

Extracellular   N   Cap   C

Rim

Membrane

Intracellular   Stem

membrane. Thus, for erythrocytes and planar membranes, the channel is 1.0 to 1.5 nm in diameter, but slightly smaller in nucleated cells carrying the toxin receptor. The pores allow the movement of ions, resulting in cell death by osmotic lysis or apoptosis. Pore formation can also trigger secondary events such as release of cytokines and increased platelet exocytosis.

Staphylococcus aureus strains may also express γ-haemolysin and leukocidin toxins. Together with the α-haemolysin, these toxins comprise a family of βCFTs that demonstrate a common amino acid sequence and three-dimensional structure.

9.4.1.1.2 Aerolysin

Aeromonas spp. cause gastrointestinal diseases and severe wound infections in humans and animals. The βCFT aerolysin, expressed by *Aeromonas hydrophila* and other *Aeromonas* spp. including the fish pathogen *Aer. salmonicida*, is one of a number of virulence factors expressed by these bacteria. Like the α-toxin of *Staph. aureus*, aerolysin permeablises susceptible membranes through the formation of a channel (approximately 1.7 nm in diameter) comprising a 14 strand β-barrel formed by the oligomerisation of seven protomers. Channel formation occurs in a similar manner, but with several important differences. Firstly, the toxin is secreted as a pro-protein that requires proteolytic nicking to remove a C-terminal fragment to allow channel formation. Secondly, the soluble form of pro-aerolysin is dimeric such that extensive hydrophobic surfaces are sequestered in the protein–protein dimer interface. The dimer binds with high affinity to receptors on susceptible membranes and several receptors have been identified including Thy-1 (CD90) on T lymphocytes and a 47 kDa erythrocyte glycoprotein. Recent evidence suggests that the glycosylphosphatidylinositol (GPI) anchor of these receptors is the binding target. Once bound on the cell surface, proteolytic nicking at the activation site is thought to induce dimer dissociation. The exposed hydrophobic regions are then available for oligomerisation and membrane insertion.

9.4.1.1.3 Anthrax toxin

It can be difficult and misleading to try to compartmentalise or subdivide toxins too rigidly according to form or function. A good example of this is provided by *Bacillus anthracis*, the causative agent of anthrax. This organism produces two toxins that function enzymically within the target cell cytoplasm (type III toxins). The toxins, lethal factor (LF) and oedema factor (EF), require a separate receptor-binding and translocation component, termed 'protective antigen (PA)', for intracellular delivery. This binary system represents an unusual variation of the AB toxin structural theme (see Section 9.5). PA is expressed as a pre-protein that binds to an as yet unidentified receptor on susceptible cells and requires proteolytic activation to permit EF or LF to bind and for oligomerisation. The complex is subsequently endocytosed, whereupon the acidic environment of the endosome promotes the insertion of a heptameric PA structure into the membrane to form a 14 strand β-barrel-defined channel with a pore size of approximately 1.2 nm. The channel is thought to permit the translocation of EF or LF into the cell cytoplasm. EF possesses a calcium- and calmodulin-dependent adenylyl cyclase activity that dramatically increases cellular c-AMP levels, activating secondary messenger pathways (see Section 4.2.3). LF, on the other hand, is a zinc-dependent protease that inactivates MAP kinase kinases, which are key components in the signal transduction pathway that regulates cell proliferation (see Section 4.2.4.2).

9.4.1.2 Channel formation involving α-helix-containing toxins

In contrast to βCFTs, α-helix-forming toxins (αCFTs) do not need to assemble into higher-ordered membrane-associated oligomers, but instead can function as monomeric proteins. The paradigm for αCFTs, colicin Ia, is a bacterial toxin that kills closely related strains of the same species. A bundle of 10 α-helices comprises the channel-forming domain. Buried within this bundle are two hydrophobic α-helices that form a hydrophobic hairpin loop similar to that proposed for signal sequences of translocated proteins. Once colicin is bound to the target membrane, it undergoes a slow conformational change where tertiary, but not secondary, structure is lost to form a so-called 'molten globule'. Slow insertion of the translocation domain into the outer leaflet of the membrane requires a transmembrane potential. Little is known about the structure of the channel, but it is likely that it is formed by a single molecule and that the transmembrane helices are highly mobile within the membrane.

9.4.1.2.1 Diphtheria toxin

Diphtheria was a major cause of death amongst children until successful mass immunisation against diphtheria toxin was introduced in the 1920s. Like anthrax toxin, DT is a type III intracellularly acting toxin that utilises a channel-forming component to facilitate translocation of the active toxin domain. DT is secreted as a soluble 58 kDa three-domain protein (see Figure 9.4a). The receptor for DT is heparin-binding epidermal growth factor (EGF)-like precursor on target cells. Once bound, the toxin is endocytosed in clathrin-coated pits (Figure 9.4b). Proteolytic nicking generates two toxin fragments that remain linked by a disulphide bridge. Fragment A (approximately 21 kDa) carries the catalytic C domain whereas fragment B (approximately 37 kDa) carries the R domain responsible for receptor binding and the T domain required for translocation of fragment A into the cytosol. Acidification of the endosome acts as the trigger for conformational changes in fragment B, promoting the membrane insertion of the α-helical T domain to form an ion-selective channel. Fragment A is 'dragged' across the membrane and into the cytosol, where the disulphide bond is reduced and the catalytic activity is released. Fragment A transfers ADP-ribose to a modified histidine, diphthamide, on elongation factor 2 (EF2), an essential co-factor for protein synthesis. ADP-ribosylation blocks protein synthesis and leads to cell death; a single toxin molecule is sufficient to kill a cell.

Diphtheria is a respiratory disease, the pathology of which is due mainly to the action of the toxin produced by the causative organism, *Corynebacterium diphtheriae*. Production of DT requires lysogenisation (insertion of a phage genome within a bacterial chromosome) of *C. diphtheriae* with a bacteriophage encoding the toxin gene. Nevertheless, non-toxigenic strains of *C. diphtheriae* are increasingly being associated with cases of pharyngitis and, more seriously, with invasive diseases

Figure 9.4a Structure of diphtheria toxin showing the three major domains.

(a)

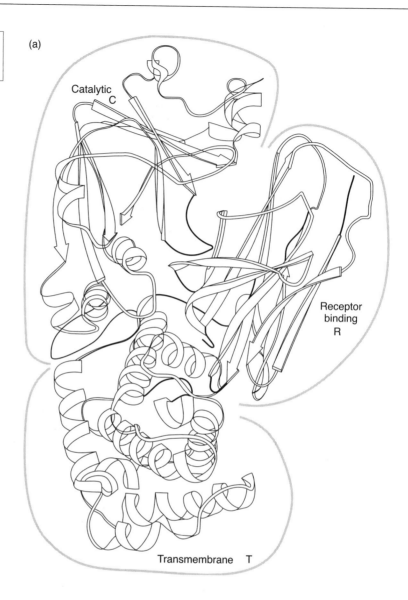

such as endocarditis and septic arthritis. In many cases, the causative organism carries a cryptic (i.e. not expressed) toxin gene and there is speculation that non-toxigenic strains can convert to toxin producers either by reversion of the disabling mutation or by lysogenisation with bacteriophage.

9.4.1.3 'Thiol-activated' cholesterol-binding cytolysins

Members of this family of toxins, expressed by four genera of Gram-positive bacteria (Table 9.3), share both antigenic and amino acid sequence similarity. Toxin monomers (50 to 60 kDa) bind to cholesterol in target membranes, insert into the membrane in a temperature-dependent manner and oligomerise to form pores. Some 25 to 80 monomers are involved to generate a transmembrane channel

(b)

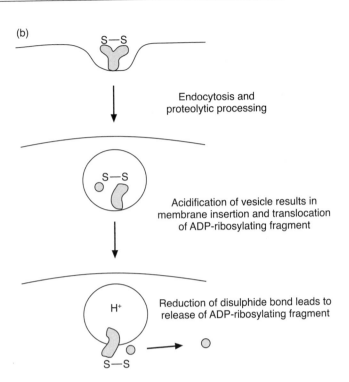

Endocytosis and
proteolytic processing

Acidification of vesicle results in
membrane insertion and translocation
of ADP-ribosylating fragment

Reduction of disulphide bond leads to
release of ADP-ribosylating fragment

Figure 9.4b Translocation of diphtheria toxin. Sequence showing the uptake of DT and the subsequent release of the ADP-ribosylating fragment.

approximately 30 nm in diameter. These are considerably larger than the channels formed by αCFTs and βCFTs (up to approximately 2 nm). Comparatively little is known about the mechanisms by which soluble toxin monomers become inserted into the membrane but, again, conformational changes and oligomerisation are thought to be important. Pore formation results in cell death.

Erythrocytes are very sensitive to these toxins, but any membrane that contains cholesterol is a target. This family of toxins is cytolytic, but at low concentrations, which have no effect on cell integrity or viability, the toxins demonstrate a range of additional effects including the induction and release of cytokines, activation of the classical complement pathway and the inhibition of polymorphonuclear neutrophils (PMN) and macrophage migration and function. The broad-spectrum activities of selected cholesterol-binding toxins (pneumolysin, perfringolysin and listeriolysin) are described below.

The term 'thiol-activated' is now known to be incorrectly applied to this family of toxins and stemmed from initial studies suggesting that the single conserved cysteine residue of each toxin was essential for cytotoxicity. Site-directed mutagenesis of this cysteine residue, which resides in an 11 amino acid residue C-terminal-located sequence conserved across the family, has demonstrated that the cysteine residue is not required for activity. In fact, changing cysteine to alanine in pneumolysin (produced by *Strep. pneumoniae*) does not affect cytotoxicity, channel formation or complement activation. Neither crude extracts of the cysteine-to-alanine mutant version of pneumolysin nor highly purified toxin require thiol activation and it is now

Table 9.3. Cholesterol-binding cytolysins. These toxins are expressed by certain genera of Gram-positive bacteria and some examples from each genus are provided

Genus	Species	Toxin
Streptococcus	Strep. pneumoniae Strep. pyogenes	Pneumolysin (intracellular) Streptolysin O
Listeria	Lis. monocytogenes Lis. seeligeri	Listeriolysin O Seeligerolysin
Clostridium	Cl. perfringens Cl. tetani Cl. sordelli	Perfringolysin O Tetanolysin Sordellilysin
Bacillus	B. cereus B. thuringiensis	Cereolysin O Thuringolysin O

thought that contaminants in early preparations were responsible for oxidative loss of activity of pneumolysin and other cholesterol-binding toxins. However, other mutations within the conserved 11 amino acid residue sequence can abolish cytotoxicity and it is now believed that the secondary structure of this region is important for cytotoxicity.

9.4.1.3.1 Pneumolysin

Pneumolysin is unusual amongst this family of toxins in that it lacks a signal sequence for directing toxin secretion from the cell. Consequently, the toxin is intracellular and is released only on cell lysis either by spontaneous autolysis or as a result of exposure to some antibiotics. The evolutionary significance of this is unclear. Nevertheless, this behaviour is thought to contribute to morbidity and mortality after antibiotic treatment of patients has eliminated live *Strep. pneumoniae.*

Besides its cytolytic activity, pneumolysin demonstrates a wide range of additional properties, many of which are thought to promote the pathogenesis of pneumococcal infection. Pneumolysin (and streptolysin O from *Strep. pyogenes*) activates the classical complement pathway, which may serve to impair complement-mediated phagocytosis. At sub-lytic levels, pneumolysin inhibits the migration and antibacterial functions of PMNs and macrophages and blocks the proliferative responses of, and immunoglobulin production by, lymphocytes. Very low doses also induce the release of TNFα and IL-1β, resulting in inflammation and tissue damage. Finally, pneumolysin is known to inhibit the normal beating of cilia on human respiratory tract epithelium and to induce significant and specific cochlear damage in an animal model of pneumococcal meningitis. Isogenic mutants of virulent pneumococci that no longer express pneumolysin (for more information on the generation of isogenic 'knockout' mutations, see Section 3.2.1) have reduced virulence in mice and are cleared

Table 9.4.	Toxins of *Clostridium perfringens*	
Toxin	Target	Action
α toxin	Membrane	Phospholipase C
β toxin	Membrane	Channel forming (β-CFT)
ε toxin	Membrane?	Channel forming?
ι toxin	G-actin	ADP-ribosylation
κ toxin	Extracellular matrix	Collagenase
λ toxin	Extracellular matrix	Protease
μ toxin	Extracellular matrix	Hyaluronidase
τ toxin, perfringolysin O	Membrane	Channel forming (RTX)
NanH, NanI	Membrane	Sialidases

Nan, neuraminidase; for other abbreviations, see the text.

more rapidly from the bloodstream than the wild-type strain. Additionally, pneumolysin-deficient mutant pneumococci do not induce cochlear damage and deafness.

Like pneumolysin, the closely related toxin streptolysin O of *Strep. pyogenes* exerts effects on cells of the immune system, amongst others, in addition to its cytolytic activity. Furthermore, secreted streptolysin O induces profound electrocardiographic changes in experimental animals and has toxic effects on pulsating heart cells in tissue culture.

9.4.1.3.2 Perfringolysin

Clostridium perfringens produces perfringolysin (theta toxin), a secreted cholesterol-binding toxin. This is only one of a barrage of toxins produced by the agent of gas gangrene, wound infections and food poisoning (Table 9.4). *Clostridium perfringens* is a soil bacterium, and it has been suggested that the most favoured outcome for the bacterium following infection of a human or animal is rapid death of the host, resulting in a bumper harvest of nutrients for the underlying soil bacteria. Nevertheless, like other cholesterol-binding toxins, perfringolysin has some subtle effects on host physiology, such as promoting anaerobiasis at the site of infection. Gas gangrene is associated with a lack of PMNs at the site of infection and this is thought to arise both from direct cytotoxic activity and from a host of effects on other cells at low concentrations (Figure 9.5). Perfringolysin induces the expression in endothelial cells of pro-inflammatory platelet-activating factor (PAF). At low levels, the toxin also affects the migration and activity of PMNs and promotes their adhesion to endothelial cells by up-regulating the expression of adhesion proteins.

Figure 9.5 Actions of perfringolysin O. High toxin concentrations (near the focus of infection) result in cytolysis. At lower toxin concentrations the effects of perfringolysin O are more subtle and include the inhibition of PMN migration and function.

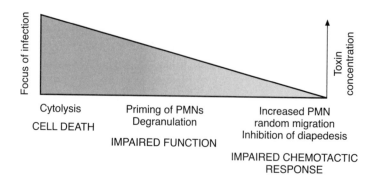

9.4.1.3.3 Listeriolysin O

Listeria monocytogenes is a food-borne pathogen that causes meningitis, meningoencephalitis and septicaemia. Listeriosis in pregnant women can result in miscarriage or septic premature birth and meningitis of the neonate. The success of *Lis. monocytogenes* as a pathogen results, in part, from its ability to grow and divide within human cells, including macrophages, and from its ability to cross barriers (intestinal, blood–brain and placental) during infection.

Once ingested in contaminated food (frequently dairy products, ready-made salads, seafood, etc.), the bacteria cross the intestinal mucosa and move via lymph and the blood to the spleen and liver. In immunocompromised individuals and pregnant women, the organism can multiply in hepatocytes, from which they disseminate haematogenously. *Listeria monocytogenes* adopts a predominantly intracellular lifestyle within the host, which offers the advantage of protection from the immune system and from antimicrobial agents. The bacteria either are phagocytosed by macrophages or actively invade non-phagocytic cells by a so-called zippering mechanism requiring the internalin proteins InlA and InlB (for a description of invasion by the organism, see Section 8.2.1.1.2). Approximately 30 minutes after uptake, the bacteria escape from their vacuoles and freely replicate within the cytosol of the host cell. The cholesterol-binding toxin listeriolysin O is essential and sufficient for this escape process, though a phosphatidylinositol-specific phospholipase C (PI-PLC) may also participate. Mutants of *Lis. monocytogenes* no longer expressing listeriolysin O are avirulent in a mouse model of infection. Unlike other members of the cholesterol-binding toxins, listeriolysin O displays optimum channel-forming activity at an acidic, rather than neutral, pH. This allows the toxin to function only once the bacterium has been internalised and the containing vacuole has been acidified. However, like other toxins in this family, listeriolysin O has additional effects including the ability to induce apoptosis, to activate MAP kinases and to stimulate PMN adhesion to infected endothelial cells by up-regulating the expression of adhesion proteins.

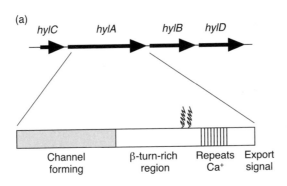

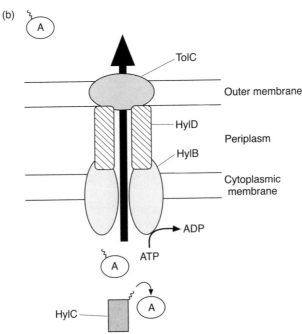

Figure 9.6 The *hylCABD* operon of *Escherichia coli* encodes HylA (an RTX toxin) and associated secretion apparatus. (a) Organisation of the RTX operon of *E. coli* and domain structure of HylA toxin. (b) HylA toxin is lipid-modified by the action of HylC. Export of modified toxin is via a type I (ABC transporter) secretion system (see Section 2.5.4) comprising HylB, HylD and the outer membrane protein TolC.

9.4.1.4 RTX toxins

The term RTX (repeat in toxin) derives from the characteristic amino acid repeats present towards the C-terminus of all members of this toxin family (Figure 9.6). Members of the RTX family of channel-forming toxins are widely distributed amongst Gram-negative bacterial pathogens of humans and animals, including *E. coli*, *Bord. pertussis*, and several *Actinobacillus* spp. (see Table 9.5). Animal studies have indicated that they play a key role in virulence. RTX toxins have channel-forming activity that results in cell death either by osmotic lysis or by the induction of apoptosis in sensitive cells. In addition, like the majority of toxins described so far, members of the RTX family may display additional activities such as inhibition of PMN migration and function and the induction of expression of pro-inflammatory mediators. Unlike the cholesterol-binding cytolysins, RTX toxins are thought to form

Table 9.5. RTX toxins are expressed by a range of Gram-negative bacteria and some examples of the family are provided. The toxins are divided into haemolysins (broad-spectrum activity) and leukotoxins (narrow cell or species range)

Organism	Toxin	Activity
Escherichia coli	HylA	Haemolysin
Actinobacillus actinomycetemcomitans	LtxA	Leukotoxin
Actinobacillus pleuropneumoniae	ApxIA–ApxIVA	Haemolysin, leukotoxin
Pasturella haemolytica	LktA	Leukotoxin
Proteus vulgaris	PvxA	Haemolysin
Bordetella pertussis	CyaA	Haemolysin-adenylyl cyclase

channels as monomers, although at higher toxin concentrations multi-merisation may occur, leading to more severe membrane disruption.

RTX toxins can be divided into two groups: the broad-spectrum haemolysins that act on a variety of cells including erythrocytes, and the leukotoxins, which display both cell-type- and species-specific activity.

9.4.1.4.1 *Escherichia coli* haemolysin

The structural gene encoding the haemolysin of *E. coli* (*hylA*) is part of an operon that also encodes a dedicated export system (HylB and HylD comprising a type I secretion system; see Section 2.5.4) and a toxin-modifying enzyme (HylC). The HylC protein is responsible for acylation of HylA, resulting in toxin activation. The product of a fifth, unlinked gene (*tolC*), is required for extracellular secretion of the toxin. The *E. coli hyl* operon (*hylCABD*; Figure 9.6) forms a template for other RTX toxin operons, though there are variations to this, as described in Section 9.4.1.4.2. The operon is transcribed as both a full-length *hylCABD* transcript and as a more abundant *hylCA* message. Depending on the organism, the *hyl* operon can be found on the chromosome, or can be present on a plasmid, as is the case for enterohaemorrhagic *E. coli* O157:H7. In other *E. coli* strains, the *hyl* operon is part of a large chromosomal element termed a pathogenicity island (see Section 2.6.4).

Escherichia coli haemolysin is a large polypeptide (110 kDa) and consists of four major domains (Figure 9.6). A hydrophobic stretch of approximately 230 amino acid residues towards the N-terminus of HylA is proposed to form eight membrane-spanning α-helices that may comprise the channel-forming domain. Adjacent to this is a β-turn-rich region, which is followed by the lysine targets for toxin modification. Acylation is essential for haemolytic activity and one or more lysine residues of HylA may be acyl modified. Beyond this region lie the characteristic nine amino acid residue repeat blocks. There are 13

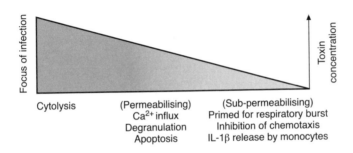

repeats (with the consensus sequence Gly-Gly-X-Gly-X-Asp-X-Leu/Ile/Val/Trp/Tyr/Phe-X-, where X can be any amino acid) in HylA, though this number varies depending on the toxin (41 repeats are present in the RTX toxin CyaA of *Bord. pertussis*). The repeats bind calcium and are essential for toxin activity. At the C-terminus is the signal that directs export of toxin by the HylB-HylD ATP-binding cassette (ABC) transporter.

Since HylA can lyse a broad range of cells and can create pores in planar lipid bilayers (see below), it is thought that binding is relatively non-specific. However, recent evidence suggests that lymphocyte-function-associated antigen (LFA) 1 is the binding target for the RTX toxins of both *E. coli* and *Actinobacillus actinomycetemcomitans*. LFA-1 is a β_2-integrin heterodimer comprising of CD11a and CD18 and is found on most circulating leukocytes, but not on cells of non-haematopoietic origin. Furthermore, the activities of some RTX toxins are exquisitely species and/or cell-type specific, as we shall see; does this arise from binding specificity or are pores formed only in susceptible cells?

It is not fully understood how HylA forms a transmembrane channel, but the eight α-helices of a single toxin molecule are predicted to insert into a planar lipid bilayer to create a channel of approximately 1 to 2 nm in diameter. Nevertheless, heterogeneous lesions are formed in erythrocyte membranes and these are time- and toxin-concentration dependent, suggesting that toxin molecules can multimerise. It is postulated that, at high toxin concentrations rapid cell death arises from osmotic lysis due to the formation of larger pores. In contrast, low doses may result in cell death via apoptosis. There is a suggestion also that bacterial lipopolysaccharide (LPS) may incorporate into aggregates of haemolysin and contribute to pore-forming activity.

In addition to lysis of target cells, HylA has numerous more subtle effects at lower concentrations, as outlined in Figure 9.7. It has been suggested that the activity of HylA at sub-lytic concentrations occurs at the level of toxin binding to membranes and is independent of pore formation.

9.4.1.4.2 RTX toxin of *Actinobacillus actinomycetemcomitans*

Actinobacillus actinomycetemcomitans is an opportunistic pathogen of humans, particularly noted for its involvement in localised juvenile periodontitis (LJP). LJP is an aggressive form of periodontitis that is associated with rapid loss of the alveolar bone supporting the affected teeth, but with only limited gingival inflammation. The organism produces a leukotoxin, Ltx, that demonstrates considerable cell specificity, and killing is limited to a subset of human leukocytes. The leukotoxic activity of Ltx, as well as its sub-lytic impairment of leukocyte function, may eliminate host immune cells from the site of infection, thus facilitating bacterial survival and proliferation, and may additionally explain the lack of gingival inflammation in LJP compared with that accompanying adult periodontitis.

Actinobacillus actinomycetemcomitans strains with low or no detectable leukotoxic activity can be readily isolated from the human oral cavity. Highly toxigenic strains, however, are associated with LJP in Europe and the USA. These strains have a deletion of approximately 500 bp within the *ltxCABD* upstream non-coding region that results in the juxtaposition of a second, stronger, promoter sequence for driving *ltx* operon transcription. Interestingly, it has recently been shown that high-level leukotoxic activity in Japanese isolates of *A. actinomycetemcomitans* correlates with the presence of an insertion sequence (see Section 2.6.3.3) within the *ltxCABD* upstream region, rather than a DNA deletion.

9.4.1.4.3 *Bordetella pertussis* haemolysin-adenylyl cyclase

The *cyaA* gene of *Bord. pertussis* encodes a larger RTX toxin that has an amino terminal extension of some 400 amino acid residues with adenylyl cyclase activity. The cytotoxicity of CyaA results from the translocation of the adenylyl cyclase domain across the cytoplasmic membrane and into the target cell cytoplasm. *Bordetella pertussis* adenylyl cyclase activity results in greatly elevated levels of c-AMP, leading to disruption of normal cellular function. It is not clear how the adenylyl cyclase domain is translocated across the membrane but, like anthrax and DT, it may involve the channel-forming component of the toxin molecule.

Fugu and Voodoo

Tetrodotoxins are potent neurotoxins that exert their effects by blocking voltage-gated sodium channels responsible for inward movement of sodium during electrical signals in the membranes of excitable cells, including neurones and heart and skeletal muscle cells. Binding to the sodium channel prevents the action potential of the cell. Tetrodotoxins are well-known marine toxins that are isolated predominantly from many species of puffer fish, including those used in Japanese Fugu cuisine, and from various other fish and molluscs. The toxins, however, are actually produced by certain species of bacteria that colonise the intestines and render their host poisonous. These include marine *Vibrio* spp. and *Pseudoalteromonas* spp. Tetrodotoxins have been immensely useful in many aspects of cell biology research, including cardiology, transduction of nervous signals and visual system development. They have also been put to nefarious use, along with other marine toxins, in the practice of Voodoo. Haitian Voodoo is a folk religion that includes in its practices both healing and poisoning. The poisoning of individuals is a means of punishment established by a clandestine justice system; tetrodotoxins are amongst the most potent Voodoo toxins and are used to induce catalepsy.

9.4.2 Toxins that damage membranes enzymically

9.4.2.1 Phospholipases

A number of Gram-positive and Gram-negative pathogens express a phospholipase C (PLC) activity that attacks phospholipids in host cell membranes. These include *Cl. perfringens* (α-toxin), *Lis. monocytogenes*, *Ps. aeruginosa* and *Staph. aureus* (β-haemolysin).

Clostridium perfringens α-toxin is extremely potent and is known to have necrotic and cytolytic activity. In contrast, the two PLCs of *Lis. monocytogenes* are important for the organism's intracellular lifestyle. PI-PLC augments listeriolysin O activity during the escape of *Lis. monocytogenes* from the primary vacuole, while the broad range PLC, or lecithinase, is required for cell–cell spread. The *Pseudomonas aeruginosa* PLC may function to damage lung surfactant and so may be important when this organism infects the lungs of cystic fibrosis patients.

9.4.2.2 Proteases

Clostridium tetani, *Cl. botulinum* and *B. anthracis* produce proteases that are intracellularly acting (type III) toxins with exquisite target specificity, as will be described in Section 9.5. However, many bacteria secrete proteases that can attack host cell membrane proteins. These destructive enzymes are undoubtedly virulence factors. Nevertheless, the term toxin has not normally been applied to them, owing to their perceived lack of specificity. Recently, however, a family of related proteases expressed by the periodontal pathogen *Porphyromonas gingivalis* has been described, each member of which has specificity for a particular integrin (for a description of integrins, see Section 6.6.2.1). These proteases may affect host cell signalling systems.

9.5 | Type III (intracellular) toxins

Type III toxins are active within the target cell. Therefore, unlike the membrane-acting (type I) and membrane-damaging (type II) toxins described in Sections 9.3 and 9.4, type III toxins must gain access to the target cell internal milieu. Consequently, the intracellular toxins have a modular design comprising receptor binding and translocation functions as well as the toxigenic component. These so-called AB toxins are widespread amongst bacteria and also offer many variations on the AB theme. Thus a single polypeptide chain may carry both A (enzymic) and B (binding and translocation) components, as is the case for *Ps. aeruginosa* **exo**toxin A (ExoA) (although proteolytic processing of the polypeptide yields separate A and B fragments), while cholera toxin comprises a single A polypeptide that complexes non-covalently with five B subunits (AB_5).

Internalisation of membrane-bound toxin involves receptor-mediated endocytosis via clathrin-coated pits. However, subsequent delivery of the toxin subunit A from the endosome to the cytoplasm is less well understood. Two AB toxins were introduced in Section 9.4 (anthrax and diphtheria toxins), since the translocation domain(s) were proposed to form transmembrane channels presumably allowing transport of the enzymic A component through the aqueous channel. Thus anthrax-protective antigen (the B component of this A_2B toxin) is thought to form a classic heptameric membrane structure generating a 14 strand transmembrane β-barrel through which both A subunits (lethal factor, a zinc-dependent protease, and EF, an adenylyl cyclase) can pass. Similarly, the translocation of DT A fragment (an ADP-ribosylating activity) may involve a transmembrane channel formed by the αCFT-like B fragment in conditions of low pH. More recent evidence suggests that many AB toxins, once endocytosed, are transported via the Golgi apparatus to the endoplasmic reticulum (ER), wherein they subvert the ER-associated protein degradation pathway to enter target cells (see Micro-aside below).

As well as the variety of AB subunit/fragment arrangements, type III toxins also demonstrate a range of enzymic activities including proteolytic, ADP ribosylation and adenylyl cyclase activity (Table 9.6) as will be discussed in this section.

Toxin translocation by subversion

DT, botulinum neurotoxin (BoNT) and tetanus toxin gain access to the cytosol directly from acidified endosomes. In contrast, other AB toxins first undergo retrograde transport (i.e. in the direction opposite to the normal movement of proteins) via the Golgi apparatus to the ER (Figure 9.8). These include CT, LT of enterotoxigenic *Escherichia coli*, pertussis toxin (PT), *Ps. aeruginosa* ExoA, Shiga toxin (Stx) and Shiga-like toxins (Stx1, Stx2) of enterohaemorrhagic *E. coli*. Once in the ER they are thought to disguise themselves as misfolded host proteins and so enter the cytoplasm via the ER-associated protein degradation (ERAD) pathway. How they do this has not been fully elucidated, but what is known reinforces, if this were needed, the mastery many bacteria and bacterial components exert over normal eukaryotic cell functions.

At the C-terminus of the A subunit or fragment of CT, LT and ExoA is an amino acid sequence that serves as an ER retrieval signal (Lys-Asp-Glu-Leu or related sequence), thus marking these proteins for retrograde transport from the Golgi apparatus to the ER. ExoA even undergoes removal of its C-terminal lysine residue to reveal an ER retrieval signal (Arg-Glu-Asp-Leu). Retrograde transport of toxins that lack this signal (e.g. Stx and Stx1, 2) may involve the Gb_3 glycolipid toxin receptor or may occur by piggyback with an as yet unidentified cellular component.

The ER has a system for recognising misfolded proteins and marking these for degradation via the ERAD pathway. Importantly for AB toxins, the proteolytic machinery resides in the cytoplasm and so misfolded proteins must first be retro-translocated from the ER to the cytoplasm. Lysine residues of misfolded proteins are covalently modified with ubiquitin and this marks the protein for subsequent proteolytic degradation by the proteasome. By pretending to be mis-folded host proteins, AB toxins may enter the ERAD pathway and thus gain access to the host cell cytoplasm. This, however, raises two obvious questions: how does the A subunit mimic the mis-folded state, and how does it avoid subsequent protein degradation once in the cytoplasm?

As to the first question, once in the ER the A and B subunits dissociate and this exposes hydrophobic, and possibly unfolded, surfaces on the A component, thus marking the toxin for translocation. The second question may be answered by the very low occurrence, or even absence, of lysine residues within the A component. Where present, lysine residues are restricted to the extreme N- or C-terminus and are thought not to be efficiently ubiquitinated. Evolutionary suppression of lysine residues within the body of the A component may thus permit avoidance of ubiquitin-mediated protein degradation. It is of note that the B components of ER-directed toxins as well as both A and B components of non-ER-directed toxins (e.g. DT) have a normal abundance and distribution of lysine residues.

9.5.1 AB$_5$ toxins

9.5.1.1 The cholera toxin paradigm

Ingestion of enterotoxigenic *V. cholerae* in contaminated water can give rise to either mild self-limiting diarrhoea or else profuse watery diarrhoea. The dehydration associated with the latter frequently leads to death in the absence of fluid and electrolyte replacement. The disease has spread throughout the world in seven major pandemics, the most recent of which started in 1961. Bacteria that survive the acid of the stomach attach to the epithelial surface of the small intestines and secrete an enterotoxin that can lead to remarkable fluid loss of up to 20 litres per day. CT is the prototype AB$_5$ toxin and comprises a single catalytic A subunit of 27.2 kDa complexed non-covalently with five B subunits of 11.6 kDa each, which are responsible for receptor binding on target cells (Figure 9.9). The receptor for CT is the glycolipid ganglioside GM1. Once bound, the six-subunit holotoxin is internalised by receptor-mediated endocytosis. The A subunit is proteolytically nicked to generate fragment A1 (22 kDa), which carries the enzymic activity, and a short A2 chain that interacts with the B pentamer. A1 and A2 fragments are held together by a disulphide bridge. The A1 fragment is subsequently delivered to the cytoplasm, possibly via the ERAD pathway. The A1 fragment is an NAD-dependent ADP ribosyltransferase and, like all ADP-ribosylating toxins (Table 9.7), the target for modification by CT is a nucleotide-binding protein. Specifically, CT adds the ADP-ribose moiety to the α-subunit of Gs, a heterotrimeric G protein involved in stimulation of adenylyl cyclase (for a summary of G proteins, see Section 4.2.2.2). ADP ribosylation of the Gs α-subunit at arginine 201 locks the protein in the activated (GTP-bound) state by blocking the natural GTP hydrolase activity of the α-subunit

Table 9.6. Range of enzyme activities expressed by type III (intracellularly acting) toxins. Some examples of each toxin type are presented; more details are given in the text

Activity	Organism/toxin (type)	Target and effect
ADP-ribosyltransferase[α]	*Vibrio cholerae* CT (AB$_5$)	Gs α-subunit. Activates adenylyl cyclase, elevated c-AMP levels
	Pseudomonas aeruginosa ExoA (AB) ExoS, ExoT (type III secreted)	
N-Glycosidase	*Escherichia coli* Stx1, Stx2 (AB$_5$)	28 S RNA. Blocks protein synthesis
Glucosyltransferase	*Clostridium difficile* TcdA, TcdB (AB)	Rho family GTPases. Disrupts cell cytoskeleton
Deamidase	*Escherichia coli* CNF1, CNF2 (AB)	Rho family GTPases. Disrupts cell cytoskeleton
Protease	*Clostridium tetani* TeNT (AB)	VAMP. Inhibits release of neurotransmitters
	Bacillus anthracis LF (A$_2$B)	MAPKK1, MAPKK2. Proteasome activation?
Adenylyl cyclase	*Bordetella pertussis* CyaA (RTX-like)	Elevated c-AMP levels
	Bacillus anthracis EF (A$_2$B)	

VAMP, vehicle-associated membrane protein; MAPKK, MAP kinase kinase. For other abbreviations, see list on p. xxii. For more details, see Section 4.2.4.2.

[α] See also Table 9.2.

(Figure 9.10). As a result of adenylyl cyclase activation, c-AMP levels increase thereby activating protein kinase A, which, in turn, leads to supranormal phosphorylation of chloride channels in the apical epithelial cell membranes. Chloride secretion is stimulated, while sodium absorbtion is inhibited, leading to a serious disturbance in fluid and electrolyte transport manifest as diarrhoea.

Enterotoxigenic *E. coli* (ETEC) produce the high molecular mass LT and low molecular mass ST toxins described earlier. LT is an AB$_5$ toxin and resembles CT structurally, functionally and immunologically. Indeed, the primary amino acid sequences of CT and LT-I are approximately 70% identical. While antibodies to CT will inactivate LT-I, they do not neutralise the less-well-conserved LT-II toxin expressed by some ETEC strains. Nevertheless, LT-II functions in a manner identical to that of CT and LT-I.

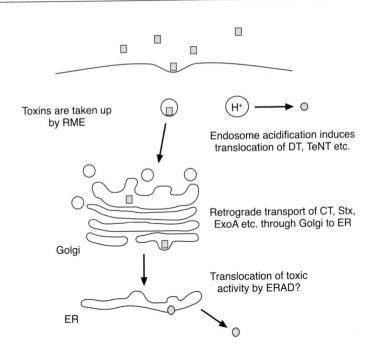

Toxins are taken up by RME

Endosome acidification induces translocation of DT, TeNT etc.

Golgi

Retrograde transport of CT, Stx, ExoA etc. through Golgi to ER

Translocation of toxic activity by ERAD?

ER

Figure 9.8 Toxin translocation. Following uptake by receptor-mediated endocytosis (RME), one group of toxins gains access to the cytoplasm directly from acidified endosomes. Members of a second group are transported through the Golgi apparatus to the endoplasmic reticulum (ER) and may gain access to the cytoplasm via the ERAD pathway (for details, see the text). For other abbreviations, see list on p. xxii.

Cholera epidemics

Cholera remains a major health problem in the 21st century, particularly so in developing countries. Cholera outbreaks also follow closely on the heels of natural and human-engendered disasters, where malnourished individuals are forced to gather in large numbers without access to clean drinking water or proper sanitation. There are some 155 serotypes of *V. cholerae*, but until very recently only strains of serotype O1 were responsible for cholera epidemics. The remainder are ubiquitous members of the microflora of aquatic environments, although they can cause sporadic outbreaks of diarrhoea. On the basis of several genotypic and phenotypic differences, O1 serotype strains can be divided into El Tor and classical biotypes. El Tor is responsible for the ongoing cholera pandemic, while the strains isolated from the previous two pandemics (1881–1896 and 1899–1923) were of the classical biotype.

In 1992 a novel epidemic strain emerged in southern India. Within months of the appearance in Madras of the epidemic O139 serotype, outbreaks of disease caused by the so-called Bengal strain had spread throughout India and into several neighbouring countries in Asia. Incidences of disease were reported also in the Western world, imported by travellers to the Indian subcontinent. The incidence of disease due to the previously endemic El Tor O1 serotype strain was greatly reduced during the Bengal rampage, which affected hundreds of thousands of people. Nevertheless, since 1994, the El Tor biotype has re-emerged as the predominant serotype associated with epidemic cholera.

So, from where did the Bengal strain emerge and how did it effectively usurp El Tor, if only briefly, as the predominant cause of cholera? The phenotypic differences between El Tor and Bengal provide the first clues to help to answer the former question. Both strains produce the major virulence factors CT and toxin co-regulated pilus (see below) but they express distinct LPS molecules that differ in the length and sugar composition of the antigenic O side chains. Additionally, strain Bengal produces a polysaccharide capsule. Either the O139 Bengal strain may have acquired the ability to cause epidemic cholera or else the pandemic O1 El Tor strain may have acquired the wherewithal to express an altered LPS and a capsule. Detailed genetic analysis including DNA fingerprinting and gene sequencing has indicated that the latter possibility is correct. Thus the O139 strain arose when an El Tor bacterium acquired, by horizontal DNA transfer, DNA involved in O antigen and capsule synthesis. Specifically, a 22 kb O1 antigen synthesis locus was replaced by a 35 kb O139 antigen-specific DNA fragment that additionally carried genes for capsule synthesis. The mechanism of transfer is not yet clear. However, an insertion sequence element within the O antigen locus may have played a role in mobilisation of genomic DNA by transposition. It is of note that Bengal also carries a conjugative transposon, although this is not closely associated with the O139 antigen genes. Bacteriophage transduction may also be involved in transfer of cell wall polysaccharide genes. Indeed, it is now clear that bacteriophages have played a crucial role in the evolution of epidemic *V. cholerae* strains. Thus the genes that encode CT, *ctxA* and *ctxB*, are carried by the filamentous phage CTXφ. The bacterial receptor for phage infection is the toxin co-regulated pilus. This is itself an important virulence factor for *V. cholerae* that allows the bacteria to colonise the intestine, and the gene encoding the toxin co-regulated pilus is, in turn, carried by another filamentous phage, VPIφ (see Micro-aside, p. 380).

Once the new strain arose, by whatever means, how did it usurp El Tor? There are a number of probable contributing factors. Firstly, the production of a polysaccharide capsule by Bengal may have offered increased resistance to complement-mediated killing and to phagocytosis (see Section 10.4.1.1). Secondly, acquired immunity to cholera in endemic areas is mediated, on the whole, by antibodies to the O antigen. A notable feature of the Bengal epidemic was the high number of adults succumbing to the disease: these individuals would normally have acquired a degree of natural immunity to cholera mediated by El Tor, but were susceptible to the Bengal strain expressing a different O antigen. The conjugative transposon of the Bengal strain carries multiple antibiotic resistance, and a third factor in the emergence of Bengal may have been the more widespread use of antibiotics in recent years. As community- or herd-immunity against the O139 Bengal strain developed, then El Tor re-emerged as the predominant disease-associated organism. Disturbingly, however, some recent Indian El Tor isolates have been shown to carry a DNA element similar to the conjugative transposon of Bengal and exhibit multiple resistance to antibiotics.

9.5.1.2 Same organisation, different function: Shiga toxins are *N*-glycosidases

Shiga toxins are produced by *Shigella dysenteriae* and Shiga-toxin producing *E. coli* (STEC) and are cytotoxic and enterotoxic. The notorious *E. coli* O157:H7 is the dominant STEC serotype and is most commonly

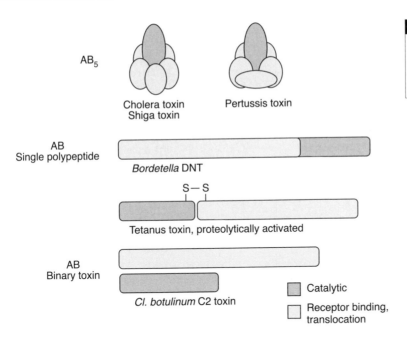

AB$_5$

Cholera toxin
Shiga toxin

Pertussis toxin

AB
Single polypeptide

Bordetella DNT

S—S

Tetanus toxin, proteolytically activated

AB
Binary toxin

Cl. botulinum C2 toxin

Catalytic

Receptor binding, translocation

Figure 9.9 Variations on a theme: molecular architecture of some AB toxins (for details, see the text). DNT, dermonecrotic toxin.

associated with large outbreaks of disease. Shigellosis is characterised by cramps, diarrhoea and dysentery, while *E. coli* O157 : H7 can cause a variety of symptoms ranging from mild diarrhoea to haemorrhagic colitis and the often-fatal **haemolytic uraemic syndrome** (HUS).

Shiga toxins of STEC were originally named Vero toxins (or Shiga-like toxins, SLTs) on account of their profound cytotoxic effects on Vero (African green monkey kidney) cells. Antibody neutralisation studies have distinguished two main types of *E. coli* Stx: Stx1 and Stx2. Stx expressed by *Shig. dysenteriae* and Stx1 of *E. coli* differ by only a single amino acid residue. In contrast, Stx2 is only approximately 60% identical in A and B subunits with either Stx or Stx1.

The Stx1 holotoxin is a 70 kDa hexamer comprising a single catalytic A subunit of 32 kDa complexed non-covalently with five B subunits, each 7.7 kDa in size. Like CT, the pentameric ring of B subunits is responsible for receptor binding. The receptor in this instance, however, is the glycolipid globotriaosylceramide Gb$_3$. Bound toxin molecules are internalised by receptor-mediated endocytosis involving clathrin-coated pits. Once internalised in a vesicle, the toxin is carried by retrograde trafficking via the Golgi apparatus to the ER (see Figure 9.8). During this process, the A subunit is proteolytically nicked by a membrane-bound host protease to generate a catalytically active 27 kDa A1 fragment and a 4 kDa C-terminal A2 fragment. While in the transport vesicle, these fragments remain attached to each other via a disulphide bond. Final delivery of the catalytic fragment to the target cell cytoplasm may involve the ERAD pathway described earlier. An alternative hypothesis is that the hydrophobic A2 fragment may function in translocation by inserting into the ER membrane. Once in the cytoplasm, the

Table 9.7. ADP-ribosylating toxins. The enzymic component of each toxin is an NAD^+-dependent ADP-ribosyltransferase that attaches an ADP-ribose moiety to the target protein (always a nucleotide-binding protein)

Toxin	Bacterium	Toxin type	Target	Action
Cholera toxin (CT)	Vibrio cholerae	AB_5	GS_α	Activates adenylyl cyclase producing elevated c-AMP levels
Heat-labile toxin (LT)	Escherichia coli	AB_5	GS_α	As above. CT and LT are functionally and immunologically very similar
Pertussis toxin	Bordetella pertussis	AB_5	Various Gs proteins	Blocks host intracellular signal transduction
Exotoxin A (ExoA)	Pseudomonas aeruginosa	AB	Elongation factor EF2	Inhibits protein synthesis
ExoenzymeS (ExoS)	Pseudomonas aeruginosa	Type III secreted	Ras proteins, also stimulates GTP hydrolysis by Rho	Disrupts cytoskeleton
ExoenzymeT (ExoT)	Pseudomonas aeruginosa	Type III secreted	Ras proteins	Disrupts cytoskeleton. Only 0.2% activity of ExoS
Diphtheria toxin (DT)	Corynebacterium diphtheriae	AB	EF2	As for ExoA
C2	Clostridium botulinum	BinaryAB	Non-muscle actin	Disrupts cytoskeleton
C3	Clostridium botulinum	–	Rho, Rac	Exoenzyme or toxin?

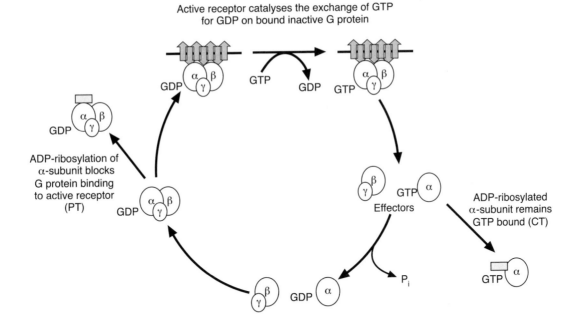

Active receptor catalyses the exchange of GTP
for GDP on bound inactive G protein

ADP-ribosylation of
α-subunit blocks
G protein binding
to active receptor
(PT)

ADP-ribosylated
α-subunit remains
GTP bound (CT)

Effectors

disulphide bond linking A1 and A2 fragments is dissociated, releasing free A1.

Stx, Stx1 and Stx2 exert their toxic effects by blocking protein synthesis, which results in cell death. The active A1 fragment is an RNA N-glycosidase that cleaves adenine from a specific adenosine residue of 28 S rRNA, a component of the eukaryotic 60 S ribosomal subunit. This depurination reaction prevents the elongation factor 1-mediated binding of amino acyl tRNAs to ribosomes, thus blocking protein synthesis.

The apparent absence of Gb$_3$ receptors on human enterocytes suggests that the gastrointestinal effects of Stx, including diarrhoea, dysentery and haemorrhagic colitis, result from systemic toxin. *Shigella dysenteriae* is invasive. STEC strains, however, are not and therefore the systemic appearance of toxin must presumably involve translocation of Stx across the gastric mucosa. Interestingly, *E. coli* O157 : H7 harbours a large virulence plasmid, pO157, that carries additional toxin genes, including a *hylCABD* operon encoding a variant RTX enterotoxin. In approximately 10% of cases, STEC infection can lead to HUS, a life-threatening sequela characterised by acute renal failure. High levels of expression of the Gb$_3$ receptor on human kidney cells may contribute to this condition.

Stx has also been shown to induce the release of pro-inflammatory cytokines (IL-6, TNFα) from macrophages.

The genes encoding Stx1 and Stx2 of *E. coli* are carried on lysogenic bacteriophages and thus are transmissible from strain to strain. In contrast, the *stx* genes of *Shig. dysenteriae* are not transmissible. However, the presence of bacteriophage-like sequences at the *stx* chromosomal locus suggests that, following acquisition of an Stx-encoding

Figure 9.10 ADP-ribosylation of heterotrimeric G proteins. ADP-ribosylation of the α-subunit by cholera toxin (CT) locks the α-subunit in the 'on' position by blocking the natural GTP hydrolase activity of the α-subunit. In contrast, ADP-ribosylation by pertussis toxin (PT) prevents G protein from binding to activated membrane receptor and thus uncouples signal reception from the appropriate effector systems.

prophage, IS element insertions and gene rearrangements effectively immobilises the defective prophage.

Antibacterials in agriculture: double trouble

The use of antibacterials in animal husbandry to promote growth is widespread. Although the European Union currently allows subinhibitory concentrations of certain growth-promoting antibacterials (which are not used in human therapy) to be included as feed additives, the practice is controversial, since it is thought to support the development of antibiotic-resistant bacteria that could subsequently be transmitted to humans.

STEC occur frequently in cattle. The genes encoding *E. coli* Shiga toxins are carried by bacteriophages that integrate into the bacterial chromosome (prophage). Stresses such as ultraviolet light or mitomycin C are known to trigger the lytic phase of the bacteriophage life cycle. Phage replication occurs, the host bacterium is lysed and mature phages are released and are able to colonise bacteria in the vicinity. Phage integration into the new host's chromosome carries with it the ability to produce Shiga toxin. It has been demonstrated recently, at least *in vitro*, that subinhibitory concentrations of some antibacterial agents induce the release of Stx phage particles and also increase the amount of Shiga toxin produced by infected bacteria.

Thus certain growth-promoting antimicrobials used in animal husbandry may simultaneously promote the development of antimicrobial resistance and also the horizontal transfer of Shiga toxin genes between bacteria.

9.5.2 ADP-ribosylation is a common toxin activity

Besides cholera toxin and the heat-labile toxin of *E. coli*, there are several other examples of bacterial toxins that possess ADP-ribosyltransferase activity and these are listed in Table 9.7.

9.5.2.1 Pertussis toxin, a variant AB₅ design

Bordetella pertussis, the causative agent of whooping cough, expresses a variant AB_5 toxin in addition to the dual RTX-adenylyl cyclase toxin (CyaA) described earlier. PT is a complex protein composed of five different subunits (S1 to S5). A single catalytic A subunit (S1) is associated with a B oligomer that is responsible for receptor binding (see Figure 9.9). The B oligomer comprises two copies of S4, plus single copies of S2, S3 and S5. S1 is an ADP-ribosyltransferase with homology to the A subunits of CT and LT. Nevertheless, the target for ADP-ribosylation differs. Thus PT modifies a cysteine residue on the α-subunits of various heterotrimeric G proteins that are important signal transduction components. GDP-bound G_α subunit interacts with β- and γ-subunits to form an active trimer that can bind to activated transmembrane signal receptors. Once bound, the α-subunit exchanges GDP for GTP and the activated (GTP-bound) α-subunit subsequently dissociates from the βγ-dimer. In some cases (e.g. the stimulation of

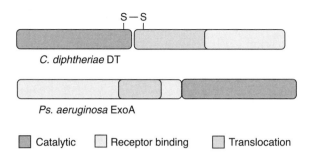

adenylyl cyclase) the activated α-subunit is responsible for effector regulation, although free βγ-dimers may also be involved in regulating effector activity (Figure 9.10; for more details of eukaryotic signal transduction, see also Chapter 4). ADP-ribosylation of GDP-bound G_α prevents the association of trimer with activated receptor, thus decoupling signal reception from the appropriate effector systems. At least seven human G protein α-subunits are sensitive to ADP-ribosylation by PT. The toxin is secreted by a type IV secretion system (see Section 2.5.3.3).

9.5.2.2 ADP-ribosyltransferases of *Pseudomonas aeruginosa*

Exotoxin A (ExoA) of *Ps. aeruginosa*, like DT (described in Section 9.4.1.2.1), transfers the ADP moiety of NAD^+ onto the diphthamide residue of eukaryotic EF2. This results in inhibition of protein synthesis and leads to cell death. ExoA is produced as an enzymically inactive pro-enzyme. The crystal structure of ExoA has revealed a modular toxin design consisting of three functional domains. Domain I mediates target cell recognition and binds a specific, but as yet unidentified, receptor. Domain II, which separates domain I into two subregions (Ia and Ib) is composed of six α-helices and is responsible for translocation of the enzymic component, domain III, into the target cell cytoplasm. The catalytic domains of ExoA (domain III) and DT (C domain) are highly conserved. However, a comparison of the primary amino acid sequence of these toxins reveals a different domain organisation (Figure 9.11).

Once bound, ExoA is taken up by receptor-mediated endocytosis via clathrin-coated pits. Acidification of the vesicle results in translocation of the ADP-ribosyltransferase activity into the cytosol, where target EF2 molecules reside. It is not yet clear whether intact ExoA or an enzymically active fragment is translocated. However, by analogy with DT, translocation would involve the formation of a transmembrane pore by the insertion of certain α-helices of domain II.

Exoenzyme S (ExoS) and exoenzyme T (ExoT) are closely related ADP-ribosyltransferases that are part of the ExoS regulon. The ExoS regulon encodes a type III secretion system that delivers ExoS and ExoT, as well as ExoU (a cytotoxin) and ExoY (an adenylyl cyclase), into target cells, as will be discussed in Section 9.5.4.2.

9.5.2.3 Binary ADP-ribosylating toxins

Certain species of clostridia produce ADP-ribosylating toxins in an AB format where the A and B components are separate polypeptides not linked by either covalent or non-covalent bonds (see Figure 9.9). Examples include toxin C2 of *Cl. botulinum*, CDT of *Cl. difficile* and iota (ι) toxin produced by *Cl. perfringens*. The target for ADP-ribosylation is monomeric G actin; modification prevents polymerisation of G actin, thus disrupting the cytoskeleton and ultimately leading to cell death. C2 toxin can also increase vascular permeability and, in a mouse intestinal loop model, can cause fluid accumulation as well as marked inflammation and necrotic effects.

C2II polypeptide, approximately 100 kDa in mass, is proteolytically processed to produce a 74 kDa active fragment that binds to specific glycoproteins on the target cell surface. Activation results in polymerisation of C2IIa proteins to form heptameric structures. Once bound, C2IIa provides a binding site for the catalytic C2I ADP-ribosyltransferase (approximately 50 kDa). The complex is subsequently internalised by receptor-mediated endocytosis. Acidification of the early endosome induces the membrane insertion of the C2IIa heptamer to generate a transmembrane channel formed by a 14-stranded β-barrel, through which the C2I polypeptide is proposed to translocate. In this respect, C2II strongly resembles protective antigen, the binding component of anthrax toxin, and indeed there is significant sequence similarity between these toxin components.

In addition to the chromosomally encoded C2 toxin, *Cl. botulinum* also expresses an ADP-ribosylating C3 exoenzyme (24 kDa) that is phage encoded. Rho and Rac GTP-binding proteins are the preferred targets for ADP-ribosylation by C3. Nevertheless, C3 is more properly labelled an exoenzyme and not an exotoxin, since there is no evidence for a binding/translocation function associated with C3. C3 cannot therefore enter most cells and so is non-toxic *per se*.

9.5.3 Other AB toxin activities

9.5.3.1 Glucosyltransferase

Enterocolitis due to *Cl. difficile* is the most common nosocomial infection of the gastrointestinal tract, particularly amongst the elderly in nursing homes. The majority of cases are associated with antibiotic therapy that alters the normal gastrointestinal microflora, allowing overgrowth of *Cl. difficile*, and can lead to the serious condition of pseudomembranous colitis. *Clostridium difficile* produces two large cytotoxins, TcdA and TcdB, that share approximately 60% amino acid sequence similarity. Nevertheless, whereas TcdA is enterotoxic, TcdB is approximately 1000 times more cytotoxic to cultured cell lines. TcdA and TcdB mediate their toxic activity by glucosylation of a specific threonine residue in members of the Rho family of GTPases (Rho, Rac, Cdc42) that regulate the actin cytoskeleton and are important signal transduction components (see Section 4.2.5.2). The source of the

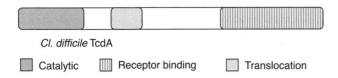

Cl. difficile TcdA

■ Catalytic ▥ Receptor binding ▨ Translocation

Figure 9.12 *Clostridium difficile* toxin A. TcdA (300 kDa) is an example of an AB^X toxin where multiple receptor-binding motifs are present within a single polypeptide chain. Large clostridial AB^X toxins are expressed also by *Cl. novyi* (Tcnα) and *Cl. sordelli* (TcsH, TcsL). The number of receptor-binding repeats varies from 14 to 30.

glucose moiety is UDP-glucose. Glucosylation within the effector domain blocks binding and hydrolysis of GTP and thus prevents Rho from binding to its effectors, resulting in the disruption of the cell cytoskeleton.

There are at least three functional domains: (i) an enzymic domain residing in the *N*-terminus of the protein, (ii) a putative translocation domain and (iii) a receptor-binding domain at the C-terminus (Figure 9.12). The receptor-binding domain comprises extensive amino acid repeats that resemble those seen in glucosyltransferase enzymes of many oral streptococci and also lytic enzymes of *Strep. pneumoniae*. Neither the receptor(s) nor the means of translocation of *Cl. difficile* cytotoxins have been fully elucidated as yet, although uptake is by receptor-mediated endocytosis via clathrin-coated pits, and acidification of endosomes is required for toxin activation.

Other large clostridial cytotoxins target Rho family GTPases. These include the functionally identical *Cl. sordellii* haemorrhagic toxin and *Cl. sordellii* lethal toxin which glucosylates Ras proteins as well as Rac from the Rho subfamily. *Clostridium novyi* α-toxin transfers *N*-acetylglucosamine to specific threonine residues on Rho, Rac and Cdc42, with toxic effects similar to those of *Cl. difficile* cytotoxins.

9.5.3.2 Deamidase

The Rho family of small GTPases is the target also for related toxins expressed by *Bordetella* spp. and by necrotoxic *E. coli* (NTEC – an animal pathogen) strains. The mode of action in this instance is deamidation of a specific glutamine residue (Gln-63 of Rho, Gln-61 of Cdc42 and Rac) to generate glutamate. Deamidation renders GTP-bound Rho proteins constitutively active by inhibition of the GTPase activity. This results in large-scale cytoskeletal rearrangements, stress fibre formation and multinucleation of cells. NTEC strains express one of two cytotoxic necrotising factors (CNF1 and CNF2) that share some 90% homology across their 1014 amino acid residue length. Both are highly lethal in mice. CNF1 is encoded on a pathogenicity island that encodes also α-haemolysin and P fimbriae (Pai II; see Section 2.6.4 and Table 2.7), while CNF2 is plasmid encoded. CNF1 induces effacement of apical microvilli on the intestinal brush border. The flat surface may provide for better adhesion of *E. coli* and possibly for increased nutrients available to the bacterium as a result of their reduced absorption by host cells. Additionally, CNF1 inhibits the transepithelial migration of PMNs.

Dermonecrotic toxin (DNT) is expressed by *Bordetella* spp. DNT shares only limited amino acid sequence homology with the CNFs of *E. coli*, and this homology is restricted to a C-terminal region responsible for enzymic activity. While CNF1 and DNT have both transglutaminase activity as well as deamidase activity, it appears from recent work that DNT is primarily a transglutamidase whereas CNF1 is primarily a deamidase. Transglutamination, nevertheless, still results in inhibition of GTPase activity.

CNF1 is unusual in that it is internalised by receptor-mediated endocytosis that does not involve clathrin-coated vesicles. Once internalised, CNF1, like DNT, requires acidification of the vesicle to induce the membrane translocation step and delivery of enzymic activity to the cytosol (see Figure 9.8).

9.5.3.3 Protease

9.5.3.3.1 Clostridial neurotoxins

Neurotoxins of *Cl. tetani* and *Cl. botulinum* constitute a family of toxins with similar structure and mode of action at the molecular level, and are the most potent protein toxins known. There is a single tetanus neurotoxin (TeNT) but seven serotypes of botulinum neurotoxin (BoNT), A to G. Tetanus is associated with wound infections whereas botulinum intoxication frequently results from ingestion of pre-formed toxin in contaminated food. *Clostridium botulinum* is also responsible for infant botulism, a mild self-limiting disease of children. The eukaryotic targets of both TeNT and BoNT are a group of proteins involved in docking and fusion of synaptic vesicles to presynaptic plasma membranes (Figure 9.13).

Both toxins are expressed as a single polypeptide chain of mass approximately 150 kDa that is released from lysed bacterial cells and subsequently undergoes proteolytic activation to generate two fragments linked by a disulphide bond (see also Section 9.4.1.2.1). The enzymically active light chain (LC; approximately 50 kDa) is a zinc metalloprotease with the consensus zinc-binding motif His-Glu-X-X-His (where X is any amino acid). The second fragment (heavy chain, HC) carries binding and translocation functions.

TeNT released in infected wounds binds first to neuromuscular terminal motor neurones and is taken up by receptor-mediated endocytosis (Figure 9.13a). TeNT is delivered by retrograde axonal transport to the motor neurone pericaryon in the central nervous system, where it is released into the synaptic cleft. TeNT is then taken up again by receptor-mediated endocytosis into the inhibitory interneurone. Following acidification of the intraneuronal compartment, the HC translocation domain is thought to insert into the endosomal membrane and to facilitate the translocation and release of enzymic LC into the cytosol. TeNT LC cleaves one of the membrane components of synaptic vesicles, preventing neurotransmitter release, which in turn causes muscle contraction, which is seen as spastic paralysis. BoNTs gain access to the bloodstream following transcytosis of the intestinal barrier (Figure 9.13b). BoNTs bind to receptors on the presynaptic membrane of motor neurones associated with the peripheral nervous system. Once taken up by receptor-mediated endocytosis, the

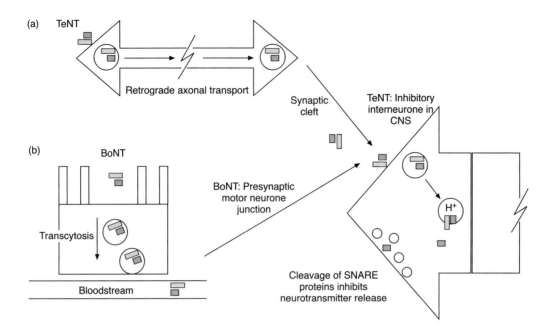

Figure 9.13 Translocation and action of clostridial neurotoxins (for details, see the text). The larger shaded rectangle represents the heavy chain and the smaller one the light chain.

enzymic LC component of BoNTs is released into the motor neurone again following vesicle acidification. BoNT LC cleaves one of the membrane components of synaptic vesicles, or else a receptor component of the target membrane, depending on the serotype. The result is again inhibition of neurotransmitter release, this time of acetylcholine, thus preventing muscle contraction and resulting in flaccid paralysis.

9.5.3.3.2 Anthrax lethal factor

This protease component of an unusual AB toxin was described in Section 9.4.1.1.3. LF is delivered to the target cell cytoplasm via a transmembrane pore formed by a heptameric PA complex. LF is a zinc protease that cleaves and inactivates MAP kinase kinases, pivotal components of the mitogen-activated protein (MAP) kinase signalling pathway that regulates, amongst other processes, cell proliferation (see Section 4.2.4.2).

9.5.3.4 Adenylyl cyclase

We have already described toxins that stimulate endogenous mammalian adenylyl cyclase activity (e.g. cholera toxin) as well as two bacterial toxins that are themselves adenylyl cyclase enzymes: EF and the unusual RTX-adenylyl cyclase CyaA of *Bord. pertussis*. It is fascinating to note that both bacterial adenylyl cyclase toxins are stimulated by calmodulin, a protein thought to be restricted to eukaryotic systems. Calmodulin is a calcium-binding protein that acts as a sensor for calcium concentration by modifying the activity of effector proteins, including some endogenous adenylyl cyclase proteins. The activation of bacterial proteins by a mammalian protein is yet another example of

Table 9.8. Toxin secretion. There are several mechanisms for bacterial release of exotoxins, and some examples are provided. A description of the secretion mechanisms is provided in Chapter 2

Secretion system	Organism/toxin (type)
None, toxin is released on bacterial lysis	*Streptococcus pneumoniae* Pneumolysin (channel-forming)

Type I	*Escherichia coli*
	    HylA (RTX, channel-forming)
	Bordetella pertussis
	    CyaA (RTX-adenylyl cyclase)

Type II (GSP)	*Pseudomonas aeruginosa*
	    ExoA (AB, ADP-ribosyltransferase)
	Vibrio cholerae
	    Cholera toxin (AB_5, ADP-ribosyltransferase)

Type III	*Yersinia pseudotuberculosis*
	    YopH (protein tyrosine phosphatase)
	Pseudomonas aeruginosa
	    ExoY (adenylyl cyclase)

Type IV	*Bordetella pertussis*
	    Pertussis toxin (AB_5, ADP-ribosyltransferase)
	Helicobacter pylori
	    CagA

For abbreviations, see text and list on p. xxii.

the intimacy of bacteria–host interactions. This may be a mechanism whereby eukaryotic cells with elevated calcium and/or calmodulin levels are specifically targeted. Another possibility is that calmodulin activation provides a mechanism to protect the producer bacterium from self-poisoning with high c-AMP levels.

9.5.4 Type III secretion and toxin 'injection'

This final grouping of intracellularly acting toxins demonstrates a novel mechanism for eukaryotic cell entry (Table 9.8). Since they have no inherent means of accessing the target cell cytoplasm, the purified proteins have no toxic activity. Instead, the toxins are transported directly from the bacterial cytoplasm to the target cytoplasm by a type III secretion apparatus (described in more detail in Section 2.5.5) that requires direct cell–cell contact. This may afford protection against degradation or inactivation by other means, although it does not seem to be a problem for the many more-conventional intracellular toxins we have seen so far. Alternatively, this could be a mechanism whereby the bacterium targets for intoxication only those cells with which it comes into intimate contact.

Regardless, the discovery of such toxins solved the conundrum posed by certain pathogenic bacteria that appeared not to produce toxins but caused cytopathic effects not attributable to LPS. The bacteria currently known to produce such toxins include species of the genera *Yersinia*, *Pseudomonas*, *Shigella*, *Salmonella* and *Escherichia*.

9.5.4.1 *Yersinia* spp.

The three *Yersinia* spp. pathogenic for humans and rodents are: Y. *pestis*, which causes plague and is transmitted by flea bites; Y. *pseudotuberculosis* responsible for mesenteric adenitis and septicaemia; and Y. *enterocolitica*, which is most commonly associated with disease in humans and causes a range of gastrointestinal syndromes. *Yersinia pseudotuberculosis* and Y. *enterocolitica* are food-borne pathogens. All species display a tropism for lymphoid tissue and are resistant to non-specific immune defences, especially phagocytosis by macrophages and PMNs (see Section 10.4.2.1). All three species carry a large (ca. 70 kb) virulence plasmid encoding some 50 gene products of the complex type III system. These proteins can be divided into three groups. One group of some 35 proteins comprises the secretion and translocation apparatus. The remainder are either effector proteins or their chaperones.

Many of the Yop effector proteins now have defined toxic activities in the target cells, including the induction of apoptosis and inhibition of PMN function (oxidative burst and phagocytosis) (Table 9.9). For example, YopE is cytotoxic and induces the depolymerisation of actin stress fibres. Like many intracellularly acting protein toxins, the targets for YopE activity are the small GTPases of the Rho family. Very recent work has demonstrated that YopE exhibits a GTPase-activating protein activity (see Section 4.2.5.2) that leads to down-regulation of Rho, Rac and Cdc42 function, resulting in cytoskeletal disruption. YopH is a protein tyrosine phosphatase that blocks phagocytosis. The dephosphorylation of target cell proteins may interfere with early tyrosine phosphorylation signals that occur in the cell during phagocytosis. Focal adhesion kinase is one target of YopH. Together, YopE and YopH provide full resistance to phagocytosis. YopP/YopJ (YopP was described first in Y. *pestis*, but is labelled YopJ in the Y. *enterocolitica* system) induces the apoptosis of macrophages by a mechanism that is not yet fully understood, although YopP can bind directly to MAP kinase kinases to modulate host cell signalling pathways. This is in contrast to the *Shigella* IpaB type III effector, which induces apoptosis by binding to, and activating, IL-1β converting enzyme (also known as caspase 1). YopP also inhibits the release of TNFα and interferon (IFN) γ, although the contribution of YopP to virulence is unclear as YopP⁻ mutants do not have reduced virulence in a mouse model of infection.

9.5.4.2 *Pseudomonas aeruginosa*

Pseudomonas aeruginosa is an opportunistic pathogen that colonises damaged epithelial tissue and causes pathology to lungs, cornea and skin. Neutropenic, cystic fibrosis and burns patients are particularly susceptible to infection with this organism. The ExoS regulon of

Table 9.9. Yop effectors. These toxins are 'injected' directly into target cells via a type III secretion system

Yop (size, kDa)	Effect on cell	Enzymic activity
YopE (22.9)	Cytotoxic, disrupts cytoskeleton	GTPase-activating protein (GAP) activity down-regulates Rho, Rac and Cdc42 function
YopH (51.0)	Inhibits phagocytosis	Protein tyrosine phosphatase activity blocks early signal transduction events
YopM (41.6)	Unknown	Unknown, though YopM is thought to traffic to the target cell nucleus
YopO (8.7)	Rounding up of cells	Serine/threonine kinase (target protein(s) not yet identified)
YopP (32.5)	Apoptosis of macrophages, inhibits TNFα release	Binds to and blocks phosphorylation of MAPKK; interferes with host signal transduction
YopT (35.5)	Cytotoxic, disrupts actin filaments	Unknown, though probably via RhoA, which is modified and redistributed in the target cell

MAPKK, MAP kinase kinase; for other abbreviations, see list on p. xxii.

Ps. aeruginosa comprises a type III secretion apparatus, similar to that of *Yersinia* spp., as well as translocated effector proteins, some of which have defined toxic enzyme functions. Expression of the ExoS regulon contributes to the dissemination of *Ps. aeruginosa* from the site of epithelial colonisation and to the inhibition of phagocyte function. ExoS and ExoT are ADP-ribosyltransferases. The cytotoxic ADP-ribosyltransferase activity of ExoS resides in the C-terminus of the protein and targets Ras GTP-binding proteins, resulting in uncoupling of Ras-mediated signal transduction. The N-terminal half of ExoS can disrupt the host cell cytoskeleton. The mechanism of action is not yet fully understood, although it probably involves small GTPases, since actin disruption by ExoS was reversed by *E. coli* CNF2 toxin, which activates Rho proteins. Once in the target cell cytoplasm, both ExoS and ExoT activities are stimulated by a eukaryotic factor that normally acts as a regulator of eukaryotic enzyme activity. ExoY is a type III system-secreted adenylyl cyclase. Like the adenylyl cyclase toxins of other pathogenic bacteria, its activity is stimulated by a host factor, although in this case the factor has not yet been identified but is not calmodulin. ExoU is a potent cytotoxin whose mode of action is not yet understood.

9.5.5 Type IV secretion and toxin 'injection'

Like the type III secretion system, the type IV system is used by some bacteria to inject proteins directly into the cytoplasm of target cells. For example, the CagA protein of *Helicobacter pylori* is translocated directly into gastric epithelial cells by the type IV secretion system (see Section 2.5.3.3). Once within the cytoplasm, CagA is phosphorylated on tyrosine residues by an as yet unidentified host cell kinase. Phosphorylated CagA induces cytoskeletal reorganisations resulting in the formation of a pedestal on which *Hel. pylori* sits in intimate contact with the host cell. CagA has been shown to activate MAP kinase cascades, leading to the induction of expression of host transcription factors that play a pivotal role in cell proliferation and neoplastic transformation.

9.6 | Toxins as therapeutic agents

9.6.1 Vaccines

Toxins produced by bacteria are among the major effectors of host damage in infection and so constitute ideal targets for vaccination to prevent infectious disease. This was recognised early, and chemically inactivated (e.g. by formaldehyde treatment) toxoid preparations provided notable success, for example in the prevention of tetanus and diphtheria. Nevertheless, there are some limitations to this approach, including short-lived immunity or unacceptable side effects.

Molecular biology, in conjunction with toxin crystallisation and structural studies, has allowed a new precise approach to vaccine development. Armed with a knowledge of the primary amino acid sequence and catalytic domain structure, toxins can now be inactivated by site-directed mutagenesis of a single or a small number of amino acid residues. The wholly inactive toxin is, nevertheless, without any gross structural distortion and, as a result, it will present the correct epitopes, which will lead to immune protection against native toxin. Furthermore, having a native structure helps to prevent its degradation in the host and also allows the first steps in intoxication (binding and uptake) to take place, enhancing the immune system processing of the antigen. Changing a single glutamate residue in the catalytic domain of many ADP-ribosyltransferase toxins, including pertussis toxin and ExoA of *Ps. aeruginosa*, is sufficient to abolish toxic activity.

9.6.2 Mucosal vaccines

The mucosal surfaces of the body often provide the first port of entry for potential pathogens and, consequently, are protected by an interconnected local immune system known overall as the mucosa-associated lymphoid tissue or MALT (see Section 5.7.1). Vaccination that targets the MALT would provide protection against mucosal (enteric, respiratory, etc.) infections. Stimulation of a protective MALT immune response, however, is problematic, owing to either poor immunogenicity or else a poor T cell memory response. CT and the closely related LT of *E. coli* are potent mucosal adjuvants that enhance greatly

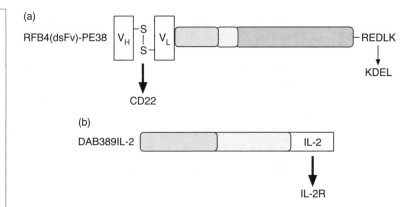

Figure 9.14 Examples of immunotoxins. (a) RFB4(dsFv)-PE38 is a truncated form of ExoA in which the receptor binding domain (component Ia (amino acid residues 1 to 252) and amino acid residues 365 to 380 of component Ib) has been replaced by V$_H$-V$_L$ of of a monoclonal antibody recognising CD22. Binding of the latter to CD22 is indicated by the bold arrow. In some ExoA-based immunotoxins, the C-terminal amino acid sequence has been modified to generate an improved endoplasmic reticulum retrieval signal. In the example given, the final five amino acid residues of ExoA (REDLK) have been changed to KDEL. (b) In DAB389IL-2, the first 388 amino acid residues of diphtheria toxin (DT) comprising enzymic (A; 1 to 193) and translocation (T; 194 to 381) domains have been fused with IL-2 to target the toxin to cells bearing the IL-2 receptor (IL-2R).

the specific sIgA response to antigens either admixed with, or covalently attached to, the toxin. Unfortunately, the ADP-ribosyltransferase activity is essential for enhancing the response to admixed, but unattached, antigens, possibly by stimulating the activity of antigen-presenting cells. Anyone with an aversion to needles finds vaccination an unpleasant experience *per se*. Imagine the levels of non-compliance if vaccination was associated, instead, with a bout of diarrhoea! Nevertheless, antigens covalently attached to the B subunit of CT – which has no toxic activity – do show enhanced uptake and presentation to the mucosal immune system, resulting in improved immunogenicity.

Other approaches to targeting the MALT include the use of live bacteria as recombinant antigen delivery systems. Being developed and tested at the moment are metabolically attenuated enteric *Salmonella* strains and wild-type commensal bacteria including oral *Strep. gordonii*.

9.6.3 Chimeric immunotoxins

The modular design of many bacterial toxins coupled with their often-remarkable potency makes them ideal tools for modern cancer therapy. Human cancers are becoming more definable by the range of proteins expressed on the surface of malignant cells. The Fv region (carrying the antigen-binding specificity) of monoclonal antibodies raised against proteins present only on the surface of particular cell types can be used to replace the binding domain of toxins such as DT or ExoA (Figure 9.14). The original toxin receptor specificity is replaced by a new, cancer-specific receptor. However, the catalytic domain of bound toxin is still translocated to the cytoplasm as normal, and the target cell is killed. In other chimeric toxins, the target recognition domain is a ligand, for example a growth hormone or cytokine that binds to a cell type-specific receptor.

Many chimeric immunotoxins have been developed and tested in clinical trials, some to phase III levels. The American Food and Drug Administration (FDA) regulatory body has recently approved for use a DT-IL-2 toxin (Figure 9.14b) for treatment of cutaneous T cell lymphoma. Problems do still exist, however, including reduced activity against solid tumours. Repeat cycles of treatment can also result in reduced efficacy, owing to the development of a neutralising antibody response, and vascular leak syndrome can arise by toxin activity on endothelial cells. Nevertheless, the future does look good for this very cell-specific cancer treatment modality.

Chimeric immunotoxins have also been developed for human immunodeficiency virus (HIV) therapy. In this instance, target cell specificity is provided by monoclonal antibodies that recognise an HIV envelope glycoprotein left on the surface of infected cells. It is hoped that such toxins, used in conjunction with highly active anti-retroviral therapy, may allow full elimination of HIV from the body.

9.6.4 Muscle spasms

The potent nerve-blocking activity of the botulinum toxins has been used to treat a variety of rare but distressing conditions where particular muscles remain permanently activated. Injection of very small doses of toxin relaxes the affected muscle and can provide relief for several months at a time. Since the dose is so low, and is applied locally, immunity to botulism toxin appears to be slow to build up. The cosmetic industry is also using this technique as an anti-wrinkle therapy.

9.7 | Concept check

- Bacterial toxins contribute significantly to the pathogenesis of many infections.
- Toxins can be classified broadly into several categories, depending on their mode or site of action. Nevertheless, there can be considerable overlap between these groups.
- Variations on a theme are common, and certain protein structures appear frequently among toxins from different bacterial species.
- Cell death may not be the only outcome of intoxication; at low toxin concentrations, target cell function may be modified or impaired.
- At low concentrations, many toxins induce the release of pro-inflammatory cytokines from host cells.
- Certain eukaryotic proteins or functions may be targets for different toxin activities.
- Many toxins subvert eukaryotic cell-signalling pathways.
- Some toxins have therapeutic uses.

9.8 | Toxins produced by the paradigm organisms

9.8.1 *Streptococcus pyogenes*

The role of toxins in diseases caused by *Strep. pyogenes* is less well defined than is the case for infections due to *E. coli*. This may be due, at least in part, to the only-recent re-emergence of serious diseases due to *Strep. pyogenes*. Additionally, there are multiple streptococcal determinants involved in disease and the host also plays a significant role.

Until the late 1800s, severe diseases due to *Strep. pyogenes* were common. These included puerperal sepsis (infection following childbirth), necrotising fasciitis, bacteraemia and septic scarlet fever, and mortality rates were quite high. The incidence of severe streptococcal diseases declined throughout the first half of the 20th century, and in some cases this decline even preceded the advent of antibiotics. The reasons for the decline are not fully understood, although factors including improved socioeconomic conditions, waning virulence of *Strep. pyogenes* strains, or increased herd immunity to virulent strains may have played a part. The latter half of the 1980s saw the re-emergence of rheumatic fever in the developed world, followed closely by fulminant *Strep. pyogenes* infections associated with shock, organ failure, necrotising fasciitis, myonecrosis, bacteraemia and death, collectively termed streptococcal **t**oxic **s**hock **s**yndrome (TSS). The incidence of TSS is sporadic (approximately 10 cases per 100 000 population) but is, nevertheless, widespread geographically, with cases reported in North America, Europe and Australasia. Certain M types (for a discussion of M types, see Section 2.9) are more frequently associated with severe disease, especially M types 1 and 3. Nevertheless, many individuals carry these M types asymptomatically. Predisposing factors may include viral infections that can disrupt or modify mucosal barriers (e.g. influenza) or impair host immunity (e.g. chickenpox), host genotype and skin or tissue damage due to trauma of some kind.

The reason for the sudden reappearance of severe diseases due to *Strep. pyogenes* is equally unclear, but there may be several contributing factors. Firstly, 'new and improved' strains may have emerged with greater virulence or fitness within the host. Acquisition of virulence by horizontal gene transfer may have occurred, and it is noteworthy that the genes encoding certain streptococcal pyrogenic exotoxins are carried by phages. Secondly, there may have been a reduction in herd immunity to the invasive strains. Whatever the reason, diseases thought to be either banished or else the curse of the poor in developing countries, are firmly back on the agenda.

Streptococcus pyogenes expresses multiple exotoxins and exoproteins that may contribute to disease. These include the haemolysins strepto-lysin-O (SLO; cholesterol-binding cytolysin) and the less-well-defined streptolysin-S (SLS), a family of pyrogenic exotoxins, some of which act as superantigens, and secreted proteases. A hypothesis has been put forward that suggests host cell and tissue damage arises from a triad of factors that may be bacterial or host in origin. Thus pore-forming components such as SLS and SLO as well as host phospholipases synergise with oxidants (e.g. hydrogen peroxide (H_2O_2) produced by the bacterium or by activated PMNs) and the third component comprising bacterial secreted proteases. The end result is a potent cytolytic and cytokine-inducing cocktail. Investigations of individual components have clearly demonstrated a role for **s**treptococcal **p**yrogenic **e**xotoxin

(SPE-B), a cysteine protease, in avoiding phagocytosis and allowing full dissemination of bacteria; knockout mutants no longer expressing SLS showed reduced virulence in a mouse model of cutaneous streptococcal infection. Similarly, purified SLO has been shown to have a range of effects, other than acting as a cytolysin, that may promote TSS. These include induction of pro-inflammatory cytokines and effects on cardiac cell function. The streptococcal superantigens also contribute to the host inflammatory response through their non-specific activation of T cells.

9.8.2 *Escherichia coli*

Escherichia coli is the predominant facultative anaerobe of the human colonic flora and is by-and-large harmless to healthy individuals. However, there are numerous highly specialised 'aggressive' *E. coli* strains that can cause a wide range of diseases, even in a healthy individual. These strains are characterised by the presence of accessory virulence factors such as toxins and invasion-associated proteins, the genes for which often reside on mobile genetic elements including plasmids and bacteriophages. Diarrhoea is one such disease and is the focus of this section. Several categories of *E. coli* are known to cause diarrhoea but we will limit the present discussion to those groups where enterotoxins are known to play an important role in pathogenesis, namely ETEC, EHEC and EAEC (Figure 9.15).

Enterotoxigenic *E. coli* express at least one member of the two families of enterotoxin, LT and ST. LTs, as we have learned, are immunologically and functionally very similar to cholera toxin, while STs are small type I membrane-acting toxins that nevertheless also cause electrolyte and fluid secretion from intoxicated cells. ETEC are responsible for diarrhoea in weaning children and for traveller's diarrhoea. Bacteria are transmitted in contaminated water or food and the onset of disease is abrupt, with an incubation period of some 14 hours. Diarrhoea is watery and usually without blood and the severity ranges from mild, self-limiting to serious purging similar to that seen in cholera.

Enterohaemorrhagic *E. coli* strains are considered a subset of STEC strains. Thus EHEC strains express one or more of the Shiga-like toxins (Stx1 and Stx2) but also possess additional virulence properties that may not be present in other STEC strains. These include the ability to form attaching and effacing (A/E) lesions on epithelial cells and the ability to cause the more severe conditions of haemorrhagic colitis and HUS. Like EPEC, EHEC have the LEE pathogenicity island (see Sections 2.6.4 and 7.5.2.1.2.1) required for formation of the A/E lesions. There is a host inflammatory response associated with the formation of these lesions. However, the major virulence factor required for bloody diarrhoea and for haemorrhagic colitis and HUS are the Shiga-like toxins. Different isolates can express either Stx1 or Stx2, both Stx1 and Stx2, or indeed multiple variants of Stx2. EAEC heat-stable enterotoxin 1 (EAST1), first described in EAEC is also found in many EHEC strains, including serotype O157 : H7 strains, and this may account for the

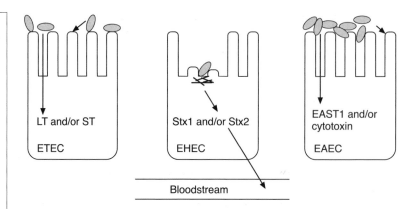

Figure 9.15 Toxins of enteropathogenic *Escherichia coli*. Heat-labile toxin (LT) and the type I (membrane-acting toxin) stable toxin (ST) are produced by enterotoxigenic *E. coli* (ETEC). Enterohaemorrhagic *E. coli* (EHEC) may express multiple Shiga toxins (Stx) and these play a major role in life-threatening haemorrhagic colitis and HUS diseases caused by these bacteria. Enteroaggregative *E.coli* (EAEC) express the type I toxin EAST1 as well as other less-well-defined cytotoxins.

non-bloody diarrhoea seen in some patients. Finally, EHEC express enterohaemolysin, although the role of this variant RTX toxin in pathogenicity is unclear.

Bacteria are transmitted in contaminated food and water. One notorious outbreak of disease that resulted in four deaths in the USA involved undercooked hamburgers, and bovine products – including unpasteurised milk – remain particularly high risk. The low infectious dose of as few as 100 cells can account for the occasional person-to-person transmission. The incubation period is some three to four days and symptoms begin with watery diarrhoea and vomiting in approximately 50% of cases. This is followed within one to two days by bloody diarrhoea that may last four to ten days. In most cases the symptoms resolve without apparent sequelae, though 10% of patients may progress to HUS.

Enteroaggregative *E. coli* are characterised by aggregative adhesion to enterocytes and are distinct from diffusely adherent *E. coli*, which may or may not be associated with diarrhoeal disease. EAEC induce copious mucus secretion such that the densely aggregated bacteria exist in a mucus layer associated with the cell surface. EAEC produce the type I (membrane-acting) EAST1 heat-stable toxin, the gene for which is plasmid encoded. EAEC also produce other less-well-defined cytotoxins.

EAEC diarrhoeal disease is characterised by mucus diarrhoea accompanied by a low grade fever but little or no vomiting. It has a short incubation period of some 8 hours, indicative of small bowel involvement. EAEC disease is common in developing countries and is frequently responsible for persistent diarrhoea.

9.9 | What's next?

In previous chapters we have dealt with many aspects of bacterial virulence, including how bacteria adhere to and invade host cells and tissues and how they can inflict damage on their host. We have also looked at the other side of the host–parasite partnership – the host defence mechanisms that operate successfully to prevent (most of the time) bacterial infections. It is now time to look at another very important aspect of bacterial virulence – the means by which they circumvent host defence systems. This will be discussed in Chapter 10.

9.10 | Questions

1. Outline the three main classes of bacterial exotoxin. Provide examples where a toxin may be included in more than one category.
2. What are the two protein-folding motifs frequently used by channel-forming toxins?
3. Describe the various mechanisms employed by β-channel-forming toxins to facilitate the soluble monomer–membrane inserted transition.
4. What are some of the sublethal effects of cholesterol-binding cytolysins and RTX toxins? How may these effects contribute to bacterial proliferation within the host?
5. The family of AB toxins comprises molecules with different molecular architecture but similar toxigenic activity, and also similar architecture but different toxigenic activity. Provide some examples of both situations.
6. What is endoplasmic reticulum-associated protein degradation and how may it be important for the activity of cholera toxin?
7. Describe the multiple ways in which toxin activity results in elevated target cell c-AMP levels.
8. Describe the beneficial uses of bacterial exotoxins.
9. Toxin genes can be transmissible. Provide examples of toxins that are encoded on mobile genetic elements.
10. What might be the advantages of type III secretion system 'injection' of toxins into host cells?

9.11 | Further reading

Books

Aktories, A. (1997). *Bacterial Toxins: Tools in Cell Biology and Pharmacology.* London: Chapman & Hall.

Moss, J., Vaughan, M., Iglewski, B. & Tu, A. (1995). *Bacterial Toxins, Virulence Factors and Disease.* New York: Marcel Dekker.

Rappuoli, R. & Montecucco, C. (1997). *Guidebook to Protein Toxins and their Use in Cell Biology.* Oxford: Oxford University Press.

Review articles

Billington, S. J., Jost, B. H. & Songer, J. G. (2000). Thiol-activated cytolysins: structure, function and role in pathogenesis. *FEMS Microbiol Lett* **182**, 197–205.

Boquet, P. (1999). Bacterial toxins inhibiting or activating small GTP-binding proteins. *Ann NY Acad Sci* **886**, 83–90.

Boquet, P., Munro, P., Fiorentini, C. & Just, I. (1998). Toxins from anaerobic bacteria: specificity and molecular mechanisms of action. *Curr Opin Microbiol* **1**, 66–74.

Dinges, M. M., Orwin, P. M. & Schlievert, P. M. (2000). Exotoxins of *Staphylococcus aureus*. *Clin Microbiol Rev* **13**, 16–34.

Ginsberg, I., Ward, P. A. & Varani, J. (1999). Can we learn from the pathogenic strategies of group A hemolytic streptococci how tissues are injured and organs fail in post-infectious and inflammatory sequelae? *FEMS Immunol Med Microbiol* **25**, 325–338.

Gouaux, E. (1997). Channel-forming toxins: tales of transformation. *Curr Opin Struct Biol* **7**, 566–573.

Johannes, L. & Goud, B. (1998). Surfing a retrograde wave: how does Shiga toxin reach the endoplasmic reticulum? *Trends Cell Biol* **8**, 158–162.

Kotb, M. (1998). Superantigens of Gram-positive bacteria: structure–function analyses and their implications for biological activity. *Curr Opin Microbiol* **1**, 56–65.

Kreitman, R. J. (1999). Immunotoxins in cancer therapy. *Curr Opin Immunol* **11**, 570–578.

Krueger, K. M. & Barbieri, J. T. (1995). The family of bacterial ADP-ribosylating exotoxins. *Clin Microbiol Rev* **8**, 34–47.

Lacy, D. B. & Stevens, R. C. (1998). Unraveling the structures and modes of action of bacterial toxins. *Curr Opin Struct Biol* **8**, 778–784.

Lally, E. T., Hill, R. B., Kieba, I. R. & Korostoff, J. (1999). The interactions between RTX toxins and target cells. *Trends Microbiol* **7**, 356–361.

Lerm, M., Schmidt, G. & Aktories, K. (2000). Bacterial protein toxins targeting Rho GTPases. *FEMS Microbiol Lett* **188**, 1–6.

McCormick, J. K., Yarwood, J. M. & Schlievert, P. M. (2002). Toxic shock syndrome and bacterial superantigens: an update. *Annu Rev Microbiol* **55**, 77–104.

Merritt, E. A. & Hol, W. G. J. (1995). AB_5 toxins. *Curr Opin Struct Biol* **5**, 165–171.

Mooi, F. R. & Bik, E. M. (1997). The evolution of epidemic *Vibrio cholerae* strains. *Trends Microbiol* **5**, 161–165.

Schmitt, C. K., Meysick, K. C. & O'Brien, A. D. (1999). Bacterial toxins: friends or foes? *Emerg Infect Dis* **5**, 224–234.

Welch, R. A., Bauer, M. E., Kent, A. D., Leeds, J. A., Moayeri, M., Regassa, L. B. & Swenson, D. L. (1995). Battling against host phagocytes: the wherefore of the RTX family of toxins? *Infect Agents Dis* **4**, 254–272.

Papers

Cossart, P. & Lecuit, M. (1998). Interactions of *Listeria monocytogenes* with mammalian cells during entry and actin-based movement: bacterial factors, cellular ligands and signaling. *EMBO J* **17**, 3797–3806.

Dubreuil, J. D. (1997). *Escherichia coli* STb enterotoxin. *Microbiology* **143**, 1783–1795.

Furlow, B. (2001). The freelance poisoner. *New Scientist* **2274**, 30–33.

Gouaux, E. (1998). α-Haemolysin from *Staphylococcus aureus*: an archetype of β-barrel, channel-forming toxins. *J Struct Biol* **121**, 110–122.

Kohler, B., Karch, H. & Schmidt, H. (2000). Antibacterials that are used as growth promoters in animal husbandry can affect the release of Shiga-toxin-2-converting bacteriophages and Shiga toxin 2 from *Escherichia coli* strains. *Microbiology* **146**, 1085–1090.

Rossjohn, J., Feil, S. C. & McKinstry, W. J. (1998). Aerolysin – a paradigm for membrane insertion of beta-sheet protein toxins? *J Struct Biol* **121**, 92–100.

Senzel, L., Huynh, P. D., Jakes, K. S, Collier, R. J. & Finkelstein, A. (1998). The diphtheria toxin channel-forming T domain translocates its own NH_2-terminal region across planar bilayers. *J Gen Physiol* **112**, 317–324.

9.12 | Internet links

1. http://www.bact.wisc.edu/Bact330/Bact330Homepage
   This web book has a section on protein exotoxins as well as sections describing diseases – including the involvement of toxins – caused by a number of bacteria including *B. anthracis*, *Bord. pertussis*, *Strep. pyogenes*, *Clostridium* spp., and *Staph. aureus*. University of Wisconsin-Madison, USA.

2. http://128.8.90.214/classroom/bsci424/HostParasiteInteractions/HostParasite.htm
   An undergraduate teaching aid that provides a nice summary of bacterial exotoxins as well as other aspects of host–parasite interactions. University of Maryland, USA.

3. http://www.kcom.edu/faculty/chamberlain/Website/Lects/Toxins.htm
   A set of lecture notes on bacterial toxins.

Bacterial evasion of host defence mechanisms

Chapter outline

Aims

The principal aims of this chapter are to describe:

- the concept of bacterial evasion of host defences
- the means by which bacteria overcome mucosal defence mechanisms
- how bacteria counteract cytokine-mediated host defences
- the mechanisms used by bacteria to evade the innate and acquired immune systems
- how bacteria use apoptosis and control the host cell cycle in order to evade host defences

10.1 | Introduction

Chapters 5 and 6 have introduced the reader to the overlapping systems of cells and molecules that mammals have evolved to defend

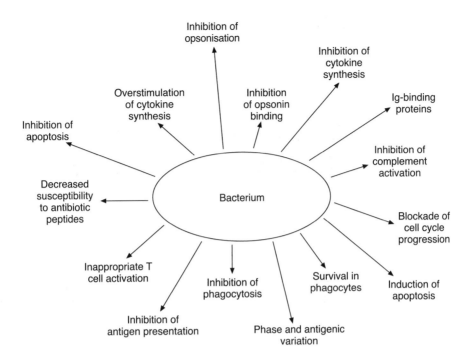

Figure 10.1 The many and varied mechanisms used by bacteria to avoid host immune defence responses.

against bacterial infection. Our protective systems of immunity and inflammation involve a range of cells (macrophages, dendritic cells, T lymphocytes, endothelial cells, etc.), cellular receptors (CD14, Toll-like receptors (TLRs), T cell receptors (TCRs), etc.), humoral effector molecules (antibacterial peptides, complement components, acute phase proteins, antibodies, etc.) and a growing array of integrating signals (cytokines and lipid mediators such as prostaglandins and lipoxins). As we learn more about the bacteria that share their lives with us, we appreciate just how many counter-measures they have evolved to help defeat our myriad systems of immunity (Figure 10.1). However, it is important to realise that the evolution of counter-measures is a two-way process, in other words it is a co-evolutionary process, which results ultimately in the survival of both the pathogen and the host.

This chapter will begin by considering the mechanisms of microbial evasion of host immunity at mucosal surfaces, and will then turn to how viruses and bacteria interfere with the integrating signals of immunity – the cytokines. Then will follow a consideration of mechanisms for evading innate immunity. This will consider the evasion of the complement pathways and of phagocytosis, and will also review recent findings suggesting that bacteria can inhibit antigen processing. The focus of the chapter will then move on to a consideration of the systems bacteria have evolved to outwit acquired immune responses. Finally, we will consider the mechanisms that bacteria have evolved to control the eukaryotic cell cycle and the process of apoptosis or programmed cell death (PCD). Again, as has been stated at various points in this book, individual bacteria may simultaneously use all of the evasion mechanisms to be described. This chapter will not attempt to be all-encompassing, but will provide examples of the main mechanisms bacteria use to evade our immune defences.

10.2 | Evasion of immune defences at mucosal surfaces

Chapter 5 has described the epithelia of the mucosal surfaces of the body and the varied mechanisms that they employ to inhibit bacterial growth or to kill bacteria. One of the spin-offs from the growing number of bacterial genomes that are being sequenced is the ability to determine what proteins bacteria produce (or can produce). This is revealing that there are great similarities between bacteria and eukaryotic cells in terms of their complement of proteins. In Section 5.4.3.1 and 5.4.3.2, the roles of host mucin and lysozyme were discussed. It has been established that bacteria also produce mucins and lysozyme-like proteins. Their role is unclear, but could be to interfere with the host defence capabilities of these proteins.

In this section, the focus will be on bacterial defences against secretory immunoglobulin (Ig) A and against antibacterial peptides. Other methods of evading mucosal defences such as the use of siderophores to compete for iron, or receptors for the iron-binding protein lactoferrin, have already been discussed in Sections 5.4.3.3 and 5.10.2 and will not be described further.

10.2.1 Evasion of secretory IgA

As has been described in Section 5.4.3.10, the major immunoglobulin at mucosal surfaces is secretory IgA. The mechanism by which IgA is secreted is an interesting example of co-operation between cells and soluble effector molecules (described in Section 5.4.3.10). It is assumed that the role of secretory IgA is to bind to pathogens in the external environment and prevent them binding to mucosal surfaces. It is likely that this is not the whole story as, for example, individuals with IgA deficiency may show no increase in susceptibility to bacterial infection. However, be that as it may, there is good evidence that bacteria have evolved mechanisms to evade secretory IgA (Figure 10.2).

Bacteria have evolved three potential mechanisms for inhibiting the activity of secretory IgA. Like most eukaryotic proteins, secretory IgA is heavily glycosylated and such glycosylation is responsible, in part, for protein conformation, net charge, hydrophilicity and susceptibility to proteases. Bacteria produce a range of glycosidases that can remove all carbohydrate side chains from IgA. Other bacteria produce sialidases, which remove the sialic acid from oligosaccharides associated with IgA. This behaviour has been reported for Gram-positive rods, streptococci and *Veillonella* spp. found in the human mouth. Removal of the sialic acid, for example, would render the glycoprotein more susceptible to proteolysis.

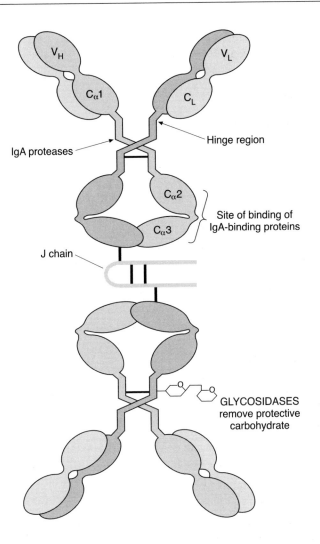

IgA proteases

Hinge region

$C_\alpha 1$

V_H

V_L

C_L

$C_\alpha 2$

$C_\alpha 3$

Site of binding of
IgA-binding proteins

J chain

GLYCOSIDASES
remove protective
carbohydrate

Figure 10.2 The three main mechanisms used by bacteria to evade secretory IgA at mucosal surfaces. These include glycosidases that can remove oligosaccharides, rendering the immunoglobulin more susceptible to proteolysis, IgA1 proteases, which cleave at the hinge region (which normally is glycosylated), and IgA-binding proteins.

Proteolysis is the second mechanism that bacteria have evolved to deal with secretory IgA. Antibodies of the α, β and γ classes have an extended peptide domain between their C_H1 and C_H2 domains, allowing the two separate antigen-binding sites some degree of flexibility (for a discussion of immunoglobulin structure, see Section 6.4.3 and Figure 6.12). This peptide is called the hinge region. Such hinge regions are susceptible to proteolytic cleavage, and in IgA1 the hinge sequence is O-glycosylated on several serine and threonine residues. This protects the region from proteolysis by conventional host or bacterial proteases. However, a number of bacteria have been found to produce proteases that specifically cleave the hinge region in IgA1 and, in consequence, these proteases have been termed IgA1 proteases. The hinge region in IgA2 is shorter and not susceptible to the majority of the IgA1 proteases. The IgA1 proteases are, surprisingly, able to cleave the hinge region by distinct enzymic mechanisms, and serine, metallo- and

Table 10.1.	Bacterial IgA1 proteases

IgA1 hinge   P^1 V P^2 S T P P^3 T P S P^4 S T P P^5 T P^6 S P S C C

1   *Clostridium ramosum*

2   *Prevotella* spp.

3   *Streptococcus mitis*
    Streptococcus oralis
    Streptococcus pneumoniae
    Streptococcus sanguis

4   *Haemophilus influenzae*
    Haemophilus aegyptius

5   *Haemophilus influenzae*
    Haemophilus parahaemolyticus
    Neisseria meningitidis
    Neisseria gonorrhoeae
    Ureaplasma urealyticum

6   *Neisseria meningitidis*
    Neisseria gonorrhoeae

Letters are the single letter code for the amino acids (C, cysteine; P, proline; S, serine; T, threonine). Letters enboldened represent amino acids that are glycosylated. The numbers represent the cleavage sites of the IgA1 proteases from the bacteria listed.

cysteine proteases have been identified. The bacteria producing such IgA1 proteases, and their specificity, are listed in Table 10.1. A number of studies examining mucosal secretions of individuals infected with bacteria producing such proteases have detected the presence of IgA1 cleavage products. By cleaving secretory IgA1 in the hinge region, these enzymes are likely to interfere with the barrier functions of mucosal IgA antibodies. Now, the message of this chapter is that we, the hosts, are locked in a never-ending evolutionary battle with the bacteria that share the planet with us. Bacterial secreted proteases with specificity for IgA1 are part of the bacterium's armamentarium to defend itself against our defences. These proteases are strong immunogens and, in turn, our own defences produce antibodies that can neutralise the activity of these IgA1 proteases and also inhibit their release. This can lead to agglutination of the bacteria, which now have the protease as a surface antigen. In turn, it has been revealed that the IgA1 proteases exhibit significant antigenic diversity. For example, more than 30 antigenically different forms of the *Haemophilus influenzae* IgA1 protease have been found. This antigenic diversity is, presumably, another evasion mechanism to escape the attention of host antibodies.

In addition to IgA1 proteases, certain bacteria, most prominently *Streptococcus pyogenes*, produce IgA-binding proteins. In the case of this organism, the proteins are called Arp and Sir and are 40 kDa surface-linked molecules. *Streptococcus agalactiae* produces a protein of molecular mass 130 kDa termed Bac. Such proteins bind to the Fc region of both subclasses of the IgA molecule and also to some IgG proteins. These proteins are members of the M protein family, which will be described in Section 10.4.1.3. Other bacteria producing IgA-binding proteins include certain strains of *Escherichia coli*, *Helicobacter pylori* and *Streptococcus pneumoniae*. With these proteins, the binding site is the Fc region of IgA. Do these bacterial IgA-binding proteins interfere with the activity of the bacterial IgA1 proteases? The answer is no. IgA1 bound to bacterial immunoglobulin-binding proteins still acts as a substrate for the IgA1 proteases. Thus these two IgA-evading mechanisms seem to act independently of each other.

10.2.2 Evasion of antibacterial peptides

The past 20 years have witnessed the discovery that all organisms produce some form of antibiotic and has opened up a new area of host immunity – the biology of antibacterial peptides. These peptides were discussed in Section 5.4.3.11. As a late discovery in the history of host immune responses, the biology of the antibacterial peptides is only now being explored in detail. The synthesis of these peptides is now known to be linked to one of the central recognition systems of innate immunity – the TLRs (see Chapters 5 and 6). These antibacterial peptides are extremely important in invertebrate immunity and so it would be expected that bacteria would have evolved systems to evade their actions. Is there any evidence yet that such systems exist?

Throughout this book, the ability of bacteria to sense their environment, and respond accordingly, has been stressed. When bacteria enter our bodies from the external environment they encounter many changes in their milieu. This will include alterations in, amongst others, growth temperature, nutrient availability, osmolarity, pH, CO_2 concentrations and metal ion content. These are cues that control specific and global transcription mechanisms in bacteria (see Chapters 2 and 4). One bacterium, which has been the subject of intensive study over the past decade or so, *Salmonella typhimurium*, has been found to be able to increase its resistance to antibacterial peptides. This bacterium was described in some detail in Chapters 7 and 8 in terms of its ability to bind to, and invade, various host cells. Entering into the host results in the activation of the sensor-linked transcription systems known as 'two-component' systems, which have been described in detail in Chapters 2 and 4. One of the most studied of such control systems in *Sal. typhimurium* is the PhoP/PhoQ system, which is regulated by a number of environmental factors, including phosphorus. One of the genes controlled by PhoP/PhoQ is *pagP*. The product of this gene is responsible for altering the lipid A of the outer membrane of the bacterium, thus rendering the organism less susceptible to the membrane-permeabilising effects of antibacterial peptides. The chemical change that the pagP gene product induces is the addition of a palmitate (acyl) group to the lipid A. This increased acylation of the outer leaflet of the outer membrane of Gram-negative bacteria is

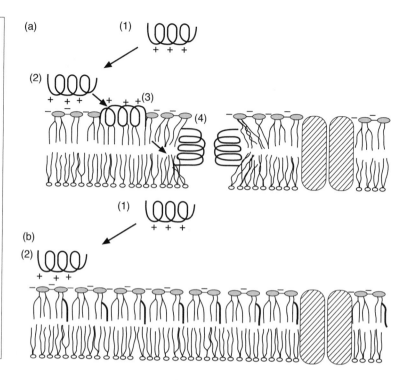

Figure 10.3 A proposed model for the role of PagP in the decreased susceptibility of *Salmonella* spp., to antibacterial peptides. In the upper diagram (a), the interaction of a cationic antibacterial peptide (1) with the outer membrane (2) leads to binding and then to the 'flipping' of the peptide (3) into the outer leaflet. This can lead to the antibacterial peptides spanning the membrane (4) and results in leakage of the bacterial contents and cell death. In (b), the introduction of palmitate (denoted by thickened lines) into the fatty acid substituents of the lipid A portion of the lipopolysaccharide molecule inhibits the ability of the antibacterial peptide to insert into the membrane and so cell leakage and death are reduced.

thought to increase the interaction between the two sides of the membrane, making it more difficult for the charged antibacterial peptides to insert into the membrane and induce membrane permeability (Figure 10.3).

Another mechanism of resistance is found in *Yersinia* spp., and this is a temperature-dependent efflux pump that is also induced by antibacterial peptides. This pump works by acidifying the cell cytoplasm, which lowers the activity of these peptides, and also ejects peptides from the cell. Organic acids such as formic acid or succinic acid have also been shown to protect cells from the potent inhibitor bacterial permeability-inducing protein (BPI).

Lactoferrin has been described in Section 5.4.3.3 as an iron-binding protein, an antibacterial protein and the precursor of the antibacterial peptide lactoferricin. A number of bacteria have evolved proteins that bind to lactoferrin and that, presumably, block its biological effects.

10.3 | Cytokines in antibacterial defence: mechanisms of microbial evasion

Cytokines play a curious role in microbial infections. As integrating signals, they are vital for the orchestration of myeloid, lymphoid and vascular cell responses that recognise infecting agents and deal with them in the most appropriate manner. This is seen most clearly in cytokine knockouts. Animals deficient in tumour necrosis factor

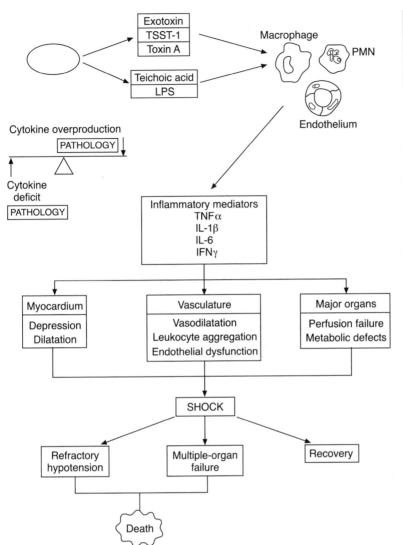

Figure 10.4 The balance of cytokines in host responses and the pathological consequences of overproduction of certain cytokines. For example, in septic and toxic shock the rapid release of various modulins (LPS, TTST-1) from Gram-negative and Gram-positive bacteria respectively results in the overproduction of pro-inflammatory cytokines such as IL-1β, TNFα, IFNγ, IL-6, IL-8 and so on. This leads to overwhelming pathology, with cardiovascular changes, inhibition of major organ perfusion and death. For abbreviations, see list on p. xxii.

(TNF) α , interferon (IFN) γ or interleukin (IL) 12, for example, are susceptible to a range of bacterial pathogens. The protective role of cytokines has been described in Sections 4.2.6 and 6.4.4. The reverse side of the coin is the pathology produced during infection as a result, largely, of the actions of these cytokines (Figure 10.4). Bacterial infections can lead to the very rapid death of the infected individual. Septic shock and toxic shock are two similar conditions, caused by Gram-negative or Gram-positive bacteria, respectively. Together, they are responsible for around 300 000 deaths each year in the USA. The problem is the rapid build-up of a number of pro-inflammatory cytokines, TNFα and IL-1β being the most pathogenic. Increased circulating levels of these very potent cytokines result in changes in cardiovascular physiology that cause lowering of blood pressure and inhibition of the perfusion of the major organs. The consequences are fatal for a large

Table 10.2.	Some examples of bacterial modulins	
Modulin	Bacterium	Minimal concentration for activity (M)
LPS	Gram-negative species	10^{-6} to $< 10^{-12}$ depending on bacterium
Peptidoglycan	Gram-positive and Gram-negative species	10^{-6} to 10^{-9}
Teichoic acids	Gram-positive species	10^{-6}
Lipoarabinomannan	Mycobacteria	10^{-6} to 10^{-9}
Porins	Gram-negative species	10^{-6} to 10^{-9}
Superantigens	Certain Gram-positive species	10^{-6} to 10^{-10}
Lipoproteins	All bacteria	10^{-6} to $< 10^{-12}$
Molecular chaperones	All bacteria	$> 10^{-9}$
Exotoxins	Certain Gram-positive and Gram-negative species	10^{-9} to 10^{-18}
Lipids	All bacteria	10^{-6} and lower
DNA	All bacteria	10^{-9}

proportion of those suffering from these shock-like conditions. Equally, in chronic infections such as tuberculosis or syphilis, the chronic inflammatory pathogenesis of these diseases, with the consequent tissue destruction, is due to the overproduction of particular cytokines.

So we have a very interesting interplay between bacteria and host cytokines, or more appropriately, host cytokine networks. Although earlier chapters have described what cytokines are, and their production in infection, the bacterial factors stimulating cytokine production have not been described and the potential role that such factors have in evading immune responses has not been emphasised.

10.3.1 Modulins

In Section 1.3.1 the concept of bacterial virulence factors was introduced. Virulence factors have been subdivided into four groups: adhesins, invasins, aggressins and impedins. By the mid 1990s it was clear from the literature that bacteria produced a wide variety of molecules, of all chemical classes, able to induce the stimulation of cells to produce cytokines (Table 10.2). Two of the authors (B.H. and M.W.) suggested that these various bacterial cytokine inducers may be an additional class of virulence factor for which the term, modulin, was suggested (Figure 10.5). The term modulin was chosen because it

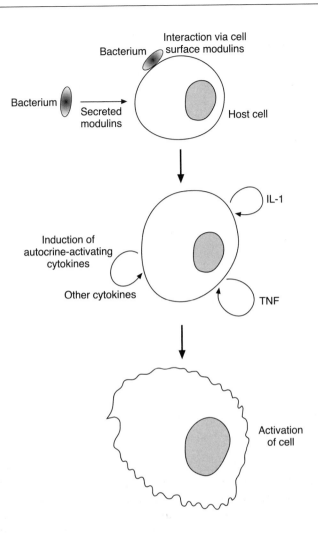

Figure 10.5 The actions of modulins. The secretion of modulins from bacteria, or their direct interaction with host cells, activates the transcription of host cell cytokines. These cytokines can act in an autocrine manner, activating the host cell. Such activation can be protective or may be the cause of pathology in infection.

described the fact that cytokines feedback onto cells and *modulate* their behaviour.

Modulins fall into at least two categories. The first are normally non-proteinaceous components of bacteria associated with the cell wall. These include lipopolysaccharide (LPS) (in Gram-negative bacteria), peptidoglycan (in Gram-positive and Gram-negative species), lipoteichoic acid (LTA) (in Gram-positive bacteria) and the lipoarabinomannan (LAM) and lipoproteins of *Mycobacterium* spp. As has been described in Section 6.6.1.1, these molecules act to stimulate cells by binding to the CD14–TLR membrane receptor system. The other group of cytokine-inducing bacterial molecules are mainly proteins and peptides that may bind to cells via the CD14–TLR receptor system, or they may act via other known, or as yet unknown, receptors. Table 10.2 offers some idea of the potency of these various bacterial molecules. To give the reader some perspective, endocrine hormones normally function to stimulate specific target cells at concentrations in the micro- to nanomolar range. Cytokines stimulate cells bearing cognate receptors at concentrations ranging from low micromolar (10^{-6} M) to high femtomolar (10^{-15} M) concentrations (normal range 100 nM to 100 pM). Thus it can be seen that some bacterial cytokine-inducing molecules are extremely active. LPS from *E. coli* and *Sal. typhimurium* can be active at concentrations in the pico- to femtomolar range. Possibly the most potent cytokine-inducing molecules are, surprisingly, the bacterial exotoxins. Many bacterial toxins have been reported to be able to stimulate cells to produce certain pro-inflammatory cytokines. In most cases, the potency of the toxin as a cytokine-inducing agonist is greater than the potency of the toxic action of the molecule. This suggests that cytokine induction may be a key attribute of bacterial exotoxins. As will be described, some bacterial toxins can also specifically inhibit the synthesis of certain cytokines, most notably TNFα.

The finding that individual bacteria, including members of the normal microflora, produce many molecules able to stimulate host cells to synthesise and secrete cytokines is, frankly, odd. Cytokines are the key signals driving inflammatory and immune responses. They are also, as has been stated, the major driving force of tissue pathology. In either case, they are dangerous both to the bacterium and to the host. It would, therefore, have been expected that evolution would have limited the numbers of components bacteria have to induce cytokine synthesis. This is clearly not the case. One possible explanation is that the many cytokine-inducing components made by bacteria may be a mechanism for evading host defence responses. These host responses can cope with only so many incoming signals to integrate appropriate cellular and humoral defences. If bacteria overcome the ability of the body to integrate its host defence mechanisms, by providing multiple signals, it may give them a better chance of surviving. Of course, this is a dangerous strategy for both sides of the bacteria–host interaction. As we shall see, certain microorganisms have 'recognised' this fact and have attempted to limit host cytokine defences.

While overstimulation of cytokine production is only a putative evasion mechanism, there is now very good evidence that microorganisms target cytokines and cytokine receptors to evade host defence mechanisms. The most persuasive evidence for this hypothesis comes from virology and a brief description of how viruses evade cytokines will be helpful for the discussion of how bacteria may also evade cytokine-driven defences.

10.3.2 Virokines and viroceptors

Although this book is devoted to bacterial pathogenesis, at certain stages in this chapter the mechanisms employed by viruses to evade immune defences will be described. The reason for this is that we know very much more about how viruses evade immunity than we do about bacteria. It is likely that viruses and bacteria will target the same immune defence mechanisms and that they may use similar mechanisms, although there is no guarantee that this will be the case.

Viruses are obligately intracellular parasites dependent on the eukaryotic cell to provide essential mechanisms such as those required for replication. The genome size of viruses will be discussed in Section 11.4. The largest viruses contain only between 100 and 200 genes. However, in spite of this small size, viruses can take over many of the functions of cells. In the late 1980s and early 1990s, reports began to appear of viral proteins with the capacity to inhibit cytokine functioning. Since these early reports, a very large number of genes encoding cytokine-modulating proteins have been found in viral genomes. These proteins fall into three main categories: (i) cytokine-like proteins or virokines; (ii) receptors for cytokines, termed viroceptors; and (iii) proteins inhibiting cytokine signalling (Table 10.3). In mechanistic terms, viruses inhibit the cytokine system by blocking cytokine synthesis, by interfering with cytokine signalling at the level of the receptor or by blocking cytokine effector functions. These proteins are made by a range of viruses such as the Epstein–Barr virus, poxviruses, cowpox virus, herpesviruses, hepatitis virus and adenoviruses.

A fascinating aspect of these viral cytokine 'inhibitors' is the fact that they appear to have been hijacked from the host genome and modified for use as an anti-host defence mechanism.

10.3.2.1 Virokines

As discussed in Sections 4.2.6 and 6.4.4, IL-10 is a macrophage-deactivating/anti-inflammatory cytokine produced by a range of cell populations including macrophages, natural killer (NK) cells, mast cells and lymphocytes. Cells binding, and responding, to IL-10 include monocytes/macrophages, dendritic cells, NK cells, mast cells, neutrophils and endothelial cells. IL-10 plays a key role in controlling inflammatory responses. Knockout of the IL-10 gene in mice gives rise to enterocolitis. If knockout mice are gnotobiotic, then no intestinal inflammation occurs, suggesting that the trigger for inflammation is the normal gut microflora. Administration of IL-10, either *in vitro* or *in vivo*, enhances the survival of a range of pathogens including yeasts and protozoans. The possession of an IL-10-like molecule would, therefore, be an advantage to a pathogen. In particular, IL-10 is an inhibitor of Th1 responses. Epstein–Barr virus possesses a gene, BCRF1, whose product both has homology to IL-10 and possesses the anti-inflammatory actions of IL-10. Other viruses (e.g. equine herpesvirus-2) also possess IL-10 homologues. A growing number of viruses have been found to contain genes encoding chemokine-like proteins. Kaposi's sarcoma-associated herpesvirus synthesises a number of chemokine homologues. For example, viral **macrophage inflammatory protein** (vMIP) II is a chemokine receptor antagonist able to bind to both CC and CXC chemokine receptors (see Section 4.2.6.5 and Tables 4.18 and 4.19) and inhibit responses to host cytokines. *Molluscum contagiosum*, a human cutaneous poxvirus, contains a homologue of an MIP-1β that, again,

Table 10.3. | Microbial cytokine evaders: some examples of virokines and viroceptors

Virus[a]	Viral gene	Cytokine function targeted[b]
Virokines		
EBV	BCRFI	IL-10 homologue/macrophage inhibition/Th1 antagonist
Equine HV-2	EHVIL-10	As above
HHV-6	U83	CC/CXC3 chemokine/evasion mechanism unclear
HHV-8	vMIP-I	CCR8 agonist/Th2 chemoattractant/immune evasion
HHV-8	vMIP-II	Chemokine receptor antagonist/Th2 chemoattractant/immune evasion
Molluscum contagiosum	MC148P	MIP-1β homologue/chemokine receptor antagonist
HHV-6	v-IL-6	Enhances angiogenesis/haematopoiesis
HSV	V-IL-17	T cell mitogen?
Viroceptors		
Vaccinia/cowpox	B15R	v-IL-1R/binds IL-1β but not IL-1α
Cowpox	CrmB	v-TNFR/binds TNFα and LTα
Shope fibroma virus	SFV T2	v-TNFR
Myxoma	MT-2	v-TNFR
Vaccinia	B18R	v-IFNR
Myxoma	M-T7	v-type II IFNR (binds IFNγ)
Vaccinia	B8R	As above
Tanapox	35 kDa protein	Binds IFNγ, IL-2 and IL-5
Measles virus	Haemagglutinin	Binds CD46 and inhibits IL-12 synthesis
Many		Chemokine receptors
Others		
Vaccinia	B13R	Inhibits several caspases including caspase 1
Cowpox	CrmA/Spi-2	As above
Adenovirus	E1A	Inhibits IFN-induced JAK/STAT signalling

[a] EBV, Epstein–Barr virus; HHV, human herpesvirus; HSV, herpes simplex virus.

[b] For abbreviations, see list on p. xxii.

has antagonistic activity. Human herpesvirus (HHV) 8 encodes an MIP-1 homologue that binds to the chemokine receptor CCR8 found preferentially on Th2 lymphocytes. This chemokine may, therefore, function to bias the Th1/Th2 balance at the site of viral infection and, by recruiting an ineffective lymphocyte population, promote its survival. Other cytokine genes possessed by viruses are homologues of growth factors and their role in viral infection is obvious.

As has been described in Sections 4.2.6.6.1, the cytokines IL-1 and IL-18 have to be proteolytically processed by an enzyme, initially termed pro-IL-1 converting enzyme (ICE) and now known as caspase 1, to form the biologically active molecule. Caspase 1 is known to be involved in the process of apoptosis. Cowpox and baculovirus encode proteins that act as inhibitors of caspase 1 and block the production of IL-1β and IL-18.

10.3.2.2 Viroceptors

Viruses encode a number of soluble forms of cytokine receptors (Table 10.3). These viroceptors include proteins such as B15R of *Vaccinia*, which has homology to the type 1 IL-1 receptor and binds IL-1β, but not IL-1α. Other soluble forms of cytokine receptors discovered include those binding to TNFα, IFNα/β, IFNγ, chemokines and colony-stimulating factor (CSF). Studies in which B15R has been deleted from the viral genome have yielded a very interesting result. The virus was more virulent when administered by an intranasal route. This, and other studies, suggest that these viral cytokine evasion mechanisms can act to inhibit damaging host responses to the viral infection. The possession of such cytokine-inhibiting genes may, therefore, be an example of co-evolution with both the host and the parasite benefiting.

10.3.2.3 Cytokine transcription

Being inside cells, viruses are in the right place to intercept signals coming from cytokine/receptor interactions. In Section 4.2.4.1, the role of the Janus kinase–signal transducers and activators of transcription (JAK-STAT) pathway in the transmission of the cell surface signal caused by binding of the haematopoietin family of cytokines (or of the interferons) to their receptors was described. A number of viruses, including murine polyoma virus and cytomegalovirus, can inhibit JAK-STAT signalling. The possibility that intracellular bacteria have similar functions needs to be examined.

10.3.3 Bacterial evasion of cytokines

We know substantially less about bacterial evasion of cytokines than we do about viruses. The first viral inhibitors of cytokines to be discovered were soluble forms of cytokine receptors. A number of reports have suggested that a range of bacteria such as *E. coli*, *Shigella* spp. and *Mycobacterium tuberculosis* have receptors for cytokines such as IL-1, TNFα and epidermal growth factor (EGF) (Table 10.4). These do not appear to be receptors captured from the host. These receptors may act to remove cytokines from sites of infection (Figure 10.6). However, there is also evidence that they act to promote the growth of bacteria. Thus virulent strains of *E. coli* appear to be able to bind to IL-1 and this stimulates cell growth. *Mycobacterium tuberculosis* binds EGF via the surface-associated glycolytic protein glyceraldehyde-3-phosphate dehydrogenase and the binding of this human growth factor stimulates growth of the bacterium. This growth-promoting action of human cytokines for certain bacteria may come under the umbrella of cytokine evasion or it may be an example of how interlinked we are with the bacteria that share our environment.

Now, a major question that has to be addressed is: do bacteria have the capacity to inhibit cytokine synthesis? The answer to this question

Table 10.4. Cytokines bound by bacteria

Cytokine bound	Bacterium	Function
IL-1	Escherichia coli	Apparent growth factor
IL-2	Escherichia coli	Apparent growth factor
IL-3	Listeria monocytogenes	Apparent growth factor (in macrophage-internalised cells)
IL-6	Mycobacterium avium	Apparent growth factor
TNFα	Escherichia coli, Salmonella typhimurium, Shigella flexneri	?
TGFβ	Mycobacterium tuberculosis	Apparent growth factor (in macrophage-internalised cells)
GM-CSF	Escherichia coli	Apparent growth factor
CSF-1	Listeria monocytogenes	Apparent growth factor (in macrophage-internalised cells)
EGF	Mycobacterium avium, Mycobacterium tuberculosis	Growth factor

GM-CSF, granulocyte-macrophage colony-stimulating factor; for other abbreviations, see the text.

is that scientists have only just started to look for bacterial molecules that inhibit cytokine synthesis. Some of the bacterial toxins which can induce the production of some cytokines have been shown to inhibit the production of others (Table 10.5).

An interesting example of a cytokine-inhibiting toxin is cholera toxin (CT), better known for its ability to cause copious diarrhoea in those infected by *Vibrio cholerae*. CT is also a powerful mucosal adjuvant able to induce Th2 (rather than Th1) responses (for a discussion of Th1/Th2 lymphocyte responses, see Section 6.7.3). As described in Section 6.7.3.3, one of the major cytokines involved in the induction of Th1 responses is IL-12. It has been found that CT is a powerful inhibitor of the secretion of IL-12 by antigen-presenting cells (APCs) (Figure 10.7) and also has the ability to block the expression of the IL-12 receptor β1- and β2-chains. CT blocks the production of IL-12 by inhibiting transcription of the two chains (p35 and p40) of this cytokine.

A number of other bacterial proteins and peptides have been reported to inhibit the synthesis of particular cytokines (Table 10.6). These range from a small peptide (nodularin) produced by cyanobacteria to the biggest protein produced by *E. coli*. The latter is a 366 kDa protein that has been named lymphostatin produced by enteropathogenic strains of *E. coli*. This protein inhibits the production of IL-2, IL-4 and IFNγ. It has homology to the catalytic domains of the large clostridial cytotoxins and so may be a toxin, possibly a new class of toxin, in its own right (for more details on clostridial toxins, see Chapter 9).

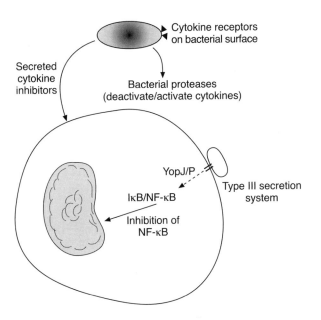

Secreted cytokine inhibitors

Cytokine receptors on bacterial surface

Bacterial proteases (deactivate/activate cytokines)

YopJ/P

IκB/NF-κB

Type III secretion system

Inhibition of NF-κB

Figure 10.6 Current thoughts on bacterial cytokine evasion tactics. Bacteria have been shown to produce a small number of soluble proteins that can inhibit the synthesis of certain cytokines. In addition, *Yersinia* spp. inject (via a type III secretion system) the protein YopJ/P, which blocks the activation of IκB/NF-κB. A less-well-established mechanism of inhibition may exist on the bacterial cell surface, with a number of groups reporting the presence of cytokine receptors. A further evasion tactic involves bacterial proteases that can hydrolyse cytokines. This can result in either the inhibition or the activation of specific cytokines, thereby disrupting cytokine networks.

The literature on cytokine-inhibiting bacterial proteins is still sparse, but it is possible to detect a pattern emerging: (i) those proteins that inhibit lymphokine production (IL-2, IL-4, etc.) and (ii) those that inhibit TNFα production. TNFα is a major defence cytokine, able to activate multiple cells to promote inflammatory responses. Abrogation of the activity of this cytokine with antibodies, or genetically by knocking out the gene, can exacerbate infection with microorganisms or increase susceptibility to infection. One example will be given of *Yersinia*, which can inhibit TNFα synthesis as part of a general mechanism to block the activity of phagocytes. As described in Section 4.2.6.6.3, the IκB/NF-κB signalling system is both involved in the synthesis of cytokines such as TNFα and also is activated by TNFα. The *Yersinia* type III secretion system (see Sections 2.5.5 and 10.4.2.1) injects a range of

Table 10.5. Bacterial exotoxins modulating cytokine synthesis

Toxin	Cytokines induced	Cytokines inhibited
Anthrax oedema toxin (EF)	IL-6	TNFα
Cholera toxin (CT)	IL-1, IL-6, IL-10	TNFα, IL-12/IL-12R
Pertussis toxin (PT)	IL-1, IL-4	IL-1
Escherichia coli heat-labile toxin (LT)	–	IL-12
Yersinia enterocolitica Yops	–	TNFα
Pseudomonas aeruginosa ADP-ribosylating toxin	–	IL-1, TNFα, IFNγ

For abbreviations, see the text.

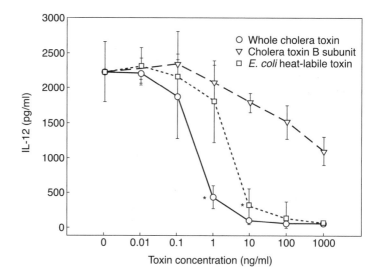

Figure 10.7 Inhibition by cholera toxin (CT) of human monocyte IL-12 secretion. Both CT and *Escherichia coli* heat-labile toxin suppress the production of IL-12 in a dose-dependent manner. Maximal inhibition is induced by between 1 and 10 ng toxin/ml . In contrast, the B subunit of CT was inactive in this respect. (Reproduced from *The Journal of Experimental Medicine*, 1999, **189**, 541–542, by copyright permission of The Rockefeller University Press.)

proteins known as Yops (*Yersinia* outer membrane proteins) into macrophages. Such injection was found to be associated with a decreased synthesis of TNFα. Using knockout mutants, the protein YopP (in *Y. enterocolitica*), also known as YopJ in *Y. pseudotuberculosis*, was identified as the inhibitor of TNFα synthesis. It turns out that YopJ/YopP blocks the inhibitory (I) κB degradation that needs to occur in order to allow nuclear factor (NF) κB to enter the nucleus and initiate gene transcription, including that of TNFα (Figure 10.8). In addition, these Yops are also able to block MAP kinases.

This brief description of the ability of bacteria to modulate cytokine synthesis is likely to be only the tip of the iceberg with regard to the manipulation of cytokine networks evoked by bacteria.

A final mechanism by which bacteria can modulate cytokine networks is by the proteolysis of cytokines. As described in Section 4.2.6.6.1, a number of cytokines such as IL-1 and TNFα are produced

Table 10.6. | Bacterial suppressors of cytokine synthesis

Organism	Cytokine(s) inhibited	Active component
Escherichia coli	IL-2, IL-4, IL-5, IFNγ	Lymphostatin (366 kDa)
Brucella suis	TNFα	40–50 kDa protein
Actinobacillus actinomycetemcomitans	IL-2, IL-4, IL-5, IFNγ	14 kDa protein
Salmonella typhimurium	IL-2	Ill-defined protein
Mycobacterium ulcerans	IL-2, IL-10, TNFα	Ill-defined protein
Cyanobacteria	IL-2	Nodularin (cyclic peptide)

For abbreviations, see the text.

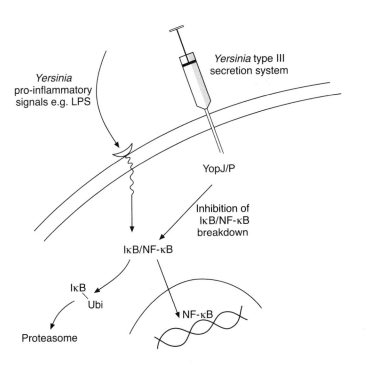

Figure 10.8 Inhibition of TNFα synthesis by YopJ/YopP. The binding of *Yersinia* spp. to cells must provide pro-inflammatory signals such as LPS, which would activate the IκB/NF-κB transcriptional control system resulting in IκB being phosphorylated and subsequently ubiquitinylated and proteolysed in the proteasome. The remaining NF-κB is then able to transfer into the nucleus and bind to cognate DNA-binding sites. YopJ/P ('injected' via a type III secretion system) acts to prevent the activation of the IκB/NFκB complex and so prevents the induction of TNFα gene transcription. For abbreviations, see the text.

as inactive pro-forms and require to be cleaved by proteases to produce an active cytokine. Many bacteria produce active proteases, and some of these enzymes have been shown to be capable of activating these pro-forms of cytokines. It is not known how important this could be in infected tissues. It would seem harmful for the bacterium to induce the formation of these pro-inflammatory cytokines. However, if these cytokines are produced inappropriately, they may not function properly. In addition to activating cytokines, bacterial proteases have also been found to inhibit the actions of cytokines. It has been reported that a major protease from the oral bacterium *Porphyromonas gingivalis* can inactivate IL-1β and TNFα while proteases from *Pseudomonas aeruginosa* can inactivate IFNγ and TNFα. Bacterial proteases have also been found to cleave off cytokine receptors such as the IL-6 receptor. This can prevent cells responding to this particular cytokine. Thus proteases can activate or inactivate cytokines and affect how cells respond to cytokines. Further work is required to establish the importance of such proteases in cytokine network control.

Don't evade it – destroy it

Bacteria produce a plethora of proteases that can cleave a large number of proteins. There is increasing evidence that such proteases can act to control the immune response. For example, as has been briefly described, there is now good evidence that bacterial proteases can both cleave and inactivate, or cleave and activate, cytokines. For example, the major pro-inflammatory cytokine IL-1β is produced as an inactive pro-form that has to be cleaved by caspase 1 to form the active 17 kDa protein. Certain bacterial proteases, such as the cysteine protease exotoxin B of

Strep. pyogenes can cleave pro-IL-1β to form an active form of the molecule. In contrast, the oral pathogen *Por. gingivalis*, which is, in part, responsible for the commonest bacterial 'infection' of humans (the periodontal diseases), produces proteases that can proteolytically cleave and inactivate IL-1β. The major lytic enzymes produced by *Por. gingivalis* are cysteine proteases called gingipains. It has recently been found that these proteases can target the LPS-binding pre-receptor CD14. When cultured monocytes were exposed to purified gingipains their ability to bind to CD14-reactive monoclonal antibodies was abolished. In contrast, TLR4 and other cell surface receptors such as CD18 and CD54, were largely unaffected. This reduction in anti-CD14 monoclonal antibody binding was due to cleavage of CD14 from the cell surface, as shown by the appearance of soluble CD14 fragments in the media supporting the cells. Loss of cell surface CD14 significantly decreased the ability of monocytes to respond to LPS. This is an interesting finding, as it suggests that this bacterium has the ability to block the stimulating activity of its own LPS. However, it has also been found that a proteolytic fragment of the gingipains can also induce cytokine synthesis and so the final picture of the control of LPS-induced inflammation has yet to be established. However, what is clear is that the induction of cytokine synthesis by bacteria cannot be explained simply as a consequence of the host's response to LPS.

10.4 | Evasion of innate immune mechanisms

It has been established that microorganisms can evade antibacterial mechanisms at mucosal surfaces and there is growing evidence that they are able to manipulate cytokine networks to their own advantage. This ability to modulate cytokine networks will influence both innate and acquired immunity. In Section 10.4, the ability of bacteria to target innate mechanisms such as complement activation and phagocytosis will be described.

10.4.1 Complement evasion

Complement is now believed to be a very ancient defence mechanism that arose some several hundred million years ago. The three pathways of complement activation were described in Sections 6.4.2 and 6.6.1.3. The classical pathway is activated by immune complexes, the alternative pathway is triggered by the binding of complement components to bacterial surfaces, and the lectin pathway is triggered by collectins, such as mannose-binding protein, or by binding to bacterial surface carbohydrates (described in detail in Section 6.6.1.3). The functions of this pathway are to provide three sets of antimicrobial substances: (i) opsonins (e.g. C3b) to bind to bacteria and enhance bacterial phagocytosis; (ii) anaphylotoxins (e.g. C5a) to enhance inflammatory events, cause leukocyte chemoattraction and enhance antibody formation; and (iii) a lytic complex to kill bacteria. The complement pathways

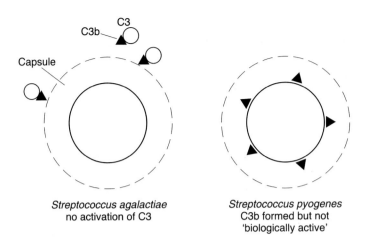

Figure 10.9 Ability of bacterial capsules to inhibit complement-mediated phagocytosis by two different mechanisms. *Streptococcus agalactiae* prevents the activation of C3 and the formation of the opsonin C3b. In contrast, in the hyaluronic acid-containing capsule of *Strep. pyogenes*, there is no inhibition of C3 activation and bacteria are opsonised with C3b. However, these covalently bound C3b proteins cannot make contact with receptors on phagocytes, owing to the masking action of the capsule.

act together with acquired immune responses to provide optimal opsonisation for subsequent phagocytosis. Complement activation also promotes antibody formation. Thus, again, the innate and acquired immune systems show co-operative responses.

Both viruses and bacteria have evolved mechanisms to evade the consequences of complement activation and Section 10.4.1 will deal only with the latter.

10.4.1.1 Bacterial capsules

Many bacteria express an outer coating of polysaccharide and/or protein termed a capsule (see Section 2.2.6). It has been known for many years that bacterial capsules can protect the organism from attack by phagocytes. Much of our understanding of the role of capsules has come from studies of streptococci. Thus, we know that the type III capsular polysaccharide of *Strep. agalactiae* contains sialic acid residues. These residues prevent the activation of the alternative pathway of complement activation and the consequent deposition of C3b on the bacterial surface (Figure 10.9). In contrast, in *Strep. pyogenes*, the capsule is composed of hyaluronic acid. This is a non-immunogenic polysaccharide that does not inhibit the generation of C3b and its binding to the bacterial surface. This has been shown by comparing C3b binding to encapsulated wild-type and unencapsulated mutants. Both bound similar amounts of C3b when incubated in serum. However, the capsulated organisms are protected from complement attack. It is thought that the mechanism of complement evasion lies in the ability of the capsule to prevent phagocytes interacting with the C3b on the cell surface (Figure 10.9).

10.4.1.2 Proteases and complement evasion

The role of proteases in evading the antibacterial actions of IgA1 has already been discussed (Section 10.2.1). There is also evidence that various bacteria produce proteases that are capable of inhibiting the activity of the complement pathways. For example, most strains of *Strep. agalactiae* encode an enzyme that proteolytically inactivates the human anaphylatoxin C5a (a major chemoattractant for neutrophils) by removing a heptapeptide from the C-terminus. This C5a-ase in soluble form is inhibited by natural IgG antibodies found in human serum or plasma. However, C5a-ase associated with the surface of encapsulated streptococci is able to inactivate C5a without being neutralised by IgG antibodies.

Another bacterium that produces proteases able to inactivate complement is the bacterium *Por. gingivalis*. This organism produces multiple forms of two cysteine proteases termed Arg-gingipain and Lys-gingipain. These proteases have been shown to be capable of potentially interacting with the complement system in a complex manner. Early reports found that the *Por. gingivalis* proteases could inactivate both C3 and C5. Later reports found that these enzymes could generate C5a activity. It was then reported that the proteases could inactivate the C5a receptor on human neutrophils. The final effects due to the release of these powerful proteases is not absolutely clear, but the evidence would suggest that inhibition of the complement system occurs.

Pseudomonas aeruginosa produces an elastase that can inactivate C3b and C5a. It is possible that proteolytic inactivation of complement activation may be a general mechanism for evading this very important host defence system. Additional studies are needed to confirm this idea.

10.4.1.3 Interference with complement regulatory proteins

As described in Section 6.6.1.3, the complement pathways are amplificatory cascades designed to produce rapidly large amounts of defence molecules such as the products of C3. Because of the 'explosive' nature of these cascades, there has evolved an extensive system of regulatory proteins to keep complement activation in check. The proteins are present both in the fluid phase and on the surfaces of cells, and are known as members of the regulators of complement activation (RCA) family. These proteins are encoded by genes found close together on chromosome 1. They are composed of short protein domains (approximately 60 residues) known either as short consensus repeats (SCRs) or complement control protein repeats (CCPs). These proteins and their interactions with bacteria are listed in Table 10.7.

The most complex of the RCA proteins is C4BP (**C4-b**inding protein), which resembles a spider in appearance, composed as it is of seven identical α-chains bound with one short β-chain. This protein was originally identified by its ability to bind to C4b and inhibit the classical pathway of complement activation. Factor **H** (FH) is a much smaller RCA and was the first of these regulatory proteins to be shown to bind to bacteria. The most recently described RCA protein is factor **H**-like protein 1 (FHL-1)/reconectin. This protein has actions similar to those of factor H.

Table 10.7. Complement control proteins and bacteria

RCA protein	Bacterium	Bacterial ligand
Fluid phase proteins		
C4BP	*Bordetella pertussis*	FHA?
	Streptococcus pyogenes	Some M proteins
	Neisseria gonorrhoeae	Porin
FH	*Neisseria gonorrhoeae*	LOS/porin
	Streptococcus pyogenes	Some M proteins
FHL-1/reconectin	*Streptococcus pyogenes*	Some M proteins
Membrane proteins		
CD46 (MCP)	*Streptococcus pyogenes*	M6 protein
	Helicobacter pylori	BabA protein
	Neisseria gonorrhoeae	Pili
	Neisseria meningitidis	Pili
CD55 (DAF)	*Escherichia coli*	Dr-like antigens
		X adhesin

FHA, filamentous haemagglutinin; LOS, lipo-oligosaccharide; BabA, blood-group antigen-binding adhesin; for other abbreviations, see list on p. xxii.

It is now established that a number of bacteria can bind to C4BP and can subvert the normal functioning of this protein for the benefit of the bacterium. The best studied of these bacterial proteins are the M proteins of *Strep. pyogenes*. These proteins are fibrillar, so-called coiled-coil dimers that are attached to the surface of the bacterium. In transmission electron micrographs, these M proteins can be seen clearly on the surface of the bacteria (see Figure 7.28, p. 398). All of the M proteins have similar overall structures, with a conserved C-terminus attached to the bacterial surface, a central repeat region and a variable N-terminus exposed to the external environment (for a more detailed description of the structure of M proteins, see Section 2.9.1). The variable region can be subdivided further into a highly variable domain of approximately 50 amino acid residues, located at the N-terminus and linked to a semi-conserved domain. The variability of the N-terminus of this protein appears to arise through the accumulation of mutations and recombination following horizontal gene transfer. The expression of M proteins is under the control of the trans-acting regulator Mga.

The M proteins of *Strep. pyogenes* bind to both C4BP (Figure 10.10) and to FHL-1/reconectin via the hypervariable N-termini. It is now thought that FHL-1, rather than FH, may be the more important bacterial M protein ligand. The binding of these RCAs allows them to retain their biological activity and so the bacterium becomes decorated with active complement inhibitors, preventing the deposition of C3b onto the bacterial surface and thus blocking the phagocytosis of the organism. The hypervariable nature of the N-terminal RCA binding site may be a

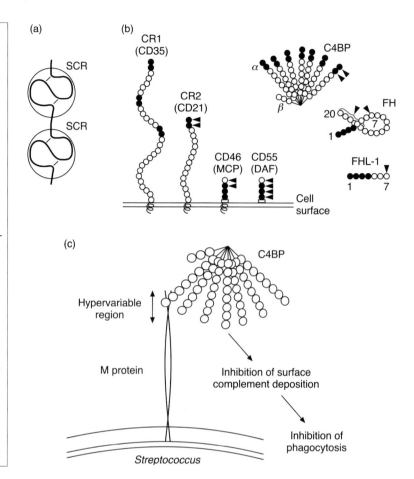

Figure 10.10 The structures of the various human complement regulatory proteins of the RCA family. In (a) there is a representation of two SCR (short concensus repeat) domains with disulphide bonds indicated. In (b) the various RCA proteins are schematically illustrated. Each circle represents one SCR domain. Filled circles indicate domains required for complement regulation and/or for binding of complement components. Arrowheads indicate SCRs implicated in the binding to microorganisms. (c) Proposed model for the interaction of C4BP with the M protein of *Strep. pyogenes*. This binding acts to down-regulate complement activation and thus inhibits opsonisation and phagocytosis. For abbreviations, see list on p. xxii. (Reprinted from *Current Opinion in Immunology* **12**, Lindahl, G., Siobring, U. & Johnsson, E., Human complement regulators: a major target for pathogenic microorganisms, pp. 44–51, Copyright 2000, with permission from Elsevier Science.)

response to the immune system targeting this site with blocking antibodies. This contention has, as yet, no experimental backing.

In addition to binding to these soluble RCAs, bacteria can also bind to the membrane-bound RCA proteins CD46 and CD55 (Figure 10.10). For example, CD46, which is widely expressed on human cells, binds to C3b and C4b monomers. It is also a receptor for *N. gonorrhoeae* and *N. meningitidis*, which bind via pili. *Helicobacter pylori* also binds to gastric tissue with the help of this receptor. It is not clear whether binding has effects on complement activation or on the activation state of the eukaryotic cell. CD55, also know as **decay-accelerating** factor (DAF), stimulates the decay of the two C3 convertases and, again, is expressed on many human cells. This protein is a receptor for many strains of *E. coli*. Again, the consequences of bacterial binding on complement activation, or on cell activation, have not been established.

Protectin – what does it protect?

The complement systems generate products that are harmful to all cells. The host has therefore to be protected from the consequence of complement activation. This is provided by the action of a range of soluble and cell surface proteins, some of which have been described in the text. One of these proteins is called protectin (CD59) and is a regulator of the terminal components of the complement system that give rise to the membrane attack complex (MAC). This is the pore system that can puncture bacterial cell membranes and cause them to lose their integrity and function. Protectin is released from cells and it has now been shown to be able to offer protection to bacteria from the effects of complement. Using radiolabelled protectin, it was found that incubation of this labelled protein with *E. coli* resulted in the binding of this host protein to the bacterium. Protectin is a glycosylphosphatidylinositol (GPI)-anchored protein (as is CD14) and it was found that the GPI moiety was required for binding of protectin to the bacteria. The bound protectin inhibited the formation of the C5b-9 membrane attack complex on the bacterial surface and this was also reflected in the failure of complement to lyse protectin-bound *E. coli*.

This shows that bacteria may be able to evade complement-induced killing by making use of the same protective system evolved to defend the host against complement activation.

10.4.2 Evasion of phagocytic killing

A major mechanism for the destruction of bacteria that have invaded the host is killing by phagocytes (macrophages and neutrophils). In Section 6.6.2, the mechanisms by which phagocytes recognise and phagocytose bacteria were described. Bacterial uptake can be by non-opsonic (e.g. by binding to scavenger receptors) or by opsonic (via C3b or Fc receptor) mechanisms. The importance of opsonisation of bacteria with C3b and/or with immunoglobulins was explained and, in Section 10.4.1.2, the evasion of opsonic phagocytosis was described. Bacteria can also evade non-opsonic phagocytosis but less is known about these mechanisms. For example, *N. gonorrhoeae* binds non-opsonically, via phase-variable outer membrane proteins (the opacity-associated (Opa) proteins) to the immunoglobulin superfamily adhesion molecule CD66 on human monocytes. This interaction inhibits the oxidative killing mechanism normally used by phagocytes to kill internalised bacteria.

Inhibition of the phagocytic uptake of bacteria, or of the process of bacterial killing, has evolved in a range of bacterial species and also in protozoans and viruses. Inhibition of phagocytic mechanisms is also related to the finding that microbial pathogens can inhibit the processing and presentation of antigens to T lymphocytes. The points at which bacteria can interact with phagocytes and inhibit phagocytic uptake or phagocytic killing are shown in Figure 10.11. The ability of bacteria to evade opsonisation, and, in so doing, to evade phagocytosis, can be described as an indirect mechanism. Bacteria can also interact directly with phagocytes and block their ability to induce phagocytosis. For example, *Yersinia* spp. can evade the attention of macrophages by injecting, via a type III secretion mechanism, proteins that inhibit phagocytosis. *Salmonella* spp. and *Shigella* spp. enter macrophages and

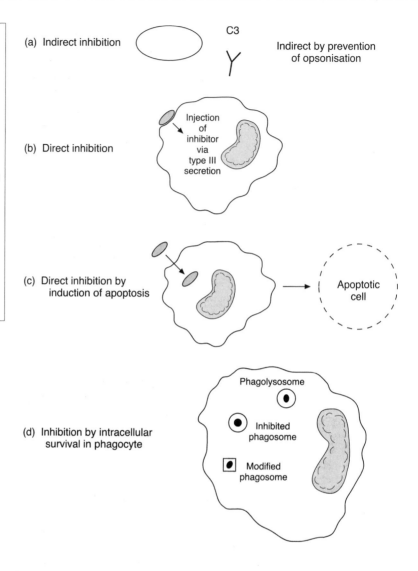

Figure 10.11 Mechanisms of the inhibition of phagocyte function by bacteria. In (a), the inhibition is of bacterial opsonisation by complement components or antibodies, which is an indirect method. In (b), inhibition is due to the action of bacterial proteins injected into the phagocyte via a type III secretion system. In (c), the bacterium prevents phagocytic killing by inducing the macrophage/neutrophil to undergo apoptosis. In (d), the strategy is to survive within the macrophage. Living within phagocytes also inhibits mechanisms of bacterial killing. Some bacteria may utilise all of these strategies to inhibit phagocytic killing.

in so doing induce apoptosis. This mechanism of evasion of phagocyte function will be described in Section 10.6.2.2. Other bacterial pathogens can enter into macrophages and can live within various locations in the endosomal compartment of these cells.

10.4.2.1 Evasion of killing by means of type III secretion systems

Yersinia enterocolitica and *Y. pseudotuberculosis* cause diarrhoeal diseases in both mice and humans. It is thought that these bacteria enter the intestine via the M cells, described in Section 5.7.1, whose function is to allow the passage of antigens to submucosal antigen-presenting cells. This brings the bacteria into direct contact with phagocytes. Interaction of the bacteria with submucosal macrophages is by direct cell–cell contact and the secretion of a group of virulence factors, the Yops. The

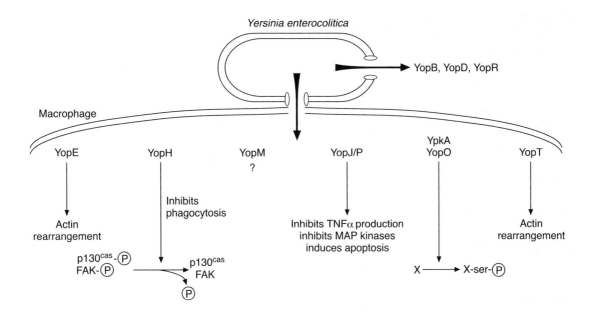

Figure 10.12 The effect of Yops on macrophage function. These include alterations in the actin cytoskeleton, inhibition of gene transcription resulting in blockade of cytokine production, and inhibition of the mechanisms leading to phagocytosis. For abbreviations, see list on p. xxii.

effects of the various Yops on macrophage functions are described in Figure 10.12. Of particular importance is YopH, a 51 kDa protein that is a potent tyrosine phosphatase with homology to eukaryotic tyrosine phosphatases (for a description of such enzymes, see Section 4.3.2.1). It has been suggested that YopH acts to dephosphorylate focal adhesion kinase (FAK) and its substrate p130^{cas}. This could cause disruption of the ability of phagocytes to adhere to the extracellular matrix, via focal adhesions, and thus compromise their ability to phagocytose. It is believed that YopE is also important in this process. In addition, as has been described earlier, YopJ/P inhibits NF-κB activation by blocking the phosphorylation of IκB, thus also blocking TNFα synthesis. This cytokine is a powerful activator of macrophage function. YopJ/P also induces apoptosis, a final mechanism for ensuring inhibition of phagocyte activity. In addition to inhibiting phagocytes, it has recently been reported that YopH can suppress the antigen-responsive activation of both B and T lymphocytes.

Pseudomonas aeruginosa can also inhibit macrophage phagocytosis by a type III secretion system that injects the protein exoenzyme S.

As will be described in Section 10.6.2.2.1, Sal. typhimurium uses the type III secretion system protein SipB (encoded by the Salmonella pathogenicity island) to induce apoptosis in macrophages. Furthermore, it has also been discovered that the second pathogenicity island in this organism (SPI-2) is involved in the inhibition of the activation of the NADPH oxidase system, as described in Section 6.6.2.3.1, and used for the oxidative killing of bacteria. This SPI-2 system appears to work intracellularly and, as will be mentioned later (Section 10.4.2.2.1), can also interfere with endosomal trafficking. Thus it is clear that Salmonella spp. have evolved multiple mechanisms for dealing with phagocytes.

10.4.2.2 Intracellular parasitism: a novel mechanism to evade phagocytosis

Once phagocytosed, bacteria are taken into the endosomal system of the cell where two main processes should occur. The internalised bacteria are killed and the bacterial antigens interact with MHC class II proteins in order to stimulate lymphocyte responses to the invading bacterium. These mechanisms are described in detail in Section 6.7.2. However, certain bacteria can evade the processes involved in phagocytic destruction, and there is growing evidence that they can inhibit the processes of antigen presentation. A few examples of how bacteria evade phagocytic killing will be given in this section.

Before dealing with the mechanisms used by certain bacterial pathogens to prevent their demise by the phagocytic killing machine, it is sensible to enquire what would be the evolutionary pressures leading to a bacterium becoming able to live in what would appear to be the most dangerous environment in the host. Partial answers to this question include the fact that being inside a cell renders the bacterium safe from 'attack' by humoral components such as complement, antibodies, antibiotic peptides, etc. Living inside a cell provides constant environmental conditions with an endless supply of nutrients. Also, living within a cell does away with the requirement for adherence to particular cell populations and the constant danger of being washed away. The obvious downside is that if the cell that the bacterium is living in is a macrophage there is the constant danger of macrophage activation leading to bacterial killing. It is the evasion of this response that will now be discussed. Section 10.4.2.2 will deal with bacteria surviving within macrophages. However, it is now known that *Ehrlichia equi* can live in the neutrophil and is the causative agent of human granulocytic ehrlichiosis. Little is known about the survival of this organism within neutrophils and it will not be discussed further.

A number of bacteria can live within phagocytes (Table 10.8). Some, like *Coxiella burnetii* and *Sal. typhimurium* (and the protozoan *Leishmania*) can actually survive within the acidified lysosomal compartment. Others, such as *Mycobacterium* spp. and *Nocardia* spp. inhibit the normal maturation of the endosomal compartment. Additional bacterial species, such as those belonging to the genera *Legionella* and *Chlamydia*, form a modified endosomal compartment that appears to exist outside of the normal endosomal trafficking system. By use of such strategies, these distinct bacteria evade the killing mechanisms of the macrophage. Most of these bacteria can induce and/or inhibit macrophage apoptosis and this will be discussed in Section 10.6.2.2.

10.4.2.2.1 Survival of bacteria within acidified lysosomes

Coxiella burnetii: This bacterium is an obligate intracellular parasite. It is the causative agent of Q (short for 'Query') fever and of a proportion of cases of infective endocarditis. The organism is phagocytosed by macrophages and the phagosome undergoes normal maturation

Table 10.8.	Bacteria living within the endosomal apparatus of macrophages
Bacterium	Survives within
Salmonella typhimurium	Early endosome, short stay passenger (can survive in lysosome)
Yersinia spp.	As above
Shigella spp.	As above
Coxiella burnetii	Mature lysosome
Mycobacterium spp.	Mature endosome (due to failure of phagosome to fuse with lysosomes)
Nocardia asteroides	As above
Legionella pneumophila	Novel compartment

to form a phagolysosome, which contains the appropriate proteins found in such organelles, such as a membrane proton ATPase, LAMPs (lysosomal-associated membrane proteins) and lysosomal enzymes (Figure 10.13). The bacterium can actually grow in the acidic environment (pH approximately 5) of the phagolysosome and, indeed, increasing the pH has a marked inhibitory effect on bacterial growth. The bacterium has recently been shown to inhibit the rate of formation of phagolysosomes but the secret to its survival at pH 5 remains to be discovered. The finding that internalisation of *Cox. burnetii* is associated with a reduction in the production of antimicrobial oxidants may be due to the acid phosphatase produced by this organism. This enzyme may act as a tyrosine phosphatase. Little is known about the effect of lysosomal survival on the processing and presentation of antigens to T cells. One must assume that there is a deficiency in the normal antigen presentation of this organism by infected macrophages.

One experimental model that is being studied in some detail is the infection of cultured macrophages with more than one intracellular pathogen. In one experiment (see Gomes *et al.* (1999); Section 10.11, Papers) murine bone marrow macrophages were infected with either *M. avium* or *M. tuberculosis* and were then co-infected with *Cox. burnetii*. The majority of the co-infected mycobacteria co-localised to the vacuole containing the *Cox. burnetii*. In such co-infected cells, the growth of *M. avium* was not impaired but that of *M. tuberculosis* was. As will be described below, both mycobacteria inhibit phagosome–lysosome fusion but, as shown by the results described, they appear to be differentially susceptible to the environment of the mature phagolysosome.

Salmonella: These organisms appear to modify the process of phagocytosis at various points. The process of phagocytosis seems to be different from that of other bacteria, being more like macropinocytosis (with substantial amounts of fluid also being ingested) and the

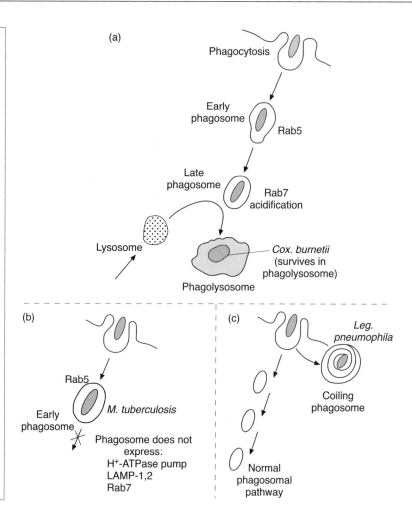

Figure 10.13 The varied mechanisms by which bacteria survive within macrophages as exemplified by *Coxiella burnetii*, *Mycobacterium tuberculosis* and *Legionella pneumophila*. In (a), the normal pathway of phagosomal maturation is shown. The early phagosome contains Rab5 and, as the phagosome matures, it loses Rab5 and gains Rab7 plus other markers (e.g. LAMPs). In addition, the phagosome acidifies. The late phagosome fuses with a lysosome to form a phagolysosome, which has an internal pH of around 5 owing to its proton-ATPase. *Coxiella burnetii* has been reported to slow the process of phagolysosome formation but can survive in the mature phagolysosome. In contrast, as seen in (b), various mycobacteria can inhibit phagosomal maturation at the early, non-acidified, Rab5 stage. Another strategy (c) has evolved with *Legionella pneumophila*, which completely bypasses the normal phagosomal maturation pathway and ends up in a specialised organelle – the coiling phagosome.

bacterium is internalised into a vacuole termed a 'spacious' phagosome. This phagosome appears to mature into a late endosome/lysosome stage with a low pH. Studies of *Sal. enterica* have shown that the second pathogenicity island (SPI-2) encoded by *Salmonella* spp. contains a gene which encodes a protein able to interfere with endosomal trafficking. This protein, SpiC, has been shown to inhibit endosome–phagosome fusion *in vitro*. Knockout of the gene encoding SpiC resulted in bacteria that were significantly less virulent. Unlike the other organisms described in this section, *Salmonella* spp. do not survive for extended periods in macrophages and, as described in Section 10.6.2.2.1 can kill the macrophage.

10.4.2.2.2 Survival of bacteria by inhibition of phagosome maturation

Before describing how *M. tuberculosis* evades phagocytic killing, it is worth while reiterating the pathway followed by phagocytosed inert particles, which have been much used to delineate the maturation of endocytotic vesicles. Following phagocytosis, there is rapid sorting of membrane proteins and recycling of incorporated plasma membrane proteins back to the plasma membrane. Early phagosomes containing inert particles rapidly acquire markers characteristic of early endosomes such as the small GTPase Rab5, and the newly discovered protein TACO (tryptophan aspartate-containing coat protein). With time, the phagosome loses Rab5, TACO and other markers of the early endosome and acquires Rab7 and markers associated with late endosomes such as LAMPs, cathepsin D and the H^+-ATPase. Indeed it is speculated that the association of Rab proteins with endosomes may constitute some form of timer function. Later, the phagosome fuses with secondary lysosomes, acquires higher concentrations of acid hydrolases and LAMPs and loses Rab7 but gains other small GTPases. It should be noted that phagosomes are extremely dynamic structures that interact with various parts of the endosomal trafficking system and exchange parts of membranes and lumenal contents. This behaviour has been termed 'kiss and run' and is part of the process of endosomal/phagosomal maturation (for more details, see Section 6.6.2.2).

Mycobacterium tuberculosis: It is almost 30 years since D'Arcy Hart and colleagues discovered that phagosomes containing live, but not dead (or non-pathogenic), mycobacteria failed to fuse with lysosomes preloaded with electron-dense colloids. It was then found that phagosomes incorporating *M. tuberculosis* lacked the H^+-ATPase in their membranes and this suggested that the presence of this bacterium prevented the fusion of the phagosome with late endosomes. This was also confirmed by the finding that phagosomes containing *M. bovis* retained Rab5, but lacked Rab7, a marker of late endosomes. The pH of the phagosomal vacuole harbouring *M. tuberculosis* is in the range 6.3 to 6.5, as compared with pH 5 in phagolysosomes. The 64 000 dollar question – how does *M. tuberculosis* control endosomal acidification – still awaits a definitive answer. The problem is not a shortage of information. However, a clear explanation for the effect still eludes researchers. For example, it has been suggested that mycobacterial urease might produce ammonia within vacuoles and prevent the lowering of the pH. Others have shown that phagocytosis of viable *M. tuberculosis* via complement receptors inhibits the normal rise in intracellular Ca^{2+} and that this somehow blocks normal endosome–lysosome fusion. Yet other workers have shown that the uptake of virulent mycobacteria is associated with disruption of the macrophage actin filament network. It is possible that all of these mechanisms are working concurrently within macrophages after ingestion of pathogenic mycobacteria. Two recent papers have shed some light on the role of one host protein in the evasion of normal endosomal trafficking by *M. tuberculosis*. This is the protein termed TACO mentioned above. It is a member of the Trp-Asp repeat protein family (see Section 6.6.2.2 and Figure 6.22) that is found on the phagosomal membrane and is normally released during the process of phagosome fusion with lysosomes. It is likely that TACO associates

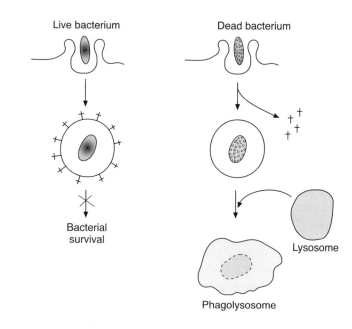

Figure 10.14 The role of TACO (tryptophan aspartate-containing coat protein) in survival of mycobacteria. This is one mechanism to explain the ability of mycobacteria to survive within phagocytes. Phagocytosis induces the recruitment of TACO onto phagosomes. When the macrophage is presented with dead mycobacteria, the association of TACO with the phagosome is transient and allows the interaction and fusion of the phagosome with the lysosome. In contrast, when the phagosome contains ingested viable mycobacteria, the TACO is actively retained on the phagosome and prevents interaction and fusion with lysosomes. †, TACO protein.

with microtubules. With the phagocytosis by macrophages of inert particles, or dead mycobacteria, the normal loss of TACO prior to lysosomal fusion was noted. In contrast, if macrophages ingested viable *M. tuberculosis*, the phagosomes failed to lose TACO (Figure 10.14). This presumably accounts, in part, for the failure of the phagosome to mature. It has subsequently been found that cholesterol is involved in the uptake of viable mycobacteria and the association of the phagosome with TACO. In tissue culture, it has been found that lowering the cholesterol content of macrophages specifically blocked their ability to phagocytose *M. tuberculosis*, while still being able to phagocytose a range of other bacteria (*E. coli*, *Y. pseudotuberculosis*, *Sal. typhimurium*, etc.). These findings raise the obvious question: why has the host retained a protein that allows mycobacteria to survive? The answer has to be that this protein provides such a useful function for cell survival that its evolutionary demise is impossible.

In Section 6.6.2.3.2, the role of Nramp1 (natural resistance-associated macrophage protein 1) in innate defence against phagocytosed bacteria was described. The *Nramp1* gene is the best characterised resistance locus for a number of intracellular bacteria, including mycobacteria. Nramp1 has been shown to be a divalent cation transporter that is found in the late endosomal/lysosomal compartment of macrophages. A second Nramp protein (Nramp2), is found in early endosomes but much less is known about it. While putatively able to transport divalent cations such as Zn^{2+}, Fe^{2+}, Mn^{2+} and Cu^{2+}, and thus act to remove these required metal ions from the phagosome, in consequence inhibiting bacterial growth, Nramp1 has also been claimed to be involved in phagosomal maturation. The role of Nramp1 in the survival of phagocytosed bacteria has been studied using macrophages from Nramp1 knockout animals and comparing them with macrophages from wild-type (Nramp-containing) mice. The uptake of latex beads (as a control), or heat-killed *M. bovis*, by macrophages from control or Nramp knockout mice, resulted in the phagosomes being fully acidified. In contrast, the uptake of live *M. bovis* was able to inhibit acidification in Nramp1 knockout animals while macrophages from wild-type mice acidified normally. The decreased acidification of the mycobacterium-containing phagosomes in the Nramp1 knockout mice was correlated with a decreased rate of fusion to late endosomes. Of course, there is a never-ending war in progress between host and pathogen and it is now known that *M. tuberculosis* (and indeed other bacteria) encodes an Nramp1-like protein that also has cation-transporting properties. The role of this protein in phagosomal survival of mycobacteria is yet to be assessed. *Nramp1* is likely to be only one of a range of genes conferring susceptibility to tuberculosis. A recently discovered genetic locus involved in susceptibility to lung infection with *M. tuberculosis* in mice has been termed **susceptibility to tuberculosis 1** (*sst1*). The function of the protein(s) encoded by this locus has yet to be determined.

Another interesting finding with regard to the mycobacterial phagosome is that it is membrane permeable. By using confocal microscopy (a technique much used by cellular microbiologists) to study the localisation of fluorescently labelled dextrans of various molecular masses, it has been found that molecules of up to 70 kDa molecular mass can gain entry into the mycobacterium-containing phagosome. This may be part of the mechanism by which mycobacteria surviving within the early phagosome can obtain nutrients. It probably also, as will be described, plays a role in the processing of mycobacterial antigens by macrophages.

10.4.2.2.3 Survival of bacteria by exiting the endolysosomal pathway

A third, and the last to be described, mechanism of avoidance of phagocytic killing is for the bacterium to establish a specialised compartment that no longer communicates with the intravacuolar apparatus. The lifestyle of *Legionella pneumophila*, a facultative intracellular pathogen that invades phagocytes, will be described. This bacterium first appeared on the world stage in 1976, when it was identified as the causative agent of a novel disease (an acute form of pneumonia) that affected attendees at an American Legion convention in Philadelphia. It infected 182 individuals, of whom 29 died. This bacterium is one of a small number, including *M. avium* and *Lis. monocytogenes*, that can survive as endosymbionts in free-living amoebae. Indeed, *Leg. pneumophila* is normally a parasite of fresh-water amoebae. *Legionella pneumophila* is taken up into macrophages via complement receptors, owing to the binding of complement components to a bacterial surface protein MOMP (major outer-membrane protein). Binding to the macrophage results in the generation of a particular form of phagosome called a

Figure 10.15 Electron micrograph of *Legionella pneumophila* and the coiling phagosome. Bar = 0.5 μm. (Reproduced with permission from Bozue, J. A. & Johnson, W. Interaction of *Legionella pneumophila* with *Acanthamoeba castellanii*: uptake by coiling phagocytosis and inhibition of phagosome-lysosome fusion. *Infection and Immunity*, 1996, **64**, 668–673.)

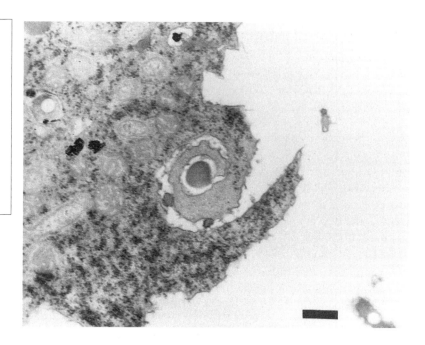

'coiling phagosome' (Figure 10.15). This coiling phagocytosis occurs with heat-killed or glutaraldehyde-killed bacteria, suggesting that the mechanism must involve some very stable surface component. The phagosomes induced by this organism are morphologically distinct and rapidly become associated with the endoplasmic reticulum and dotted with ribosomes. It is only within the past two years that clues have begun to emerge to explain how *Leg. pneumophila* manages to wrest control of the endosome from the host phagocyte. This information has come from genetic screens of the organism to detect mutants that failed to grow within macrophages. This has identified two loci (*dot* and *icm*) that consist of 24 genes and are believed to encode a secretion machine that rapidly exports bacterial virulence proteins that cause remodelling of the phagosome as it forms. Of interest is the finding, made by using *dot-icm* mutants, that these mutants can survive within the specialised vacuole made by wild-type bacteria. To determine whether growth of the wild-type organisms was necessary for survival of the *dot-icm* mutants, use has been made of a thymine auxotroph in which restriction of thymidine blocks growth of the mutant. This had no effect on the growth of the non-auxotrophic *dot-icm* mutant. The *dot-icm* genes are therefore not necessary for intracellular growth but function only to enable the bacterium to alter host phagosome biogenesis. As with all of the bacteria described, further work is necessary to establish the details of the genesis of the specialised phagosome generated by the co-operation between *Leg. pneumophila* and host phagocytes.

10.4.3 Evasion of antigen processing

This final subsection on the evasion of phagocytosis spans the divide between innate and acquired immunity and will describe briefly the very recent findings that bacteria can evade the key process of antigen presentation. The reader is reminded that effector lymphocytes are required to deal with cytosolic, intravacuolar and free-living pathogens ranging from viruses to multicellular worms (Figure 10.16; for details see Section 6.7.3). For an appropriate lymphocyte to be activated, it must be exposed to its cognate antigen in the correct context on the surface of an APC. This requires that the antigens, protein or otherwise, are processed and presented on major histocompatibility complex (MHC) or MHC-like proteins (CD1) on the surface of the APC (described in detail in Section 6.7.2).

Our understanding of the mechanism used by pathogens to evade antigen processing is, again, more developed in virology than in bacteriology and many mechanisms have been identified that block MHC class I/II-dependent antigen processing (Table 10.9). For example, several herpesviruses produce proteins that inhibit the TAP (transporter with associated antigen processing) complex involved in the transport of peptides from the cytoplasm into the endoplasmic reticulum. Adenoviruses produce a 19 kDa protein, E3/19, that retains class I molecules in the endoplasmic reticulum and prevents their display on the cell surface. Viruses can also interfere with endocytic pathways.

Extracellular bacteria can have effects on antigen processing by modulating cytokine synthesis (as described earlier) and the case of cholera toxin has been explained. *Helicobacter pylori* produces a toxin, VacA, that has a direct effect on vacuolar traffic, disrupting the late endosomes and inhibiting the MHC class II presentation pathway.

The effects of intracellular bacteria on intracellular antigen processing are only now being studied in any detail. *Coxiella burnetii* infection has been shown to be associated with an increased interaction of human leukocyte antigen (HLA) DR with HL-DM proteins. As described in Section 6.7.2.2.2, DM proteins stabilise empty class II proteins. Indeed, an increase in unstable HLA-DR proteins has been found in *Cox. burnetii*-infected cells. These findings suggest that there is a deficiency in adding bacterial antigens to MHC class II molecules in *Cox. burnetii*-infected cells. *Mycobacterium tuberculosis* is believed to block antigen processing by a variety of mechanisms. Inhibition of macrophage activation is a well-known effect of internalised *M. tuberculosis* and is the reason that antigen-activated T cells are required to provide key activating signals for infected cells. There are also contradictory reports that *M. tuberculosis* inhibits MHC II expression on infected phagocytes. It has been proposed that *Listeria monocytogenes* can impair antigen processing through the haemolytic action of the toxin, listeriolysin. The other intracellular bacterium reported to modulate antigen processing is *Chlamydia trachomatis*. This organism has been reported to decrease levels of HLA-DR, HLA-DM and invariant

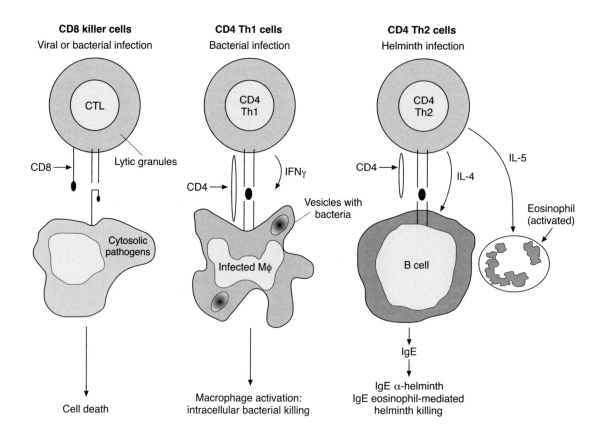

CD8 killer cells
Viral or bacterial infection

CD4 Th1 cells
Bacterial infection

CD4 Th2 cells
Helminth infection

Cell death

Macrophage activation:
intracellular bacterial killing

IgE α-helminth
IgE eosinophil-mediated
helminth killing

Figure 10.16 The role of effector T lymphocytes. CD8 cells are cytotoxic cells that recognise major histocompatibility complex (MHC) class I-bearing cells infected with cytosolic pathogens (viruses or some bacteria). CD4 Th1 lymphocytes recognise MHC class II-bearing cells (antigen-presenting cells mainly) containing intravacuolar pathogens (bacteria and protozoans) and activate them by various means. CD4 Th2 lymphocytes have evolved to deal with helminthic infections and induce reaginic antibodies. Mφ, macrophage.

chain in infected cells. There were no alterations in the levels of class I proteins in infected cells.

Again, these results are all very preliminary and further investigations are needed to understand how bacteria can control the process of antigen presentation.

10.5 | Evasion of acquired immunity

The key to the working of the acquired immune system has been described in Chapter 6 – the ability to make an enormous number of different recognition molecules. It comprises the antibody molecule and the TCR. Both molecules are encoded by multiple genes and there is a system for combining these individual gene segments in a manner that generates enormous diversity of protein sequence and, by definition, structure. This clonal system with its recombinase genes and associated MHC proteins appeared late in evolutionary development and exists only in the vertebrates. Viruses have evolved a wide range of strategies for avoiding detection by the immune response. These include, among others: (i) inhibition of the intracellular pathways by which viral antigens are linked onto MHC class I proteins (described above), (ii) inhibition of cytokines, (iii) blockade of NK cell

Table 10.9. Mechanisms of pathogen evasion of antigen processing

Pathogen	Gene product	MHC class	Phenotype blocked
EBV	EBNA1	I	Blocks proteasome processing
HCMV	US6	I	TAP binding/blocks peptide import
	US3	I	Retains class I molecules in endoplasmic reticulum
	Pp65	I	Inhibition of proteolysis
	US11	I	Targets MHC I heavy chains for degradation
HSV	IPC47	I	TAP binding/blocks peptide binding
Adenovirus	E3/19K	I	TAP binding/tapasin inhibitor
	E1A	II	Interferes with IFNγ-induced MHC II up-regulation
HIV	Nef	I	Down-regulation of class I/CD4
	Vpu	I	Destabilises newly synthesised class I proteins
Coxiella burnetii	Unknown	II	Phagosome alteration
Mycobacterium tuberculosis	Unknown	II	Inhibition of lysosomal fusion, etc.
Salmonella typhimurium	Unknown	II	Altered phagosome processing
Listeria monocytogenes	Listeriolysin	II	Altered endocytic trafficking
Yersinia spp.	Yops	II	Inhibition of phagocytosis
Helicobacter pylori	VacA	II	Disrupts late endosomes
Escherichia coli	LT	II	Inhibits phagocytic processing
Vibrio cholerae	CT	II	Inhibits phagocytic processing

EBV, Epstein–Barr virus; HCMV, human cytomegalovirus; HSV, herpes simplex virus; HIV, human immunodeficiency virus; EBNA, Epstein–Barr nuclear antigen; Vac, vacuolating toxin. Where available, other abbreviations are listed on pp. xxii–xxix.

Table 10.10.	Bacterial immunoglobulin-binding proteins	
Protein	Bacteria producing	Immunoglobulin binding
Protein A	*Staphylococcus aureus*	IgG Fc
Sbi	*Staphylococcus aureus*	IgG Fc and β_2-glycoprotein
Protein G	Some streptococci	IgM
Protein H	*Streptococcus pyogenes*	IgG
Protein L	*Peptostreptococcus magnus*	Variable domain of κ light chains
Protein P	?	Immunoglobulin polymers
Ig-binding proteins	*Haemophilus somnus*	IgG

?, unknown.

function, (iv) interference with endocytotic trafficking, and (v) inhibition of apoptosis. We do not know quite as much about the mechanisms of bacterial evasion of acquired immunity. In Section 10.5, the evasion mechanisms discussed will be: (i) evasion of antibodies, (ii) bacterial superantigens and evasion of T cell responses and (iii) modulation of apoptosis.

10.5.1 Evasion of antibodies

What are the functions of antibodies raised against bacteria? As discussed in Section 6.4.3, they can neutralise the actions of secreted virulence factors. Antibodies are also required for the processes of opsonisation, activation of the classical pathway of complement activation and enhancement of phagocytosis. Antibodies are important in host antibacterial defences and thus it would be expected that bacteria would have evolved mechanisms to avoid the actions of these multifunctional host proteins. The actions of anti-IgA defence systems at mucosal surfaces have already been described. In Section 10.5.1, two additional antibody-evading mechanisms will be described. These are the role of IgG/IgM binding proteins and the capacity that bacteria have to alter their surface antigens so as to confuse immune recognition systems.

10.5.1.1 Bacterial immunoglobulin-binding proteins

Bacteria have evolved a number of genes encoding proteins with the ability to bind to host antibodies (Table 10.10). The best known of these proteins is protein A, produced by *Staphylococcus aureus*.

- Protein A is a cell-wall-anchored protein with the ability to bind to the Fc domain of the IgG molecule. This results in the bacterium being coated in antibody (Figure 10.17). The antibody, being bound via its Fc region, is unable to interact with the Fc receptors on phagocytes. This helps to inhibit the phagocytosis of *Staph. aureus*.

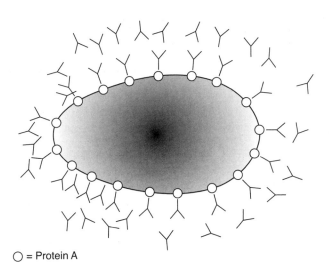

○ = Protein A

Knocking out the gene encoding protein A is associated with a decrease in virulence. A second staphylococcal IgG-binding protein has recently been discovered. This protein, termed Sbi, binds both to IgG and also to β_2-glycoprotein I (also known as apolipoprotein H).

- Protein G is a surface-bound protein, produced by streptococci, that binds to the Fc region of IgG. Protein H is another streptococcal surface IgG-binding protein. Proteins G and H appear to interfere with antibody-mediated complement activation at the bacterial surface. This obviously plays a role in immune evasion.

- Protein L is a multidomain cell wall protein of *Peptostreptococcus magnus*. There is strain variation in the structure of this protein and it can contain up to five repeating immunoglobulin-binding domains. Unlike proteins A and G, this bacterial protein binds to the framework region of the variable domain of the κ light chains. It is reported that such binding does not interfere with the binding of antibodies to their antigens. It is not clear what role protein L plays in bacterial virulence but the presence of this protein in strains of *Pep. magnus* has been correlated with the virulence of the bacterium.

- Protein P is another reported immunoglobulin-binding protein for which little information is available. A growing number of bacteria are being found to encode proteins that bind to components of the extracellular matrix such as the multivalent protein fibronectin. The fibronectin-binding protein, SfbI, of *Strep. pyogenes* has been shown to also bind to the Fc region of human IgG and to be involved in the evasion of antibody/complement-mediated attack.

10.5.1.2 Antigenic variation

Antibodies generally recognise molecules present on the surfaces of bacteria. It is the binding to these molecules that opsonises the bacteria. As the bacterial cell surface molecules are antigenic, an obvious strategy to evade the effects of antibodies towards these molecules would be to change them. This is what is now known to happen to a growing number of bacterial surface components. This phenomenon is known as antigenic variation and is related, in certain aspects, to the process of phase variation described in Section 2.6.2.

The best-studied example of antigenic variation is found in the African trypanosomes, responsible for sleeping sickness. This is a protozoal disease in which the infectious organism has a glycoprotein coat composed of a molecule termed variant surface glycoprotein (VSG). This protein is immunogenic and sufferers raise an effective antibody response to it, which leads to the opsonisation of the cells by antibody and complement, and killing of the parasite by phagocytes. Unfortunately, the trypanosome has more than 1000 VSG genes. Only one VSG gene is expressed by the majority of the population of trypanosomes at any one time, although a small percentage of the population will express other genes. One of these 'variant gene'-expressing populations escapes the initial antibody-mediated killing, begins to proliferate and produces a new wave of parasitaemia. This disease therefore has a cyclic pathology with wave after wave of new VSG variants appearing in the patient's bloodstream.

A number of examples of antigenic variation are seen in bacteria. The M protein of *Strep. pyogenes* has already been described and its hypervariable N-terminus has been highlighted. Over 100 different M protein serotypes have been recognised and it is believed that this is due to reiterated regions in the encoding DNA that could lead to homologous recombination across the repeated sequences. The details of this mechanism have not yet been elucidated.

A well-understood example of antigenic variation in bacteria is the variation in the pili of *Neisseria gonorrhoeae*. This organism, which causes gonorrhoea, uses pili to attach to host cells. These pili are among the immunodominant proteins of the bacterium and thus should be neutralised by host antibodies. The reason that they are not, and that individuals cured of gonorrhoea can be reinfected, is because of antigenic variation. The pili are formed from the transcription of the *N. gonorrhoeae* pilin gene and the complex structure of the protein so formed is shown in Figure 10.18. It is the polymerisation of this 17 kDa pilin protein that produce the pilus. The pilin protein contains a constant N-terminal region and a hypervariable C-terminal region, which is immunodominant, and several semi-variable regions; known as minicassettes which vary in DNA sequence from one gene to another. *Neisseria gonorrhoeae* has only one copy of the complete pilin gene, known as *pilE*. However, in addition to this one complete gene, the bacterium also encodes 10 to 15 copies of variant-encoding genes for pilin (cf. the trypanosome above). These variant genes are truncated at the 5' end, do not have the sequences encoding the N-terminal part of the pilin and are lacking the transcriptional promoter elements (Figure 10.19). These copies of the pilin gene are known as the *pilS* locus, the S denoting that the locus is 'silent' or non-functional. Variation in the

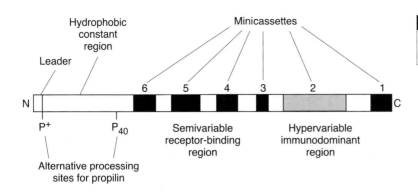

Figure 10.18 Structure of the pilin protein of *Neisseria gonorrhoeae*.

pilin genes involves the recombination of small stretches of the *pilS* gene sequences into the *pilE* expression locus, thus replacing the same minicassette DNA sequence. This process can give rise to an extremely large number of antigenic variants and presumably accounts for the chronicity of gonorrhoea.

Other examples of antigenic variation in bacteria are provided in Table 10.11.

Protein A – more than an IgG-binding protein

Staphylococcal protein A (SpA) is a well-known product of *Staph. aureus* and is much used by industry for purifying antibodies. It is also a well-established immune evasion factor owing to its ability to bind to IgG and thus coat the bacterium in immunoglobulin molecules, thereby rendering it 'invisible' to the immune system. However, in recent years SpA has been shown to be a representative of a new class of antigen, the B cell superantigens (SAgs). These SAgs, unlike conventional antigens, bind to the Fab regions of immunoglobulin molecules outside their complementarity-determining regions. In addition, B cell SAgs can react with a substantial amount of a host's serum immunoglobulins by virtue of their ability to interact with many members of an entire variable heavy (V_H) chain or variable light (V_L) chain gene family. For example, SpA reacts with the Fabs of most human immunoglobulins using heavy chains from the V_H3 gene family (V_H3^+). What then are the possible consequences of this activity of the SAgs?

The immune system is a continuum between innate systems that are immediately responsive and acquired systems that take some time to act. Part of this continuum is the population of B lymphocytes found in the body cavities and termed B-1 lymphocytes. These cells produce clonally restricted IgM antibodies that recognise bacterial components, particularly carbohydrates and lipids. These are called natural antibodies and serve as a pre-formed (innate) defence mechanism against bacteria. It has now been found that treatment of human neonates or adults with SpA induces a T cell-independent deletion of a large supraclonal set of susceptible B cells that includes clan III/V_H S107 family-expressing lymphocytes. In studies of different SpA forms, the magnitude of the induced deletion directly correlated with the V_H-specific binding affinity/avidity. When exposure of animals to SpA ceased, the conventional, splenic (B-2 subset) lymphocytes recovered their normal numbers. In contrast, the V_H family-restricted decrease in peritoneal B-1 cells was found to persist. SpA treatment also induced a persistent loss of splenic S107-mu transcripts, with a loss of certain natural antibodies and specific tolerance to phosphorylcholine immunogens that normally recruit protective antimicrobial responses dominated by the S107-expressing B-1 clone T15. Thus it appears that SpA, a B cell superantigen, has evolved to exploit an 'Achilles heel' in the immune system, for which B-1 cells are especially sensitive.

10.5.2 Superantigens and evasion of T lymphocyte responses

As described in Section 6.3.4, TCR is composed of two polypeptide chains (termed either α and β or γ and δ) and each polypeptide chain, like the immunoglobulin molecule, has a variable (V_α/V_β) and constant (C_α/C_β) domain. In the human genome there are 25 genes encoding the V_β chains of the TCR. Certain bacteria synthesise proteins which recognise both the V_β domains of the TCR and the MHC class II proteins, which form the complexes with antigenic peptides recognised by the TCR. These bacterial proteins are called superantigens and have become the subject of intense study because of their ability to cause human pathology. The best known superantigen is toxic shock syndrome toxin (TSST) 1, which came to worldwide attention in the late 1970s as the causative agent of a new condition, toxic shock, that resembles Gram-negative septic shock. These superantigens are now known to cause a large amount of human pathology. However, it is likely that superantigens, which are increasingly recognised as being produced by a growing number of bacteria (Table 10.12), have evolved to evade the acquired immune response and that tissue pathology is only a side effect of this process.

T lymphocytes can be activated by three distinct classes of molecules

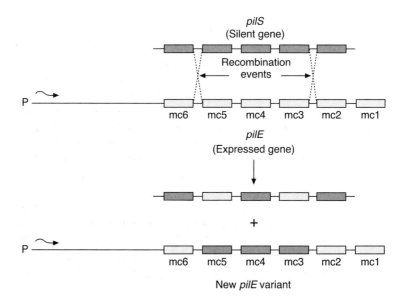

Figure 10.19 The mechanism of antigenic variation in *Neisseria gonorrhoeae*. The bacterium contains the gene *pilE* and up to 20 copies of *pilS* or so-called silent genes. These are partial genes lacking the conserved N-terminal sequence but containing minicassettes (mc) of variant sequences. Bacterial pilus variation involves a transfer of minicassette DNA from a silent locus to *pilE*, where a homologous recombination event replaces the existing *pilE* minicassette DNA by material from one of the silent loci. P, promoter.

Table 10.11.	Examples of bacterial antigenic variation	
Bacterium	Gene	Protein
Neisseria gonorrhoeae	*pilE/pilS*	Pilin
Streptococcus pyogenes	*emm*	M protein
Borrelia hermsii	*vmp* genes	Membrane proteins

Table 10.12.	Bacterial superantigens	
Bacterium	Superantigen	Human V_β specificity[a]
Staphylococcus aureus	SEA	1.1, 5.3, 6.3, 6.4, 6.9, 7.3, 7.4, 9.1
	SEB	3, 12, 14, 15, 17, 20
	SEC1	12
	SEC2	12, 13.2, 14, 15, 17, 20
	SEC3	5, 12, 13.2
	SED	5, 12
	SEE	5.1, 6.1–6.4
	TSST-1	2
	Exfoliative cytotoxin	2
Streptococcus pyogenes	SPE-A	2, 12, 14, 15
	SPE-C	1, 2, 5.1, 19
Mycoplasma arthritidis	MAM	3, 5.1, 7.1, 8.2, 10, 12.2, 16, 17.1, 20
Yersinia pseudotuberculosis	?	3, 9, 13

SE, staphylococcal enterotoxin; TSST-1, toxic shock syndrome toxin 1; SPE, streptococcal pyrogenic exotoxin; MAM, *Mycoplasma arthritidis* mitogen; ?, unknown.

[a] The numbers represent specific V_β chains that are recognised by the particular antigen.

(Table 10.13). Antigens, for example protein components of viruses, bacteria and protozoa, are normally the major stimulants of T cells. Each antigen will be recognised by only a tiny fraction of the T cells circulating in the body at any one time. As described in Section 6.7.2.2.2, protein antigens are proteolytically processed within APCs and peptides derived from the antigens are expressed on the surface of these cells in a complex with an MHC class II protein. It is this complex of peptide and MHC class II that is recognised by cognate T cells and the process of antigen presentation is therefore termed MHC-restricted (Figure 10.20). In contrast, there are certain molecules, such as **phytohaemagglutinin** (PHA) and **con**canavalin A (ConA), which can activate the majority of T cells. These molecules are termed mitogens and they have been used extensively by immunologists as probes of T lymphocyte function. Activation of T cells by such mitogens is not MHC restricted. Superantigens interact with both the MHC class II molecule on the surface of the APC and selected V_β domains of the TCR. This

Table 10.13. | Superantigens: comparison with mitogens and antigens

Feature	Antigen	Superantigen	Mitogen
Responsive T cells (%)	10^{-4}–10^{-6}	5–20	80–90
Requirement for APC	Yes	Yes	Yes
Requirement for MHC class II	Yes	Generally	No
Antigen processing required	Yes	No	No
Restricted V_β usage	Possible	Yes	No

APC, antigen-presenting cell.

process activates both the APCs and the T lymphocytes (Figure 10.20). As stated above, there are 25 genes encoding V_β domains. Thus, on average, each V_β domain is expressed on 4–5% of the circulating T lymphocyte pool. Superantigens, depending on how many V_β domains they recognise, can therefore activate from 4–5% to >50% of the T cell repertoire of the human body. Table 10.12 shows that the *Staph. aureus* exotoxin staphylococcal enterotoxin A (SEA) can bind to eight different V_β domains, and can thus theoretically activate 30–40% of the T cells of a human. As staphylococci can make a large number of these superantigenic exotoxins, a large fraction of the T cell pool could become activated if it had access to these superantigens. Although these bacterial superantigens bind to MHC class II proteins, the activation of the T lymphocytes is not MHC restricted in the classical sense (as described in Section 6.7.2.1) and most MHC class II alleles, and even xenogenic MHC class II proteins, can present superantigens to a given T lymphocyte.

What is the consequence to the acquired immune system of the introduction of a bacterial superantigen? When a T cell meets its

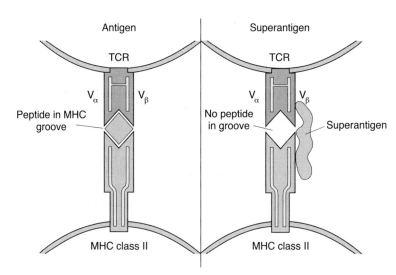

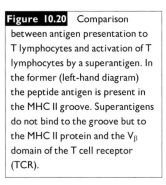

Figure 10.20 Comparison between antigen presentation to T lymphocytes and activation of T lymphocytes by a superantigen. In the former (left-hand diagram) the peptide antigen is present in the MHC II groove. Superantigens do not bind to the groove but to the MHC II protein and the V_β domain of the T cell receptor (TCR).

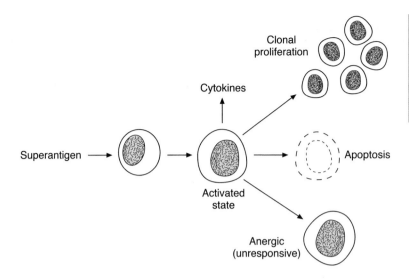

Figure 10.21 Fate of T lymphocytes stimulated by a superantigen. This can be: (i) proliferation of the multiple clones of cells, (ii) apoptosis or (iii) induction of the non-responsive state known as anergy.

cognate antigen it is stimulated into the process known as clonal proliferation. This results in an increase in the numbers of this particular antigen-specific T lymphocyte. This process, as described in Section 6.7.2.2.5 and Figure 6.30, requires the antigenic peptide, the MHC class II molecule, TCR and additional co-stimulatory molecules such as CD80 and CD86. When a T lymphocyte is stimulated by a superantigen, it can undergo one of three fates (Figure 10.21). The first is that the T cells become stimulated. It is this initial stimulation of T cells and APCs by certain superantigens, such as TSST-1, that results in tissue pathology. Following stimulation, these activated T cells can enter a state of chronic unresponsiveness known as T cell anergy. This is the second fate awaiting a superantigen-stimulated T cell. The third path the T lymphocyte can travel following encounter with a bacterial superantigen is apoptosis (elimination). The end result of these three pathways tends to be the loss of (or loss of function of) large numbers of the T cells that are encountering the bacterial infection.

The consequences of producing superantigens for bacteria is not absolutely clear. Staphylococci and streptococci produce large numbers of superantigens, and much of the toxicity that is produced by these bacteria can be due to such toxins. The ability to evade a substantial proportion of the antigen-reactive T lymphocytes of the host is an obvious advantage for the bacterium. However, it could be assumed that, if this superantigenic behaviour were to be a general activity of these bacteria, it would lead to some form of chronic disease. In fact, most staphylococcal and streptococcal infections are not chronic. However, as described in Section 11.5.3, there is the possibility that the production of bacterial superantigens may be associated with chronic idiopathic diseases such as psoriasis and arthritis.

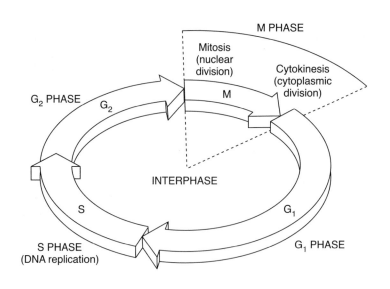

Figure 10.22 The conventional way in which the cell cycle is represented. Following mitosis and cytokinesis (cell division), the only visible signs of the cell cycle, daughter cells enter G_1. The duration in this phase of the cell cycle depends on the presence of growth factors in the cell's milieu. In the presence of growth factors, cells transit through G_1 and enter the S phase, at which time DNA and other components are replicated. At the end of the S phase, cells enter a checking phase called G_2. If the quality control checks are satisfactory, cells will then enter mitosis.

10.6 Bacterial control of the cell cycle and apoptosis as evasion mechanisms

The human body contains cells that are constantly dividing (e.g. epithelial cells in the gut, leukocyte precursor cell populations in the bone marrow), cells that can be stimulated to divide (e.g. lymphocytes) and cells that never divide (neurones). Inhibition of cell division or induction of the process of apoptosis (by which cells undergo a programmed set of responses leading to death) can be used to evade host defence responses and these will be described in turn.

10.6.1 Bacterial inhibition of cell cycle progression

10.6.1.1 The eukaryotic cell cycle

The cell cycle is the term given to describe the complex cellular processes by which cells replicate their cellular DNA and cytoplasmic contents and then partition them into two daughter cells (Figure 10.22). The process of mitosis and cytokinesis are the only visible signs of the cell cycle and these are, in fact, the shortest part of this process. Following mitosis, a cell enters a quiescent period termed G_1. The letter G simply stands for 'gap'. The duration of G_1 depends on the presence of appropriate growth factors in the cell's external environment. If such factors are absent, the G_1 period can be very long indeed and, in this state, the cell is said to be in G_0. If growth factors are produced, the cell passes through a restriction, or checkpoint, some 14 to 16 hours later and becomes committed to enter the S (synthesis) phase. In the S phase, the cell replicates its nuclear DNA. Once the DNA has been replicated, cells go through another checkpoint to enter the G_2 phase which can be thought of as a period in which quality control procedures are instituted in order to check the quality of the newly produced DNA.

If the quality of the DNA is satisfactory, the cell can go through the third checkpoint in the cell cycle and enter mitosis. It is in mitosis that the chromosomes condense, the DNA is partitioned and this is followed by partitioning of the cell cytoplasm to form two viable daughter cells.

The reader can envisage that the cell cycle is a complex process and will be carefully controlled. Indeed, failure to control the cell cycle is, ultimately, the cause of cancer. Over the past 30 years it has been established that three groups of proteins control the cell cycle. These are (i) the cyclins, (ii) the cyclin-dependent kinases (Cdks) and (iii) the cyclin-dependent kinase inhibitors (CKIs) (Figure 10.23). In mammalian cells, a small number of related Cdks (numbered on the basis of their discovery – Cdk1, 2, 4, 6) control cell cycle progression. Mammalian cells also express multiple cyclins – A and B function in S phase, G_2 and early mitosis. Cyclins D and E are expressed in G_1. Cyclin kinase inhibitors are recently discovered proteins that fall into two families. The Cdk-inhibitory protein (CIP) family bind, and inhibit, all Cdk1, 2, 4 and 6 cyclin complexes. The inhibitors of the kinase 4 family inhibit Cdk4 cyclin D and Cdk6 cyclin D complexes.

The discussion in the following sections will concentrate on bacterial inhibitors that block cell cycle progression in G_2. The principal Cdk involved in the passage of cells through G_2 is Cdk1 (often termed Cdc2 in older literature), which associates with cyclins A and B. In turn, the activity of the Cdk1–cyclin complexes is regulated by proteins homologous to the yeast kinases Wee1 and CAK (Cdc2-activating kinase) and the protein phosphatase Cdc25 (Figure 10.24).

10.6.1.2 Bacterial proteins with the ability to control cell cycle progression

It was only during the 1980s that the first tentative reports appeared suggesting that bacteria could produce molecules able to interfere with cell cycle progression. Only a few of the cell cycle-modulating proteins that bacteria produce have been identified and their genes cloned and they fall into two categories: (i) inhibitors of cell cycle progression and (ii) stimulators of cell proliferation. Our current understanding of the bacterial cell cycle-modulating proteins is presented in Table 10.14. The most prevalent of the bacterial cell cycle inhibitors is cytolethal distending toxin (CDT), which is now found to be produced by a growing number of Gram-negative bacteria. The mechanism of action of this toxin will be described below. CDT has been shown to inhibit T lymphocyte cell cycle progression and so may act as an immunosuppressant. Other proteins that block cell cycle progression (and most work by blocking cells at the G_2/mitosis transition) could also have immunosuppressant actions if released in the presence of lymphocytes. Indeed, *Sal. typhimurium* produces STI (*Sal. typhimurium* inhibitor of T cell proliferation), a protein discovered because it inhibited T cell proliferation.

A number of bacteria produce proteins that can induce cell proliferation. The most interesting is *Pasteurella multocida* toxin (PMT), which has been shown to be the most potent mitogen for the Swiss 3T3 mouse

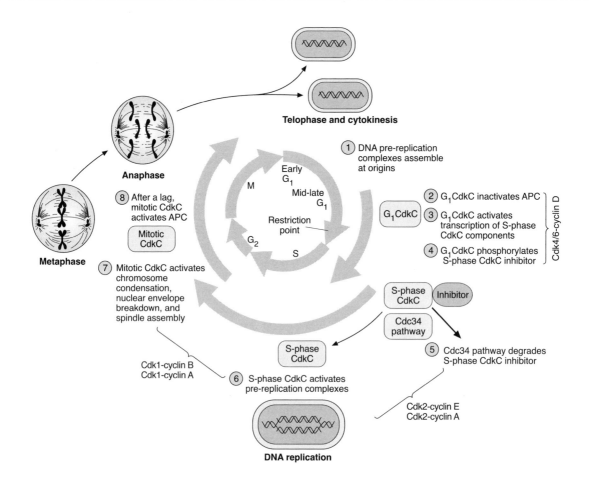

Telophase and cytokinesis

Anaphase

(8) After a lag, mitotic CdkC activates APC

Mitotic CdkC

Metaphase

(7) Mitotic CdkC activates chromosome condensation, nuclear envelope breakdown, and spindle assembly

M

Early G₁

Mid-late G₁

Restriction point

G₂

S

Cdk1-cyclin B
Cdk1-cyclin A

(6) S-phase CdkC activates pre-replication complexes

S-phase CdkC

(1) DNA pre-replication complexes assemble at origins

G₁CdkC

(2) G₁CdkC inactivates APC

(3) G₁CdkC activates transcription of S-phase CdkC components

(4) G₁CdkC phosphorylates S-phase CdkC inhibitor

Cdk4/6-cyclin D

S-phase CdkC | Inhibitor

Cdc34 pathway

(5) Cdc34 pathway degrades S-phase CdkC inhibitor

Cdk2-cyclin E
Cdk2-cyclin A

DNA replication

Figure 10.23 Schematic diagram outlining the regulation of the eukaryotic cell cycle. Traversing the cycle depends on the interaction of a range of proteins, primarily cyclin-dependent kinase complexes (CdkCs). Such complexes consist of a regulatory cyclin (e.g. cyclin A, B, E, etc.) and a catalytic cyclin-dependent kinase (e.g. Cdk1) subunit. The various Cdk–cyclin complexes involved in the various parts of the cell cycle are shown in the diagram. Cyclin kinase inhibitors (CKIs) are also involved in control of the cell cycle. APC, anaphase-promoting complex; Cdc, cell division cycle; for other abbreviations, see the text.

cell line much used in mitogen research. Surprisingly, this protein appears to be able to dysregulate the myeloid and mesenchymal cell differentiation pathways that are important in normal bone remodelling. In *in vitro* assays of bone resorption, PMT is one of the most potent inducers of bone destruction. Thus it comes as no surprise to learn that the major symptom associated with *P. multocida* infection, which occurs in pigs and other domestic farm animals, is atrophic rhinitis, a condition involving destruction of the bones of the snout. This leads to collapse of the snout and problems in breathing, etc.

The CDTs, first discovered in *E. coli*, are a family of protein toxins now known to be produced by a range of bacteria including *Shigella* spp., *Campylobacter* spp, *H. ducreyi* and *Actinobacillus actinomycetemcomitans* (see Table 10.14). These toxins initially came to light by the observations that culture supernatants from *E. coli* 0128 caused Chinese hamster ovary (CHO) cells to become distended and then, over a prolonged period, to die. Analysis of the DNA of cells exposed to this toxic material revealed that cells contained twice the normal DNA content, suggesting that cells were blocked either in G₂ or early M phase. The effect of CDT on HeLa cells is shown in Figure 10.25.

Initial genetic analysis of CDT production by *E. coli* strain 6468/62

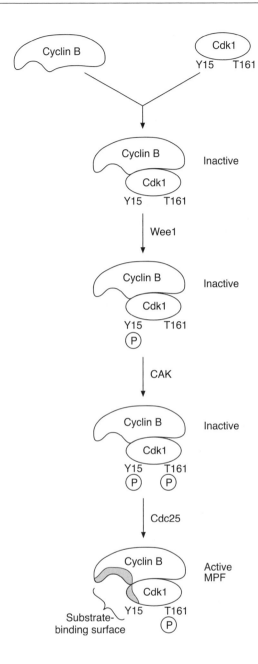

Figure 10.24 Control of the activation of the Cdk1 cyclin B complex, which controls the transit of cells from G_2 into mitosis. The example given here is for the yeast enzymes about which most is known. The inactive complex is first phosphorylated by Wee1 on tyrosine (Y) at position 15 and then by Cdc2-activating kinase (CAK) at threonine (T), position 161. The phosphate at position 15 then has to be removed by the tyrosine phosphatase, Cdc25, to produce the active complex. MPF, maturation-promoting factor.

(O86 : H34) revealed the presence of three contiguous genes, *cdtA*, *cdtB*, *cdtC*, which coded for proteins of molecular mass 26, 30 and 20 kDa, respectively. A surprising finding was that the *cdt* genes from *E. coli* strain 9142-88 (O128 : H⁻) showed significant sequence divergence, with the CdtA proteins being only 40% identical between these two strains. Comparison between the identities of the three proteins in *E. coli* (O86 : H34) and the oral bacterium *A. actinomycetemcomitans* reveals 34% (CdtA), 50% (CdtB) and 27% (CdtC) identity. It is not yet clear whether all three proteins are required for activity. Most workers claim that the active component is CdtC but that this must be

Table 10.14. Cell cycle-modulating bacterial proteins

Bacterium	Protein	Mechanism
Escherichia coli	CDT	Blocks cells in G_2 by inhibition of Cdk I /cyclin B complex / nuclease activity
Campylobacter spp.	CDT	As above
Haemophilus ducreyi	CDT	As above
Shigella spp.	CDT	As above
Actinobacillus actinomycetemcomitans	CDT	As above
Helicobacter hepaticus	CDT	As above
Escherichia coli	CNF	Deamidates Gln-63 in Rho and blocks cytokinesis
Bordetella bronchiseptica	DNT	As above
Actinobacillus actinomycetemcomitans	Gapstatin	Inhibits cyclin B synthesis and blocks cells in G_2
Actinobacillus actinomycetemcomitans	80 kDa protein	Blocks cells in G_2
Pneumocystis carinii	Unknown	Blocks cells in G_2 by blocking CdkI–cyclin B complex
Fusobacterium nucleatum	95 kDa protein	Inhibits G_1 cyclin production/blocks cell in G_1
Helicobacter pylori	100 kDa protein	Reversible inhibition of cell proliferation
Campylobacter rectus	48 kDa protein	Inhibits cell division
Prevotella intermedia	50 kDa protein	Inhibits cell division
Salmonella typhimurium	STI	Inhibits T cell proliferation
Vibrio cholerae	CT	Stimulates lymphocyte proliferation
Porphyromonas gingivalis	FAF	Stimulates fibroblast proliferation
Pasteurella multocida	PMT	Most potent mitogen reported
Bartonella henselae	Protein	Endothelial cell mitogen

CDT, cytolethal distending toxin; CNF, cytotoxic necrotising factor; DNT, dermonecrotic toxin; STI, *Salmonella typhimurium* inhibitor of T cell proliferation; FAF, fibroblast-activating factor; PMT, *Pasteurella multocida* toxin.

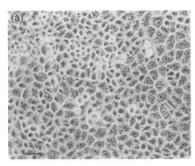
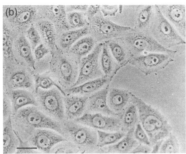

Figure 10.25 The effect of CDT on HeLa cells. In (a) we have normal HeLa cells photographed 72 hours after adding control supernatants. In (b), cells have been exposed to *Campylobacter jejuni* CDT for 72 hours. In the presence of the toxin the cells are grossly distended and some show nuclei that are starting to fragment. Bar = 36 μm. (Reprinted from *Trends in Microbiology*, **7**, Pickett, C. L. & Whitehouse, C. A. The cytolethal distending toxin family, 292–297. Copyright 1999, with permission from Elsevier Science.)

'activated' by interactions with the other two proteins. However, other groups have reported that CdtB is active alone.

What is the mechanism of action of CDT? The first studies to concentrate on the mechanism of action concluded that this toxin inhibited the Cdk1–cyclin B complex by preventing the dephosphorylation of the complex at Tyr-15. This dephosphorylation is carried out by Cdc25, a tryosine phosphatase. It has subsequently been shown that overexpression of Cdc25 will overcome the G_2 block induced by CDT. In a more recent study, it has been reported that CdtB has deoxyribonuclease activity and that this causes breakdown of DNA. As described above, G_2 is a phase of the cell cycle where the cellular DNA undergoes quality control checks. If there is DNA damage, cells remain in G2. Thus CdtB may be a unique DNA-cleaving cell cycle-inhibiting toxin.

Coming back to the subject of immune evasion, it is suggested that CDT can act to inhibit lymphocyte proliferation. Experiments have shown that the CdtB from *A. actinomycetemcomitans* can arrest the cell cycle progression of human CD4 and CD8 T lymphocytes and that these cells are significantly more sensitive to this toxin than are HeLa cells.

The study of bacterial cell cycle-modulating proteins is just in its infancy and it is a safe prediction that many unexpected findings will be made during the next decade. This prediction is based on the enormous advances in our understanding of how bacteria can control another vital cell mechanism – the process known as apoptosis.

10.6.2 Bacterial control of apoptosis

As long-lived creatures with memories, we think of death in wholly negative terms. However, cell death is a vital part of life in a multicellular organism. To give an example, during embryonic development, the formation of tissues such as the fingers and toes requires that certain cells die to allow the digits to form. The working of the immune system requires that many of the putative T cells entering the thymus die. In these examples, the process by which cells die is termed apoptosis. An alternative name often used by immunologists is programmed **c**ell **d**eath (PCD). This process allows the body to remove cells without inducing any untoward responses. The alternative to apoptosis is cellular necrosis. Necrosis is a process that can induce local pathology. Apoptosis was only discovered in the 1970s and the mechanisms of apoptosis are still the subject of active study. Perhaps one of the most unexpected findings in recent years is that CD14, a receptor introduced in this, and earlier, chapters as the LPS receptor, is also involved in allowing macrophages to recognise apoptotic cells. In this context, binding to the CD14 receptor does not induce inflammation.

It has become clear that pathogenic microorganisms target the mechanisms of apoptosis either to stimulate the process or to inhibit it. For example, viruses live within cells and the natural response of virally infected cells is to apoptose. To evade this particular host defence, viruses have evolved mechanisms to block apoptosis. Bacteria were initially found to promote the apoptotic process. Now it is clear that bacteria can both stimulate or inhibit apoptosis as part of evasion strategies to defeat host protective defences.

10.6.2.1 The mechanism of apoptosis

Before discussing the ability of bacteria to manipulate apoptosis, the process needs to be described. The description will be abbreviated and the reader is referred to Section 10.11 for more details.

Apoptosis, from the Greek for dropping/falling off (as in leaves on a tree), is regulated in one of two ways. It appears that most, if not all, cells in multicellular organisms are programmed to die and are kept alive by particular factors called 'trophic factors'. In the absence of such factors, cells die. The second method of regulation occurs in developmental and immune contexts in which specific signals induce apoptosis. Whether induced directly by 'murder signals' (via so-called death receptors) or by lack of survival signals, the process of apoptosis appears to have a final common pathway involving: (i) a group of specialised proteases called caspases; (ii) a family of proteins that control mitochondrial permeability; and (iii) certain mitochondrial proteins, most notably the electron-transport chain component beloved of all students of biochemistry, cytochrome c. These individual components will be described in turn, and then their interactions with each other, and with the cell surface death receptors (members of the TNF/nerve growth factor family, e.g. Fas/APO-1/CD95, TNF-RI), to cause apoptosis will be reviewed.

A number of enzymes (known as caspases) are involved in apoptosis. The term caspase arises from the enzymic specificity of these proteins. They are cysteine proteases with an absolute specificity for **asp**artic acid in the P1 position and contain a conserved Gln-Ala-Cys-X-Gly (where X is any amino acid residue) pentapeptide active site motif. They are produced as inactive pro-enzymes, comprising a pro-domain with one large and one small subunit; activation requires proteolytic processing to remove the regulatory (pro-) domain. At least 14 caspases have now been cloned. The first caspase to be discovered (caspase 1 or IL-1β-converting enzyme), and caspase 11, mainly function as cytokine-processing enzymes. Caspases 2, 3, 6, 7, 8, 9 and 10 are involved in the apoptotic process. The roles of the other caspases remain to be deter-

Table 10.15.	Mammalian caspases	
Caspase	Synonym	Major phenotype of knockout
1	ICE	Resistance to LPS, IL-1β production inhibited
2	Nedd2, ICH1	Female germ cell hyperplasia
3	CPP32, YAMA	Neuronal hyperplasia, thymocytes resistant to pro-apoptotic agents
4	TX, ICH2, ICErII	?
5	TY, ICErIII	?
6	MCH2	?
7	MCH3, CMH	?
8	FLICE, MACH	Knockout is lethal
9	MCH6, ICELAP6	Similar to caspase 3 knockout
10	FLICE2 MCH4	?
11		Similar to caspase 1 knockout
12		?
13	ERICE	?
14	MICE	?

?, unknown.

mined (Table 10.15). The functions of the caspases are to break down the major structural proteins of the cells, including the nuclear lamins and the cytoskeletal proteins in the cytoplasm.

Other molecules involved in controlling apoptosis include members of the Bcl-2 family of proteins. The first of these, Bcl-2, was identified as the product of a proto-oncogene in follicular B cell lymphomas in which it was overexpressed. It was then recognised as being the mammalian homologue of the *Caenorhabditis elegans* apoptosis repressor Ced-9. Thus this protein was identified as having the ability to repress apoptosis. The next member of this family of proteins to be identified was Bax (**Bcl-2** associated protein **X**), which is found to co-precipitate with Bcl-2 from the cytoplasm of cells. In contrast to Bcl-2, Bax promotes apoptosis. A large number of members of the Bcl-2 family have now been identified (Table 10.16) and they can have pro- or anti-apoptotic activity. It is curious to find a 'family' of proteins in which members can have two diametrically opposed activities. The three-dimensional structures of one pro-apoptotic (Bid) and one anti-apoptotic (Bcl-X_L) member of the Bcl-2 family have been elucidated. These proteins demonstrate structural homology to the pore-forming domains of certain bacterial toxins and in particular the colicins A and E1 and the neisserial pore-forming toxin PorB. Indeed, as will be described, these

Table 10.16.	Members of the Bc1-2 family of proteins
Anti-apoptotic	**Pro-apoptotic**
Bcl-2	Bax
Bcl-x_L	Bak
Bcl-w	Bok
Mcl-1	Bcl-X_S
A1	Bad
	Bid
	Bik/Nbk
	Bim
	Krk
	Mtd

proteins form channels in lipid bilayers. However, it appears that the characteristics of the channels formed by pro- or by anti-apoptotic proteins are different.

One of the most fascinating findings in Cell Biology in recent years is the role of the mitochondrion in the control of apoptosis. This organelle (which has evolved from an endosymbiotic bacterium) now appears to be charged with deciding the apoptotic fate of cells. Moreover, this fate appears to be controlled, in part, by proteins that look like bacterial toxins. One of the key processes in apoptosis is the release of cytochrome c from mitochondria through pores made by proapoptotic Bcl-2 family members. The cytochrome c forms a complex with apoptotic protease-activating factor (APAF-1) and pro-caspase 9 and activates this caspase.

Having introduced the major players in the apoptotic process, the remainder of Section 10.6.2.1 will try to explain the interactions between these various proteins. It should be realised that we do not yet have all the answers, and gaps still exist in our understanding. The reader has our apologies for the equally large number of acronyms to be introduced. This simply underlines the large number of interacting proteins involved in controlling apoptosis.

10.6.2.1.1 Apoptosis via cellular death receptors

One pathway that initiates apoptosis is caused by binding of the cognate ligand to one of the cell surface proteins that are known as death receptors (Table 10.17). The death receptors (receptors containing a death domain (DD) in their cytoplasmic tails) are members of the TNF/ nerve growth factor receptor superfamily (see Section 4.2.6.6.3 and

Table 10.17.	TNF superfamily members with death domains	
Receptor	Cell expression	Intracellular interacting proteins
TNFR-1 (CD120a)	Ubiquitous	TRADD
CD95 (Fas/APO-1)	Ubiquitous	FADD, FAP, Daxx
DR3	Spleen, thymus, PBL	TRADD
DR4	Most tissues	TRADD, FADD, RIP
DR5	Ubiquitous	TRADD, FADD, RIP
DR6	Ubiquitous	TRADD, FADD, RIP

FAP, Fas-associated phosphate; Daxx, death-associated protein; for other abbreviations, see the text.

Table 4.21) and include TNFR1, CD95 (alternative names: Fas/APO), DR3 (death receptor 3, also called APO-3/WSL-1/TRAMP) and the TRAIL (TNF-related apoptosis-inducing ligand) receptors DR4 (TRAIL-R1), DR5 (TRAIL-R2/KILLER/APO-2) and DR6. Not all TNF/NGF receptor family members contain death domains. For example, the second TNF receptor, TNF-RII, lacks a DD. The best studied of these death receptors is CD95 (Fas). When the Fas (CD95) receptor binds its ligand, FasL, the receptor undergoes trimerisation, and the cytoplasmic domain of the receptor, which contains a DD, interacts with and binds a DD-containing adaptor molecule designated FADD (Fas-associated death domain) (Figure 10.26). This system has also been described in Section 4.2.6.3.3. FADD and Fas interact by homotypic interactions of their DDs. Other proteins containing homologous death domains have been identified including TRADD (TNFR-associated death domain), RIP (receptor-interacting protein), RAIDD (RIP-associated ICH-1/CED-3-homologous protein with a death domain) and MADD (MAP kinase-activating death domain) (see Table 10.17). In addition to the death domain, FADD also contains a separate N-terminal domain known as a death effector domain (DED) which can interact with the DED of pro-caspase 8 and probably of pro-caspase 10. The binding of FADD and then pro-caspase 8 to the trimeric Fas results in the formation of what is termed the death-inducing signalling complex (DISC) and the activation of pro-caspase 8 to form enzymically active caspase 8. Caspase 8 can activate all known pro-caspases *in vitro*, and so is a prime candidate for being the initiator caspase in many forms of death-receptor-induced apoptosis. Signalling via the TNF receptor (CD 120a) recruits not FADD, but TRADD, which then binds to FADD and, as described, activates pro-caspase 8 (Figure 10.26).

Caspase 8 can also proteolytically activate one of the pro-apoptotic Bcl-2 proteins – Bid. The product is a 15 kDa C-terminal fragment (tBid) which interacts with Bax and results in it interacting with the mitochondrial membrane to release cytochrome *c* (see Section 10.6.2.12).

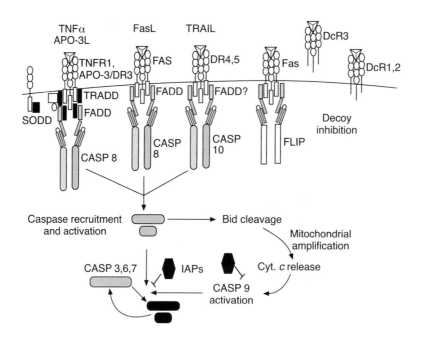

Figure 10.26 Caspase activation via cell surface death receptors. This diagram shows a number of the potential death receptors (TNFR-1, APO-3/DR3, FAS/Fas, DRs) on the surface of a leukocyte. Binding of a ligand to these receptors results in the association of three of the receptor proteins to form a trimeric complex. This results in the recruitment of proteins such as FADD or TRADD, which contain death domains (DD) that allow them to bind to the activated receptor. These DD-containing proteins have the ability to interact with pro-caspases such as pro-caspase-8 (CASP 8) or CASP 10. Binding of pro-caspase to the DD-containing proteins results in the autoproteolytic cleavage to form the active caspase. The caspases can then undergo recruitment and cause the cleavage of BID, which enables signals from the cell surface death receptors to interact with the mitochondria. The role of BID in controlling the mitochondrial mechanisms of the apoptotic process is detailed in Section 10.6.2.1.2 and Figure 10.27.

It is obviously important to be able to control the apoptotic process induced by death receptors. A number of mechanisms exist. These include the soluble forms of death receptors (decoy receptors, DcR) which bind the apoptosis-inducing ligand in the extracellular fluid and thus prevent it acting at the cell surface. The second is to prevent the recruitment of pro-caspases to the liganded receptor. Proteins known as FADD-like ICE inhibitory proteins (FLIPs) can bind to FADD but fail to recruit the pro-caspase and thus block subsequent caspase-mediated events. Another protein, silencer of death domains (SODD), is found in association with the DD of TNFR-1, preventing signalling through this receptor. Other points of regulation of apoptosis occur at the level of the mitochondrion. One key group of such control proteins are the inhibitors of apoptosis (IAPs), which appear to act at the level of the activation or action of certain caspases such as caspase 9. (Reproduced from Budihardjo *et al.*, 1999 (see Section 10.11, Review articles) Copyright 1999 by Annual Reviews.)

As would be expected with so important a signalling pathway, a large number of safety circuits have evolved to control death receptor-mediated apoptosis. Without going into detail, there are three basic inhibitory mechanisms: (i) prevention of pro-caspase recruitment or activation at the DISC, (ii) decoy receptors for TRAIL and (iii) direct inhibition of proteolytic activation of pro-caspases or inhibition of specific caspases. This is the function of a group of proteins known as IAPs (**i**nhibitors of **ap**optosis). These proteins have been discovered in baculoviruses as well as in humans. Another host inhibitor has been discovered and termed **s**ilencer **o**f **d**eath **d**omains (SODD). In unstimulated cells, SODD is associated with the death domain of TNFR-1 preventing the spontaneous signalling by this receptor and the initiation of DISC formation (Figure 10.26).

10.6.2.1.2 The mitochondrial pathway to apoptosis

At the time of writing, it is only four to five years since the discovery that mitochondria are involved in controlling apoptosis. The original discovery was made using cell-free extracts where it was found that adding mitochondria resulted in changes in nuclear chromatin characteristic of apoptosis. The current working model to explain how mitochondria control apoptosis is based on the finding that cytochrome c released from these organelles can form a death-inducing signal complex (DISC – see similar complex described earlier) by complexing with APAF-1 – a 130 kDa protein with three distinct domains – and with pro-caspase 9. In the presence of ATP or dATP, the complex activates pro-caspase 9 and the active caspase 9 can, in turn, activate pro-caspases 3 and 7 (Figure 10.27). The release of cytochrome c from the mitochondria is controlled by the actions of the Bcl-2 family, which contains proteins that are either pro-apoptotic or anti-apoptotic (Table 10.16). Cell stress alters the cytosolic binding of pro-apoptotic Bcl-2 family members such as Bax, Bad or Bid. Thus dephosporylation of Bad results in its release from the phosphoserine-binding protein 14-3-3 (see Section 4.2.4.2) and proteolytic cleavage of Bid by caspase 8 results in the translocation of these proteins onto the mitochondrial membrane (Figure 10.27). These proteins, as described above, have structures similar to bacterial pore-forming toxins and can interact with mitochondrial proteins involved in altering permeability such as the adenine nucleotide transporter, which is part of the permeability transition pore. This effect of Bax and activity-related proteins can be inhibited by Bcl-2. The exact mechanism by which Bax and Bcl-2, and the other Bcl-2 family members, control cytochrome c release from mitochondria still remains to be worked out in detail, but the simple description given should convince the reader of the role of the mitochondrion in cellular apoptosis and of the interaction between the two pathways of programmed cell death.

These recent discoveries indicate clearly that the mitochondria act to control apoptosis. Now, mitochondria have evolved from bacterial endosymbionts and thus the control of apoptosis in eukaryotic cells is actually under the control of what was formerly a bacterium. Over the past six years it has become clear that bacteria can exert control over the process of apoptosis and this will be described in the next few sections.

10.6.2.2 Bacterial control of the apoptotic process

Bacteria have two reasons to modulate the apoptotic process. The first, which involves activating apoptosis, is to eliminate leukocytes involved in host defence. The second, which involves blocking apoptosis, is to survive within cells. Before discussing the mechanisms used by bacteria

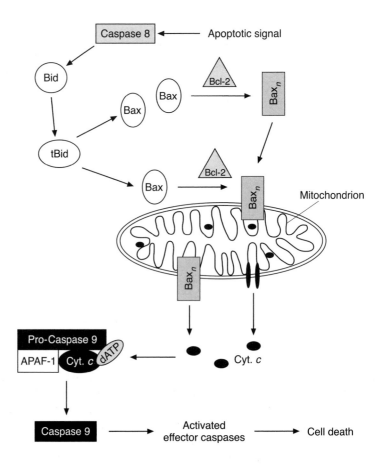

Figure 10.27 Role of mitochondria in apoptosis. Apoptotic signals can activate pro-caspase 8, which, in turn, can cleave the Bcl-2 family member Bid to form a 15 kDa C-terminal fragment (tBid). This cleavage product can induce a change in Bax conformation in the cytoplasm, leading to Bax translocation to the mitochondrion. Further conformation changes and oligomerisation result in Bax inserting into the mitochondrial membrane triggering cytochrome c release. The cytochrome c forms a complex with pro-caspase 9, APAF-1 and dATP, resulting in activation of the associated pro-caspase, which activates additional caspases, leading to cell death.

to induce or block apoptosis, it is worth looking briefly at the mechanisms employed by viruses to combat apoptosis.

Viruses have evolved mechanisms for blocking all the aspects of the apoptotic pathways (Table 10.18). The encoding of soluble forms of cytokine receptors such as the TNFR has been described in earlier chapters. These can block the interactions of ligands with death receptors. Adenovirus encodes a number of proteins that can induce the internalisation and degradation of DD-containing receptors such as the Fas receptor (CD95). A number of the herpesviruses encode proteins that interfere directly with FADD−caspase 8/10 interactions and several viral proteins have caspase-inhibitory actions. The importance of the Bcl-2 family in apoptosis control has been emphasised. Numerous viruses encode either anti-apoptotic Bcl-2 homologues or proteins that bind and inhibit the pro-apoptotic members of the Bcl-2 family. For details of these viral anti-apoptotic mechanisms the reader is referred to Tortorella *et al.* (2000) in Section 10.11, Review articles.

10.6.2.2.1 Bacterial induction of apoptosis

Bacteria are more concerned with inducing apoptosis of cells but it is only in very recent years that evidence has begun to emerge that

Table 10.18. Viral anti-apoptotic strategies

Cellular target	Virus	Gene	Anti-apoptotic mechanism
Death receptor	Many Adenovirus	Many E3-10.4/14.5K	Inhibition of binding of ligand to death receptor complex, which causes internalisation/destruction of Fas
DED-containing proteins	Many	e.g. K13	Prevent caspase activation via death receptors
Caspases	Cowpox Vaccinia Baculovirus	crmASP1-2 SPI-2/B13R2 p35	Inhibits caspase 1, 8 and granzyme B Homologous to CrmA Inhibits multiple caspases
Bcl-2	EBV EBV HHV-8	BHRF1 LMP-1 kcbcl-2	Bcl-2 homologue Up-regulates Bcl-2 Bcl-2 homologue
Cell cycle	HPV Adenovirus	E6 E1B 55K	Targets p53 for degradation Binds and inactivates p53

EBV, Epstein–Barr virus; HHV, human herpes virus; HPV, human papilloma virus.

bacteria may be involved in inhibiting this process. A wide range of proteins (toxins) have been demonstrated to be involved in the killing of cells and it is assumed that this is very often due to activating the apoptotic process (Table 10.19). However, the mechanisms by which bacteria induce apoptosis are often unclear. An obvious means of inducing apoptosis would be to engage the normal physiological mechanisms that cause apoptosis. The actions of bacterial superantigens are presumed to replicate the physiological processes by which T cells undergo apoptosis, say in the thymus. Thus it has been demonstrated that SEB, which is associated with the killing of specific V_β-expressing T lymphocytes, induces Fas expression on normal human T lymphocytes. *Helicobacter pylori* has been shown to induce the formation of Fas on peripheral blood mononuclear cells by an indirect mechanism involving the synthesis of pro-inflammatory cytokines such as TNFα and IL-1β.

A number of bacteria kill cells by the use of toxins active at the plasma (or other) membranes. Pore-forming toxins of bacteria such as *E. coli*, *Staph. aureus*, *A. actinomycetemcomitans* and *Lis. monocytogenes* are associated with the apoptotic death of cells. It is not clear whether apoptosis is due to activation of death receptors or alterations in mitochondrial permeability. The AB toxins of *Corynebacterium diphtheriae*, *E. coli*, *Ps. aeruginosa* and *Shigella dysenteriae* inhibit protein synthesis and are associated with apoptotic cell death. Again, the mechanisms associated with the death of these cells have not been established.

At the time of writing, the best-understood mechanism of bacterial apoptosis induction is that used by *Shig. flexneri*, an organism causing shigellosis or bacillary dysentery. The pathology associated with this

Table 10.19. Bacterial induction of apoptosis

Mechanism	Bacteria utilising mechanism
Pore-forming toxins	*Staphylococcus aureus, Listeria monocytogenes, Actinobacillus actinomycetemcomitans, Escherichia coli*
Toxins inhibiting protein synthesis	*Corynebacterium diphtheriae, Pseudomonas aeruginosa, Shigella* spp., *Escherichia coli*
Type III secretion	*Shigella* spp., *Salmonella typhimurium, Yersinia* spp.
Superantigens	*Staphylococcus aureus, Streptococcus pyogenes*
Other toxins	*Clostridium difficile, Bordetella pertussis*

infection involves bloody diarrhoeae and bacterial invasion of the intestines (see Section 8.2.1.1.3). This organism is unable to invade through the apical surface of intestinal epithelial cells and invades the body through the intestinal M cells. The macrophages underlying the M cells engulf the organism and are killed. The bacteria can then enter into epithelial intestinal cells through their basolateral surface. The ability to invade requires the genes encoding invasion plasmid antigens (Ipas) that are carried on a 200 kb virulence plasmid. The *ipa* operon encodes four secreted proteins IpaA, IpaB, IpaC and IpaD. Studies of these four proteins have identified IpaB as the protein responsible for inducing macrophage apoptosis. IpaB is released from phagocytosed bacteria and binds to, and activates, pro-caspase 1. Activation of this caspase then appears to set in train the whole apoptotic process. Apoptosis is generally held to be a safe process with few sequelae. However, it is now established that caspase 1-negative animals do not develop inflammation, neither do animals in which IL-1 or IL-18 have been knocked-out. Thus the process of macrophage apoptosis results in the release of pro-inflammatory cytokines such as IL-1 and IL-18 and this then gives rise to an inflammatory response. *Salmonella typhimurium*, which causes a pathology similar to that of *Shigella* spp. also causes macrophages to undergo apoptosis. This requires the type III secretion system of this organism. One of the proteins translocated by this type III secretion system is *Salmonella* invasion protein (Sip)B, which has significant sequence homology to the IpaB of *Shigella* spp.

Other bacteria whose mechanisms of action have been partially uncovered include *Lis. monocytogenes* and *Yersinia* spp. *Listeria monocytogenes* can escape from the phagosome and enter into the cell cytoplasm. It can induce apoptosis in lymphocytes and dendritic cells by means of its potent toxin, listeriolysin O (see Section 9.4.1.3.3). It is speculated that one means by which the organism is able to induce apoptosis involves listeriolysin O-mediated release of cytochrome *c* from mitochondria. *Yersinia* spp., as described above, can induce apoptosis through injection of YopJ/P into macrophages.

Less is known about the mechanisms of apoptosis induced by the chronic intracellular pathogen *M. tuberculosis*. It has been suggested that binding of mycobacterial cell wall components such as lipoproteins via the TLR-2 receptor on macrophages can induce apoptosis via a caspase 1-dependent pathway and/or by controlling cellular levels of Bcl-2. However, there is also evidence that mycobacteria can inhibit apoptosis and this will be discussed in the Section 10.6.2.2.2.

10.6.2.2.2 Bacterial inhibition of apoptosis

Many bacteria live within the cytosol or vacuolar apparatus of cells. It would therefore be expected that these organisms would inhibit apoptosis in order to survive. It is only within the past three years that evidence for bacterial inhibition of apoptosis has come to light.

The first report of the blockade of apoptosis by bacteria revealed that cells infected with the intracellular bacterium *Chl. trachomatis* were resistant to a range of inducers of apoptosis. This included the kinase inhibitor staurosporine, the DNA-damaging compound etoposide, TNFα, Fas antibodies and granzyme. The mechanism of inhibition appeared to involve blockade of cytochrome *c* release from mitochondria and inhibition of caspases. A related organism, *Chl. pneumoniae*, has also been reported to inhibit apoptosis. In this system, the role of cytokines has been explored and it has been proposed that inhibition of apoptosis is due to the induction of IL-10 by the infected cells down-regulating the synthesis of pro-apoptotic cytokines such as TNFα.

Virulent strains of *M. tuberculosis* have been reported to induce less macrophage apoptosis than avirulent strains. This failure to induce apoptosis may be due to the induction of IL-10 synthesis and concomitant release of the TNFR-2, which could neutralise the pro-apoptotic signal coming from induced TNFα. Another proposed mechanism for inhibiting apoptosis is through induction of TLR-2-induced activation of NF-κB.

It is therefore clear that we are witnessing just the tip of the iceberg in terms of the numbers of bacteria able to inhibit apoptosis and the variety of mechanisms by which they block this key cellular process. Our current understanding of the richness of these mechanisms can be seen in Figure 10.28.

Evasion of Toll: lessons from viruses

Where will the next bacterial evasion mechanisms be found? As has been suggested in the text, it is important to look at viruses for clues to the potential evasion mechanisms evolved by bacteria. One of the main growth areas in host–bacteria signalling is the TLRs. These are being shown to be important cell surface receptors involved in the recognition and discrimination of cellular pathogens including bacteria and fungi. It is likely, therefore, that bacteria and fungi will have developed mechanisms to abrogate TLR recognition.

This statement can now be made with much greater confidence because of a recent finding with poxviruses. These viruses, as has been described, employ many strategies to evade and neutralise the host immune response. It has now been

Figure 10.28 A schematic diagram illustrating the diverse mechanisms by which bacteria can modulate the apoptotic process. Details of the various mechanisms are given in the text. MAPKK, MAP kinase kinases; CAD, caspase-activated DNase; ICAD, inhibitor of CAD; for other abbreviations, see list on p. xxii. (Reprinted from *Trends in Microbiology* **8**, Gao, L.-Y. & Kwaik, Y. A., The modulation of host apoptosis by intracellular bacterial pathogens, pp. 806–313, Copyright 2000, with permission from Elsevier Science.)

demonstrated that vaccinia virus contains two open reading frames (ORFs), *A46R* and *A52R*, that are homologous to the Toll/IL-1 receptor (TIR) domain, a motif that defines the IL-1/TLR superfamily of receptors (see Section 4.2.6.6.2). When expressed in mammalian cells, the protein products of both ORFs were shown to interfere specifically with IL-1 signal transduction. A46R partially inhibited IL-1-mediated activation of the transcription factor NF-κB, and A52R potently blocked both IL-1- and TLR4-mediated NF-κB activation. MyD88 is a TIR domain-containing adapter molecule known to have a central role in both IL-1 and TLR4 signalling. A52R mimicked the dominant-negative effect of a truncated version of MyD88 on IL-1, TLR4, and IL-18 signalling but had no effect on MyD88-independent signalling pathways. Therefore, A46R and A52R are likely to represent a mechanism used by vaccinia virus to suppress TIR domain-dependent intracellular signalling. In consequence, it can be predicted that bacteria will also have evolved mechanisms to evade TLR signalling.

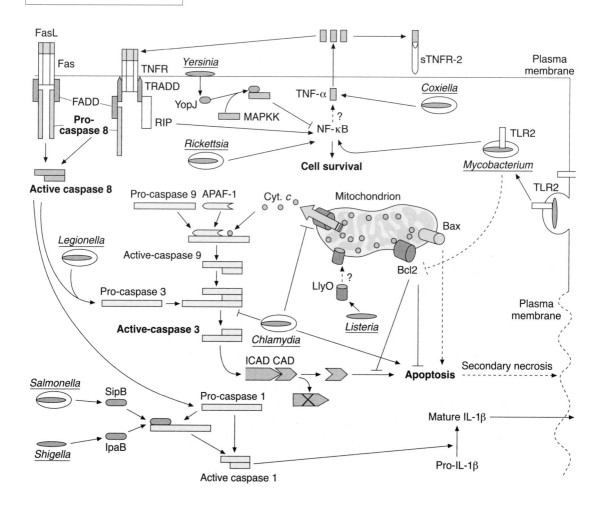

10.7 | Concept check

- Bacteria have co-evolved with the host and with host defence mechanisms.
- The host has co-evolved with bacteria to generate a complex series of immune defences.
- Bacteria have evolved mechanisms for evading both innate and acquired immune defences.
- Evasion occurs at all the major sites of host–bacteria interactions.
- On mucosal surfaces, bacteria have evolved mechanisms for inactivating IgA antibodies and for preventing the antibacterial actions of host defensive peptides.
- Bacteria can evade innate defence mechanisms such as complement activation by producing capsules, by binding complement control proteins or by use of proteases.
- The evasion of complement also prevents bacterial opsonisation and phagocytic uptake.
- Many intracellular bacteria have evolved complex means of living within phagocytes and thus evading the killing consequences of phagocytosis.
- Microorganisms in general have developed means of controlling the cytokine communication of the host and in doing so can evade host immunity.
- Many mechanisms have evolved to defeat acquired immune defences.
- Evasion of antibody-mediated defences involves immunoglobulin-binding proteins and antigenic variation.
- Superantigens and cell cycle toxins can interfere with T cell-mediated responses.
- There is a never-ending evolutionary war between bacteria and their hosts.

10.8 | Evasion of host defences by the paradigm organisms

10.8.1 *Streptococcus pyogenes*

Infections due to this organism declined in their morbidity and mortality up until the 1980s, when they began to increase and it has been during this period that clinicians recognised the condition known as streptococcal toxic shock syndrome. It is not known what has changed to enable this organism to become more pathogenic and it is important to find out. A number of the evasion mechanisms utilised by this bacterium have been described in this book (and are summarised in Table 10.20) and the role of the M proteins in evading

Table 10.20. Strategies used by *Streptococcus pyogenes* to evade host defence systems

Impedin	Function
Hyaluronic acid capsule	Anti-phagocytic, mimics a major constituent of host connective tissue, therefore poorly antigenic
M protein	Anti-phagocytic, inhibits complement activation, binds to Fc region of antibodies, antigenic variation
G protein	Binds to Fc region of antibodies
C5a peptidase	Inhibits recruitment of phagocytic cells
Streptococcal inhibitor of complement-mediated lysis (SIC)	Inhibits the complement membrane attack complex
Superantigens	Prevent a co-ordinated immune response to the organism, induce apoptosis
Heparin-binding protein	Mimicry of host molecules, antigen masking
Protein H	Inhibits complement activation
Streptolysin O, streptolysin S	Kill leukocytes, thereby interfering with immune response

complement-mediated opsonisation has already been discussed. These proteins bind C4BP and prevent opsonisation. However, they are also immunogenic and induce bactericidal antibodies. A recent study has identified a new, and unexpected, surface protein in *Strep. pyogenes* that may contribute to bacterial virulence and that may also be a novel target for vaccine development. Using an M protein-negative strain of *Strep. pyogenes*, it was found, contrary to expectations, that this bacterium was fully virulent. This suggested that some additional factor was responsible. Using antibodies raised to a crude extract of the M protein mutant, a new cell surface protein was identified. This has been termed streptococcal protective antigen. This is a 24 kDa protein with a novel sequence. Antibodies raised to the protein were protective, suggesting that this new protein may be a useful target for vaccine development. It also raises the key question of the role of this protein in bacterial evasion of host defences. This can be defined only by making isogenic mutants that lack the streptococcal protective antigen and others that lack both this protein and M protein.

10.8.2 *Escherichia coli*

The means by which *E. coli* is known to evade host defences are summarised in Table 10.21. Over the past decade, *E. coli* strains have been found to produce a growing number of toxins that have profound effects on leukocyte, and particularly lymphocyte, function. For example, the non-toxic B subunit of the heat-labile toxin (LT) is a poly-

Table 10.21.	Strategies used by *Escherichia coli* to evade host defence systems
Impedin (or evasion strategy)	Function
Capsule	Anti-phagocytic, anti-opsonic, certain types mimic host tissue components and so are non-antigenic
IgA-binding proteins	Render IgA ineffective by binding to Fc region
Lipopolysaccharide	Long antigenic side chain prevents access of the complement membrane attack complex to the outer membrane
Unidentified	Binds factor H thereby inhibiting complement activation
Cytokine-binding receptors	Remove cytokines from site of infection
Lymphostatin	Inhibits production of IL-2, IL-4 and IFNγ
Cytolethal distending toxin (CDT)	Induces apoptosis, inhibits lymphocyte proliferation
Heat-labile toxin (LT)	Induces apoptosis, polyclonal B cell activator
Phase variation in type I fimbriae	Undermines the effectiveness of the immune response
Phase variation in antigen 43 (an outer membrane protein)	Undermines the effectiveness of the immune response

clonal B cell activator and induces apoptosis of CD8 lymphocytes. In contrast, this toxin has minimal effects on CD4 lymphocytes. In an animal model of arthritis, the B subunit of LT was found to have significant inhibitory activity on disease induction and progression, suggesting that the subunit has the potential for therapeutic use.

As has been described, *E. coli* also makes a cell cycle inhibitory toxin, CDT. As a consequence of the binding of this toxin, cells undergo apoptosis. While the *E. coli* toxin has not been shown to be immunosuppressive, the homologous toxin from *A. actinomycetemcomitans* has been reported to block T cell proliferation. This toxin could also inhibit the proliferation of a range of cells and so could be an extremely effective way of dealing with the complete armamentarium of host defence cells, most of which have to divide to produce their effects.

Enteropathogenic *E. coli* (EPEC) have been described in detail in Section 7.5.2.1.2.1 and their ability to modulate the epithelial cell actin cytoskeleton has been described. EPEC encode an enormous toxin, which has been termed lymphostatin. The predicted molecular mass of this toxin is 366 kDa and the gene encoding it spans 9669 bp and is the largest reported gene in *E. coli*. Lymphostatin is one of the largest bacterial toxins produced. The first evidence that such a toxin existed was the finding that EPEC strains blocked the synthesis of IL-2, IL-4, IL-5 and IFNγ by human lymphocytes and also blocked lymphocyte proliferation.

Surprisingly, in contrast to the actions of CDT, lymphocytes exposed to lymphostatin survived and did not become apoptotic. The N-terminus of lymphostatin has homology to the large clostridial cytotoxins, discussed in Sections 4.2.2.3 and 9.5.3.1, that can block the activity of Rho, Rac and Cdc42. It was possible therefore that inhibition of lymphocyte function was due to an effect on these cytoskeletal-controlling proteins. However, no changes in cell shape were noted when lymphostatin-containing strains of non-EPEC were incubated with epithelial cells. The mechanism of action of this toxin therefore awaits discovery.

It therefore appears that pathogenic strains of *E. coli* are rich in proteins that can modulate leukocyte function and induce a decreased immune response to this organism.

10.9 | What's next?

In the final chapter a number of concepts are discussed including: (i) the future of bacterial pathogenesis research in the context of advances in genomics, (ii) the role of bacteria in idiopathic diseases, (iii) the role of bacteria in health, and (iv) some unusual bacteria–host interactions.

10.10 | Questions

1. Compare and contrast antigens, mitogens and superantigens and explain the consequences to the host of infection by bacteria producing superantigens.
2. Describe the process of apoptosis and explain with examples how bacteria can induce this process.
3. What strategies have bacteria evolved to cope with phagocytes?
4. How do bacteria evade attention from host antibodies?
5. Explain how type III secretion systems contribute to evasion of immune responses.
6. Compare and contrast viral and bacterial evasion of host immunity.
7. Compare and contrast bacterial evasion of immunity at mucosal surfaces and after invasion of the tissues of the body.
8. What biological activities of the complement system do bacteria avoid?
9. Explain the importance of the bacterial capsule in the evasion of host defence mechanisms.
10. What role does antigenic variation play in bacterial survival?

10.11 | Further reading

Books
Baumberg, S. (1999). *Prokaryotic Gene Expression*. Oxford: Oxford University Press.

Brogden, K., Roth, J., Stanton, T., Bolin, C., Minion, C. & Wannemuehler, M. (2000). *Virulence Mechanisms of Bacterial Pathogens*. Washington: ASM Press.

Henderson, B., Poole, S. & Wilson, M. (1998). *Bacteria–Cytokine Interactions in Health and Disease*. London: Portland Press.

Henderson, B., Wilson, M., McNab, R. & Lax, A. (1999). *Cellular Microbiology: Bacteria–Host Interactions in Health and Disease*. Chichester: John Wiley Ltd.

Higgs, G. A. & Henderson, B. (eds.) (2000). *Novel Cytokine Inhibitors*. Basel: Birkhauser.

Review articles

Abbas, A. K. & Janeway, C. A (2000). Immunology: improving on nature in the twenty-first century. *Cell* **100**, 129–138.

Aepfelbacher, M., Zumbihl, R., Ruckdeschel, K., Jacobi, C. A., Barz, C. & Heesemann, J. (1999). The tranquilizing injection of *Yersinia* proteins: a pathogen's strategy to resist host defense. *Biol Chem* **380**, 795–802.

Bermudez, L. E. & Sangari, F. J. (2001). Cellular and molecular mechanisms of internalization of mycobacteria by host cells. *Microbes Infection* **3**, 37–42.

Brodsky, F. M., Lem, L., Solache, A. & Bennett, E. M. (1999). Human pathogen subversion of antigen presentation. *Immunol Rev* **168**, 199–215.

Budihardjo, I., Oliver, H., Lutter, M., Luo, X. & Wang, X. (1999). Biochemical pathways of caspase activation during apoptosis. *Annu Rev Cell Dev Biol* **15**, 269–290.

Cornelis, G. R. (2000). Molecular and cell biology aspects of plague. *Proc Natl Acad Sci USA* **97**, 8778–8783.

Deretic, V. & Fratti, R. A. (1999). *Mycobacterium tuberculosis* phagosome. *Mol Microbiol* **31**, 1603–1609.

Dinges, M. M., Orwin, P.M. & Schlievert, P. M. (2000). Exotoxins of *Staphylococcus aureus*. *Clin Microbiol Rev* **13**, 16–34.

Fraser, J., Arcus, V., Kong, P., Baker, E. & Proft, T. (2000). Superantigens – powerful modifiers of the immune system. *Mol Med Today* **6**, 125–132.

Gao, L.-Y. & Kwaik, Y. A. (2000) The modulation of host cell apoptosis by intracellular bacterial pathogens. *Trends Microbiol* **8**, 306–313.

Gao, L.-Y. & Kwaik, Y. A. (2000). Hijacking of apoptotic pathways by bacterial pathogens. *Microbes Infect* **2**, 1705–1719.

Goebel, W. & Gross, R. (2001). Intracellular survival strategies of mutualistic and parasitic prokaryotes. *Trends Microbiol* **9**, 267–273.

Hackstadt, T. (1998). The diverse habitats of obligate intracellular parasites. *Curr Opin Microbiol* **1**, 82–87.

Henderson, B., Wilson, M. & Hyams, J. (1998). Cellular microbiology: cycling into the millennium. *Trends Cell Biol* **8**, 384–387.

Henderson, B., Wilson, M. & Wren, B. (1997). Are bacterial exotoxins cytokine network regulators? *Trends Microbiol* **5**, 454–458.

Lee, V. T. & Schneewind, O. (1999). Type III secretion machines and the pathogenesis of enteric infections caused by *Yersinia* and *Salmonella* spp. *Immunol Revs* **168**, 241–245.

Lindahl, G., Sjöbring, U. & Johnsson, E. (2000). Human complement regulators: a major target for pathogenic microorganisms. *Curr Opin Immunol* **12**, 44–51.

Maksymowych, W. P. & Kane, K. P. (2000). Bacterial modulation of antigen processing and presentation. *Microbes Infec* **2**, 199–211.

Marrack, P. & Kappler, J. (1994). Subversion of the immune system by pathogens. *Cell* **76**, 323–332.

O'Garra, A. & Arai, N. (2000). The molecular basis of T helper 1 and T helper 2 cell differentiation. *Trends Cell Biol* **10**, 542–550.

Papageorgiou, A. C. & Acharya, K. R. (2000). Microbial superantigens: from structure to function. *Trends Microbiol* **8**, 369–375.

Pickett, C. L. & Whitehouse, C. A. (1999). The cytolethal distending toxin family. *Trends Immunol* **7**, 292–297.

Pieters, J. (2001). Evasion of host cell defence mechanisms by pathogenic bacteria. *Curr Opin Immunol* **13**, 37–44.

Rhen, M., Eriksson, S. & Pettersson, S. (2000). Bacterial adaptation to host innate immunity responses. *Curr Opin Microbiol* **3**, 60–64.

Tortorella, D., Gewurz, B. E., Furman, M. H., Schust, D. J. & Ploegh, H. L. (2000). Viral subversion of the immune system. *Annu Rev Immunol* **18**, 861–926.

Wick, M. J. & Ljunggren, H.-G. (1999). Processing of bacterial antigens for peptide presentation on MHC class I molecules. *Immunol Revs* **172**, 153–162.

Papers

Agranoff, D., Monahan, I. M., Mangan, J. A., Butcher, P.D. & Krishna, S. (1999). *Mycobacterium tuberculosis* expresses a novel pH-dependent divalent cation transporter belonging to the Nramp family. *J Exp Med* **190**, 717–724.

Barker, H., Kinsella, N., Jaspe, A., Friedrich, T. & O'Connor, C.D. (2000). Formate protects stationary-phase *Escherichia coli* and *Salmonella* cells from killing by cationic antimicrobial peptide. *Mol Microbiol* **35**, 1518–1529.

Bengoechea, J. A. & Skurnik, M. (2000). Temperature-regulated efflux pump/potassium antiporter system mediates resistance to cationic antimicrobial peptides in *Yersinia*. *Mol Microbiol* **37**, 67–80.

Braun, M. C., He, J., Wu, C.-Y. & Kelsall, B. L. (1999). Cholera toxin suppresses interleukin (IL)-12 production and IL-12 receptor β1 and β2 chain expression. *J Exp Med* **189**, 541–552.

Escalas, N., Davezac, N., De Rycke, J., Baldin, V., Mazars, R. & Ducommun, B. (2000). Study of the cytolethal distending toxin-induced cell cycle arrest in HeLa cells: involvement of the CDC25 phosphatase. *Exp Cell Res* **257**, 206–212.

Fan, T., Lu, H., Lu, H., Shi, L., McClarty, G. A., Nance, D. M., Greenberg, A. H. & Zhong G. (1998). Inhibition of apoptosis in *Chlamydia*-infected cells: blockade of mitochondrial cytochrome *c* release and caspase activation. *J Exp Med* **197**, 487–496.

Ferrari, G., Langen, H., Naito, M. & Pieters, J. (1999). A coat protein on phagosomes involved in the intracellular survival of mycobacteria. *Cell* **97**, 435–447.

Fletcher, J., Nair, S., Poole, S., Henderson, B. & Wilson, M. (1998). Cytokine degradation by biofilms of *Porphyromonas gingivalis*. *Curr Microbiol* **36**, 216–219.

Gatfield, J. & Pieters, J. (2000). Essential role for cholesterol in entry of mycobacteria into macrophages. *Science* **288**, 1647–1650.

Gomes, S. M., Paul, S., Moreira, A. I., Appelberg, R., Rabinovitch, M. & Kaplan, G. (1999). Survival of *Mycobacterium avium* and *Mycobacterium tuberculosis* in acidified vacuoles of murine macrophages. *Infect Immun* **67**, 3199–3206.

Gruenheid, S. & Gros, P. (2000). Genetic susceptibility to intracellular infections: Nramp1, macrophage function and divalent cations transport. *Curr Opin Microbiol* **3**, 43–48.

Guerin, I. & de Chastellier, C. (2000). Pathogenic mycobacteria disrupt the macrophage actin filament network. *Infect Immun* **68**, 2655–2662.

Guo, L., Lim, K. B., Poduje, C. M., Daniel, M., Gunn, J. S., Hackett, M. & Miller, S. I. (1998). Lipid A acylation and bacterial resistance against vertebrate antimicrobial peptides. *Cell* **95**, 189–198.

Keane, J., Remold, H. G. & Kornfield, H. (2000). Virulent *Mycobacterium tuberculosis* strains evade apoptosis of infected alveolar macrophages. *J Immunol* **164**, 2016–2020.

Klapproth, J.-M. A., Scaletsky, I. C. A., McNamara, B. P., Lai, L.-C., Malstrom, C., James, S. P. & Donnenberg, M. S. (2000). A large toxin from pathogenic *Escherichia coli* strains that inhibit lymphocyte activation. *Infect Immun* **68**, 2148–2155.

Lara-Tejero, M. & Galan, J. E. (2000). A bacterial toxin that controls cell cycle progression as a deoxyribonuclease I-like protein. *Science* **290**, 354–357.

Malik, Z. A., Denning, G. M. & Kusner, D. J. (2000). Inhibition of Ca^{2+} signalling by *Mycobacterium tuberculosis* is associated with reduced phagosome–lysosome fusion and increased survival within human macrophages. *J Exp Med* **191**, 287–302.

Marques, M., Kasper, D. L., Pangburn, M. K. & Wessels, M. R. (1992). Prevention of C3 deposition is a virulence mechanism of type III group B streptococcus capsular polysaccharide. *Infect Immun* **60**, 3986–3993.

Nashar, T. O., Webb, H. M., Eaglestone, S., Williams, N. A. & Hirst, T. R. (1996). Potent immunogenicity of the B subunits of *Escherichia coli* heat-labile enterotoxin: receptor binding is essential and induces differential modulation of lymphocyte subsets. *Proc Natl Acad Sci USA* **93**, 226–230.

Perez-Caballero, D., Albert, S., Vivanco, F., Sanchez-Corral, P. & Rodriguez de Cordoba, S. (2000). Assessment of the interaction of human complement regulatory proteins with group A streptococcus. Identification of a high-affinity group A streptococcus binding site in FHL-1. *Eur J Immunol* **30**, 1243–1253.

Sandt, C. H. & Hill, C. W. (2000). Four different genes responsible for nonimmune immunoglobulin-binding activities within a single strain of *Escherichia coli*. *Infect Immun* **68**, 2205–2214.

Sansonetti, P. J., Phalipon, A., Arondel, J., Thirumalai, K., Banerjee, S., Akira, S., Takeda, K. & Zychlinsky, A. (2000). Caspase-1 activation of IL-1β and IL-18 are essential for *Shigella flexneri*-induced inflammation. *Immunity* **12**, 581–590.

Sharp, L., Reddi, K., Fletcher, J., Nair, S., Wilson, M., Curtis, M., Poole, S., Shepherd, P., Henderson, B. & Tabona, P. (1998). A lipid A-associated protein of *Porphyromonas gingivalis*, derived from the haemagglutinating domain of the R1 protease gene family, is a potent stimulator of interleukin-6 synthesis. *Microbiology* **144**, 3019–3026.

Shenker, B. J., McKay, T., Datar, S., Miller, M., Chowhan, R. & Demuth, D. (1999). *Actinobacillus actinomycetemcomitans* immunosuppressive-protein is a member of the family of cytolethal distending toxins capable of causing a G_2 arrest in human T cells. *J Immunol* **162**, 4773–4780.

Teitelbaum, R., Cammer, M., Maitland, M. L., Freitag, N. E., Condeelis, J. & Bloom, B. R. (1999). Mycobacterial infection of macrophages results in membrane-permeable phagosomes. *Proc Natl Acad Sci USA* **96**, 15190–15195.

Uchiya, K.-I., Barbieri, M. A., Funato, K., Shah, A. H., Stahl, P. D. & Groisman, E. A. (1999). A *Salmonella* virulence protein that inhibits cellular trafficking. *EMBO J* **18**, 3924–3933.

Vazquez-Torres, A., Xu, Y., Jone-Carson, J., Holden, D. W., Lucia, S. M., Dinauer, M. C., Mastroeni, P. & Fang, F. C. (2000). *Salmonella* pathogenicity island 2-dependent evasion of the phagocyte NADPH oxidase. *Science* **287**, 1655–1658.

Wurzner, R. (1999). Evasion of pathogens by avoiding recognition or eradication by complement, in part via molecular mimicry. *Mol Immunol* **36**, 249–260.

Yao, T., Mecsas, J., Healy, J. I., Falkow, S. & Chien, Y.-H. (1999). Suppression of T and B lymphocyte activation by a *Yersinia pseudotuberculosis* virulence factor, YopH. *J Exp Med* **190**, 1343–1350.

Zilber, M. T., Gregory, S., Mallone, R., Deaglio, S., Malavasi, F., Charron, D. & Gelin, C. (2000). CD38 expressed on human monocytes: a coaccessory molecule in the superantigen-induced proliferation. *Proc Natl Acad Sci USA* **97**, 2840–2845.

10.12 | Internet links

1. http://mercury.faseb.org/aai/default.asp
   The website of the American Association of Immunologists.
2. http://vlib.org/Science/Cell_Biology/apoptosis.shtml
   List of links to apoptosis-related sites.
3. http://www.edc.com/%7Ekimball/BiologyPages/A/Apoptosis.html#The_Mechanisms_of_Apoptosis
   Basic description of apoptosis with links to related sites.
4. www.bact.wisc.edu/Bact330/lectureai
   The section of an on-line lecture course (from the University of Wisconsin-Madison, USA) that deals with bacterial evasion of immune responses.
5. http://www.bact.wisc.edu/Bact330/lecturead
   The section of an on-line lecture course (from the University of Wisconsin-Madison, USA) that deals with bacterial evasion of phagocytes.
6. http://dunn1.path.ox.ac.uk/~cholt/Index.html
   Lots of useful information about macrophages.

Bacteria in human health and disease: the future?

Chapter outline

Aims

To provide the reader with a glimpse into the future with regard to:

- new approaches to studying bacterial pathogenesis
- the development of new methods of treating diseases caused by bacteria
- the involvement of bacteria in human diseases of unknown causation
- the role of the normal microflora in human development

11.1 | Introduction

The 1950s and 60s heralded what was assumed to be the pinnacle of our control over bacterial diseases. Antibiotics could treat most bacterial infections, and, in time, would treat all bacterial infections, and that was all there was to it. If you could eradicate the bacterium, why bother understanding how it caused disease? While there was a certain logic to this viewpoint, it went against our fundamental human need to understand how things work. In retrospect, it also slowed down progress in our understanding of the workings of the cell and of immunity. The renaissance in the study of the biology of bacteria, which has been forced upon us by the rapid rise in antibiotic resistance and the discovery of 'new' bacterial diseases, has revealed a whole new world of cellular interactions between prokaryotes and eukaryotes, and between prokaryotes and prokaryotes, that were undreamed of only 10 to 20 years ago. The rapid progress in Cellular and Molecular Microbiology made in the past 10 to 15 years suggests that the next decade will see major advances in our understanding of the cellular and molecular microbiology of prokaryotic/eukaryotic interactions. Our final chapter aims to give an insight into what the reader may expect to see in this period.

So, at the end of a book on the mechanisms used by bacteria to cause disease we need to ask: where do we go from here? It is clear that at the beginning of the 21st century we have a growing number of cellular and molecular tools for dissecting the interactions between bacteria and host cells in order to define the processes responsible for disease. Many of these techniques have been described in Chapter 3 and in later chapters and they include *in vivo* expression technology (IVET), signature-tagged mutagenesis (STM), genomic analysis and mapping by *in vitro* transposition (GAMBIT), various forms of polymerase chain reaction (PCR), gridded DNA libraries, proteomics, etc. These techniques are all underpinned by having the complete genomic DNA sequences of a growing number of bacterial species. For an increasing number of bacteria there are also proteomic databases. A key requirement is that we understand what happens to bacteria *in vivo* and there is increasing pressure to focus only on bacteria–host interactions *in vivo*. This involves the use of experimental animals. The use of humans in such studies would be ideal, but this is ethically difficult. A major spin-off of our knowledge about the mechanisms of bacterial virulence is that it enables novel antibacterial therapeutic targets to be identified. There is now a major drive by the pharmaceutical industry to identify bacterial virulence factors, or virulence mechanisms, that can be the target of new therapeutics, including drugs and vaccines.

Genomics is beginning to provide sufficient numbers of complete bacterial genomic sequences to allow sophisticated comparisons of different bacteria or groups of bacteria to be made. This will become increasingly important in tracing phylogenetic relationships between organisms and in determining from where they derived their virulence genes.

A prize that genomics is offering the biologist is the understanding of the minimal set of genes required for life. Another fascinating area of bacteriology, from which the veil is only beginning to be removed, is the role that bacteria may play in the induction of idiopathic diseases, i.e. diseases of unknown origin. Common diseases such as atherosclerosis, rheumatoid arthritis, psoriasis and even cancer are currently being re-evaluated to determine whether they could be caused by bacteria. Furthermore, interest is being shown in using living bacteria to treat cancer.

In addition to these relatively obvious areas, what else will the future bring for our understanding of bacterial pathogenesis and host–bacteria interactions? There is increasing interest in how commensal bacteria interact with their hosts. An obvious question is: how do we survive with bodies full of bacteria? Recent work is suggesting that our microflora not only lives in harmony with us but may even play a role in shaping our bodies.

These various areas of research will be described briefly to give an idea of the excitement that is currently pervading the world of Cellular Microbiology. To provide a starting point to the discussion in this chapter, in May 2000 it was estimated that 4473 bacteria in 905 genera had been described validly. Of these bacteria, 29 species in 21 genera had been sequenced. Considering that the number of species of bacteria on our planet could be in the millions, we have a very long way to go before we have a full understanding of the bacteria that live with us on Earth. Having said this, we are currently advancing our under-standing at an enormous rate, a rate with which the reader, if s/he decides to go into scientific research, will have to cope.

11.2 | Identification of bacterial virulence genes and virulence mechanisms *in vivo*

Chapter 3 was devoted to describing the techniques that are currently being used to investigate the mechanisms of bacterial virulence at the cellular and molecular level. Much of the short history of the study of bacterial virulence has involved using bacteria cultured on agar or in a liquid medium. While there are sensible reasons for continuing such studies, the general feeling within the microbiological community is that to fully comprehend how bacteria cause disease it is important to identify those genes that are switched on when the organism enters the host. These are the population of genes known as *in vivo*-induced (ivi) genes. The reader has been introduced to a number of these tech-niques such as IVET, STM and DFI (differential fluorescence induction) in Chapter 3; these techniques have been used successfully for a number of years and have undergone substantial development (Table 11.1).

Chapter 8 described the cellular and molecular biology of cell inva-sion by bacteria. This aspect is now recognised to be a common feature of bacteria, or at least of those that interact with mammals. To survive inside cells, bacteria secrete specific proteins and it is important to identify what these proteins are. A recently described technique, known as disseminated insertion of class I epitopes (DICE), involves the random insertion of a DNA sequence, encoding a major histocom-patibility complex (MHC) peptide epitope (see Section 6.7.2.1), into the genome of the bacterium of interest. The library of organisms with the randomly inserted peptides is cultured with macrophages, and intra-cellular bacteria that secrete the epitope are recognised and isolated by flow cytometry. The gene to which the epitope was tagged can then be

Table 11.1.	Methods for assessing bacterial gene expression *in vivo*
Technique	**Results obtained**
IVET	Identifies genes required for survival in animal host
RIVET	Modified methodology using recombinase
STM	Complementary approach to IVET
DFI	Uses GFP reporter plasmids to identify genes induced under specific host-like conditions
DICE	See text, p. 585
Differential display	PCR-based technique to identify genes induced under specific conditions relative to control
IVIAT	Uses human serum absorbed with agar-grown bacteria to screen λ libraries for *in vivo* expressed genes
IVIF	Uses plasmid-based GFP reporter to monitor gene expression in infected tissues
DNA microarrays	Uses annotated arrays of cDNA probes on solid support to define genes induced by *in vivo* conditions using hybridisation format

For abbreviations, see list on p. xxii.

sequenced and the gene ascribed the property of being transcribed when the bacterium enters macrophages. This technique is actually an *in vitro* method, but allows the dissection of those genes that are switched on when the bacterium enters macrophages *in vivo*. Increasingly, there is a move towards *in vivo* systems for studying bacterial virulence mechanisms. Two polarised examples of such techniques will be described.

11.2.1 *Caenorhabditis elegans* and bacterial virulence

Techniques such as IVET use the mouse as a 'filter' to determine which bacterial genes are involved in pathogenicity. With the growing realisation that bacteria–host virulence/defence mechanisms are common across species barriers there is a increasing interest in using simpler organisms for exploring bacteria–host interactions. One of the first multicellular organisms to be targeted for genome sequencing was the nematode worm *Caenorhabditis elegans*. This soil organism was chosen for study because it is a multicellular organism, but a relatively simple creature, which is composed of approximately 1000 cells. The 20 Mb genome of this organism was sequenced (completed in 1998) and contains approximately 19 000 genes. This is more than half of the estimated number of genes in the human genome (approximately 32 000). In the past few years, a number of groups have begun to study the interactions of bacteria with this worm and have identified the usefulness of this organism in bacteriological research.

The use of *Caenorhabditis elegans* for studying bacterial virulence is still in its infancy and only a brief description will be provided. *Pseudomonas aeruginosa* is an opportunistic pathogen that often infects human individuals who are immunosuppressed. It has a surprisingly broad host range, being able to infect animals as well as plants. One strain of *Ps. aeruginosa* has been reported to kill the worm by two distinct mechanisms. The first, so-called 'fast killing', is due to a low molecular mass pigment/toxin called pyocyanin, which can catalyse the NADPH-dependent conversion of oxygen to superoxide and hydrogen peroxide and is known to damage mammalian cells *in vitro* through oxidative stress. *Caenorhabditis elegans* mutants with known altered sensitivity to oxidative stress demonstrate altered responses to this fast killing mechanism. Mutants of the worm that lack toxin effusion pumps were also shown to be more sensitive to fast killing. A second mechanism – 'slow killing' – appears to be due to infection of the nematode and takes days rather than hours. Using mutants of *Ps. aeruginosa*, it has been possible to identify the genes responsible for virulence in the nematode and they are, remarkably, similar to those identified as being important in mammals.

In a separate study, an undefined toxin, controlled by a quorum sensing system (see Section 4.3.2.1) in *Ps. aeruginosa*, was found to cause lethal paralysis of the nematode. The use of random mutagenesis of *Caen. elegans* to select for resistant mutants has identified the gene encoding the protein EGL-9, currently of unknown function, as being important for susceptibility to bacterial paralysis. The fact that the complete genome sequences are known for both *Ps. aeruginosa* and *Caen. elegans*, makes it relatively easy to identify genes in either organism involved in the induction of bacterial pathology.

Caenorhabditis elegans appears to be susceptible to infection by a number of human pathogens and is going to be extremely useful in the study of bacterial pathogenesis in the future.

11.2.2 Study of bacterial pathogenic mechanisms in humans

There is a pressing need to identify bacterial genes expressed *in vivo* in the human host. However, it is not ethical to use techniques such as IVET or STM in *Homo sapiens*. Fortunately, a technique has recently been described to detect only those bacterial genes that are being transcribed when the bacterium is in its host and causing disease. This technique utilises the ability of the immune system to recognise protein immunogens. This method is called *in vivo*-induced antigen technology (IVIAT) and is shown schematically in Figure 11.1. The advantage of IVIAT is that it utilises the human immune system to interrogate bacterial virulence genes and is a remarkably simple, and potentially powerful, method for identifying bacterial genes involved

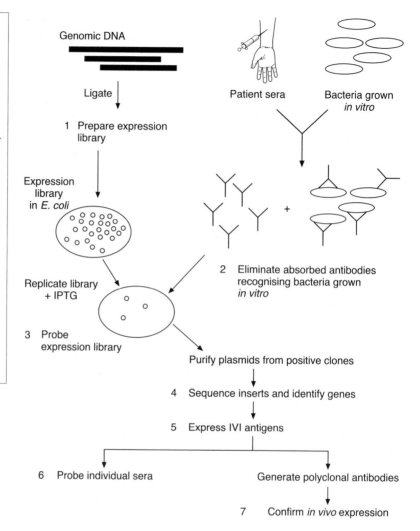

Figure 11.1 A schematic description of *in vivo*-induced antibody technology (IVIAT). An expression library of the organism of interest is generated. Blood is taken from patients who are infected with the bacterium under study and the sera prepared. These sera are immunoadsorbed with bacteria grown under conventional *in vitro* conditions. It is assumed that such conditions do not induce virulence genes invoked when the organism invades the human host. The immunoadsorbed sera are used to immunoscreen the genomic library. It is envisaged that the only antibodies remaining in the sera are those recognising the proteins encoded by the bacterium that are switched on as part of the organism's virulence strategy. IPTG, isopropylthiogalactoside.

in disease pathogenesis. All that is required for IVIAT is sera from patients who have been infected with the bacterium of interest. As has been described in Section 6.4.3, individuals with bacterial infections raise antibodies to many of the protein components of the infecting cells. The sera from infected patients are incubated with bacteria that have been grown by conventional bacteriological culture. The assumption is that under these conditions of culture the bacterium does not express those proteins that allow the organism to induce disease. Incubating the laboratory-grown bacterium with the patient's sera removes all the antibodies recognising the gene products produced by the organism under the *in vitro* conditions. This removal of antibodies is called immunoadsorption. This leaves, at least in theory, only those antibodies recognising the bacterial gene products that are produced specifically when the bacterium is infecting the human host. To identify those genes, the immunoadsorbed antisera are used to immunoscreen a genomic library of the organism under study in an appropriate expression vector. The use of immunoscreening to identify bacterial genes has been described in Section 3.3.1. In this way, bacterial genes that are switched on only when the organism is infecting the human host can be identified. Such genes are likely to be related to disease pathology. At the time of writing, this methodology is in its infancy and its potential usefulness was still the subject of debate because of the problems of defining the conditions under which bacteria are grown *in vitro*. However, IVIAT may allow the dissection of the bacterium–host interactions in terms of the bacterial virulence genes employed in the human host. Of course, as we have seen in Chapter 10, the evolution of bacterial immune evasion strategies may limit IVIAT.

11.3 | Development of new antibacterials

The pharmaceutical industry is one of the major global players and hardly a day goes by without some announcement appearing about a breakthrough in drug discovery or a mega-merger taking place. It is therefore chilling to learn, in a world running out of antibiotics, that no new chemical classes of antibiotics have been introduced successfully into clinics for over 30 years. This means that the antibiotic administered to any reader unfortunate enough to have a bacterial infection was developed when his or her parents were children. Moreover, these antibiotics are based on natural compounds whose evolutionary history is unknown. This failure to introduce new antibiotics, linked with the massive overuse of these compounds, has brought us to a point in the 21st century where many of the bacterial pathogens with which we share this planet are resistant to most, if not all, antibiotics. A good example is *Staphylococcus aureus*, which rapidly exhibited resistance to penicillin and has, in the past few years, thrown up strains that are insensitive to the last-ditch compound – vancomycin.

What has gone wrong? A general answer to this question is that it is due to a combination of greed, inertia and lack of imagination. The pharmaceutical industry was making billions from antibiotics and apparently could not see that any more money would be made from newer classes of compounds. More than 150 antibacterial drugs have been approved for clinical use in the USA. These compounds are all based on the original antibiotic classes and, in total, only target 15 different bacterial mechanisms (protein synthesis, DNA synthesis, cell wall synthesis, etc.). Only the oxazolidinones (which bind to the bacterial 50 S ribosomal subunit and inhibit protein synthesis initiation) have novel therapeutic targets.

Antibiotics are only one approach to dealing with bacterial infections but, until very recent years, novel approaches to developing antibacterials were not actively pursued. However, with the rapidly emerging discoveries about bacterial virulence mechanisms linked with the enormous outpouring of information about bacterial genomes, the situation is changing. Use is now being made of the array of new technologies such as genome sequencing, transposon mutagenesis, proteomics, DNA arrays (or DNA microchips) to define new therapeutic targets, and such targets are emerging. For example, transposon mutagenesis (described in Section 3.2.2) is a key tool in determining gene function. This technology had its limitations because of the low frequency of insertions and their essentially random nature. A new development is GAMBIT. Here the secret is to cover the bacterial genome with overlapping sets of long-range PCR products of approximately 10^4 bp. These PCR products undergo *in vitro* transposition leading to random and high frequency insertion into the DNA of the bacterial target species. The bacterium can then be transformed by the mutagenised PCR products and the location of transposon insertions can be mapped by PCR. Genes essential for survival under given conditions can be recognised by their lack of transposon insertions. This is a rapid method for detecting specific populations of genes.

Comparative genomics now allows the searching of multiple bacterial genomes for orthologous genes, i.e. genes encoding proteins that have the same function in different organisms. Such information could be used to develop broad-spectrum antibacterial compounds. Alternatively, a search for genes that are restricted to certain species could be utilised for the development of narrow spectrum drugs – these are particularly useful as they have only a limited effect on the composition of the normal microflora. In the past, in order to identify a target for novel drug discovery, the activity of the target molecule (mainly proteins) had to be known. However, there are now computational methods that can infer protein function from sequence information. Another aid in deciphering protein function is coming from the increasing rapidity with which it is possible to gain the crystal- or nuclear magnetic resonance (NMR)-based three-dimensional structures of proteins. Such bioinformatic strategies are now central to drug design.

Identification of molecular targets is just the first step in developing novel antibacterials. The next step is to determine what classes of chemicals can inhibit the activity of the specific molecular target. Major advances in pharmaceutical chemistry such as combinatorial chemistry and phage display (techniques allowing enormous numbers of molecular variants to be produced), coupled with robotic screening, now allow tens to hundreds of thousands of compounds to be screened each day.

A brief description of some of the novel therapeutic targets that pharmaceutical companies are using to develop new antibacterial compounds will now be given (Figure 11.2).

11.3.1 Inhibition of bacterial adhesion

Chapters 1 and 7 have highlighted the importance to bacteria of being able to adhere to specific host cells. If bacteria cannot adhere, they cannot infect. Therefore adhesion is a major therapeutic target.

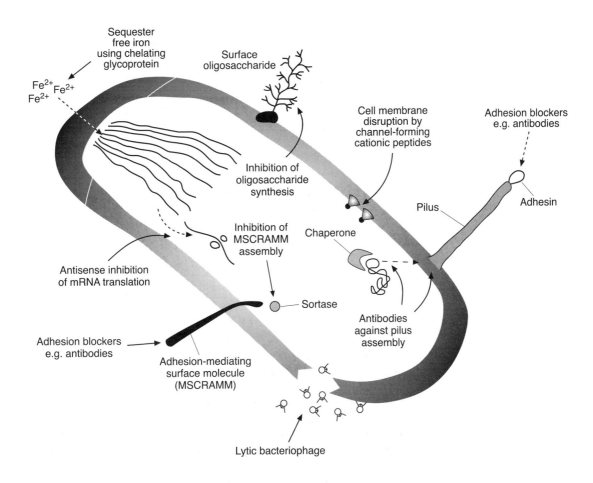

Figure 11.2 The new 'antibiotics'. A schematic view of the some of the newer targets for antibacterial compounds. MSCRAMM, microbial surface components recognising an adhesive matrix molecule. (Modified from Breithaupt, H. (1999), The new antibiotics. *Nature Biotechnology* **17**, 1165–1169, with permission from *Nature* Publishing Group.)

Bacteria use a large number of surface components such as lipopolysaccharide (LPS), lipoteichoic acid, various proteins (e.g. microbial surface components recognising adhesive matrix macromolecules (MSCRAMMs)) and complex adhesive structures (fibrils, pili, fimbriae) that recognise specific components on host cell surfaces. Host cell glycoproteins and glycolipids are used by many bacteria as receptors and one approach is to design synthetic oligosaccharides that mimic host receptors to antagonise bacterial adhesion. Such oligosaccharides are in clinical trial for blocking infections with *Helicobacter pylori*, *Strep. pneumoniae* and *Haemophilus influenzae*. Another approach being taken is to target the MSCRAMMs – the key adhesins on *Staph. aureus* and *Staph. epidermidis*. Here one strategy is to produce blocking humanised monoclonal antibodies for passive immunisation. It is also proposed that the MSCRAMMs are targets for vaccination. Other adhesive structures are being examined for use in vaccines. These include the FimH adhesin found on the pili of uropathogenic *E. coli*. An alternative approach to blocking the adhesins of bacteria is to inhibit their production. One method of doing this would be to target the molecular chaperone required for pilus assembly in bacteria such as *Escherichia coli*, *H. influenzae* and *Klebsiella pneumoniae*.

11.3.2 Antibacterial peptides

The fact that all of the higher organisms tested have been found to produce antibacterial peptides that target the bacterial membrane was discussed in Section 5.4.3.11. The actions of these peptides are being investigated by a number of companies and novel compounds are under test. For example, Magainin Pharmaceuticals (now Geneara Corporation) whose founder, Michael Zasloff, discovered the magainin antibacterial peptides secreted by frog skin, are synthesising compounds based on natural peptide compounds. One of these compounds, pexiganan acetate, has been shown to be topically effective in treating diabetic foot ulcers. The antibacterial peptides suffer from one major problem. Being peptides with α-amino bonds, they are targets for proteases and can be rapidly broken down when administered. However, it has now proved possible to produce magainin peptides with a β-amino acid backbone that retain their antibacterial activity but are resistant to proteolytic breakdown. Bactericidal/permeability-increasing protein (BPI, commercial name Neuprex), a neutrophil protein that binds LPS and kills bacteria (see Section 5.4.3.6), is in clinical trial for *Neisseria meningitidis* infections in young children. A range of other peptides are in clinical trial and the results, while generally encouraging, suggest that more information is required before the full potential of antibacterial peptides can be ascertained.

11.3.3 Antisense

The initiation of specific protein translation in cells can be blocked by small antisense oligonucleotides targeted to the mRNA. Using Watson–Crick base-pairing interactions, such oligonucleotides can be synthesised to bind to one specific mRNA and inhibit only the translation of that protein. Antisense therapeutics have been the subject of enormous interest for idiopathic diseases. The results, it should be said, have not lived up to the original hype. However, in principle, the technology should yield therapeutic agents, provided the basic problems of delivering the drugs can be solved. It may be easier to deliver an antisense compound to a bacterium, and at least one company is exploring the possibility of utilising antisense oligonucleotides as antibacterials.

11.3.4 Other approaches

Humans are limited only by their imagination and that can be no limitation at all. One fascinating idea is to use bacteriophages to target bacteria. This is an idea that first surfaced in the 1920s and relies on the fact that bacteriophages kill bacteria through their lytic cycle. Using an *in vivo* screening strategy for isolating phages that persist in the circulation, it has been possible to develop a phage with persistence and limited host activation. It is hoped to test the clinical efficacy of such phage particles in the near future.

Another interesting approach currently being developed is the use of light-activated antimicrobial agents. These are drugs which have no antimicrobial effect in the dark (or under normal light conditions) but are activated by light of a particular wavelength (matching that of the absorption maximum of the drug) to produce potent antimicrobial molecules. The latter include free radicals and singlet oxygen which kill bacteria by damaging their cytoplasmic membranes. These drugs could be used clinically to treat burn and wound infections or other infectious conditions confined to the skin or mucosal surfaces. Treatment would involve the application of the drug to the disease lesion, followed by short-term irradiation (usually less than 1 minute) with light of an appropriate wavelength from a low-power laser.

Thus far, the targeting of therapies has been to the bacterium. However, it may be possible to provide the host with the genes required to deal with a particular infection. For example, would it be possible to deliver to cells extra copies of genes required to produce antimicrobial cytokines or antibacterial peptides? The delivery of such genes could boost the natural defence systems of the body and would be free from the side effects of synthetic compounds.

11.3.5 New targets from genomics

The size of the genomes of pathogenic bacteria range from 0.6 to 6.3 Mb. Surprisingly, in each bacterial genome, a large fraction of the open reading frames (ORFs) encode as yet uncharacterised proteins unique to a given species. It is likely that many of these proteins are involved in the virulence characteristics of these organisms and are therefore therapeutic targets. The disadvantage of such therapeutic targets is that they are not generic; thus only single pathogenic species could be targeted. This, of course, as has been stated, may not be a problem as compounds selective for one bacterial species would not necessarily interfere with the normal microflora of mammals. Genomics is being used in conjunction with DNA microarrays (see Section 3.8) to identify the patterns of genes switched on by a bacterium under given sets of conditions. The genes switched on can then be complemented by proteomic technology to identify the proteins being transcribed and what happens to them (are they exported, broken down, etc.). Proteomics can also define what post-translation modifications proteins undergo. These chemical changes in proteins may be crucial to protein function and may enable further therapeutic avenues to be explored.

The discovery of new 'antibiotics' has now started and is gaining pace rapidly. There will be major discoveries made along the way and we can confidently predict that new drugs will eventually emerge. This will be an exciting time and one in which the reader of this book may be able to participate.

11.3.6 Using genomics to identify vaccine candidates

The twin foundations of antibacterial treatment are antibiotics and vaccines. One major problem with regard to the latter is identifying proteins that are vaccine targets. A new approach, developed in the past year or so, is to utilise genomics to identify vaccine candidates and then to clone and express those genes, identified *in silico*, and use the encoded protein(s) to vaccinate animals and determine whether there is protection. The method used to identify vaccine targets is simple: those proteins that are exported or are surface expressed are identified as vaccine candidates. Such identification can now be done easily using algorithms such as SignalP (see Section 11.9) that can identify signal sequences of exported proteins. In a recent study of *N. meningitidis*, 570 candidate ORFs were identified and 350 of these genes were able to be expressed. The expressed proteins were purified and used to immunise mice. Of these proteins, seven were found to raise antibodies that both recognised the surfaces of the bacterium and had serum bactericidal activity. These were mainly lipoproteins. Similar genomic screens have also been carried out for organisms such as *Hel. pylori* and *H. influenzae*. Such screening exercises are extremely time consuming and labour intensive and so can be done only by large pharmaceutical companies. The paper that describes this exercise quoted in Section 11.8, has 36 authors (Pizza *et al.*, 2000).

11.4 | Genomics to identify the basics of life

One of the most exciting problems to which genomics may provide an answer is the question of the minimum set of genes required for 'life'. This is almost approaching the ultimate question: what is life?

Viruses, which exist at the borderline between the living and the non-living, have genome sizes ranging from 3.5 to 235 kb. The smallest bacterial genome discovered so far is that of *Mycoplasma genitalium*, which is 580 kb, just over twice the size of the largest viral genome. Mycoplasmas are the smallest free-living bacteria, and are related to Gram-positive bacteria. They are unique among bacteria in that they have no rigid cell wall and their cell membranes contain sterols, which must be obtained from the environment. Mycoplasmas are common in nature but only three species are known to cause disease in humans: *Myc. pneumoniae*, *Myc. hominis* and *Myc. urealyticum*, which cause inflammatory disease of the lungs or genitourinary tract. *Mycoplasma genitalium* may also cause genitourinary inflammation.

Analysis of the genome of *Myc. genitalium* has identified 480 protein-coding genes and 37 RNA-coding genes. Of these 517 genes, almost one fifth encode proteins of unknown function. Compare this with the largest known viruses, which cannot survive without the aid of a host cell. Viruses such as poxvirus have 200 to 250 kb of DNA, which means that these life forms have approximately 150 to 200 or so genes.

The big question is: how many genes could this mycoplasma lose before it became non-viable? It is possible that the answer is none. The genome of another mycoplasma, *Myc. pneumoniae*, has also been sequenced and is a genome size of 816 kb in size. Comparison of these two mycoplasma genomes reveals that *Myc. pneumoniae* has orthologues of virtually every one of the *Myc. genitalium* protein-coding genes. Experimental studies with the bacterium *Bacillus subtilis* suggested that it could survive with only 9% of its genome, or some 562 kb of DNA. This is very similar to the size of the genome of *Myc. genitalium*.

To explore the minimum gene set required for the survival of *Myc. genitalium*, use has been made of global transposon mutagenesis to identify those genes that are not essential (Figure 11.3). The methodology of transposon mutagenesis has been described in Section 3.2.2. The insertion of a transposon within an ORF alters the protein being encoded and renders it, in most cases, functionless. In using transposon mutagenesis, the investigators were looking for those recombinant organisms in which the insertion of the transposon was not lethal. Provided the transposon had inserted within a functional gene, and active protein was no longer being produced, then it could be deduced that this particular gene was not essential for survival. The conclusions from this study (Hutchison *et al.*, 1999; see Section 11.8, Papers) suggested that 265 to 350 of the 480 protein-encoding genes of *Myc. genitalium* are essential for the survival of the bacterium when it is grown under laboratory conditions. At the lower limit, the number of genes that may be compatible with the term 'living' is very close to that of the largest virus. Thus the border between living and non-living is being reached (Figure 11.4). This is going to be an area of intense research over the next decade and will redefine our concepts of life and establish it clearly at the molecular level.

11.5 | Bacteria and idiopathic diseases

That bacteria cause bacterial diseases is obvious – or is it? It was not until the latter half of the 19th century that the organisms causing the common bacterial diseases were identified. This created a paradigm for medicine, and during the first half of the 20th century many attempts were made to identify the bacteria responsible for diseases of unknown causation. Such diseases are known as idiopathic diseases and examples include arthritis, psoriasis, certain cancers, certain forms of epilepsy, etc. Many diseases of unknown origin were treated with antibacterial compounds. For example, among the standard drugs used for the treatment of rheumatoid arthritis, an autoimmune disease of the joints, are compounds containing gold. The reason why gold compounds were originally tested in rheumatoid arthritis was because such compounds killed *M. tuberculosis* and were used to treat tuberculosis. At the time, rheumatoid arthritis was thought to be caused by an organism such as *M. tuberculosis*.

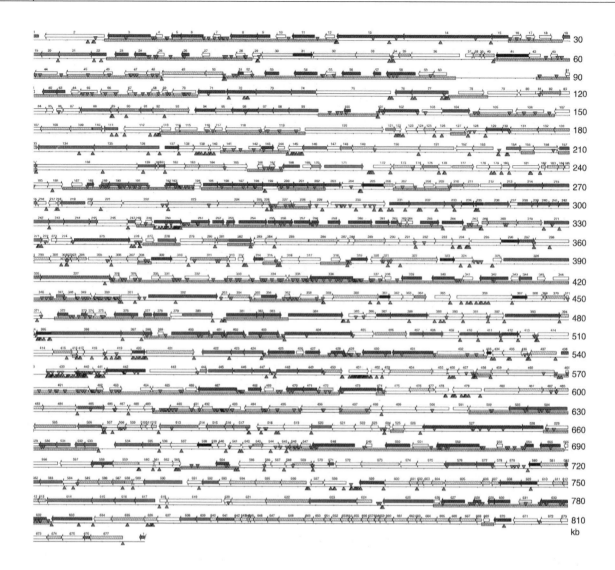

Figure 11.3 Global transposon mutagenesis of *Mycoplasma genitalium*. This diagram shows the *Mycoplasma pneumoniae* genome at a scale of 30 kb per line. The genes are numbered sequentially from 1 to 677. Triangles below the line indicate positions of transposon insertions documented in *Mycoplasma genitalium* mapped onto the *Myc. pneumoniae* genome. Triangles above the line indicate positions of transposon insertions documented in *Myc. pneumoniae*. (Reprinted with permission from Hutchison, C. A., Peterson, S. N., Gill, S. R., Cline, R. T., White, O., Fraser, C. M., Smith, H. O. & Venter, J. C. Global transposon mutagenesis and a minimal mycoplasma genome. *Science*, 1999, **286**, 2165–2169. Copyright 1999 American Association for the Advancement of Science.)

The first few decades of the 20th century saw the popularisation of the idea of focal infection as a cause of disease. It was believed that pockets of bacteria in the mouth, gut and elsewhere could give rise to the symptoms of a variety of human diseases. The obvious way to treat such diseases was, therefore, to remove the bacteria or, failing that, the

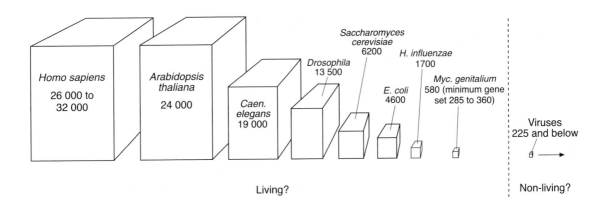

Figure 11.4 Schematic representation (not to scale) of the numbers of genes in the genomes of organisms ranging from the human (approximately 30 000) down to viruses. The minimum genome size compatible with life is estimated to be between 285 and 350 genes, which is very close to the numbers of genes in the largest virus. *Caen. elegans*, *Caenorhabditis elegans*.

tissues containing the bacteria. Thus patients with idiopathic disease had their teeth and portions of gut removed to treat their symptoms. As one would expect, the treatment was often worse than the disease. With the introduction of antibiotics, the focal infection concept had to change organisms and moved from viewing 'straightforward' bacteria as the causative agents of disease to mycoplasmas, bacterial L forms and, eventually, to viruses.

By the 1970s, those researchers hunting an infectious causation of idiopathic diseases were few and far between – largely because they could not get funding. However, for a number of diseases, such as the various forms of arthritis which mysteriously affect the joints and immune system, infection was one of the few likely explanations for the causation of the disease. The past 20 years have been more kind to those who thought that idiopathic diseases were caused by infectious agents. The breakthrough came with a re-evaluation of the causes of a very common human condition – the stomach ulcer.

11.5.1 *Helicobacter pylori*, ulcers and cancer

Would you drink a culture of an unknown bacterium? This is in fact what Barry Marshall, a young Australian medical doctor, did to test the hypothesis that a newly discovered bacterium, *Hel. pylori* was the cause of stomach ulcers (see Section 1.4.1.1.1). Until this experiment, it was dogma that ulcers of the stomach, a very common condition world-wide, were caused by stress. Indeed, the world's best selling drug at the time of this experiment was Zantac™, used to treat stomach ulcers. This targeted a particular histamine receptor, and raised the pH in the stomach. Marshall and his supervisor, J. R. Warren, had discovered curved or spiral Gram-negative bacteria in the stomachs of humans in 1983 and had pioneered the hypothesis that this organism caused stomach ulcers (see Micro-aside, p. 32). However, direct evidence that this bacterium caused stomach ulcers was lacking, owing largely to a failure to infect animals with *Hel. pylori*. To overcome these problems, and to directly test the hypothesis (and to satisfy Koch's postulates; see Table 1.5) Marshall used himself as a guinea pig and carried out the experiment described in Chapter 1. The results of this experiment began to convince the sceptics and, at the time of writing, it is now accepted that stomach ulcers are indeed caused by the bacterium *Hel. pylori*. Treatment of *Hel. pylori*-induced stomach ulceration consists of a proton pump inhibitor, which raises the stomach pH (*Hel. pylori* has evolved to live at low pH) and two antibiotics.

Studies on *Hel. pylori*, stimulated by Marshall's experiment on himself, have revealed that over half the world's population have their stomachs colonised by *Hel. pylori*. Yet, in spite of this, only a proportion of those colonised have any signs of stomach pathology. Thus *Hel. pylori* appears to be a commensal bacterium for a proportion of the population. Their failure to develop pathology may be related to the strain of bacterium colonising the stomach, the genetics of the individual, or a combination of both factors.

More surprising still is the hypothesis that *Hel. pylori* can cause cancer of the stomach (see Micro-aside, p. 32). In addition to the established association between *Hel. pylori* and stomach ulceration there is epidemiological evidence that infection with this bacterium is associated with the development of adenocarcinoma and Hodgkin's lymphoma of the stomach. Many hypotheses have been formulated to explain how the chronic inflammation in the stomach induced by this bacterium can lead to the development of cancer. However, the mechanism or mechanisms responsible have not yet been elucidated. This is one of the most fascinating challenges in Microbiology and will assuredly change our perceptions of bacteria/host interactions.

More recently, evidence has been presented that incriminates another bacterium, *Chlamydia trachomatis*, with cancer at a different anatomical site – the cervix. Analysis of blood samples taken from 530 000 women in Scandinavian countries has shown that those with antibodies to the serotype G strain of *Chl. trachomatis* had an increased risk of developing cervical squamous cell carcinoma.

11.5.2 Bacteria and heart disease

The finding that a bacterium could cause a chronic human disease (in this case stomach ulcers) and, moreover, a human disease that had never been associated with having a bacterial cause, has made clinicians and scientists think again about the relationship between bacteria and idiopathic human diseases. Over the past 5 to 10 years, evidence has accumulated to suggest that there is an association between infection with certain bacteria and coronary heart disease (CHD). This is one of the consequences of the major cardiovascular disease, atherosclerosis (hardening of the arteries). The major arteries (coronary arteries) supplying the heart can become blocked by a fatty deposition on the walls of the blood vessels. This deposition is in fact an inflammatory focus containing macrophages, differentiated macrophages called foam cells, B lymphocytes and T lymphocytes. The occlusion of these coronary arteries limits blood supply to the heart and gives rise to the symptoms of angina. Complete blockage of one of

Table 11.2.	Bacteria implicated in the pathogenesis of coronary heart disease

- *Helicobacter pylori*
- *Chlamydia pneumoniae*
- *Porphyromonas gingivalis*
- *Prevotella intermedia*
- *Actinobacillus actinomycetemcomitans*

the arteries cuts off the supply of blood to one part of the heart and is known as myocardial infarction or, in common parlance, a heart attack.

The causation of CHD has been under active investigation for more than 40 years. It is now clear that atherosclerosis is a very complex (so-called multifactorial) disease and a large number of risk factors have been identified over the past few decades. These include cigarette smoking, obesity, levels of circulating lipoproteins and other diseases such as diabetes.

During the past decade, epidemiological studies have suggested that a number of bacteria may predispose to CHD (Table 11.2). These include the intracellular bacterium *Chl. pneumoniae*, *Hel. pylori* and various oral bacteria implicated as causative agents of periodontitis. This is still a controversial area and the next decade should reveal whether or not the hypothesis that bacteria predispose to atherosclerosis is valid. If bacteria are implicated in this disease, a major question would be: what is the mechanism by which bacteria cause atherosclerosis? There is currently much research aimed at identifying whether these various organisms can be found in diseased blood vessels. However, a very persuasive hypothesis relates the causation of atherosclerosis to the immune response to heat shock proteins (Hsps) or molecular chaperones.

Proteins need to take up a particular three-dimensional configuration in order to be functional. Many proteins within cells require help with their folding. Such help is provided by families of proteins known as molecular chaperones. These proteins increase in concentration when cells are stressed (e.g. by heat) and, in consequence, are known as heat shock proteins or stress proteins. One of the best-studied molecular chaperones is the oligomeric protein known as chaperonin 60 (or heat shock protein (Hsp) 60) (Figure 11.5). This is composed of two rings of seven subunits stacked back-to-back. As well as being a molecular chaperone, chaperonin 60 is also a powerful immunogen. Each time we have a bacterial infection we raise antibodies to chaperonin 60. In our own bodies, chaperonin 60 is found within the mitochondria of cells. Mitochondria have evolved from bacteria captured by eukaryotic cells billions of years ago. The chaperonin 60 protein is an example of an evolutionarily conserved protein and the antibodies raised to bacterial

Figure 11.5 Function and structure of chaperonin 60 (Hsp60). (a) Schematically represents the mechanism by which chaperonin 60, a multimeric protein composed of two stacked rings of seven 60 kDa subunits, folds proteins within its central cavity. In (b) a cryoelectron image of chaperonin 60 is shown. (Reproduced with permission from Ranson, N. A., White, H. E. & Saibil, H. R., 1998, *Biochemical Journal*, **333**, 233–242). © the Biochemical Society.)

(a)

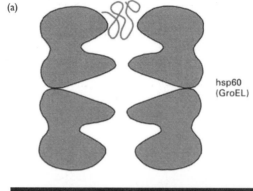

hsp60 (GroEL)

(b)

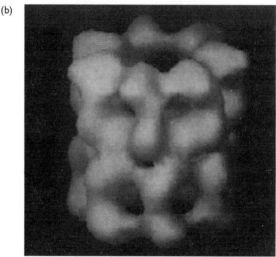

chaperonin 60 could also cross-react with the human chaperonin 60 if the latter protein were to be released from human cells.

It is this piece of information that has been taken by George Wick in Austria and turned into a hypothesis (Figure 11.6). In an attempt to show that atherosclerosis was an autoimmune disease, Wick and colleagues immunised rabbits with components of blood vessels. However, it was found that animals immunised only with the adjuvant (which is composed of freeze-dried *M. tuberculosis*; see Section 6.7.2.2.5) had an increased incidence of experimental atherosclerosis. These animals had raised antibodies to the chaperonin 60 of the *M. tuberculosis* present in the adjuvant. It was then shown that stressed blood vessels would express the mitochondrial chaperonin 60 on the surface of the endothelial cells and that, if antibodies to mycobacterial chaperonin 60 were added to these cells, then they were killed by a complement-mediated cytotoxic mechanism.

This has led Wick to propose the hypothesis that bacterial infections give rise to increased concentrations of antibodies to bacterial chaperonin 60. Blood vessels have areas of high hydrodynamic stress at regions where vessels bifurcate. This stress can lead to the over-expression of chaperonin 60 and its presence on the outer surface of

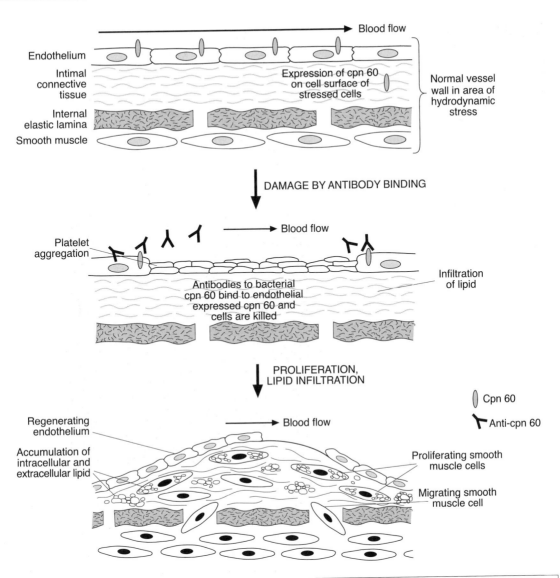

Figure 11.6 Explanation for the role of chaperonin (cpn) 60 in the pathogenesis of atherosclerosis. In areas of high hydrodynamic flow the vascular endothelial cells are stressed and express cpn 60 on their surface. This cpn 60 is recognised by antibodies raised by the individual to bacterial cpn 60. Due to the significant sequence homology, these antibodies to bacterial cpn 60 cross-react with self-cpn 60, causing complement-mediated killing of the endothelial cells. This leads to denudation of the endothelium and starts the process of atherosclerosis.

endothelial cells. Where this happens, antibody to the chaperonin can bind to the vascular endothelial cells and cause activation of complement and complement-mediated cell killing. Such denuded areas of the vasculature are then subject to the process of atherogenesis, with the build-up of atherosclerotic plaque.

This is a very interesting hypothesis, which, if it can be supported by scientific testing, should allow novel therapeutics to be developed to inhibit the process of atherosclerosis.

11.5.3 Other diseases possibly caused by bacteria

11.5.3.1 Psoriasis

Psoriasis is an inflammatory and proliferative disease of the skin, the major symptoms being red plaques on the skin covered by a silvery scaling. The disease consists both of an inflammatory component and an overgrowth of the keratinocytes that make up our skin (see Sections 1.2.1 and 5.3.1). Many explanations have been pursued in the study of the causation of psoriasis and at the present time it is thought to be an immunological disease in which T lymphocytes play a key role in perpetuating inflammation and stimulating the hyperproliferation of keratinocytes. In fact, evidence for the role of T cells has come from experiments in which an experimental drug composed of interleukin (IL) 2 linked to diphtheria toxin has been shown to be beneficial in psoriasis patients. This cytokine-toxin conjugate delivers the killing power of the toxin only to T cells that express the IL-2 receptor, providing a specific way of targeting the effector cells in this disease. This is also an example of the growing clinical applications of bacterial exotoxins (discussed in Section 9.6).

The nature of the antigens driving the T cell response in psoriasis is unclear. Studies of the nature of the T lymphocyte infiltrate in psoriatic skin has utilised methods of identifying the V_β chain of the T cell receptor. As discussed in Section 10.5.2, bacterial superantigens have the property of activating all T cells with a particular V_β chain. As there are only about 25 genes coding for V_β segments, then superantigens tend to activate at least 5% of available T cells. This so-called bias in the V_β repertoire can be recognised using various techniques. There are now a number of studies suggesting that the T cell repertoire in patients with psoriasis shows evidence of the action of a bacterial superantigen. Moreover, the T lymphocytes of patients with psoriasis react to a range of antigens from *Strep. pyogenes* and injection of small numbers of these bacteria into the clinically normal skin of patients with psoriasis induces the classic psoriatic lesion.

These findings suggest that streptococcal (or staphylococcal) infection of the skin of individuals of the correct genetic background can lead to aberrant T cell responses and hyperproliferation of the keratinocyte population. It is not clear whether the continued presence of the bacteria is required, or whether patients develop cross-reactive autoantibodies that recognise both the bacterium and some component of the skin, thus keeping the lesion going.

11.5.3.2 Kidney stones

In Section 11.4, the mycoplasmas were discussed and it was stated that they were the smallest free-living bacteria. In the 1990s even smaller bacteria were found. These cells are as small as 0.05 μm in diameter and have been termed nanobacteria or ultramicrobacteria. Almost nothing is known about these bacteria. The nanobacteria are about the size of the supposed bacteria found in a Martian meteorite recovered from the South Pole that became a *cause célèbre* a few years ago.

One of the most painful conditions known to medicine is the kidney stone. These are formed from calcium salts. In two recent studies of kidney stones, nanobacteria were found in the majority of stones. The reason for this may be the fact that these bacteria incorporate calcium salts, such as hydroxyapatite, on their cell surface and thus the bacteria may act as nucleation centres for ectopic calcification. Studies of nano-bacteria and calcium stones have also revealed that nanobacteria are present in normal human and bovine blood. If they are normally present in blood, these organisms may play roles in other pathological calcification mechanisms. The question of the role of nanobacteria in biology, and specifically in human health, is just beginning to be con-sidered and the next decade should see exciting answers and more questions about these very intriguing microbacteria.

11.5.3.3 Asthma and the hygiene hypothesis

The final example of the proposed role of bacteria in idiopathic human disease is both counterintuitive and fascinating as it concerns the allergic diseases, of which the most severe is asthma. This is a chronic inflammatory disease of the airways and one that is clinically serious and difficult to treat. There has been a remarkable increase in the incidence of allergic diseases, including asthma, in the past few decades. Initially, this was believed to be related to the enormous increase in atmospheric pollution, due to industry and cars. However, this hypothesis has been refuted by the finding that the incidence of asthma in the former Soviet block countries has not increased in line with that in Western Europe. In these communist countries the levels of industrial/transport pollution is substantially greater than in Western Europe and, therefore, they should have been experiencing an even larger rise in asthma incidence if the pollution hypothesis was correct. If this is not the explanation, then what is? A fascinating, but still controversial, hypothesis (the hygiene hypothesis) has been sug-gested to account for this increase in allergic disease. It proposes that in present-day Western society, with its small family sizes, high standards of living and hygiene and mass vaccination programmes, the chances of catching infectious diseases are limited. As described in Chapter 6, our immune systems appear to have evolved to recognise bacteria, particularly intracellular bacteria such as *M. tuberculosis*, and deal with them via Th1 lymphocyte responses that activate macrophages both directly and indirectly. The alternative Th2 lymphocyte response is focused on larger parasites, such as filarial worms, and requires activation of eosinophils and mast cells. It is the products of these eosinophils and mast cells that are responsible for the signs of allergy and asthma. From *in vitro* experiments, and studies in the intact mouse, these Th1 and Th2 responses exhibit reciprocal regula-tion. It is therefore contended that if early infections are not occurring, then Th1 responses do not fully develop and are unable to exert control over the development of Th2 responses. These Th2 responses may therefore, in individuals with the correct genetic background, become dominant and induce pathology.

Evidence supporting this hypothesis has come from a number of international epidemiological studies and infections with pathogens such as *M. tuberculosis*, hepatitis A and measles virus have been suggested to prevent the induction of allergic responses. In experimental studies, infection of mice with *M. bovis*-BCG (bacillus Calmette–Guérin) was found to suppress experimental asthma. Of course, while these early studies look extremely interesting, much more work is required to fully test the hygiene hypothesis and to identify the potential mechanisms by which microbial-induced development of our immune responses prepares us for living in our environment.

11.6 | Conversations with the normal microflora

The last section of this glimpse into the crystal ball is focused on one of the most mysterious aspects of microbiology – the microflora of *Homo sapiens*. As has been explained in various part of this book, we pick up these microscopic partners at birth and live with them until the day we die. It is worth while reminding the reader, yet again, that there are approximately 10 bacteria in our bodies for every one of our own cells. The ability of members of the normal microflora to cause disease has been described in Section 1.3.2. The possibility that they may be able to down-regulate pro-inflammatory responses has been described (see Micro-aside, p. 531). In this section of the final chapter, the possibility that the normal microflora have additional functions, such as controlling body shape and patterning, will be discussed briefly.

11.6.1 *Vibrio fischeri* and the bobtail squid

Little swirls of sand make way for bigger ones and suddenly, as if from nowhere, appears the body of the squid. Up, up and away, leaving the sea floor and making for the surface is *Euprymna scolopes*, otherwise known as the bobtail squid. This nightly occurrence is only found in the Hawaiian archipelago where the dozen or so species of this family of squid live.

The bobtail squid has come to microbiological prominence in the past decade because of the fascinating association this organism has with one particular bacterium, *Vibrio fischeri*. This bacterium lives in the light organ of the squid where, each day, it reaches a very high cellular density and because of this is able to switch on the *lux* operon and emit light. It is the emission of this light that allows the bobtail squid to avoid predators. The strategy is called counter-illumination and it is hypothesised that the production of light by the ventral surface prevents the animal casting a shadow, which could be seen by predators. The mechanism by which the *lux* operon is switched on, the process

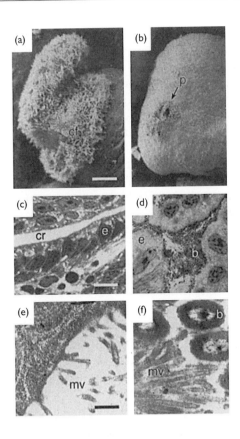

Figure 11.7 Comparison of developmental changes in the juvenile light organ of the bobtail squid over the first four days from hatching. In the left panel of micrographs (A, C and E) the animals have not been exposed to *Vibrio fischeri*, while in the right panel (B, D and F) the animals have been exposed to this bacterium. cr, crypt; e, epithelial cell; cf, ciliated field; p, pores; mm, microvillus; b, bacteria. Bars: (a), (b) 50 μm; (c), (d) 20 μm; (e), (f), 0.5 μm. (Reproduced with permission from Hirsch, A. M & McFall-Ngai, M. J. Fundamental concepts in symbiotic interactions: light and dark, day and night, squid and legume. *Journal of Plant Growth Regulation*, 2000, **19**, 113–130. Copyright 2000 Springer-Verlag.)

known as quorum sensing, has been described in Section 4.3.2.1 and will not be discussed further. In this section the relationship between *V. fischeri* and the generation of the squid's light organ will be described.

Extensive experimental studies of the colonisation of the light organ of the bobtail squid have been conducted over the past decade. Newly hatched bobtail squids have a nascent light organ composed of crypts lined with epithelial cells and there are three pores leading from the light organ to the external environment. Of importance is the fact that this nascent light organ is free from bacteria. Animals can be kept in germ-free environments or various bacteria or groups of bacteria can be added to their water to study the process of colonisation. If newly hatched squids are kept in water containing *V. fischeri*, then the light organ is colonised within hours by this bacterium. Indeed, upon hatching of the squid, the epithelial cells of the light organ appear to function to capture *V. fischeri*. With the capture of the bacteria, the light organ undergoes extensive morphogenesis (Figure 11.7). Animals kept in germ-free conditions, or cured before 8 hours after infection with bacteria (by using antibiotics), do not undergo this extensive morphogenetic programme, which lasts four days. There is, as far as we are aware, no human analogy of the light organ of the bobtail squid. But it is equivalent to growing one's spleen or kidneys only after one had been born and had been colonised by one's microflora.

The interaction of *V. fischeri* with the light organ of the bobtail squid appears to induce a two-way process in both organisms. This is obviously due to the passage of signals between bacteria and light organ and *vice versa*. Little is currently known about the nature of such signals. The bacteria, upon colonisation, speed up their rate of growth until their numbers reach a plateau. They also become smaller in size and lose their characteristic flagella. In the adult squid, the light organ contains 10^9 *V. fischeri*. This is an extremely large number of bacteria for a small squid to contain. In spite of these huge numbers, the bacteria fail to invade adjacent tissues, suggesting very tight mutual control of bacterial behaviour. Another striking finding is that the light organ, which is open to the environment through pores, is never colonised by the many other bacteria present in the ocean.

Colonisation of the light organ triggers a developmental programme in the squid (Figure 11.7). The interaction of *V. fischeri* with the cells lining the nascent light organ causes these cells to undergo terminal differentiation – a process in which they become more cuboidal and undergo a four-fold increase in volume. These cells contain microvilli and the cell extensions also change and become more extensive and branched, thus covering more of the surface of the cells, and allowing greater contact with the bacteria. These changes in the lining cells are reflected by changes in the crypts themselves, which expand. The ciliated microvillus epithelial structures that line the pores of the light organ, and aid the initial colonisation, then undergo a process of regression.

Here is a wonderful example of a symbiosis that involves major changes both in the bacterium and in the host, with the ultimate aim of fooling predators by allowing the squid to not cast a shadow. The rough details of the molecular mechanisms that have to be invoked to begin and sustain this symbiosis are known but much more work needs to be done to fully map out what is happening in this fascinating prokaryotic–eukaryotic marriage. The use of natural and transposon-generated mutants will allow the genes in the bacterium required for colonisation to be determined. It will be much more difficult to address the nature of the genes in the bobtail squid required to enable the light organ to be colonised and those involved in the morphogenesis of the adult light organ. However, understanding this fascinating symbiosis is important for our understanding of bacteria–host interactions generally.

11.6.2 The mammalian intestine: another host–microbe interface

The bobtail squid light organ enlarges and undergoes significant cellular changes upon the introduction of *V. fischeri*. Is this simply an isolated, and therefore unique, example of the outer limits of bacteria–host symbiosis, or is it part of the normal distribution of such

Table 11.3.	Changes in physiology in germ-free rodents
Physiological characteristic	**Alteration in gnotobiotic animals**
Physiological	
Basal metabolic rate	Decreased
Cardiac output	Decreased
Blood volume	Decreased
Small bowel motility	Slower
Weight of caecum	Greater
Urea in caecal contents	Present (but none in conventionalised animals)[a]
Ammonia in caecal contents	Less
Ammonia in portal blood	Less
Bile acid transformation in intestine	None (occurs in conventionalised animals)[a]
Trypsin activity in large bowel contents	Present (none in conventionalised animals)[a]
Immunological	
Mucosal surface area in small bowel	Decreased
Lamina propria	Thinner
Myeloid cells in bowel	Less
Ileocaecal lymph nodes	Smaller
Serum immunoglobulins	Less

[a] Conventionalised animals were gnotobiotic but have been recolonised with their normal microflora.

interactions? We do not have to look too far to see that other bacteria–host systems can involve the alteration of tissue behaviour. The example that will be given is the intestinal microflora of the gnotobiotic mouse. Gnotobiotic (or 'germ-free') animals were first produced at the turn of the 19th century, but the technology only became established for producing large numbers of germ-free animals in the 1950s and 60s. Animals are delivered by aseptic caesarean delivery, kept in isolators and fed only germ-free food. It has been noted over the past 30 years that many physiological and immunological changes occur in these gnotobiotic animals (see Table 11.3).

Louis Pasteur, the great French pioneer of Bacteriology, believed that we could not live without our normal microflora. As far as we can tell, at least with certain animals, he was wrong. There have also been examples, although only a few, of humans having to be kept under sterile conditions, although it is not clear that they were totally germ-free. Having said this, it is always good in science to have doubts about any technique. It is now clear that so many different bacterial and archaeal species exist and can survive under extremes of conditions that they may be able to populate so-called germ-free animals. For example, are germ-free animals free from nanobacteria? However, in Section 11.6 we will assume that the germ-free mouse is free from bacteria.

11.6.2.1 Alterations in intestinal anatomy/renewal in gnotobiotic mice

The intestine of the germ-free mouse has been studied in some detail. What are the major changes that occur in this tissue, which normally contains huge numbers of diverse bacteria interacting with host cells? The most visible change in the intestine of the germ-free animal is a dramatic enlargement of the caecum. The caecum is the first 5 to 8 cm of the large intestine and is in the form of a blind pouch located, in the human, in the lower right quadrant of the abdomen. It is a thin-walled, highly vascularised organ with a muscular wall enabling mixing of the caecal contents. The characteristic caecal enlargement in germ-free animals can be reversed by mono-infection with the bacterium *Peptostreptococcus micros*. Gnotobiotic rodents also show changes in gut motility. This appears to be due to a slowing in the spread of motor signals in the germ-free intestine. These differences can be eliminated by reintroducing the normal gut microflora to rodents. Another change is in the length of the villi of the small intestine. In germ-free mice the villi (finger-like projections that increase the surface area so aiding resorption of dietary components, as discussed in earlier chapters) are longer than in matched non-gnotobiotic controls. They consist of four major cell populations (absorptive enterocytes, goblet cells, enteroendocrine cells and Paneth cells; see Sections 5.3.1 and 5.7) but this discussion will be limited to the enterocytes. These cells arise in the crypts of Lieberkühn from multipotent stem cells located near the base of each crypt. Each crypt in the mouse produces, on average, 12 cells per hour. Each enterocyte migrates to the top of the villus in a period of two to five days. Once they reach the top of the villus, cells apoptose and exfoliate. Thus there is constant renewal of the cell populations of the villi of the intestines. The rate of renewal of these cell populations is very impressive. For example, it is estimated that the mouse will turn over a mass of intestinal epithelial cells equal to the body mass (around 25 g) in four months. Similar estimates of cell turnover in humans come up with figures of 250 g of intestinal epithelial cells being produced each day of our lives. In the absence of intestinal bacteria, the process of villus formation continues, but at a slower rate than in conventional animals.

11.6.2.2 Alterations in the GALT of gnotobiotic mice

In Section 5.7.1, the gut-associated lymphoid tissue (GALT) was described and the role of M cells in sampling gut antigens introduced. If the gut does not contain bacteria, what effect does this have on the GALT and the immune system generally? It has been known for years that germ-free animals have less circulating immunoglobulin, suggesting that much of our circulating antibody recognises components of our microflora. In the intestine, reintroduction of bacteria into gnotobiotic mice results in an increase in the number of intra-epithelial lymphocytes. However, much needs to be done to address the question of how the gut microflora controls the GALT.

11.6.2.3 Alteration in epithelial cell differentiation in gnotobiotic mice

In the last few years, work from Jeffrey Gordon's laboratory in Washington University School of Medicine in St Louis, Missouri, has opened up new avenues for the exploration of the bacteria–intestine symbiosis. This work involves the glycosylation patterns of intestinal epithelial cells in gnotobiotic and normal mice. Many proteins in eukaryotes have oligosaccharides or glycolipids attached to them and this process is termed glycosylation. It is now possible to recognise specific glycosylation patterns on cells by using proteins called lectins that bind only to specific oligosaccharides. In the 1980s it was found that colonisation of germ-free mice with the normal gut microflora led to changes in the glycosylation pattern of intestinal epithelial cells. In the presence of the normal gut microflora, these cells produced an enzyme called GDP-fucose asialo-GM1α1,2-fucosyltransferase. This caused an increased expression of fucosyl asialo-GM1, a neutral glycolipid, and was associated with a decrease in expression of asialo-GM1 on intestinal epithelial cells. Gordon and colleagues have followed up these initial observations using *Ulex europeaeus* agglutinin (UEA-1) lectin, which binds to fucose α-1,2-galactose (Fucα1,2Gal). Following groups of gnotobiotic and conventional postnatal mice from birth for binding of this lectin to epithelial cells, it was found that there was no detectable binding before day 17. Between days 17 and 21, scattered cells in both groups bound the lectin. However, after postnatal day 21, the germ-free animals failed to express Fucα1,2Gal while the normal animals expressed Fucα1,2Gal epitopes in all enterocytes in the villi. Expression is found only in the distal part of the small intestine. Thus there is a major difference in the expression of this Fucα1,2Gal glycoconjugate in germ-free animals. However, if animals are colonised by the normal microflora of mice, the normal glycosylation pattern occurs and is associated with the transcription of the gene encoding GDP-L-fucose:β-D-galactosidase 2-α-L-fucosyltransferase in epithelial cells. Surprisingly, reintroduction of one single bacterial species, *Bacteroides thetaiotamicron*, can reproduce this reinitiation of glycosylation. This organism is an abundant member of the gut microflora of the mouse and also of humans. It has the capacity to utilise a wide variety of host cell glycans both in culture and *in vivo*. Thus it appears that, in the presence of this bacterium, the host epithelial cells undergo specific gene transcription in order to provide the appropriate nutrients. The effect does not seem to occur by direct cell–cell contact.

How does this bacterium alter gene expression in intestinal epithelial cells? For a start, the change in glycosylation depends on there being sufficient numbers of *Bac. thetaiotamicron* present in the gut. To explore the interaction of this bacterium with the host, Gordon and colleagues resorted to the use of transposon mutagenesis (a technique that has been used in many other experimental studies described in earlier chapters) to identify the bacterial genes involved in the process. One mutant, termed Fu-4, had the transposon inserted in a

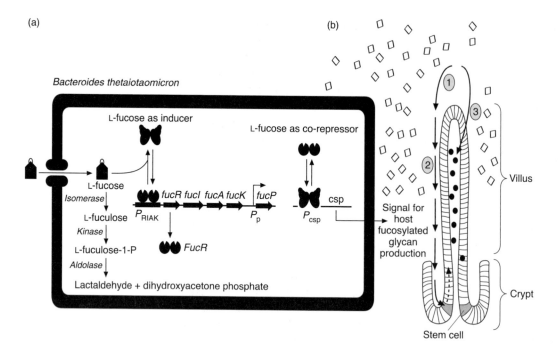

Figure 11.8 Model of the interaction of *Bacteroides thetaio-tamicron* and the intestinal entero-cyte. For fuller explantation, refer to text. In (a) L-fucose is proposed to act as an inducer of the fucose utilisation operon and as a co-repressor of a second locus termed control of signal production (*csp*). The *csp* locus produces a signal (◇) whose target is currently unkown, but may be one or more of the following: (i) enterocytes in the villus epithe-lium, (ii) stem cells in the crypt or (iii) components of the GALT. P_{RIAK}, promoter for the operon containing the genes *fucR*, *fucI*, *fucA* and *fucK*; P_{csp}, promoter for the *csp* operon; P_p promoter for *fucP*. (Reproduced from Hooper, L. V., Falk, P. G. & Gordon, J. I. (2000) Analysing the molecular foundations of commensalism in the mouse intestine. *Current Opinion in Microbiology* **3**, 79–85 with permission.)

gene identified as *fucI* which encodes L-fucose isomerase, an enzyme which can convert L-fucose to L-fuculose. This mutant proved to be unable to signal to the host to produce appropriate fucosylated glycans on the surface of intestinal epithelial cells (Figure 11.8). Mutants of *fucA* and *fucK* were not significantly different from wild-type bacteria. This suggested that the product of L-fucose isomerase is a signal that triggers host enterocytes. However, when the gene *fucR* was disrupted, which would eliminate production of FucR, FucI, FucA and FucK (but not FucP, see Figure 11.8) this mutant behaved like the wild type. This rules out L-fuculose as a signal and suggests that there is a second genetic locus, which Lora Hooper and colleagues have termed *csp* (control of signal production), that signals to the host. Hooper and colleagues have put forward the model to explain signalling to the host. Thus L-fucose acts through FucR as an inducer of the transcription of the fucose metabolism operon and as a repressor of *csp*. Knockout of *fucI* leads to a build-up of L-fucose, increases the proportion of FucR with bound L-fucose, and results in the silencing of *csp*. A prediction from the model is that lowering L-fucose levels will relieve *csp* repression and increase host signalling. This has been found with mutants of the L-fucose permease gene (*fucP*) which reduces import of this sugar and increases host signalling. Thus we have a nice feedback system from bacteria to host. When L-fucose is abundant, the *csp* is repressed and no signalling occurs. When levels of L-fucose decrease, repression of *csp* is lifted and host signalling molecules can be produced.

What is the benefit to the host of providing nutrients for bacteria? An obvious one is that it can exert control over the microflora, in terms of its composition and the number of bacteria present.

These studies are showing how one single member of the normal microflora of the mouse is interacting with the adjacent host epithelial cells to produce a situation that is, presumably, beneficial to both species. It is likely that this is just one of many sets of interactions that must be going on between the host and its microflora to ensure that both the prokaryotic and the eukaryotic cells of the composite organism (e.g. the reader) are satisfied. This area of host–bacteria interaction is going to be one of the most complex and exciting areas of research and we expect many illuminating results to appear during the next decade.

11.6.3 *Wolbachia pipientis* and insect reproduction

The final example of bacteria–host interactions that is generating great excitement at the present time is the interaction of the intracellular bacterium *Wolbachia pipientis* with insects. *Wolbachia* spp. were first reported in the 1920s as an unnamed rickettsia in the ovaries of the mosquito *Culex pipiens*. However, it was not until the 1970s that it was discovered that the organism affected the breeding of these insects: uninfected eggs from this mosquito, if fertilised with sperm infected with *W. pipientis*, died. It was then found that other insects, including the fruitfly much beloved of experimenters, *Drosophila*, were also infected by this bacterium and suffered the same problems. The discoveries that have made the study of this bacterium imperative are that infection with the organism can cause many alterations in insect reproduction, including inducing parthenogenesis (virgin birth), feminisation and male killing (infected male embryos die, while infected female embryos turn into infected female adults).

Wolbachia pipientis is an obligate intracellular bacterium that makes study of its mechanisms of 'virulence' difficult. The genome of this organism has just been sequenced, which will make study of bacteria–host interactions more tractable. No further information on this fascinating organism and its interactions with insects will be provided and the interested reader should read the review by Stouthamer and colleagues (1999) (see Section 11.8, Review articles).

Wolbachia spp. also infect filarial nematodes, including those responsible for two horrendous diseases – river blindness and elephantiasis. These are severe diseases endemic in Africa. There is now preliminary evidence to suggest that the inflammatory pathology of these infections is driven by the *Wolbachia* and that these diseases may be cured by treatment with antibiotics such as tetracycline.

11.7 | Summary

The major advances in bacterial pathogenesis have been presented in the preceding chapters of this book. In this chapter a number of areas where rapid advances are likely to be made, or where the area is likely to produce revolutionary ideas in the future, have been discussed. The authors hope that the readers have shared in their excitement about the new science of Cellular Microbiology and that they have a chance to involve themselves in its future development.

11.8 | Further reading

Books

Higgs, G. A. & Henderson, B. (eds.) (2000). *Novel Cytokine Inhibitors*. Basel: Birkhauser.

Review articles

Bax, R., Mullan, N. & Verhoef, J. (2000). The millennium bugs – the need for and development of new antibacterials. *Int J Antimicrob Agents* **16**, 51–59.

Binder, S., Levitt, A. M., Sacks, J. J. & Hughes, J. M. (1999). Emerging infectious diseases: public health issues for the 21st century. *Science* **284**, 1311–1313.

Bush, K. & Macielag, M. (2000). New approaches in the treatment of bacterial infections. *Curr Opin Chem Biol* **4**, 433–439.

Cohen, M. L. (2000). Changing patterns of infectious disease. *Nature* **406**, 762–767.

de Boer, O. J., van der Wal, A. C. & Becker, A. E. (2000). Atherosclerosis, inflammation, and infection. *J Pathol* **190**, 237–43.

Handfield, M., Brady, L. J., Progulske-Fox, A. & Hillman, J. D. (2000). IVIAT: a novel method to identify microbial genes expressed specifically during human infections. *Trends Microbiol* **8**, 336–338.

Hooper, L. V., Bry, L., Falk, P. G. & Gordon, J. I. (1998). Host–microbial symbiosis in the mammalian intestine: exploring an internal ecosystem. *BioEssays* **20**, 336–343.

Hooper, L. V., Falk, P. G. & Gordon, J. I. (2000). Analysing the molecular foundations of commensalism in the mouse intestine. *Curr Opin Microbiol* **3**, 79–85.

Kotra, L. P., Vakulenko, S. & Mobashery, S. (2000). From genes to sequences to antibiotics: prospects for future developments from microbial genomics. *Microbes Infec* **2**, 651–658.

Kurz, C. L. & Ewbank, J. J. (2000). *Caenorhabditis elegans* for the study of host-pathogen interactions. *Trends Microbiol* **8**, 142–144.

Loferer, H., Jacobi, A., Posch, A., Gauss, C., Meier-Ewert, S. & Seizinger, B. (2000). Integrated bacterial genomics for the discovery of novel antimicrobials. *Drug Discovery Today* **5**, 107–114.

Mäkelä, P. H. (2000). Vaccines, coming of age after 200 years. *FEMS Microbiol Rev* **24**, 9–20.

Muhlestein, J.B. (2000). Chronic infection and coronary artery disease. *Med Clin North Am* **84**, 123–48.

Relman, D. A. (1999). The search for unrecognised pathogens. *Science* **284**, 1308–1310.

Rosamond, J. & Allsop, A. (2000). Harnessing the power of the genome in the search for new antibiotics. *Science* **287**, 1973–1976.

Skov, L. & Baadsgaard, O. (2000). Bacterial superantigens and inflammatory skin diseases. *Clin Exp Dermatol* **25**, 57–61.

Stouthamer, R., Breeuwer, J. A. J. & Hurst, G. D. D. (1999). *Wolbachia pipientis*: microbial manipulator of arthropod reproduction. *Annu Rev Microbiol* **53**, 71–102.

Sulakvelidze, A., Alavidze, Z. & Morris, J. G. (2001). Bacteriophage therapy. *Antimicrob. Agents Chemother* **45**, 649–659.

Summers, W. C. (2001). Bacteriophage therapy. *Annu Rev Microbiol* **55**, 437–451.

Tan, Y.-T., Tillett, D. J. & McKay, I. A. (2000). Molecular strategies for overcoming antibiotic resistance in bacteria. *Molecular Medicine Today* **6**, 309–314.

Von Hertzen, L. C. (1998). The hygiene hypothesis in the development of atopy and asthma – still a matter of controversy? *Q J Med* **91**, 767–771.

Walsh, C. (2000). Molecular mechanisms that confer antibacterial drug resistance. *Nature* **406**, 775–781.

Washburn, M. P. & Yates, J. R. (2000). Analysis of the microbial proteome. *Curr Opin Microbiol* **3**, 292–297.

Papers

Anttila, T., Saikku, P., Koskela, P., Bloigu, A., Dillner, J., Ikaheimo, I., Jellum, E., Lehtinen, M., Lenner, P., Hakulinen, T., Narvanen, A., Pukkala, E., Thoresen, S., Youngman, L. & Paavonen, J. (2001). Serotypes of *Chlamydia trachomatis* and risk for development of cervical squamous cell carcinoma. *J Am Med Assoc* **285**, 47–51.

Darby, C. M., Cosma, C.L., Thomas, J.H. & Manoil, C. (1999). Lethal paralysis of *Caenorhabditis elegans* by *Pseudomonas aeruginosa*. *Proc Natl Acad Sci USA* **96**, 15202–15207.

Hooper, L. V., Xu, J., Falk, P. G., Midtvedt, T. & Gordon, J. I. (1999). A molecular sensor that allows a gut commensal to control its nutrient foundation in a competitive ecosystem. *Proc Natl Acad Sci USA* **96**, 9833–9838.

Hutchison, C. A., Peterson, S. N, Gill, S. R., Cline, R. T., White, O., Fraser, C. M., Smith, H. O. & Venter, J. C. (1999). Global transposon mutagenesis and a minimal mycoplasma genome. *Science* **286**, 2165–2169.

International Human Genome Sequencing Consortium. (2001). Initial sequencing and analysis of the human genome. *Nature* **409**, 860–921.

Kajander, E. O. & Ciftcioglu, N. (1998). Nanobacteria: an alternative mechanism for pathogenic intra- and extra-cellular calcification and stone formation. *Proc Natl Acad Sci USA* **95**, 8274–8279.

Langworthy, N. G., Renz, A., Mackenstedt, U., Henkle-Duhrsen, K., de Bronsvoort, M. B., Tanya, V. N., Donnelly, M. J. & Trees, A. J. (2000). Macrofilaricidal activity of tetracycline against the filarial nematode *Onchocerca ochengi*: elimination of *Wolbachia* precedes worm death and suggests a dependent relationship. *Proc R Soc Lond B Biol Sci* **267**, 1063–1069.

Levy, O. (2000). A neutrophil-derived anti-infective molecule: bactericidal/permeability-increasing protein. *Antimicrob Agents Chemother* **44**, 2925–2931.

Pizza, M., Scarlato, V., Masignani, V., Guilianic, M. M., Arico, B., Comanducci, M., Jennings, G. T., Baldi, L., Bartolini, E. *et al.* (2000). Identification of vaccine candidates against serogroup B meningococcus by whole genome sequencing. *Science* **287**, 1816–1820.

Ryby, E. G. & McFall-Ngai, M. J. (1999). Oxygen-utilising reactions and symbiotic colonisation of the squid light organ by *Vibrio fischeri*. *Trends Microbiol* **7**, 414–420.

Taylor, M. J., Cross, H. F. & Biol, K. (2000). Inflammatory responses induced by the filarial nematode *Brugia malayi* are mediated by lipopolysaccharide-like activity from endosymbiotic *Wolbachia* bacteria. *J Exp Med* **191**, 1429–1436.

Venter, J.C., Adams, M.D., Myers, E. W., Li, P. W., Mural, R. J., Sutton, G. G., Smith, H. O., Yandall, M., Evans, C. A. *et al.* (2001). The sequence of the human genome. *Science* **291**, 1304–1351.

11.9 | Internet links

1. www.sanger.ac.uk/Projects/C_elegans
   A comprehensive database of genetic and molecular data on *Caenorhabditis elegans*.

2. http://www.niaid.nih.gov/dmid/genomes
   Latest information on the genome sequencing projects supported by the National Institute of Allergy and Infectious Diseases, Maryland, USA.

3. http://www.cdc.gov/ncidod/eid/about.htm
   Free access to the journal *Emerging Infectious Diseases* published by the National Center for Infectious Diseases, Georgia, USA.

4. http://www.healthsci.tufts.edu/apua/apua.html
   The website of the Alliance for the Prudent Use of Antibiotics. Up-to-date information on antibiotics and bacterial resistance to them. Many useful links to other websites.

Appendix A

Glossary of terms used

ABC transporter. One of a family of transporter systems with common molecular architecture. The hydrolysis of ATP drives the import/export of substrates.

Accessory cells. Immune cells that form part of immune defences but are not antigen specific. Examples include mast cells, natural killer cells and phagocytes.

Acidophile. An organism that prefers an acidic environment for growth.

Aciduric. An organism that can survive exposure to low pH values.

Acquired immunity. As opposed to innate immunity. The immunity generated only after an individual has been exposed to a pathogen (or specific immunogen).

Actin filaments. Key components of the cytoskeleton of eukaryotic cells.

Acute disease (infection). A disease (infection) that lasts only a short period of time.

Acute phase proteins. A group of proteins including C-reactive protein (CRP) and serum amyloid protein (SAP) that have opsonic and other actions against bacteria and protozoans.

Acute phase response. Part of the innate defence mechanism against pathogens. Refers largely to the induction of a group of proteins that can increase in concentration, in the blood, by up to three log orders.

Acylation. The attachment of a lipid molecule to a protein.

Adapter proteins. In cell signalling – proteins that act to connect one cell signalling protein with another, thus allowing passage of the signal.

Adenocarcinoma. Malignant growth of endocrine (glandular) tissue.

Adenylyl cyclase. Enzyme responsible for synthesis of cyclic AMP (c-AMP).

Adhesin. The particular molecule (or part of the molecule) on a bacterium (usually on one of its organelles) that is responsible for mediating adhesion of the organism to a host cell, to another bacterium or to an inanimate surface.

Adhesion molecules. Present in eukaryotic organisms and evolved to aid cell–cell interactions. Examples of these are integrins, selectins, CD44, CD4, etc.

Adhesiotope. The region of a receptor molecule that interacts with a complementary structure (epitope) on the adhesin molecule thereby enabling specific binding of the adhesin to its receptor.

Adjuvant. Material added to an immunogen to increase its ability to induce the immunised species to raise a specific immune response. Common adjuvants include bacterial components and alum.

Aerobe. An organism that needs oxygen for growth.

Agonist. Any molecule that binds to a receptor and produces a biological effect.

Alleles. Variants of a single genetic locus.

Allergens. Antigens that elicit hypersensitivity or allergic reactions.

Allergy. Inappropriate immune response to environmental antigens (pollen, dander, etc.).

Alternative pathway. In complement activation, the pathway that is triggered by binding of C3b to the pathogen.

Alveoli. The terminal, sack-like structures at the end of bronchioles that are responsible for gaseous exchange in animals.

Anaerobe. An organism that does not need oxygen for growth and is killed by it.

Anaphylatoxins. Specifically C3a, C3a and C5a – products of the process of complement activation that have pro-inflammatory actions.

Anergy. A state of non-responsiveness to antigen.

Angiogenesis. The origin and development of blood and lymphatic vessels.

Anoxic. Without oxygen.

Antagonist. Molecule that prevents an agonist binding to its receptor.

Anterior nares. The nostrils.

Antibody. Protein present in blood that binds specifically to a cognate molecule known as an antigen. Antibodies are produced in response to immunisation with antigen.

Antigen. Molecule that reacts with its corresponding antibody.

Antigen presentation. The display of an antigen, in the form of peptide fragments (or non-peptidic moieties) linked to MHC proteins, on the surface of a cell.

Antigen-presenting cell (APC). Specialised cells that process antigens intracellularly and present them on their surface in combination with MHC class II proteins.

Antigenic variation. Alteration of bacterial surface antigenic properties to avoid host immune responses.

Antisense. In generic terms, a sequence of DNA or RNA complementary to another segment of DNA/RNA.

Antisense therapy. Treatment of a disease state by using specific antisense oligonucleotides to block the translation of a particular mRNA.

Apnoea. The cessation of breathing.

Apoptosis. Also know as programmed cell death (PCD) – a natural programme of cell manipulation that leads to the death of the cell with few sequelae.

Apoptosome. A complex of cytochrome *c*, APAF-1 and caspase 9 that is important in the control of apoptosis.

Archea. Single-celled organisms that are bacteria-like but have cell walls consisting of pseudopeptidoglycan or protein only. Most are anaerobes and thrive under unusual growth conditions e.g. hot springs, extremely salty water, highly acid or alkaline soils.

Arachidonic acid. Also known as eicosatetraenoic acid. It can form side chains of membrane phospholipids and be a substrate for cyclo-oxygenases and lipoxygenases that produce prostanoids and leukotrienes. The latter substances are important control factors in inflammation.

Arthropods. The largest group of invertebrates – includes insects and spiders.

Asthma. Inflammatory disease of the lungs due to some form of allergic reaction leading to problems in breathing.

Atherogenesis. Process of development of atherosclerosis.

Atherosclerosis. Thickening, and loss of elasticity, of arterial walls resulting in interference with blood supply.

Attenuation. (i) A process used to produce non-virulent strains of a pathogenic organism so that they can be used for immunisation purposes. (ii) A means of controlling gene expression in which transcription is terminated before a full-length mRNA transcript is produced.

Autoantibodies. Antibodies reactive against self-antigens.

Autoantigens. Self-antigens that are recognised by the individual's own immune system.

Autoimmunity. The process by which the immune system recognises self-antigens. This occurs normally in most individuals but can give rise to pathology.

Auto-inducer. A molecule, secreted by bacteria, that is involved in quorum sensing. On reaching a critical concentration, it can induce the activation (or repression) of gene transcription.

Autolysin. An enzyme able to hydrolyse the peptidoglycan of the organism producing it. Such enzymes are involved in the formation of new cell wall material in the growing bacterium.

Autophosphorylation. A process in which an enzyme (termed a 'kinase') undergoes self-phosphorylation, i.e. it adds a phosphate group to itself.

Autotransporter. A protein secreted by a Gram-negative bacterium which itself contains all of the information required to ensure its translocation across the outer membrane.

Axenic. A pure culture of an organism.

Axillae. The armpits

B cells. *See* B lymphocytes.

B lymphocytes. One of the two major classes of lymphocyte. They are responsible for the synthesis of antibodies.

Bacteraemia. The transient presence of bacteria in the bloodstream, cf. septicaemia.

Bacteriophage. A virus that infects a bacterium.

β-Barrel. Protein tertiary structure motif formed by the organisation of multiple β-sheets.

Basophil. Leukocytes in blood which contain granules that stain with basic dyes.

Biofilm. An accumulation of microbes and their extracellular polymers on a surface.

Bioinformatics. The study of the management of biological information. Often implies use of biological databases of sequences and structures.

Biotechnology. The use of living organisms in industrial processes.

Bone marrow. The site of haematopoiesis, i.e. the place where all the cells of the blood originate. One of the major primary lymphoid organs.

Cadherins. Adhesion molecules that act to glue epithelial cells together.

Caecum. The blind sac-like tissue that is present at the commencement of the colon.

Cancer. A general term describing any malignant growth in any part of the body. It is due to dysregulation of the cell cycle or apoptotic mechanisms.

Capsule. Polysaccharide or protein matrix associated with the bacterial cell surface. Functions in attachment and in resistance to phagocytosis.

Carrier. An individual who is colonised by a pathogenic organism but who is not suffering from any disease caused by that organism.

Caspase. Protease involved in the process of apoptosis.

Catalase. An enzyme capable of cleaving the damaging chemical hydrogen peroxide to water and oxygen.

Cathepsins. A family of proteases.

Catheter. An indwelling surgical instrument for withdrawing or administering fluids through blood vessels or the urethra.

Caveolae. Flask-like indentations in the cytoplasmic membrane of a wide range of eukaryotic cells that are rich in the cholesterol-binding protein caveolin. They are important in signal transduction and in transport as they are capable of vesicle formation.

CD. *See* Cluster of differentiation.

cDNA. A double-stranded DNA copy of an mRNA molecule.

cDNA arrays. Collection of cDNAs used for determining the transcription of genes with cells under given conditions.

Cecropins. Insect antibacterial peptides.

Cellular Immunology. The study of the cellular basis of immunology.

Cellulitis. Inflammation of the connective tissues – particularly the subcutaneous tissues.

Centrosome. An organelle found in most animal cells, usually situated near the nucleus. Microtubules of the cytoskeleton are attached to this structure and it functions as a microtubule-organising centre.

Chancroid. A sexually transmitted disease caused by *Haemophilus ducreyi*.

Chaperone. A protein involved in the correct folding of other proteins.

Chemoattractant. Molecule that stimulates a cell to move towards it (examples include chemokines, C5a, *N*-formyl-methionyl-leucyl-phenylalanine (FMLP)).

Chemokine. Large group of small cytokines involved in the migration and activation of cells, principally leukocytes.

Chemotaxis. The movement of a cell towards, or away from, a chemical.

Chondrocyte. Cell found in cartilage and responsible for cartilage matrix synthesis.

Chronic disease (infection). A disease (infection) that lasts a long time.

Chronic granulomatous disease (CGD). Immunodeficiency disease in which chronic granulomas form as a result of defective killing of bacteria. Basic defect is in the NADPH oxidase system that generates superoxide radical.

Clone/clonal. A clone is a population of cells derived from a single precursor cell.

Cluster of differentiation (CD). Refers to groups of monoclonal antibodies that recognise the same cell surface molecules.

Coagulase-negative staphylococci. All species belonging to the genus *Staphylococcus* with the exception of *Staph. aureus*. The term refers to the fact that *Staph. aureus* is distinguishable from other members of the genus because it can produce coagulase – an enzyme that clots plasma.

Coagulation. Clumping together as in the process of blood coagulation.

Cognate. Its own, belonging to.

Collectins. Structurally related family of calcium-dependent lectins that have collagen-like domains. Mannose-binding protein is a collectin.

Colonisation. The multiplication of an organism once it has adhered to a host tissue or structure.

Combinatorial chemistry. A process by which a large range of compounds, differing around a central structural theme, are generated. The process is much used in the pharmaceutical industry to generate new therapeutic compounds.

Commensalism. An interaction between two organisms in which one member derives benefit while the other is unaffected.

Competent. Able to take up DNA and be genetically transformed.

Complement. A large group of interacting plasma proteins evolved to deal with invading pathogens.

Complementarity-determining regions (CDRs). Part of the antibody molecule that makes contact with the specific ligand and determines specificity.

Conjugation. Bacterial fornication. Gene transfer in bacteria that requires direct cell–cell contact.

Conjugative pilus. Bacterial surface appendage involved in conjugation. May also serve as a receptor for bacteriophage infection.

Conjugative transposon. A mobile genetic element that can transfer between bacteria by direct cell–cell contact.

Co-stimulator. Proteins required for the induction of T cell activation by antigen-presenting cells. These molecules include CD28, CD80 and CD86.

Constitutive expression. Genes that are continually expressed regardless of the environmental conditions.

Counter-illumination. Mechanism used by various oceanic creatures for avoiding detection by predators and involving the generation of light from light organs.

C-reactive protein. Acute phase protein that binds to phosphatidylcholine.

Cryptic gene. A gene that is not expressed, because of (often multiple) mutations. Also known as a pseudogene.

Cryptidins. Antibiotic peptides produced in the intestines of mammals.

Cystic fibrosis (CF). Commonest genetic disease of Caucasians involving defect in chloride channel and resulting in multiple lung infections and pathological changes in the pancreas.

Cytokine. Protein or glycoprotein that binds to high affinity cell surface receptors and acts as an intercellular signal.

Cytokine network. The basic unit of cytokine biology. An interacting group of cytokines (e.g. Th1 or Th2 networks) leading to a specific biological response.

Cytoskeleton. The collection of intracellular tube-forming proteins that enables a cell to maintain or change its shape and is involved in the intracellular transport of organelles and molecules.

Cytotoxic T lymphocyte (CTL). A T cell that can kill other cells.

Cytotoxicity. The process of exogenous killing of cells.

Dalton. Unit of molecular mass.

Danger model. An alternative to the self-non-self model of immunity that proposes that the immune system recognises those signals that cause cytotoxic death of cells.

Defensins. Mammalian antibiotic peptides.

Delayed-type hypersensitivity (DTH). A form of cell-mediated immunity elicited by antigen in the skin. It is controlled by CD4 Th1 cells and is so named because the skin inflammation takes some time to arise (i.e. is delayed).

Dendritic cells. Cells with a branch-like morphology to enable them to make contact with T lymphocytes and pathogens. They are the most powerful inducers of T cell responsiveness.

Dental caries. A disease involving destruction of the tooth due to acid production by certain bacteria.

Desmosomes/hemidesmosomes. Specialised region of the eukaryotic plasma membrane consisting of dense protein plaques connected to intermediate filaments and evolved to 'glue' cells together.

Diacylglycerol. Product of the breakdown of inositol trisphosphates that binds to protein kinase C and activates this kinase family.

Diapedesis. The movement of leukocytes across the vascular endothelial barrier from the blood and into tissues.

Differential fluorescence induction (DFI). A technique, similar to IVET, that utilises green fluorescent protein as a promoter trap to identify genes expressed *in vivo*.

Disulphide bridge. Cross-link between polypeptide chains formed by the oxidation of cysteine residues.

DNA library. Collection of cloned DNA molecules consisting of fragments of the complete genome (or of cDNA copies of mRNA) inserted into a suitable cloning vector.

Domain. Region of a protein with a distinct tertiary structure (e.g. rod-like, globular, etc.) and characteristic activity.

Dorsal. Relating to the back of an animal or the upper surface of a structure.

Downstream. For a gene, this refers to the direction the RNA polymerase moves during transcription, which is toward the $3'$ end.

Dynein. A eukaryotic motor protein (see also kinesin) that can propel organelles along microtubules.

Effector cell. Any cell capable of mediating an immune function. The term normally refers to lymphocytes and macrophages.

Electroporation. Introduction of DNA into a bacterium using a high voltage electrical pulse.

Embryogenesis. The formation and development of the embryo.

Embryonic stem (ES) cells. Continuously growing cells that retain the ability to contribute to all cell lineages in developing mouse embryos. Such cells can be genetically manipulated in tissue culture and then inserted into mouse blastocysts to generate mutant lines of mice. Such cells can also be used to generate gene knockout mice.

Enamel. The extremely hard external layer of that portion of a tooth that protrudes from the gum.

Endocrine. Relating to glandular tissues that secrete hormones into the blood.

Endocytosis. Uptake of material from exterior of cell by production of invaginations in the plasma membrane to form small membrane-bound vesicles known as early endosomes.

Endogenous pyrogens. Cytokines that induce fever. Also the original name for interleukin 1.

Endoplasmic reticulum (ER). The network of interconnected membranous structures within the cytoplasm of eukaryotic cells.

Endoplasmic reticulum-associated protein degradation pathway (ERAD). Eukaryotic system for the elimination of misfolded proteins. Subverted by a number of bacterial exotoxins to facilitate access to the target cell cytoplasm.

Endoscope. Instrument for visualising body cavities or organs.

Endosome. A cell vesicle with an acidic interior.

Endothelium. A layer of cells that lines the inside of blood and lymph vessels as well as body cavities such as the chambers of the heart.

Endotoxin. Bacterial toxin released only when the cell is damaged or dead. The classic endotoxin is lipopolysaccharide.

Enterocolitis. Inflammation of the small intestine and the colon.

Enterocytes. Absorptive epithelial cells of the intestine.

Enteroendocrine cells. Endocrine cells within the intestine.

Enterovirulent. An organism able to cause disease in the intestinal tract.

Eosinophil. Blood leukocytes involved in our defence against parasitic infestations.

Epidermis. A multilayered structure forming the outer, protective layer of the skin.

Epigenetic variation. Alteration of the antigenic nature of bacterial surface components in the absence of genetic variation.

Epiglottitis. Inflammation of the epiglottis – a cartilaginous layer that prevents food, etc. from entering the larynx during swallowing.

Epithelia. Sheets of cells comprising one of more cell layers covering an external or internal body surface and having a protective function.

Epitope (of a receptor). The particular molecular structure of a receptor molecule to which a bacterial adhesin binds.

Erysipelas. An acute inflammation of the skin and underlying tissues accompanied by fever.

Erythrasma. An infection of the horny layer of the epidermis.

Eukaryote. An organism whose chromosomal DNA is enclosed within a membrane thereby forming a distinct entity – the nucleus.

Exocytotic vesicles. Small vesicles that are produced within eukaryotic cells and designed for export of material from the cell.

Exogenous infection. An infection produced by an organism that is not normally found in the healthy host, for example anthrax, diphtheria, tuberculosis.

Exon. Those parts of an interrupted eukaryotic gene that are represented in the mature transcription product (mRNA).

Exotoxin. Bacterial toxins that are actively exported from the cell.

Expression vector. A modified plasmid or virus that is able to carry a gene or cDNA into a suitable host cell to allow synthesis of the encoded protein.

Extracellular matrix (ECM). Matrix material formed between eukaryotic cells and composed of numerous macromolecules such as collagens, proteoglycans and glycoproteins.

Facultative anaerobe. An organism that can grow in the presence or absence of oxygen.

Fas/Fas ligand. Members of the TNF gene family. Fas is expressed on certain cells that are susceptible to killing by cells expressing Fas ligand on their cell surface.

Fc receptors. Receptors for the Fc region of immunoglobulins when they form immune complexes.

Fibril. Surface appendage involved in bacterial adhesion.

Fibrinolysis. Breakdown of fibrin.

Fibroblast. A major cell of connective tissue involved in the synthesis of the extracellular matrix as well as able to secrete cytokines in response to bacteria and their products.

Filopodia. Long (can be up to 50 μm), narrow (0.1 μm) protrusions of a eukaryotic cell.

Fimbria. Multisubunit surface appendage involved in adhesion. Also known as a pilus.

Flagellum. Complex surface appendage that enables a bacterium to move.

Focal adhesions. Contact points between cells and the extracellular matrix.

Fold (protein). A topographical description of the shapes that amino acid sequences can take up within a protein molecule.

Functional genomics. The assignation of functions to genes.

G protein. Protein involved in intracellular signalling that binds GTP and converts it to GDP.

G protein-coupled receptor. Receptor coupled to G proteins, allowing signal to be transduced into cell (e.g. chemokine receptors).

Genome analysis and mapping by *in vitro* transposition (GAMBIT). A technique for the stepwise saturation mutagenesis of a bacterial chromosome.

GDP dissociation inhibitor (GDI). Protein controlling the activity of G proteins.

Gene. Physical and functional unit of heredity encoding a protein or RNA molecule.

Gene knockout. Common parlance for the disruption of a gene by the process of homologous recombination.

Generation of diversity (GOD). The system of genetic manipulations that enables B and T lymphocytes to generate the enormous diversity of antibodies and T cell receptors that enables the immune system to recognise pathogens.

Genome. Total genetic information carried by a cell or organism.

Genomic library. Collection of cloned DNA molecules consisting of fragments of the complete genome inserted into a suitable cloning vector.

Genomics. Comparative analysis of the genomes of different organisms using bioinformatics tools.

Genotype. The genetic complement of an organism, i.e. the sum total of the genes it contains – many of these may not be expressed by the organism under a particular set of environmental conditions: cf. Phenotype.

Germinal centres. Sites in secondary lymphoid organs of intense B cell activity (proliferation, selection, maturation, death) stimulated during antibody responses.

Gingiva. The mucosa adjacent to the teeth – the gums.

Gingivitis. Inflammation of the gingiva (gums).

Glomerulonephritis. Inflammation of the glomeruli of the kidneys.

Glycolipid. Lipid covalently modified with one or more sugar residues.

Glycoprotein. Protein containing oligosaccharides covalently linked via serine or threonine residues.

Gnotobiotic. A term used to describe an animal that has been born and reared in a microbe-free environment.

Goblet cell. Cell responsible for production of mucin.

GPI-anchored proteins. Proteins anchored to plasma membranes of cells via glycosylphosphatidylinositol (GPI) moiety (e.g. CD14).

Granzymes. Serine proteases found in the granules of cytotoxic lymphocytes.

Group A streptococci. *Streptococcus pyogenes.*

Growth factor. Large family of cytokines involved in stimulating growth of mesenchymal cells.

GTPase. Enzyme that converts GTP to GDP. *See also* Small GTPases.

GTPase-activating protein (GAP). Protein involved in control of G proteins.

Guanine nucleotide exchange factor (GEF). Protein involved in control of G proteins.

Guanylyl cyclase. Enzyme converting GMP to cyclic GMP (c-GMP).

Haematopoiesis. The generation of the cellular elements of the blood.

Haemolysin. An agent that lyses erythrocytes.

α-helix. Protein secondary structure motif.

Helminth. A worm. There are three major groups of helminths containing members that have humans as their main host – digenean flukes, tapeworms and roundworms (nematodes).

Helper T cells. CD4 T lymphocytes with effector functions. They include Th1 and Th2 lymphocytes.

Hemidesmosome. A region of the plasma membrane responsible for anchoring an epithelial cell to the extracellular matrix.

High endothelial venules (HEV). Specialised blood vessels found in lymphoid tissues and allowing lymphocyte migration into these tissues.

Histidyl-aspartyl phosphorelay. An alternative (and more correct) name for the two-component signal transduction system used by bacteria to detect, and respond to, environmental signals.

Histocompatibility. Immunological term to describe the genetic systems that determine the rejection of tissues.

Homeostasis. In physiological terms, the dynamic situation in which the organism, as a system, remains the same despite changing external conditions.

Homologous recombination. Reciprocal exchange between a pair of homologous DNA sequences.

Homoserine lactones. Signalling molecules that are involved in quorum sensing in certain bacteria.

Housekeeping genes. Genes that encode molecules needed for the normal, day-to-day metabolic processes of the cell.

Human leukocyte antigen (HLA). The genetic designation for the major histocompatibility complex in humans.

Humoral immunity. The soluble factors (antibodies, complement, cytokines) that drive the immune system.

Hydroxyapatite. The non-cellular, inorganic component of bone and teeth.

Hypoglycaemia. A low concentration of sugar in the blood.

Iatrogenic. A disease which results from some medical procedure, e.g. insertion of a catheter.

Idiopathic. A disease in which the causative agent has not yet been identified.

Immune system. A collective term for the cells and mediators involved in innate and acquired immunity.

Immunity. The ability to resist infection.

Immunoadsorption. Removal of specific components from a mixture by use of an antiserum or monoclonal antibodies.

Immunoassay. An assay that is based on the ability of an antibody molecule to specifically recognise its complementary antigen.

Immunobiology. The study of the biological basis of host defence against infectious agents.

Immunodeficiency. A situation in which an individual has one aspect of their immune response impaired.

Immunogen. Any molecule that can induce an adaptive immune response.

Immunoglobulin. An antibody.

Impetigo. Inflammation of the skin accompanied by the formation of discrete round blisters.

In silico. By computer.

Inducible expression. Genes that are expressed only in response to a particular environmental stimulus.

Inflammation. The general term for the vascular-, leukocyte- and mediator-based responses that lead to local accumulation of leukocytes and plasma at sites of infection or injury.

Injectisome. The secretory apparatus of type III secretion systems – also known as a secreton.

Innate immunity. A system of immunity active all the time in organisms and able to deal with pathogens. It includes the actions of phagocytes and complement but not those of lymphocytes.

Insertion sequence. A small mobile genetic element that carries only the genes needed for its own transposition.

Integrins. Molecules on the surface of mammalian cells that are important in mediating adhesion of the cell to the extracellular matrix and to other cells.

Interleukin (IL). One of the six families of cytokines.

Intron. A segment of an interrupted eukaryotic gene that is transcribed but subsequently removed during mRNA processing.

In vivo **expression technology (IVET).** A series of promoter trap techniques for the identification of *in vivo*-expressed genes.

In vivo-**induced antigen technology (IVIAT).** A screening method using human serum antibodies to identify bacterial genes expressed during infection.

Ion channel receptor. A cell surface receptor that, when it binds to its specific agonist, causes an ion channel to open.

Isoforms. One of several forms of the same protein whose amino acid sequence varies slightly but whose general function is unchanged.

Isogenic mutant. A mutant, usually created by insertional inactivation, that differs from the parental strain only at the mutant locus.

Isotype. One of the five types of antibody.

Intercellular adhesion molecules (ICAMs). Ligands on cell surfaces for integrins.

Intra-epithelial lymphocytes (IELs). T cells found within the epithelial cells of the mucosae.

Invariant chain. A non-polymorphic protein that binds to newly synthesised MHC class II protein in the endoplasmic reticulum and prevents peptide binding.

Keratin. Family of proteins associated with certain epithelia, particularly the keratinocytes of the skin.

Keratinocytes. Skin cells that produce keratin.

Kinase. Enzyme that can phosphorylate specific residues within distinct proteins.

Kinesin. A protein (motor protein) found in a eukaryotic cell that is able to move organelles along the microtubules of the cytoskeleton.

Lactoferrin. An iron-binding protein found on mucosal surfaces.

Lamellipodium. A thin, broad, sheet-like extension of a eukaryotic cell. Often forms at the leading edge of a moving cell – when not at the leading edge it is often called a 'ruffle'.

Lectin. A protein that binds specifically to sugars.

Lectin pathway of complement activation. Triggered by binding to circulating lectins such as mannose-binding protein.

Leukocyte. A white blood cell.

Leukotrienes. Lipid mediators of inflammation produced by the action of 5-lipoxygenase.

Lewis antigens. Cell surface carbohydrates used as a basis for distinguishing between red blood cells. These carbohydrate groups are ligands for adhesion receptors (selectins) on vascular endothelial cells. The Lewis antigen system is used as a basis for blood group classification.

Ligand. Any molecule that can bind to a receptor.

Lipopolysaccharide (LPS). A complex glycolipid found on the surface of Gram-negative bacteria that is a powerful inflammatory stimulus – it is also often referred to as 'endotoxin'.

Lipoxygenase. An enzyme that oxygenates arachidonic acid to form pro-inflammatory lipid mediators.

Lyme disease. A chronic disease caused by *Borrelia burgdorferi*, a spirochaete that is transmitted to humans by a tick.

Lymph. The fluid (usually pale yellow) that flows in the lymphatic vessels. Its composition is similar to that of plasma.

Lymphadenopathy. Any disease of the lymph glands. Inflammation of the lymphatic vessels.

Lymphangitis. Inflammation of the lymphatic vessels.

Lymphatics. A system of vessels permeating the body that collects intercellular fluid called lymph and returns it, via the thoracic duct, to the circulation.

Lymphocytes. Antigen-specific cells of the immune system.

Lymphokine. Older term for cytokines that are produced by, or have effects on, lymphocytes.

Lysis. The rupture of a cell.

Lysosome. Small vesicular organelle with an acidic pH and a complement of hydrolytic enzymes.

Lysozyme. Enzyme capable of cleaving bacterial peptidoglycan.

M (microfold) cells. Specialised epithelial cells overlying Peyer's patches in the gut and through which many pathogens can pass and gain access to underlying tissue.

Macrophage. Tissue-based phagocytic cell derived from blood monocytes.

Macropinocytosis. A process by which a eukaryotic cell takes in soluble and particulate matter from its external environment. It involves invagination of the cytoplasmic membrane and the formation of internal vesicles.

Major histocompatibility complex (MHC). A cluster of genes in a mammalian cell that encodes cell surface proteins important for antigen presentation to T cells.

MALT. Mucosa-associated lymphoid tissue. This comprises all of the lymphoid cells present in the epithelia and lamina propria below the mucosal surfaces.

MAP kinases. Mitogen-activated protein kinases. A group of important kinases that transduce signals from growth factors, bacterial products and inflammatory molecules.

Mast cell. Bone marrow-derived cell recognised by its large intracellular granules and involved in hypersensitivity (allergic) reactions and in antibacterial defence.

Membrane attack complex (MAC). Terminal stage of complement activation, which forms pores in target cells.

Memory (immunological). The ability of acquired immunity to respond more rapidly, and with greater effect, to subsequent exposure to any particular pathogen.

Meninges. The membranes covering the brain and spinal cord of vertebrates.

Meningitis. Inflammation of the membranes of the brain or spinal cord.

MHC restriction. The characteristics of T cells by which the cells recognise a peptide antigen only when it is bound to a particular allelic form of an MHC molecule.

Microarray. A collection of DNA molecules (genes or their fragments) arrayed on a surface (e.g. a glass slide) in a form suitable for probing with cDNA. These are used for identifying patterns of gene expression.

Microflora. Those microbes found at a particular anatomical site – this may be a disease-free site or a lesion produced during the cause of a disease (e.g. a boil) or by an injury (e.g. a burn or wound).

β_2-Microglobulin. Small protein associated with MHC class I proteins.

Microtubule. One of the key proteinaceous filaments that comprise the cytoskeleton of the eukaryotic cell.

Microvillus. A finger-like projection (approximately 1.0 μm long and 0.1 μm thick) present on certain cells – particularly epithelial cells. A single cell will have very large numbers of microvilli – at least along one of its surfaces.

Mitochondrion. Organelle responsible for energy generation and for controlling apoptosis in eukaryotic cells.

Mitogen. Any extracellular substance that promotes cell proliferation.

Molecular chaperone. A protein involved in catalysing the correct folding of proteins.

Monoclonal antibody. An antibody that is specific for one antigen and is the product of one hybridoma cell.

Monocyte. Bone marrow-derived precursor of tissue macrophages.

Mononuclear phagocytes (or monocytes). Bone marrow-derived leukocytes involved in phagocytosis but distinct from polymorphonuclear cells (neutrophils).

Morphogenesis. The process of generating the cellular structure of the adult animal during embryogenesis.

Motif. In proteins – a structural unit exhibiting a particular three-dimensional architecture that is found in a variety of proteins and has a particular function.

Mucociliary escalator. A cleansing mechanism that employs ciliated epithelial cells and mucus-secreting cells to remove particulate matter (including bacteria) from the respiratory tract.

Mucosa. The moist membrane that lines the respiratory, genitourinary and intestinal tracts.

Multifactorial (disease). Complex conditions such as rheumatoid arthritis and atherosclerosis that are due to multiple causes – both genetic and environmental.

Mutualism. An association between two organisms from which both partners derive some benefit.

Mutagenesis. A process that generates any change in the sequence of the DNA of an organism.

Myeloid cells. Cells derived from myeloid tissue, which is the main site of haematopoiesis. Such cells include monocytes, macrophages, neutrophils, eosinophils and mast cells.

Myositis. Inflammation of the muscles.

Nanobacteria. Very small bacteria – also called ultramicrobacteria.

Nasopharynx. The nasal region of the pharynx.

Natural antibodies. IgM antibodies, largely produced by B-1 B lymphocytes, that are specific for bacteria found on mucosal surfaces.

Natural killer (NK) cells. A subset of bone marrow-derived lymphocytes, distinct from T and B cells, that function as part of the innate immune responses to pathogens.

Necrotising fasciitis. A disease caused by *Streptococcus pyogenes* that involves inflammation and killing of the fascii, i.e. sheets of connective tissue surrounding muscles and other organs.

Nephritogenic. Organisms able to damage the kidneys.

Neuropeptides. Peptides with actions on neurones.

Neutrophil (also polymorphonuclear neutrophil (PMN)). The most abundant circulating leukocyte recruited to inflammatory sites, where it phagocytoses and kills pathogens.

NF-κB. A family of transcription factors important in innate and acquired immunity.

Nitric oxide synthase. A group of enzymes that function to produce the potent mediator, nitric oxide (NO), from the amino acid L-arginine.

Normal (endogenous) microflora. Those microbes usually present on an animal which, usually, do not cause diseases in an otherwise healthy individual. It is usual to speak of the normal microflora of a particular anatomical site such as the mouth, the skin, etc.

Nosocomial infection. An infection acquired in hospital.

Obligate. An adjective used to indicate that a particular environmental factor is always required for growth of the organism.

Oligosaccharide. A molecule composed of a number of sugar residues.

Oncogene. A gene whose product is involved in transforming cells and which may be involved in the process of carcinogenesis.

Open reading frame (ORF). The part of a protein-coding gene that is transcribed and translated into a protein molecule.

Operon. Organisation of several genes into a single transcriptional unit.

Opportunistic infection. An infection caused by an organism that is a member of the normal microflora. The term is also applied to infections caused by organisms (e.g. *Pseudomonas* spp.) that normally inhabit the external environment (e.g. water, soil).

Opsonin/opsonise. An opsonin is a macromolecule that becomes attached to a pathogen and can be recognised by surface receptors on phagocytes, thus enhancing phagocytosis of the pathogen through 'opsonisation'.

Orphan gene. A gene for which a nucleotide sequence is available but whose product has, as yet, no ascribed function.

Oropharynx. That part of the pharynx behind the mouth and below the soft palate.

Orthologous. With regard to genes, refers to homologous genes in the genomes of different organisms.

Otitis media. Inflammation of the middle ear.

Outer membrane proteins (OMPs). Proteins found in the outer membrane (as opposed to the cytoplasmic membrane) of Gram-negative bacteria.

Paneth cell. A cell of the intestine involved in antibacterial defence.

Parasitism. An interaction between two organisms in which one member derives benefit (the parasite) while the other is harmed in some way.

Parologous. Two or more homologous genes located within the same genome.

Pathogen. A microbe that is able to damage its host.

Pathogen-associated molecular pattern (PAMP). Evolutionarily conserved molecules that can be recognised by receptors on phagocytes and thus allow innate recognition of infection.

Pathogenicity. The ability of a microorganism to cause disease.

Pathogenicity island (PAI). Large chromosomal element encoding virulence determinant(s) that is thought to have been acquired by horizontal gene transfer.

Pattern recognition receptor (PRR). The phagocyte receptors that recognise PAMPs. CD14 is one of the best examples.

PCR (polymerase chain reaction). A technique for amplifying specific nucleic acid sequences.

Peptidoglycan. The main structural component of the cell wall of most bacteria.

Perforin. A pore-forming protein, homologous to the C9 component of complement and produced by cytolytic T cells.

Perineal. Belonging to the perineum, i.e. the region between the thighs.

Periodontitis. A disease involving inflammation of the periodontal (i.e. tooth-supporting) tissues accompanied by destruction of the alveolar (i.e. jaw) bone.

Periplasm. Fluid-filled region between the inner and outer membrane of Gram-negative bacteria.

Peyer's patch. Organised lymphoid tissues in the lamina propria of the small intestine. Immune responses to bacterial and ingested antigens may be initiated here.

Phage display. A genetic screening system for identifying genes encoding proteins that interact with other proteins.

Phagocyte. A cell that can ingest particles such as bacteria, for example neutrophils, macrophages, eosinophils and mast cells.

Phagocytosis. The process by which certain large particles (e.g. bacteria) are ingested by certain eukaryotic cells.

Phagosome. A membrane-bound intracellular vesicle that contains bacteria (or other particles) that have been ingested from the cell's external environment.

Phase variation. The switching off or on of expression of surface components such as fimbriae.

Pharynx. The portion of the alimentary canal that extends from the base of the skull to the beginning of the oesophagus. It is subdivided into a number of distinct regions.

Phenotype. The observable characteristics of an organism (e.g. colonial morphology, staining characteristics, production of certain enzymes); cf. Genotype.

Pheromone. A chemical produced by an organism which, when released into the environment, can influence the behaviour or development of other members of the same species.

Phorbol esters. Synthetic compounds that can mimic the action of diacylgylcerol.

Phosphatase. An enzyme able to remove phosphate groups from proteins.

Phosphoinositides. A family of membrane-bound lipids containing phosphorylated inositol derivatives involved in intracellular signalling.

Phospholipase. An enzyme capable of cleaving acyl groups from phospholipids.

Phospholipid. A major class of lipid found in cell membranes and normally composed of two fatty acid chains esterified to glycerol with the phosphate esterified to one of various polar groups (e.g. phosphatidylcholine).

Phosphorelay. A process involving a number of proteins in which a phosphate group is transferred sequentially from one protein to another.

Phosphorylation. Process of adding a phosphate group to a molecule.

Phylogenetics. The study of phylogeny – the science that describes the evolutionary relationship between organisms.

Pilus. *See* Fimbria.

Pinocytosis. Uptake of extracellular fluid by a cell as a result of invagination of the cytoplasmic membrane followed by vesicle formation.

Planktonic bacteria. Bacteria that are suspended in a fluid environment (usually water) as opposed to those attached to a surface.

Plasma cell. A terminally differentiated antibody-secreting B lymphocyte.

Plasmid. A piece of DNA (usually circular) capable of independent replication, often found in bacteria (and some other types of cell) and much used in molecular biology for gene cloning and expression.

Platelet. A small, non-nucleated, cellular element found in blood and essential for blood clotting.

Pleiotropy. Having multiple functions (as applied to cytokines).

Polymorphic genes. Those genes that can exist in a number of different alleles or haplotypes in the population.

Porins. Proteins in the outer membrane of Gram-negative bacteria that group together to form channels through which small molecules can pass.

Post-capillary venules (PCVs). Part of the vasculature within which inflammatory events, with leukocyte migration, take place.

Prebiotic. A substance (often an oligosaccharide) that selectively stimulates the growth and/or activity of one or several members of the normal microflora.

Probiotic. An organism used to alter the microflora of a particular site in the host so that the latter benefits in some way, for example by having an increased resistance to infection, by eliminating a harmful organism, by boosting nutrition.

Programmed cell death (PCD). *See* Apoptosis.

Prokaryote. An organism that does not have a membrane-bound nucleus – its chromosomal DNA is usually present in the cytoplasm as a tightly coiled aggregate.

Proliferation. As applied to cells – to increase in number.

Prostaglandins. Pro-inflammatory products produced from arachidonic acid by the action of cyclo-oxygenases.

Prosthesis. A manufactured device meant to replace some structure or system that has been damaged or destroyed by disease, for example heart valves, limbs, joints, etc.

Protease/proteinase. An enzyme capable of cleaving peptide bonds.

Proteasome. Large multifunctional protease complex found in the cytosol of cells that functions to degrade ubiquinylated proteins.

Protein. Polymeric molecules composed of amino acids.

Proteoglycans. Acidic glycoproteins that contain a greater proportion of carbohydrate than protein. Present in the extracellular matrix of many animal tissues.

Proteome. The total protein complement of a cell under a defined set of environmental conditions. Confusingly, the term is sometimes used to refer to the total complement of proteins that the cell has the potential to produce.

Proteomics. The study of the proteome.

Protozoa. A group of single-celled, eukaryotic, motile organisms that do not have a cell wall.

Pseudogene. *See* Cryptic gene.

Psoriasis. Hyperproliferative/inflammatory disease of the skin.

Puerperal fever. An infection of the uterus during childbirth – caused by *Streptococcus pyogenes*.

Pyogenic bacteria. Pus-forming bacteria, i.e. bacteria able to kill neutrophils and produce a localised disease lesion containing pus.

Quorum sensing. A process by which bacteria regulate gene expression on the basis of their population density. Once a critical population density has been reached, the level of gene expression is altered.

Reaginic. Pertaining to IgE antibodies that mediate type I immediate hypersensitivity.

Receptor. Generically, any molecule that binds to another such that a cellular response is initiated. Also, the complementary molecule on a surface (cellular or inanimate) to which an adhesin on a bacterium binds, thereby enabling adhesion of the bacterium to that surface.

Receptor tyrosine kinase (RTK). A class of cell surface receptors whose internal domain contains a tyrosine kinase active site.

Redox potential (E_h). A measure of the reducing power of a system. A high, negative value (expressed in millivolts) means that the system is strongly reducing, while a high, positive value indicates that the system is highly oxidising.

Regulon. A group of unlinked genes and/or operons subject to co-ordinated regulation.

Response regulator. A DNA-binding protein that can be phosphorylated by a sensor kinase. This phosphorylation alters the DNA-binding properties of the molecule and results in either activation or repression of gene transcription. It is an essential element of a two-component signal transduction system.

Reverse transcription. Synthesis of DNA from an RNA template. Accomplished by the enzyme reverse transcriptase.

Rheumatoid arthritis. An autoimmune disease of the joints.

Rocky Mountain spotted fever. An infection caused by *Rickettsia rickettsii*.

S layer. Crystalline protein coat found on the surface of some bacterial species.

Scavenger receptors. A family of cell surface receptors on macrophages that mediate the phagocytosis of pathogens.

Second messenger. An intracellular signalling molecule (e.g. c-AMP) whose concentration increases, or decreases, in response to the binding of an extracellular ligand.

Secretin. One of a family of outer membrane proteins that functions in bacterial protein translocation and the assembly of pili and filamentous phage.

Secreton. The secretory apparatus of type III secretion systems – also known as an injectisome.

Secretory component. The proteolytically cleaved component of the extracellular domain of the poly(Ig) receptor that remains bound to the IgA molecule at mucosal surfaces.

Secretory leukocyte protease inhibitor (SLPI). A protease inhibitor that can also block the actions of lipopolysaccharide.

Selectin. A group of cell surface carbohydrate-binding proteins involved in leukocyte-vascular endothelial cell interactions.

Sensor kinase. A protein that is able to detect a signal (e.g. the presence of a nutrient, a change in pH, etc.) in the external environment and then responds by undergoing auto-phosphorylation at a specific histidine residue. It is an essential element of a two-component signal transduction system.

Septicaemia. The persistent presence of bacteria in the bloodstream; cf. Bacteraemia.

Serotype. Different strains of a species may be subdivided on the basis of their antigenic composition – these strains are known as 'serotypes' of the species.

β-Sheet. Protein secondary structure motif.

Siderophore. A protein that can bind iron at very low concentrations.

Sigma factor. A prokaryotic protein factor that directs RNA polymerase binding to specific promoter sequences.

Signal sequence. Short sequence of amino acid residues that directs the protein to the secretory apparatus. Normally found at the N-terminus of proteins.

Signal transduction. Conversion of the binding of a protein at the cell surface into an intracellular signal that alters cell shape, function or gene transcription.

Signature-tagged mutagenesis (STM). A technique to detect genes required for *in vivo* survival and growth that uses modified transposons to allow high throughput screening of random mutants.

Sinusitis. Inflammation of the epithelium lining the paranasal sinuses.

Slime layer. Similar to a capsule but less well organised.

Small GTPases. Proteins such as Ras and Raf.

Splicing. A process that removes introns and joins exons post gene transcription.

Sporulation. The process by which certain organisms (e.g. *Bacillus* spp., *Clostridium* spp.) produce resistant structures known as spores.

Squames. Skin scales shed from the outer, keratinised layers of the epidermis.

STATs. Signal transducers and activators of transcription. Proteins that act as cytoplasmic transcription factors requiring phosphorylation and dimerisation to allow them to enter the nucleus.

Stratum corneum. Outermost layer of skin.

Subarachnoid space. Membrane-bound region of the brain and spinal cord that contains cerebrospinal fluid.

Submucosa. Tissue underlying the mucosa.

Superantigens. Bacterial or viral proteins that bind to, and activate, all the T cells expressing a particular set or family of V_β T cell receptor genes.

Superoxide. A damaging oxygen radical.

Symbiosis. A state involving two organisms living together. Each member of the symbiosis is termed a 'symbiont'. *See also*: Mutualism, Commensalism, Parasitism.

Systemic infection. An infection that spreads throughout the body of the infected individual.

T lymphocyte. One of the two major classes of lymphocytes and responsible for cell-mediated immunity.

Thymus. One of the primary lymphoid organs involved in T cell 'education'.

Tight junctions. Ribbon-like bands that connect adjacent epithelial cells and function to prevent leakage across epithelial barriers.

Toll. *Drosophila* gene involved in dorso-ventral patterning.

Toll-like receptors (TLRs). Homologues of Toll involved in recognition of pathogens.

Transcription. Synthesis of an RNA molecule from a DNA template. Catalysed by RNA polymerase.

Transcription factor. Protein required to initiate/regulate the transcription of a gene.

Transduction. Transfer of bacterial DNA from one bacterium to another by a phage.

Transferrin. An iron-binding protein in serum.

Transformation (bacterial). Uptake of free DNA by competent bacteria and its incorporation into the genome.

Transformation (eukaryotic). Conversion of eukaryotic cells to a state of uncontrolled growth.

Transforming growth factor β (TGFβ) superfamily. A large family of growth-promoting cytokines.

Transgene. A cloned gene that is introduced and stably incorporated into a plant or animal.

Transgenic. Referring to any plant or animal carrying a transgene.

Translation. Synthesis of a polypeptide from a mRNA template, utilising ribosomes and other cellular constituents.

Translocase. The secretory apparatus of the general secretory pathway.

Transporter associated with antigen processing (TAP). An ATP-dependent peptide transporter involved in transferring peptides from the cytosol to the site of assembly of class I MHC molecules in the endoplasmic reticulum.

Transposon. A mobile genetic element that may carry additional genes over and above those required for its own transposition.

Transposon mutagenesis. The use of transposons to generate random mutants.

Trefoil peptides. Polypeptides with protective actions on epithelia.

Tularaemia. An infection caused by *Francisella tularensis*.

Tumour necrosis factor. An important 'early response' cytokine with many functions in the immune response. It is produced by macrophages and T cells.

Trypanosome. A protozoal parasite belonging to the genus *Trypanosoma*. Members of the genus are responsible for a number of human infections including sleeping sickness and Chaga's disease.

Ubiquitin. A small and highly conserved protein that becomes covalently linked to lysine residues in intracellular proteins, thus marking proteins for proteolytic breakdown by the proteasome.

Ultramicrobacteria. *See* Nanobacteria.

Uropathogenic. Organisms able to cause infections of the urinary tract.

Usher. A secretin protein involved in pilus assembly.

Vacuolar system. Intracellular system of organelles that moves proteins around the cell.

Virulence. The degree of pathogenicity of a disease-causing organism.

Zinc finger. This is a structure formed when a zinc atom binds to two amino acid residues (usually cysteine or histidine) in a protein chain. The resulting intervening span of amino acid residues (usually between 15 and 30 in number) is known as a zinc finger.

Zoonosis. A disease of animals that can be transmitted to humans.

Brief descriptions of bacteria frequently mentioned

Organism	Morphology	Other distinguishing characteristics	Atmospheric requirements	Normal habitat	Diseases caused
Actinobacillus actinomycetemcomitans	Gram −ve cocco-bacilli	–	Facultative anaerobe, prefers increased CO_2 levels	Human oral cavity	Periodontitis
Bacillus anthracis	Gram +ve bacilli	Spore former	Obligate aerobe	Sheep, cattle, soil	Anthrax
Bartonella henselae	Gram −ve bacilli	Pleomorphic cells	Aerobe, prefers increased CO_2 levels	Cats	Cat-scratch disease, bacillary angiomatosis
Bordetella pertussis	Gram −ve cocco-bacilli	–	Obligate aerobe	Human upper respiratory tract	Pertussis
Borrelia burgdorferi	Long, slender Gram −ve spiral cells	Motile	Microaerophilic	Rodents and deer	Lyme disease
Brucella spp.	Gram −ve cocco-bacilli	–	Obligate aerobe, prefers elevated levels of CO_2	Domesticated mammals	Brucellosis
Chlamydia pneumoniae	Gram −ve cocci	Obligate intracellular parasite	Cannot be grown in cell-free media	Human respiratory tract	Atypical pneumonia
Chlamydia trachomatis	Gram −ve cocci	Obligate intracellular parasite	Cannot be grown in cell-free media	Human genital tract, eye	Trachoma, genital infections
Citrobacter freundii	Gram −ve bacilli	Motile	Facultative anaerobe	Human intestinal tract	Meningitis, urinary tract infections
Clostridium difficile	Gram +ve bacilli	Motile, spore former	Obligate anaerobe	Human intestinal tract	Antibiotic-associated diarrhoea, pseudomembranous colitis
Clostridium tetani	Gram +ve bacilli	Motile, spore former	Obligate anaerobe	Soil and faeces	Tetanus

Organism	Gram	Morphology	Oxygen requirement	Habitat	Diseases
Clostridium perfringens	Gram +ve bacilli	Non-motile, spore former	Obligate anaerobe	Soil and faeces	Gangrene, food poisoning, cellulitis
Corynebacterium diphtheriae	Gram +ve bacilli	Irregularly shaped rods	Obligate aerobe	Human upper respiratory tract	Diphtheria
Escherichia coli	Gram −ve bacilli	Motile	Facultative anaerobe	Human intestinal tract	Diarrhoea, urinary tract infections
Francisella tularensis	Gram −ve cocco-bacilli	Pleomorphic, slow growing	Facultative anaerobe	Rabbits, ticks	Tularaemia
Haemophilus ducreyi	Gram −ve cocco-bacilli	—	Aerobe, prefers elevated levels of CO_2	Human genitourinary tract	Chancroid
Haemophilus influenzae	Gram −ve cocco-bacilli	—	Aerobe, prefers elevated levels of CO_2	Human upper respiratory tract	Meningitis, pneumonia, epiglottitis, bacteraemia, cellulitis, otitis media, sinusitis, bronchitis
Helicobacter pylori	Gram −ve spiral-shaped bacilli	Motile	Aerobe but prefers reduced oxygen and increased CO_2	Human gastric mucosa	Ulcers, carcinomas
Lactobacillus spp.	Gram +ve bacilli	Non-motile	Microaerophilic	Human oral cavity, vagina, intestinal tract	Dental caries
Legionella pneumophila	Gram −ve bacilli	Motile, long cells	Obligate aerobe	Warm, moist environments	Pneumonia
Listeria monocytogenes	Gram +ve bacilli	Motile	Facultative anaerobe	Widely distributed in environment	Meningitis, bacteraemia, perinatal sepsis

(Continued)

Organism	Morphology	Other distinguishing characteristics	Atmospheric requirements	Normal habitat	Diseases caused
Mycobacterium tuberculosis	Acid-fast bacilli	Difficult to stain	Obligate aerobe	Human lungs	Tuberculosis
Mycoplasma spp.	Gram −ve	Pleomorphic, do not have a cell wall	Aerobe, prefers elevated levels of CO_2	Human mucous membranes	Pneumonia Genital infections
Neisseria gonorrhoeae	Gram −ve kidney-shaped cocci	Cells are in pairs	Aerobe, prefers elevated levels of CO_2	Human genitourinary tract	Gonorrhoea
Neisseria meningitidis	Gram −ve kidney-shaped cocci	Cells are in pairs	Aerobe, prefers elevated levels of CO_2	Human nasopharynx	Meningitis, bacteraemia
Porphyromonas gingivalis	Gram −ve bacilli	—	Obligate anaerobe	Human oral cavity	Periodontitis
Propionibacterium acnes	Gram +ve bacilli	Pleomorphic bacilli	Facultative anaerobe grows best anaerobically	Human skin	Acne
Pseudomonas aeruginosa	Gram −ve bacilli	Motile	Obligate aerobe	Widely distributed in environment	Infections of burns and wounds, systemic infections in immunocompromised patients, meningitis, infections of prosthetic devices, lung infections in cystic fibrosis patients
Rickettsia rickettsii	Gram −ve bacilli	Obligate intracellular parasite	Cannot be grown in cell-free media	Rodents, dogs	Rocky Mountain spotted fever
Salmonella typhi	Gram −ve bacilli	Motile	Facultative anaerobe	Human intestinal tract, water	Typhoid fever
Salmonella typhimurium	Gram −ve bacilli	Motile	Facultative anaerobe	Animals and environment	Gastroenteritis

Organism	Morphology	Characteristics	Oxygen requirement	Habitat	Diseases
Shigella spp.	Gram −ve bacilli	−	Facultative anaerobe	Human intestinal tract, water	Dysentery
Staphylococcus aureus	Gram +ve cocci	Cells are in clusters produce coagulase	Facultative anaerobe	Human skin	Abscesses, wound infections, impetigo, cellulitis, osteomyelitis, toxic shock syndrome, food poisoning
Staphylococcus epidermidis	Gram +ve cocci	Cells are in clusters; do not produce coagulase	Facultative anaerobe	Human skin	Wound infections, prosthetic device-associated infections, infective endocarditis
Streptococcus agalactiae	Gram +ve oval cocci	Cells are in pairs or chains	Facultative anaerobe	Human female genital tract	Neonatal meningitis
Streptococcus mutans	Gram +ve oval cocci	Cells are in chains	Facultative anaerobe	Human oral cavity	Dental caries
Streptococcus pneumoniae	Gram +ve oval cocci	Cells are in pairs	Facultative anaerobe	Human throat	Meningitis, pneumonia, bacteraemia, otitis media, sinusitis
Streptococcus pyogenes	Gram +ve oval cocci	Cells are in pairs or chains	Facultative anaerobe	Human throat	Pharyngitis, scarlet fever, impetigo, cellulitis, necrotising fasciitis, pneumonia, erysipelas, puerperal fever
Vibrio cholerae	Gram −ve curved bacilli	Motile	Obligate aerobe	River estuaries, human intestinal tract	Cholera

(Continued)

Organism	Morphology	Other distinguishing characteristics	Atmospheric requirements	Normal habitat	Diseases caused
Viridans streptococci	Gram +ve oval cocci	Cells are in pairs or chains	Facultative anaerobe	Human mouth	Dental caries, infective endocarditis
Yersinia enterocolitica	Gram −ve bacilli	Motile	Facultative anaerobe	Rodents, farm animals	Diarrhoea
Yersinia pestis	Gram −ve bacilli	Exhibit bi-polar staining	Facultative anaerobe	Rodents	Plague

+ve, positive; −ve, negative.

Index